W9-CHH-140

Dr. Andrew M. Gordon, R.P.F.
Dept. of Environmental Biology
University of Guelph
Guelph, Ontario N1G 2W1
Canada

The Forest Alternative for Treatment and Utilization of Municipal and Industrial Wastes

The *FOREST ALTERNATIVE*

for Treatment and Utilization
of Municipal
and Industrial Wastes

Edited by DALE W. COLE,
CHARLES L. HENRY,
and WADE L. NUTTER

University of Washington Press
SEATTLE AND LONDON

The following chapters were written and prepared by U.S. government employees on official time, and are therefore in the public domain: "Forest Land Applications of Sludge and Wastewater" by Robert S. Burd; "Overview on Sludge Utilization" by Robert K. Bastian; "Nitrification and Leaching of Forest Soil in Relation to Application of Sewage Sludge Treated with Sulfuric Acid and Nitrification Inhibitor" by Carol G. Wells, Debra Carey, and R. C. McNeil; "Effect of Sewage Sludge from Two Sources on Element Flux in Soil Solution of Loblolly Pine Plantations" by C. G. Wells, C. E. Murphy, C. Davis, D. M. Stone, and G. J. Hollod; "Influence of Municipal Sludge on Forest Soil Mesofauna" by Gary S. MacConnell, Carol G. Wells, and Louis J. Metz; "Growth Response of Loblolly Pine to Municipal and Industrial Sewage Sludge Applied at Four Ages on Upper Coastal Plain Sites" by W. H. McKee, Jr., K. W. McLeod, C. E. Davis, M. R. McKevlin, and H. A. Thomas; "Policies and Guidelines: The Development of Technical Sludge Regulations" by Elliot D. Lomnitz; "Forest Land Treatment with Municipal Wastewater in New England" by Sherwood C. Reed and Ronald W. Crites; and "Reclamation of Severely Devastated Sites with Dried Sewage Sludge in the Southeast" by Charles R. Berry.

The following chapters were prepared in connection with work done under contract with the U.S. Department of Energy. The U.S. government retains a nonexclusive, royalty-free license in and to any copyright covering these chapters, along with the right to reproduce and to authorize others to reproduce all or part of the copyrighted material: "Response of Loblolly Pine to Sewage Sludge Application: Water Relations" by Gail L. Ridgeway, Lisa A. Donovan, and Kenneth W. McLeod; "Understory Response to Sewage Sludge Fertilization of Loblolly Pine Plantations" by K. W. McLeod, C. E. Davis, K. C. Sherrod, and C. G. Wells; and "The Sludge Application Program at the Savannah River Plant" by J. C. Corey, M. W. Lower, and C. E. Davis.

Library of Congress Cataloging-in-Publication Data

The Forest alternative for treatment and utilization of municipal
    and industrial wastes.

    Includes index.
    1. Factory and trade waste—Environmental aspects.
2. Refuse and refuse disposal—Environmental aspects.
3. Sewage sludge—Environmental aspects.
4. Forest soils—Fertilization. I. Cole, Dale W. II. Henry, Charles L. III. Nutter, Wade L.
TD897.F65 1986   363.7'28   86-7759
ISBN 0-295-96392-1

Proceedings of the Forest Land Applications Symposium, held at the University of Washington, June 25–28, 1985, and hosted by the University of Washington College of Forest Resources, the University of Georgia School of Forest Resources, and Washington State University Cooperative Extension. Contributing sponsors include the Washington State Department of Ecology, the U.S. Environmental Protection Agency, the Municipality of Metropolitan Seattle, the USDA Forest Service, the Georgia Forestry Commission, and the Society of American Foresters.

Institute of Forest Resources Contribution No. 56

Contents

Preface   ix

Acknowledgments   x

PROGRAM OVERVIEWS

1   Forest Land Applications of Sludge and Wastewater   3
    Robert S. Burd

2   Overview on Sludge Utilization   7
    Robert K. Bastian

3   The Status of Wastewater Irrigation of Forests, 1985   26
    Dean H. Urie

4   Silvicultural Land Application of Wastewater and Sludge from the
    Pulp and Paper Industry   41
    William E. Thacker

5   Future Directions: Forest Wastewater Application   55
    Wade L. Nutter and Jane T. Red

6   Future Directions: Forest Sludge Application   62
    Dale W. Cole and Charles L. Henry

ENVIRONMENTAL EFFECTS

7   Microbiological Aspects of Forest Application of Wastewater
    and Sludge   73
    Charles A. Sorber and Barbara E. Moore

8   Trace Element Considerations of Forest Land Applications
    of Municipal Sludges   85
    A. C. Chang, T. J. Logan, and A. L. Page

9   Water Quality in Relation to Sludge and Wastewater Applications
    to Forest Land   100
    R. J. Zasoski and R. L. Edmonds

10   Wildlife Responses to Forest Application of Sewage Sludge   110
    Jonathan B. Haufler and Stephen D. West

11   Forest Land Application of Municipal Sludge: The Risk
    Assessment Process   117
    Sydney Munger

12   Behavior of Organic Compounds in Land Treatment Systems
    with the Presence of Municipal Sludge   125
    Michael R. Overcash and Jerome B. Weber

13 Nitrification and Leaching of Forest Soil in Relation to
Application of Sewage Sludge Treated with Sulfuric Acid and
Nitrification Inhibitor   132
Carol G. Wells, Debra Carey, and R. C. McNeil

14 Nitrogen Transformations in Four Sludge-amended Michigan
Forest Types   142
Andrew J. Burton, Dean H. Urie, and James B. Hart, Jr.

15 Effect of Sewage Sludge from Two Sources on Element Flux in
Soil Solution of Loblolly Pine Plantations   154
C. G. Wells, C. E. Murphy, C. Davis, D. M. Stone, G. J. Hollod

16 Heavy Metal Storage in Soils of an Aspen Forest Fertilized
with Municipal Sludge   168
A. Ray Harris and Dean H. Urie

17 Influence of Municipal Sludge on Forest Soil Mesofauna   177
Gary S. MacConnell, Carol G. Wells, and Louis J. Metz

18 Deer and Elk Use of Forages Treated with Municipal Sewage Sludge   188
Henry Campa III, David K. Woodyard, and Jonathan B. Haufler

19 The Influence of Forest Application of Sewage Sludge on the
Concentration of Metals in Vegetation and Small Mammals   199
David K. Woodyard, Henry Campa III, and Jonathan B. Haufler

FOREST RESPONSE

20 Growth Response of Forest Trees to Wastewater and Sludge
Application   209
Roberta Chapman-King, Thomas M. Hinckley, and Charles C. Grier

21 Wastewater and Sludge Nutrient Utilization in Forest Ecosystems   221
Dale G. Brockway, Dean H. Urie, Phu V. Nguyen, James B. Hart

22 Effect of Sludge on Wood Properties: A Conceptual Review with
Results from a Sixty-year-old Douglas-fir Stand   246
D. G. Briggs, F. Mecifi, and W. R. Smith

23 Growth Response, Mortality, and Foliar Nitrogen Concentrations
of Four Tree Species Treated with Pulp and Paper and Municipal
Sludges   258
Charles L. Henry

24 Aspen Mortality Following Sludge Application in Michigan   266
John H. Hart, James B. Hart, and Phu V. Nguyen

25 Growth Response of Loblolly Pine to Municipal and Industrial
Sewage Sludge Applied at Four Ages on Upper Coastal Plain Sites   272
W. H. McKee, Jr., K. W. McLeod, C. E. Davis, M. R. McKevlin,
H. A. Thomas

26 Municipal Sludge Fertilization on Oak Forests in Michigan:
Short-term Nutrient Changes and Growth Responses   282
Phu V. Nguyen, James B. Hart, Jr., and Dennis M. Merkel

27 Municipal Sludge Fertilization on Oak Forests in Michigan:
Estimations of Long-term Growth Responses   292
Dennis M. Merkel, J. B. Hart, Jr., Phu V. Nguyen, Carl W. Ramm

28 Response of Loblolly Pine to Sewage Sludge Application:
   Water Relations   301
   Gail L. Ridgeway, Lisa A. Donovan, and Kenneth W. McLeod

29 Understory Response to Sewage Sludge Fertilization of
   Loblolly Pine Plantations   308
   K. W. McLeod, C. E. Davis, K. C. Sherrod, C. G. Wells

30 Remote Sensing Forest Biomass for Loblolly Pine Using
   High Resolution Airborne Remote Sensor Data   324
   John R. Jensen, Michael E. Hodgson, Halkard E. Mackey, Jr.

PROGRAM IMPLEMENTATION

31 Policies and Guidelines: The Development of Technical
   Sludge Regulations   337
   Elliot D. Lomnitz

32 Planning for the Public Dimension in Forest Sludge and
   Wastewater Application Projects   341
   R. Ben Peyton and Larry M. Gigliotti

33 Technology and Costs of Wastewater Application to
   Forest Systems   349
   Ronald W. Crites and Sherwood C. Reed

34 Technology and Costs of Forest Sludge Applications   356
   Charles L. Henry, Charles G. Nichols, and Terrill J. Chang

35 Utility of a Public Acceptance Survey for Forest
   Application Planning: A Case Study   367
   Larry M. Gigliotti and R. Ben Peyton

36 Relating Research Results to Sludge Guidelines for
   Michigan's Forests   383
   Dean H. Urie and Dale G. Brockway

CASE STUDIES: MUNICIPAL WASTEWATER

37 Forest Land Treatment of Wastewater in Clayton County, Georgia:
   A Case Study   393
   Wade L. Nutter

38 Penn State's "Living Filter": Twenty-three Years of Operation   406
   William E. Sopper

39 Forest Land Treatment with Municipal Wastewater in New England   420
   Sherwood C. Reed and Ronald W. Crites

40 Irrigation of Tree Plantations with Recycled Water in Australia:
   Research Developments and Case Studies   431
   H. T. L. Stewart, E. Allender, P. Sandell, P. Kube

41 Municipal Wastewater Renovation on a Coastal Plain, Slash Pine
   Land Treatment System   442
   Jane T. Red and Wade L. Nutter

42 Fourteen Years of Wastewater Irrigation at Bennett Spring
State Park   452
Don Barnett and Ken Arnold

CASE STUDIES: MUNICIPAL SLUDGE

43 Pack Forest Sludge Demonstration Program: History and Current
Activities   461
Charles L. Henry and Dale W. Cole

44 Silvigrow: Metro's Forest Sludge Application Program   472
John Spencer and Peter S. Machno

45 Municipal Sludge Application in Forests of Northern Michigan:
A Case Study   477
Dale G. Brockway and Phu V. Nguyen

46 Reclamation of Severely Devastated Sites with Dried Sewage
Sludge in the Southeast   497
Charles R. Berry

47 Land Treatment of Sludge in a Tropical Forest in Puerto Rico   508
Wade L. Nutter and Roberto Torres

48 The Sludge Application Program at the Savannah River Plant   515
J. C. Corey, M. W. Lower, and C. E. Davis

CASE STUDIES: INDUSTRIAL AND PULP AND PAPER

49 Pulp Mill Sludge Application to a Cottonwood Plantation   533
Walter J. Shields, Jr., Michael D. Huddy, and Sheldon G. Somers

50 Land Treatment of Chemical Manufacturing Wastewater   549
R. L. Kendall, W. G. Algiere, and W. L. Nutter

51 Applying a Resin Based Sludge to Forested Areas: A Case Study   556
Glenn L. Taylor and Ron D. Presley

Contributors   563

Index   567

Preface

The 1970s marked an important decade as worldwide attention was focused on water pollution problems. Innovative solutions were sought to recycle and utilize wastes instead of disposing of them in waterways and landfills. In the United States, the Federal Water Pollution Control Act of 1972 required both municipalities and industries to implement a high degree of wastewater treatment to abate stream pollution. The action resulted in a corresponding increase in the quantity of wastewater residues produced. The Act required serious consideration of land treatment as an alternative to dumping or disposing of wastewater and sludge materials. Today we find an increasing amount of these products effectively and safely recycled into the environment for the purposes of providing further treatment in a cost-effective manner, increasing site productivity, and for further uses such as groundwater recharge or streamflow augmentation. These programs have included application to agricultural crops, reclamation of disturbed sites, composting, and application to forests. The use of forest land as an alternative to other options in the reuse of municipal and industrial wastewaters and sludges has not received the research attention or public acceptance received by other land application alternatives. This is somewhat surprising in that 40% of the land area of the United States is forested. In addition, since forests are a step removed from the human food chain, theoretically there should be less risk to the public health in a forest application program compared with a similar program in agriculture.

To address the role that the forest can play in a land application program, a symposium was organized by scientists at the College of Forest Resources, University of Washington, and the School of Forest Resources, University of Georgia. This four-day symposium was held in June 1985 on the campus of the University of Washington as well as the Charles Lathrop Pack Experimental Forest, a field research site managed by the College of Forest Resources. Over 125 scientists, consulting engineers, regulators, and municipality and industry personnel attended the symposium, which was sponsored by the Washington Department of Ecology, the U.S. Environmental Protection Agency, the Municipality of Metropolitan Seattle, the USDA Forest Service, the Georgia Forestry Commission, and the Society of American Foresters. The program was a part of the continuing education program of the University of Washington College of Forest Resources and the Cooperative Extension Service, Washington State University.

It was the purpose of the symposium to examine thoroughly the forest alternative in a land treatment program. Every aspect of this complex program was addressed, including the regulatory constraints and opportunities, current state of knowledge on the risks involved, response of forest systems to application programs, and the response of the public to this use of a forested area. Case studies of forest land treatment programs throughout the United States were also reported.

This book contains the papers presented as well as papers from posters at the symposium. All papers were subjected to peer review and technical editing before publishing to make this book truly a contribution to the body of scientific knowledge on treatment of wastes in forest ecosystems.

Acknowledgments

On June 25-28, 1985, Forest Land Applications: An International Symposium on Forest Utilization of Municipal and Industrial Wastewater and Sludge was held at the University of Washington and Pack Forest. As an outgrowth of the symposium, this book defining the forest alternative for the treatment and utilization of municipal and industrial wastes has evolved. The symposium and this book have been supported by the Washington Department of Ecology, the U.S. Environmental Protection Agency, the Municipality of Metropolitan Seattle, the USDA Forest Service, the Georgia Forestry Commission, and the Society of American Foresters. The editors would specially like to acknowledge the Washington Department of Ecology for its contribution making the publication of this book possible.

A number of individuals have been responsible in making this symposium and book possible. Special thanks are given to the speakers and poster presenters who are authors of chapters in the book; the symposium planning committee:

Symposium Directors:	Dale W. Cole, University of Washington
	Wade L. Nutter, University of Georgia
Symposium Organizers:	Charles L. Henry, University of Washington
	Donald Hanley, Washington State University
	Roberta Chapman-King, University of Washington
	Jane T. Red, University of Georgia
Regulations and Acceptance:	Robert K. Bastian, U.S. EPA
Municipal Sludges:	Charles Berry, USDA Forest Service
Wastewater Effluent:	Wade L. Nutter, University of Georgia
Pulp and Paper Wastes:	Walter J. Shields, CH2M-Hill
Industrial Wastes:	Robert L. Kendall, Earth Systems Assoc., Ltd.

and especially the technical editor: Leila Charbonneau.

Program Overviews

Chapter 1

Forest Land Applications of Sludge and Wastewater

ROBERT S. BURD

ABSTRACT   Congress in passing the Clean Water Act intended that beneficial uses of sludges and liquid wastewater be promoted. In recent years the Congress, as well as the public, has become increasingly concerned about toxic substances. As a result, waste recycling is under much scrutiny. It is therefore necessary for government agencies to do a better job of convincing the public that their health can be protected while wastes are recycled. Part of the program of convincing the public to accept recycling of wastes on land should include a risk management component.

To paraphrase Mark Antony's funeral oration for Julius Caesar: "I have come to praise sludge not to bury it." In preparing to write this paper, I reread the oldest book on the subject of waste disposal that I could find in my personal library; it dates back over seventy years (Elsner 1912). The author wrote: "It seems an opportune time to present a discussion on utilizing sludge. In many cases, the question of utilizing sludge, even in thoroughly worked-out projects, has been left open." The book reveals that sludge became an issue in 1857 when sanitation authorities in England first proposed sewage treatment. Land application of untreated wastewater was common in those days. The author goes on to say, "The principal field for the use of settled sludge is as a fertilizer. In the future, the greater part will also be utilized in this way." He quotes a 1909 letter from Baltimore, Maryland, describing the use of sewage sludge to fertilize vegetable farms: "The odors in the vicinity of the sludge lagoons are very offensive, but, as far as known, they have not had an unfavorable effect on the health of those living on the farms. However, the nuisance from flies is considerable."

An even earlier history from Brockton, Massachusetts, identified in Elsner's book, also may have been prophetic. In 1890, the sludge was sold to farmers for $125 per year; in 1901 to 1906, $150 per year; but after 1909 it was given away. Today, we realize that there is economic value in waste products but the price must be competitive with other sources of nutrients and water.

Elsner concluded what we know today: the use of sludge for land application depends on (1) the cost of land, (2) the character of the sludge, and (3) the proximity of the disposal area to dwellings and the general character of the actual and prospective development in the neighborhood.

When Congress passed the Clean Water Act, it very much intended that the beneficial uses of sludges and liquid wastewater be promoted. The act has a number of references giving encouragement and financial incentives for recycling these substances to the land. That is an example of the legislation passed in the 1970s. In the 1980s we find that Congress, as well as the public, has become very concerned about toxic substances—wherever they may appear. As a result, waste recycling is more difficult because it has fallen in with bad company, such as Love Canal and other hazardous waste disposal sites. Therefore, the public demands to know that their health can be protected while wastes are recycled.

In reviewing land application projects, we must concentrate on the reduction of important environmental risk. But we should not get overzealous. Nothing erodes the public's tolerance of a regulatory agency more than the imposition of burdens that appear to have only petty results in terms of some substantive public benefit. At the same time, nothing erodes the public's faith in a regulatory agency more than the appearance that the agency, for whatever reason, is not acting aggressively in the public interest.

The Environmental Protection Agency has been discussing risk management a lot lately, and it certainly is a topic closely related to this symposium. Reducing risks—to human health and environmental values—is the basic reason we remove pollutants from the environment. By closely watching the movement of pollutants that result from regulatory options and calculating the resultant risks for each, we can ensure that our actions are indeed connected with a measurable, permanent good.

The risk management approach has some obvious problems. It is relatively easy to compare the risk of a single public health effect delivered via two different media. We can agree that a one in a million chance of getting cancer from drinking water is pretty much equal to the same chance of getting it through breathing something in the air. But what about comparing the chance of human disease with the chance of harming the marine environment? For example, if particularly toxic wastes are incinerated on land, there is always some residual risk to the surrounding human population. If they are incinerated at sea, that risk virtually disappears. But there is a quite small, though still calculable, possibility that something could happen to the incineration ship, thus causing unpredictable effects on marine organisms.

Dilemmas such as this cannot be solved without considerable information on risks, costs, and probabilities, and without the ability to respond flexibly, depending on what that information yields. Most important, that kind of information is needed in order to communicate to the public how the decision was made, and how all the factors involved were balanced.

Up to now, EPA's approach to land application of wastes has been conservative—involving little or no risk. But, this can be changed to a different risk level as long as the public judges the risks to be acceptable. But the public will not be satisfied, as they were in the past, with assurances that odors will be suppressed and traffic along the dusty road leading to the land application area will be controlled.

Land application projects must deal with three major issues: (1) Is the project environmentally safe? (2) Is the cost reasonable? (3) Can public acceptance be gained? Public acceptance, because of the toxics concern, is generally the most difficult of the three issues.

Those of us in governmental agencies need to gain public understanding of these difficult issues. Risk management decisions are ultimately very tough to make—they are decisions about "how safe is safe enough?" And in an environment of potential carcinogens, it is difficult for anyone to decide that a little bit of a carcinogen may be acceptable.

The temptation is to push for perfection—that is, as low as reasonably achievable—or not to accept any risk at all.

Ultimately, then, a government agency must decide on some course of action that will produce less than perfect control of a hazardous pollutant. Will the public then accept that decision? The answer will depend in part on how good a job is done to involve the public in the decision and how good a data base is presented to them. We need good monitoring, good siting of projects, and effective industrial waste pretreatment programs to remove toxics. We in the government agencies need to open our decision making to public scrutiny and to respond to questions and criticism openly. Such openness overcomes suspicions the public may have about the integrity of the agency's actions.

Land application decisions sometimes are complex and involve some uncertainties, but we can assist public involvement and understanding by (1) publicizing all relevant data through the media, public meetings, and availability of government officials, (2) developing easily understood informational materials and distributing them as widely as possible, and (3) providing opportunities for oral presentations at workshops and before other groups.

EPA has been part of the public acceptance problem. A few years ago, we publicly debated whether sewage sludge should be classified as a hazardous waste. The Resource Conservation and Recovery Act of 1976 (RCRA) fanned some of the emotionalism surrounding land application of any material considered as waste. Today, I think we're on track that only if the sludge fails the RCRA toxics test would it be regulated as a hazardous substance.

In Washington State, 85% of the sewage sludge is applied to the land. In Idaho and Oregon, a high percentage is also recycled. But if the different levels of government get too conservative, this percentage will drop and a good conservation practice will be discouraged.

Since forests are not a food-chain crop, there are fewer public health concerns with plant uptake of contaminants in sludges and wastewater. In addition, research indicates that some tree species are very tolerant to impurities that may present problems for some agricultural crops.

With our many cities located near forest lands, forest application may often be less expensive than other means of waste disposal. The recycling of nutrients and water will benefit the growers. Since forests are perennial, the scheduling of waste application is not as complex as it may be for agricultural uses. And the city may be able to get the land from the federal government.

The total commercial forest land in EPA Region 10 (including coastal Alaska) is 65 million acres. Of the total, 59% is federally owned, 34% is privately owned, and 7% is owned by the state, county, or municipality. In the state of Washington, 29% is under federal land ownership, in Idaho 64%, and in Oregon 52%. Therefore, there are many potential land application sites—particularly if you have any influence with the USFS or BLM. If forest land application can be successful anywhere, it's in the Northwest.

In conclusion, remember the old Chinese proverb: "One generation plants the trees, another gets the shade." But we have the opportunity to rewrite history. The proper use of sludges and wastewater in the forest will allow the same generation to plant trees and sit in their shade.

REFERENCE

Elsner, A. 1912. Sewage sludge: Treatment and utilization of sludge. McGraw-Hill, New York.

Chapter 2

Overview on Sludge Utilization

ROBERT K. BASTIAN

ABSTRACT   Utilization of the residual sludges produced as a result of treating municipal and certain industrial wastewaters is strongly encouraged by both Congress and EPA. The Clean Water Act (CWA) directs the Agency administrator to encourage waste treatment management that results in facilities for (1) the recycling of potential pollutants through the production of agricultural, silvicultural, and aquacultural products, (2) the reclamation of wastewater, and (3) the elimination of the discharge of pollutants. Regulations and guidance issued by EPA to implement provisions of CWA have attempted to encourage these practices and to identify good management practices to be followed. A formal Policy on Municipal Sludge Management issued by the Agency in June 1984 stated that EPA will actively promote "sludge management practices that provide for the beneficial use of sewage sludge while maintaining or improving environmental quality and protecting public health." Many sludge management practices have been investigated and employed on full-scale projects. In addition to efforts to recover energy from sludge through combustion and anaerobic digestion and methane recovery processes, sludge use includes a wide range of land application practices for recycling the nutrient and soil amendments, including land application to forest lands as well as to cropland, rangeland, parks, golf courses, and a variety of disturbed areas and marginally productive lands. Sludge reuse projects are under way in many parts of the country and involve sludges produced by both large and small POTWs and a variety of industries. Problems associated with forest land application projects are discussed in this paper, along with steps to ensure that good sludge management practices are followed.

Concern over environmental problems has led Congress to pass many laws that direct the U.S. Environmental Protection Agency (EPA) in dealing with waste management problems, including the control of sludge disposal and utilization practices (see Appendix). While EPA has actively encouraged the beneficial reuse of waste materials (including sewage sludge) as mandated by the Clean Water Act (CWA), the Resource Conservation and Recovery Act (RCRA), and other legislation, both Congress and EPA have recognized that certain waste materials and waste management practices have potential for causing environmental or public health problems. In developing guidelines and regulations to help control waste management practices, the Agency has actively sought acceptable levels of risk while encouraging the maximum recycling of wastes as a resource.

Currently the major sludge management alternatives for most municipalities are incineration (thermal conversion), landfill, and land application. Various techniques for processing sludge into products (e.g., soil amendments, organic fertilizer, dried bulking agents, fuel materials, aggregate, clay tile, and bricks) are available, but most of them eventually involve some form or combination of land disposal, land application, thermal combustion, or conversion of the end products. Ocean disposal of sewage sludge (both ocean dumping from vessels and discharges from ocean outfalls) has been phased out by many municipalities in recent years, in part because of secondary treatment provisions in the CWA and the ocean dumping requirements under the Marine Protection, Research and Sanctuaries Act (MPRSA) and its amendments. Although interest in ocean disposal practices has increased somewhat as a result of recent studies and assessments of deep ocean sites, court rulings, and ocean dumping site designation decisions, the issuance of future ocean dumping permits for sludge disposal will most likely require a clear demonstration of need and unavailability of environmentally sound alternatives.

Where possible, current EPA policy encourages the use of sludge management practices involving beneficial use in preference to strictly disposal options such as conventional landfilling and ocean disposal practices (EPA 1984). In addition to supporting a number of long-term research and demonstration projects and the development of detailed design guidance for various land application practices, EPA has supported and promoted pretreatment and source-control programs to improve sludge quality and such technologies as composting and lime stabilization. The Agency has also supported the development of dewatering systems, pyrolysis, chemical fixation, digestion, and other technologies to help improve energy recovery from thermal conversion systems, methane recovery from anaerobic stabilization systems, and the recovery of various potentially marketable by-products from sludge.

Well over 26 billion gallons of wastes are flushed daily into the nation's sewers by the 150 million Americans and at least 87,000 industrial contributors serviced through centralized wastewater collection and treatment facilities. The treatment of this wastewater results in the production of nearly 7 million dry tons of raw sewage sludge each year as well as treated wastewater effluent for disposal or reuse. Nearly 20% of this volume is produced by the POTWs (publicly owned and operated wastewater treatment works) in the country's ten largest urban centers. In addition, it is estimated that nearly 700,000 dry tons of septage are produced annually from the septic tanks that treat wastes from about 25% of the U.S. homes and well over 100 million dry tons of "other pollution control sludges" are produced each year as a result of controlling air emissions and treating both drinking water and industrial wastewaters (Table 1).

TABLE 1. Approximate mass of sludge generated by pollution control activities.

Type of Sludge	Total Sludge (dry Mg/yr x 10^6)
Air pollution control	
Electric utilities (fly ash and scrubber sludges)	50
Other	43
Drinking water treatment (surface and groundwater)	4
Industrial wastewater treatment*	16
Municipal wastewater treatment	7

Source: Based on data from JRB, Associates, Report to U.S. EPA, "Inventory of Air Pollution Control, Industrial Wastewater Treatment and Water Treatment Sludges," December 1983.

*Iron and steel, inorganic chemicals, food processing, and pulp and paper manufacturing account for 90% of the total.

The 1982 Needs Survey (EPA 1983a) identified over 15,500 existing POTWs, more than 78% of which treat flows of 1 mgd or less while fewer than 400 treat flows of more than 10 mgd (Table 2). Over 6,600 of the POTWs consist simply of wastewater treatment ponds, while some of the others involve rather sophisticated treatment processes.

Efforts to estimate what is currently being done nationwide with the sludge produced by POTWs are greatly hampered by the type of data available. For example, although the 1982 Needs Survey contains extensive information on sludge processing used by POTWs (Table 3), complete information for the end sludge use and disposal practices is included for only about half of the POTWs (Table 4). However, data generated from a random survey of 1,011 POTWs (Table 5) conducted for EPA in 1980 do account for all the sludge end-use and disposal practices of about 6.5% of the POTWs nationwide, which produced about 35% of the total municipal sludge generated at that time. This survey found that 33% of the POTWs surveyed used two or more sludge use and disposal practices.

Although comprehensive survey data are not available, it is estimated that as much as 40% of the sewage sludge currently produced is applied to the land as a soil amendment or fertilizer in one form or another and the remainder incinerated, used as landfill, or disposed of in the ocean. Large volumes have also been stored for long periods in la-

TABLE 2. Estimated municipal sludge production in 1982 by POTW size.

POTW Size (mgd)	Number of POTWs	Sludge Produced (dry Mg/yr)	% of Total
0-2.5	14,168	1,070,829	17
2.5-5	631	463,954	8
5-10	352	529,601	9
10-20	187	560,230	9
20-50	125	832,406	14
50-100	40	610,282	10
>100	41	2,091,847	34
Totals	15,544	6,159,149	

Source: U.S. EPA 1983a.

TABLE 3. Status of major sludge processing methods at municipal wastewater treatment facilities.

Sludge Handling Methods	Now in Use		Under Construction	
	Number	mgd*	Number	mgd*
Aerobic digestion	3,400	28,662	134	452
Anaerobic digestion	4,189	22,449	90	480
Composting	28	1,050	10	51
Sludge lagoons	611	4,399	26	102
Heat treatment	163	3,659	10	690
Chlorine oxidation	40	402	5	23
Lime stabilization	74	978	9	63
Wet air oxidation	51	856	1	10
Air drying	6,728	13,566	216	301

Source: U.S. EPA 1983a.

*Projected design flow of wastewater treatment plants using each sludge method.

TABLE 4. Major municipal sludge disposal and utilization practices.

Practice	Number of POTWs	Estimated Sludge Produced (dry Mg/yr)	% of Total Sludge	% of Total Facilities
Landfill	5,613	2,219,430	32	36
Incineration	306	1,207,734	18	2
Land application	1,722	873,151	13	11
Ocean disposal	41	483,488	7	1
Other	183	305,330	4	1
Missing data	7,679	1,754,360	26	49
Totals	15,544	6,843,493		

Source: U.S. EPA 1983a.

TABLE 5. Estimated distribution of municipal sewage sludge in 1980, by management method and treatment plant size.

Practice	Small POTWs (1 mgd or less)	Medium POTWs (1 to 10 mgd)	Large POTWs (over 10 mgd)	Percent of Total
Landfill	31	35	12	15
Thermal processing (e.g., incineration)	1	1	32	27
Land application	39	39	21	24
To food-chain crops	(31)	(22)	(10)	(12)
To nonfood-chain crops	(8)	(17)	(11)	(12)
Distribution and marketing	11	18	19	18
Ocean disposal	1	--	4	4
Other (e.g., lagoons, stockpiles)	17	12	12	12

Source: Based on data collected for the U.S. EPA, Office of Solid Waste, in a random survey of 1,011 POTWs conducted in 1980, covering 6.5% of all POTWs and 35% of total sewage sludge mass generated.

goons, but eventually most of these lagoon sludges are applied to the land or otherwise disposed.

LAND APPLICATION PRACTICES FOR SLUDGE RECYCLING

Many land application practices for recycling sewage sludge have been investigated and employed to date, including application to highly productive agricultural and forest lands, marginally productive areas, parks, golf courses, and a variety of disturbed areas. Sludge reuse projects are under way in many large metropolitan areas, including Washington, D.C., Philadelphia, Chicago, Milwaukee, Denver, San Diego, and Seattle, as well as in thousands of smaller cities and towns across the country, especially in the Midwest. Such systems are prime examples of the basic land treatment, recycle-reuse concepts that have been strongly encouraged by Congress and EPA. The research and demonstration experience and guidance available covering such practices (Hornick et al. 1984, U.S. EPA, 1983b, Page et al. 1983, Sopper et al. 1982, Bledsoe 1981) and the growing number of successful projects across the country clearly indicate the potential benefits of land application of sludge.

Land application practices have been implemented by many rural communities with adequate available land, and agreeable landowners and neighbors. In some cases "dedicated" or publicly owned and controlled sites have been used, but a more common practice is application to privately owned and managed farmland, strip mines, and so forth. Nationwide, well over 30% of the smaller communities have applied their sludge to the land for over forty years.

For years many communities simply stockpiled dried sludge and allowed the public to haul it away for their own use. Now there is a growing interest in the potential for marketing sludge to farmers or others as an organic fertilizer and soil amendment. A number of communities have been heat drying, composting, or otherwise processing their sludge prior to marketing or giving it away as a soil amendment for many years. Others have established programs for applying liquid sludge to cropland (and in the case of Seattle to forest land) at carefully predetermined rates, and in at least some cases farmers pay for the sludge or sludge application service.

The potential for overseas shipment of municipal sewage sludge materials has also been extensively examined. Proposals include sludge shipment in ore boats, oil tankers, or sludge ships to such areas as the Bahamas, Haiti, and other Caribbean areas, several Central and South American countries, Africa's Gold Coast, Egypt, the Middle East, and Asian countries for use as soil conditioners. To date no such projects for large-scale overseas shipment of sewage sludge have been implemented, although negotiations between certain project promoters and municipalities are continuing.

NUTRIENT AND SOIL AMENDMENT BENEFITS OF SLUDGE

Although their total nutrient content is substantial (Table 6), municipal wastewater, sewage sludge, and septage currently supply a very small amount of the nitrogen, phosphorus, and potassium applied to cropland. Total reuse of the nutrients calculated as present in these organic wastes would amount to about 1.4 million tons of the nitrogen, 300,000 tons of the phosphorus, and 600,000 tons of the potassium—or about $950 million worth of nutrients—currently supplied by commercial fertilizer sources (assumes

TABLE 6. Estimated total amounts of primary plant nutrients* in municipal wastewater, sludge, and septage.

Nutrient	Commercial Fertilizer Use in 1978	Raw Waste-water**	Treated Effluent	Sewage Sludge	Septage
		(1,000 Mg/yr)			
Nitrogen	10,642	1,534	1,012	399	18
Phosphorus	2,453	275	204	72	11
Potassium	4,844	590	572	18	3

Source: Bastian et al. 1982.

*Assumes 66% total nitrogen in raw wastewater remains in treated effluent and ends up in sewage sludge, 74% total phosphorus remains in treated effluent and 26% in sewage sludge, 97% potassium remains in treated effluent and 3% in sewage sludge.

**Assumes 39 mg/l nitrogen, 7 mg/l phosphorus, 15 mg/l potassium.

the price of these nutrients at $300/ton for nitrogen, $800/ton for phosphorus, and $200/ton for potassium). About 35% of this total is associated with the sludge and septage. In addition, treated effluents have considerable value as a source of irrigation water in areas of limited water supplies; both sludges and septage can be converted into excellent organic soil amendment materials by composting or other processing; and municipal wastewater, sewage sludge, and septage all contain many micronutrients in addition to nitrogen, phosphorus, and potassium.

The potential benefits from recycling the organic matter and nutrient resources in sewage sludge through various land application practices have been well demonstrated and have led to an increased use of these practices in many parts of the country. Not only can land application help the municipalities by serving as a cost-effective sludge "disposal" technique, it can also serve the farmer or other landowners by improving soil characteristics, reducing fertilizer costs, and increasing productivity. Although sludge is not a high grade fertilizer, the $30 to $60/dry ton worth of organic nitrogen and phosphorus in typical sewage sludges alone can make land application worthwhile as a partial replacement for commercial inorganic fertilizers in certain cases. Sludge compost has been effectively used as a substitute for topsoil and peat for certain horticultural applications, showing yield improvement valued at $35 to $50/dry ton over the usual potting media (Hornick et al. 1983). The potential value of sludge-related materials (liquid sludge, sludge cake, compost, etc.) in the production of field crops, vegetables, nursery crops and ornamentals, forages, and sod production and maintenance has been clearly demonstrated in many locations across the country.

Generally, the greatest cost savings may be realized as a result of using the nutrients from sludge on less productive soils. However, the increase in operational costs associated with sludge use in agriculture can be more than offset, especially for crop rotations requiring relatively large amounts of nitrogen, if site conditions do not require additional contouring or other expensive measures for runoff control or other substantial investments as a result of sludge use. A study of the utilization of sewage sludge and effluent on selected agricultural crops in one area of Oregon found that the return per acre from sludge use compared with traditional fertilizer sources ranged from a loss of $6 to an increase of $15 per acre, depending on the crop rotation involved, previous soil management practices, soil type, and level of sludge application (Schotzko et al. 1977). These calculated economic results from sludge use were limited to the fertilizer savings after subtracting costs for new production practices when the sludge was available to the farmer with no cost for application.

Municipal sludges are also being used as an effective topsoil substitute, soil conditioner, and organic fertilizer in forestry production and in reclamation, revegetation, and stabilization of strip mines, mine spoils, and tailings piles, construction sites, quarries, dredge spoils, borrow pits, and other drastically disturbed areas (Sopper et al. 1982, Bledsoe 1981). Municipal sludges have even been used to help stabilize shifting beaches and sand dunes, and to help create near-shore islands for recreational uses. Although limited to areas where transport to such sites is cost effective and locally acceptable, reclamation practices offer an opportunity to solve several environmental problems at one time. Both experimental and full-scale projects using liquid, dewatered, or composted sludges for reclamation uses have been undertaken in Illinois, Ohio, Pennsylvania, Maryland, Virginia, West Virginia, South Carolina, Alabama, Florida, Wisconsin, Washington, Colorado, and Montana.

FOREST LAND APPLICATION POTENTIALS

In recent years there has been a growing interest in the potential for increasing productivity in managed forest through fertilization, irrigation, and other more intensive tree farming practices. The use of sludge to help shorten wood production cycles and increase production (especially on marginally productive soils), as well as revegetate and stabilize areas that have been clearcut or devastated by forest fires, could play an important role in achieving increased forest productivity, at least in certain locations (Henry and Cole 1983, Bledsoe 1981).

The area classified by the U.S. Department of Agriculture as forest land occupies about 662 million acres (265 million ha): over 285 million acres (114 million ha) of federally controlled and 376 million acres (150 million ha) of nonfederal forest land (Table 7). About 75% of the commercial nonfederal forest land is primarily under control of farmers and other private owners (commercial forest land is forest land that produces or can produce more than 20 cubic feet per acre per year of industrial wood under proper management and has not been withdrawn from timber production). Industry owns nearly 18% of this land while state, county, and municipal governments own about 8% (Table 8). Nearly 30% of the noncommerial forest land is in nonfederal ownership. The largest part of this is held by farmers and other private owners. There are an estimated 4 million private owners of noncommercial forest land, with 72% of the holdings at 500 acres or less (USDA, Soil and Water Resource Conservation Act, Part 1, 1980).

Large acreages of forest are harvested or devastated by forest fires, landslides, or other natural disasters each year and require reforestation if full production or recovery is desired within a reasonably short time. The National Forest System, which occupies some 197 million acres (79 million ha) and harvested well over 10 million board feet of lumber in fiscal year 1979 alone, had over 1.6 million acres (0.6 million ha) requiring reforestation. It would appear that a continuing supply of forested areas disturbed by harvesting and natural disasters will be available as possible sites where sludge could be beneficially used as part of forest stabilization and reforestation activities as well as to increase forest productivity and shorten wood production cycles.

A wide range of land reclamation and biomass production projects (Sopper et al. 1982), many involving either direct forest land application or reforestation efforts, have been investigated to date (see Figure 1). Also, a growing amount of research and demonstration experience and guidance information addresses various forest land application practices. In their recent review of the role of forests in sludge and wastewater utilization programs, Cole et al. (1983) noted the following reasons for considering forested sites as potential candidates for the disposal and reuse of sewage from a municipal treatment plant: "(1) In many regions of the U.S. extensive acreages of forest land are potentially available for such a program. Forests occupy 40 percent of the landscape in the contiguous 48 states. (2) Forests are typically located in the better drained sites and not subject to the periodic flooding of alluvial agricultural areas. (3) Many of the forests of the U.S. are markedly deficient in major nutrients that are found in municipal sludge and wastewater, especially nitrogen and phosphorus. A lack of adequate nutrition is the main factor limiting forest productivity in this country. (4) Since forests are not food chain crops, many of the public health concerns and land application regulations should not be as critical as those associated with agricultural sites. (5) Forest soils theoretically have properties well suited to receive sludge and wastewater additions, including a great deal of organic carbon which will immobilize available nitrogen, a high infiltration rate which should minimize the potential for surface runoff, and a perennial root system which should allow for year around uptake of available nutrients."

A considerable amount of research has been undertaken in the United States and elsewhere for over a decade examining the effects of both wastewater and sludge applications to forested sites (Cole et al. 1983, Bledsoe 1981). The results suggest that when proper management practices are followed, sludge can be effectively used to increase forest productivity without causing significant environmental problems. By limiting ap-

TABLE 7. Federal and nonfederal forest land in 1977, by regions, subregions, and states.

	Federal	Nonfederal	Total
	(1,000 acres)		
North			
Northeast			
Connecticut	2	1,418	1,420
Delaware	5	360	365
Maine	228	16,520	16,748
Maryland	155	2,160	2,315
Massachusetts	60	2,756	2,816
New Hampshire	694	3,976	4,670
New Jersey	94	1,967	2,061
New York	200	15,445	15,645
Pennsylvania	557	14,349	14,906
Rhode Island	7	303	310
Vermont	275	3,931	4,206
West Virginia	964	9,805	10,769
Total	3,241	72,990	76,231
North Central			
Illinois	340	3,026	3,366
Indiana	364	3,533	3,897
Iowa	26	1,483	1,509
Michigan	3,358	15,322	18,680
Minnesota	2,980	13,807	16,787
Missouri	1,414	10,829	12,243
Ohio	206	5,860	6,066
Wisconsin	1,642	13,252	14,894
Total	10,330	67,112	77,442
Total for North	13,571	140,102	153,673
South			
Southeast			
Florida	2,319	12,146	14,465
Georgia	1,498	21,567	23,065
North Carolina	1,825	16,818	18,643
South Carolina	757	10,770	11,527
Virginia	2,081	13,237	15,318
Total	8,480	74,538	83,018
South Central			
Alabama	840	19,792	20,632
Arkansas	2,661	14,069	16,730
Kentucky	936	10,645	11,581
Louisiana	740	12,594	13,334
Mississippi	1,299	14,416	15,715
Oklahoma	323	4,933	5,256
Tennessee	1,061	11,639	12,700
Texas	807	9,240	10,047
Total	8,667	97,328	105,995
Total for South	17,147	171,866	189,013

TABLE 7. Continued

	Federal	Nonfederal	Total
		(1,000 acres)	
Rocky Mountain and Great Plains			
Great Plains			
Kansas	786	857	857
Nebraska	45	439	484
North Dakota	35	368	403
South Dakota	1,057	333	1,390
Total	1,208	1,926	3,134
Rocky Mountains			
Arizona	9,853	1,804	11,657
Colorado	14,961	3,343	18,304
Idaho	16,978	4,229	21,207
Montana	16,324	6,343	22,667
Nevada	5,352	229	5,581
New Mexico	10,559	3,426	13,985
Utah	11,446	1,066	12,512
Wyoming	8,523	1,163	9,686
Total	93,996	21,603	115,599
Total for Rocky Mountain and Great Plains	95,204	23,529	118,733
Pacific Coast			
Pacific Northwest			
Alaska	112,245	6,900	119,145
Oregon	18,698	10,062	28,760
Washington	9,474	12,413	21,887
Total	140,417	29,375	169,792
Pacific Southwest			
California	18,819	9,857	28,676
Hawaii	0	1,443	1,443
Total	18,819	11,300	30,119
Total for Pacific Coast	159,236	40,675	199,911
Caribbean	28	428	456
Total for United States and Caribbean	285,186	376,600	661,786

Source: USDA, Soil and Water Resource Conservation Act, Part 1, 1980.

TABLE 8. Commercial forest land acreage and ownership, 1974.

Area	Acreage	Land Ownership (10^6 acres)				
		Federal	State and local	Industry	Farm	Private
Northwest	177.9	12.3	19.6	17.6	51.0	77.4
Southwest	192.5	14.6	3.0	35.3	65.1	74.8
West	129.3	80.6	6.4	14.4	15.0	12.8
Totals	449.5	107.5 (21.4%)	29.0 (5.8%)	67.3 (13.5)	131.1 (26.2%)	165.0 (33.0%)

Source: U.S. Forest Service 1974 and Burwell 1978.

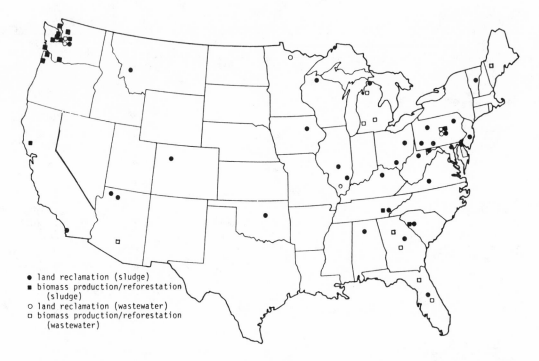

Figure 1. Location of some reclamation and biomass production/reforestation projects using municipal wastewater or sewage sludge (from Bastian et al. 1982).

plication rates and slopes to reasonable levels (typically application rates to 20 dry tons per acre and slopes to 30%) and taking steps to control other important siting factors, problems such as excessive nitrate leaching and movement of pathogens or other sludge constituents into surface waters or groundwaters can be avoided. Although significant differences in plant uptake and accumulation of heavy metals have been documented under experimental conditions, the uptake levels observed under field conditions have been rather small and apparently have not resulted in phytotoxic conditions. While not all tree species have responded well under all test conditions, excellent growth responses—even exceeding those achieved by commercial forest fertilization trials—have been noted for Douglas-fir in response to sludge applications to established production sites. It appears that, at least in the Pacific Northwest, the increased growth responses for Douglas-fir from a single sludge application at the rate of 20 dry tons per acre can be maintained for at least five years, while minimizing the potential for nitrate leachate to groundwater to exceed drinking water standards. Although not all the results from research studies involving forest land application of sludge have been favorable, it is clear that the problems identified early in these studies (many apparently associated with excessive applications of sludge), such as significant nitrate leaching, excessive competition by grass and weed species, browsing and vole damage, differential response to sludge nutrients by various tree species, and inadequate equipment for operating in many forested sites, have now been largely resolved.

Major Problems Facing Forest Land Application Projects

Municipal sewage sludge constituents can vary widely (Table 9), depending on such factors as the type and amount of industrial discharges to the POTW and the sludge

TABLE 9. Common sludge constituents.

Type of Sludge	Conventional Organics	Conventional Inorganics	Toxic Metals	Toxic Organics	Nutrients	Hazardous Waste
Air pollution control	None	High	Medium	None	Low	Rarely
Drinking water treatment	Low	High	Low	Low	Low	No
Industrial wastewater	Low to high	Low to high	Low to high	Low to high	Low to high	Several catagories listed
Municipal wastewater treatment	High	Low to high	Low to high	Low to high	Low to high	Rarely

Source: Based on data from JRB, Associates, Report to U.S. EPA, "Inventory of Air Pollution Control, Industrial Wastewater Treatment and Water Treatment Sludges," December 1983.

treatment processes used, making it difficult to generalize about the physical, chemical, and biological properties of sewage sludge. In reality almost anything can be found in sludge. Some of the physical, chemical, and biological characteristics and constituents of municipal sludge that are of potential importance or concern regarding sludge management are listed in Table 10. Ranges and typical values for a number of these properties, as reported or often cited in the literature, are shown in Table 11 for several types of sewage sludge. Note especially the wide ranges shown for most chemical constituents. The variability in content of heavy metals, toxic organics, and pathogens reported in sludges has been a major drawback in allowing greater use of these materials as a source of nutrients, organic soil amendments, or feed supplements.

Although the potential benefits from recycling the organic matter and nutrient resources in sewage sludge through various land application practices have been well demonstrated and have led to an increased use of these practices in many parts of the country, questions are frequently raised when projects involving land application of sewage sludge are proposed. The issues generally center on concerns that pathogens or toxic chemicals, which may be present in the sludge at varying levels, may contaminate the soil, nearby surface waters or groundwaters, or plant and animal life on sites that have received sludge. Differential responses of various pathogens to specific stabilization processes, toxic organic compounds to breakdown by soil microbes, and plants to uptake of heavy metals applied with sludge clearly add to the concerns over the potential for harm to public health or the environment.

Clearly, the doubts of the public and the concerns of officials about adding potentially toxic substances and pathogens to productive farmlands, watersheds, or other land application sites must be closely examined in terms of protecting human health, crop quality, environmental impacts, and future land productivity. Concerns are also often expressed about potential odors, aesthetic problems, increased traffic and noise, and their potential impact on neighboring property values. Also, there is often a general psychological opposition to the idea of recycling human wastes on land, since many people have developed an out-of-sight attitude to waste management problems which generally favors the more highly engineered approaches to sludge disposal.

While many persons consider forests as being removed from the concerns typically associated with sludge application to farmland, the application of sludge to forest lands raises its own special concerns. In addition to the usual concerns over effects on water quality, the question of adverse impacts on wildlife as well as heavy metal uptake by edible mushrooms and berries is often raised. The possibility of limitations being placed

TABLE 10. Sludge characteristics and constituents of importance or potential concern.

Physical Characteristics

Volume	Dissolved solids
Density	Volatile solids
Particle size	Heat value (heat of combustion)
Total solids	

Biological Constituents

Bacteria	Fungi
Salmonella spp.	Aspergillus spp.
Escherichia coli	
Pseudomonas spp.	

Virus	Parasites and Protozoa
Enteroviruses	E. histolytica
Polioviruses	Ascaris spp.
Coxsackie viruses	Toxacara spp.
Echoviruses	Trichluris spp.
Hepatitis viruses	Hymenolepis spp.
Adenoviruses	
Reoviruses	

Chemical Constituents

Organic carbon	DDT	Benzene
Total nitrogen	DDD	Chlorobenzene
Ammonia-N	DDE	1,4-dichlorobenzene
Nitrate-N	Dieldrin	1,2-dichlorobenzene
Total phosphorus	Aldrin	1,3-dichlorobenzene
Total potassium	Chlordane	1,2,4-dichlorobenzene
Total sulfur	Heptachlor	1,2,3-trichlorobenzene
Sodium	Lindane	1,3,5-trichlorobenzene
Calcium	Toxophene	Hexachlorobenzene
Aluminum	Endrin	Benzidine
Arsenic	Bis(2-ethylhexyl)phthalate	Phenol
Beryllium	Di-n-butyl phthalate	Pentachlorophenol
Boron	Chloroethane (ethyl chloride)	2,4-dichlorophenol
Cadmium	Methyl chloride	4-chloro-m-cresol
Chromium	Methyl bromide	(4-chloro-3-methylphenol)
Copper	Vinyl chloride	Toluene
Cyanide	1,1-dichloroethylene	Naphthalene
Fluorine	1,2-dichloropropane	2-chloronapthalene
Gold	1,3-dichloropropane	Acenaphthylene
Iron	Pentachloroethane	Anthracene
Lead	Hexachloroethane	Benzo(a)anthracene
Magnesium	Carbon tetrachloride	(1,2-benzanthracene)
Manganese	Dichlorodifluoromethane	Indeno(1,2,3-c,d)pyrene
Mercury	Dichlorobromomethane	Benzo(a)fluoranthene
Molybdenum	Trichlorofluoromethane	PCBs
Nickel	Tetrachloroethylene	(Polychlorinated biphenyls)
Palladium	Trichloroethylene	
Platinum	Hexachlorobutadiene	
Selenium	Chloroform	
Silver	Bromoform	
Tin	1,1,2,2-tetrachloroethane	
Vanadium	1,1,1-trichloroethane	
Zinc	1,1,2-trichloroethane	

TABLE 11. Properties of various sludges.*

	Raw Primary Sludge		Raw Activated Sludge		Trickling Filter Humus		Digested Sludge	
	Range	Median	Range	Median	Range	Median	Range	Median
Total solids (TS) (%)	3-7	5	1-2	1	2-7	4	6-12	10
Volatile solids (% TS)	60-80	70	60-80		50-80		30-60	40
Thermal content (kJ/kg) x 10⁴	1.6-2.3						0.72-1.6	
Nutrients (% dry wt)								
Nitrogen	1.5-8	3	4.8-6	5.6	1.5-5	3	1.6-6	3.7
Phosphorus	0.8-2.7	1.6	3.1-7.4	5.7	1.4-4	3	0.9-6.1	1.7
Potassium	0-1	0.4	0.3-0.6		0-1		0.1-0.7	
pH	5-8	6					6.5-7.5	7
Alkalinity (ppm CaCO₃)	500-1,500	600					2,500-3,500	3,000
Metals (ppm dry wt)								
Arsenic							3-30	14
Cadmium							5-2,000	15
Chromium							50-30,000	1,000
Copper			385-1,500	916			250-17,000	1,000
Lead							136-7,600	1,500
Mercury							3.4-18	6.9
Nickel							25-8,000	200
Selenium							1.7-8.7	
Zinc			950-3,650	2,500			500-50,000	2,000
Persistent Organics (ppm dry wt)								
PCBs							1.2-105	3.2
Chlordane							3-30	
Dieldrin							0.3-2.2	0.16
Pathogens								
Virus (PFU/100 ml)	11.0-11.4	7.9				11.5		0.85
Coliform (10⁶/100 ml)	2-2.8							0.4
Salmonella (per 100 ml)	74-23,000	460				93		29
Pseudomonas (per 100 ml)	1,100-24,000	46,000				11,000		34

Source: Adapted from National Academy of Sciences 1977.

*Raw primary sludge results from sedimentation of wastewater solids, activated sludge from biomass of suspended microorganisms, and trickling filter humus from biomass of attached microorganisms. Stabilization of organic matter in these sludges by aerobic or anaerobic biological processes produces digested sludge. Levels of metals, persistent organics, and pathogens affect the reuse of sludges. Digestion reduces virus and bacteria from levels found in raw sludges, but does not affect metals or persistent organics. Nutrient content of digested sludge can be used by crops. Thermal content determines how easily sludge can be oxidized after sufficient dewatering.

on public access to areas that receive sludge is also seen as being in conflict with the traditional recreational uses of forests. The fact that not all tree species respond favorably to sludge applications and that much less field experience and research dealing with forest land application exists than for agricultural uses of sludge suggests to some that sludge use in forests is more speculative.

Public acceptance by landowners and their neighbors is a key to successful projects involving land application of municipal sludge. Odors from poorly managed sludge management systems and perceived odors from anything that has to do with sewage and sludge may well be the biggest problem facing the successful establishment and operation of land application projects. General reluctance of rural areas to receive urban waste (at least until adequate economic or other incentives are offered) can also be a significant factor. "What do I get out of it" and "Not in my backyard" (NIMBY) attitudes often become apparent when land application projects are proposed.

Such factors can lead to the passage of what appear to be overly restrictive regulations, special land use ordinances or zoning restrictions, and extensive monitoring and record keeping requirements. Concerns that additional restrictions may be placed on land application practices once they begin, excessive nuisance claims may be filed, contractor performance may lapse, or other factors may lead to increased operating costs and a loss of public and regulatory acceptance have often prevented serious consideration of land application practices in areas where they were clearly appropriate.

Land application systems are also faced with the usual array of equipment and operating problems associated with hauling and spreading large volumes of liquid or dewatered organic materials—not unlike those faced by livestock producers or food processors involved in land application of their wastes.

Actions Taken to Deal with Land Application Problems

As a sludge disposal technique, land application of sewage sludge has been practiced for many years in this country and overseas. Only in recent years, however, have the necessary research and monitoring studies been undertaken to develop sound design guidelines for recycling sludge on the land. Appropriate management practices have been developed so that land application systems will be properly designed and operated from the standpoint of public health and environmental impacts, and this also helps protect the long-term productivity of the sites to which the sludge is applied. Many of these practices have been closely monitored and carefully evaluated for extended periods. The results have led to the publication of numerous scientific reports and state-of-the-art symposium proceedings (Page et al. 1983, Sopper et al. 1982, Bledsoe 1981). Several universities, states, and federal agencies have issued detailed guidelines for the proper use of municipal sludge as a soil conditioner and fertilizer (Pennsylvania State University 1985, Hornick et al. 1984, U.S. EPA 1983b, WDOE 1982).

Concerns over the levels of contaminants in sludges led to the issuance of regulations by EPA in 1979 (40 CFR 257), and by many states, which establish loading limits for certain contaminants and require specific levels of treatment to ensure adequate pathogen reduction in sludges that are land applied. Probably further regulations will be established in this area. However, through recent efforts to improve source control and industrial pretreatment, many POTWs have been able to improve greatly the quality of their sewage sludge and as a result its acceptability for utilization. For the most part, the highly contaminated sludges often cited in the literature, and represented by the high end of the range for many of the chemical contaminants shown in Table 11, are now a thing of the past. Design guidance and acceptable management practices have also been developed that will allow most sludges to be land applied in one manner or another if proper controls are implemented (Reed and Crites 1984, U.S. EPA 1983b, Parr et al. 1983, Overcash and Pal 1979).

Sludge usually must be stabilized before application to the land in order to reduce the numbers of pathogens and amount of odor-causing volatile organic matter present in raw sludge. Many processes have been successfully used, including composting, heat treatment, digestion, long-term storage, and chemical stabilization. However, efforts undertaken to date to develop processes that effectively remove chemical contaminants from sludges have generally proved to be very expensive, leaving source control and pretreatment as the most effective means of controlling the levels of most chemical contaminants.

Stabilized liquid sludge can be sprayed directly on the soil surface, incorporated into the upper layer by plowing or disking, or injected beneath the soil surface with specially desired injection systems. The equipment to apply sludge by such methods is now generally available even to the smallest communities. However, when there are considerable distances between the treatment plant and application sites, sludges are often dewatered in drying beds or with the aid of vacuum filters, centrifuges, presses, or other devices to concentrate the sludge solids and lower sludge transportation costs. Specialized equipment and management procedures have even been developed to help make forest land application of sludge more efficient and easier to control.

Many communities rely on interested farmers or other landowners as a continuing source of land application sites, while others have purchased land in order to ensure that application sites are available when needed. In some areas the treatment plant operators or their contractors must seek out cooperative landowners, while in others ample potential land application sites have been found simply by answering inquiries that have come to the treatment plant in response to local advertisements of a planned or ongoing land application program. Written agreements or contracts between the sludge applier (the city or its contractor) and the landowners are often used to help avoid uncertainties about future responsibilities, liability, and record keeping. Public education programs and the use of local advisory panels, demonstration projects, and so forth, aimed at providing factual information to help counteract rumors and vague apprehensions can result in the identification of interested landowners as well as lead to improvements in public acceptance. Special demonstration projects, participation in local fairs, university research programs and farm science reviews, and hosting or participating in farm tours set up through local soil and water conservation districts, granges, farm bureaus, or other farm organizations can also go a long way toward gaining greater farmer and landowner interest in sludge use. In some cases, communities or private companies add nonsludge sources of nutrients in order to enrich the nutrient content of their sludge and to attract greater interest in the use of their high analysis organic fertilizer "products."

If efforts to prevent impacts (e.g., careful siting of projects, truck routing, odor and noise control measures, spill prevention, etc.) fail, various means of contingency planning as well as mitigation measures can be employed to help deal with problems that may occur as a result of land application operations. In some cases compensation for unavoidable impacts may be appropriate (e.g., payments for correcting any damage to local highways, culverts, bridges, etc.); in other cases offsetting benefits may be feasible (e.g., creating public parks and recreation areas, generating local jobs and business opportunities, paying local taxes, etc.). Some other contingency management measures that have been effective include the posting of performance bonds, funding of independent project monitoring, the use of liability insurance, and setting up "hotlines" to facilitate access to project information.

CONCLUSIONS

It appears that certain types of forested lands can be used safely and effectively in land application of sludge programs. Such projects have been implemented in a number of locations using management techniques that have adequately protected public health while dramatically increasing tree growth rates. The potential silvicultural benefits appear to be greatest in areas of naturally poor production—such as highly porous and relatively infertile soils, or areas disturbed by construction or mining activities—where dramatic tree growth improvements have been demonstrated as a result of single applications of sewage sludge.

To encourage greater use of forested systems for sludge management, the results of research and development efforts, as well as past experience gained from operating projects, need to be made more widely available and applied to the design and implementation of future projects. Because many of the early problems faced by researchers undertaking projects involving sludge application to forest areas have not been resolved and good management techniques developed, it also appears to be time to reach out for greater involvement by private forestry companies. Such involvement also should strengthen the acceptability of—if not develop a demand for—sludge use in forestry and might help avoid some of the political, institutional, and public acceptance problems often faced by other land application practices.

Finally, in an effort to guide the future implementation of the Agency's sewage sludge regulatory and management programs, EPA recently issued a Policy on Municipal Sludge Management which states that the Agency will "actively promote those municipal sludge management practices that provide for the beneficial use of sludge while maintaining or improving environmental quality and protecting public health." It is hoped that this policy will help set the stage for greater EPA assistance to both state and local authorities in creating a climate that can assist in the establishment of land application projects that are well designed and operated.

APPENDIX

Major Federal Legislation Concerning Municipal
Wastewater Treatment and Sludge Management

FEDERAL WATER POLLUTION CONTROL ACT (FWPCA), as amended in 1972 (PL 92-500), 1977 (PL 95-217; the *Clean Water Act*), and 1981 (PL 97-117; the *Municipal Wastewater Treatment Construction Grant Amendments*), focuses on the restoration and maintenance of the chemical, physical, and biological integrity of the nation's waters. Research, standards and enforcement, water quality planning, and construction grants program authorities are included. They cover control of both point and nonpoint sources of water pollution. These laws authorize federal funding for the planning, design, and construction of publicly owned wastewater treatment works (POTWs), including sludge management facilities. They also authorize the issuance of comprehensive sewage sludge management guidelines and regulations, the issuance of National Pollution Discharge Elimination System (NPDES) permits for point source discharges, and the development of areawide waste treatment management plans, including best management practices (BMPs) for nonpoint sources of water pollution, and require the development and implementation of pretreatment standards for industrial discharges into POTWs. The 1977 and 1981 amendments added several important waste management provisions, including special incentives for greater use of innovative and alternative waste treatment technologies and methodologies, broad authority to regulate sewage sludge management practices, pretreatment credits for industrial dischargers to POTWs, and the opportunity for coastal communities to apply for modified discharge permits which would allow for less than secondary treatment for discharges into marine waters.

THE SOLID WASTE DISPOSAL ACT, as amended in 1976 (PL 94-580; the *Resource Conservation and Recovery Act* [RCRA]) and 1984 (PL 98-616; the *Hazardous and Solid Waste Amendments*), focuses on the regulation of solid waste management practices to protect human health and the environment while promoting the conservation and recovery of resources from solid wastes. Technical and financial assistance, training grants, solid waste planning, resource recovery demonstration assistance, and hazardous waste regulatory program authorities are included. The key aspect of RCRA is the comprehensive regulatory system to ensure the proper management of hazardous waste. RCRA also provides for technical and financial assistance to state, local, and interstate agencies for the development of solid waste agencies and solid waste management plans. In addition, it prohibits open dumping of wastes; promotes a national R&D program for improving solid waste management practices; and calls for a cooperative effort among federal, state, and local governments and private enterprise to recover valuable materials and energy from solid wastes.

THE CLEAN AIR ACT AMENDMENTS (CAA) of 1970 (PL 91-604) and 1977 (PL 95-95) focus on the protection and enhancement of the quality of the nation's air resources in order to protect public health and welfare and the productive capacity of the country. A national R&D program, technical and financial assistance, emission standards, and air quality planning assistance program authorities are included. CAA provides for technical and financial assistance to state and local governments for the development and execution of their air pollution control programs, encourages and assists the development and operation of regional air pollution control programs, and initiates an accelerated national R&D program to achieve the prevention and control of air pollution. It authorizes the development of state implementation plans (SIPs) for the purpose of meeting minimum federal ambient air quality standards. It also authorizes issuance of regulations to control hazardous air pollutants and new source performance standards (i.e., emission standards).

THE MARINE PROTECTION, RESEARCH AND SANCTUARIES ACT (MPRSA) of 1972 (PL 92-532) and its amendments provide for regulating the dumping of all types of materials into ocean waters and limiting the ocean dumping of materials that would adversely affect human health and welfare of the marine environment and its commercial values. The MPRSA prohibits ocean dumping and transportation from the United States for purposes of dumping except pursuant to permit; and permits are not to be issued where dumping would "unreasonably degrade" the marine environment. Permitting regulations, marine research, and provisions for establishment of marine sanctuaries are included. EPA is required to establish criteria for evaluating ocean dumping permit applications, applying specific statutory factors. A 1977 Amendment (PL 95-153) effectively established December 31, 1981, as the deadline for terminating ocean dumping of "sewage sludges," defined as municipal waste "which may unreasonably degrade or endanger human health, welfare, amenities, or the marine environment, ecological systems, or economic potentialities." A similar amendment was enacted in 1980 for "industrial wastes."

THE TOXIC SUBSTANCES CONTROL ACT (TSCA) of 1976 (PL 94-469) provides for the testing and premanufacture notification of chemical substances and mixtures, and the regulation of production or use of certain ones which present an unreasonable risk of injury to health or the environment. Along with its many regulatory, testing, and reporting requirements, EPA is also required under Section 9 of TSCA to coordinate actions taken under TSCA with actions taken under other federal laws. In addition, TSCA requires that EPA issue rules respecting the manufacturing, processing, distribution in commerce, and disposal of PCBs.

REFERENCES

Bastian, R. K., A. Montague, and T. Numbers. 1982. The potential for using municipal wastewater and sludge in land reclamation and biomass production as an I/A technology: An overview. In W. E. Sopper, E. M. Seaker, and R. K. Bastian (eds.) 1982. Land reclamation and biomass production with municipal wastewater and sludge. Pennsylvania State University Press, University Park. 524 p.

Bledsoe, C. S. (ed.) 1981. Municipal sludge application to Pacific Northwest forest lands. Institute of Forest Resources Contribution 41. College of Forest Resources, University of Washington, Seattle. 155 p.

Burwell, C. C. 1978. Solar biomass energy: An overview of U.S. potential. Science 199:1041–1048.

Cole, D. W., C. L. Henry, P. Schiess, and R. J. Zasoski. 1983. Forest systems. p. 125–143. In A. L. Page, T. L. Gleason III, J. E. Smith, Jr., I. K. Iskandar, and L. E. Sommers (eds.) 1983. Proceedings of the 1983 Workshop on Utilization of Wastewater and Sludge on Land. University of California, Riverside. 480 p.

Henry, C. L., and D. W. Cole (eds.) 1983. Use of dewatered sludge as an amendment for forest growth. Vol. 4. Institute of Forest Resources, University of Washington, Seattle. 110 p.

Hornick, S. B., L. J. Sikora, S. B. Sterrett, J. J. Murray, P. D. Miller, W. D. Burge, D. Calacicco, J. F. Parr, R. L. Chaney, and G. B. Willson. 1984. Utilization of sewage sludge compost as a soil conditioner and fertilizer for plant growth. USDA/ARS Agric. Information Bulletin 464. 32 p.

National Academy of Sciences. 1977. Multimedia management of municipal sludge. Analytical Studies for the U.S. Environmental Protection Agency. Vol. 9.

Overcash, M. R., and D. Pal. 1979. Design of land treatment systems for industrial wastes. Ann Arbor Science Publishers, Ann Arbor, Michigan.

Page, A. L., T. L. Gleason III, J. E. Smith, Jr., I. K. Iskandar, and L. E. Sommers (eds.) 1983. Proceedings of the 1983 Workshop on Utilization of Wastewater and Sludge on Land. University of California, Riverside. 480 p.

Parr, J. F., et al. (eds.) Land treatment of hazardous wastes. Noyes Data Corp., Park Ridge, New Jersey.

Pennsylvania State University. 1985. Criteria and recommendations for land application of sludges in the Northeast. Pennsylvania State University Agriculture Experiment Station Bulletin 851. Pennsylvania State University, University Park. 94 p.

Reed, S. C., and R. W. Crites. 1984. Handbook of land treatment systems for industrial and municipal wastes. Noyes Publications, Park Ridge, New Jersey. 427 p.

Schotzko, R. T., C. Allison, V. V. Volk, and A. G. Nelson. 1977. Projecting farm income effects of sewage sludge utilization in the Tualatin Basin of Oregon. Special Report 498. Agriculture Experiment Station, Oregon State University, Corvallis.

Sopper, W. E., E. M. Seaker, and R. K. Bastian (eds.) 1982. Land reclamation and biomass production with municipal wastewater and sludge. Pennsylvania State University Press, University Park. 524 p.

U.S. Environmental Protection Agency. 1983a. The 1982 needs survey: Conveyance, treatment, and control of municipal wastewater, combined sewer overflows, and stormwater runoff. Summaries of technical data. EPA 430/9-81-002. Office of Water Program Operations. 192 p.

———. 1983b. Process design manual: Land application of municipal sludge. EPA 625/1-83-016. CERI, Cincinnati, Ohio. 434 p.

———. 1984. Municipal sludge management policy. Federal Register 49(144):24358–24359. June 21.

U.S. Forest Service. 1974. Forest resources report.

Washington Department of Ecology. 1982. Municipal and domestic sludge utilization guidelines. WDOE 82-11.

———. Best management practices for use of municipal sewage sludge. Draft.

Chapter 3

The Status of Wastewater Irrigation of Forests, 1985

DEAN H. URIE

ABSTRACT   After more than twenty years of scientific study, irrigation of forests with waste-water is only beginning to be considered seriously as an optional method of sewage treatment. The major growth is occurring in the southeastern United States, with its combination of demonstration projects, enthusiastic designers, and forest lands available at reasonable cost. A 1983 review of national research and design needs has identified inadequate nutrient cycling definition as a major problem. Wastewater irrigation has been tested over a range of application rates on about a dozen forest types and stages of growth. In general, juvenile stands and open forests with vigorous herbaceous understories have provided the highest nitrogen assimilation, sometimes equaling that found in agricultural crops. Irrigation in older forests has provided less reliable nitrogen recycling, especially when wastewater with high concentrations of inorganic nitrogen is applied. Pretreatment, especially pond storage, is well suited to forest irrigation, providing both storage and nitrogen reduction, through ammonia volatilization. Problems in managing wastewater irrigated forests have arisen from understory competition, rodent damage, toxic chemicals from industrial sources, and boron in domestic wastewater. Wildlife benefit from improved forage, but nesting habitat can be seasonally degraded. Earthworms are particularly vulnerable to trace elements accumulated in detritus. However, no documented wildlife health hazard or reproductive abnormality has been associated with forest irrigation. As more demonstrations of environmentally harmonious wastewater irrigation systems are built in forested regions, commercial projects can be expected to increase accordingly.

Sewage is an unpleasant detail of everyday life that we try to ignore. As a society, we pay someone else to take care of collection and treatment. We hope that they will do so in environmentally safe ways and that they will not charge us too much. Ultimately, these wastes must either be placed in some final disposal site or be reassimilated into our productive cycles. Energy and economic pressures tend to favor recycling, if environmental needs can be met. For nearly thirty years, major efforts to increase reuse of our wastes have been under way. Since 1977, treatment methods featuring reuse of sewage have received special financial incentive through preferential funding by the federal government. Wastewater irrigation of forests has been found to be an economical and safe sewage effluent reuse method in comprehensive research studies conducted at four locations in the United States: Pennsylvania, Georgia, Michigan, and Washington. These findings have augmented the results from research on use of agricultural crops to recycle waste nutrients.

Land treatment of sewage effluent has been widely publicized over the past twenty years, with optimistic predictions of application in a significant proportion of the new waste treatment plants built in the past decade. Now that the construction of new waste treatment facilities has slowed down, we can examine how widely the land treatment of forest and cropland irrigation has been applied.

Out of more than a thousand land treatment systems in total, only a few are devoted to forest production. The 1981 EPA Design Manual (EPA 1981) lists seven operational land treatment systems in the United States. Nutter and Red (1985) have described several additional projects in the southeastern United States, and noted that more are in the beginning stage. This list of forest wastewater systems is still not very long (Table 1). These projects tend to be small, with only two exceeding one million gallons a day (3,785 m³/day) in flow (Figure 1). From this list of operational forest land treatment systems we can look for patterns of growth.

According to Nutter and Red (1985), in the southeastern United States the technology developed at the Helen, Georgia site is being applied to numerous municipal wastewater systems. In Georgia alone, systems operating and planned will use more than 4,000 ha of forest to treat 3.5×10^6 m³/day (80 mgd) of sewage. Projects in the Southeast benefit from a climate that permits use of year-round irrigation, minimizing storage costs. The forest soils of this region can denitrify excess nitrate in the soil solution, helping to reduce leachate to acceptable concentrations. Pond pretreatment on several projects also serves to dilute nitrogen with rainfall and to reduce ammonia-N through volatilization losses.

Elsewhere in the United States few, if any, new forest land projects are being built. We may ask whether this is because the design information is inadequate in other regions (as suggested from the Page et al. 1983 review). If we compare progress in the use of

TABLE 1. Wastewater irrigation of forests.

Project Type and Location		Sewage Type	Forest Type
R	Detroit Lakes, MN	M	Mature hardwoods
R/O	Seabrook Farms, NJ	I	Hardwoods and grass
R/O	Fremont, MI	I	Hardwood seedlings and saplings
R/O	Pennsylvania State University	M	Red pine, old fields-spruce, hardwoods, oak
R/O	Unicoi State Park, GA	M	Mixed pine and hardwoods
O	Helen, GA	M	Mixed pine and hardwoods
R	Middleville, MI	M	Red pine, poplars
R	Harbor Springs, MI	M	Christmas trees, northern hardwoods
R	Pack Forest, WA	M	Douglas-fir seedlings and old growth, poplars
O	West Dover, VT	M	Hardwoods and conifers
O	Mount Sunapee, NH	M	Mixed hardwoods
O	Mackinaw City, MI	M	Aspen, white pine, birch
O	Kings Bay Base, GA	M	Pine
O	Bennett Spring State Park, MO	M	Pine, hardwoods
R/O	Clayton County, GA	M	Pine, hardwoods
O	Round Spring Park, MO	M	Pine
O	Mount Lemmon, AZ	M	Pine
O	Killington, VT	M	Northern hardwoods (two systems)
O	Sugarloaf Mountain, ME	M	Hardwoods
O	Wolfeboro, NH	M	Hardwoods
O	Covington, GA	M	Hardwoods and pine
O	Dalton, GA	M	Hardwoods and pine

R = research study. O = operational wastewater treatment system.
M = municipal wastewater. I = industrial wastewater.

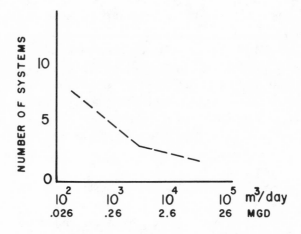

Figure 1. Size distribution of forest land wastewater treatment systems in the United States in 1981.

forest lands with the number of new irrigation projects being installed on agricultural lands in forested regions, we can test whether land treatment designs are popular in general. For if cropland systems have adequate design data available and are growing in number, it is likely that we just do not yet know how to design forest systems adequately.

In general, land application of sewage was judged by a national review panel to be both environmentally acceptable and cost effective as of 1983 (Page et al. 1983). (More than a thousand land treatment systems for wastewater were in operation at that time.) Information was judged as adequate for designing wastewater systems for all climates and for all geographical areas in the United States. However, an obvious gap in knowledge exists in nutrient assimilation criteria for scheduling long-term rates of wastewater application. Specifically, this review concludes, we need long-term information on the behavior of forest ecosystems treated with wastewater, so that site aging can be recognized before ecosystem collapse.

This lack of adequate design criteria may be a significant factor in delaying acceptance of silvicultural treatment. Another reason may be a lack of adequate forest land in the vicinity of municipal treatment plants. Most agricultural cropland management systems are located where there is demand for irrigation water. There are few slow-rate agricultural sewage systems being built in the humid zone. A reason so few forest land systems are built could be a lack of any significant demand by managers for forest irrigation. A count of all slow-rate wastewater treatment systems in six states with important forest resources illustrated the preponderance of agricultural land use (Table 2). Out of seventy-five systems listed in the 1981 EPA Design Manual, only four involved significant forest crops.

Wastewater irrigation of forests in other countries has received some scientific attention. Studies in Poland have included irrigation of pine plantations and poplar forests with both domestic and industrial sewage (Bialkiewicz 1978). In Australia, current studies include combining tree growth with crop production under sewage effluent treatments.

TABLE 2. Number of slow-rate land treatment systems in selected humid zone states, total and forest land (U.S. EPA 1981).

State	Total Slow-rate Systems	Forest Land Systems
Washington	12	0
Montana	8	0
Minnesota	25	0
Florida	23	1 (in part)
New Hampshire	2	2
Pennsylvania	5	1
Total	75	4

RESEARCH AND DEMONSTRATION TESTS

We will never have long-term data for every growth stage in every forest type eligible for sewage irrigation. As with other practical uses of science, good judgment will be required in applying biological skills and engineering skills to extrapolate to specific sites from the relatively small number of trials we will have available. It is worthwhile at this time to examine where we do have strong research data.

The longest continuous studies of forest irrigation have been made at Pennsylvania State University (Sopper and Kardos 1973, Sopper and Kerr 1979). Performance data are available for northern hardwoods, red pine, white spruce, old fields, and a mixed oak type game management area. Irrigation schedules have ranged from 25 to 100 mm (1 to 4 inches) per week with growing season and year-round schedules. The wastewater used in these experiments has been a secondary effluent from a municipal biological treatment plant. Prescriptions from these Penn State studies should apply to use of effluents with high levels of inorganic nitrogen through much of the north central and northeastern forest region.

The study at the Unicoi State Park in Georgia has provided field performance data over a twelve-year period for a mixed pine and hardwood forest irrigated year-round with pond-treated, municipal type wastewater (Nutter and Red 1985). The wastewater at Unicoi is relatively dilute in mineralized forms of nitrogen and high in the organic forms. Since 1979, when the site was harvested, the vegetation has been in the regeneration stage. This experimental system has provided design data for many operational and proposed wastewater irrigation systems throughout the southeastern United States, illustrating useful extrapolation to a range of soils and forests within a climatic region.

Demonstrations of forest irrigation with effluent from sewage ponds were made in Michigan on pole-size red pine plantations, juvenile plantings, pole-size northern hardwoods, and Christmas trees (Urie et al. 1984). The wastewater used in these studies contained less than 10 mg per liter of total nitrogen, most of this in organic forms. Ammonia volatilization from storage ponds reduced inorganic nitrogen to very low levels during the growing season irrigation periods. Growth responses were measured in the young plantations. In the older stands growth changes over the five to ten years of the studies were modest at best.

These studies demonstrated the principals advocated by King (1979) and Burton (1982) regarding pond pretreatment for nitrogen removal, followed by land irrigation for phosphorus reduction. These principles are probably more broadly applicable than just in the Lake State where these studies were conducted. Wetlands forest irrigation offers a special case, as defined in cypress domes by Ewel and Odum (1978).

A five-year wastewater irrigation study (Schiess and Cole 1981) in Washington provided data on Douglas-fir seedlings and a juvenile plantation of Lombardy poplar. An established Douglas-fir forest was also irrigated for four years. Year-round irrigation at 50 mm/wk resulted in a very high rate of plant and soil assimilation of nitrogen. The wastewater was of typical biological secondary plant quality, relatively high in inorganic nitrogen. Biomass measurements and soil samples verified that as much as 50% of the nitrogen applied was being assimilated in the young plantations. Difficulties with weed competition in the juvenile stands suggested that the least troublesome method for forest land treatment of a secondary wastewater in this region might be year-round application in established Douglas-fir forests on well-aerated glacial outwash soils. Low concentrations of leached nitrogen were also measured beneath the established Douglas-fir over a four-year period, indicating high site retention of applied nitrogen.

In summarizing these results Cole et al. (1983) listed these important criteria for a successful sewage irrigation project using a tree crop: (1) tree species with a high growth and nitrogen assimilation potential, (2) well-drained, well-aerated soils, and (3) use of year-round irrigation only where freezing conditions are not a problem.

Most of these projects are in experimental or demonstration stages. We have yet to see routine adoption of forest irrigation as a wastewater treatment process anywhere in the world.

The forest irrigation systems that use sewage effluent have developed under one or more of the following circumstances: (1) they were the outgrowth of nearby experimental studies; (2) forests provided the only available lands near sewered areas; (3) low quality, cheap land was chosen for a land treatment system and found best suited to tree production; (4) an evaluation of treatment alternatives showed forest irrigation to be the cheapest acceptable method.

An examination of the progress that has been made in the last twenty-five years in forest land irrigation indicates that we are technically capable of a lot more sewage effluent treatment on forests than we are now doing. Lack of technical knowledge seems not to be the principal barrier. Let us examine some other problems and guess the future of the use of forest irrigation with sewage effluent.

The research background available as a base for enlarging this list of projects shown in Table 1 has reached a significant size. I have chosen to outline the result of these studies under three principal categories of effects: (1) effects on tree production, (2) effects on water quality and public health, and (3) effects on the forest as a wildlife habitat.

Tree Production under Wastewater Irrigation

When forest managers consider effluent irrigation, they expect improved tree growth. The growth data available for young plantations are quite promising, as shown in Table 3. Doubling of height growth is common, especially on sites where unirrigated trees grow slowly. Diameter growth increases are even more impressive, a result of extending the growing season and the combined effects of supplemental nutrients and water.

In older trees, growth has generally increased, although effects have been variable. At Penn State, red pine poles grew faster for a few years with 25 and 50 mm (1 and 2 inches) per week of effluent. Subsequently, restricted root penetration resulted in windfall, and

TABLE 3. Height and diameter growth in trees receiving wastewater irrigation (McKim et al. 1982).

Region	Species	Growth Response (% above controls)	
		Height	Diameter
Northeast			
Seedlings	White pine	167	
	Red pine	80	
	White spruce	79	
	Pitch pine	200	
Established forests	Red pine	38	187
	White spruce	140	122
	Sugar maple		
Lake States			
Seedlings	Eastern cottonwood	100	
	Hybrid cottonwoods	65-169	
	European larch	76	
	Red oak	67-83	
Established forest	Sugar maple		400
Pacific Northwest			
Seedlings	Lombardy poplar	93	210
	Douglas-fir	41	77

there was suspected toxicity from boron (Sopper and Kardos 1973). Cooley (1978) and Neary et al. (1975) also measured elevated boron concentrations in foliage of red pine trees in Michigan after wastewater irrigation. Foliar necrosis was evident, but ten-year diameter growth was not reduced below that measured in controls in pole-size red pine trees (Urie et al. 1984).

About 11 to 12 tonnes (Mg) per ha per year is the practical maximum dry-matter production rate reached with sewage or with mineral fertilization and irrigation (Schiess and Cole 1981). These dry-matter accumulation rates represent a nitrogen assimilation of 500 to 1,000 kg/ha per year at an average nitrogen content of 0.4 to 0.8%. The maximum rate of nitrogen accumulation, measured under year-round irrigation for five years of hybrid poplars in Washington, was 1,247 kg/ha, which represents about 250 kg/ha per year of nitrogen assimilation (Table 4). Somewhat lower rates have been measured in other studies.

Douglas-fir seedlings with associated understory accumulated 873 kg/ha over the same period. Comparative values from the Lake States indicate only 400 to 500 kg/ha in four years of growth for a hybrid poplar irrigated with a dilute (<10 mg/l of N) sewage pond effluent (Cooley 1978). Older forests of mixed pine and hardwoods in Georgia accumulated about 150 kg/ha per year of nitrogen under year-round irrigation (Nutter and Red 1985).

In the juvenile stands, understory grasses and herbs constituted an important nitrogen sink until crown closure. In the Washington poplar study, 30% of the total nitrogen was incorporated in the grass component (Table 5). Under the Douglas-fir plantation, grass accounted for 49% of the nitrogen uptake. With crown closure, nitrogen in decomposed herbs is transferred to tree uptake or leached to groundwater.

TABLE 4. Accumulation of nitrogen in forest vegetation irrigated with sewage effluent.

Location and Type	Nitrogen Applied (kg/ha)	Net Nitrogen Stored (kg/ha·yr)	Reference
Pacific Northwest (juvenile stands)			
Poplar (5 yr) with understory	2,171	250	Schiess and Cole 1981
Douglas-fir (5 yr) with understory	1,811	190	
Lake States (juvenile stands)			
Raverdeaux poplar	412	+100	Cooley 1979
Southeast (older forests)			
Mixed hardwoods and pine (includes denitrification)		150	Nutter and Red 1985

TABLE 5. Nitrogen content of understory (percentage of nitrogen added to forest plantations by irrigation with wastewater).

	Nitrogen Added (kg/ha)	Nitrogen Uptake (%)
Pacific Northwest (fourth year)		
Poplar, grass	400	30
Douglas-fir, grass	350	49
Northeast (fifth year)		
White spruce, old field	310	39

Soil accumulation of nitrogen has been measured in a few irrigated forests. After five years of irrigation with facultative pond wastewater, Boyer loamy sand (Typic Hapludalf) at Middleville, Michigan, accumulated about 600 kg/ha of nitrogen in the 0 to 10 cm of mineral soil (Harris 1979). This increase was related to an increase of 50 to 100% in organic matter, presumably including decomposed forest floor materials, since the increase in nitrogen was independent of irrigation rates from 25 to 88 mm/wk (180 to 412 kg/ha of total nitrogen over five years). Thus, in these tests, the added nitrogen could be more than accounted for in added soil nitrogen.

Wastewater irrigation allows us to create nearly ideal growth conditions for any plant crop, because we can regulate both water and nutrient supply. Problems do arise in natural forests when we try to add too much water or too many nutrients. If we consider that most wastewater use on land occurs where reuse of the water is essential because of the climate, we might expect that the opportunity to raise more productive, more water demanding forest species on sites not currently suitable for these species could lead to acceptance of forest land treatment. In the humid region, short-rotation energy-wood plantations appear to be the best choice. We already have a few examples in the drier regions of the United States of the use of wastewater to establish forest greenbelts where trees would not exist without supplemental irrigation (Younger et al. 1973).

In most research studies, only the nitrogen applied to the forest in wastewater and the leached nitrogen have been measured directly. The nitrogen balance also includes plant-assimilated nitrogen, accumulated soil nitrogen, and volatilization losses. Table 4 illustrates the highest rates of plant assimilation measured so far. In older forests, plant assimilation is less, so the other pathways become critical.

Inorganic nitrogen loading rates are a major factor in the effectiveness of the forest in assimilating a large enough portion of the added nitrogen to produce acceptable leachate for groundwater recharge. Established hardwood forests in Pennsylvania and Michigan were not effective in lowering the concentrations of inorganic nitrogen leached to groundwater (Hook and Kardos 1978, Burton and Hook 1979). Juvenile forests, on the other hand, have shown uniformly high nitrogen renovation capability, owing to high plant uptake in trees and understory and, in some locations, to denitrification in residual grass sod.

Illustrations of the role of understory vegetation in nutrient (nitrogen) cycling have been derived from biomass accumulations in poplars and conifer plantings having luxuriant understory (Figure 2).

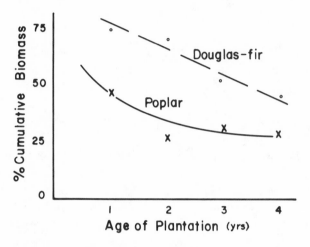

Figure 2. Cumulative biomass in understory plants in young forest plantations as a percentage of total cumulative biomass (after Schiess and Cole 1981).

Managing the nitrogen renovation is usually based on removal from the site of those succulent plant tissues that contain the highest proportion of stored nitrogen. In short-rotation hybrid poplars, about 75% of the aboveground assimilated nitrogen was in limbs and foliage (Urie et al. 1978) (Figure 3). Leaving the foliage and twigs on the site in a harvesting operation is not unlike mowing a herbaceous cover and letting it lie on the ground. Maybe it is not necessary to carry off these residuals. Burton (1982) and Sopper and Kerr (1979) have measured nitrogen balances in old field vegetation systems that assimilated inorganic nitrogen loadings of 200 to 400 kg/ha per year. Nitrate concentrations in leachate at both these sites were reduced to acceptable levels without any vegetation being removed for several years.

When the old fields in Michigan were mowed, there was little change in the leachate concentrations over a two-year period. Burton did report that there was evidence that nitrate concentrations were increasing after several years without crop removal—a trend that was also evident in the Pennsylvania old field experiments.

At present, the short rotation forests that best suit irrigation needs have limited markets for their product. Christmas tree production provides short rotations and, with her-

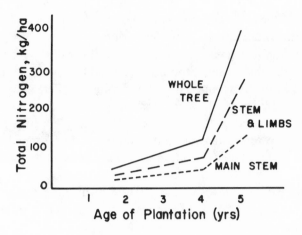

Figure 3. Nitrogen content of harvestable portions of a young poplar plantation (after Cooley 1979).

baceous vegetation growth between the tree rows, could maintain high rates of nitrogen assimilation. Regular mowing would be needed, however, to control competition.

Irrigation with nutrient-rich effluents can result in poor crown density (Sagmuller and Sopper 1967). Use of dilute wastewater from sewage storage ponds at Harbor Springs, Michigan, improved growth of Scotch pine, white spruce, and balsam fir on a droughty soil without severely lowering Christmas tree quality (Cooley 1980). Sprout growth of aspen and of other hardwoods offers a brief period when nutrient uptake is high. The design life for a fixed slow-rate irrigation treatment system, commonly twenty to thirty years, usually outlasts the period of rapid growth for Christmas trees and sprout forests grown on short rotations. A revised loading rate could be needed in the second and third decades if intermediate harvests are not scheduled.

King (1978) and Burton (1982) have described forest irrigation using a low maintenance, high recycling system with a long-term life. By combining wastewater pretreatment with storage in ponds, the ammonia volatilization capability of facultative ponds permits irrigation with an effluent that has a very low concentration of mineralized nitrogen. The nitrogen loading rate can then be held well within the assimilation capability of most forests.

Nurtients applied in irrigation wastewater accumulate in trees, understory vegetation, and soils. The organic material in the forest floor and in surface mineral soils represents the largest potential sink. Because of litter decomposition, accumulation of nitrogen in the forest floor under a sewage irrigation regime may not occur, as Richenderfer and Sopper (1979) illustrated in Pennsylvania.

Effects of Wastewater Irrigation on Water Quality

The quality of water leaving the forest site provides the measure of treatment effectiveness. Nitrogen removal from wastewater is the criterion by which most forest irrigation projects are judged. Although the mechanisms that make one system satisfactory while another fails are not clearly understood, enough research information is available to construct a satisfactory system given almost any effluent and any forest. That is not to say, however, that all forests can be used in a cost-effective system. Old-growth forests irrigated with a wastewater high in mineral forms of nitrogen pose the most difficult problem. The forest can assimilate only small amounts of added nitrate and ammonia. In fact, in the early stages, irrigation may release excess nitrogen from the forest floor and from soil organic matter, discharging more nitrogen from the system than is added by irrigation. Such a forest much either be irrigated at a very low rate or the effluent must be pretreated to reduce the nitrogen concentration.

Nitrogen management is based on plant assimilation and atmospheric release of nitrogen through volatilization and denitrification. Juvenile forest ecosystems and fast-growing, intermediate-aged stands provide the best opportunities for assimilation. Agroforestry promises to maximize the process, although it has not received enough study. The research results from several plantation experiments that emphasize the importance of nitrogen taken up by ground cover species illustrate how important understory vegetation can be.

Harvesting a wastewater-irrigated forest can result in short-term flushes of nitrate until a vigorous ground cover is reestablished. A red pine harvest following a blow-down in Pennsylvania and a 50% partial cutting in Michigan produced such pulses of nitrate in leachate. In both cases, concentrations of nitrate returned to precutting concentrations within eighteen months (Figure 4).

Phosphorus renovation has been no problem in the short-term studies for which records are available. Exhaustion of phosphorus retention was suggested in a sequence of measurements in a sandy loam (Hook et al. 1973) and loamy sand (Harris 1979). Phos-

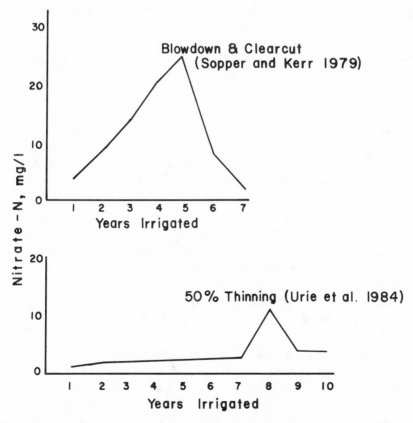

Figure 4. Nitrate concentrations in soil leachate after clearcutting and thinning in pole-size red pine plantations.

phorus was significantly increased in mineral soils, especially in the surface horizon. After ten years, concentrations of total phosphorus at 120 cm depths were increased, indicating that leachate below the root zone was enriched with phosphorus.

Besides the nitrate hazard, three major pathogen or toxic chemical pathways must be considered in evaluating the risks to human health resulting from forest land irrigation with sewage effluent. The most serious influence is the quality of renovated water leaving the treatment site. Nitrate concentration in groundwater is so critical as to constitute the major criterion for system design. Movement of trace chemicals and pathogens in groundwater from land treatment systems to points of human use has not been reported. We can assume at this stage that a system designed to achieve nitrogen renovation of wastewater will not pose an unacceptable risk in these other groundwater quality hazards. Wildlife and wild plants harvested for food from sewage-irrigated forests could convey chemicals or pathogens to humans. Research to date shows such low accumulations of chemicals in wild animal body tissues that human risks are trivial. However, harvest of wild food plants could be a low-frequency risk.

Studies have consistently failed to show increased disease in domestic animals raised on sewage-irrigated farmland. Wildlife is likely to be even less threatened because of free movement to and from irrigated sites.

Direct human exposure to the irrigated sewage or by aerosol drift to places of human activity is controlled by isolation strips and by limiting public access. Workers on the site have the exposures common to sewage plant workers throughout the industry. Many studies of this population have confirmed that sickness frequencies in that work force are the same as for the general public (Page et al. 1983).

In summary, we would expect that human health risks from a forest irrigation system designed to achieve adequate nitrogen renovation from wastewater will be lower than that of a similar project using agricultural crops.

Effects of Wastewater Irrigation on Wildlife

Two avenues of wildlife effects have been evaluated: (1) changes in the vegetation and other habitat alterations affecting food supply, cover, and nesting environment, (2) transfer of waste chemicals or biological agents from the sewage effluent to wildlife by direct ingestion or through food-chain accumulation.

Changes in cover distribution, production of food plants, and nutritional quality of food plants have been identified in forests treated with sewage sludge. Fertilization and irrigation combined in an effluent irrigated forest would be expected to show even greater changes. More nitrogen should result in greater protein content, but rapid growth means that the early stages of plant growth quickly pass out of reach of ground-dwelling animals.

Trace chemicals and pathogenic organisms have received little attention in wastewater irrigated forests, largely because similar research in croplands and with farm animals has shown little cause for concern.

In general, wildlife collected from sewage irrigated forests have shown little evidence of higher accumulations of trace metals. Major changes in vegetation due to irrigation may benefit or detract from habitat value for specific wildlife species. For example, Anthony and Wood (1979) found songbird populations to be less diverse on sewage irrigated lands, while forage production for deer and rabbits improved in irrigated aspen, pine, and shrub types. Introduction of unpalatable succulent species, such as pokeweed

(*Phytolacca americana*), into irrigated forest understory detracted from summer forage values.

Wildlife benefits probably outweigh negative aspects, although irrigation tends to favor food production over breeding habitat. Trace chemical increases in wild animal tissues have not been related to impaired health of animals. Earthworms are particularly vulnerable to trace elements in surface mineral soil, and, because of their importance in woodcock diets, there may be reason to avoid siting forest irrigation projects in prime woodcock habitat.

MANAGEMENT NEEDS

One of the advantages of land treatment systems that is often cited is the relative simplicity of management. Small municipal systems can be operated without highly trained, high salaried, technical workers. Land treatment systems also have the ability to absorb shocks and upsets and to buffer changes in waste quality that would, and frequently do, render biological-chemical sewage treatment inoperative for considerable periods of time. Because of the resiliency of the system, untrained or careless operators have been reported to abuse land treatment systems. Some regulatory agencies expect land treatment systems to gravitate toward sloppy maintenance and careless operation. To reduce costs, needed maintenance is put off. This is nothing new. At a 1972 symposium in Philadelphia, Maurice Goddard, then secretary for Pennsylvania's Department of Environmental Resources, stated that in Pennsylvania the record of management of land treatment systems had been uniformly poor (Goddard 1973). He hoped that improved permitting procedures would cure this problem. We seem to be still pursuing that hope. The urge to economize by attending only to crisis maintenance is still a major obstacle to proper operation, and thus to acceptance of land treatment systems for wastewater. To the extent that regulatory agencies become skeptical of land treatments as a result of bad experiences, future growth will be seriously jeopardized.

In the seclusion of the forest, the clogged sprinkler and the broken distributor line can be hidden for months before any regulator or supervisor is likely to insist on its repair. Forest land treatment systems are probably even more susceptible to sloppy maintenance than cropland systems are, because growth is not easily evaluated. Unfortunately, the good intentions and the high environmental goals of the designers cannot be passed on to the worker who spends his career amid the countless chores of maintaining a sewage collection and treatment system funded by a rate system that may have been barely adequate years ago, when the system was built.

I see two avenues for improvement. First, regulatory agencies must recognize the propensity for using shortcuts and be equally thorough in demanding adequate maintenance of land treatment systems as they are for systems discharging to surface water. To accomplish this, a penalty-free system seems to be a useful tool in overcoming both misguided thrift and indolence. The second way, especially pertinent to forest lands, is use of proprietary interests to protect the land and forest resource. Whenever possible the crop management should be independent of the sewage treatment process, thus ensuring an independent land management which demands high quality wastewater distribution so as to protect the plants, soil, and groundwater.

Weed and rodent problems can be serious barriers to plantation growth when understory grasses are abundant. Cooley (1982) also found that some clones of poplar exhibited disease problems but not demonstrably as a result of wastewater irrigation.

If the goal of a forest land irrigation system is simply to accomplish a high level of wastewater renovation without important tree growth benefits, then irrigation at a rate of 25 mm per week or less during the frost-free season will be possible on all established forests tested so far. Higher rates have been possible when the effluent was low in nitrogen.

THE FUTURE

Wastewater irrigation of forest lands has enough demonstrated successes to ensure that future planners should consider it as an option. What is needed is more good demonstrations in local conditions to illustrate benefits and the lack of nuisances.

As long as wood supplies are abundant we are not going to find much demand from foresters for more rapid wood production. Should we enter a period when short-rotation energy plantations for fuel are in great demand, wastewater irrigation will provide an ideal source of the low-cost water and the fertilizer needed for their growth. Energy forests will also provide a nutrient-demanding crop that is outside the human food chain.

Until then, we can anticipate slow acceptance where forest lands are available and where competition for their use is not restrictive. We need to accumulate histories of trouble-free operation with those systems that have been in operation for many years. When I pass the Mackinaw City facility, I wish I could compare its benefits and reliability with alternatives that could have been used when it was built in the mid-1970s. The list of operating projects includes several such opportunities for long-term studies.

Various reasons have been advanced for the slow acceptance of land treatment. One of these, the need for large land area, is partly alleviated when forest land is used, for forest lands are more likely to be held in large blocks and there are few restrictions on the land's use in ongoing forestry enterprises. Ownership and management of the forest crops can easily be left to the land's owners without transfer of these functions to the' managers of the wastewater system.

Concerns over lowered land values when sewage treatment systems move in next door are real and important. Nuisances such as odors, unsightliness, and noise are so uncharacteristic of a properly operated land treatment system that research scientists tend to overlook their importance as a factor in the acceptance of the technology. As Forster and Southgate (1983) pointed out in a review of constraints, protests about public health concerns may be a rationalization of these economic fears.

Reluctance to accept the land treatment concept is more the result of attitudes than a lack of information. Attitudes are slow to change. Established procedures are easier to follow than new ones, especially when there is little support from the public. The sewage treatment industry has a tradition of technological remedies for sewage treatment problems. Lack of enthusiasm for the innovative technology of land treatment is perhaps due to the uncertainty of unfamiliar methods. Also, the relatively low capital investment required for land treatment systems has prompted a low level of industry interest.

We do not have to oversell forest irrigation as a method of wastewater treatment. The facts should be allowed to speak for themselves. There is ample evidence that, with good maintenance and design for the forest's renovation capabilities, forest irrigation can offer the best alternative method for sewage treatment in a given locality and can provide these benefits over a long site life.

REFERENCES

Anthony, R. G., and G. W. Wood. 1979. Effects of municipal wastewater irrigation on wildlife and wildlife habitat. p. 213–223. In W. E. Sopper and S. N. Kerr (eds.) Utilization of municipal sewage effluent and sludge on forest and disturbed land. Pennsylvania State University Press, University Park.

Bialkiewicz, F. 1978. Lysimetric and forest studies on cleaning and productivity utilization of municipal sewage. File report. Project 521-FS-59. U.S. Forest Service.

Burton, T. M. 1982. Studies of land application in old growth forests in southern Michigan. p. 181–193. In F. M. D'Itri (ed.) Land treatment of municipal wastewater. Ann Arbor Science Publishers, Ann Arbor, Michigan.

Burton, T. M., and J. E. Hook. 1979. A mass balance study of application of municipal wastewater to forests in Michigan. J. Environ. Qual. 8:489–596.

——. 1982. Old field management studies on water quality management facility at Michigan State University. p. 107–134. In F. M. D'Itri (ed.) Land treatment of municipal wastewater. Ann Arbor Science Publishers, Ann Arbor, Michigan.

Cole, D. W., C. L. Henry, P. Schiess, and R. J. Zasoski. 1983. Forest systems. p. 125–146. In A. L. Page, T. L. Gleason III, J. E. Smith, Jr., I. K. Iskandar, and L. E. Sommers (eds.) Proceedings of the 1983 Workshop on Utilization of Municipal Wastewater and Sludge on Land. University of California, Riverside.

Cooley, J. H. 1978. Nutrient assimilation in trees irrigated with sewage oxidation pond effluent. p. 328–340. In Proceedings of the Central Hardwoods Forest Conference. Purdue University, West Lafayette, Indiana.

——. 1980. Christmas trees enhanced by sewage effluent. Compost Science/Land Utilization 21(6):28–30.

——. 1982. Growing trees on effluent irrigation sites with sand soils in the upper Midwest. p. 155–164. In F. M. D'Itri (ed.) Land treatment of municipal wastewater. Ann Arbor Science Publishers, Ann Arbor, Michigan.

Ewel, K. C., and H. T. Odum. 1979. Cypress domes: Nature's tertiary treatment filter. p. 103–114. In W. E. Sopper and S. N. Kerr (eds.) Utilization of municipal sewage effluent and sludge on forest and disturbed land. Pennsylvania State University Press, University Park.

Forster, D. L., and D. D. Southgate. 1983. Institutions constraining the utilization of municipal wastewater and sludge on land. p. 19–49. In A. L. Page, T. L. Gleason III, J. E. Smith, Jr., I. K. Iskandar, and L. E. Sommers (eds.) Proceedings of the 1983 Workshop on Utilization of Municipal Wastewater and Sludge on Land. University of California, Riverside.

Goddard, M. K. 1973. Needed directions in land disposal. p. 1–5. In W. E. Sopper and L. T. Kardos (eds.) Recycling treated municipal wastewater and sludge through forest and cropland. Pennsylvania State University Press, University Park.

Harris, A. R. 1979. Physical and chemical changes in forested Michigan sand soils fertilized with effluent and sludge. p. 155–161. In W. E. Sopper and S. N. Kerr (eds.) Utilization of municipal sewage effluent and sludge on forest and disturbed land. Pennsylvania State University Press, University Park.

Hook, J. E., and L. T. Kardos. 1978. Nitrate leaching during long-term spray irrigation for treatment of secondary sewage effluent at woodland sites. J. Environ. Qual. 7:30–34.

Hook, J. E., L. T. Kardos, and W. E. Sopper. 1973. Effects of land disposal of wastewaters on soil phosphorus relations. p. 200–217. In W. E. Sopper and L. T. Kardos (eds.) Recycling treated municipal wastewater and sludge through forest and cropland. Pennsylvania State University Press, University Park.

King, D. L. 1978. The role of ponds in land treatment of wastewater. p. 191–198. In State of knowledge of land treatment of wastewater. Vol. 2. U.S. Army Corps of Engineers, Cold Regions Research and Engineering Laboratory (CRREL), Hanover, New Hampshire.

McKim, H. L., W. E. Sopper, D. W. Cole, W. L. Nutter, D. H. Urie, P. Schiess, S. N. Kerr, and H. Farquhar. 1982. Wastewater applications in forest ecosystems. CRREL Report 82–19. CRREL, Hanover, New Hampshire.

Neary, D. G., G. Schneider, and D. P. White. 1975. Boron toxicity in red pine following municipal wastewater irrigation. Soil Sci. Soc. Am. Proc. 30:981–982.

Nutter, W. L., and J. T. Red. 1985. Treatment of wastewater by application to forest land. TAPPI J. 68(6):114–117.

Page, A. L., T. L. Gleason III, J. E. Smith, Jr., I. K. Iskandar, and L. E. Sommers (eds.) 1983. Proceedings of the 1983 Workshop on Utilization of Municipal Wastewater and Sludge on Land. University of California, Riverside.

Richenderfer, J. L., and W. E. Sopper. 1979. Effect of spray irrigation of treated municipal sewage effluent on the accumulation and decomposition of the forest floor. p. 163–177. *In* W. E. Sopper and S. N. Kerr (eds.) Utilization of municipal sewage effluent and sludge on forest and disturbed land. Pennsylvania State University Press, University Park.

Sagmuller, C. J., and W. E. Sopper. 1967. Effect of municipal sewage effluent irrigation on height growth of white spruce. J. For. 65:822–823.

Schiess, P., and D. W. Cole. 1981. Renovation of wastewater by forest stands. p. 131–147. *In* C. S. Bledsoe (ed.) Municipal sludge application to Pacific Northwest forest lands. Institute of Forest Resources Contribution 41. College of Forest Resources, University of Washington, Seattle.

Sopper, W. E., and L. T. Kardos. 1973. Vegetation responses to irrigation with treated municipal wastewater. p. 271–294. *In* W. E. Sopper and L. T. Kardos (eds.) Recycling treated municipal wastewater and sludge through forest and cropland. Pennsylvania State University Press, University Park.

Sopper, W. E., and S. N. Kerr. 1979. Renovation of municipal wastewater in eastern forest ecosystems. p. 61–76. *In* W. E. Sopper and S. N. Kerr (eds.) Utilization of municipal sewage effluent and sludge on forest and disturbed land. Pennsylvania State University Press, University Park.

Urie, D. H., J. H. Cooley, and A. R. Harris. 1978. Irrigation of forest plantations with sewage lagoon effluents. p. 207–214. *In* Proceedings of the International Symposium on the State of Knowledge in Land Treatment of Wastewater. CRREL, Hanover, New Hampshire.

Urie, D. H., A. R. Harris, and J. H. Cooley. 1984. Forest land treatment of sewage wastewater and sludge in the Lake States. p. 101–110. *In* 1984 TAPPI Research and Development Conference, Appleton, Wisconsin. Technical Association of the Pulp and Paper Industry, Technology Park, Atlanta, Georgia.

U.S. Environmental Protection Agency. 1981. Process design manual: Land treatment of municipal wastewater. EPA 625/1–81–013. CERI, Cincinnati, Ohio.

Younger, V. B., W. D. Kesner, A. R. Berg, and L. R. Green. 1973. Ecological and physiological implications of greenbelt irrigation with reclaimed wastewater. p. 396–407. *In* W. E. Sopper and L. T. Kardos (eds.) Recycling treated municipal wastewater and sludge through forest and cropland. Pennsylvania State University Press, University Park.

Chapter 4

Silvicultural Land Application of Wastewater and Sludge from the Pulp and Paper Industry

WILLIAM E. THACKER

ABSTRACT   The pulp and paper industry in the United States generates considerable quantities of process wastewater and sludge resulting from the treatment of that wastewater. One option for the handling of these wastes is to apply them to forest land and to tree nurseries. NCASI (National Council of the Paper Industry for Air and Stream Improvement, Inc.) is unaware of any mill in the United States that so handles its wastewater, but there is published research on the subject. The silvicultural land application of sludge is practiced by a few mills, and several more are actively investigating this option. Characteristics of mill wastewaters and sludges, with an emphasis on their significance in land application, are discussed in this paper. Literature on the application of these materials to forest land and tree nurseries is reviewed, and the current activity in the silvicultural land application of mill sludges is summarized.

The pulp and paper industry in the United States generates considerable quantities of wastewater and wastewater treatment sludge. At most mills, wastewater is discharged to surface water after primary (sedimentation) and secondary (biological) treatment, and sludge is used as landfill. An alternative is to land apply these materials, as is the practice at a small percentage of mills.

This paper is an overview on the subject of applying wastewater and sludge from the pulp and paper industry to land, especially land occupied by a forest or utilized as a tree nursery. Following a general description of the industry, characteristics of mill wastewaters and sludges are described. Literature on the application of these materials to silvicultural land is reviewed, and the current status of land application in the industry is summarized.

PROFILE OF THE PULP AND PAPER INDUSTRY

The U.S. pulp and paper industry comprises nearly 700 mills distributed among 42 states that collectively produce about 6×10^7 tonnes (7×10^7 tons) of paper and paperboard each year. Mills differ in size, raw materials, processes, and products. In terms of production capacity, mills range in size from under 45 tonnes/day (50 tons/day) to more than 1,360 tonnes/day (1,500 tons/day). Processes for the pulping of wood can be categorized as mechanical (groundwood), semimechanical and chemimechanical, and chemical (kraft, soda, and sulfite). In addition, wastepaper may be deinked or otherwise prepared for recycling into new products. In the United States, approximately 70% of the pulp is produced by the kraft or sulfate process, whereby sodium hydroxide and sodium sulfide aid in the separation of wood into fibers. Products of the industry include bleached and unbleached pulp, newsprint, tissue and toweling, writing and printing (fine) paper, bag and other coarse paper, paperboard such as boxboard, and numerous specialty grades of paper. For the purpose of establishing effluent standards, the U.S. EPA (1982) has divided the industry into approximately two dozen categories.

Various steps in the production of pulp and the manufacture of paper generate wastewater. Examples of specific waste streams include spent pulping liquor, pulp wash water, bleach plant effluent, and white water from papermaking. Recycling reduces the quantity discharged, and about 5% of the mills, mostly those producing paperboard from wastepaper, are described as "closed" or "self-contained" because they reuse essentially all of their process wastewater. Nonetheless, the industry as a whole discharges roughly 5.7×10^{12} liters per year (1.5×10^{12} gal/yr) of wastewater (NCASI 1983). Usually the individual waste streams are combined prior to treatment, yielding "whole mill" or "combined" wastewater. A slight majority of the mills have their own treatment plants from which effluent is discharged to surface water, and most of the others discharge wastewater, possibly after sedimentation, to municipal treatment works.

The removal of solids from wastewater during conventional treatment creates a residue commonly known as sludge. The annual production of wastewater sludge from the U.S. pulp and paper industry is approximately 2.2×10^6 dry Mg (2.4×10^6 dry tons), or 4.5×10^7 wet Mg (5.0×10^7 wet tons) prior to possible dewatering (NCASI 1979). About 1.9×10^6 dry Mg/yr (2.1×10^6 dry tons/yr) are primary sludge (i.e., from sedimentation of raw wastewater), and the remainder is secondary sludge (i.e., from sedimentation of wastewater following biological treatment). Some mills that produce paperboard from wastepaper recycle their primary sludge back into the manufacturing process.

WASTEWATER AND SLUDGE CHARACTERISTICS AND THEIR SIGNIFICANCE IN LAND APPLICATION

Wastewater

Quantity (Flow) and Conventional Pollutants. The quantity and strength of raw whole mill wastewater is a function of mill type and mill operating practices such as the degree of water reuse. The flow for mills that discharge process wastewater may vary from much less than 2.1×10^4 to about 4.2×10^5 liters per Mg (tonne) of production ($\leqslant 5 \times 10^3$ to 1×10^5 gal/ton) (U.S. EPA 1982, NCASI 1983). The BOD_5 (5-day biochemical oxygen demand) concentration in untreated mill wastewater tends to fall within the range of 50 to 1,500 mg/l (U.S. EPA 1982). The bottom and top of this range are populated by nonintegrated (paper production without on-site pulping) and deink mills, respectively. Total suspended solids (TSS) tend to vary from about 100 to 2,500 mg/l, with the higher concentrations also generally associated with deink mills (U.S. EPA 1982). Hence pulp and paper mill wastewater can be more concentrated than typical municipal wastewater. Moreover, some individual mill waste streams can be highly concentrated. Spent liquor from chemical or semichemical pulping, for example, can have BOD_5 levels on the order of 20,000 to 30,000 mg/l.

Conventional treatment can substantially reduce the level of organic and suspended matter. Primary treatment generally removes more than 80% of the suspended solids; BOD_5 reduction depends on the fraction of the organic matter that is in suspended form and varies from about 15% for newsprint mills to 90% for tissue mills (NCASI 1964).

Biological treatment units generally are capable of reducing the BOD_5 of the primary effluent by 80 to 90%.

The parameters of BOD_5 and TSS are of interest because paper companies have turned to land application often to avoid overloading a receiving stream with these substances. In addition, it has been common in land application to evaluate BOD_5 loadings and, to a lesser extent, suspended solids loadings. An estimated upper limit on the assimilative capacity of soils, in terms of high percolate quality and maintenance of cover vegetation and of good soil permeability, is about 225 kg/ha (200 lb/acre) per day of BOD_5 for mill wastewaters (Blosser and Owens 1964). There is at least one instance of a company with a land application system resorting to the installation of primary clarification to reduce ponding and odors caused by suspended solids.

The pH of untreated whole mill wastewaters tends to be between 6 and 9. Individual waste streams such as spent pulping liquor or bleach plant effluent may be highly acidic or alkaline, however. In some cases the pH of a wastewater applied to land may be a concern relative to detriments to vegetation or soil.

Process wastewaters of the pulp and paper industry normally do not contain sanitary wastes and thus are free of fecal coliform and enteric pathogens. The microbial quality of mill wastewaters has been discussed in detail elsewhere (NCASI 1972).

Color. Whole mill wastewaters may be highly colored, on the order of 1,000 or more platinum-cobalt units, and the color may not be well removed in primary and secondary treatment (U.S. EPA 1982). Strong color is frequently ascribed to dissolved lignin derivatives present in waste streams from chemical pulping and bleaching operations, but it may result from other sources as well. Color is mentioned here primarily because it too has been an impetus for mills to consider land application in lieu of discharging to surface water. The migration of color toward groundwater could be an aesthetic concern in some land application situations.

Nitrogen and Phosphorus. Process wastewaters from pulp and paper mills are prone to be deficient in the nutrients nitrogen and phosphorus. The exception is nitrogen in wastewater from mills with the variant of sulfite or semichemical pulping that is ammonium based. Nitrogen and phosphorus frequently are added before biological treatment to enhance biodegradability, but secondary effluents still tend to be lower in these nutrients than similarly treated municipal wastewater (NCASI 1977). Nitrate leaching is normally much less of a concern in the land application of mill wastewater compared with municipal wastewater. Paper companies as a rule do not add nutrients to wastewater prior to land application, yet poor vegetation response due to a nitrogen or phosphorous deficiency has not been reported.

Sodium and Salinity. The most commonly raised concern about the quality of pulp and paper mill wastewaters in a land application setting is probably salinity—high sodium concentrations in particular. Sodium compounds are employed as pulping chemicals in certain processes, namely kraft, soda, chemimechanical, and some forms of sulfite and semichemical pulping. Pulping liquors can be high in both sodium and dissolved solids. Vegetation damage and reduced soil permeability due to sodium or total salts in land-applied mill wastewaters have been reported (Blosser and Caron 1965, Guerri 1973, Jorgensen 1970). It has been suggested that the sodium adsorption ratio (SAR) of mill effluents should be less than 8 for application to permeable soils, and possibly much less than 8 for soils high in clay, in order to avoid either a severe reduction in soil permeability or

TABLE 1. Reported SAR (sodium adsorption ratio) for pulp and paper mill wastewaters.

Mill Type	Wastewater Type	SAR	Reference
Kraft	Raw whole mill	2.5-13*	NCASI 1959
Kraft	Whole mill	5	Vomocil 1974
Kraft	Pulp wash water	46	NCASI 1959
Kraft	Bleach plant	17-23**	Hayman and Smith 1979
Deink	Primary effluent	6	Flower 1969

*A survey of ten mills. Bleach mills tended to have higher SAR values than nonbleach mills. Total dissolved solids ranged from 270 to 1,250 mg/l.

**This wastewater had an electrical conductivity of 3.3 to 4.2 mmhos/cm.

the need for gypsum addition to the soil (Blosser and Owens 1964). Some reported SAR values for mill wastewaters are given in Table 1.

Priority Pollutants and Miscellaneous Constituents. Pulp and paper mill wastewaters have been well characterized in terms of heavy metals and other trace elements and specific organic compounds (U.S. EPA 1982). Typically, most of the priority pollutants in raw whole mill wastewaters are either undetectable or found at low concentrations, and substantial reductions in the concentrations of many organic species can be achieved in secondary treatment. A dependency of raw wastewater quality on mill type and practices can be seen. Reviews on the presence and toxicity of chlorinated compounds in waste streams from the chlorine bleaching of pulp are available (NCASI 1980, Kringstad and Lindstrom 1984). The behavior and significance of trace organic compounds during the land application of mill wastewaters have been given little study.

Sludge

Organic Matter and Clay. Because pulp and paper mill sludges contain organic matter as a significant portion of their dry solids, they should be able to condition the soil. The organic fraction of solids in primary sludge is predominantly wood fiber, whereas the dry matter in secondary sludge tends to be mostly microbial biomass.

Improvements in soil properties also may occur because clay minerals are present in mill sludges. High ash sludges, those with about 50 to 70% ash on a dry weight basis, are primary sludges associated principally with deink pulping and the manufacture of printing and writing papers and coated paperboard. The ash in such sludges should be mainly koalin clay, which should be particularly beneficial to sandy soils.

Studies have demonstrated noticeable improvements in soil properties of plant available water, organic matter content, and cation exchange capacity, following the application of mill sludge to agricultural land (Simpson et al. 1983, Thiel 1984, Einspahr et al. 1984). Mill sludge has been successful in promoting vegetation establishment on strip mine spoil (Shoemaker and Dickinson 1979, Hoitink and Watson 1980).

Nitrogen and Other Macronutrients. Data on the plant macronutrient content of pulp and paper mill sludges are displayed in Table 2. Mill sludges typically contain a lower concentration of each of the six elements than their municipal counterparts, but they possess measurable levels such that placement on the land could be beneficial in some cases.

Mill sludges exhibit great variability in macronutrient composition, a consequence of differences in the type and operation of the mill and in the wastewater and sludge treatment provided. Rock and Beyer (1982) observed that high sulfur sludges were associated with sulfite pulping. Large calcium concentrations are most likely to be found in primary and combined sludges from kraft pulp mills (NCASI 1984a). Secondary sludges are usually much higher in nitrogen and phosphorus than primary ones, because of the addition of these two nutrients to wastewater prior to biological treatment. Some mill secondary sludges are comparable to municipal sludges in terms of these two nutrients. The macronutrient content of a combined sludge is, of course, a function of the ratio and composition of the primary and secondary components.

TABLE 2. Elemental content of pulp and paper industry sludges.

Element	54 Pulp and Paper Mill Sludges* (NCASI 1984a)		Municipal Sewage Sludges (Chaney 1980, Sommers 1977)	
	Range	Median**	Range	Median
Macronutrients (% dry weight)				
Nitrogen	0.051-8.75	0.898	<0.1-17.6	3.3
Phosphorus	0.001-2.54	0.235	<0.1-14.3	2.3
Potassium	0.012-1.0	0.22	0.02-2.64	0.30
Calcium	0.028-21.0	1.4	0.10-25.0	3.9
Magnesium	0.02-1.9	0.155	0.03-1.97	0.45
Sulfur	0.020-2.00	0.468	0.6-1.5	1.1
Heavy Metals (mg/dry kg)				
Cadmium	<0.09-56	1.2	1-3,410	15
Chromium	3.0-2,250	42	10-99,000	500
Cobalt	ND -9.7	--†	1-260	10
Copper	3.9-1,590	52	84-17,000	800
Iron	97.1-10,800	1,540	1,000-15,400	1,700
Lead	<0.05-880	28	13-26,000	500
Manganese	13-2,200	155	32-9,870	260
Mercury	0.0009-3.52	0.35	0.6-56	6
Molybdenum	<2.5-14	--†	1.2-40	10
Nickel	1.3-133	18.3	2-5,300	80
Silver	<0.1-<11	0.55	--	--
Tin	<70.6	--††	40-700	150
Zinc	13-3,780	188	101-49,000	1,700
Aluminum	590-89,000	13,400	1,000-13,500	4,000
Arsenic	<0.07-8.3	1.2	1.1-230	10
Barium	17.9-1,800	160	<0.01-9,000	200
Boron	<1-491	25	4-1,000	33
Chlorine	<0.06-8,500	383	--	--
Selenium	<0.01-<31	0.21	1.7-17.2	5
Sodium	300-66,700	2,200	100-30,700	2,400
Titanium	3,100-76,000	--	--	--

NOTE: Observed range in pH for pulp and paper mill sludges was 5.0 to 9.9 (NCASI 1984a).

ND = Not detected, detection limit not stated.

*An assortment of primary, secondary, and combined (mixed primary and secondary) sludges.

**May be based on one or more "less than" values. Not given for elements analyzed for in four or fewer sludges.

†A survey of seven sludges from mills in Wisconsin found the cobalt and molybdenum contents to range from 1.8 to 4.5 and 5.2 to 8.9 mg/dry kg, respectively, and to have means of 2.5 and 6.7 mg/dry kg (McGovern et al. 1983).

††A survey of twelve mills in Maine revealed the tin content to range from 5.07 to 135 mg/dry kg and to have a median of 10.7 mg/dry kg (Rock and Beyer 1982).

The majority of the total nitrogen in mill sludges, usually 65 to essentially 100%, is in organic compounds. In general, the organic fraction is above 90%, with lower percentages associated most often with liquid or aged (lagooned) sludge.

Nitrogen mineralization is consequently an important process in land application. Moreover, mill sludges can be low enough in nitrogen that their ratio of organic carbon to total nitrogen (C:N) is greater than 20-30:1, and therefore the potential exists for the temporary microbial immobilization of nitrogen. Primary, combined, and secondary mill sludges have exhibited a range in C:N values of 32-930:1, 13-81:1, and 6-115:1, respectively (NCASI 1984a).

Numerous reports, among them those by Hermann and Gilbert (1982), Simpson et al. (1983), Yerkes (1971), and Smith (1980), have shown that land application of mill sludges has created plant yields of agricultural crops that are (1) either greater than those from control areas receiving no sludge or commercial fertilizer or (2) equal to or greater than areas receiving fertilizer alone. Of course, enhanced soil productivity can result not only through the addition of nitrogen but also by the addition of other nutrients and of soil conditioning substances.

Poor growth of agricultural crops has been attributed, on the other hand, to the microbial immobilization of nitrogen from the application of mill sludges with high C:N ratios (Dolar et al. 1972, Huettl 1981, Aspitarte et al. 1973). Fortunately, this problem can be handled by a number of possible management techniques related to crop selection and the timing and magnitude of the application (NCASI 1984a). Presumably forests should be more tolerant of short periods of immobilization than typical agricultural crops.

Heavy Metals, Soluble Salts, and Miscellaneous Elements. Data on the level of elements in pulp and paper mill sludges, in addition to macronutrients, are displayed in Table 2. One can conclude that mill sludges tend to be relatively clean in terms of heavy metals and other trace elements of potential concern. The median or mean levels of arsenic, cadmium, chromium, cobalt, copper, mercury, molybdenum, nickel, lead, selenium, tin, and zinc are substantially lower in mill sludges than in their municipal counterparts. Only for aluminum is the median greater in the case of the mill sludges. It is noteworthy that results of EP toxicity tests indicate that classification of a mill sludge as hazardous due to heavy metals would be a highly atypical circumstance (NCASI 1979).

As with the macronutrients, other elements in a sludge are a consequence of the type of mill and treatment plant and the specifics of their operation. Sludges found to be relatively high in heavy metals have often been ones from deink and wastepaper mills, but certainly not all sludges from such mills are high in these elements. Sludges from semichemical pulp mills may be high in sodium. Rock and Beyer (1982) associated high levels of aluminum with the use of clays in papermaking and the use of aluminum salts in the coagulation of wastewater solids.

Several studies, including those by Simpson et al. (1983), Hermann and Gilbert (1982), Jacobs (1978), and Huettl (1981), have examined the uptake of heavy metals and other elements into agricultural crops grown in soils amended with mill sludges. Though short term, these studies clearly demonstrate that mill sludges can be applied to land without adverse levels of such elements being absorbed into plant tissue.

Some pulp and paper mill sludges are high in sodium or total salts. Buchanan (1978) reported the sealing of a clayey soil that received a liquid mill sludge having a sodium concentration of 50,000 mg/dry kg, findings suggestive of salt phytotoxicity have been observed by Huettl (1981), Dolar et al. (1972), and Wilde (1979). Nonetheless, experience indicates that neither greatly reduced soil permeability nor salt phytotoxicity is a common problem. Furthermore, the potential for these problems can be minimized, if necessary, through monitoring or other management practices.

Specific Organic Compounds. Information on specific organic compounds in pulp and paper industry sludges is limited, but data have been presented recently (NCASI 1984a, NCASI 1984b). Mill sludges appear to be fairly clean in terms of priority pollutants and other compounds of possible concern; concentrations are typically much lower than those in the more contaminated municipal sludges. Compounds that have been found in mill sludges at concentrations above 10 mg/dry kg include naphthalene, some phthalates, chloroform, PCBs, and wood extractives or derivatives (abietic acid, dehydroabietic acid, norabietetriene, tetrahydroretene, and retene). Detectable TOX (total organic halogen) levels, 600 and 1,900 mg/dry kg, have been measured in two of four sludges examined (NCASI 1984b). Chlorinated lignin derivatives of large molecular weight are surmised to be responsible for the TOX readings.

The composition of organic matter in mill sludges should vary with mill type and practices, as it does with wastewater. Deink and other wastepaper mills may have PCBs in their sludge as a result of older carbonless copy paper being a component of the wastepaper recycled for pulp. Because PCBs have deliberately not been added to paper for several years, however, their level in sludges should be decreasing. Facilities using chlorine for bleaching pulp or chlorophenolic compounds for slime control also would be more likely to have detectable levels of chlorinated organics. The two sludges mentioned previously that contained detectable TOX were from mills with bleach plants. Chlorophenolic slimicides are no longer commonly used.

Limited research (NCASI 1984a, Thiel 1984, Thiel 1985) suggests that organic compounds in mill sludges pose minimal risk in land application. The issue of TOX in sludge is currently under examination by the industry.

Pathogens. It was noted earlier that process wastewaters from pulp and paper mills do not contain human waste and are free of enteric pathogens. Accordingly, sludges derived from these wastewaters are also free of pathogens.

LITERATURE REVIEW ON SILVICULTURAL APPLICATION

Literature on the land application of wastewater and sludge from pulp and paper mills mainly involves grasses and agricultural crops as the cover vegetation (NCASI 1984a, NCASI 1985). There is literature dealing with silvicultural aspects, however, and it is summarized below.

Wastewater
In an early report on applying pulp and paper mill wastewater to silvicultural land, Crawford (1958) noted that preliminary observation of a wooded area sprayed with a lagooned mixture of individual kraft mill wastes—evaporator condensate, digester condensate, and dregs washings—indicated no noticeable tree injury. Tree growth was to be monitored in the future. The woodland spraying was part of a pilot project whose main thrust was the application of wastewater to agricultural land.

Gellman and Blosser (1960) presented a survey of eighteen mills with full-scale or pilot programs for the land application of wastewater. Four mills reported experimental irrigation of wooded areas. It was noted that benefits to tree growth from wastewater application had not been documented as yet.

Barrett (1962) investigated the effects of irrigating slash pine grown in fine-grained soil with various kraft mill wastewaters. In a study using large, undisturbed soil cores, the wastes (pulp wash water, lagooned whole mill effluent, and two blends of wash water and black liquor—one 10% black liquor and the other 20%) were applied during two growing seasons at rates of 1.8 and 4.6 cm/wk (0.7 and 1.8 in./wk). Compared with controls receiving either no irrigation or irrigation with groundwater, wastewater application had no effect on seedling survival, which was excellent at 97%. Growth trends were not particularly distinct as to type of irrigant or application rate, possibly owing in part to a high level of rainfall, but the 20% black liquor mixture did depress tree growth.

A study using small pots employed the wash water and black liquor in five different blends (Barrett 1962). The application was made at a rate of 3.6 cm/wk (1.4 in./wk) for a period of five months, and water also was added to simulate rainfall. After two years, pots receiving 0 or 10% black liquor mixtures had 100% survival, a 20% black liquor mixture produced 82% survival, and treatments with higher proportions of the liquor caused 100% mortality. Black liquor was found to increase the pH, salinity, and sodium content of the soil; it was assumed that mortality resulted from elevated levels of soil salts.

The aforementioned study using large, undisturbed soil cores was continued for a total of five years, as reported by Jorgensen (1970). The higher irrigation rate produced waterlogged soil and reduced seedling growth regardless of irrigant. The black liquor mixtures at either hydraulic loading tended to reduce growth more than other treatments, however, and by the fourth year they caused some tree mortality. Poor tree growth was due to two factors in addition to excessive water: (1) the buildup of soluble soil salts and (2) reduced plant uptake of the nutrients potassium, calcium, and magnesium due to excessive soil sodium.

Flower (1969) presented pilot and full-scale experience with spraying primary effluent from a deink mill on a sandy, wooded hillside having a slope of 15 to 25%. The BOD_5 level in runoff, which constituted approximately 10% of the applied effluent, and in springs indicated good ($\geq$85%) removal of organics. Casual observation indicated no adverse effects to vegetation, and a detailed study of the vegetation and soil was noted to be in the planning stage.

A study of sodium accumulation in potted soils receiving various segregated wastewaters from a pulp and paper mill has been described briefly by Kadambi (1971). The wastes were introduced at a hydraulic application rate approximately equivalent to 2 cm/wk (0.8 in./wk) for a two-year period. Apparently no water was added to simulate leaching from rainfall. Plants grown in the pots were loblolly pine, cottonwood, sweetgum, sycamore, elm, and four species of eucalyptus. Instances of increased levels of soil sodium, up to as high as 1,000 mg/kg, were observed, as were large reductions in soil magnesium and calcium. The health of vegetation was not described.

Research with the surface irrigation of a mill secondary effluent on a short rotation, intensively managed tree plantation has been described (Hansen et al. 1980). Wastewater was applied during three consecutive summers to highly permeable sand planted with cuttings of poplar hybrids and black willow. For the first two summers the hydraulic loading was 28 cm/wk (11 in./wk), which corresponded to an annual application of about 3.6 m (11 ft); for the final summer the hydraulic loading was 71 cm/wk (28 in./wk), corresponding to an annual rate of 11 m (36 ft).

Groundwater quality, soil chemistry, and tree growth were monitored during the investigation (Hansen et al. 1980). Analysis of the shallow groundwater, present at depths of 2.7 to 4.2 m (9 to 14 ft), showed phosphorus and nitrogen in the wastewater to be essentially removed during soil percolation. Sodium, chloride, and sulfate were, however, among those substances whose groundwater concentrations rose significantly. It was noted that systems allowing substantial passage of wastewater constituents are situated best near a groundwater discharge area to restrict the affected portion of an aquifer. The nutrients phosphorus, calcium, and magnesium, as well as organic carbon, increased noticeably in the soil surface as a result of wastewater application. Tree growth varied because of uneven wastewater distribution, but was greatest where the wastewater was applied most heavily.

Sludge

Experimentation with a secondary sludge (N = 6%, dry weight basis) from a NSSC (neutral sulfite semichemical) mill as a fertilizer on intensive forestry sites has been reviewed by Smithe and Morin (1977). The sludge was incorporated into test plots at two sites, one of sandy and the other of loamy soil. The timing or method of application varied among the plots (i.e., first or second year of growth only, each of the first three years of growth, or sludge placement in trenches), but eventually all plots receiving sludge had loadings of total nitrogen amounting to 560 kg/ha (500 lb/acre). Compared with no-sludge controls, a 25 to 30% increase in the height of hardwood seedlings (species not stated) was attained after one year of growth. Groundwater monitoring revealed no noticeable increase in nitrate levels. Results were so encouraging that a full-scale program was initiated whereby liquid sludge was applied, usually by subsurface injection, to company tree farms at 560 kg N/ha (500 lb N/acre).

Buchanan (1978) discussed the development and operation of a program for applying mill secondary sludge (N = 1.9%, dry weight basis) to both agricultural and nursery land. The latter was leased from a company growing evergreen trees, and nursery stock occupied about one-third of the site at any one time. The nursery site was equipped with groundwater monitoring wells. It received sludge when nearby farmland was unavailable for application. Sludge was drawn from tanks through a hose to a subsurface injection implement drawn by a tractor. The cost of the program at that time was estimated to be $110/dry Mg ($100/dry ton) inclusive of capital and operation/maintenance.

A research project investigating the efficacy of a lagooned primary mill sludge (C:N = 40:1) as a soil conditioner has been described by Magnuson (1978), Wilde (1979), and Sopcich and Giesfeldt (1980). The sludge was sprayed on clearcut, sandy forest land at rates of 5 to 12 dry Mg/ha (2.2 to 5.4 dry tons/acre) with and without 340 kg/ha (304 lb/acre) of ammonium sulfate, and then incorporated into the soil. Red pine seedlings were planted and then monitored for three years. Sludge treatment was determined to have no influence, either beneficial or detrimental, on seedling development. The monitoring of groundwater, which was found at depths of 3 and 5 m (11 to 17 ft), indicated no influence on water quality as a result of the operation of the test plots.

In a companion study to the preceding research, a combined mill sludge (N = 1.7%, dry weight basis) added at rates of 11 and 22 dry Mg/ha (5 and 10 dry tons/acre) caused poor growth and high mortality of the pine seedlings (Wilde 1979). Planting was conducted soon after sludge incorporation, and the adverse reaction of the seedlings was thought to be due possibly to harmful levels of either ammonia or soluble salts. Analysis of fresh sludge sampled at a much later data revealed electrical conductivity values of 7 to 12 mmhos/cm, suggesting that salt toxicity may have been the cause of the poor plant response (NCASI 1984a).

An investigation into the influence of a mill combined (80% primary on a dry solids basis) sludge on the growth of Douglas-fir seedlings has been summarized (Aspitarte 1980). Some field plots received initial treatments of either sludge incorporated at rates of 56 to 448 dry Mg/ha (25 to 200 dry tons/acre), ammonium nitrate, or Milorganite (a composted municipal sludge). Other plots received no treatment at all (controls). Some of the treated and untreated areas were given surface ("mulch") applications of sludge approximately 30 days after seedling emergence and also during the second year of growth. Color, survival, and height of the trees were recorded at 4.5 and 22 months after planting. The various initial treatments improved seedling performance compared with controls. In addition, surface applications of sludge enhanced growth compared with similar plots without this surface addition.

Aspitarte (1980) noted an additional experiment, which evaluated the growth and survival of red alder seedlings and black cottonwood cuttings grown in riverbed sand with and without mill primary sludge amendment. Survival in amended plots was much greater than in unamended controls, and this was attributed to improved water retention by the sand.

White (1981) briefly reviewed the initiation of a program in which dewatered primary sludge from a municipal treatment plant was applied to cleared forest land owned by a paper company. The sludge was in essence one from a pulp and paper mill because the company contributed about 97% of the flow and waste load (BOD_5 and TSS) to the treatment system. The site had been instrumented for groundwater monitoring, but no seedlings (pine) had been replanted at that time.

The same municipal plant contracted to have sludge, which had accumulated in a biological treatment lagoon, spread on sandy soil vegetated with palmetto trees and young pines (White 1984). The sludge was dredged from the lagoon, pumped to the application site through a portable pipeline, and spread by traveling gun irrigators.

The effects of the surface application of a mill liquid secondary sludge on a 40-year-old red pine plantation situated in sandy soil were examined in detail (Brockway 1983). The sludge had total nitrogen and phosphorus levels of 7.6 (C:N = 9:1) and 1.0% dry weight, respectively, and was low in heavy metals and other trace elements of possible concern. It was applied at rates of 0 (control), 4, 8, 16, and 32 dry Mg/ha (0, 1.8, 3.6, 7.1, and 14.3 dry tons/acre). Sludge application significantly increased concentrations of nitrogen and phosphorus in the forest floor, but not those for heavy metals and other trace elements. Litter pH and salinity were elevated. Movement of chemical species into the surface soil was limited mainly to nitrate, ammonia, and phosphorus. In related work, Brockway and Urie (1983) estimated that the application rate should not exceed 9.5 dry Mg/ha (4.2 dry tons/acre) in order for groundwater immediately below the site to remain within the drinking water standard of 10 mg/1 nitrate-N. Understory vegetation showed raised levels of nitrogen and phosphorus, and its biomass was increased by as much as 92%, compared with controls. The uptake of other elements could not be related to sludge treatment, and no metal toxicity symptoms were observed. Improvements to overstory trees were manifested as increases in fasicle dry weight and needle length, and needles were darker in color. Significant changes in other tree growth parameters were not found. Based on other research, however, it was felt that sufficient time had not passed for the full benefits of the sludge to be seen.

Finally, two research programs discussed elsewhere in this volume that deal with the silvicultural land application of mill sludge, namely those by Crown Zellerbach Corporation and the University of Washington, have been given short descriptions previously (Silvicycle 1984a, Silvicycle 1984b, Amberg 1984).

CURRENT STATUS OF LAND APPLICATION IN THE PULP AND PAPER INDUSTRY

Wastewater

Seventeen pulp and paper mills have been identified recently as routinely applying to land all or part of their wastewater (NCASI 1985). The number increases to twenty-five if one includes fiberboard (insulation board and hardboard) mills as part of the paper industry. None of the mills irrigate wooded areas. It is still worthwhile to note that (1) with the exception of landfill leachates, whole mill wastewaters rather than segregated waste streams (e.g., spent pulping liquor, bleach plant effluent) are currently applied to land, (2) at a majority of mills the wastewater receives at least primary treatment prior to land application, and (3) for the most part, land application is restricted to smaller mills or to only part of the discharge of larger mills, so that the flow to land is less than or equal to 3.8×10^6 l/day (1×10^6 gal/day). As to the latter point, land requirements have been a factor in limiting the number of mills practicing the land application of wastewater.

It is unclear why none of the mills with land application systems irrigate trees. Early land application philosophy was prone to view trees as a nuisance, however, and in a few instances trees were removed from areas being prepared for application. Also, a surprising number of mills are probably too far from sufficient tracts of forests to make silvicultural land application economically attractive.

Sludge

Landfilling is the predominant means for managing mill sludges, but landfills are becoming more difficult to site and more costly to construct and operate because of increasingly stringent regulations, diminishing land availability, and public opposition. The most notable alternative receiving the attention of the pulp and paper industry is land application, as shown by a recent survey (NCASI 1984a). Forty-five mills with full-scale programs for the land application of sludge were identified. Nearly twenty of the programs were temporary, such as would occur with the dredging of wastewater or sludge lagoons; the majority involved the routine application of all or a portion of the daily sludge production. Thirty-five mills were found to be seriously considering one or more types of land application. Most of the exploratory programs included field demonstration projects.

Industry activity is mostly in agricultural utilization, but there are examples of silvicultural utilization. Six mills have full-scale programs in silvicultural land application, and sixteen have exploratory programs. The emphasis has been on application to nurseries and land cleared for afforestation or reforestation, but because of problems with weed growth, the interest in applying sludge to established forests has increased. Based on the design and conduct of full-scale silvicultural programs, application rates for mill sludges range from 9 to 224 dry Mg/ha (4 to 100 dry tons/acre) with 45 dry Mg/ha (20 tons/acre) being "typical." Application frequency to any given area varies from a one-time act to that resembling the application of wastewater.

The application of sludge to forest land can be attractive to paper companies for the same reason that it can be for municipalities. These reasons include the avoidance of food-chain crops and the use of isolated sites. In addition, silvicultural sites for mill sludges are usually on land owned by paper companies. This is desirable for program stability and means that the company receives the benefits of any enhanced plant growth. As with wastewater, a significant factor in limiting the use of forest land is transportation distance.

SUMMARY AND CONCLUSIONS

Land application is of increasing importance as a method of managing sludges from the pulp and paper industry. A few mills have full-scale silvicultural land application programs, and several more are seriously examining this option. In contrast, of the several mills applying wastewater to land, none are presently irrigating wooded land.

Mill wastewaters and sludges are both similar and different in quality compared with their respective municipal counterparts. Although concerns with the quality of mill wastewaters have been raised and problems in wastewater land application have been described, it must be stressed that there are successful full-scale programs with grasses and agricultural crops. Most mill sludges have certain properties, such as low heavy metal concentrations, that are desirable in land application, and they may serve as fertilizers or soil conditioners. Many mill sludges are low in nitrogen; this can be a source of concern with agricultural crops but may be acceptable for forest application.

Problems with the land application of wastewater or sludge can be minimized through a careful assessment of waste quality and site characteristics. Paper companies commonly perform laboratory and field studies to improve the design and operation of full-scale land application programs.

REFERENCES

Amberg, H. R. 1984. Sludge dewatering and disposal in the pulp and paper industry. J. Water Poll. Control Fed. 56:962.

Aspitarte, T. R. 1980. Agricultural utilization of wastewater treatment sludges. In Proceedings of the 1979 NCASI Central–Lake States Regional Meeting. NCASI Special Report 80–02. National Council of the Paper Industry for Air and Stream Improvement, New York.

Aspitarte, T. R., A. S. Rosenfield, B. C. Smale, and H. R. Amberg. 1973. Methods for pulp and paper mill sludge utilization and disposal. EPA R2-73-232. U.S. Environmental Protection Agency, Washington, D.C.

Barrett, J. P. 1962. The effect of paper mill effluents on slash pine seedlings. Progress report, Project AL 14.1. Southern Forest Experiment Station, USDA Forest Service, Pineville, Louisiana.

Blosser, R. O., and A. L. Caron. 1965. Recent progress in land disposal of mill effluents. TAPPI 48(5):43A–46A.

Blosser, R. O., and E. L. Owens. 1964. Irrigation and land disposal. Pulp Paper Mag. Canada 65(6):15–19.

Brockway, D. G. 1983. Forest floor, soil, and vegetation responses to sludge fertilization in red and white pine plantations. Soil Sci. Soc. Am. J. 47:776–784.

Brockway, D. G., and D. H. Urie. 1983. Determining sludge fertilization rates for forests from nitrate-N in leachate and groundwater. J. Environ. Qual. 12:487–492.

Buchanan, B. 1978. Land application of secondary sludge. In Proceedings of the 1975 NCASI Central–Lake States Regional Meeting. NCASI Special Report 78–06.

Chaney, R. L. 1980. Agents of health significance: Toxic metals. *In* G. Bitton, B. L. Damron, G. T. Edds, and J. M. Davidson (eds.) Sludge: Health risks of land application. Ann Arbor Science Publishers, Ann Arbor, Michigan.

Crawford, S. C. 1958. Spray irrigation of certain sulfate pulp mill wastes. Sewage Ind. Wastes 30(20):2306.

Dolar, S. G., J. R. Boyle, and D. R. Keeney. 1972. Papermill sludge disposal on soils: Effects on the yield and mineral nutrition of oats. J. Environ. Qual. 1:405–409.

Einspahr, D., M. Fiscus, and K. Gargan. 1984. Paper mill sludge as a soil amendment. p. 253–257. *In* TAPPI Proceedings, 1984 Environmental Conference. TAPPI Press, Atlanta, Georgia.

Flower, W. A. 1969. Spray irrigation for the disposal of effluents containing deinking waste. TAPPI 52(7):1267–1269.

Gellman, I., and R. O. Blosser. 1960. Disposal of pulp and papermill waste by land application and irrigation use. *In* Proceedings of the Fourteenth Industrial Waste Conference. Purdue University, Lafayette, Indiana. Extension Series 104. Engineering Bulletin 104(5):479.

Guerri, E. A. 1973. Industrial land application sprayfield disposal and wildlife management can coexist. TAPPI 56(4):95–97.

Hansen, E. A., D. H. Dawson, and D. N. Tolsted. 1980. Irrigation of intensively cultured plantations with paper mill effluent. TAPPI 63(11):139–143.

Hayman, J. P., and L. Smith. 1979. Disposal of saline effluent by controlled-spray irrigation. J. Water Poll. Control Fed. 51(3):526.

Hermann, D. J., and F. A. Gilbert, Jr. 1982. Prospects for land application of wastewater sludges from the paper industry. *In* TAPPI Proceedings, 1982 Environmental Conference. TAPPI Press, Atlanta, Georgia.

Hoitink, H. A. J., and M. E. Watson. 1980. Reclamation of acidic stripmine spoil with papermill sludge. *In* Utilization of municipal wastewater and sludge for land reclamation and biomass production. EPA 439/9–81–012. U.S. Environmental Protection Agency, Washington, D.C.

Huettl, P. J. 1981. Disposal of a primary papermill sludge on cropland soil. Ph.D. diss., University of Wisconsin, Madison.

Jacobs, L. W. 1978. Utilizing paperboard wastewater sludge on agricultural soils. *In* Proceedings of the First Annual Madison Conference of Applied Research and Practice on Municipal and Industrial Waste. University of Wisconsin–Extension, Madison.

Jorgensen, J. R. 1970. Growth and survival of slash pines irrigated by effluents. World Irrigation 20(3):16.

Kadambi, K. 1971. Accumulation of sodium in potted soil irrigated with papermill effluents. Consultant 93.

Kringstad, K. P., and K. Lindstrom. 1984. Spent liquors from pulp bleaching. Environ. Sci. Tech. 18(8):236A.

Magnuson, C. 1978. Options for solid waste utilization: Soil amendment. *In* Proceedings of the 1977 NCASI Central–Lake States Regional Meeting. Special Report 78–01.

McGovern, J. N., J. G. Berbee, J. G. Bockheim, and A. J. Baker. 1983. Characteristics of combined effluent treatment sludges from several types of pulp and paper mills. TAPPI J. 66(3):115–118.

National Council of the Paper Industry for Air and Stream Improvement (NCASI). 1959. Pulp and paper waste disposal by irrigation and land application. Stream Improvement Tech. Bulletin 124.

————. 1964. Settleable solids removal practices in the pulp and paper industry. Stream Improvement Tech. Bulletin 178.

————. 1972. *Klebsiella pneumoniae* infection: A review with reference to the water-borne epidemiological significance of *K. pneumoniae* presence in the natural environment. Stream Improvement Tech. Bulletin 254.

————. 1977. Pulp and papermill effluent nitrogen and phosphorus requirements for biological treatment and residuals after treatment. Stream Improvement Tech. Bulletin 296.

————. 1979. Nature and environmental behavior of manufacturing-derived solid wastes of pulp and paper origin. Stream Improvement Tech. Bulletin 319.

————. 1980. Chlorinated organics in bleach plant effluents of pulp and paper mills. Stream Improvement Tech. Bulletin 332.

————. 1983. A compilation of data on the nature and performance of wastewater management systems in the pulp and paper industry. Special Report 83–09.

————. 1984a. The land application and related utilization of pulp and paper mill sludges. Tech. Bulletin 439.

————. 1984b. Pulp and paper sludges in Maine: A characterization study. Tech. Bulletin 447.

————. 1985. The land application of wastewater in the forest products industry. Tech. Bulletin 459.

Rock, C. A., and B. S. Beyer. 1982. Pulp and paper wastewater sludges: The Maine perspective. *In* C. A. Rock and J. A. Alexander (eds.) Long range alternatives for pulp and paper sludges. University of Maine, Orono.

Shoemaker, G. H., and R. H. Dickinson. 1979. Disposal of sludge from pulp and paper mills: A case history. TAPPI 62(10):53–55.

Silvicycle. 1984a. Crown Zellerbach RENU project. Silvicycle 2(2):2.

————. 1984b. Pulp and paper/municipal sludge nursery demonstration. Silvicycle 2(2):1.

Simpson, G. G., L. D. King, B. L. Carlile, and P. S. Blickensderfer. 1983. Paper mill sludges, coal fly ash, and surplus lime mud as soil amendments in crop production. TAPPI J. 66(7):71–74.

Smith, K. E. 1980. Sludge-disposal costs reduced from Appleton's land-application program. Pulp and Paper 54(11):151.

Smithe, R. J., and M. Morin. 1977. The use of secondary treatment plant sludge for fertilizing intensive forestry tree farms at Packaging Corporation of America, Filer City, Michigan. *In* Proceedings, 1976 NCASI Central–Lake States Regional Meeting. Special Report 77–02.

Sommers, L. E. 1977. Chemical composition of sewage sludges and analysis of their potential use as fertilizers. J. Environ. Qual. 6:225–232.

Sopcich, D. J., and M. G. Giesfeldt. 1980. Land-spreading of pulp and paper mill sludges. Wisconsin Department of Natural Resources, Bureau of Solid Waste, Madison.

Thiel, D. A. 1984. Sweet corn grown on land treated with combined primary/secondary sludge. p. 93–102. *In* Proceedings, 1984 TAPPI Environmental Conference. TAPPI Press, Atlanta, Georgia.

——. 1985. Combined primary/secondary sluge proves beneficial for potatoes and corn. *In* Proceedings, 1985 TAPPI Environmental Conference. TAPPI Press, Atlanta, Georgia.

U.S. Environmental Protection Agency. 1982. Development document for effluent limitations guidelines and standards for the pulp, paper and paperboard point source category. EPA 440/1–82–025. U.S. EPA, Washington, D.C.

Vomocil, J. A. 1974. Irrigated farming utilizing pulp mill wastes. Preprint, 41st Annual Meeting Pacific Northwest Pollution Control Association, Richland, Washington.

White, P. 1981. Utilization of paper mill sludge in woodlands. *In* Proceedings, 1981 NCASI Southern Regional Meeting. Special Report 81–11. NCASI, New York.

White, T. E. 1984. Moving sludge from lagoon to land. BioCycle 25(3):41.

Wilde, S. A. 1979. Unpublished data summarized in P. J. Huettl, Disposal of primary papermill sludge on cropland soil. Ph.D. diss., University of Wisconsin, Madison (1981).

Yerkes, W. D., Jr. 1971. Secondary sludge as soil amendment. *In* Proceedings, 8th TAPPI Air and Water Conference. TAPPI Press, Atlanta, Georgia.

Chapter 5

Future Directions: Forest Wastewater Application

WADE L. NUTTER and JANE T. RED

ABSTRACT   Considerable advances have been made in the past decade in knowledge and design of forest land treatment systems using wastewater. Today there are numerous examples of the application and treatment of both municipal and industrial wastewater on forest land. However, much of the research on forest systems has been in only four regions of the United States, and as a result those are the areas where most of the operating systems have been implemented. More regional demonstration/prototype systems are needed in order to improve local technology and encourage acceptance by the public and regulatory agencies. To advance the knowledge of forest land treatment, to make the technology more widely applicable, and to improve the design of systems, the following issues have been identified as future needs: improved design criteria, assessment of risk, evaluation of economic factors, development of operational procedures, public and regulatory acceptance, and development of improved design and operational guidelines.

In the past decade a number of wastewater land treatment systems have been implemented on forest land—far fewer, however, than on agricultural land. Municipal and industrial systems have been put in operation on land areas ranging from a few hectares to over 2,000. Forest systems have been most widely used in regions where supplemental irrigation of crops is not generally practiced, land slopes are steep, inexpensive land is readily available, intensive management is undesirable, and a crop is required that is not part of the human food chain. Although a few forest land treatment systems are implemented each year, the selection of forest vegetation (and often the consideration of land treatment itself) occurs only because someone on the study team is aware that forests are a viable alternative and can be cost effective.

The 1973 Conference on Recycling Municipal Sludges and Effluents on Land (National Association of State Universities and Land Grant Colleges 1973), a workshop to evaluate the state of knowledge on land treatment, did not address the potential of forests for land treatment of wastewaters, nor did the first U.S. Environmental Protection Agency design manual for application of municipal wastewaters (U.S. EPA 1977). Thus little attention has been focused on the potential of forests for treatment of wastewater except for research at a few locations around the United States, most notably in Pennsylvania, Georgia, Michigan, and Washington. It is generally in these regions that most forest systems are found today. Local demonstration or prototype systems appear to have been important in leading the way to acceptance and implementation of forest land treatment systems.

The 1983 follow-up Workshop on Utilization of Municipal Wastewater and Sludge on Land (Page et al. 1983) recognized the role of forests in land treatment, as did the revised U.S. Environmental Protection Agency design manual (U.S. EPA 1981). The 1983 conference also identified a few areas of additional forest system research that were needed, many of them common to both forest and agricultural systems (such as public acceptance and health issues). However, solutions for agricultural systems will not necessarily provide valid answers for forest systems.

A number of land treatment design, operation, and risk issues are unique to forests. Some of these are: (1) the land area requirement is generally greater than for agriculture, because of lower nutrient uptake; (2) the presence of forest floor and soil organic matter improves potential for hydraulic management and is important to the fate of pathogens, organics, and heavy metals; (3) year-round operation is possible without risk of soil erosion; (4) a low level of management is required; (5) forests offer a flexibility in design and operation that permits easy modification to accommodate improved knowledge or site specific operational experience; (6) frequent access for harvest or crop management is not required; and (7) forest land is removed from the food chain.

The following general topics are those encompassing issues and gaps in knowledge specific to forest systems, as identified at the symposium on which this volume is based, the 1983 conference (Page et al. 1983), and from experiences of researchers, designers, and regulators. The topics are listed in order of relative importance, although all issues must be addressed in the near future.

1. *Design criteria*. Better design criteria for forest systems, including hydraulic loading, nutrient assimilation, growth response, and ecosystem simulation or response models.

2. *Risk assessment*. Procedures for evaluating the risk from pathogens and organics to humans and to wildlife.

3. *Economic factors*. Assessment of economic factors such as expected short- and long-term growth response and wood quality, potential opportunities for utilization, and cost-benefit analyses.

4. *Operational procedures*. Better information about system installation and operation to include application systems, year-round operation, site rehabilitation after construction, and harvesting techniques.

5. *Public and regulatory acceptance*. Improved means of communicating technology to the public and regulatory agencies to gain acceptance.

6. *Design and operational guidelines*. Improved guidelines at the federal and state levels for design and operation of forest systems to include technology transfer to and training of forest system operators.

Each of these topics will be discussed separately, with particular attention to the specific information needed to advance the design and implementation of wastewater application to forests. Land treatment of wastewater in forests has been successful, as demonstrated by the descriptions of a variety of systems reported in this volume. It is important to note that the systems described were designed to provide additional treatment to the applied wastewater and not merely to dispose of the wastewater. Improved tree growth was a secondary objective in some of the systems.

The future needs that are discussed below are directed at enhancing forest land treatment technology to make forest systems more widely used by designers and accepted by the public as an application of cost-effective and beneficial technology. As this information becomes available (some of which can be gained by better documentation of the performance of existing systems), the options to the designer for broader application of the technology will be expanded. In the meantime, the designs must follow the relatively conservative path used to date.

DESIGN CRITERIA

There are several important design criteria unique to forests that must be researched to improve our understanding—and subsequently the feasibility—of forest land treatment.

Perhaps the greatest gap in development of adequate design criteria for forest systems is the lack of knowledge of nutrient cycles, particularly nitrogen, in the forest ecosystem. Nitrogen loading is often the critical design factor controlling land area requirements and vegetation selection and management needs. Because each forest ecosystem is unique, nutrient cycling information must be developed for each forest type. Unfortunately, research to date in natural forest ecosystems is of little value in developing design criteria for wastewater application, because the addition of water and nutrients markedly alters the nutrient pathways in the normally nutrient-poor cycle.

With respect to nitrogen, the principal concern is maintenance of drinking water standards for nitrate-N in the groundwater. Thus net nitrogen storage in the harvestable biomass, the role of understory vegetation in the annual uptake and storage cycle, and occurrence and management of denitrification are important in developing site specific design criteria for nitrogen. This is not to imply that nitrogen assimilation design criteria suggested for many parts of the United States (Table 4-12, U.S. EPA 1981) are not suitable for current design. Improved knowledge of nitrogen assimilation design criteria will result in better long-term management and economic criteria. In addition, other forest systems for the production of biomass must be considered to improve the efficiency of nutrient storage and possibly reduce the land area required.

The expected growth response in the harvestable and nonharvestable biomass is directly related to nutrient cycling. Although economic return from the harvestable products is important, the increment of biomass that can be expected and its nutrient storage capacity are of greater significance. Growth response data are vital in determining nutrient assimilation design criteria as well as species selection and management techniques.

Forest systems generally have a greater hydraulic loading capacity than agricultural systems because of steeper slopes, the presence of the forest floor, less frequent access, higher evapotranspiration losses, and good soil hydraulic properties. But there is no widely accepted methodology for determining the hydraulic loading capacity. Programs developed by the U.S. Weather Service for the U.S. EPA that determine storage capacities based on climatic factors (see U.S. EPA 1981) grossly underestimate the hydraulic capacity of forests. Programs specific to forest conditions (including winter operation) need to be developed for use by the designer and regulatory agencies.

Other critical design criteria for which information specific to forests is required are the depth to water table and the slope. In many locations, significant areas with high water tables are occupied by forests. The treatment efficiency under controlled water table conditions is not known for many soil and forest types. In contrast to the flat-land systems, application of wastewater has been practiced successfully on slopes in excess of

35%. As slope increases, what critical design factors must be considered to ensure adequate treatment of the wastewater?

Improved knowledge of each of these design factors will improve efficiency of current designs and most likely expand the land and forest ecosystem base for the implementation of additional systems. At present, most systems account for these factors by providing a conservative design, which may not be the most economical.

RISK ASSESSMENT

Page et al. (1983) identified risk assessment of land treatment systems as a high priority need. Because of the unique characteristics of the forest (remoteness, presence of organic matter as forest floor and in the surface soil, removal from the human food chain, retention of biomass on site, etc.), a risk assessment addressing these factors must be forthcoming. (Many of the risk assessments of agricultural lands will also be useful for forest systems.)

The fate of organics, heavy metals, and pathogens must be better defined for forest systems. The importance of soil pH for metals retention is one area that needs further definition for forest systems. These improvements in knowledge about basic processes will provide the designer, regulatory agency, and public with better information for decision making and will serve to make designs more efficient and expand the land base considered suitable for land treatment.

The risk to wildlife living or transiting a forest land treatment system has been addressed in several studies. Results have shown no adverse effect. However, as forest systems are implemented in different regions and ecosystems, a survey of wildlife impacts will be necessary to assess the feasibility of land treatment, especially on large-scale sites.

ECONOMIC FACTORS

Little information exists about the cost of irrigating forests, since irrigation is not a common practice as it is in agriculture. Detailed analysis of existing forest irrigation systems and evaluation of alternatives would provide better information to the planner for developing cost-effective comparisons. Forest systems are frequently eliminated from consideration—in comparison with conventional treatment or other land treatment alternatives—because inadequate cost information is available.

Information about forest tree growth response and potential markets of the wood products is also necessary to prepare adequate cost comparisons of alternative treatment systems as well as to select the optimum forest system management scheme. Growth response models for the major forest types where land treatment is most likely to occur would be important tools for the economic planner and the designer who must consider nutrient budgets. The economics (and suitability for wastewater treatment) of other forest systems, such as short rotation hardwoods for biomass or energy production, must be considered. Quality of the wood and its intended use must also be considered. Little work has been done on wood quality associated with land treatment systems.

Several land treatment systems, two of which are reviewed by Nutter (Chapter 37) and Sopper (Chapter 38), employ recycling of drinking water and/or other resources (woody biomass energy, etc.). There is a potential at many locations to incorporate recycling into the design. Improved understanding of the economics of recycling is likely to result in greater application of this technology.

OPERATIONAL PROCEDURES

Experience at a number of forest land treatment systems has shown that improved procedures must be developed for installing an irrigation system with minimum site disturbance. Site disturbance becomes the limiting factor controlling application of wastewater. The site cannot be fully utilized until complete rehabilitation occurs. Regulatory agencies must also develop realistic time schedules for implementing application of wastewater following construction, recognizing that a construction rehabilitation period is necessary before the system can come on line.

An important consideration to the long-term operation of a forest land treatment system is harvesting of the biomass and removal of the stored nutrients. Harvesting must be followed by some type of regeneration scheme if the land treatment operation is to be continued. Wastewater application must be reduced or discontinued during the harvesting and regeneration period. Too often designs have failed to incorporate adequate land area for use during harvesting and regeneration, or have failed to consider the need to harvest the biomass at all. The efficiency with which the harvesting and regeneration program is carried out and the rate at which the site is rehabilitated to receive the design water and nutrient loading are important for successful system operation. Harvesting and regeneration techniques that preserve the qualities necessary to achieve successful treatment of the wastewater must be developed and tested.

PUBLIC AND REGULATORY ACCEPTANCE

Public acceptance of wastewater irrigation has greatly improved in the past decade. However, because of the lack of operating forest systems and demonstration sites, there tends to be a lower acceptance level of forest systems. Three principal issues must be addressed to gain improved public and regulatory agency perceptions of forest systems: public education, agency technical staff education, and improved communication between the research and design communities on the one hand and the public and agencies on the other. Public acceptance was an issue given high priority at the 1973 conference (National Association of State Universities and Land Grant Colleges 1973), and although gains have been made, it was still identified in 1983 (Page et al. 1983) as a high priority topic.

Forests present several unique issues concerning public and regulatory acceptance, many of them related to misconceptions the public has about forest management practices in general. Experience at many locations around the country has shown that a well-founded public awareness program initiated early in the project-planning stage elicits a positive response when it is time to compare alternatives. This same approach has also been successful with regulatory personnel (who frequently have not been exposed to land treatment technology). Documentation of these successful programs as case studies and assessments of attitudes, such as that described by Peyton and Gigliotti (Chapter 32), must be undertaken not only at completed projects but also at those in the planning stage to provide the planner and designer with the opportunity for public input.

An integral part of public and regulatory agency education is communication and technology transfer between the scientific, design, public, and regulatory communities. In the United States, federally funded technology transfer programs have been cut and the responsibility must be assumed by other interested parties. Technology transfer through professional groups, specially arranged seminars, and public meetings is necessary to keep the rapidly emerging technology before all parties. This can be achieved through improved communication within the technical community and between it and the public. In all likelihood, the technical community must lead the way.

DESIGN AND OPERATIONAL GUIDELINES

There is a continuing need for new information to be translated into working design and operational guidelines that regulatory agencies and the public can use to judge the worthiness of projects. After the technology is accepted, there must be an orderly means of incorporating it into existing or new statutes and regulations. This issue must be addressed at both the national and state levels. The incorporation of emerging technology into design and operational guidelines ensures application of the latest technology to achieve a cost-effective design and system operation within prescribed environmental constraints.

Vital to proper implementation of the technology is development of guidelines for system operation and certification of trained operators. Most states currently require an operator to maintain a wastewater treatment plant operator's certification, a situation wholly inadequate to manage most land treatment systems—forest or agriculture. To ensure that environmental and public goals are met, operator training programs must be developed. These should be established on a regional basis, since most states do not have an adequate base for such certification programs. The training programs should be designed to include a continuing education component.

CONCLUSIONS

Six general issues have been identified as future needs to improve the state of knowledge and implementation of wastewater treatment by forest systems: improved design criteria, assessment of risk, evaluation of economic factors, development of operational procedures, public and regulatory acceptance, and development of improved design and operational guidelines. These six issues can be divided into two general categories: design/operation and public acceptance.

The principal issues that would provide the greatest step forward in improving our knowledge and implementation of land treatment systems are: (1) improved understanding of the nitrogen cycle, including improved estimates of net annual nitrogen storage in the biomass, the role of understory vegetation, encouragement and management of denitrification, and procedures to manage the vegetation to maximize nitrogen assimilation; (2) development of growth response models for use in nutrient assimilation assessments as well as economic forecasting; (3) assessment of alternative forest vegetation systems (i.e., short rotation hardwoods) to minimize land area requirements and maximize biomass production without compromising wastewater treatment goals; (4) improvement in installation and operation procedures for forest irrigation systems; (5) development of forest management techniques, including harvesting and regeneration procedures, that recognize the unique features and goals of a forest land treatment system; (6) improved understanding of the relative risks associated with forest wastewater applications of pathogens, heavy metals, and organics compared with other wastewater treatment and disposal alternatives; (7) improved procedures to transfer information to the public and regulatory agencies to gain acceptance of the appropriate technology; and (8) implementation of regional demonstration/prototype systems for designers, regulatory agencies, and the public to observe and use to improve communications.

Specification of knowledge gaps and future needs does not imply that systems designed today can't meet environmental goals or regulatory intent. Quite the contrary. Forest systems have been designed and are operated successfully, as evidenced by the number of systems described in this volume. Expansion of the knowledge base will serve to fine-tune design and operation procedures as well as expand the base of potential forest land areas available for land treatment. It will also serve to aid in developing realistic economic comparisons with other vegetation systems and other treatment alternatives.

REFERENCES

National Association of State Universities and Land Grant Colleges. 1973. Proceedings of the Joint Conference on Recycling Municipal Sludges and Effluents on Land. NASULGC, Washington, D.C. 244 p.

Page, A. L., T. L. Gleason III, J. E. Smith, Jr., I. K. Iskandar, and L. E. Sommers (eds.) 1983. Proceedings of the 1983 Workshop on Utilization of Wastewater and Sludge on Land. University of California, Riverside. 480 p.

U.S. Environmental Protection Agency. 1977. Process design manual for land treatment of municipal wastewater. EPA 625/1-77-008.

——. 1981. Process design manual for land treatment of municipal wastewater. EPA 625/1-81-013.

Chapter 6

Future Directions: Forest Sludge Application

DALE W. COLE and CHARLES L. HENRY

ABSTRACT As illustrated in the numerous research papers and case studies included in this volume as well as the other published literature, there is a general conclusion that environmentally sound applications of municipal and industrial sludge can be made to forest systems. This confidence in forest applications has lead to a number of demonstration as well as operational sludge application programs. But there remain unanswered questions and fine-tuning efforts that must be addressed in order to provide maximum economic benefits and minimize the risk to the environment and public health. This paper addresses these needs, with specific reference to the contributions and discussions of the symposium papers in this volume.

To a large extent, the 1983 Workshop on Utilization of Municipal Wastewater and Sludge on Land (Page et al. 1983) identified major needs for a forest sludge application program. Expanding on this list, the following needs are seen as the most critical to forest systems.

1. *Public and regulatory acceptance.* Better communication is needed between researchers, municipalities, regulators, and the general public.

2. *Economics.* Costs of forest application, as well as the benefits received through increased forest production, must be determined.

3. *Design criteria for applying sludge.* A sound basis for calculating nitrogen application rates must be established, primarily through a better understanding of the pathways and rates of nitrogen accumulation and loss; appropriate application rates of heavy metals for the broad range of forest soil conditions must be determined; buffer requirements and slope limitations must be developed to ensure protection of drainage waters; and the importance of organics as a potential design limitation must be resolved.

4. *Improved application and management techniques.* Better application and management techniques, including frequency of application, are essential to improve the efficiency and reliability of forest application programs. Management of the forest for sludge application needs to be integrated with traditional forest management practices.

5. *Risk Assessment.* Risk assessment procedures useful for decision making must be developed. It is particularly important to study heavy metals, pathogens including viruses and parasites, and organic substances.

Each of these categories will be discussed with specific attention to the type of information needed before there will be full acceptance of forest application of sludge. This is not to suggest that operational use of sludge in forestry must wait until all the answers are available. This has not proved to be necessary nor desirable in either agriculture or land reclamation programs. In forestry, there are numerous examples, as we have seen from the case studies described in this publication, where sludge has been successfully applied to forested sites. Rather, the following suggestions for future needs are of the type designed to promote public and regulatory acceptance of this concept, improve the efficiency of application, target forest application areas that will maximize benefits while minimizing costs, and develop application prescriptions tailored to specific sites.

In discussing these needs, it must be recognized that application technology is still much in its infancy and future advancements can be expected. The added economic benefits that a sludge application program in forestry could provide from timber production must wait until additional and longer term forest growth information is available. Although we believe sufficient information is available to establish design criteria for applying sludge to forest sites, there remains much work before we can identify the best candidate sites and prescribe optimum application rates. The uncertainties associated with the fate of nitrogen, the stability of sludge on slopes, especially during periods of heavy rain, the mobility of heavy metals in low pH forest soils, and the fate of organics all suggest that the research phase of this program is not over.

PUBLIC AND REGULATORY ACCEPTANCE

This is, in all probability, the most critical area that must receive attention before there is wide acceptance of forest sludge application. No matter how sound a program might be from a scientific or engineering perspective, if it has not been accepted by the general public and the regulatory agencies, for whatever reasons, it clearly will not be implemented. The kind of systematic attention that has been given to the underlying science and application technology has not been applied to the problems associated with acceptance. Public acceptance is typically addressed only after there has been an acceptance problem. A systematic approach to public information as described by Peyton and Gigliotti (Chapter 32) is seldom undertaken. To add to the acceptance difficulties, the public, and in many cases the regulatory agencies, have higher standards and expectations for a forest application program than for alternative programs for sludge disposal or utilization. Examples one could mention include public "giveaway" programs, incineration, agricultural applications, and the marketing of sludge to the general public.

There are several reasons for this dual standard, some of which are emotional and others stem from a lack of operational experience with forest systems. The major reason, we suspect, is that forest application is a new concept with little precedence for acceptance. It should not surprise us that forest application is viewed by many with suspicion. Without doubt, the application of sludge to a forest will temporarily change some public activities that the forests can provide. For example, hiking and the gathering of edible forest products must be temporarily curtailed. The fact that a forest area may also be part of a defined watershed unit providing water for recreational or domestic purposes is a cause for additional concern in a sludge application program. Forests are viewed by most as a pristine environment. Sludge hardly adds to this image.

A major problem facing a regulatory agency in issuing an application permit may be a lack of experience and knowledge in forest sludge application. The agency faced with a permit request may have no previous experience with this alternative. In some cases a process for permitting such a practice is not available. Few state or local health districts have design manuals or operational procedures for forest application of sludge. Consequently, sludge application may be denied for procedural reasons without regard for the merits of the case.

It is clear that there is a need for better information transfer to the regulatory officials.

In part, symposiums and workshops can help to correct this information gap. Before we can expect wide acceptance of programs applying sludge to forested sites, it will be necessary to develop design manuals specific to given areas as well as training courses for regulatory officials. This will not happen rapidly, nor will the acceptance of forests as application sites.

In order to implement a sludge application program, we need to develop a systematic program addressing the process of public and regulatory acceptance. This process has be be carried out with the same commitment given the assessment of the other aspects of the program, including the environmental risks, growth response, public health concerns, and operational technology. Public acceptance research, typically little more than opinion surveys and public meetings, seldom involves the public in a meaningful way. The regulatory officials, especially at the local level, need to become far more knowledgeable about the risks and benefits of forest land application and not focus so narrowly on permit processes that may or may not be appropriate to a forest land application program.

The key to both public and regulatory acceptance lies in better communication between those doing the research, the treatment plant operators and administrators, the regulatory agencies involved in the permit and monitoring process, and the public and industry, who are the primary sources of sewage materials. Unfortunately those who generate the raw sewage do not usually involve themselves with its disposal or reuse.

ECONOMICS

The economic feasibility of a forest application program depends on the cost of the operation minus any values that might be derived. In forests, these added values are measured by the quantity and quality of the wood produced due to the sludge addition. The total costs for such a program, including any benefits that may be derived, must then be compared with other alternatives available to a treatment plant. A number of uncertainties enter into this calculation, making clear comparisons between alternatives difficult. For example, the application technology of placing sludge into forests is only now at the stage of development where large-scale operations are possible, as suggested by Henry et al. (Chapter 34) and Nutter and Torres (Chapter 47). Consequently, costs of such operations are not firmly established. Considering the rapid development of this technology during the past few years, one would have to conclude that more efficient equipment will be available in the future. With our current limited experience, the maintenance and life expectancy of such equipment is uncertain at this time.

More important, the economic value of the forest response is only now being recognized. We still do not know how long a forest will respond to a single application of sludge. The growth response at Pack Forest, which now covers a period of seven years from a single application, has not shown any sign of diminishing (Henry and Cole, Chapter 43). It is conceivable from our current evidence that a single application of sludge can result in the permanent enhancement of the forest site and the productivity of the forest stand. Should this prove to be the case, we will have achieved an accomplishment rarely obtained with chemical fertilizers. Current studies in the Pacific Northwest with Douglas-fir indicate that the growth response from sludge is at least twice as great and has lasted at least twice as long as that obtained when this species is fertilized with urea, a common management practice in the Pacific Northwest (Cole et al. 1984).

This is not to suggest that forest lands should be limited to a single application. Depending on the local conditions, it is entirely possible that a site should receive continuous or periodic applications for optimum response. To maximize economic benefits we must be able to recognize those sites where a high growth response is expected. Current studies indicate a great deal of variation in growth response between sites and tree species receiving sludge. Unfortunately, we do not know with certainty how to identify sites with a high growth response potential.

We must also be able to establish better prescriptions for the quantity and timing of the application. As is true of agricultural crops, the best growth response will probably be obtained with application rates exceeding the fertilizer needs of the vegetation. The ideal prescription rate should maximize growth response while at the same time minimizing the possibility of any adverse environmental or public health consequences (Brockway et al., Chapter 21; Zasoski and Edmonds, Chapter 9). This optimal application rate has not been established.

In order to assess the economic value of the wood produced, we must also have information on changes in wood quality caused by the sludge addition. As discussed by Briggs et al. (Chapter 22), we can expect to see a shift in some wood properties. What remain to be established, however, are the significance of this shift and if the changes in wood properties will detract from the value of the product. Clearly, the accelerated rate of growth will increase the log diameter and thus the grade for sawlogs. Where trees are grown for biomass or pulpwood, changes in wood quality may not be an issue.

When growth and wood quality information is available, perhaps in three to four years, there will still be uncertainty in calculating the economic value of sludge. There are too many alternative ways to consider costs, enter discount rates, project growth, and assume future values of wood products for any common agreement on the projected costs or values of such a program.

It is clear from the studies in the Pacific Northwest that we can get very close to achieving a balance between the cost of applying sludge (approximately $150/dry Mg) and the potential benefits from the increased volume of wood produced. Preliminary calculations based on a 10% discount rate carried forward for forty years project the value of sludge from $50 to $300/dry Mg.

DESIGN CRITERIA FOR APPLYING SLUDGE

To establish an appropriate application rate, it is necessary to take into consideration the response of the forest as well as the potential risks to the public. If tree growth is the only consideration, it is difficult to apply too much sludge to an established forest. For example, at Pack Forest we have applied levels in excess of 450 dry Mg/ha—approximately 25 cm of 18% solid sludge—with excellent response to forest growth. However, other aspects of the environment were significantly changed. Peak nitrate levels in the groundwater, 10 to 15 meters deep, exceeded drinking water standards by nearly 20-fold (Reikerk and Zasoski 1979). Fortunately, we do not have to add sludge at these quantities to receive acceptable growth response. As stated earlier, application prescriptions must be developed that will allow an excellent growth response without compromising public health and environmental considerations. This can only be done by (1) understanding and managing the nitrogen associated with a given application treatment, (2) determining the fate of heavy metals associated with a sludge addition, with special consideration to the acidic soils associated with most forest sites, (3) establishing appropriate buffer zones to protect drainages and wetlands that could potentially be affected, and (4) determining maximum slopes that should receive sludge for given application rates and weather conditions. These four considerations are discussed below.

Nitrogen

A sludge application of 90 dry Mg/ha typically includes enough nitrogen (assuming a 3.5% N content) to double the total nitrogen level of many forest ecosystems. To develop an application prescription it is essential to know the fate of this nitrogen, including how much is utilized by the forest. Unfortunately, there is insufficient information, in spite of many years of research, to define clearly the pathways and transformation rates of nitrogen following application. Most critically we need to know how much nitrogen under what conditions is converted to an available form (ammonium or nitrate) in the soil. An excess of nitrogen in the available form can result in nitrate leaching (Henry 1985). A shortage of available nitrogen will result in a poor growth response. The rate at which the nitrogen is transformed to available forms varies widely with the type of sludge as well as the environmental conditions, making it difficult to establish an appropriate application rate and adding to the uncertainty of the results. It is clear we need a better understanding of the pathways and rates of nitrogen transformation following a sludge application if we want to optimize growth response and minimize risks.

Heavy Metals and Organics

Heavy metal loadings are typically not limiting factors for most forest applications as long as the sludge is primarily from a "clean" or nonindustrial municipal source. There have not been any reported cases of phytotoxicities apparently due to heavy metals even when excessive loading rates have been applied. The heavy metals do not appear to leach in forest soil, despite the relatively low pH (Zasoski et al. 1984). In addition, since the forest is typically a step removed from the human food chain, there is little in the way of public risk.

Questions have been raised, however, that do need additional attention (Chang et al., Chapter 8). Many forest soils, especially in the surface horizons, have soil pH values below 5. There has been only limited research on heavy metal mobility in the soil and availability to plants under these acidic conditions. Theoretically, we would expect the mobility and availability of heavy metals to increase substantially under these conditions. However, whether this increase could cause a problem in uptake or leaching is an unanswered question. Although the likelihood of such problems occurring is reasonably low, it must be recognized that definitive research on this issue has not been completed.

The same point could well be made regarding organics. Although there is no evidence to suggest that organics associated with most municipal and industrial sludges or decomposition products from sludge pose a public health risk (Overcash, Chapter 12), it must be recognized that research regarding this issue is limited. The large number of organic compounds found in sludge adds to this confusion. Hopefully it will be possible to select a limited number of these compounds for screening purposes. Research is obviously needed to identify which compounds could best be used for this purpose.

Buffer Requirements and Slope Limitations

A forest site offers several challenges to those involved in application design. Unlike most agricultural sites, a forested area is often located on sloped terrain divided by permanent or intermittent streams. It is obviously necessary in designing a forest application area to avoid any condition that might allow the sludge to move overland into a water course. A successful design depends on stabilizing the sludge to minimize sludge movement. To avoid the possibility of stream contamination, buffer zones ranging from 15 to 60 meters are recommended. The amount of buffer needed is in turn dependent on a long list of variables, including slope steepness, surface roughness, ground vegetation cover, amount of sludge applied, fluid characteristics of the sludge, and climatic conditions such as the amount of rain the site might receive before the sludge becomes stabilized.

Establishing such buffer zones is more of an art than a science. We clearly need a better understanding of the conditions leading to potential overland flow of sludge or any of its components before precise recommendations can be made for buffer areas or specific slope limitations. This will require additional research as well as direct experience through demonstration projects. Until we have this additional information and research results, it is essential to approach application design in a conservative way. The application of more that 1 cm of dewatered sludge on sloped, forested terrain might require a series of separate applications, making certain that the previous application is fully stabilized before any additional sludge is applied. Stabilization will obviously take longer in winter than in summer, when drying and decomposition are far more rapid. Unfortunately, sludge cannot be injected into the soil of established stands as a means of increasing its stability—a common practice in agriculture—because of the presence of the forest and its rooting system. This, however, could be an option for recently cleared areas prior to planting, if the terrain is sufficiently level.

It is essential to gain such an understanding in order to maximize the forest land area available for sludge. If the buffer zones become too large or slope limitations too limiting, the efficiency of a sludge application operation could be affected, as could the amount of area benefiting from the addition. General design manuals and "rules of thumb" are often inadequate when one is designing an application program for a specific site. A strong case could be made for site-specific design criteria as a means of minimizing environmental problems while maximizing the available space.

IMPROVED APPLICATION AND MANAGEMENT TECHNIQUES

It has only been in recent years that sludge has been applied to forest sites of sufficient land area that this phase of the operation could be addressed. The early application technologies were designed to service research plots and not operational sites. The application vehicles were prone to frequent breakdown and performance problems. The application system and management program developed by Seattle Metro and the University of Washington (Henry and Cole, Chapter 43) for their demonstration programs has moved us very close to a system that is fully operational. The cost of forest application with this latest technology is approximately $150/dry Mg. With increased experience and the further modification of equipment, it is hoped to reduce this cost even further. The information that will aid in equipment improvement can be derived from demonstration level studies and operational programs.

Demonstration studies play an additional role in a sludge application program by providing a link between research studies and full-scale operational programs. Without demonstration studies, it is highly speculative to go from research to operations. These studies offer a logical means by which research information can be scaled up to provide the bases for a program at the operational level. They also provide a means to fine-tune application technology before making major commitments for capital equipment. Through a demonstration program, it should be possible to develop application manuals as well as cost-benefit information necessary before sludge can be applied as a standard silviculture prescription. Thus a well-designed demonstration program is an obvious step that should help any large-scale sludge generator in formulating a full-scale sludge application program.

RISK ASSESSMENT

An important part of the planning process for a land application program is establishing risk assessment of the hazards involved. The major concerns mentioned throughout this symposium are nitrates (Zasoski and Edmonds, Chapter 9), heavy metals (Chang et al., Chapter 8), pathogens including viruses and parasites (Sorber and Moore, Chapter 7), and organic substances (Overcash, Chapter 12). It is generally recognized by the regulatory agencies that an adequate risk assessment procedure does not exist for a sludge land application program, and that what is available in not readily usable by decision makers. This deficiency in the planning process has an obvious effect on implementing new sludge application programs and clearly adds to the difficulties and uncertainties that already exist in establishing a program in forestry.

Seattle Metro (1983) has made an excellent start in developing a risk assessment document, the basis of which is reported by Munger in Chapter 11. It must be recognized, however, that such a document has to be developed through a reiterative process including systematic research as well as experience associated with demonstration land application programs. Risk assessment must be achieved through a collective and integrative analysis of the total system. Addressing only a specific condition without understanding how this condition interacts with other components of the system leads to a misunderstanding of the risks involved. As a case example, sampling nitrate levels in the soil will provide an assessment of nitrate risks only if one knows the relation between such concentrations and the movement of nitrate from the application site, as well as the dynamics of the receiving groundwater aquifer.

There is no such thing as zero risk. It is important to decide on an acceptable level of risk and design into the risk assessment the variable factors such as frequency of flooding, storm intensities, and so forth. To do so does not require more research on risk assessment as such. Rather research is needed on understanding forest ecosystem behavior in relation to the risk in question.

CONCLUSIONS

1. Public and regulatory education remains an area that has received far too little attention. Without acceptance from the public and especially the regulatory agencies, a forest application program cannot be implemented.

2. The economic viability of forest sludge application compared with other alternatives is an unanswered question and is likely to remain so until long-term growth response information is available, as well as costs of application.

3. Preliminary design criteria for a forest application program have now been developed. Additional experience and carefully directed research are needed in fine-tuning design criteria to target the best sites and develop the best application prescription and application techniques for each site.

4. The uncertainty associated with fate of nitrogen, sludge stability on slopes, and mobility of heavy metals, especially in low pH forest soils, points to the need for additional research in these areas.

5. A more complete and definitive risk assessment document should be developed that will be accepted by the regulatory agencies, and that can be used as a tool by municipalities.

REFERENCES

Cole, D. W., M. L. Rinehart, D. G. Briggs, C. L. Henry, and F. Mecifi. 1984. Response of Douglas-fir to sludge application: Volume growth and specific gravity. p. 77–84. *In* 1984 TAPPI Research and Development Conference, Appleton, Wisconsin. Technical Association of the Pulp and Paper Industry, Technology Park, Atlanta, Georgia.

Henry, C. L. 1985. Nitrate leaching from fertilization of three Douglas-fir stands with municipal sludge. Unpublished report.

Municipality of Metropolitan Seattle. 1983. Health effects of sludge land application: A risk assessment. Metro Report.

Page, A. L., T. L. Gleason III, J. E. Smith, Jr., I. K. Iskandar, and L. E. Sommers (eds.) 1983. Proceedings of the 1983 Workshop on Utilization of Municipal Wastewater and Sludge on Land. University of California, Riverside. 480 p.

Riekerk, H., and R. J. Zasoski. 1979. Effects of dewatered sludge applications to a Douglas fir forest soil on the soil, leachate, and groundwater composition. p. 35–45. *In* W. E. Sopper and S. N. Kerr (eds.) Utilization of municipal sewage effluent and sludge on forest and disturbed land. Pennsylvania State University Press, University Park.

Zasoski, R. J., R. L. Edmonds, C. S. Bledsoe, C. L. Henry, D. J. Vogt, K. A. Vogt, and D. W. Cole. 1984. Municipal sewage sludge use in forests of the Pacific Northwest, U.S.A.: Environmental concerns. Waste Management and Research 2:227–246.

Environmental Effects

Microbiological Aspects of Forest Application of Wastewater and Sludge

CHARLES A. SORBER and BARBARA E. MOORE

ABSTRACT   For many years there has been considerable discussion regarding the potential microbiological health effects with the land application of wastewater and sludge. During this period, research has focused on defining a scientific basis for organism survival and transport during wastewater treatment and in soils. Such data are essential to the formulation of a consensus on risk associated with the use of wastewater and residuals on forest land. This paper attempts to bring together the factors important in the use of wastewaters and sludges in forest applications. The microorganisms of importance are identified, as are the most important potential pathways of these organisms to humans. Mitigating factors such as various types of wastewater and sludge treatment, pathogen survival in this environment, and pathogen transport along the pathways are elucidated. Some fundamental conclusions regarding the forest application of wastewater and sludge are drawn. In general, risks to human health by microorganisms are judged to be minimal under conditions of well-defined treatment, transport, and management of forest application.

Increased quantities of municipal wastewaters and sludges are being generated annually in the United States. With this increased production comes the obvious need for disposal. However, it has long been accepted that treated municipal wastewaters and stabilized sludges have value. They contain nutrients, particularly nitrogen and phosphorus, as well as organic matter. Furthermore, the use of wastewater for irrigation can be invaluable, especially in arid and water-scarce regions of the world. These attributes suggest that the beneficial use of treated municipal wastewaters and sludges would be more prudent than direct, ultimate disposal.

While several beneficial uses of treated municipal wastewater and sludges have been identified, demonstrated, and initiated, their application in silviculture holds particular promise. This judgment is based on the relative isolation of forested areas, the amount of land dedicated to silviculture, the documented increase in forest productivity, and the inherent value in the reuse of any resource. Nevertheless, this paper is not intended to extol the virtues of municipal wastewater and sludge use. Rather, it will focus on the microbiological aspects of this practice as they relate to public health.

By and large, microorganisms of public health consequence occur in wastewater, owing to the existence of infection or disease in the community served. In fact, it has been suggested that in a normal, healthy community as many as 0.1 to 10% of the population may be shedding an infectious agent at any time (Sobsey and Olson 1983). Obviously, during epidemic disease periods, the shedding rate increases substantially. In addition, a variety of microorganisms can be introduced into domestic wastewater by animals.

The type of organisms present in wastewater is a function of a number of factors. These include the age and socioeconomic status of the community, the geographic location, and the health practices such as immunization. Once organisms leave their human or animal hosts they enter a hostile environment. Under unfavorable environmental conditions such as limiting nutrients, elevated temperature, desiccation, or pH extremes, living organisms will gradually decrease in numbers. Thus, from a public health standpoint, the issue becomes one of understanding what conditions affect the decline in microbial numbers and being able to predict the time required to achieve acceptable organism levels in either wastewater or sludges. Then it is necessary to consider the factors that control the transmission of microorganisms from wastewater or sludges applied to forest lands to a susceptible human population.

The objective of this paper is to assess the microbiological health effects of forest application of municipal wastewater and sludges, to identify the factors that mitigate the risk to humans along various pathways of potential contact, and to draw some fundamental conclusions in regard to these practices.

POTENTIAL MICROBIOLOGICAL HEALTH EFFECTS

In attempting to describe the potential for microbiological health effects associated with the land application of wastewater and sludges, the understanding of a number of interrelated functions is extremely important. Of particular importance are the specific organisms of concern, the critical pathways by which these organisms can come in contact with humans, and the relation between the presence of an organism and the potential for human infection.

Microorganisms of Concern

The microorganisms of public health concern generally fall into three broad categories: bacteria, parasites, and viruses. Parasites are often further differentiated into helminths and protozoans. There are hundreds of organisms in these categories that may be present in domestic wastewaters.

Available detailed reviews describe a wide variety of disease causing microorganisms that may be transmitted through environmental exposure (Akin et al. 1978, Sobsey and Olson 1983, Bitton and Gerba 1984, Gerba 1983). Generally, such lists include most infectious agents known to be transmitted by the fecal-oral route. A more focused list of microbial agents can be prepared, however, if additional criteria such as demonstrable presence in wastewaters or sludges or documented environmental transmission of disease are applied. Table 1 provides such a listing. Note that while some of the organisms listed are overt pathogens, reports of their occurrence in wastewater and sludges in the United States are quite rare.

The concentrations of some of these organisms isolated from raw wastewater are shown in Table 2. Similarly, concentrations of these organisms found in raw municipal sludges are shown in Table 3. Ranges are a function of the population demography as well as regional and social issues such as water use in the home and sanitary practices.

Bacteria. During 1983 over 44,000 cases of salmonellosis and almost 20,000 cases of shigellosis in the United States were reported to the Center for Disease Control (CDC) (1984). This is in contrast with one case of cholera and 852 cases of legionellosis reported during the same period. While such cumulative data highlight the relative prevalence of

TABLE 1. Organisms of major concern in land application of municipal wastewater and sludges.

Group and Name of Organism	Primary Disease	Remarks
Bacteria		
Legionella pneumophila	Acute respiratory disease	Aerosol transmission documented but no cases linked to waste-water exposure to date
Salmonella sp.	Gastroenteritis, typhoid and paratyphoid fever	
Shigella sp.	Bacillary dysentery	Overt pathogens but low proba-bility of occurrence in waste-water in the United States
Vibrio cholerae	Cholera	
Helminths		
Ascaris sp.	Ascariasis	
Protozoans		
Giardia lamblia	Giardiasis (gastroenteritis)	
Viruses		
Hepatitis A virus	Infectious hepatitis	
Non-A, Non-B hepatitis	Hepatitis	Preliminary evidence for water-borne transmission
Norwalk-like agents	Gastroenteritis	
Rotavirus	Gastroenteritis	

TABLE 2. Ranges of selected organisms found in raw municipal wastewater.

Organism	Concentration (organisms/100 ml)	Reference
Salmonella sp.	10^2-10^6	Foster and Engelbrecht 1973
Shigella sp.	$<10^0$->10^1	Camann et al. 1983
Ascaris sp.	10^0-10^1	Larkin et al. 1978
Giardia lamblia	10^2-10^3	Gerba 1983
Total enteroviruses	10^1-10^3	Gerba 1983, Camann et al. 1983

TABLE 3. Ranges of selected organisms found in untreated municipal wastewater sludges.

Organism	Concentration (organisms/gram dry weight)	Reference
Salmonella sp.	10^2-10^3	Ward et al. 1984
Shigella sp.	$<10^1$	Camann et al. 1983
Ascaris sp.	10^1-10^3	Reimers et al. 1981
Giardia lamblia	Unknown*	
Viruses	10^2-10^4	Ward et al. 1984

* Limited data available; however, low concentrations expected.

these diseases in the United States, the routes of infection are seldom documented. Presumably, few if any of these cases were associated with wastewater. Nevertheless, both *Salmonella* sp. and *Shigella* sp. remain very important in the United States. Generally, pathogenic bacteria are fairly stable in raw wastewater, especially during the relatively short time required to reach the treatment works. Although the wastewater environment is hostile to most bacteria, there have been reports of replication of some species in wastewater sludges (Yeager and Ward 1981).

Parasites. Most of the helminths in wastewater and sludges are only incidentally considered a potential problem to humans in most areas of the United States. More often than not, they are involved in the infection of animals. Nevertheless, *Ascaris* is considered to be the most important of the helminths and *Giardia lamblia* is considered to be the most important protozoan as far as infection of humans in the United States is concerned. *G. lamblia* is particularly important in those areas where outbreaks are common, owing largely to the natural occurrence of this protozoan in wild animal populations (i.e., the Pacific Northwest, the Rocky Mountain region, and the higher elevations of New England).

Viruses. During 1983, more than 21,500 cases of hepatitis were reported to the CDC (1984). A limited number of cases were known to be waterborne, as were cases of Norwalk type viral gastroenteritis and rotavirus gastroenteritis (Sobsey and Olson 1983). Furthermore, almost 3,500 cases of non-A, non-B hepatitis were reported to the CDC in 1983 (1984). While there have been no documented cases of environmental transmission of this agent in the United States, such transmission has been reported in foreign countries (Nouasria et al. 1984, Tabor 1985).

Although certain human viruses have been shown to be fairly stable in raw wastewater, they cannot replicate outside a susceptible host. Thus their numbers will decrease with time as they are exposed to an unfavorable environment. It has been shown, however, that protection from inactivation can be afforded by viral adsorption or occlusion in solids.

Pathways to Humans

There are many pathways whereby microorganisms can infect humans. Generally, these pathways involve ingestion, inhalation, entry through abrasions in the skin, entry through orifices other than the nose or mouth, and, rarely, diffusion through the pores of the skin. Of primary importance in the land application of wastewater and sludges are ingestion and inhalation. To be important, a pathway must involve a susceptible population. Finally, it must be understood that few, if any, environmental risks can be reduced to zero. In fact, absolute zero risk may not be a rational goal (Lowrance 1976).

Figure 1 provides a schematic of the various potential pathways by which microorganisms in wastewater and sludges applied to forests can travel to humans. Among these pathways, the potential for contamination of groundwater used as a source of potable water is considered to be most sensitive.

The aerosol pathway is mitigated by the lack of a substantial susceptible population during application of wastewater or sludges. The potential for exposure through this pathway is further reduced by impaction and interception of aerosols in the forest.

The importance of surface water as a pathway can be substantially reduced by restricting the slope of the application area and by application control within 50 to 100 m of surface waters. Under these conditions, available data suggest less contamination of surface waters than observed from direct discharge of treated wastewater (Lue-Hing et al. 1979).

The potential for human disease acquired through the food pathway is essentially nonexistent owing to the limited amount of potential food products found in forested areas. Furthermore, the extent of the susceptible population subject to either the food or vector pathway is minimal.

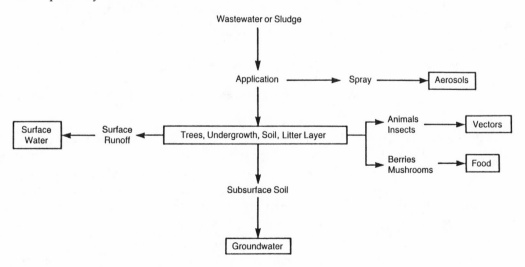

Figure 1. Pathways of microbial transport from wastewater and sludge in forest application.

Minimal Infectious Dose

The infectious dose is generally reported as that number of organisms required to infect 50% of the test population. Infection is defined not as disease but rather as a response in humans as measured by fecal shedding of infecting organisms or an increase in specific antibodies in the blood.

The minimal infectious dose can be described at any arbitrary point along a gradation of infection response frequencies which are dependent on dose. Obviously, the actual response measured will depend on the pathogen species. It is not uncommon for the minimal infectious dose to be defined as that dose required to infect 1% of the test population.

Table 4 provides available information on the infectious dose required for infection of 50% of the test population by selected organisms of primary interest in the context of this paper. The 1% infectious dose may be somewhat lower, perhaps as much as an order of magnitude in susceptible individuals.

TABLE 4. Infectious dose for a 50% infection rate for selected organisms in humans (after Akin 1983, and Ward and Akin 1984).

Organism	50% Infectious Dose	Mode of Administration
Salmonella sp.	10^5-10^8	Oral
Shigella	10^1-10^2	Oral
Giardia lamblia	10	Oral
Ascaris sp.	1-10	Oral
Poliovirus 1 (Sabin)*	72 ($TCID_{50}$)	Oral
Echovirus 12	919 (PFU)	Oral
Coxsackievirus A21	28 ($TCID_{50}$)**	Nasal
Adenovirus 4	1 ($TCID_{50}$)	Nasal

*Study conducted in healthy infants.

**Infectious dose was found to approximate the illness dose.

$TCID_{50}$ = 50% tissue culture infectious dose. PFU = plaque forming unit.

While minimal infectious doses are important, most of the population recognizes illness as the absolute criterion for problems of public health significance. To effect illness, as determined by observable symptoms, a dose of some organisms higher than even the 50% infectious dose may be required. In addition, the virulence of a given organism and the susceptibility of the host play important roles in the initiation of disease. Nevertheless, since an infectious dose results in fecal shedding, it will be used as the level to attain to minimize the risk of disease and infection to the general population.

MITIGATION OF HEALTH RISKS TO HUMANS

As with any nonconservative pollutant in the environment, the concentrations of microorganisms will be reduced with time. The more hostile the environment, the more rapid will be their inactivation. There are a number of factors that contribute to this environmental die-off or decay. Not least of these are the specific wastewater treatment processes, the environmental conditions at the application site, and the topography and hydrogeology of the application site.

Wastewater Treatment

Wastewater treatment affects the various organism types in different ways. In general, it has been observed that microbial segregation occurs during conventional wastewater treatment. Bacteria and viruses tend to become associated with the sludge component, as do the heavier eggs of certain parasites such as *Ascaris*. On the other hand, available information suggests that a significant number of parasitic cysts and ova may exit a wastewater treatment facility in the liquid component (effluent).

A typical treatment train for conventional wastewater treatment is shown in Figure 2. It should be noted that none of the processes shown (with the exception of disinfection) are designed for the reduction of pathogens. Thus the greater part of pathogen removal from wastewater is fortuitous. Table 5 is a summary of the effectiveness of selected unit processes in the reduction of pathogen types. A range is provided for the various combinations of conventional wastewater treatment that are most often encountered.

The reduction of various pathogen groups associated with typical sludge treatment processes is shown in Table 6. It can be seen that parasites are least affected by most of the processes, although composting is superior to the others in that regard.

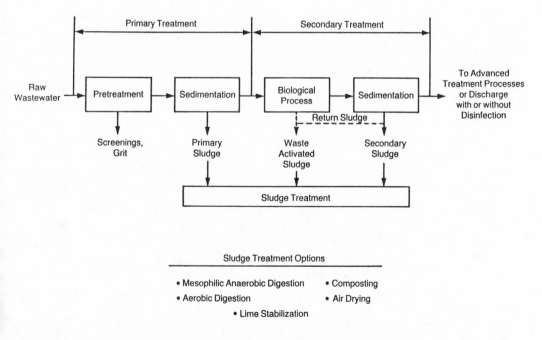

Figure 2. Selected conventional wastewater treatment options.

TABLE 5. The effectiveness of conventional treatment processes in the reduction of pathogens in wastewater (after Sproul 1978, Foster and Engelbrecht 1973, Sorber et al. 1974, Moore et al. 1981, and Sorber 1983).

Treatment Process	Log_{10} Reductions		
	Bacteria	Parasites	Viruses
Primary alone	0.2-0.5	0.1-1.0	0.1-0.5
Secondary (including primary)			
Trickling filtration	0.1-1.5	0-0.3	0.1-1.3
Activated sludge	1.0-2.0	0-1.5	0.4-2.9
Disinfection	0.5-3.7	Unknown	0.3-1.0
Total, all processes	0.8-5.7	0.1-1.5	0.4-3.9

TABLE 6. Effectiveness of various sludge treatment processes in the reduction of pathogens (after Ward et al. 1984).

Sludge Treatment Process	Log_{10} Reductions		
	Bacteria	Parasites	Viruses
Mesophilic anaerobic digestion	0.5-4	<0.5	0.5-2
Aerobic digestion	0.5-4	<0.5	0.5-2
Composting	2->4	2->4	2->4
Air drying	0.5->4	0.5->4	0.5->4
Lime stabilization	2->4	<0.5	>4

Clearly, whether intentional or not, significant reductions in the number of pathogens in raw sewage occur as a result of efficiently operated wastewater treatment. In fact, activated sludge wastewater treatment followed by effective disinfection can reduce the concentrations of bacteria and viruses in 100 ml of wastewater to near or below the 50% infectious dose. Similarly, well-treated wastewater sludges can reach 50% infectious dose levels per gram dry weight. On the other hand, the concentrations of parasites can remain above the 50% infectious dose in either treated wastewater or sludges.

Survival of Microorganisms

Survival of bacterial pathogens in wastewater or sludge-amended soil applied in the forest will depend on a number of factors, including the soil moisture content, temperature, sunlight, pH, organic matter, and antagonistic soil microflora. A rather extensive discussion of this topic has been presented by the authors and their coworker elsewhere (Sagik et al. 1979). There is great variability in the survival of different microorganisms in soil systems (see Table 7). For example, *Salmonella typhosa* has been recovered from loam and peat soils for periods up to eighty-five days, while survival in drying sand was four to seven days. Additionally, *S. typhosa* may survive as long as two years at freezing temperatures.

Survival of viruses in soils is influenced by many of the same parameters described above, although at this time little direct evidence supports viral inactivation by antago-

TABLE 7. Half-life for selected microorganisms in soil systems (after Reddy et al. 1981).

| Microorganisms | Half-life (hours) | | | Number of Observations |
	Minimum	Average	Maximum	
Fecal coliforms	1.8	20.9	237.6	46
Salmonella sp.	2.4	37.1	184.8	16
Shigella sp.	22.5	24.5	26.8	3
Viruses	4.5	11.5	415.9	11

nistic microorganisms. The effect of temperature on the survival of viruses is well documented. Lower temperatures favor longer survival times. In one study, observation of a 1 $\log_{10}$ loss of poliovirus titer required approximately three months at 4°C, one month at 20°C, and less than one week at 30°C (Duboise et al. 1976). An optimal soil moisture content favors poliovirus survival in soil, while desiccation results in a more rapid loss of virus recoverability. Another study involving a wide variety of human enteroviruses, including polioviruses, Coxsackieviruses, and echoviruses, reported survival times ranging from 110 to 170 days at a soil pH of 7.5 and a soil temperature of 3 to 10°C (Bagdasaryan 1964).

Less quantitative data are available for parasites. Protozoa seem to be very sensitive to drying, and under these conditions survival rates are usually short. Kowal (1983) suggested maximum survival times for protozoa in soil to be ten days, with two days being more common. On the other hand, helminths appear to be quite resistant to environmental stress. For example, *Ascaris* eggs have been observed to remain viable in soil for up to fifteen years (Krasnonos 1978, Jackson et al. 1977). However, drying reduced survival substantially (Smith et al. 1980), and a more practical maximum survival time appears to be two years (Kowal 1983).

Microbial Transport

In addition to survival of pathogens in wastewater and sludge-soil systems, consideration must be given to their ability to move in this environment, particularly to groundwater. As in the case of pathogen survival, microbial transport is a complex issue.

Removal of bacteria from wastewater percolating through a soil is due to both mechanical removal (i.e., straining or sieving at the soil surface) and adsorption to soil particles. Bacterial movement through soils has been demonstrated at several field sites. Reporting from the available literature, Gerba et al. (1975) noted movement of bacteria in a variety of soils for distances ranging from 0.9 to 456 m. Release and movement of microorganisms would be expected, since physical adsorption to particulates is a reversible phenomenon and, in part, ion dependent.

The phenomenon of adsorption as a mechanism for the retention of viruses in soil systems has been demonstrated by a number of investigators (Duboise et al. 1976, Lance et al. 1976, and Funderberg et al. 1981). In general, virus adsorption by soils increases with increasing ion exchange capacity, clay content, and organic carbon. Movement of viruses through soils in field studies has been demonstrated to depths of 67 m (Keswick 1984).

In the case of viruses in sludge, several studies have demonstrated that viruses are tightly bound to sludges (Damgaard-Larsen et al. 1977). This phenomenon appears to restrict the transport of viruses to groundwater.

The transport of protozoa and helminths in soils appears to be more limited than for bacteria or viruses (Metro 1983). This may be the result of the considerable size difference between viruses, bacteria, and parasites (see Table 8). For example, protozoa are up to 20 times larger than bacteria and up to 2,000 times larger than enteroviruses. *Ascaris* eggs are even larger. Clearly, mechanical straining may be the most important factor in parasite transport.

TABLE 8. Sizes of selected viruses, bacteria, and parasites (after Bitton and Gerba 1984).

Microorganism	Size (um)
Viruses Hepatitis A Rotavirus Norwalk-like agents	0.02-0.08
Bacteria Salmonella sp. Shigella sp. Vibrio cholerae	1-10
Protozoans Giardia lamblia	5-20
Helminths (eggs) Ascaris sp.	25-38

TABLE 9. Die-off rate constants for viruses and bacteria in groundwater (after Gerba and Bitton 1984, Yates et al. 1985).

Microorganism	Die-off Rate (per day)*
Poliovirus 1	0.035-0.77
Coxsackievirus	0.19
Echovirus 1	0.051-0.63
Rotavirus SA-11	0.36
Salmonella typhimurium	0.22

*As $\log_{10} N_r/N_o$, where N_r equals concentration of organisms after 24 hours and N_o equals the initial concentration of organisms.

Once pathogens reach groundwater their inactivation is reduced. Examples of die-off rates for selected microorganisms in groundwater can be found in Table 9. This situation suggests that distance of groundwater travel to a surface discharge or a potential well site is a critical factor in site selection. Fortunately, groundwater travel is quite slow, generally ranging from 1.5 m/year to 150 m/year, depending on the hydraulic conductivity of the aquifer. Consequently, significant die-off of microorganisms will take place should groundwater contamination occur. Nevertheless, groundwater contamination can be avoided if distance to groundwater and the existence of confining strata at a forest application site are used as important site selection criteria.

CONCLUSIONS

1. Groundwater used as a source of potable water is considered to be the most sensitive pathway for the infection of humans from the forest application of wastewater or sludges.

2. Considerable numbers of pathogens of environmental significance can be found in municipal wastewater and sludges.

3. Well-designed and -operated conventional wastewater and sludge treatment processes can reduce the concentration of pathogens appreciably.

4. Parasites may be least affected by conventional wastewater treatment processes.

5. Reduction of pathogens in wastewater or sludge-amended soils is accelerated by high temperatures, drying, and time.

6. The most important criterion for selection of forest wastewater or sludge application sites is time of travel to groundwater that may be used as a source of potable water.

REFERENCES

Akin, E. W. 1983. Infective dose of wastewater pathogens. p. 24–39. *In* A. D. Venosa and E. W. Akin (eds.) Municipal wastewater disinfection. EPA 600/9-83-009. U.S. Environmental Protection Agency, Cincinnati, Ohio.

Akin, E. W., W. Jakubowski, J. B. Lucas, and H. R. Pahren. 1978. Health hazards associated with wastewater effluents and sludge: Microbiological considerations. p. 9–25. *In* B. P. Sagik and C. A. Sorber (eds.) Proceedings of the Conference on Risk Assessment and Health Effects of Land Application of Municipal Wastewater and Sludges. University of Texas, San Antonio.

Bagdasaryan, G. A. 1964. Survival of viruses of the enterovirus group (poliomyelitis, echo, Coxsackie) in soil and on vegetables. Hyg. Epidemiol. Microbiol. Immunol. 8:497–505.

Bitton, G., and C. P. Gerba (eds.) 1984. Groundwater pollution microbiology. John Wiley and Sons, New York.

Camann, D. E., R. L. Northrop, P. J. Graham, M. N. Guentzel, H. J. Harding, K. T. Kimball, R. L. Mason, B. E. Moore, C. A. Sorber, C. M. Becker, and W. Jakubowski. 1983. An evaluation of potential infectious health effects from sprinkler application of wastewater to land: Lubbock, Texas, second interim report. Southwest Research Institute, San Antonio, Texas.

Center for Disease Control. 1984. Reported morbidity and mortality in the U.S. Mobidity and Mortality Weekly Report, Annual Review, 1983. U.S. Dept. of H.E.W./Public Health Service. 33:38:542–547.

Damgaard-Larsen, S., K. O. Jensen, E. Lund, and B. Nissen. 1977. Survival and movement of enterovirus in connection with land disposal of sludges. Water Res. 11:503–508.

Duboise, S. M., B. E. Moore, B. P. Sagik, and C. A. Sorber. 1976. The effects of temperature and specific conductance on poliovirus survival and transport in soil. p. 31–36. *In* National Conference on Environmental Research, Development, and Design. American Society of Civil Engineers, University of Washington, Seattle.

Foster, D. H., and R. S. Engelbrecht. 1973. Microbial hazards in disposing of wastewater on soil. p. 247–270. *In* W. T. Sopper and L. T. Kardos (eds.) Recycling treated municipal wastewater and sludge through forest and cropland. Pennsylvania State University Press, University Park.

Funderburg, S. W., B. E. Moore, B. P. Sagik, and C. A. Sorber. 1981. Viral transport through soil columns under conditions of saturated flow. Water Res. 15:703–711.

Gerba, C. P. 1983. Pathogens. p. 147–187. *In* A. L. Page, T. L. Gleason III, J. E. Smith, Jr., I. K. Iskandar, and L. E. Sommers (eds.) Proceedings of the 1983 Workshop on Utilization of Municipal Wastewater and Sludge on Land. University of California, Riverside.

Gerba, C. P., and G. Bitton. 1984. Microbial pollutants: Their survival and transport pattern to groundwater. p. 65–88. *In* G. Bitton and C. P. Gerba (eds.) Groundwater pollution microbiology. John Wiley and Sons, New York.

Gerba, C. P., C. Wallis, and J. L. Melnick. 1975. Fate of wastewater bacteria and viruses in soil. J. Irrig. Drain. Div., Am. Soc. Civil Eng. 101:157–174.

Jackson, G. T., J. W. Bier, and R. A. Rude. 1977. Recycling of refuse into the food chain: The parasite problem. p. 116–127. *In* B. P. Sagik and C. A. Sorber (eds.) Proceedings on the Conference on Risk Assessment and Health Effects of Land Application of Municipal Wastewater and Sludges. University of Texas, San Antonio.

Keswick, B. 1984. Sources of groundwater pollution.

p. 39–64. *In* G. Bitton and C. P. Gerba (eds.) Groundwater pollution microbiology. John Wiley and Sons, New York.

Kowal, N. 1983. An overview of public health effects. p. 329–394. *In* A. L. Page, T. L. Gleason III, J. E. Smith, Jr., I. K. Iskandar, and L. E. Sommers (eds.) Proceedings of the 1983 Workshop on Utilization of Municipal Wastewater and Sludge on Land. University of California, Riverside.

Krasnonos, L. I. 1978. Many-year viability of *Ascaris* eggs (*Ascaris lumbricoides*) in soil of Samarkand (Abstract). Trop. Disease Bull. 75:991–992.

Lance, J. C., C. P. Gerba, and J. L. Melnick. 1976. Virus movement in soil columns flooded with secondary sewage effluent. Appl. Environ. Microbiol. 32:520–526.

Larkin, E. P., J. T. Tierney, J. Lovett, D. Van Donsel, and D. W. Francis. 1978. Land application of sewage wastes: Potential for contamination of foodstuffs and agricultural soils by viruses, bacterial pathogens. p. 102–115. *In* B. P. Sagik and C. A. Sorber (eds.) Proceedings on the Conference on Risk Assessment and Health Effects of Land Application of Municipal Wastewater and Sludges. University of Texas, San Antonio.

Lowrance, W. W. 1976. Of acceptable risk: Science and the determination of safety. William Kaufmann, Inc., Los Altos, California.

Lue-Hing, C., S. J. Sedita, and K. C. Rao. 1979. Viral and bacterial levels resulting from the land application of digested sludge. p. 445–462. *In* W. E. Sopper and S. N. Kerr (eds.) Utilization of municipal sewage effluent and sludge on forest and disturbed land. Pennsylvania State University Press, University Park.

Metro (Municipality of Metropolitan Seattle). 1983. Metro sludge quality: Monitoring report and literature review. Municipality of Metropolitan Seattle.

Moore, B. E., B. P. Sagik, and C. A. Sorber. 1981. Viral transport to groundwater at a wastewater land application site. J. Water Poll. Control Fed. 53:1492–1502.

Nouasria, B., B. Larouze, M. C. Dazza, C. Gaudebout, A. G. Saimot, and A. Aouati. 1984. Direct evidence that non-A, non-B hepatitis is a waterborne disease. The Lancet II:8394:94.

Reddy, K. R., R. Khaleel, and M. R. Overcash. 1981. Behavior and transport of microbial pathogens and indicator organisms in soils treated with organic wastes. J. Environ. Qual. 10:255–266.

Reimers, R. S., M. D. Little, A. J. Englande, D. B. Leftwich, D. D. Bowman, and R. F. Wilkinson. 1981. Parasites in southern sludges and disinfection by standard sludge treatment. EPA 600/52-81-

166. U.S. Environmental Protection Agency, Cincinnati, Ohio.

Sagik, B. P., B. E. Moore, and C. A. Sorber. 1979. Public health aspects related to the land application of municipal sewage effluents and sludges. p. 241–253. *In* W. E. Sopper and S. N. Kerr (eds.) Utilization of municipal sewage effluent and sludge on forest and disturbed land. Pennsylvania State University Press, University Park.

Smith, G. S., H. E. Kiesling, E. E. Ray, D. M. Hallford, and C. H. Herbel. 1980. Fate of parasites in drying bed studies (Abstract). p. 357. *In* G. Bitton, B. L. Damron, G. T. Edds, and J. M. Davidson (eds.) Sludge: Health risks of land application. Ann Arbor Science Publishers, Ann Arbor, Michigan.

Sobsey, M., and B. Olson. 1983. Microbial agents of waterborne disease. *In* Assessment of microbiology and turbidity standards for drinking water. EPA 570/9-83-001. U.S. Environmental Protection Agency, Washington, D.C.

Sorber, C. A. 1983. Removal of viruses from wastewater and effluents by treatment processes. p. 39–52. *In* G. Berg (ed.) Viral pollution of the environment. CRC Press, Inc., Boca Raton, Florida.

Sorber, C. A., S. A. Schaub, and H. T. Bausam. 1974. An assessment of a potential virus hazard associated with spray irrigation of domestic wastewaters. p. 241–252. *In* J. F. Malina, Jr., and B. P. Sagik (eds.) Virus survival in water and wastewater systems. Center for Research in Water Resources, University of Texas, Austin.

Sproul, O. J. 1978. The efficiency of wastewater unit processes in risk reduction. p. 282–296. *In* B. P. Sagik and C. A. Sorber (eds.) Proceedings on the Conference on Risk Assessment and Health Effects of Land Application of Municipal Wastewater and Sludges. University of Texas, San Antonio.

Tabor, E. 1985. The three viruses of non-A, non-B hepatitis. The Lancet I:8431:743–745.

Ward, R. L., and E. W. Akin. 1984. Minimum infective dose of animal viruses. CRC Critical Reviews in Environmental Control 14:297–310.

Ward, R. L., G. A. McFeters, and J. G. Yeager. 1984. Pathogens in sludge: Occurrence, inactivation, and potential for regrowth. Sandia National Laboratories, Albuquerque, New Mexico.

Yates, M. V., C. P. Gerba, and L. M. Kelley. 1985. Virus persistence in groundwater. Appl. Environ. Microbiol. 49:4:778–781.

Yeager, J. G., and R. L. Ward. 1981. Effects of moisture content on long-term survival and regrowth of bacteria in wastewater sludge. Appl. Environ. Microbiol. 41:5:1117–1122.

Chapter 8

Trace Element Considerations of Forest Land Applications of Municipal Sludges

A. C. CHANG, T. J. LOGAN, and A. L. PAGE

ABSTRACT   The ultimate concern over trace elements in municipal wastewater and sludge is their possible entry into the food chain at levels harmful to humans or animals, as well as their possible adverse effects on the capacity of soils to sustain the growth of native vegetation and commercial crops. Trace elements in sludges and wastewaters may enter the food chain via uptake and accumulation by vegetation, through contamination of vegetation by surface deposition, by direct ingestion of sludge or sludge-contaminated soil, and by contamination of surface water and groundwater through runoff and leaching. Present knowledge concerning the mode of entry of trace elements into the food chain from the use of sludge and treated wastewater on agricultural lands is quite extensive. Also, the conditions under which trace elements applied to cropland in the form of sludge become phytotoxic are reasonably well understood. The fate of trace elements applied in the form of sludge or wastewater to forest lands, however, is not nearly as well documented. The physical and biological mechanisms that control the transformation of sludge-borne metals on forest land, however, are the same as those on cropland. This paper reviews what is known about the levels of trace elements in municipal wastewater and sludge, their effects on the growth and chemical composition of crops, their mobility in soil, and their entry into the food chain via direct ingestion of sludge-treated soil and sludge deposited on the surface of vegetation. Information obtained from applying sludge to agricultural lands is evaluated in terms of the possible effects when sludge or wastewater is applied to forests.

During the past decade much progress has been made in our knowledge of the chemical elements in municipal sewage sludges. The potential detriments and benefits of applying the trace elements and metals latent in sludges to cropland have been investigated extensively. However, our understanding of the effects of such sludge-borne elements on forest ecosystems is not nearly as complete. To delineate the forest ecosystem's responses to sludge application, it is necessary to reference the data derived from studying sludge-treated cropland. This paper outlines the effects of trace elements from sludge application on soils and crops, points out how croplands differ from forests, and reviews effects of trace elements on forest stands from sources other than sludges.

TRACE ELEMENTS IN MUNICIPAL SEWAGE SLUDGE

Trace elements are ubiquitous in municipal wastewater. The concentrations of trace elements in sewage from residential communities usually are less than those from industrial sources. During the wastewater treatment process, most trace elements tend to concentrate in the sludge. Trace element concentrations in sludge are the result of concentrations in the domestic water supply, elemental addition during treatment, storage, and conveyance (e.g., zinc from galvanized pipe), the wastewater treatment process employed, and the nature of waste discharges. As a result, the trace element composition of sludges varies considerably.

Several studies have summarized data on the chemical composition of municipal sewage sludges from treatment plants in cities throughout the United States and other parts of the world (Page 1974, Furr et al. 1976, Sommers 1977, Doty et al. 1978, Tabatabai and Frankenberger 1979, Beckett 1980, Logan and Chaney 1983). In general, data presented by these sources show that the chemical composition of sludges from the wastewater treatment plants that were surveyed varies substantially from one plant to another, as well as with time. The data presented by Sommers (1977) are outlined in Table 1 as an example. The concentrations of trace elements in sludges, with a few exceptions, appear to follow a log-normal distribution, indicating that the overwhelming majority of the sludges have trace element contents much less than the maximum concentrations observed. A sludge of unusually high metal concentration often is the result of industrial waste discharge. Consequently, the geometric mean is a better representation of the typical elemental content of sludge than the arithmetic mean. Industrial waste pretreatment is effective in reducing unusually high levels of trace elements in sludges. For example, a pretreatment program adopted by the wastewater treatment plant at Defiance, Ohio reduced the concentrations of cadmium and nickel in the sludge by as much as 90% (Miller and Logan 1979). Likewise, Nelson (1981) reported substantial reduction in the concentrations of cadmium, chromium, copper, lead, nickel, and zinc in sludge from wastewater treatment plants in Philadelphia following the implementation of a pretreatment program.

A comparison of the median concentrations of trace elements in sewage sludges with typical concentrations in soils shows that concentrations for most trace elements are substantially greater in sludges than in typical soils (Table 2). This means that sludge applications will increase the trace element concentrations of soils. Since most trace elements (boron is an exception) are quite immobile in soil, and the amounts removed by crops are small compared with the amounts present (see Page et al. 1981), the trace elements introduced with repeated sludge applications tend to accumulate in the surface soil horizons.

TABLE 1. Concentrations of trace elements in municipal sewage sludges (Sommers 1977, Logan and Chaney 1983).

Element	Concentration (mg/kg, dry weight basis)			
	Minimum	Maximum	Mean	Median
Arsenic	6	230	43	10
Boron	4	760	77	33
Cadmium	3	3,410	110	16
Chromium	10	99,000	2,620	890
Cobalt	1	18	5.3	4
Copper	84	10,400	1,210	850
Fluorine	80	33,500	--	260
Lead	13	19,700	1,360	500
Mercury	0.2	10,600	733	5
Molybdenum	5	39	28	30
Nickel	2	3,520	320	82
Selenium	1.7	17	--	5
Zinc	101	27,800	2,790	1,740

TABLE 2. Comparison of median concentrations of trace elements in sludge with typical concentrations for soils.

| Element | Median Concentration in Sludge (mg/kg) | Amount Present in Soil | |
		Normal Range (mg/kg)	Typical Level (mg/kg)
Arsenic	10	0.1-40	6
Boron	33	2-100	10
Cadmium	16	0.05-1.0	0.3
Chromium	890	5-3,000	100
Cobalt	4	1-40	8
Copper	850	2-100	20
Fluorine	260	20-700	200
Lead	500	20-200	15
Mercury	5	0.01-0.3	0.03
Molybdenum	30	0.2-5.0	2
Nickel	82	10-1,000	40
Selenium	5	0.01-2.0	0.2
Zinc	1,740	10-300	50

TRACE ELEMENTS OF CONCERN IN LAND APPLICATION

When used as a fertilizer to satisfy nitrogen requirement of plants, sludge usually will also supply adequate amounts of all other essential plant nutrients (Page and Chang 1983). Aside from the nutrient elements, the organic matter in sludges acts as a soil conditioner and as such improves the agronomic properties of soils. But a variety of trace elements, when they are present in soil at concentrations in excess of the critical levels, are undesirable. The trace elements of concern in municipal sewage sludges include arsenic, boron, cadmium, chromium, cobalt, copper, fluorine, lead, mercury, molybdenum, nickel, selenium, and zinc.

The elements arsenic, chromium, lead, and mercury, although acutely toxic to all biological systems when present in available forms, are not absorbed by crops in amounts considered to be harmful to consumers. Under ordinary chemical conditions they are present in soils in rather inert chemical forms or are converted subsequently and do not affect the growth of plants. Ingesting these elements in the form of either sludge-treated soils or sludges adhering to foliage, however, can be harmful to livestock animals (Logan and Chaney 1983). Besides arsenic, chromium, lead, and mercury, high levels of iron, copper, fluorine, cobalt, molybdenum, and selenium in sludge or sludge-treated soil could also present a direct ingestion hazard to grazing animals (Chaney 1980).

Concentrations of copper, nickel, and zinc are high in all municipal sludges. These metals will be phytotoxic if their concentrations in soils, especially acid soils, exceed the critical levels. The phytotoxic thresholds of these metals, however, are highly dependent on the plant species (e.g., see Sommers 1980). Even in heavily contaminated soil, amounts of copper, nickel, and zinc absorbed by crops usually do not present a health hazard to consumers (Logan and Chaney 1983).

Unlike copper, nickel, and zinc, cobalt may be absorbed by crops in amounts potentially harmful to the health of ruminant animals (Logan and Chaney 1983). Given the ideal soil conditions, most plants will accumulate cobalt in the foliage to levels higher than ruminants may tolerate in their diets. Unless levels of cobalt in the sludge were considerably higher than commonly observed (Table 1), it is not likely that cobalt concentrations in plants grown on sludge-amended soils would become excessive (Chaney 1980).

Until recently, chromium in soil and in sludge-amended soils was thought to be present as or rapidly converted to the Cr(III) oxidation state, which would be chemically inert and quite immobile. Bartlett and James (1979) and James and Bartlett (1983a, 1983b) show that Cr(VI) may be present in soils for extended periods or that Cr(III) may be subsequently oxidized in soils to Cr(VI), and may, through leaching, escape the active oxidation-reduction zone in soil. Based on these findings, the U.S. Department of Agriculture (USDA) and the Cooperative State Research Service (CSRS) technical committee of Northeastern Regional Research Project NE-96 (Pennsylvania State University 1985) has recommended limiting the amounts of chromium that may be added to soils in the form of sludge. Except for those contaminated by significant amounts of tannery or chrome plating wastes, municipal sludges do not contain sufficient chromium to pollute the groundwater during land application (Pennsylvania State University 1985).

Molybdenum and selenium may be accumulated by crops in amounts sufficient to cause either acute toxicity or metabolic imbalance in animals that consume the crops. Although molybdenum is an essential element in animal diets, particularly ruminant animals, concentrations as low as 5 mg per kilogram of dietary intake may be harmful (Allaway 1968). The occurrence and severity of the toxicity is influenced by amounts of molybdenum relative to copper and sulfate ingested. High molybdenum and low copper levels in the animal feed are the worst possible combination. Molybdenum absorption by crops is influenced by the soil's chemical properties. Generally, the availability of molybdenum increases and that of copper decreases as soil pH rises. Molybdenum toxicity to animals, therefore, is commonly associated with forage grown on alkaline soils. Although selenium is not considered to be an essential element for the growth of plants, it is required in the diets of animals and avian species. Like molybdenum, selenium may accumulate in plants to levels considered unsafe for consumption by animals (National Academy of Sciences 1983). Most plants are capable of absorbing large amounts of molybdenum and selenium without exhibiting phytotoxic symptoms. In sludge-amended soils, neither molybdenum nor selenium is expected to reach phytotoxic levels.

From a nutritional standpoint boron is unique. It is an essential element for plant growth, but is considered nonessential for animals and human beings. Although boron is not harmful to animals in a wide range of concentrations, it is toxic to plants at levels only slightly above the optimum. Plants grown on soils whose water-soluble boron is less than 0.04 mg per liter often exhibit symptoms of boron deficiency. At concentrations greater than 0.5 mg/l boron becomes toxic to sensitive plant species (Bingham 1973). In contrast to other trace elements reviewed so far, boron is only weakly adsorbed and readily passes through soils with leaching water. Since boron does not accumulate in soils, problems associated with boron phytotoxicity occur only in arid and semiarid regions where there is inadequate leaching, and rarely manifest themselves in semihumid and humid regions.

In land application of sewage sludges, the trace element attracting the most attention is cadmium. The concern stems from the discovery in Japan that cadmium poisoning of human beings resulted from the consumption of rice grown on soils irrigated with cadmium-contaminated water (Tsuchiya 1978). A number of studies in the United States and other parts of the world have shown that when sewage sludges containing cadmium are applied to agricultural land, the concentrations of cadmium in the harvested crops become elevated (see Council for Agricultural Science and Technology 1980, Logan and Chaney 1983, U.S. EPA 1983). However, there is no documented clinical case of cadmium poisoning related to the consumption of foods grown on sludge-amended soils. Since the percentage of agricultural land in the United States that has received sludge is extremely small, the probability for one individual consuming foods with elevated cadmium from the marketplace over an extended period sufficient to cause chronic cadmium poisoning is rather remote (Council for Agricultural Science and Technology 1976). In theory, it is possible for a person who derives a substantial portion of his or her food requirements from cadmium-affected sources and maintains this diet daily for a number of years to develop symptoms of cadmium toxicity. The concern over the potential health effects of excessive cadmium has prompted the U.S. Environmental Protection Agency to regulate the annual and cumulative amounts of cadmium that can be applied to cropland in the form of sludge (Code of Federal Regulations, Title 40, Protection of the Environment, Part 257: Criteria for classification of solid waste disposal facilities and practices).

LIMITS ON QUANTITIES OF TRACE ELEMENTS APPLIED

The U.S. Environmental Protection Agency at present regulates only the amount of cadmium that may be applied to land in the form of sewage sludges. Briefly, the regulations state that the resulting pH of the soil and sludge mixture at the time of application must be equal to or greater than 6.5 wherever food-chain crops are grown. If the indigenous soil pH is less than 6.5, the maximum annual and cumulative inputs of cadmium shall not exceed 0.5 and 5.0 kg/ha, respectively. For soils with pH greater than 6.5, annual cadmium applications will be gradually reduced from 1.25 kg/ha to 0.5 kg/ha according to a set timetable (after January 1, 1987). Cumulative cadmium inputs to land used for food-chain crops are based on the cation exchange capacity (CEC) of the soil. For soils with low CEC (<5 c mol/kg) the limit is 5 kg/ha; for soils with a CEC from 5 to 15 c mol/kg the limit is 10 kg/ha; and the cadmium limits for soils with a CEC greater than 15 c mol/kg is 20 kg/ha. A provision in the federal regulations exempts land application of sludges from the above limits if the following conditions are met: (1) the pH of the soil-sludge mixture is 6.5 or greater where food-chain crops are grown, (2) the food-chain crops grown are to be used exclusively for animal feed, (3) a facility management plan to preclude the use of the crops by humans is implemented, and (4) notification is given to future property owners that the land should not be used for food-chain crops.

Although the U.S. Environmental Protection Agency has no specified limits for cropland land application of trace elements other than cadmium, a number of states and two USDA regional technical committees recommended regulating application rates of other metals as well. In 1976 the USDA-CSRS North Central (NC) regional technical committee NC-118 suggested input limits for cadmium, copper, nickel, zinc, and lead (Knezek and Miller 1976) (Table 3). Indiana, Wisconsin, and Ohio have adopted metal loading guidelines similar to levels specified by NC-118. Recently, the USDA-CSRS NE-96 has proposed the upper limits of trace element applications to soil in the northeastern United States (Pennsylvania State University 1985). The limits suggested by NE-96 for agricultural land (Table 4) are more stringent than those recommended by NC-118. Members of NE-96 have proposed more restrictive guidelines, since most soils in the northeastern region are acid and thus thought to be more susceptible to trace element induced phyto-toxicity and leaching. Annual limits, not exceeding 20% of the cumulative limits speci-fied in Table 4, are suggested by NE-96.

No guidelines or criteria have been developed by the federal regulating agencies to limit amounts of trace elements applied to forest production systems in the form of sludge. Since trace elements applied to forest systems would not directly enter the hu-man food chain, the considerations in developing such guidelines no doubt will be dif-ferent from those for cropland systems.

TABLE 3. Recommended cumulative limits for trace elements applied to agricultural land (Knezek and Miller 1976).

Trace Element	Cation Exchange Capacity of Soil (c mol/kg)		
	<5	5 to 15	>15
Cadmium	5	10	20
Copper	140	280	560
Nickel	140	280	560
Lead	560	1,120	2,240
Zinc	280	560	1,120

TABLE 4. Recommended maximum cumulative metal loading for soils in the Northeast (Pennsylvania State University 1985).

Metal	Textural Classes		
	Loamy sand, sandy loam (kg/ha)	Fine and very fine sandy loam, loam, silt loam (kg/ha)	Silt, clay loam, sandy and silty loam, sandy clay, silty clay, clay (kg/ha)
Cadmium	2.24	3.36	5.0
Zinc	56	168	336
Copper	28	84	168
Nickel	11.2	33.6	67.2
Lead	112	336	672
Chromium	112	336	672

CONTRASTS BETWEEN CROPLAND AND FOREST LAND

The direct research on trace element effects in forest systems from sludge and waste-water application is limited (Lepp and Eardley 1978, Anderson et al. 1982, Sopper and Kardos 1973, Grant and Olesen 1984, Bramryd 1984, Cole 1982, Urie et al. 1982, Zasoski et al. 1984, Harris and Urie 1985), and few of these works are based on long-term field studies. On the other hand, effects of sludge and wastewater application of trace ele-ments on cropland have been extensively studied in the last decade, and this research has been the subject of numerous extensive reviews (Chang and Page 1983, Chaney

1980, Logan and Chaney 1983). Because of this disparity, it becomes necessary to extrapolate the results of cropland-related sludge research to forest systems. In doing so, it is important to point out major differences in cropland and forest ecosystems.

Soil

Soils owe many of their properties to the dominant vegetation on the site during soil formation, but present-day land use can greatly modify the original soil characteristics. Conversion of forest to cropland has been widespread in the United States in the last two centuries or more, but there has also been significant reversion of cropland to "old field," hence forest, through abandonment of eroded or worn-out cropland (Foth and Schafer 1980). While these processes tend to reduce the differences between cropland and forest soil properties, some general distinctions can be made.

The profile of a cultivated soil commonly consists of a plow layer (Ap), transition horizons (A3, B1, or both), a zone of clay enrichment (B2), and one or more horizons of weathered parent material (C) (Brady 1984). The thickness of the Ap horizon is determined by depth of plowing and erosion and is often in the range of 8 to 25 cm. The pH of this horizon varies with soil mineralogy, organic matter accumulation, and lime additions to meet specific crop pH requirements (often in the range of pH 5.5 to 7.5). Soil structure is greatly affected by type and content of clay and organic matter content, as well as type and degree of tillage; Ap horizons of cultivated soils probably contain fewer large continuous pores than those of forest soils.

A forest soil typically has one or more organic litter layers (L, F, H), an Al horizon of organic matter accumulation generally less than 10 to 15 cm, a zone of removal of weathering products (AE), transition horizons, a zone of clay enrichment (B2), and weathered parent material (C). Most decomposing leaf litter is acidic (pH <4.5), with that from conifers most acidic (Williams and Gray 1974). Acid soils have a pH similar to that of the litter; only alkaline soils have a pH higher than that of the litter (Williams and Gray 1974). Most tree species (except conifers) take up more calcium than any other mineral elements, and if calcium is largely retained in the biomass, the result will be lowered soil pH. However, this effect is ameliorated somewhat by species that release calcium back to the soil through litter decomposition (Spurr and Barnes 1980). (Addition of nitrogen to forest ecosystems can also contribute to the acidification of forest soils.)

Soil CEC is a function of type and concentration of clay minerals, organic matter content, and soil pH. With the same mineralogy, forest soils would usually have lower total CEC in the surface 30 cm or so as a consequence of their shallow A horizons and low pH. Forest soils tend to have more large, continuous, water-conducting pores than cultivated soils formed from the same parent material as a consequence of the undisturbed profile and distribution of large tree roots.

The combination of low pH, low organic matter content with depth, low CEC, and greater content of large, water-conducting pores could represent an increased hazard of trace element leaching in forest soils if hydraulic loading through precipitation or sludge application were to exceed evapotranspiration. However, subsurface (B2 and C horizons) properties of cropland and forest soils would be more similar than surface soil properties and could reduce any differences in trace element leaching from surface application of sludge.

Hydrology

Cultivated soils, although they generally have more level slopes than forest soils, are more susceptible to surface runoff because of the greatly reduced surface cover of cropland versus forest. This is particularly true of temperate regions where cropland may be bare from late fall to early spring. While surface runoff from cropland is common during spring rains on medium to fine-textured soils, runoff from forest land is rare, except where sites have been disturbed. Trails, logging, and other disturbances reduce soil cover, compact the soil, and reduce infiltration (Brown 1980).

In contrast to cultivated soils where evaporative water loss is important, most of the evapotranspiration from forests is by transpiration (Spurr and Barnes 1980). Tree growth is greatest in the spring when available soil moisture is highest (Spurr and Barnes 1980) and rains of less than 4 cm rarely penetrate the surface horizons of forest soil if roots are active (Coile 1952).

Differences in hydrology of cultivated and forest soils would indicate that trace element losses by runoff or erosion are probably negligible on forest soils if disturbances due to sludge applications are minimized. Considering the difference in forest types and climate, the potential for trace element leaching in forest soil is limited by high transpiration rates in the spring and would be greatest during the fall and winter.

Soil Organisms

Cultivated soils, especially those in which chemical weed and insect control is practiced, have a much lower diversity of soil organisms than forest soils do. Undisturbed forest soils contain a wide range of fauna—from microorganisms to earthworms, insects, arthropods, small mammals, and rodents. The activity of these organisms is primarily concentrated in the surface meter or less of soil, with the highest activity in the litter and A1 horizon; and this zone will be directly affected by surface applications of sludge or wastewater. Also, some decomposition of litter will return to the surface trace elements taken up by trees from sludge applications.

Trace elements may be biomagnified in the food chain depending on the type of tissue in which they are found and the extent to which uptake of the metal is regulated by the organism. Lead is found in calcareous tissues and is often biomagnified in carnivores (Hughes 1981, Martin and Coughtrey 1981), while cadmium is subject to accelerated plant uptake and shows greater concentration in herbivores than in carnivores. Zinc uptake, retention, and loss is regulated in a large number of species, including invertebrates; and zinc concentration in animals may vary less than in their food (Hughes 1981).

Direct study of sludge sources of trace elements on soil organisms is limited. Studies (Helmke et al. 1979) have reported biomagnification of trace elements by earthworms; and Anderson et al. (1982) have shown that a 9 Mg/ha application of a digested sewage sludge (Milorganite) containing 60 mg Cd/kg increased the cadmium content of liver and kidneys of meadow voles (*Microtus pennsylvanicus*) compared with a control. The animals were kept in enclosures in the field on which the treatments were applied. There was some detectable increase in tissue cadmium, but sludge application did not result in increased tissue lead or zinc concentrations. Metal uptake had no short-term effects on the voles. Additional information on the effects of trace elements on wildlife is presented by Haufler and West in this volume.

Biomass

In cropping systems, biomass is produced over a relatively short period (one growing season for annual crops), and part of that biomass is harvested. Where crops are cultivated (excludes no-till systems), the remaining biomass of roots and aboveground plant parts is incorporated into the soil, where it undergoes fairly rapid decomposition. Cultivated soils maintain a microfauna and microflora adapted to rapid decomposition of biomass. Only small amounts of trace elements are taken up by crops, and much of this is returned to the soil from biomass decomposition. In cropped, cultivated systems, therefore, trace elements applied with sludge are almost entirely found uniformly mixed in the plow layer.

In forest systems, annual turnover of biomass is restricted to litter deposited on the soil surface and small roots that are concentrated near the surface. Litter addition to soil is 1.5 to 5 Mg/ha per year, of which 70% is leaf litter (Spurr and Barnes 1980). In open canopies, understory litter may contribute up to 28% of the total litter, and roots may represent up to 75% of the total biomass returned annually to the forest (Spurr and Barnes 1980). Decomposition of forest litter is dependent on the activity of macroorganisms such as isopods, millipedes, earthworms, and larval insects, as well as soil microorganisms, dominated by fungi (Southwick 1972). The macroorganisms play an essential role in reducing the size of surface litter and incorporating it into the surface soil, where it can be further decomposed by soil microorganisms.

In the early growth of forests, the rate of litter production may exceed decomposition, then steady state is reached, and, finally, in old open forests, litter decomposition may exceed litterfall. Organic matter accumulates in forest soils when decomposition is inhibited by moisture saturation or increased acidity (Spurr and Barnes 1980). Based on data from smelter sources, trace element accumulation at the surface of forest soils has been shown to reduce litter decomposition (Martin and Coughtrey 1981). This result cannot be extrapolated to sludge sources of trace elements, however, because (1) sludge additions tend to raise pH of acid forest soils (Lepp and Eardley 1978, Kardos and Sopper 1973) and (2) nutrients applied with the sludge may stimulate litter decomposition. Most of the trace elements in the forest ecosystem are contained in the soil (Hughes 1981, Martin and Coughtrey 1981), with litter containing the second largest quantity compared with other compartments. This is true for natural ecosystems and those with surface contamination of trace elements. Hughes (1981) indicated that trace element concentrations in forest soil are correlated with soil organic matter, and suggested that organic matter serves to immobilize metals (such as copper) that form strong complexes with humic materials and has less effect on metals (such as cadmium and zinc) that form weaker complexes with organic matter.

Unlike the cropland where nutrients are added to produce harvestable biomass annually, cycling of mineral elements in forest ecosystems is critical to their long-term productivities. This process involves the decomposition of organic matter, the weathering and transport of minerals in soils, and the uptake and storage of nutrients by plants. Although the fundamental mechanisms of nutrient cycling are the same in all forests, the rates and reaction kinetics will change with climate zones, forest types, and soil types. Cole and Rapp (1981) examined thirty-two sets of elemental cycling data representing fourteen forest stands around the world. They observed (1) significant changes in the rate of elemental cycling over the life of forest ecosystems, (2) marked differences in the cycling of different nutrient elements (generally, elements that were deficient cycled more efficiently than those in excess), and (3) in terms of nutrient cycling, coniferous forest ecosystems were more efficient in producing biomass than deciduous forests (under comparable conditions, the rate of nutrient uptake and the amounts required were both higher in deciduous forests). The differences in the cycling of nutrients in forest ecosystems, in turn, account for the distinctive characteristics of each forest stand.

Differences in biomass production, organic matter decomposition, and elemental cycling in cultivated versus forest soils would suggest that, in both systems, trace elements are primarily retained in the soil. In the case of forest systems, most of the trace element would be concentrated very near the surface and in litter, the environment primarily in contact with forest organisms and wildlife.

EFFECTS OF ANTHROPOGENIC TRACE ELEMENTS

By nature, forest ecosystems have tendencies to capture air-borne and precipitation-associated chemical constituents. Even for forests located in relatively unpolluted areas, the inflows of chemical elements frequently exceed the outflows (Heinrichs and Mayer 1977). Forests and wooded areas near urban and industrial complexes are especially susceptible to environmental contamination. When forests are affected by anthropogenic emissions, the influxes of chemicals become substantially greater than those of the unpolluted areas. Although the airborne and precipitation-associated pollutants are often adsorbed by the foliage, the forest floor remains the ultimate repository for trapped chemical constituents. The long-term impacts of trace elements emitted by stationary sources (such as primary smelters) on the forest ecosystems have been investigated extensively.

From the numerous studies of trace element-contaminated forests, a consistent pattern of elemental deposition on the forest floor has emerged (Little and Martin 1972, Buchauer 1973, Martin and Coughtrey 1981). Data obtained by Martin et al. (1982) at a forest that has been subjected to the fallout of a primary smelter for over fifty years may be used as an illustration (Table 5). The concentrations of cadmium, lead, zinc, and cop-

TABLE 5. Accumulation of trace elements in a forest downwind from a primary lead-zinc smelter (Martin et al. 1982).

Depth	Trace Element Concentration (mg/kg)			
	Cadmium	Lead	Zinc	Copper
L horizon	6.24	293	644	44
F horizon	22.3	1,027	1,450	67
H horizon	52.1	1,808	2,348	83
Depth below H horizon (cm)				
0-1	17.1	203	944	35
1-3.5	8.2	111	623	37
3.5-6	4.1	65	434	35
6-8.5	2.7	47	345	29
8.5-11	2.1	41	236	43
11-13.5	1.3	35	201	44
13.5-16	1.1	29	202	48
16-18.5	0.8	24	110	39
18.5-21	2.6	25	135	43
21-30	0.7	48	145	60

per in the forest litter were higher than other components of the ecosystem. Within the litter layer, the metal contents rose with the increased organic matter decomposition. In mineral soil, the concentrations of cadmium, lead, zinc, and copper decreased with depth (Table 5). For this forest, concentrations of cadmium, zinc, and copper at 30 cm depth were still above those of the control. Most trace elements, however, are relatively immobile in soil. They will be strongly complexed by organic matter and/or adsorbed by various soil minerals (Bloomfield et al. 1976). Among the metal elements commonly studied, copper always exhibited the greatest potential of movement in the soil profile. With 9.5, 168, 934, and 130 kg/ha of cadmium, lead, zinc, and copper, in the 0 to 30 cm soil depth (Table 5), over 60% of the cadmium, lead, and zinc, respectively, were deposited in the top 6 cm and only 25% of the copper. Harris and Urie (this volume) found sludge-borne metals in an aspen forest also were immobilized by the humus layer and confined within the upper 5 cm soil layer three to five years following sludge application.

In general, plants are sensitive to trace elements. For plants subjected to aerial fallout, the trace element contamination of the vegetation is largely superficial (MacLean et al. 1969). Metal particulates lodged on the leaf's surface are relatively inert and seldom result in phytotoxicity. On the other hand, trace elements in the soil often affect plant growth (Carlson and Bazzaz 1977). Concentrations in plant leaves appear to be an accepted yardstick to assess the phytotoxic effects of trace elements (Beckett and Davis 1977, Burton et al. 1983). Although trees have differential susceptibility to metals, 8, 11, 35, 20, and 200 mg/kg (dry weight), respectively, are the generalized upper critical leaf tissue concentrations of cadmium, nickel, lead, copper, and zinc (Burton et al. 1983) beyond which tree growth would be affected by the presence of these metals (Carlson and Bazzaz 1977, Lamoreaux and Chaney 1977, Lozano and Morrison 1982). Even for trees that were grown in polluted forests, such levels were rarely found in the plant tissue (Martin et al. 1982, Burton et al. 1983). Some deforestation was reported, however, when 30,000 and 1,500 mg/kg of zinc and cadmium, respectively, in a contaminated forest soil resulted in 4,500 mg/kg of zinc and 70 mg/kg cadmium in the leaf tissue (Buchauer 1973). Rooting predominantly in the surface soil layer, the under-vegetation of the forests frequently contained higher levels of metals than the established trees and was more severely affected by the metals (Coughtrey and Martin 1977). When municipal wastewater or sewage sludge is applied on forest land, considerably greater amounts of plant nutrients than the potentially harmful trace elements are introduced. As a result, the trees appear to respond well to the nutrient inputs (Sidle and Sopper 1976, Lepp and Eardley 1978, McIntosh et al. 1984). Results of these land application experiments with short-term and low metal inputs indicated few harmful effects due to the sludge-applied trace elements. It was clear, however, that concentrations of cadmium, chromium, copper, nickel, lead, and zinc in the forest soils were increased by the waste applications (Sidle and Kardos 1977).

Many investigators have reported the slowdown of litter decomposition and reduction of soil microbial populations in forests contaminated by trace elements (Tyler 1972, Jordan and Lechevalier 1975). In turn, the productivity of the forest could be affected by reduction of nutrient mineralization and cycling. It was also found that saprophagous invertebrates and small mammals that inhabit the contaminated woodlands carried significantly greater body burdens of trace elements (Hopkins and Martin 1983, Scanlon et al. 1983). The long-term ecological consequences of accumulating heavy metals in forests are nevertheless difficult to gauge.

In conceptualizing the relation between environmental pollution and forest ecosystems, Smith (1974) grouped the effects of pollution on forests into three classes. In Class I, no detectable physiological effects appear to take place. The vegetation and the soils of the forest ecosystem function as a sink for pollutants. With intermediate pollutant inputs (Class II) the subtle changes in microbial and entomological population, nutrient cycling, photosynthesis, and water consumption start to affect individual (weaker) trees and individual (less competitive) species. The overall integrity of the forest has not yet been challenged. Exposure to high rates of pollutants (Class III) finally will take its toll. The rapid deforestation will be coupled with other environmental consequences. Applying municipal wastes on forest land obviously will not result in Class III effects. However, considering the potential for trace element accumulation by the forest ecosystem and the transfer of elements through the forest ecological chain, there is undoubtedly a need to set an upper limit on trace element inputs to forest land.

CONTRASTS BETWEEN SOURCES OF TRACE ELEMENTS AND EFFECTS

As previously indicated, there is little specific research on the effects of sludge applications of trace elements on forest ecosystems. However, there is an extensive and well-developed literature on the effects of trace elements from smelting and mining operations on natural systems (e.g., Hutchinson and Whitby 1974). There have also been studies on ecosystem effects of trace elements such as lead and cadmium from automobile emissions and tire wear near highways (e.g., Lagerwerff 1967). These studies provide valuable insights on anthropogenic impacts and are extensively cited in this paper; however, it is important to distinguish between the chemistry and bioavailability of trace elements from these sources and from sludge applications.

Trace element additions to soil from mining and smelting operations, from automobile emissions, and from other anthropogenic sources primarily occur as inorganic compounds. These may include oxides, sulfides, silicates, and carbonates from mining operations, or base metal particulates from smelters. The solubility and bioavailability of these sources in soil are a function of the thermodynamic stabilities of the inorganic compounds in a particular soil environment and the kinetics of their reactions in soils. For example, trace elements added to soil as base metal or sulfide (e.g., galena PbS) may be thermodynamically unstable in well-aerated soil but may persist in these forms in soil if particle size inhibits the kinetics of dissolution. Ultimate solubility and bioavailability of the trace element will be determined by its reactions with the soil system following dissolution of the original chemical compound.

Until recently, forms of trace elements in sludges and wastewaters were poorly understood (Page 1974, Logan and Chaney 1983, Chang and Page 1983), and it was commonly accepted that, particularly in the case of trace metals such as copper, cadmium, chromium, nickel, lead, and zinc, the element was primarily associated with the organic fraction through complexation. A consequence of this assumption was the often-stated concern that trace metal solubility and bioavailability might increase in soil with sludge mineralization (Logan and Chaney 1983). Recent characterization of sludges by fractionation (Lake et al. 1984), equilibrium solution modeling (Lake et al. 1984), and solid-phase mineralogy (Essington 1985) has indicated that while some elements such as copper have a strong affinity for the organic fraction, other elements primarily occur in solid-phase minerals or adsorbed to various solid phases in the sludge. Of particular importance is the work of Essington (1985), who has shown that considerable amounts of the trace elements in sludge may occur in forms determined by their original source. He found the trace metals in a wastewater sludge to be frequently associated with mineral oxides, silicates, sulfides, or carbonates, and has suggested that these may represent the original industrial or detrital sources.

Extensive field studies in agricultural (Logan and Chaney 1983) and forest (Kardos and Sopper 1973, Lepp and Eardley 1978) systems have shown that sludge and wastewater applications tend to shift soil pH to near 7. This is considerably higher than the pH of most forest soils, and should, at least initially, result in reduced trace element solubility, since solubilities of most trace elements in soil decrease with increasing pH. In the absence of long-term field studies of sludge effects on forest soils, it is not clear how long soil pH will remain higher after a single sludge application.

Because of the differences between the chemistry of trace elements in sludges and that from other anthropogenic sources, extrapolation to sludge systems of results from other studies must be done with considerable caution.

REFERENCES

Allaway, W. H. 1968. Agronomic controls over the environmental cycling of trace elements. Adv. Agron. 20:235–274.

Anderson, T. J., G. W. Barrett, C. S. Clark, V. J. Elia, and V. A. Majeti. 1982. Metal concentrations in tissues of meadow voles from sewage sludge-treated fields. J. Environ. Qual. 11:272–277.

Bartlett, R. J., and B. R. James. 1979. Behavior of chromium in soils: III. Oxidation. J. Environ. Qual. 8:31–35.

Beckett, P. H. T. 1980. The statistical distribution of sewage and sludge analysis. Environ. Poll. (Ser. B) 3:27–35.

Beckett, P. H. T., and R. D. Davis. 1977. Upper critical levels of toxic elements in plants. New Phytol. 79:95–106.

Bingham, F. T. 1973. Boron in cultivated soils and irrigation waters. In Trace Elements in the Environment. Adv. Chem. Ser. 123:130–138.

Bloomfield, C., W. I. Kelso, and G. Pruden. 1976. Reactions between metals and humified organic matter. Soil Sci. 27:16–31.

Brady, N. C. 1984. Nature and properties of soils. 10th ed. Macmillan, New York.

Bramryd, T. 1984. Uptake of heavy metals in pine forest vegetation fertilized with sewage sludge. p. 423–425. In P. L'Hermite and H. Ott (eds.) Processing and use of sewage sludge. D. Reidel, Dordrecht, Holland.

Brown, G. W. 1980. Forestry and water quality. Oregon State University, Corvallis. 124 p.

Buchauer, M. J. 1973. Contamination of soil and vegetation near a zinc smelter by zinc, cadmium, copper and lead. Environ. Sci. Technol. 7:131–135.

Burton, K. W., E. Morgan, and A. Roig. 1983. The influence of heavy metals upon the growth of Sitka spruce in South Wales forests. I. Upper critical and foliar concentrations. Plant Soil 73:327–336.

Carlson, R. W., and F. A. Bazzaz. 1977. Growth reduction in American sycamore (Plantanus occidentalis L.) caused by Pb-Cd interaction. Environ. Poll. 12:243–253.

Chaney, R. L. 1980. Health risks associated with toxic metals in municipal sludges. p. 59–83. In G. Bitton, B. L. Damron, G. T. Edds, and J. M. Davidson (eds.) Sludge: Health risks of land application. Ann Arbor Science Publishers, Ann Arbor, Michigan.

Chang, A. C., and A. L. Page. 1983. Hydrologic management: Trace metals. p. 107–124. In A. L. Page, T. L. Gleason III, J. E. Smith, Jr., I. K. Iskandar, and L. E. Sommers (eds.) Proceedings of the 1983 Workshop on Utilization of Municipal Wastewater and Sludge on Land. University of California, Riverside.

Coile, T. S. 1952. Soil and the growth of forests. Adv. Agron. 4:329–398.

Cole, D. W. 1982. Response of forest ecosystems to sludge and wastewater applications: A case study in western Washington. p. 274–291. In W. E. Sopper, E. M. Seaker, and R. K. Bastian (eds.) Land reclamation and biomass production with municipal wastewater and sludge. Pennsylvania State University Press, University Park.

Cole, D. W., and M. Rapp. 1981. Elemental cycling in forest ecosystems. p. 341–409. In D. E. Reichle (ed.) Dynamic properties of forest ecosystems. Internal Biological Programme 23, Cambridge University Press.

Coughtrey, P. J., and M. H. Martin. 1977. Cadmium tolerance of *Holcus lanatas* from a site contaminated by aerial fallout. New Phytol. 79:273–280.

Council for Agricultural Science and Technology. 1976. Application of sewage sludge to cropland: Appraisal of potential hazards of the heavy metals to plants and animals. Office of Water Program Operations, U.S. Environmental Protection Agency, Washington, D.C.

———. 1980. Effects of sewage sludge on the cadmium and zinc content of crops. Report 83. 250 Memorial Union, Ames, Iowa.

Doty, W. T., D. E. Baker, and R. F. Shipp. 1978. Heavy metals in Pennsylvania sewage sludge. Compost Sci. 19:26–29.

Essington, M. E. 1985. Characterization of inorganic solid phases in density-fractionated sewage sludge and sewage sludge-amended soil. Ph.D. diss., University of California, Riverside. 345 p.

Foth, H. D., and J. W. Schafer. 1980. Soil geography and land use. John Wiley and Sons, New York. 484 p.

Furr, A. K., W. C. Kelly, C. A. Bache, W. H. Guttenmann, and D. J. Lisk. 1976. Multielement uptake by vegetables and millet grown in pots on fly ash-amended soil. J. Agric. Food Chem. 24:885–888.

Grant, R. O., and S. E. Olesen. 1984. Sludge utilization in spruce plantations and sand soils. p. 79–90. *In* S. Berglund, R. D. Davis, and P. L'Hermite (eds.) Utilisation of sewage sludge on land: Rates of application and long-term effects. D. Reidel, Dordrecht, Holland.

Heinrichs, H., and R. Mayer. 1977. Distribution and cycling of major and trace elements in two central European forest ecosystems. J. Environ. Qual. 6:402–407.

Helmke, P. A., W. P. Robarge, R. L. Koroter, and P. J. Schomberg. 1979. Effects of soil-applied sewage sludge on concentrations of elements in earthworms. J. Environ. Qual. 8:322–327.

Hopkins, S. P., and M. H. Martin. 1983. Heavy metals in the centipede *Lithobius variegatus* (Chilopoda). Environ. Poll. (Ser. B) 6:309–318.

Hughes, M. K. 1981. Cycling of trace metals in ecosystems. p. 95–118. *In* N. W. Lepp (ed.) Effects of heavy metal pollution on plants. Vol. 2. Metals in the environment. Pollution Monitoring Series. Applied Sci. Pub., London and New Jersey.

Hutchinson, T. C., and L. M. Whitby. 1974. Heavy-metal pollution in the Sudbury mining and smelting region of Canada. I. Soils and vegetation contamination by nickel, copper and other metals. Environ. Conserv. 1:123–132.

James, B. R., and R. J. Bartlett. 1983a. Behavior of chromium in soils. VI. Interactions between oxidation-reduction and organic complexation. J. Environ. Qual. 12:173–176.

———. 1983b. Behavior of chromium in soils. VII. Adsorption and reduction of hexavalent forms. J. Environ. Qual. 12:177–181.

Jordan, M. J., and M. P. Lechevalier. 1975. Effects of zinc-smelter emissions on forest soil microflora. Can. J. Microbiol. 21:1855–1865.

Kardos, L. T., and W. E. Sopper. 1973. Effects of land disposal of wastewater on exchangeable cations and other chemical elements in the soil. p. 220–229. *In* W. E. Sopper and L. T. Kardos (eds.) Recycling treated municipal wastewater and sludge through forest and cropland. Pennsylvania State University Press, University Park.

Knezek, B. D., and R. H. Miller. 1976. Application of sludges and wastewaters to agricultural land: A planning and educational guide. North Central Regional Research Publ. 235. Ohio Agricultural Research and Development Center, Wooster. 88 p.

Lagerwerff, J. V. 1967. Heavy-metal contamination of soils. *In* N. C. Brady (ed.) Agriculture and the quality of our environment. Am. Assoc. Adv. Sci. Publ. 85:343–364.

Lake, D. L., P. W. W. Kirk, and J. N. Lester. 1984. Fractionation, characterization and speciation of heavy metals in sewage sludge and sludge-amended soils: A review. J. Environ. Qual. 13:175–183.

Lamoreaux, R. J., and W. R. Chaney. 1977. Growth and water movement of silver maple seedlings affected by cadmium. J. Environ. Qual. 6:201–205.

Lepp, N. W., and G. T. Eardley. 1978. Growth and trace metal content of European sycamore seedlings grown in soil amended with sewage sludge. J. Environ. Qual. 7:413–421.

Little, P., and M. H. Martin. 1972. A survey of zinc, lead, and cadmium in soil and vegetation around a smelting complex. Environ. Poll. 3:241–254.

Logan, T. J., and R. L. Chaney. 1983. Utilization of municipal wastewater and sludge on land—metals. p. 235–323. *In* A. L. Page, T. L. Gleason III, J. E. Smith, Jr., I. K. Iskandar, and L. E. Sommers (eds.) Proceedings of the 1983 Workshop on Utilization of Municipal Wastewater and Sludge on Land. University of California, Riverside.

Lozano, F. C., and I. K. Morrison. 1982. Growth and nutrition of white pine and white spruce seedlings in solutions of various nickel and copper concentrations. J. Environ. Qual. 11:437–441.

MacLean, A. J., R. L. Halstead, and B. J. Finn. 1969. Extractability of added lead in soils and its concentration in plants. Can. J. Soil Sci. 49:327–334.

Martin, M. H., and P. J. Coughtrey. 1981. Impact of metals on ecosystem function and productivity. p. 119–158. *In* N. W. Lepp (ed.) Effects of heavy metal pollution on plants. Vol. 2. Metals in the environment. Pollution Monitoring Series. Applied Sci. Pub., London and New Jersey.

Martin, M. H., E. M. Duncan, and P. J. Coughtrey. 1982. The distribution of heavy metals in a contaminated woodland ecosystem. Environ. Poll. (Ser. B) 3:147–157.

McIntosh, M. S., J. E. Foss, D. C. Wolf, K. R. Brandt, and R. Darmody. 1984. Effect of composted municipal sewage on growth and elemental composition on white pine and hybrid poplar. J. Environ. Qual. 13:60–62.

Miller, R. A., and T. J. Logan. 1979. Temporal variation in sewage sludge composition from six Ohio treatment plants. Agron. Abs. 1979:34.

National Academy of Sciences. 1983. Selenium in nutrition. National Academy Press, Washington, D.C. 174 p.

Nelson, M. D. 1981. Philadelphia's sludge marketing program. p. 148–153. In The sludge management syndrome: Burn, barge, bury or bust. Assoc. Metro. Sewage Agencies, Washington, D.C.

Page, A. L. 1974. Fate and effects of trace elements in sewage sludge when applied to agricultural land. EPA 670/2-74-005. 108 p.

Page, A. L., and A. C. Chang. 1983. Nutritional supplements to plants: Sewage sludge origin, composition and uses. p. 377–387. In Miloslav Richagl, Jr. (ed.) Handbook of nutritional supplements. Vol. 2. Agricultural use. CRC Press, Boca Raton, Florida.

Page, A. L., A. C. Chang, G. Sposito, and S. Mattigod. 1981. Trace elements in wastewater: Their effects on plant growth and composition and their behavior in soils. p. 182–222. In I. K. Iskandar (ed.) Modeling wastewater renovation land treatment. John Wiley and Sons, New York.

Pennsylvania State University. 1985. Criteria and recommendations for land application of sludges in the Northeast. Pennsylvania State University, Agricultural Experiment Station Bulletin 851.

Scanlon, P. F., R. J. Kendall, R. L. Lochmiller II, and R. L. Kirkpatrick. 1983. Lead concentrations in pince voles from two Virginia orchards. Environ. Poll. (Ser. B) 6:157–160.

Sidle, R. C., and L. T. Kardos. 1977. Transport of heavy metal in a sludge-treated forested area. J. Environ. Qual. 6:431–437.

Sidle, R. C., and W. E. Sopper. 1976. Cadmium distribution in forest ecosystems irrigated with treated wastewater and sludge. J. Environ. Qual. 5:419–422.

Smith, W. H. 1974. Air pollution effects on the structure and function of the temperate forest ecosystem. Environ. Poll. 6:111–129.

Sommers, L. E. 1977. Chemical composition of sewage sludges and analysis of their potential use as fertilizers. J. Environ. Qual. 6:225–232.

——. 1980. Toxic metals in agricultural crops. p. 105–140. In G. Bitton, B. L. Damron, G. T. Edds, and J. M. Davidson (eds.) Sludge: Health risks of land application. Ann Arbor Science Publishers, Ann Arbor, Michigan.

Sopper, W. E., and L. T. Kardos. 1973. Vegetation responses to irrigation with treated municipal wastewater. p. 271–294. In W. E. Sopper and L. T. Kardos (eds.) Recycling treated municipal wastewater and sludge through forest and cropland. Pennsylvania State University Press, University Park.

Southwick, C. H. 1972. Ecology and the quality of our environment. Van Nostrand Reinhold, New York. 247 p.

Spurr, S. H., and B. V. Barnes. 1980. Forest ecology. 3d ed. John Wiley and Sons, New York. 687 p.

Tabatabai, M. A., and W. T. Frankenberger. 1979. Chemical composition of sewage sludge in Iowa. Research Bulletin 586, Agriculture and Home Economic Experiment Station. Iowa State University of Science and Technology, Ames, Iowa.

Tsuchiya, K. 1978. History of itai-itai disease. p. 1–6. In K. Tsuchiya (ed.) Cadmium studies in Japan: A review. Elsevier/North-Holland Biomedical Press, New York.

Tyler, G. 1972. Heavy metals pollute nature, may reduce productivity. Ambio 1:51–59.

Urie, D. H., C. K. Losche, and F. D. McBride. 1982. Leachate quality in acid mine-spoil columns and field plots treated with municipal sewage sludge. p. 386–398. In W. E. Sopper, E. M. Seaker, and R. K. Bastian (eds.) Land reclamation and biomass production with municipal wastewater and sludge. Pennsylvania State University Press, University Park.

U.S. Environmental Protection Agency. 1983. Process design manual: Land application of municipal sewage sludge. EPA 625/1-83-016. CERI, Cincinnati, Ohio.

Williams, S. T., and T. R. G. Gray. 1974. Litter on the soil surface. p. 611–622. In C. H. Dickinson and G. J. F. Pugh (eds.) Biology of plant litter decomposition. Vol. 2. Academic Press, London and New York.

Zasoski, R. J., R. L. Edmonds, C. S. Bledsoe, C. L. Henry, D. J. Vogt, K. A. Vogt, and D. W. Cole. 1984. Municipal sewage sludge use in forests of the Pacific Northwest, U.S.A.: Environmental concerns. Waste Management and Research 2:227–246.

Chapter 9

Water Quality in Relation to Sludge and Wastewater Applications to Forest Land

R. J. ZASOSKI and R. L. EDMONDS

ABSTRACT   Because forests occupy areas where rainfall exceeds evapotranspiration, the potential for mass transfer of pollutants in water exists. Both surface runoff and deep leaching can influence water quality. These processes may transport suspended and dissolved materials including microorganisms, organics, heavy metals, and nitrates. The potential for affecting water quality depends on site hydrology and sludge or wastewater characteristics. Since site design criteria restrict surface runoff, leaching of sludge components is the major water quality concern, and among these components nitrate is the primary concern. Deep leaching of nitrate can be expected whenever the supply of mineralized nitrate exceeds forest demands and water is available for transport. This will vary considerably with forest type, climatic conditions, and loading rates. While the potential for heavy metal movement exists, adsorption by the soil and limited release from the sludge matrix limits leaching. Pathogen movement also appears to be limited. Few viable bacteria reach groundwater, but data for virus mobility are less complete. Although organic toxins can reach groundwater, the movement of organics from sludge or wastewater is limited. Organics can be absorbed on soil colloids or removed by degradation; however, information from field studies of organics is limited. Available data for wastewater and sludge applications to forests show nitrate leaching to be the major problem relative to water quality.

Water quality deals with the concentration of suspended and dissolved substances in water sources, and the end use of the water determines acceptable standards. Since forests are sources of high quality water, this paper will consider water quality as applied to drinking water. In this context, leachates from sludge and wastewater that enter groundwater may significantly alter water quality, as will overland flow and direct input into surface water. Direct movement of sludges or wastewaters into surface water supplies is unacceptable, and the consequences are easily predicted. Appropriate site design and proper management can avoid overland flow, and for this reason overland flow will not be considered in this paper.

From a conceptual standpoint, the soil on which sewage applications occur is a buffer that moderates input from sludges or wastewater into water supplies. A general scheme of how water quality is related to land application is shown in Figure 1. Sludge and wastewater can be generated from several sources and are treated by many different methods. In Figure 1 soil is interposed between the applied material and the groundwater. When levels of nitrogen, cadmium, and fecal coliforms in typical sludge and wastewater are compared with U.S. EPA drinking water guidelines (Table 1), it is obvious that sludge and wastewater are several orders of magnitude higher. Thus the soil to be effective must render a significant change in sludge and wastewater leachates.

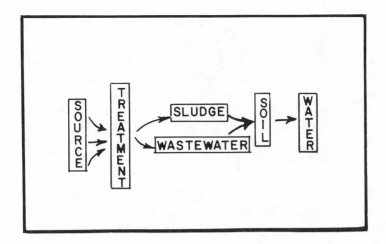

Figure 1. A conceptual scheme of sludge and wastewater application to soil in relation to water quality.

It must be recognized that sludge components are not all water soluble, and therefore the leachates from surface-applied sludge are not as concentrated as Table 1 seems to suggest. Nonetheless, land application of sludges and wastewater has the potential to modify water quality significantly, depending on the quantity of water moving through a soil. In arid regions leaching may be limited, but in forests the precipitation usually substantially exceeds evapotranspiration, making the water necessary for leaching soluble sludge components available during at least some portion of the year. Wastewater applications, at rates generally used, furnish this potential year around. Special considerations in forest applications relative to agricultural use are the perennial nature of vegetation in forests, a generally acidic soil, and an accumulation of organic material at the soil surface. Both constraints and benefits are the result of use of forest soils for sludge and wastewater recycling.

Chemical compounds and elements, microbes, and organic chemicals are present in both sludges and wastewaters. Many representatives from each group pose potential environmental and public health risks. Discussions of these components are well presented in papers of the recent Denver conference (Page et al. 1983) and in earlier symposia (Sopper and Kardos 1973, and Sopper and Kerr 1979). These publications generally recognize nitrates, heavy metals, pathogenic organisms, and toxic organics as the princi-

TABLE 1. U.S. EPA drinking water guidelines compared with the content of nitrogen, cadmium, and fecal coliforms in sludge and wastewater.

Component	Content in Wastewater	Content in Sludge	EPA Guidelines	Source
Total N (mg/kg)	20-40	30,000-50,000	10	Linden et al. 1983
Cadmium (mg/kg)	0.02-0.4	--	0.01	Chang and Page 1983
Cadmium (mg/kg)	--	10 (1-3,400)	0.01	Logan and Chaney 1983
Fecal coliforms (no./ml)	$3-18 \times 10^6$	--	<0.02*	Hunter and Kotalik 1973

*EPA guidelines are for total coliforms.

pal potential contaminants in sludges and wastewaters. Large differences in contaminant loading exist among sludges and between sludges and wastewaters. Therefore, in assessing the application of these materials to lands, a case-by-case analysis should be conducted. For example, heavy metal burdens in wastewaters are low compared with sludges because the solid phase concentrates metals from the liquid phase.

Soluble materials added with wastewater or leached from sludges are acted on by a number of processes, which may modify or remove components from the percolate. These processes include decomposition or transformations of toxic and nontoxic organic compounds; filtration of particulates and microbes from the leachate; precipitation reactions that remove metals, phosphates, and nutrients; immobilization reactions that convert inorganic soluble compounds into microbial tissue; and adsorption-exchange reactions that retain metals and other nutrient cations. These processes vary in effectiveness with soil properties, soil depth, and environmental conditions. The relative importance of both the input to these processes and the process itself may also vary with time.

This temporal variation is shown conceptually in Figure 2, which indicates that while pathogens are a major concern following application, they soon die off and other factors

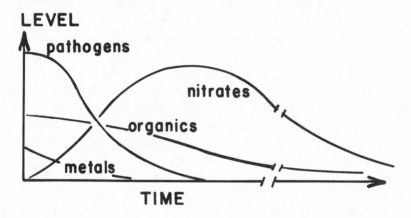

Figure 2. The relationship of sludge and wastewater components to their relative levels over time.

become relatively more important. While the trends shown in Figure 2 for nitrates, pathogens, and metals are generalized from field data, the trend for organics is mostly supposition. Some organics are readily decomposed, while others take longer to be metabolized or altered. Some components in sludges and wastewaters are conservative and will always be associated with the sludge unless removed by leaching; others decompose, die, or are transformed into other compounds. The specifics of these processes are covered in other papers in this volume. Regardless of the process involved, research studies have documented that the soil forms an effective filter for potentially toxic materials introduced in sludge or wastewater.

The extent of areal coverage that an application site occupies in relation to the source area of the receiving water body is also important. In the situation shown in Figure 3, a portion of the landscape above an aquifer has been treated with sludge. Leachate from the sludge (or wastewater) mixes with leachate from untreated areas, and a certain amount of dilution takes place. Sampling and interpretation in this situation must be

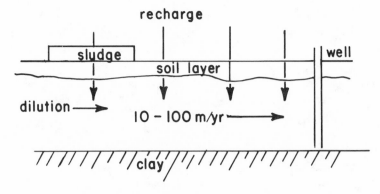

Figure 3. The relationship of sludge occupancy to leachate dilution and monitoring.

carefully considered, since samples from the lower soil profile may not reflect conditions in the aquifer. MacKay et al. (1985) suggest that a reasonable flow rate for water in an aquifer is 10 to 100 m per year. If a sampling well is 100 m from the source, then a contaminant, if present, may not be detected for one to ten years. When the possibility of dilution is also considered, the question of when and where to sample is complex indeed.

Contaminants from a sewage treatment area that enter an aquifer or other water source are not in steady state and may appear as pulses. How should these pulses be interpreted? Is a daily, yearly, or monthly average the appropriate sampling period? Does a one-time pulse that exceeds published guidelines mean the project is not in compliance? These questions may pertain differently to the individual components.

In the following section, nitrates, metals, organics, and pathogens will be considered as potential contaminants of groundwater systems. This section is not intended to be inclusive, since other papers in this volume review the status of wastewater and sludge applications. The objective here is to suggest why each of these components may or may not be a concern.

PATHOGENS

Pathogens are nonconservative: they will die off with time. During the initial phase of application, restricted public access is desirable until pathogen populations are lowered. Water quality concerns are directed mostly at organisms leaching into water supplies. Although pathogens have been known to move rather large distances through soils in some circumstances (Gerba 1983), there is evidence that movement is limited in most soil conditions (Zasoski et al. 1984). Figure 4 shows the time course of fecal coliform populations in sludge and the soil beneath sludge applied to a Douglas-fir forest. In this gravelly, loamy sand the soil acted as a very efficient biological filter. Gerba (1983) provided an exhaustive review of the subject. Additional information is also available in this volume. The authors of this paper conclude that unless the soil is extremely coarse in texture and the rate of water flow in the system is rapid, as in rapid infiltration systems, the movement of organisms to groundwaters is sufficiently limited. Kowal (1983) considered rapid infiltration systems to pose a threat from bacterial contamination and concluded that the question of viruses in groundwater was unsettled. He also concluded, however, that the threat was likely to be low at slow infiltration sites.

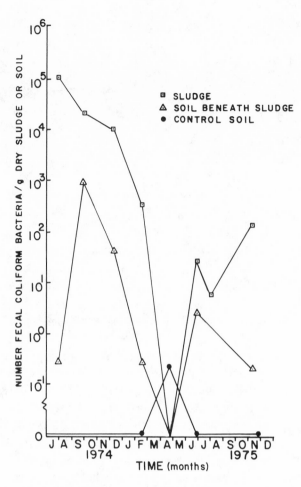

Figure 4. Numbers of fecal coliform bacteria in sludge, soil 5 cm beneath the sludge, and control soil after sludge application to a clearcut over time.

ORGANICS

The quantity and type of organics applied in wastewater and sludges are usually different from those in the soil before application. Like pathogens, organics are not conservative and can be chemically or photochemically degraded, decomposed, or altered by organisms in the soil or by organisms imported with the sludge or wastewater. In addition, many of the imported organics can be adsorbed and immobilized by the soil (Kowal 1983). The large number of organic compounds that can occur in sludges and wastewaters does not allow for generalizations about their degradation and transport in soils. Kowal (1983), however, suggests that the more complex the molecule, the greater the number of condensed rings, and the more halogenated, the more difficult is the biodegradation. The organics found in sludge and wastewater generally increase soluble carbon in the soil and add to the native organics.

Table 2 shows the total organic carbon (TOC) in the litter and A, B, and C horizons of a site irrigated with wastewater and sludge (Hanson 1978). TOC in the wastewater before application (11 mg/l) is less concentrated than that found in the A horizon of the control

TABLE 2. Total organic carbon (mg/l) in leachates from a site treated with sludge and wastewater at Pack Forest (Hanson 1978).

Plot	Horizon			
	Litter	A	B	C
Control	34	25	17	14
Wastewater*	--	22	8	6
Sludge	81	72	36	34

*Total organic carbon in wastewater before application = 11 mg/l.

plot. Sludge addition increased the TOC down to the C horizon (160 cm), and also changed the character of the organics. Figure 5 shows the molecular weight distribution of organics from a control site and a sludge-treated site following sludge application (Hanson 1978). Retention time is a measure of molecular weight, with longer retention times indicating lower molecular weights. The sludge treatment had more low molecular weight organics in both the surface and subsurface horizons, and apparently had a major effect on decomposition in the litter layer. Thus more and different types of organics are reaching the C horizon, although their identity is unknown and little research has been done to quantify the reactions of the myriad organics found in sludges and wastewater. Jenkins and Palazzo (1981) suggested that toxic organics can be transported into groundwater. Overcash (1983) reviewed the fate of organics in soils and concluded that because of adsorption by the soil and decomposition reactions, little leaching would occur. Lands receiving sludges or wastewater with high concentrations of potentially toxic organics should be closely monitored, especially sites with shallow soils or rapid infiltration characteristics.

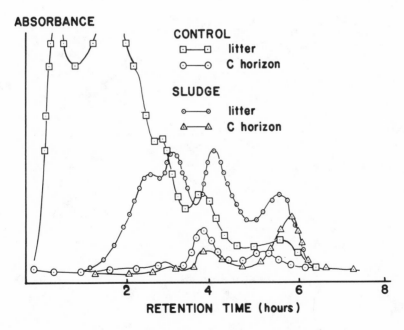

Figure 5. Gel filtration chromotograms of organics from a sludge-treated site and a control site in the litter layer and the C horizon (data from Hanson 1978).

METALS

Unlike pathogens and organics, metals imported with wastewater and sludge applications are conservative and will remain at the application site unless they are leached from the system. In order to be leached, sludge-borne metals must first be released from the sludge matrix. Figure 6 shows the loss of nickel and cadmium from sludge over a year of continuous monitoring at a clearcut site. In the data shown (McKane 1984), the two metals are behaving differently. The original nickel content of the sludge was about twice as high as the cadmium level, yet in solution the cadmium concentration was several times lower than nickel. Individual metals behave differently, and each must be considered separately; in this instance, neither metal was exiting the sludge at high levels. These solutions were collected as they exited directly from the sludge interacting with the soil. Solution concentrations would be expected to decrease even further after percolating through a soil.

Questions of analytical techniques and background levels must be considered when solution concentrations are very low, as is the case for most metals. Table 3 shows data from a forest in Germany that had not been treated with sludge. Average cadmium levels in the seepage water at a depth of 80 cm varied from 2.9 to 5.1 ppb (µg/l). For comparison, at a sludge-treated site, Sidle and Kardos (1977) found cadmium in leachates to range from 0.2 ppb in the control to 1.8 ppb at the highest sludge application rate. Even though cadmium and other metals can be leached through forest soils treated with sludge and wastewater, concentrations in leachates are very low and often similar

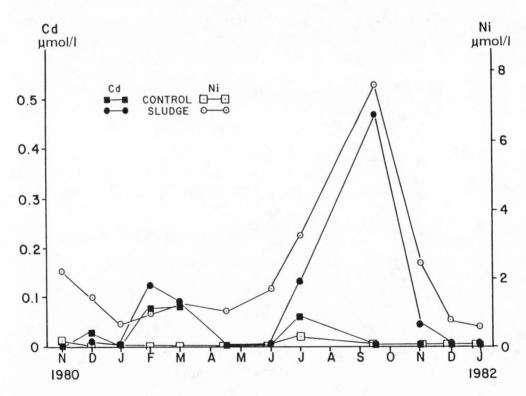

Figure 6. Solution concentration of cadmium and nickel in leachates collected directly beneath a 10 cm deep sludge layer over time (data from McKane 1984).

TABLE 3. Dissolved metal levels in water moving through a forest site (not treated with sludge) in West Germany (Heinrichs and Mayer 1980).

	Dissolved Metal Level (ug/1)	
Water Source	Zinc	Cadmium
Precipitation	180	3.4
Stemflow	1,720	2.0
Seepage at 80 cm depth	190-570	2.9-5.1

to or below native levels at other sites. Metal concentrations reported in leachate from forested sites are near the detection limits for many analytical techniques, and great care must be exercised in the sampling, storage, and analysis of these very low level solutions.

The role of organic matter and pH appears to be particularly important in forest soils. Tyler (1978) has shown that metals will leach from the forest floor if the pH is reduced significantly. But in the case of lead (Figure 7) pH must be below 3.2 before significant metal movement is observed. Other metals were somewhat more mobile than lead, but show the same trends (Tyler 1978).

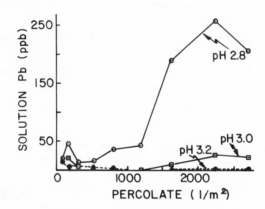

Figure 7. Solution lead concentration leached from a litter layer in relation to volume and imposed pH (after Tyler 1978).

It appears that unless a soil has limited adsorption capacity (low organic matter and colloid content), metals will not leach into groundwater in significant amounts, extremely acid systems and very high metal loading rates notwithstanding.

NITRATES

Nitrates are a major concern in forests receiving sewage applications because they are not adsorbed by the soil in significant quantities. Both sludge and wastewater applications import large quantities of nitrogen to a site. Over time some of the organic nitrogen and ammonium in the sludge is converted to nitrate, which is either taken up by the vegetation and microbes or is leached. There is ample research evidence that nitrate can be leached and that the leachate concentration can exceed drinking water standards

(Brockway and Urie 1983, Kardos and Sopper 1973, Riekerk and Zasoski 1979, Zasoski et al. 1984). The question of what application rate is appropriate for a given forest type is critical to water quality.

One major difference between forest applications and those in an agricultural setting is the large amount of organic material in the forest floor and upper soil horizons. This large organic pool typically has a high C:N ratio and can immobilize a significant amount of nitrogen. Forest vegetation can also accumulate soluble nitrogen. Resolving these factors into a predictive model is a pressing need. In general, nitrate leaching can be minimized by matching application rates to plant uptake and microbial immobilization rates.

CONCLUSIONS

Applications of wastewaters and sludges to forest sites generate concerns about leaching of nitrates, heavy metals, pathogens, and organics into groundwater. Nitrate leaching is the major concern. Immobilization and uptake will limit leaching of applied nitrogen, but these aspects are not well quantified. Further definition of the nitrogen cycle in the forest environment is needed in relation to application, decomposition, and leaching rates.

Metals do not appear to leach downward in significant quantities from sludge applications in the short run because of the strong interactions with the soil and because they are slowly released from the sludge matrix. However, in the long run, acidification from nitrification and other oxidative processes may mobilize a larger fraction of the metals. Long-term data are needed. Wastewater applications raise the pH of forest soils, making the long-term mobility of the metals unlikely under these circumstances as long as the irrigation continues.

Decomposition and adsorption reactions would appear to lower the risk of organics leaching into groundwater. However, the decomposition rates of specific compounds in the sludge are largely unknown. In acid forest soils, toxicity from aluminum or from the large metal content imported with sludge may affect the microbes involved in decomposition.

Although pathogens are not known to be a problem in forest applications, there is a need to define the conditions where this is true, especially since movement of viruses to depth has been demonstrated in some systems.

It is also important to know where and when to monitor water quality in relation to sludge and wastewater applications. Is daily, monthly, or yearly averaging the appropriate criterion for monitoring water quality? Also, the role of dilution in modifying leachate concentrations beyond the confines of the application site needs attention. All water quality criteria have a temporal component. Both short-term and long-term consequences must be considered.

REFERENCES

Brockway, D. G., and D. H. Urie. 1983. Determining sludge fertilization rates for forests from nitrate-N in leachate and groundwater. J. Environ. Qual. 12:487–492.

Chang, A., and A. L. Page. 1983. Hydrologic management: Trace metals. p. 107–122. *In* A. L. Page,

T. L. Gleason III, J. E. Smith, Jr., I. K. Iskandar, and L. E. Sommers (eds.) Proceedings of the 1983 Workshop on Utilization of Municipal Wastewater and Sludge on Land. University of California, Riverside.

Gerba, C. P. 1983. Pathogens. p. 147–187. *In* A. L.

Page, T. L. Gleason III, J. E. Smith, Jr., I. K. Iskandar, and L. E. Sommers (eds.) Proceedings of the 1983 Workshop on Utilization of Municipal Wastewater and Sludge on Land. University of California, Riverside.

Hanson, D. M. 1978. Effects of the land disposal of treated sewage wastewater and sludge on groundwater organics. M.S. thesis, College of Forest Resources, University of Washington, Seattle.

Heinrichs, H., and R. Mayer. 1980. The role of forest vegetation in the biogeochemical cycle of heavy metals. J. Environ. Qual. 9:111–118.

Hunter, J. V., and T. A. Kotalik. 1973. Chemical and biological quality of treated sewage effluents. p. 6–25. In W. E. Sopper and L. T. Kardos (eds.) Recycling treated municipal wastewater and sludge through forest and cropland. Pennsylvania State University Press, University Park.

Jenkins, T. F., and A. J. Palazzo. 1981. Wastewater treatment by a prototype slow rate land treatment system. Report 81–14. U.S. Army Corps of Engineers, Cold Regions Research and Engineering Laboratory (CRREL), Hanover, New Hampshire.

Kardos, L. T., and W. E. Sopper. 1973. Renovation of municipal wastewater through land disposal by spray irrigation. p. 148–163. In W. E. Sopper and L. T. Kardos (eds.) Recycling treated municipal wastewater and sludge through forest and cropland. Pennsylvania State University Press, University Park.

Kowal, N. E. 1983. An overview of public health effects. p. 329–394. In A. L. Page, T. L. Gleason III, J. E. Smith, Jr., I. K. Iskandar, and L. E. Sommers (eds.) Proceedings of the 1983 Workshop on Utilization of Municipal Wastewater and Sludge on Land. University of California, Riverside.

Linden, D. R., C. E. Clap, and R. H. Dowdy. 1983. Hydrologic management: Nutrients. p. 79–101. In A. L. Page, T. L. Gleason III, J. E. Smith, Jr., I. K. Iskandar, and L. E. Sommers (eds.) Proceedings of the 1983 Workshop on Utilization of Municipal Wastewater and Sludge on Land. University of California, Riverside.

Logan, T. J., and R. Chaney. 1983. Metals. p. 235–323. In A. L. Page, T. L. Gleason III, J. E. Smith, Jr., I. K. Iskandar, and L. E. Sommers (eds.) Proceedings of the 1983 Workshop on Utilization of Municipal Wastewater and Sludge on Land. University of California, Riverside.

McKane, R. B. 1984. Effects of sludge redox potential on the leaching of nutrients and heavy metals from surface applications of sewage sludge. M.S. thesis, College of Forest Resources, University of Washington, Seattle.

Mackay, D. M., P. V. Roberts, and J. A. Cherry. 1985. Transport of organic contaminants in groundwater. Environ. Sci. and Tech. 19:384–392.

Overcash, M. R. 1983. Land treatment of municipal effluent and sludge: Specific organic compounds. p. 199–277. In A. L. Page, T. L. Gleason III, J. E. Smith, Jr., I. K. Iskandar, and L. E. Sommers (eds.) Proceedings of the 1983 Workshop on Utilization of Municipal Wastewater and Sludge on Land. University of California, Riverside.

Page, A. L., T. L. Gleason III, J. E. Smith, Jr., I. K. Iskandar, and L. E. Sommers (eds.) 1983. Proceedings of the 1983 Workshop on Utilization of Municipal Wastewater and Sludge on Land. University of California, Riverside.

Riekerk, H., and R. J. Zasoski. 1979. Effects of dewatered sludge applications to a Douglas fir forest soil on the soil, leachate, and groundwater composition. p. 35–45. In W. E. Sopper and S. N. Kerr (eds.) Utilization of municipal sewage effluent and sludge on forest and disturbed land. Pennsylvania State University Press, University Park.

Sidle, R. C., and L. T. Kardos. 1977. Transport of heavy metals in a sludge-treated forested area. J. Environ. Qual. 6:431–437.

Sopper, W. E., and L. T. Kardos (eds.) 1973. Recycling treated municipal wastewater and sludge through forest and cropland. Pennsylvania State University Press, University Park.

Sopper, W. E., and S. N. Kerr (eds.) 1979. Utilization of municipal sewage effluent and sludge on forest and disturbed land. Pennsylvania State University Press, University Park.

Tyler, G. 1978. Leaching rates of heavy metal ions in forest soil. Water, Air and Soil Poll. 9:137–148.

Zasoski, R. J., R. L. Edmonds, C. S. Bledsoe, C. L. Henry, D. J. Vogt, K. A. Vogt, and D. W. Cole. 1984. Municipal sewage sludge use in forests of the Pacific Northwest, U.S.A.: Environmental concerns. Waste Management and Research 2:227–246.

Chapter 10

Wildlife Responses to Forest Application of Sewage Sludge

JONATHAN B. HAUFLER and STEPHEN D. WEST

ABSTRACT   Application of municipal sewage sludge to forest vegetation can have beneficial or detrimental effects on wildlife populations. Direct effects from contact with sludge appear to be minimal. Indirect effects can result from changes in composition, structure, or productivity of vegetation, changes in nutritional quality of forages, or toxicities from elements or compounds entering food chains. Sludge application produces a fertilizing effect, increasing site productivity in many types of vegetation. This increase, along with corresponding changes in plant protein levels, causes an increase in populations of many wildlife species. Those dependent on more open, poorer quality sites may decline in numbers. Toxicity problems from trace organic compounds or metals in sludges could have an impact on wildlife populations. No problems have been found using municipal sewage sludge from areas not supporting major industrial complexes. Industrial sludges, with higher concentrations of metals such as cadmium, need to be investigated further before they are applied to forest vegetation. Although some questions remain to be answered, sludge application can offer valuable benefits to wildlife. Wildlife agencies should be encouraged to consider sludge application to forest lands as a method of wildlife habitat management.

Land application has received increased attention in recent years as an alternative means for disposing of sewage sludge while recycling nutrients contained in the sludge. Sludge has been applied to agricultural lands in many areas, and silvicultural application has been proposed as an option for communities near forests. Forest ecosystems are complex, and the effects of sludge application on wildlife communities must also be considered. This paper summarizes known responses of wildlife communities to forest application of sewage sludge.

Wildlife populations may be affected either beneficially or detrimentally. The sludge can affect wildlife through direct contact or through alterations in the vegetation, such as by clearing certain areas to provide application trails. Wildlife populations may increase or decrease as a result of vegetative changes caused by the nutrient enrichment from sludge. Wildlife may also be affected by toxicities from trace organic compounds or metals contained in the sludge.

DIRECT EFFECTS ON WILDLIFE

Sewage sludge can be applied to forests in many ways, including spraying dissolved sludge from irrigation type systems or applying dried or dissolved sludge by using application vehicles. Irrigation systems produce little site disturbance other than that of the sludge itself. Application vehicles, unless application is to be made in a forest opening or recent clearcut, require access trails, with some resultant site disturbance and associated vegetation and soil changes.

Application of sewage sludge to forest vegetation does not appear to affect wildlife populations significantly. Small mammal populations were monitored on control and sludge-treated plots in a four-year-old jack pine clearcut one month after dissolved sludge application, with no reduction in numbers exhibited (Woodyard 1982). Small mammal populations monitored within two weeks after dissolved sludge was applied to plots in a northern hardwood forest (Thomas 1983) and in a jack pine, red pine forest (Seon 1984) showed no significant difference from control plots in each type. Sludge applied as a dried, commercial fertilizer called Milorganite did not cause any reductions of small mammals in an old-field stand (Anderson and Barrett 1982).

Clearing areas to provide trails for access of sludge application vehicles will cause vegetation changes that can influence wildlife populations (Thomas 1983, Seon 1984, Campa et al., this volume). These changes are similar to those produced by other forest thinning practices or trail construction for other purposes, and are not felt to be a major consideration affecting sludge application.

EFFECTS ON WILDLIFE HABITAT

Vegetative Composition, Structure, and Productivity

Application of sludge to forest vegetation can produce a number of changes in the plant community. Changes in the composition of plant species are possible through introduction of new species from seeds contained in the sludge or from shifts in competition of plant species favored by nutrient enrichment. Few changes in the composition of plant species were noted in any of the five forest types that have been investigated for sludge application in Michigan (Woodyard 1982, Thomas 1983, Seon 1984, Haufler, unpublished data). This lack of change in species composition may be partly due to the sludge being applied to relatively undisturbed forest stands. In contrast, sludge application to recent clearcuts in Washington was found to reduce or eliminate regeneration of woody species (Edmonds and Cole 1977, Taber and West 1982), owing to plant competition that resulted from the rapid growth of herbaceous species and animal damage caused primarily by the increased numbers of voles (*Microtus townsendii*) that girdled seedlings.

Application of sewage sludge has affected the structure and productivity of plant communities. Woodyard (1982) found that the application of municipal sewage sludge to a four-year-old jack pine clearcut altered the vegetation by increasing both the vertical and horizontal cover and by increasing annual productivity. Understory vegetation has also been monitored following sludge application to four forest types in Michigan (Thomas 1983, Seon 1984, Campa et al., this volume, Haufler, unpublished data). Numerous significant differences were found in the vegetation on sludge-treated and untreated plots. Annual productivity was increased the first summer after a fall application of sludge to aspen and oak stands, especially for herbaceous vegetation. A corresponding increase in vertical cover in lower height strata of vegetation was also observed.

These same changes in vegetative productivity and structure were found in sludge-treated pine and northern hardwood stands, but not until the second summer after sludge application, since the sludge was not applied until late June. Although sludge application to the hardwood stand did cause some mortality of small seedlings, it was felt that this would not have occurred if the sludge had been applied before the active growing season. The northern hardwood stand, having the highest quality soil of the four forest types, exhibited the least vegetative response to sludge application.

Plant Nutritive Quality

Sludge application to forest vegetation has been found to alter the chemical composition of plant species. This is of greatest interest in plant species that are important forages for herbivores. The most consistent and significant change has been in protein levels. Working in Washington, Anderson (1983) found that sludge application increased crude protein content of Italian rye grass (*Lolium multiflorum*) 160% in December samples and 176% in March samples. Elevated nitrogen levels were also found in current annual growth for seven understory plants common in Douglas-fir forests (West, Zasoski, Taber, unpublished data). Campa (1982) found that spring sludge application to a four-year-old jack pine clearcut in Michigan significantly increased crude protein levels in all six plant species sampled that summer and in all three plant species sampled in late fall.

In studies of sludge application to other Michigan forest types, crude protein levels were consistently higher in spring and summer vegetation from an aspen study area the first year after treatment (Campa et al., this volume). The only significant increase in crude protein in winter vegetation was found in bigtooth aspen (*Populus grandidentata*). The protein levels of vegetation from the sludge-treated plots had declined to levels not significantly different from control plot samples by the second growing season. Of the three woody species sampled from an oak study area, only red maple (*Acer rubrum*) was found to have significant increases in protein levels on sludge-treated plots, being greater in spring, summer, and winter samples (Haufler et al., unpublished data). This increase, as in the aspen vegetation, was not observable the second year after treatment. Sludge application increased protein levels in summer samples of sedge (*Carex* spp.) and red oak (*Quercus rubra*) from a jack pine and red pine study area (Seon 1984), and in red maple in winter. Sedge protein levels remained significantly greater the next spring and summer, while red maple levels remained higher just for spring (Haufler et al., unpublished data). Sludge application did not produce significant increases in plant protein levels in vegetation collected during the summer immediately following the June application to the northern hardwood study area (Haufler et al., unpublished data). However, protein levels in sugar maple (*Acer saccharum*) were significantly greater the following winter and spring, but were at control levels by summer. The limited response of plant protein levels in the hardwood study area was probably due to its higher site quality.

The only other plant nutrient that showed significant trends following sludge application is phosphorus. Campa (1982) reported higher phosphorus levels in summer samples of herbaceous species following sludge treatment. Woody species were not higher in phosphorus in summer, but all species sampled were significantly higher in winter. Campa et al. (this volume) have also reported higher phosphorus levels in herbaceous vegetation sampled in spring and summer from an aspen study area in Michigan. No significant differences were found in winter or second year posttreatment samples. Red maple, sampled from the oak study area in Michigan, had significantly higher phosphorus levels in the first spring, summer, and winter following sludge application, but not thereafter (Haufler et al., unpublished data). Seon (1984) found higher phosphorus levels in sedges (*Carex* spp.) from a pine study area the summer after a June application of sludge. No other increases in phosphorus from the pine study area were found. No increases in phosphorus content of plants were found following sludge application of northern hardwood plots in Michigan, nor were increases seen in understory plants in Douglas-fir forests in Washington (West, Zasoski, Taber, unpublished data).

Other nutrients have not shown the consistent or significant trends following sludge treatment that have been observed for protein and phosphorus. An additional indicator of forage nutritive quality, in vitro digestibility, has been found to increase in some species on low quality sites (Campa 1982, Seon 1984), but no significant differences have been found in other cases.

WILDLIFE RESPONSES TO HABITAT CHANGES

Good indicators of habitat change are small mammal populations: they have small home ranges and tend to be specific habitat selectors. Several studies have been made of the response of small mammals to forest application of sewage sludge, such as Wood-yard's (1982) investigation of small mammal responses to sludge applied to a four-year-old jack pine clearcut in Michigan. He found that meadow voles (*Microtus pennsylvanicus*) responded to the increased plant productivity on the sludge-treated plots, and appeared only on these plots. Meadow jumping mice (*Zapus hudsonius*) and eastern chipmunks (*Tamias striatus*) were found in greater numbers on sludge plots than on control plots. Sludge plots had a correspondingly higher number of species captured, greater number of individuals, and a higher species diversity.

Small mammal populations were studied in the other four forest types investigated in Michigan. In general, sludge plots, compared with control plots, contained a greater number of species, a greater number of individuals, and a higher species diversity of small mammals (Thomas 1983, Seon 1984, Haufler et al., unpublished data). Species that tended to increase on most sludge plots were deer mice (*Peromyscus* spp.), meadow voles or red-backed voles (*Clethrionomys gapperi*), and meadow jumping mice. Thirteen-lined ground squirrels (*Citellus tridecemlineatus*) increased on sludge plots in the pine area the second year after sludge application. This species declined in numbers on sludge plots in the aspen area, apparently a response to the increase in ground-level vegetation, which became too dense for suitable habitat for this species.

West et al. (1981) reported small mammal responses one to two years after sludge application on a forty-year-old Douglas-fir stand in Washington. Results from four paired sites sampled in one year suggested that herbivore numbers declined on sludge plots, although samples from additional years needed to confirm this pattern were not obtained. More recent work in Washington on earlier successional stages has shown positive responses by herbivores, especially voles, to increased herbaceous vegetation resulting from sludge application (West, unpublished data).

Large herbivores have also responded to sludge application to forest lands. Woodyard (1982) found significantly higher rates of browsing by white-tailed deer (*Odocoileus virginianus*) on sludge-treated plots than on control plots; he attributed this to higher plant protein levels. Campa et al. (this volume) also found increased rates of browsing by deer and elk (*Cervus elaphus canadensis*) on sludge plots, which they felt to be associated with the higher nutritive quality of forages on these plots. Anderson (1985) found that Columbian black-tailed deer (*O. hemionus columbianus*) altered their food habits to take advantage of higher protein levels of grasses in sludge-treated areas. They were able to feed on these grasses through the winter, when deer on untreated areas had switched their foraging primarily to browse species. Anderson (1983) also found significantly higher productivity among the does that used the sludged areas than among those that did not use them. He attributed the higher productivity to the greater protein availability.

These studies indicate that wildlife generally respond favorably to the increased productivity and nutritive quality of forages produced by sludge treatment. Only species that are favored by open or thin ground-level vegetation appear to be negatively influenced.

POTENTIAL TOXICANTS IN SLUDGE

Although wildlife and wildlife habitat have generally responded favorably to sludge application to forest lands, negative effects could result from releases of metals or trace organic compounds into wildlife habitats or food chains. The metal of greatest concern is cadmium (Wade et al. 1982), but other elements with maximum recommended concentrations in sludge for land application include zinc, lead, copper, nickel, boron, and mercury (Logan and Chaney 1983). Although numerous trace organic compounds can occur in sludges (Overcash 1983), no potential toxicities to wildlife have been identified.

Several studies have investigated the effects of sludge application on element levels in wildlife or wildlife foods. West et al. (1981) reported higher levels of cadmium in pooled samples of livers and kidneys in deer mice from sludge-treated areas than from control areas. Hegstrom (unpublished data) also found significantly higher concentrations of cadmium in insectivores from sludge-treated areas in Washington, with kidney levels as high as 200 ppm. No kidney, liver, or testis lesions were observed in these animals, but a laboratory study is needed to confirm these results. The sludge applied in these studies contained approximately 50 ppm cadmium, dry weight.

Williams et al. (1978) reported an increase in cadmium in tissues of meadow voles fed an experimental diet from sludge-treated forages. Similarly, Anderson et al. (1982) reported increases in cadmium levels in vole tissues following exposure to sludge plots, but Anderson and Barrett (1982) reported no toxic effects on vole population densities, survival, or reproductive parameters. Woodyard et al. (this volume) found no increases in levels of cadmium in small mammal tissues or wildlife forages from sludge-treated areas having cadmium loading rates of 0.08 to 0.42 kg/ha.

Earthworms have been found to concentrate some elements, particularly cadmium, from sludge-treated soils (Beyer et al. 1982, Wade et al. 1982). If additional accumulation were to occur in higher trophic levels of these food chains, potential toxicities might develop.

No accumulation of heavy metals in deer organs was reported by West et al. (1981) and Anderson (1981) in sludge-treated Douglas-fir stands in Washington. Campa et al. (this volume) reported a cadmium level of 31 ppm in kidneys of deer collected from sludge plots in Michigan, but felt that this level was within the normal range.

These studies indicate that accumulation of metals in animal tissues following sludge applications can occur. Metal concentrations tend to be low in herbivores and granivore-omnivores, but can be higher in insectivores. No studies have found levels high enough to produce any measurable effects on populations. However, sludges are quite variable in levels of different elements, and application of sludges with high cadmium levels could produce different responses. Additional research is warranted on food-chain accumulation of elements from sludges with high levels of metals such as cadmium, particularly in food chains containing earthworms. At present, if guidelines are followed for levels of elements considered acceptable for agricultural land application of sludge (Logan and Chaney 1983), no toxicity problems for wildlife or health risks to human consumption of game species are anticipated.

CONCLUSIONS AND RECOMMENDATIONS

Application of sewage sludge to forest lands will produce changes in wildlife habitat and wildlife populations. Direct effects of sludge application appear minimal. Sludge will increase the productivity of vegetation on poor quality sites, particularly herbaceous vegetation. Sludge application will also increase the quantity and quality of forages for herbivorous animals. Protein and phosphorus levels in forages will increase on poor quality sites, especially in herbaceous species. Animal populations favored by greater quantities of ground-level vegetation will increase in numbers, while those with habitat requirements for more open understories will decrease. Herbivores will selectively forage more heavily on sludge-treated areas, with probable increases in the productivity rates of their populations.

No toxicity problems have been found from use of nonindustrial, municipal sewage sludge. However, more research on sludges containing higher levels of cadmium or other toxicants is needed before they can be applied to forest ecosystems.

Sludge application to forest lands is a viable disposal option, generally improving wildlife habitat. Sites with poor quality soils are the best candidates. Sludge application to wildlife openings should improve their value and use by wildlife, and may help maintain these areas in herbaceous vegetation. Sludge treatment of aspen clearcuts would appear to have advantages of easy access coupled with a very favorable vegetative response. The black sludge, darkening the soil surface, stimulates growth of greater numbers of aspen shoots; and the nutrient addition should increase aspen growth rates, especially on poor quality sites. Increased production of herbaceous species should also be readily available for wildlife use on these areas. However, sludge application to recent clearcuts in areas with high herbivore populations could lead to regeneration problems if browsing or girdling pressures are great enough.

The timing of sludge application influences the response by wildlife and vegetation. Application during the growing season will not produce sizable responses until the next growing season, and may kill some seedlings. Application of sludge prior to the active growing season is recommended.

Although more questions remain to be answered, sludge application can offer valuable benefits to wildlife. Wildlife agencies should be encouraged to consider sludge application to forest lands as a method of wildlife habitat management.

Michigan Agricultural Experiment Station journal article number 11820.

REFERENCES

Anderson, D. A. 1981. Response of the Columbian black-tailed deer to fertilization of Douglas-fir forests with municipal sewage sludge. Ph.D. diss., University of Washington, Seattle. 176 p.

_____. 1983. Reproductive success of Columbian black-tailed deer in a sewage-fertilized forest in western Washington. J. Wildl. Manage. 47:243–247.

_____. 1985. Influence of sewage sludge fertilization on food habits of deer in western Washington. J. Wildl. Manage. 49:91–95.

Anderson, T. J., and G. W. Barrett. 1982. Effects of dried sludge on meadow vole (*Microtus pennsylvanicus*) populations in two grassland communities. J. Appl. Ecol. 19:759–772.

Anderson, T. J., G. W. Barrett, C. S. Clark, V. J. Elia, and V. A. Majeti. 1982. Metal concentrations in tissues of meadow voles from sewage sludge-treated fields. J. Environ. Qual. 11:272–277.

Beyer, W. N., R. L. Chaney, and B. M. Mulhein. 1982. Heavy metal concentrations in earthworms from soil amended with sewage sludge. J. Environ. Qual. 11:381–385.

Campa, H., III. 1982. Nutritional responses of wildlife forages to municipal sludge application. M.S. thesis, Michigan State University, East Lansing. 88 p.

Edmonds, R. L., and D. W. Cole. 1977. Use of dewatered sludge as an amendment for forest growth. Vol. 2. Center for Ecosystem Studies, College of Forest Resources, University of Washington, Seattle. 120 p.

Logan, T. J., and R. L. Chaney. 1983. Utilization of municipal wastewater and sludge on land—metals. p. 235–323. *In* A. L. Page, T. L. Gleason III, J. E. Smith, Jr., I. K. Iskandar, and L. E. Sommers (eds.) Proceedings of the 1983 Workshop on Utilization of Municipal Wastewater and Sludge on Land. University of California, Riverside.

Overcash, M. R. 1983. Land treatment of municipal effluent and sludge: Specific organic compounds. p. 199–231. *In* A. L. Page, T. L. Gleason III, J. E. Smith, Jr., I. K. Iskandar, and L. E. Sommers (eds.) Proceedings of the 1983 Workshop on Utilization of Municipal Wastewater and Sludge on Land. University of California, Riverside.

Seon, E. M. 1984. Nutritional, wildlife, and vegetative community response to municipal sludge application of a jack pine/red pine forest. M.S. thesis, Michigan State University, East Lansing. 75 p.

Taber, R. D., and S. D. West. 1982. Biological processes in the control of risk tree species on rights-of-way in forested mountains: Pacific Northwest. p. 410–415. *In* A. Crabtree (ed.) Third Symposium on Environmental Concerns in Rights-of-way Management. Mississippi State University, Mississippi State.

Thomas, A. H. 1983. First-year responses of wildlife and wildlife habitat to sewage sludge application in a northern hardwoods forest. M.S. thesis, Michigan State University, East Lansing. 81 p.

Wade, S. E., C. A. Bache, and D. J. Lisk. 1982. Cadmium accumulation by earthworms inhabiting municipal sludge-amended soil. Bull. Environ. Contam. Toxicol. 28:557–560.

West, S. D., R. D. Taber, and D. A. Anderson. 1981. Wildlife in sludge-treated plantations. p. 115–122. *In* C. S. Bledsoe (ed.) Municipal sludge application to Pacific Northwest forest lands. Institute of Forest Resources Contribution 41. College of Forest Resources, University of Washington, Seattle.

Williams, P. H., J. S. Shenk, and D. E. Baker. 1978. Cadmium accumulation by meadow voles (*Microtus pennsylvanicus*) from crops grown on sludge-treated soil. J. Environ. Qual. 7:450–454.

Woodyard, D. K. 1982. Response of wildlife to land application of sewage sludge. M.S. thesis, Michigan State University, East Lansing. 64 p.

Chapter 11

Forest Land Application of Municipal Sludge: The Risk Assessment Process

SYDNEY MUNGER

ABSTRACT   The concept of risk assessment was used to define the probability of harm occurring to human health as a result of the forest application of sludge. The methodology required that the toxic or disease-causing substances in the sludge be identified and quantified and all potential pathways of human exposure to sludge-borne constituents be defined. The outcome, in terms of migration of these constituents along the exposure pathways to environmental compartments, was estimated based on an extensive literature review. The consequences of exposure to one or more of these environmental compartments was determined based on minimal infectious dose information or dose-response curves for individual toxic chemicals. The conclusion of this assessment was that normal human exposure in and around a sludge forest land application site would result in no observable increased risk. This conclusion is based on several assumptions and estimates of sludge constituent degradation and migration rates. Information collected from future large-scale demonstration projects will allow for further refinement of the risk assessment.

Scientific presentations at symposia typically address what toxic chemicals and pathogenic microorganisms are present in sludge and what various studies have shown about environmental and health-related effects of sludge land application. That is usually the point at which most conferences end. The scientists generally feel fairly comfortable with the proceedings. They have gained new insights and are perhaps excited by new questions raised by the work just presented. But the policy makers—those persons who must decide which projects, if any, to fund or which processes to regulate—are still waiting for the "bottom line." Their typical reaction is, "So what if there are human viruses in the sludge and what does it mean if we find PAHs and PCBs in the sludge?" As an environmental health scientist in a public agency, the Municipality of Metropolitan Seattle (Metro), the author likes to think of the risk assessment process as a tool to answer these questions—as a framework for decision making.

THE PROCESS OF ASSESSING RISK

Risk is defined in this presentation as a measurement of the probability of harm occurring to human health as a result of the land application of sludge. It is important to understand that this paper will not be making judgments as to the absolute safety of land application of municipal waste but rather analyzing the data available to determine the level of risk. In his book *Of Acceptable Risk*, Lowrance (1976) stresses the idea that nothing we do or have imposed on us is absolutely free of risk. The concept of safety should be viewed as a judgment concerning the acceptability of the risk involved. The scientist can, using assumptions, measure the risks. It is the role of public and private institutions and the general public to form judgments as to the acceptability of the risks.

It is the scientist's responsibility to translate available data into risk factors and make a determination as to level of risk. This is not a clearcut process, although many of the regulatory agencies have prepared guidelines to assist the scientist. Key to the outcome of the risk assessment are the assumptions that must be made to fill in the data gaps. For example, in writing a risk assessment concerning the effects of organic priority pollutants in sludge, one must assume that migration of these compounds in the environment will occur at the same rate and to the same extent as when tested as pure compounds, since data on organics in sludge are limited. One must also assume that the toxicity of the mixture of compounds found in sludge is not different from that of each of the single compounds. Assumptions are an accepted part of the risk assessment process as long as they are clearly stated in the text of the assessment.

After the risks are defined, it is the job of the scientist to communicate the conclusions of the risk assessment to (1) the generators of the condition causing the risk—in this case the policy makers of the municipal agencies responsible for sludge reuse, (2) the regulatory agencies responsible for control of risk to public health (Environmental Protection Agency, Washington Department of Ecology, and the county health departments), and (3) the population at risk—those persons living close enough to a sludge application site to be potentially affected.

It is the responsibility of institutions or agencies to work with the general public to determine levels of acceptable risk. This is probably best done on a regional basis so that factors important to a certain locale can be considered. Once the level of acceptable risk is defined, decisions must be made on how to mitigate excess risk. The regulatory agencies must determine appropriate regulations to ensure that the acceptable risk level is maintained and enforcement mechanisms to ensure that regulatory levels or procedures are met by local agencies.

METHODOLOGY FOR DETERMINING RISK

Several versions of risk assessment methodology have appeared in the past few years. The steps outlined by Rowe (1977) are (1) causative event, (2) outcome, (3) exposure probability, and (4) consequences. Using these steps, an example of the risk assessment process as it relates to the forest application of municipal sludge follows. The Metro sludge risk assessment (Munger 1983) contains a thorough evaluation of potential concerns. This example is therefore limited to a discussion of the organic priority pollutant risk assessment findings since that publication was completed.

Causative Event

The causative event includes the conditions of sludge application and the levels of organic priority pollutants present in the sludge. Management guidelines are defined in Washington State by the Department of Ecology, and operating practices are presented in the Metro Sludge Management plan (Metro 1983). The following criteria are assumed in this assessment: (1) an application rate of 20 dry tons per acre, (2) buffer zones of 200 feet between a sludge-applied area and any type of surface water, (3) a minimum of 5 vertical feet between sludge applied to soil and any groundwater aquifer, (4) site access regulated for at least one year, and (5) no sludge applied on a slope greater than 30%.

The organic priority pollutants as defined by the U.S. EPA will be considered to represent the organic toxicants of greatest concern in this analysis. The concentration of organic priority pollutants measured in Seattle Metro dewatered digested sludge are given in Tables 1 to 4 and can be classified in the following groups: polychlorinated biphenyls (PCBs), polynuclear aromatic hydrocarbons (PAHs), phthalic acid esters and substituted monocyclic aromatics, and halogenated short chain aliphatics.

For the purpose of this risk assessment it was necessary to select a few chemicals that could serve as representatives of larger groups of compounds with similar physical and chemical characteristics. The criteria considered when selecting these representative chemicals related to transport through environmental compartments and the toxicity of the compounds. Specifically these criteria, as developed by Overcash (1983), were: prevalence in Metro sludge, vapor pressure to indicate volatilization loss, aqueous solubility, octanol/water partition coefficient to indicate sorption to soil, rate of decomposition, and toxic dose and availability of dose response studies.

Based on these criteria, Overcash (1983) recommended to Metro that the following compounds act as representatives of the priority pollutants in Metro sludge: *Polyaromatic*

TABLE 1. Polychlorinated biphenyls (PCBs) in Seattle Metro dewatered sludge.

Compound	Concentration (ug/kg dry weight)
Aroclor 1242	180
Aroclor 1248	180
Aroclor 1254	400
Aroclor 1260	325
Total	1,085

TABLE 2. Polynuclear aromatic hydrocarbons (PAHs) in Seattle Metro dewatered sludge.

Compound	Concentration (ug/kg)
Noncarcinogenic	
Naphthalene	3,600
Acenaphthene	2,800
Fluorene	3,400
Phenanthrene	8,500
Anthracene	1,950
Fluoranthene	4,250
Pyrene	4,150
Benzo(k)fluoranthene	1,600
Benzo(g,h,i)perylene	250
Carcinogenic	
Benzo(a)anthracene	1,950
Chrysene	2,850
Benzo(b)fluoranthene	1,600
Benzo(a)pyrene	2,600
Indeno(1,2,3-c,d)pyrene	420
Dibenzo(a,h)anthracene	100

Concentrations based on one sample of digested dewatered sludge.

TABLE 3. Phthalic acid esters and substituted monocyclic aromatics in Seattle Metro sludge.

Compound	Concentration (ug/kg)
Phthalates	
Dibutyl phthalate	1,950
Butylbenzyl phthalate	1,950
Diethylhexyl phthalate and dioctyl phthalate	170,000
Substituted Monocyclic Aromatics	
Dichlorobenzenes	1,750
Phenol	1,000
4-methyl phenol	1,000
2,4-dichlorophenol	5
Pentachlorophenol	60
Hexachlorobutadiene	60
N-nitrosodiphenylamine	950

TABLE 4. Halogenated short chain aliphatics in Seattle Metro sludge.

Compound	Concentration (ug/kg)
1,1-dichloroethylene	200
Methylene chloride	350
Chloroform	15
Trichloroethylene	5
Toluene	22
Tetrachloroethylene	5
Ethylbenzene	50

hydrocarbons: phenanthrene, benzo(a)anthracene, and benzo(a)pyrene (BaP). *Polychlorinated biphenyls*: PCB mixture. *Phthalates*: di-n-butyl phthalate and bis(2-ethylhexyl)phthalate. *Volatiles*: methylene chloride, chloroform, and tetrachloroethylene.

From this list four compounds were selected with which to assess the risk of toxic effects due to organic priority pollutants in land-applied sludge: Benzo(a)pyrene, PCBs, bis(2-ethylhexyl)phthalate, and chloroform. These selections were based primarily on the toxicity of the compounds and availability of adequate animal or human studies demonstrating a dose response. This paper will be limited to BaP, the most toxic of the PAHs in Metro sludge and total PCBs. The assumption is that these compounds or groups of compounds will serve as worst-case models for the other toxicants.

Outcome

The land application of sludge can potentially result in the transfer of organic toxicants from the sludge to air, soil, surface water, groundwater, and edible animals and plants. Using the known concentrations of representative compounds in Metro sludge, and literature on environmental fate and physical and chemical characteristics of these compounds, it is possible to estimate their fate in forest land application projects. Tables 5 and 6 give the estimated concentrations for BaP and PCBs in environmental compartments as a result of sludge application (Overcash 1983).

The levels of BaP and PCBs contributed to the environment by the land application of sludge should be compared with background levels of these compounds to gain per-

TABLE 5. Estimated concentrations of benzo(a)pyrene (BaP) in selected environmental compartments resulting from Silvigrow application (Overcash 1983).

Environmental Compartment	Concentration (ppb)				
	0 Time	3 Months	6 Months	12 Months	24 Months
Sludge	2,600	NA	NA	NA	NA
Aerosols	Unknown	NA	NA	NA	NA
Sludge-soil litter (top 15 cm)	60	25	10	1-10	1-10
Control soil	1-10	1-10	1-10	1-10	1-10
Blackberries	0.02-0.2	0.6-6	0.3-3	0.02-0.2	0.02-0.2
Control blackberries	0.02-0.2	0.02-0.2	0.02-0.2	0.02-0.2	0.02-0.2
Deer tissue	Unknown				
Groundwater (>5 feet)	<0.1	<0.1	<0.1	<0.1	<0.1
Surface water	<0.1	<0.1	<0.1	<0.1	<0.1

NA = Not applicable.

TABLE 6. Estimated concentrations of PCBs in selected environmental compartments resulting from Silvigrow application (Overcash 1983).

Environmental Compartment	Concentration (ppb)				
	0 Time	3 Months	6 Months	12 Months	24 Months
Sludge	1.1	NA	NA	NA	NA
Aerosols	Unknown	NA	NA	NA	NA
Sludge-soil litter	26	24	22	18	18
Control soil	<10	NA	NA	NA	NA
Blackberries	<0.1	0.2-1	0.2-1	0.2-1	0.2-1
Control blackberries	<0.1	NA	NA	NA	NA
Deer fat	10-12	10-12	10-12	10-12	10-12
Control deer fat	Unknown				
Groundwater (>5 feet)	<0.01	<0.01	<0.01	<0.01	<0.01
Surface water	<0.01	<0.01	<0.01	<0.01	<0.01

NA = Not applicable.

spective of any increased risks. PCBs have been observed in water, food, soils, sediments, and animal and human tissue sampled throughout the world. PCB levels in food are gradually declining as a result of restrictions on their use, but as of 1975 were commonly in the parts per million range. The teenage male in 1975 was estimated to consume 8.7 micrograms of PCBs per day (Jelinek and Corneliussen 1976, as reported in U.S. EPA 1980a). BaP is a natural as well as an anthropogenic component of the environment. The major sources are related to combustion products created during power generation, refuse burning, coke production, and motor vehicle travel. The levels of BaP found in food and water are directly related to their proximity to sources of BaP. In 1980, the U.S. EPA estimated the average human consumption of BaP per day to be 160 to 1,600 ng (nanograms).

Exposure Probability

To illustrate the probability of exposure to BaP or PCBs in the environment, the pathways for transport are diagrammed in Figure 1.

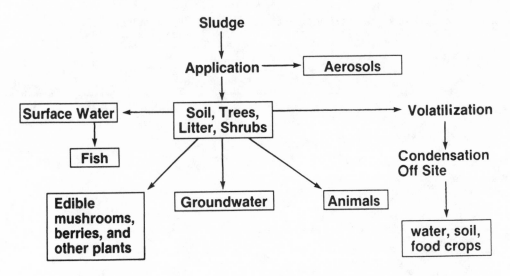

Figure 1. Pathways of organic transport from sludge in silvicultural application.

Consequences

PCBs. The International Agency for Research on Cancer (IARC) and the U.S. EPA have concluded that PCBs are animal, and potentially human, carcinogens. A study by Kimbrough et al. (1975) was used by the EPA (1980a) to calculate the level of risk associated with the ingestion of PCBs. Using the linearized multistage extrapolation model developed by Crump (Federal Register 45 [230], 1980), it was concluded that ingestion of 204 ng PCBs per day by a 70 kg (154 lb) person was equivalent to a lifetime cancer risk of one per 100,000 persons. In 1975 the average daily consumption was 8,700 ng. This risk estimation, combined with predicted amounts of PCBs in selected environmental compartments as a result of forest sludge application, can be used to calculate quantities of food, soil, and water that can be consumed without exceeding an excess lifetime cancer risk of 1/100,000 (Table 7).

It can be seen in Table 7 that surface and groundwater consumption will not be limited by PCBs from a sludge application. Blackberries and other edible forest plants can take up PCBs, but 200 to 1,000 grams per day can still be ingested if the plants are growing in a sludge-amended forest. Deer can accumulate PCBs in fat tissue, leading to elevated

TABLE 7. Maximum daily dose of PCBs (excess lifetime cancer risk 1/100,000).

Environmental Compartment	0 Time	3 Months	6 Months	12 Months	24 Months
Aerosols	Unknown				
Sludge-soil litter (kg)	0.008	0.008	0.009	0.01	0.01
Blackberries (kg)		0.2-1	0.2-1	0.2-1	0.2-1
Control blackberries (kg)	>2	>2	>2	>2	>2
Deer fat (kg)		0.02	0.02	0.02	0.02
Control deer fat	Unknown				
Groundwater (>5 feet)	>20.4 liters				
Surface water	>20.4 liters				

levels. Intake of deer fat should be limited to 20 grams per day to keep the excess lifetime cancer risk at or below 1/100,000.

PCB levels in the soil will be elevated as a result of sludge application, and after an initial decline remain elevated for many years. Children with a tendency to eat dirt may be at increased risk if they play daily in a sludge-amended forest. However, in such as area they would have to eat 8 to 10 grams of soil per day to increase the level of risk above 1/100,000.

Benzo(a)pyrene. Short-term mutagen assays and long-term animal testing support the conclusion the BaP is a potent mammalian carcinogen. Epidemiological studies of coke oven and coal tar industry employees implicate BaP as a human carcinogen. A study by Neal and Rigdon (1967) was used by the EPA to estimate the risk associated with ingestion of BaP. As with PCBs, the Crump model was used to extrapolate the observed risk to low-dose human exposure levels. It was concluded that consumption of 61 ng per day was equivalent to a lifetime cancer risk of 1/100,000. The average daily consumption is estimated at 160 to 1,600 ng.

The probability that BaP in a sludge forest application project could cause a cancer risk greater than 1/100,000 can be estimated by the quantities of environmental compartments that can be ingested without exceeding 61 ng per day of BaP (Table 8). Neither surface water nor groundwater will be affected by a detectable increase in BaP concentrations.

TABLE 8. Maximum daily dose of BaP (excess lifetime cancer risk 1/100,000).

Environmental Compartment	0 Time	3 Months	6 Months	12 Months	24 Months
Sludge-soil litter (kg)	0.001	0.002	0.006	0.006-0.06	0.006-0.06
Blackberries (kg)	0.3-3.0	0.01-0.1	0.02-0.2	0.3-3.0	0.3-3.0
Control blackberries (kg)	0.3-3.0	0.3-3.0	0.3-3.0	0.3-3.0	0.3-3.0
Deer tissue	Unknown				
Groundwater (1)	>6.0	>6.0	>6.0	>6.0	>6.0
Surface water (1)	>6.0	>6.0	>6.0	>6.0	>6.0

The addition of sludge to forest soils will cause an initial increase in the concentration of BaP in the surface soil. During the 6 to 12 months following a silvicultural sludge application, children with a tendency to eat dirt are at an increased risk of exceeding 61 ng per day of BaP if they eat from 1 to 60 grams of soil per day. After one year the levels of BaP contributed by the sludge are estimated to have returned to background concentrations.

Silvicultural applications can have an impact on the human food chain and increase risks if wild berries and game animals bioconcentrate BaP in edible tissues. If the growing season started within a few months of sludge application, the blackberries could contain elevated BaP levels. Limiting consumption to 10 to 100 grams per day of sludge-grown blackberries compared with 300 to 3,000 grams per day of control blackberries would be required to remain at or below a lifetime cancer risk of 1/100,000.

Estimation of risk due to consumption of deer tissue is not possible because of a lack of information on uptake in animals. However, the BaP in the raw meat will be significantly less than in the same cooked meat. In other words, the cooking process will determine the BaP content of the meat to a greater extent than the quantity of BaP ingested by the deer or cow.

PRODUCTS OF THIS RISK ASSESSMENT

This risk assessment process has provided the following information: (1) The local risks that could occur from exposure to a forest sludge site can be defined as small and controllable with proper site management. (2) Data gaps have been defined for which assumptions were made in this risk assessment. Estimates for the quantity of organics migrating from sludge to surface water, groundwater, plants, and animals were based on studies using pure compounds, not on the complex mixture of sludge. Data should be collected that measure the fate of organic priority pollutants on a silvicultural site. (3) The probability of risk was communicated to policy makers, regulatory agencies, and the public.

The next step in the risk assessment process is to determine the level of acceptable risk. This should be done as a cooperative effort between government agencies and the affected public. After the level of acceptable risk has been defined, appropriate site management guidelines can be derived to meet acceptable risk levels.

REFERENCES

Jelinek, C. E., and P. E. Corneliussen. 1976. Levels of PCBs in the U.S. food supply. *In* National Conference, Polychlorinated Biphenyls, Chicago, Illinois. EPA 560/6-75-004. Office of Toxic Substances, U.S. Environmental Protection Agency, Washington, D.C.

Kimbrough, R. D., et al. 1975. Induction of liver tumors in Sherman strain female rats by polychlorinated biphenyl Aroclor 1260. J. National Cancer Inst. 25:553.

Lowrance, W. W. 1976. Of acceptable risk: Science and the determination of safety. William Kaufmann, Inc., Los Altos, California.

Munger, S. F. 1983. Health effects of sludge land application: A risk assessment. Metro Report. Municipality of Metropolitan Seattle.

Municipality of Metropolitan Seattle (Metro). 1983. Sludge management plan. Metro Report. Municipality of Metropolitan Seattle.

Neal, J., and R. H. Rigdon. 1967. Gastric tumors in mice fed benzo(a)pyrene: A quantitative study. Texas Reports Biol. Med. 25:553.

Overcash, M. 1983. Identification and selection of organic priority pollutants in Pilchuck Demonstration Site. Report to Metro. Municipality of Metropolitan Seattle. 10 p.

Pucknat, A. W. 1981. Health impacts of polynuclear aromatic hydrocarbons. NDC, Park Ridge, New Jersey.

Rowe, W. D. 1977. An anatomy of risk. John Wiley and Sons, New York.

U.S. Environmental Protection Agency. 1980a. Ambient water quality criteria for polychlorinated biphenyls (PCBs). Washington, D.C.

_____. 1980b. Ambient water quality criteria for polychlorinated aromatic hydrocarbons (PAHs). Washington, D.C.

Chapter 12

Behavior of Organic Compounds in Land Treatment Systems with the Presence of Municipal Sludge

MICHAEL R. OVERCASH and JEROME B. WEBER

ABSTRACT   Municipal sludge land treatment in forest and agricultural areas involves the assimilation of a variety of inorganic and organic constituents. While forest areas offer certain short-term advantages for the treatment of organics, the issue remains the continued ability to assimilate compounds, thus ensuring long-term success. This study has examined the uptake by four plant species and decomposition in soil of two organic priority pollutants (di-n-butyl phthalate ester and paradichlorobenzene). These assimilative pathways were measured with and without the presence of municipal sludge generally at two rates (0.36 and 3.6 dry Mg/ha). The uptake of carbon 14 was found not to be statistically different with the presence of sludge. Likewise, no significant sludge effect was found for the rate an organic compound disappeared from the soil, although the trend was to accelerate the decomposition with increasing sludge rate. These data put in perspective the possible order of magnitude for changes in pure compound information when used in systems containing sludge. However, further compound information is needed to expand this limited comparison. In addition, this information is expected to be qualitatively correct for forest herbaceous growth, but it is unclear whether organic chemical uptake by trees is a relevant issue.

The use of forest systems in the land treatment of municipal sludge is increasing. Considering all the constituents in sludge and the various terrestrial pathways for assimilating waste components, forested sites offer advantages, but also retain certain limitations. In general, the pathways or mechanisms for treating the various classes of municipal sludge constituents (anions, metals, and organics) are essentially the same in forest and agricultural systems (Overcash and Pal 1979). However, trees, as a land treatment system vegetation, offer the advantage of being less subject to phytotoxicity related to sludge organic constituents than certain sensitive agricultural crops. The magnitude of this advantage is unknown, since few, if any, phytotoxic effects of chemicals have been found for municipal sludge land treatment areas.

A major advantage of forest systems lies in the implications of chemical uptake of forests versus agricultural crops. If a tree were to take up an organic or inorganic compound from applied sludge, the end use of the wood would probably preclude any realistic concern over the constituent getting into the food chain. Also, the growth cycle of the forest is sufficient to provide a time buffer against organic uptake. Since most organics may undergo decomposition or loss, any subsequent crop, particularly agricultural vegetation should a site be cleared and used for that purpose, would not be likely to show any significant uptake. However, if an organic or inorganic constituent (such as metals) is not lost from the soil, then the level present might affect any subsequent crop.

There are many factors involved in the design and evaluation of a municipal sludge land treatment system, in forests or elsewhere. The research reported herein focuses on only one component of this larger, complex entity: the behavior of organic compounds in light of the presence or absence of municipal sludge—that is, whether the presence of the organic or inorganic materials in sludge alters the uptake of specific organics by vegetation or the decomposition and loss of those organics from the soil.

The experimental difficulties and lack of any criteria for concern with organic uptake ruled out the use of tree species as the vegetation for research purposes. Since agricultural crops are the vegetation of possible concern, they were selected for study. The implications of an effect of sludge on organic compound fate are of interest in absolute terms, and also as this information relates to the ability to use existing data, generally available in the literature. That is, if the presence of sludge has a small or negligible effect on the assimilative pathways of organic constituents, then significant available data would be of direct use.

Thus the objective of this research was to select two organic priority pollutants as an initial examination of the effect of sludge on plant uptake and chemical decomposition or loss when applied to soil. One sludge loading rate was zero so that the behavior of the compounds could be directly studied.

MATERIALS AND METHODS

The greenhouse and analytical facilities were located on the North Carolina State University campus, at approximately 35.8°N latitude, 78.6°W longitude. The temperature and relative humidity of the greenhouse were controlled at about 80°F and 60%, respectively.

The organic chemicals were from the list of priority pollutants (U.S. EPA 1979). The two organic chemicals were di-n-butyl phthalate ester (DnBP) and paradichlorobenzene (pDCB). These chemicals were applied separately to pots of soil and uniformly mixed into the top 15 cm of soil in each pot to achieve loading rates of 0.22 and 0.20 mg/kg (dry soil basis), respectively. The compound loading rate was developed to approximate that received by a soil area over an annual cycle of municipal sludge loading based on the average concentration of the organic constituent in sludge and a sludge nitrogen loading of 560 kg N/ha (500 lbs N/acre) per year. The carbon 14 labeled compounds were used with nonlabeled compounds to achieve about 4,500 disintegrations per minute (dpm) per gram soil and the desired soil levels of total organic compounds.

Each pot was 21 cm deep and 20 cm in diameter, containing 7.8 kg of soil. Since four plant species were to be grown, two pots were needed for each condition studied. In one pot, corn and soybeans were grown, with the aboveground corn completely removed after about thirty days and the soybeans allowed to grow to maturity. In the companion pot, wheat and fescue were grown, with aboveground fescue completely removed after about thirty days and wheat grown to maturity. For the species carried to maturity, seeds and plants were harvested separately. The crops grown were Kentucky 31 tall fescue (*Festuca arundinacea* Schreb.), Pioneer 3368A corn (*Zea mays* L.), Altona soybeans (*Glycine max* [L.] Merr.), and Butte wheat (*Triticum aestiuum* L.). Triplicates were used in all cases.

The soil selected was Norfolk sandy loam, used in a number of previous land treatment studies (Overcash et al. 1985). The bulk density in the pots was 1.2 g/cc. The soil organic matter is about 1.5%, with about half being humic matter. Soil pH is typically 5.0 and the cation exchange capacity about 3.5 meq/100 g. Fertilization was achieved by the time-released product Osmocote®. Insect and disease pests were controlled with routine chemicals. Sludge was obtained at the Raleigh, North Carolina wastewater treatment plant from sludge aeration basins and incorporated simultaneously with the DnBP or pDCB throughout the upper 15 cm of soil.

In addition to the harvesting of plant material, the soil was sampled periodically during the experiment. These samples were used to assess the decomposition of the study chemicals. The experiments were done in triplicate.

Extraction methods were modified from Erickson et al. (1981) and Harrold and Young (1982). Plant and soil samples were homogenized for one minute with a 45 Virtis homogenizer at 39,000 rpm in methylene chloride for pDCB and hexane:acetone (1:1) for DnBP. After extract concentrating, the extracted liquid was analyzed on a scintillation spectrometer. Detailed conditions are given in Overcash et al. (1985). In addition, total carbon 14 was determined in a Harvey Biological Oxidizer, with trapped carbon dioxide analyzed in the scintillation spectrometer. Data were analyzed with the General Linear Model (GLM) program from SAS (Barr et al. 1979). Variables used in the ANOVA were replicate and sludge rate.

RESULTS

Plant Growth

The average height of all plants was measured during the growth cycle (Table 1). Comparison of the plant growth with zero sludge to that at 0.36 and 3.6 dry Mg/ha (0.16 and 1.6 dry tons/acre)—whether with chemical present or not—indicates that only a few species were affected. Corn height was increased, while immature wheat was decreased at the 0.36 dry Mg/ha rate. The remainder of the young vegetation and the mature crops were unaffected by the sludge applied. While not reported in Table 1, sludge loading of 7.2 and 14.4 dry Mg/ha (3.2 and 6.4 dry tons/acre) severely depressed soybean growth. Likewise the comparison of pots with DnBP or pDCB (in the absence of sludge) indicated little influence on crop heights. An exception was mature wheat versus the control pots, which showed some phytotoxicity.

Plant Uptake

The most complete evaluation of soil chemical levels was for the pDCB (Table 2). The results of uptake are on the basis of extractable carbon 14, thus leaving unresolved the total uptake (extractable plus that from oxidizing extracted residues). The percentage extractable, however, was measured for DnBP and was in the range of 25 to 40%. In addition, the carbon 14 analysis includes both parent and metabolites retaining the radiolabel. Thus the preliminary character of these results must be kept in mind.

For all four crop species, in the absence of sludge, there was carbon 14 taken up into the plant when pDCB was applied to the soil (Table 2). However, none of the radiolabel was detected in the seed portion (wheat or soybean). The results are on a dry basis with a typical moisture content of 80 to 90%. Thus on a fresh weight, or as harvested, basis these concentrations are reduced by a factor of five to ten. For DnBP, the single crop

TABLE 1. Average height and standard deviation (in parentheses) of plants grown in greenhouse with municipal sludge and specific organic priority pollutants (cm).

Sludge Rate (dry Mg/ha)	Corn	Fescue	Immature Soybeans	Mature Soybeans	Immature Wheat	Mature Wheat
Controls without chemical addition						
0	62 (2)	24 (6)	39 (2)	66 (8)	30 (4)	52 (4)
0.36	76 (2)	32 (4)	33 (4)	70 (.)	18 (4)	43 (11)
3.6	78 (2)	34 (5)	35 (4)	64 (6)	34 (6)	46 (1)
Paradichlorobenzene (200 ppb in soil)						
0	60 (0)	24 (4)	35 (1)	51 (13)	30 (0)	37 (8)
0.36	71 (8)	24 (6)	38 (0)	66 (6)	22 (6)	43 (1)
3.6	71 (8)	28 (4)	32 (6)	77 (4)	36 (6)	37 (3)
Statistics						
L.S.D. (p≤0.05)	12	11	10	26	9	15
F	5.6*	1.5	0.4	0.8	8.2*	1.8
Controls without chemical						
0	-	-	3.2 (0.4)	23 (5)	-	-
Di-n-butyl phthalate ester (220 ppb in soil)						
0	-	-	5.2 (1.1)	23 (3)	-	-
0.36	-	-	3.5 (2.1)	22 (3)	-	-
3.6	-	-	8 (.)	16 (2)	-	-
Statistics						
F	-	-	1.1	0.98	-	-

*Significant at $p \leq 0.05$.
(.) Only one plant remaining.

studied was also found to contain the radiolabel in the absence of applied sludge (Table 2). The extractable carbon 14 DnBP in soybeans was about twice that of extractable carbon 14 pDCB, even though both were applied to the soil at approximately equal loadings. Again, on a fresh weight basis, there is a similar reduction in actual chemical concentration.

Since the greenhouse environment and pot experimentation are not expected to give absolute quantitative agreement with field studies, the main objective was to determine the relative effect of sludge. That is, the uptake by plants of these radiolabeled organic compounds is evaluated in the absence and presence of the sludge organic matrix. This comparison is thus a preliminary assessment of the behavior of specific organics with municipal sludge land treatment.

For pDCB, the presence of sludge over a tenfold range (0.36 and 3.6 dry Mg/ha) had no significant effect (at the 5% level) on the uptake by all plants (Table 2). The seeds remained without detectable levels of pDCB. Mature plants and those harvested after about thirty days did not appear to increase or decrease in uptake with the various sludge additions. The average plant uptake of pDCB was between 68 and 1,400 ppb on a dry matter basis and between 14 and 210 ppb on a fresh weight basis (average moisture assumed 15%). These uptakes as a ratio to that applied to the soil approach 1.0, on a fresh weight bioaccumulation basis. Compared with other organic priority pollutants,

TABLE 2. Average uptake and standard deviation (in parentheses) as ppb extractable carbon 14 in plants grown in greenhouse (dry basis).

Sludge Rate (dry Mg/ha)	Corn	Fescue	Immature Soybeans	Mature Plants	Soybeans Seeds	Immature Wheat	Mature Plants	Wheat Seeds
Paradichlorobenzene (200 ppb in soil)								
0	86	1,400	550	100	0	1,000	94	0
	(91)	(180)	(11)	(30)		(.)	(14)	
0.36	68	1,200	280	500	0	-	150	0
	(52)	(.)	(110)	(500)			(70)	
3.6	120	1,500	210	430	0	1,100	220	0
	(50)	(500)	(4)	(11)		(.)	(120)	
Statistics								
L.S.D. (p$\leq$0.05)	160	2,000	540	650	-	1,700	260	-
F	0.21	0.16	0.41	0.5	-	0.17	0.23	-
Di-n-butyl phthalate (220 ppb in soil)								
0	-	-	1,100	280	-	-	-	-
0.36	-	-	150	200	-	-	-	-
0.72	-	-	250	50	-	-	-	-
1.08	-	-	200	80	-	-	-	-
2.2	-	-	150	230	-	-	-	-
3.6	-	-	50	60	-	-	-	-

(.) Only one plant remaining.

this ratio for pDCB may be one of the highest detected, although compound verification and complete extraction percentages would have to be determined to verify this relative rank of terrestrial bioaccumulation.

For DnBP, the effect of sludge on soybean plant uptake was also not significant. In the case of immature soybeans, there appeared to be a trend of lowered compound uptake when sludge was present. This might be related to competitive absorption of DnBP by the sludge matrix to reduce that translocated into the plant. As before, these experiments were aimed at determining major effects related to the presence of municipal sludge. It appears that larger plant populations are needed to improve the statistical aspects of organic compound uptake experiments so that the effects of sludge can be verified. In addition, sludges with specific organics should be studied to assess the effect on uptake of the sludge chemical origin.

Behavior in Soil

The phthalate ester was first examined to determine whether the extractability of DnBP from soil remained constant over the 73 day investigation. No significant alteration in extractable versus total oxidizable carbon 14 was found, indicating only minor changes in the compound as present on the soil and sludge organic matrices (Table 3).

A progressive loss of the phthalate ester concentration in the soil occurred over the experiment period (Table 4). Assuming a first order loss mechanism, the loss coefficient was subsequently calculated. For DnBP in the absence of sludge, the loss coefficient was 0.10/day. Similar calculations at 0.36 and 3.6 dry Mg of sludge/ha yielded loss coefficients of 0.11/day, and 0.25/day, respectively.

Soil levels of DnBP at the various sludge loading rates were measured at several time intervals. These showed a slight but perceptible decrease with increasing sludge concen-

TABLE 3. Soil extraction of di-n-butyl phthalate ester over experiment period.

Sludge Rate (dry Mg/ha)	Period after Loading (days)			
	7	14	28	73
	(extraction percentage*)			
0	45	44	44	36
0.36	39	45	43	35
0.72	43	46	42	36
1.08	41	44	45	35
2.2	40	46	40	32
3.6	43	45	43	29
7.2	32	45	44	35
14.4	38	40	48	28
Average	40	44	44	33

*Extractable concentration/(extractable concentration + oxidized residue concentration).

TABLE 4. Soil concentrations based on carbon 14 over experiment period.

Sludge Rate (dry Mg/ha)	Time after Loading (days)	Concentration (ppb)
Paradichlorobenzene (200 ppb in soil)		
0	49	9.0
0.36		6.5
3.6		10
0	90	6.0
0.36		5.8
3.6		7.4
0	151	4.8
0.36		4.6
3.6		6.2
Di-n-butyl phthalate ester (220 ppb in soil)		
0	7	90
0.36		90
3.6		77
0	14	85
0.36		76
3.6		65
0	28	55
0.36		62
3.6		20
0	73	41
0.36		43
3.6		28

trations. However, this trend is minor and proved not to be statistically significant. Thus the remaining DnBP in the soil is reasonably extractable, is undergoing loss, and is not generally sensitive to the presence of municipal sludge.

Soil sampling for the paradichlorobenzene experiments was first done after 49 days (Table 4). At that stage, over 95% of the carbon 14 had been lost from the soil. No kinetic coefficients for pDCB were calculated, since short time intervals were not sampled. In regard to the effect of sludge, no differences in soil level were detected in comparison to the pots without sludge. This situation would represent the long-term status in the land treatment site; however, the form of the carbon 14 was unknown. At comparable times, there was a much greater soil loss of pDCB than for DnBP.

CONCLUSIONS

There has been virtually no published information on crop uptake and loss (microbial and physical processes) from soil of organic compounds in the presence of municipal sludge. This is the case for both agricultural and forest vegetation systems. Far more is known of the terrestrial behavior of such organics in the absence of sludge, hence the importance of assessing relative behavior in these separate situations. The findings in this preliminary study were as follows:

1. For two organic priority pollutants (paradichlorobenzene and di-n-butyl phthalate) the uptake by plants was statistically similar for sludge at 0, 0.36, and 3.6 dry Mg/ha. The crops were immature corn, fescue, immature and mature soybeans, and immature and mature wheat.

2. With or without sludge, the mature wheat and soybean seeds did not show uptake of paradichlorobenzene.

3. Over the study period, the soil levels of paradichlorobenzene and di-n-butyl phthalate ester decreased with no statistically significant effect of the presence of municipal sludge. Although not statistically significant, there appeared to be a slight acceleration of DnBP loss with increasing sludge level.

4. Loss of DnBP from the soil was slower than that of pDCB.

5. Tree species offer advantages in regard to specific organic uptake, in that the end use of wood probably precludes any realistic concern with level of chemical constituent.

6. In a forest receiver system for municipal sludge, it will remain a critical factor that specific organics continue to be assimilated or lost from the forest soil system.

REFERENCES

Barr, A. J., J. H. Goodnight, and J. P. Sall. 1979. SAS user's guide. SAS Institute, Cary, North Carolina.

Erickson, M. D., M. T. Giguere, and D. A. Whitaker. 1981. Comparison of common solvent evaporation techniques in organic analysis. Anal. Letters 14(A11):841–857.

Harrold, D. E., and J. C. Young. 1982. Extraction of priority pollutants from solids. J. Env. Eng. Div., Proc. Am. Soc. of Civil Eng. 108(EE6):1211–1227.

Overcash, M. R., and D. Pal. 1979. Design of land treatment systems for industrial wastes. Ann Arbor Science Publishers, Ann Arbor, Michigan (now Technomics Publishers, Lancaster, Pennsylvania). 684 p.

Overcash, M. R., J. B. Weber, and W. P. Tucker. 1985. Organic priority pollutants in plant-soil systems. U.S. Environmental Protection Agency report. MERL, Cincinnati, Ohio. 135 p.

U.S. Environment Protection Agency. 1979. Fate of priority pollutants in publicly owned treatment works: Pilot study. EPA 440/1-79-300. Effluent Guidelines Division, Washington, D.C.

Chapter 13

Nitrification and Leaching of Forest Soil in Relation to Application of Sewage Sludge Treated with Sulfuric Acid and Nitrification Inhibitor

CAROL G. WELLS, DEBRA CAREY, and R. C. McNEIL

ABSTRACT   A reconstructed profile of forest floor layers 0i (pH 3.9) and 0e (pH 4.2), and part of the sandy A layer (pH 4.8), from a 28-year-old loblolly pine stand was treated with sludge amended in factorial combination with sulfuric acid (to pH 6.6) and nitrapyrin (6.7 µg/l), and incubated at 16 and 32°C for five months. Sludge had two rates of nitrogen equivalent to 300 and 600 kg N/ha from liquid sludge with 2.5% solids and 7% N oven-dry basis. The simulated profile was 7 cm in diameter and 10 cm deep over a fiberglass filter pad fashioned in a 500 ml bottle with holes drilled in the bottom for application of suction to facilitate leaching. An acid-saturated fiberglass filter was periodically placed in the bottle cap to trap ammonia. The profiles were leached with 150 ml of deionized water monthly for five months, and nitrate, ammonium, potassium, calcium, magnesium, manganese, zinc, and copper ions were determined in the leachate. The acid treatment decreased ammonia volatilization; however, estimated volatilizations from treatments without acid were less than 2% of the nitrogen added in the sludge. The nitrification inhibitor failed to decrease nitrate leaching significantly. Acidification of sludge effectively controlled nitrate leaching. Leaching of ammonium and other cations was increased by acidification; however, the low concentration of nitrate as a result of acidification would probably decrease the depth of cation leaching in the field. Acidification of the sludge had no effect on forest floor (0i, 0e) and mineral soil (A) pH at the 300 N rate at 16 and 32°C or the 600 N rate at 16°C. At 32°C, acid treatment of sludge increased forest floor and mineral soil pH at five months after treatment.

Ammonium and nitrate ions are the major source of nitrogen (N) for plants; however, excessive amounts of either cause an imbalance in plant nutrition and alter soil processes that are related to nitrogen, including nutrient loss. Because of the changes that can occur in the soil due to quality and quantity of nitrogen, nitrogen loading is often the controlling factor in waste product application to soil (Breuer et al. 1979). The leaching loss of nitrate from soil systems into the groundwater can cause water sources to exceed safe public health levels for nitrate.

Altering the nitrogen transformation processes by amending an ammonium-rich material with the inhibitor nitrapyrin can delay nitrification until agricultural crops reach a stage of development where nitrates can be utilized. Terry et al. (1981) reported that nitrapyrin delayed nitrification in a sludge-amended soil for 112 days in laboratory incubation, and McClung et al. (1983) reported 85% inhibition of nitrification by nitrapyrin and etridiazol in soil-compost combinations following incubation for 18 weeks at 25°C.

This information indicates nitrification inhibitors could be used to regulate nitrate levels when it is desirable to apply large amounts of ammonium or readily available organic nitrogen to forest soils.

Nitrification is reduced in soils by a low pH, a property of most soils producing loblolly pine (*Pinus taeda* L.) in the Southeast; however, high concentrations of nitrate were found in forest soils treated with sewage sludge at the Savannah River Plant, Aiken, South Carolina (Wells et al. 1984). The near- or above-neutral pH of the added nitrogen-rich material can contribute to nitrification and increase volatilization of ammonia. The addition of acid or acidic materials to sludge could possibly prevent ammonia volatilization (Terry et al. 1978), improve nitrogen utilization by trees, and reduce nitrate concentrations in drainage water. If acidification prevents nitrate production, the overall effect on soil acidity may be approximately equivalent to acidification through nitrification.

The objective of this study was to determine the effect of temperature, pH control, and a nitrification inhibitor on ammonia volatilization and on nitrate and ammonium leaching from sludge-amended soil.

METHODS

Soil Preparation and Incubation Leaching

A 6 to 8 mm thick filter was prepared from fiberglass filter paper macerated to a slurry and drawn by vacuum against holes in 500 ml Nalgene™ bottles. There were approximately sixty 2 mm holes distributed over the bottom of each bottle. Soil A horizon and forest floor (0i and 0e) samples were collected from a 28-year-old loblolly pine stand in the upper coastal plain at the Savannah River Plant, Aiken, South Carolina. The soil was Lucy loamy sand, loamy, siliceous, thermic Arenic Paleudults. The soil, stand, and results of field studies are more completely described in Corey et al., McKee et al., and Wells et al. (this volume). Approximately 360 grams of air-dried A horizon were put over the filter in each bottle and tamped by dropping on the table to a bulk density of about 1.2 g/cc. Then 0e (14.4 g) and 0i (3.4 g) forest floor layers approximating those found in the forest were placed on top. The 0i layer was cut into 1 to 2 cm lengths to facilitate placement. This produced a 7 cm diameter reconstructed profile of 0 horizon and part of the A horizon totaling 10 cm. Bottles were leached with 150 ml of water, which was sampled for analysis.

Sludge was composited from samples collected when installing a field study and stored in the refrigerator about a year. Tests for nitrate and ammonium indicated no change during storage. One portion of sludge was treated with 6.7 µg nitrapyrin per liter to yield 4.4 kg/ha active ingredient. Another portion was treated with 1.67 ml/l sulfuric acid to obtain a pH of about 6.6. A portion was also treated with the same amounts of both nitrapyrin and sulfuric acid. Water was applied as the control treatment. Soils were amended with 66 ml of treated or untreated sludge to obtain approximately 300 kg N/ha (low level). Treatments were added without sludge for the 0 kg N/ha samples. Percolating water was collected for analysis of ammonium. Bottles with the caps loose were incubated at either 16 or 32°C. These temperatures approximate the 31.8°C average for July at 7.5 cm in the soil and a lower temperature representative of November to March. An additional 6 ml of sludge were added to some bottles for the 600 kg N/ha treatment (high rate) after the first leaching, one month after the original application. This addition of sludge also doubled the nitrapyrin and sulfuric acid application. Three replicates were prepared for each sludge treatment and incubation combination. Since evaporation from the bottles between monthly leaching was small, addition of water was not required.

Water was applied to the soil in each bottle at approximately one-month intervals for five months. The first month, 125 ml were applied. Thereafter, 150 ml equivalent to 4 cm were applied. The bottles were returned to incubation, allowed to drain into beakers overnight, then vacuum pumped at 20 to 25 cm Hg to recover additional leachate. The leachate was analyzed for potassium, calcium, magnesium, manganese, zinc, and copper by atomic absorption, nitrate and ammonium by autoanalyzer, and pH by glass electrode.

Ammonia Volatilization

Three to five days after treatment of soils, a fiberglass filter saturated with 0.5 M sulfuric acid was placed in the cap of each bottle. The filter was removed to a vial after three days and the cap rinsed with four 5 ml portions of water. This solution was refrigerated for ammonium analysis. The procedure was repeated at the end of the first and third months of incubation.

Sludge and Soil Properties

The sludge material was 7.2% nitrogen, of which 42% was ammonium (Table 1) but very low in nitrate. Other elements were below the average found in sludge (Table 2).

TABLE 1. Properties of materials applied, forest floor, and mineral soil in leaching study.

Material	pH	Ammonium-N (mg/kg)	Organic N (mg/kg)	Ammonium-N (mg/bottle)	Organic N (mg/bottle)	Amount of Sludge Materials (g/bottle)
Sludge, low N	8.6	810	998	53.5	65.9	66
Sludge, high N	8.6	810	998	107.0	131.8	132
Oi layer	3.9	50.5	6,034	0.2	20.5	3.4
Oe layer	4.2	95.5	3,174	1.4	45.7	14.4
Mineral soil, A	4.8	6.4	341	2.3	122.8	360

RESULTS AND DISCUSSION

Ammonium and Nitrate Ions in Leachate

Acid treatments increased ammonium and decreased nitrate concentrations for the first three months in all application rates (Table 3). These effects were generally statistically significant at the 0.05 level by analysis of variance and Duncan's multiple range test. In the fourth and fifth months, ammonium concentration decreased and nitrification increased sharply at 32°C but not at 16°C. It is interesting that nitrification occurred in the fourth month in the nonacid treatment (control and nitrapyrin) without any sludge. This is in agreement with the moderate amount of nitrate found in lysimeters in the control plots of an age 1 loblolly pine stand in South Carolina (Wells et al., this volume) and in oak and pine in Indiana (Vitousek and Matson 1985). Little or no nitrate has been detected in lysimeters of the control plots of the 28-year-old stand from which the soil was obtained. Rapid nitrification in the field (Wells et al., this volume) and in the

TABLE 2. Sludge properties and application of nutrients and heavy metals in sludge treatments with low or high N.

Sludge Property	Sludge (mg/kg, oven-dry)	Low N (mg/bottle)	High N (mg/bottle)
Sludge, oven-dry	24,800	1,637	3,274
Ash	11,408	753	1,506
Phosphorus	16,200	26.6	53.2
Potassium	2,360	3.9	7.8
Calcium	14,550	23.8	47.6
Magnesium	2,460	4.0	8.0
Manganese	137	0.22	0.44
Zinc	1,330	2.2	4.4
Copper	318	0.52	1.04
Sodium	20,900	34.2	68.4
Lead	238	0.39	0.78
Nickel	44	0.07	0.14
Cadmium	44	0.07	0.14
Chromium	173	0.28	0.56
Sulfur	5,967	9.8	19.6
Boron	254	0.42	0.84

low and high rates with incubation without acid indicate low ammonium concentrations, acidity, and an inhibitor may have caused a nitrification delay in loblolly pine stands.

The nitrification inhibitor, nitrapyrin, failed to delay nitrification at the rate applied except possibly for the low sludge rate at 16°C. Results for effective inhibition of nitrification in sludge composts and sludge soil mixtures (Terry et al. 1981, McClung et al. 1983) cannot be extrapolated to applications of sludge to the surface of forest soils. This study's rates of nitrapyrin at approximately 1 µg/g of oven-dry sludge were low compared with 3 to 50 µg/g of soil or compost sludge mixture in the other studies. In addition to the low application rate, organic matter of the sludge and 0 horizon contributed to the ineffectiveness of nitrapyrin in this study (McClung et al. 1983).

At 16°C, 20 to 38 mg N per bottle as ammonium and nitrate were leached at the low sludge rate and 50 to 77 at the high rate, with about 50% more leached with the acid treatment (Table 4). Data for 32°C show there was no effect of acid on total leaching of nitrogen. Approximately 25 and 40% of the nitrogen leached at 16 and 32°C, respectively.

Acidification effectively controlled nitrification over a five-month period, and it is expected to have similar effects in forests. Actual nitrate leaching was less with the high rate of treatment with acidification than in the control (0 rate) without acidification. Although ammonium was leached with acidification, very small quantities would move below the root zone in this forest as indicated by lysimeter studies (Wells et al., this volume).

Potassium, Calcium, and Magnesium Leaching

As for ammonium, leaching of potassium, calcium, and magnesium was increased by acid treatment (Table 4). Only for potassium was the additional leaching from acid greater than application in the sludge. Any increased leaching of cations must be considered an adverse effect; however, cations would probably not move through the profile in the absence of the highly soluble nitrate anion.

Calcium in the leachate was nearly 250 mg/l at 32°C with acid application without sludge (Table 3). Compared with acid application, calcium leaching was low without acid; however, when nitrification in the control sharply increased in the fourth and fifth months, calcium in the leachate also increased. In the 600 N sludge treatments, the effect of acid addition was greater than was the nitrification effect.

Microelements and Heavy Metals in Leachate

The application of sludge increased leaching of zinc and copper without any acid treatment. The amount leached in relation to that added was less than 1% for zinc and less than 3% for copper without acid (Table 4). Leaching of zinc and copper was increased slightly by acid at 32°C for the high rate, but at the low rate at 16°C, acid decreased leaching of both elements. Lead, cadmium, and nickel were not quantitatively measurable in the leachate by flame direct atomic absorption. These heavy metals were immobilized in the organic matter and surface soil in agreement with results from a soil core leaching study of a sandy soil in Indiana (Miller et al. 1983). In this soil, which is high in manganese, leaching exceeded the small amounts in sludge application for several treatments. The data for manganese are not shown because inconsistent results preclude explanation except for the strong effect of acid without sludge application.

Forest Floor and Mineral Soil pH Five Months after Treatment

Acidification of sludge had no significant ($p < 0.05$) effect on forest floor and mineral soil pH at the low sludge rate at 16 or 32°C or the high rate at 16°C (Table 5). At 32°C, sludge acid increased forest floor and mineral soil pH at five months after treatment. Although this seems to be an unexpected reaction, it can be explained by the action of sulfuric acid preventing mobile nitrate formation from ammonium (Van Miegroet and Cole 1984). The sulfate ion was adsorbed by the soil, and the anion concentration was lowered, thus decreasing the leaching of cations. The pH of the leachate was between 4 and 5 throughout the five months.

Ammonia Volatilization

Acid reduced ammonia volatilization for the first month after treatment; however, after three months the effect seems to be very small, with the exception of the low sludge rate at 16°C (Table 6). The volatilization test was not designed to be quantitative. Since the measurements were made at the end of months 1 and 3, when moisture was lowest, the estimates are probably high. Estimated volatilization without acid was less than 2% of the added nitrogen—a small loss compared with estimates of 50% for field conditions (Beauchamp et al. 1978). The effectiveness of the acid treatments in forest applications cannot be determined from these data. If plants had been growing in the soil, much of the ammonium would have been taken up before the third month, when volatilization showed the increase for acid treatments; with normal air movement, ammonia volatilization would be greater without than with acid application.

SUMMARY

Acidification of sludge to pH 6.6 before application controlled nitrification and decreased ammonia volatilization. Leaching of cations was increased, but the amount leached did not exceed the application in sludge except for potassium and manganese,

TABLE 3. Average concentration of leachate for one through five months after sludge treatment at low and high rates of N and incubation at 16 and 32°C.

Month	Temp. (°C)	0 Sludge C	I	H	IH	Low Sludge C	I	H	IH	High Sludge C	I	H	IH
						Nitrate-N (mg/liter)							
1	16	0.11a	0.12a	0.10a	0.10a	4.89a	3.93a	0.20b	0.18b	6.12a	3.84a	0.23b	0.20b
2	16	0.03a	0.05a	0.08a	0.03a	6.77a	5.97a	0.13b	0.13b	46.7a	37.0a	0.32ab	0.22ab
3	16	0.22a	0.27a	0.06a	0.06a	9.59a	8.97a	0.10b	0.08b	62.3a	55.8a	0.16b	0.18b
4	16	0.26a	0.12ab	0.03b	0.05b	25.0a	14.4ab	0.11b	0.39b	54.2a	52.1a	0.25b	0.55b
5	16	0.35a	0.18ab	0.07b	0.05b	26.3a	4.20a	0.29a	0.25a	43.4a	37.2a	0.55b	0.83b
1	32	0.10a	0.10a	0.09a	0.11a	13.5a	12.0a	0.26b	0.22b	10.2a	9.27a	0.19b	0.20b
2	32	0.69a	0.42ab	0.08b	0.07b	13.8a	14.7a	0.22b	0.20b	51.3a	68.0a	0.38b	0.33b
3	32	2.93a	3.61a	0.06b	0.06b	30.8a	31.2a	0.16a	0.15a	82.6a	81.4a	1.92b	0.18b
4	32	49.9a	61.5a	0.95b	0.37b	73.9a	86.7a	4.25b	3.89b	112.0a	106.0a	11.7b	3.86b
5	32	64.1b	33.0b	2.14a	0.61a	100.0a	70.1ab	34.8ab	34.1b	111.0a	93.1a	21.8b	15.6b
						Ammonium-N (mg/liter)							
1	16	2.0c	2.0c	13.2b	14.9a	49.3ab	24.6b	104a	92a	20.2b	22.7b	126a	113a
2	16	2.0b	2.1b	16.3a	15.3a	36.3b	36.7b	105a	101a	89b	78.7b	270a	290a
3	16	2.2b	2.5b	10.7a	9.2a	32.0b	33.6b	61.1a	67.2a	97.3b	79.7b	174a	178a
4	16	2.1b	2.1b	8.6a	8.0a	28.4a	29.0a	30.5a	30.6a	57.2b	57.2b	75.9ab	88.8a
5	16	2.1b	2.1b	6.6a	6.9a	25.6a	22.4a	28.5a	30.3a	59.3a	53.0a	56.0a	58.3a
1	32	3.5b	4.2b	30.4a	24.8a	41.5b	42.1b	134a	141a	34.7b	35.3b	138a	148a
2	32	7.1b	6.3b	27.7a	23.0a	56.7b	55.3b	123a	137a	120.0b	143.0b	360a	3.40a
3	32	9.4b	8.2b	24.2a	15.7a	52.5b	50.3b	75.1a	71.3ab	106.0b	119.0b	190a	191a
4	32	8.3a	6.8a	15.3a	13.1a	52.5a	57.8a	34.1a	38.7a	74.3a	77.4a	72.4a	93.0a
5	32	5.1b	3.8b	10.9a	8.6b	51.4a	46.8a	47.0a	43.2a	70.8a	68.5a	80.5a	70.2a

TABLE 3. Continued

Calcium (mg/liter)

Month	Temp. (°C)	0 Sludge				Low Sludge				High Sludge			
		C	I	H	IH	C	I	H	IH	C	I	H	IH
1	16	11.6c	14.8c	178.0b	294.0a	33.6b	36.6b	148.0a	189.0a	39.0b	29.6b	178.0a	186.0a
2	16	2.54c	2.29c	39.4b	52.6a	9.17b	8.85b	42.8a	51.3a	23.2b	19.9b	91.5a	121.0a
3	16	1.53c	1.83c	18.7a	16.7b	4.13c	4.57c	10.3b	14.7a	13.0b	12.7b	35.7a	40.0a
4	16	1.28b	1.16b	9.74a	8.74a	4.18a	3.83a	4.05a	4.89a	5.31b	6.28b	11.7a	13.6a
5	16	1.13b	1.05b	5.67a	5.13a	5.70a	2.26a	2.41a	2.24a	3.03a	4.06a	4.1a	4.8a
1	32	12.4a	12.3a	244.0a	146.0ab	34.0a	30.6a	178.0a	157.0a	32.2a	25.6a	139.0a	162.0a
2	32	2.45b	2.05b	39.7a	33.5a	7.98b	8.20b	40.2a	37.7a	18.8b	22.6b	117.0a	109.0a
3	32	3.80b	4.83b	16.0a	12.3a	7.07a	7.40a	9.80a	9.30a	13.7b	4.2b	31.7a	27.0a
4	32	21.6ab	26.2a	10.0bc	7.90c	14.0a	16.1a	5.63b	5.19b	19.8a	14.3a	13.6a	10.7a
5	32	34.2a	18.4a	6.05b	4.24b	22.4a	15.1a	8.30a	9.42a	23.4a	14.7ab	8.66ab	6.73b

C = control. I = nitrapyrin nitrification inhibitor. H = acid.

Values within sludge rate and row followed by the same letter are not significantly different at the 0.05 level.

TABLE 4. Total nutrients leached over five months, by sludge N rate, acid and nitrapyrin treatment, and temperature.

Temperature (°C)	O Sludge				Low Sludge				High Sludge			
	C	I	H	IH	C	I	H	IH	C	I	H	IH
Nitrate-N (mg/bottle)												
16	0.13a	0.09ab	0.04ab	0.03b	7.80a	3.80ab	0.09b	0.11b	22.60a	19.00a	0.14b	0.20b
32	10.77a	10.53a	0.33b	0.13b	26.73a	23.77a	4.50b	4.50b	42.00a	39.33a	3.87b	2.19a
Ammonium-N (mg/bottle)												
16	1.28b	1.31b	6.83a	6.57a	19.00b	15.77b	38.00a	35.00a	34.67b	29.77b	74.00a	76.67a
32	3.77b	3.19b	12.07a	8.97a	28.83b	28.53b	46.67a	48.00a	45.00b	48.33b	88.00a	91.33a
Nitrate + Ammonium-N (mg/bottle)												
16	1.29b	1.40b	6.88a	6.59a	26.83ab	19.58b	38.11a	35.21a	57.18b	48.76b	74.13a	76.84a
32	14.54a	13.72a	12.41a	9.10a	55.55a	52.29a	51.05a	52.41a	86.98a	87.75a	91.85a	93.47a
Potassium (mg/bottle)												
16	1.72b	1.66b	4.83a	6.37a	4.13b	3.80b	9.27a	8.97a	5.43b	5.30b	12.57a	11.97a
32	3.05b	3.47ab	7.13a	5.13ab	5.93b	5.87b	11.70a	11.03a	8.63b	8.03b	14.03a	14.40a
Calcium (mg/bottle)												
16	2.12b	2.50b	29.10a	40.33a	5.97b	5.67b	23.20a	26.77a	8.00b	7.43b	32.73a	38.67a
32	6.93b	6.97b	33.13a	22.03ab	8.93a	8.10a	25.03a	22.40a	11.70a	9.30a	32.20a	29.87a
Magnesium (mg/bottle)												
16	0.49b	0.46b	5.53a	3.91a	1.33b	1.27b	3.59a	4.01a	2.15b	2.17b	6.40a	6.53a
32	2.14a	2.78a	3.05a	2.84a	2.50ab	2.18b	3.90ab	4.01a	3.25ab	2.62b	5.05a	4.76ab
Zinc (g/bottle)												
16	13.60b	27.60ab	56.60a	51.30a	39.00a	40.30a	30.30a	28.00a	37.30a	27.60a	44.60a	38.30a
32	33.30a	24.60a	37.30a	41.30a	31.00a	36.60a	33.30a	38.60a	27.30a	31.30a	59.60a	62.60a
Copper (g/bottle)												
16	6.70a	5.30a	8.60a	5.10a	24.20ab	30.60a	10.80a	10.00b	18.20a	15.60a	14.40a	10.40a
32	13.10a	4.30a	6.50a	3.70a	23.50a	20.10a	22.00a	20.20a	11.90b	18.20ab	48.30a	50.00a

C = control. I = nitrapyrin nitrification inhibitor. H = acid.
Values within sludge rate and row followed by the same letter are not significantly different at the 0.05 level.

TABLE 5. Effect of sludge N rate, acid and nitrapyrin treatments, and temperature on mineral soil and forest floor pH after five weeks of incubation.

Layer	Temperature	O Sludge				Low Sludge				High Sludge			
		C	I	H	IH	C	I	H	IH	C	I	H	IH
Soil	16°C	5.0b	4.9b	4.3c	4.3c	5.2ab	5.5a	5.3ab	5.2ab	5.6a	5.7a	5.4ab	5.2ab
	32°C	4.0e	4.2cde	4.4bcd	4.5bc	4.1de	4.2cde	4.4bcde	4.4bcde	4.0e	4.1de	4.6ab	4.9a
Forest floor	16°C	4.3c	4.3c	4.0c	4.0c	6.2a	6.3a	5.9b	6.1ab	6.4a	6.3a	6.4a	6.5a
	32°C	4.7efg	4.7ef	4.3fg	4.3g	5.0de	5.2d	5.8bc	5.7c	4.9de	5.9bc	6.2ab	6.3a

C = control. I = nitrapyrin nitrification inhibitor. H = acid.

Means in the same row followed by the same letter are not significantly different at the 0.05 level.
Temperature effect was significantly different for mineral soil and forest floor by analysis of variance test.

TABLE 6. Ammonia volatilization one day (0), one month, and three months after sludge treatment at low and high rates of N.

Month	Temperature (°C)	O Sludge				Low Sludge				High Sludge			
		C	I	H	IH	C	I	H	IH	C	I	H	IH
		(milligrams/bottle per month)											
0	16	0.048	0.050	0.057	0.042	0.077	0.088	0.061	0.045	0.068	0.077	0.058	0.063
1	16	0.036	0.077	0.017	0.027	0.113	0.095	0.047	0.080	0.208	0.158	0.043	0.054
3	16	0.059	0.105	0.059	0.051	0.138	0.164	0.055	0.046	0.198	0.300	0.092	0.239
0	32	0.062	0.051	0.066	0.059	0.132	0.370	0.098	0.141	0.533	0.288	0.068	0.082
1	32	0.017	0.027	0.021	0.021	0.560	0.648	0.167	0.185	0.556	0.553	0.239	0.168
3	32	0.067	0.149	0.055	0.133	0.360	0.353	0.280	0.300	0.413	0.380	0.400	0.700

C = control. I = nitrapyrin nitrification inhibitor. H = acid.

the latter of which was in good supply in the soil. The cations leached through the 10 cm column of forest floor and surface mineral soil would be mostly retained within the root zone under field conditions. Although it could not be tested in the study, there were indications from laboratory and field studies that the very low nitrate concentration in the acid treatments would prevent leaching of cations to as great a depth as would occur in the nonacid treatments, where nitrate levels were high. Acid had little effect on or increased soil pH after five months of incubation.

Nitrapyrin at approximately 1 mg/kg of oven-dry sludge did not effectively control nitrification. Leaching of nitrate and ammonium was considerably less at 16 than 32°C, indicating that sludge application in fall and winter months may reduce nitrate leaching in coniferous forests in the South compared with spring and summer application.

ACKNOWLEDGMENTS

This work was supported in part by contract DE-A109-80SR10711 with the U.S. Department of Energy.

The use of trade, firm, or corporation names in this paper is for the information and convenience of the reader and does not constitute an official endorsement or approval by the U.S. Department of Agriculture or the Forest Service of any product or service to the exclusion of others that may be suitable.

REFERENCES

Breuer, D. W., D. W. Cole, and P. Schiess. 1979. Nitrogen transformation and leaching associated with wastewater irrigation in Douglas fir, poplar, grass, and unvegetated systems. p. 19–33. In W. E. Sopper and S. N. Kerr (eds.) Utilization of municipal sewage effluent and sludge on forest and disturbed land. Pennsylvania State University Press, University Park.

Beauchamp, E. G., G. E. Kidd, and G. Thurtell. 1978. Ammonia volatilization from sewage sludge applied in the field. J. Environ. Qual. 7(1):141–146.

McClung, G., D. C. Wolf, and J. E. Foss. 1983. Nitrification by nitrapyrin and etridiazol in soils amended with sewage sludge compost. Soil Sci. Soc. Am. J. 47:75–80.

Miller, W. P., W. W. McFee, and J. M. Kelly. 1983. Mobility and retention of heavy metals in sandy soils. J. Environ. Qual. 12:579–584.

Terry, R. E., D. W. Nelson, and L. E. Sommers. 1981. Nitrogen transformations in sewage sludge-amended soils as affected by soil environmental factors. Soil Sci. Soc. Am. J. 45:506–513.

Terry, R. E., D. W. Nelson, L. E. Sommers, and G. J. Meyer. 1978. Ammonia volatilization from wastewater sludge applied to soils. J. Water Poll. Control Fed. 2657–2664.

Van Miegroet, H., and D. W. Cole. 1984. The impact of nitrification on soil acidification and cation leaching in a red alder ecosystem. J. Environ. Qual. 13:586–590.

Vitousek, P. M., and P. A. Matson. 1985. Causes of delayed nitrate production on two Indiana forests. For. Sci. 31:122–131.

Wells, C. G., K. W. McLeod, C. E. Murphy, J. R. Jensen, J. C. Corey, W. H. McKee, and E. J. Christensen. 1984. Response of loblolly pine plantations on two sources of sewage sludge. p. 85–94. In 1984 TAPPI Research and Development Conference, Appleton, Wisconsin. Technical Association of the Pulp and Paper Industry, Technology Park, Atlanta, Georgia.

Chapter 14

Nitrogen Transformations in Four Sludge-amended Michigan Forest Types

ANDREW J. BURTON, DEAN H. URIE, and JAMES B. HART, JR.

ABSTRACT   Anaerobically digested municipal sludge was added in liquid form to the surface of intact soil cores from pine, hardwood, oak, and aspen forest types. Cores contained the forest floor and first 10 cm of mineral soil. Pairs of sludge-treated and control cores were incubated in the laboratory at 20°C and 80% relative humidity, with randomly selected pairs destructively sampled and analyzed for nitrate-N and ammonium-N at 0, 2, 4, and 8 weeks. Additional paired cores from the oak and pine sites were placed in polyethylene bags and incubated in situ. Nitrification in sludge-treated pine cores incubated in the laboratory was approximately half that of the other sites. Sludge-treated pine cores also retained more ammonium-N. Control cores did not nitrify even though large pools of ammonium-N were present after eight weeks. Forest floor ammonium-N concentrations in sludge-treated cores decreased over time. Soil ammonium-N varied but tended to increase in similar amounts in both control and treated cores. Total inorganic N differences between treated and control cores decreased over time. As a result, mineralization of sludge organic N could not be estimated. Nitrification in sludge-treated cores and mineralization in control cores were less in field incubations than in lab incubations. Nitrate-N in soil leachate and groundwater following sludge application at the four sites could not be predicted by incubation results alone.

Knowledge of nitrogen transformations that may occur following land application of sludge can be used in determining loading rates that provide the maximum available N for plant growth without causing nitrate-N pollution of groundwater. Liquid sludges vary considerably in composition, but N content is typically about half organic N and half ammonium-N with only small amounts of nitrate-N present (Sommers 1977, Sabey et al. 1975). Following application, the ammonium-N can be volatilized as ammonia-N, immobilized by soil microorganisms, taken up by plants, converted to nitrate-N through nitrification, or held on soil exchange sites. Ammonium-N derived from mineralization of sludge organic N is subject to the processes listed above. Nitrate-N can be utilized by vegetation, lost to the atmosphere through denitrification, or leached from the soil profile to groundwater, where it presents a potential pollution hazard.

Many environmental factors can affect nitrogen transformations. Forest species and stocking will influence plant uptake. Temperature and moisture condition have been shown to influence mineralization (Cassman and Munns 1980), volatilization (Sommers and Nelson 1981), nitrification (Gilmour 1984, Terry et al. 1981), denitrification (Focht and Verstraete 1977), and immobilization (Terry et al. 1981). Slightly alkaline pH optimizes volatilization (Donovan and Logan 1983), as well as nitrification and denitrification (Focht and Verstraete 1977), and immobilization and mineralization are affected by the C:N ratio of the organic material involved. Where the C:N ratio exceeds 30:1, immobilization proceeds faster than mineralization, but if the C:N ratio drops below 15:1 mineralization will dominate (Foth 1978). Forest ecosystems that differ in these properties can be expected to show differences in the rates of the processes involved in the nitrogen cycle and thus will require different sludge application rates in order to maximize growth while avoiding nitrate-N leaching to groundwater.

One method that has been used to study nitrogen transformations involves aerobic incubation of samples and subsequent analysis for nitrogen forms (Bremmer 1965, Stanford and Smith 1972). In general, these experiments have involved agricultural soils in which dried sludge and soil have been mixed. Samples were either periodically leached with 0.01 M calcium chloride (Epstein et al. 1978, Parker and Sommers 1983) or replicate samples were periodically destructively sampled and extracted with potassium chloride solutions (Magdoff and Chromec 1977, Lindemann and Cardenas 1984, Parker and Sommers 1983). Incubations can occur in the laboratory, under controlled temperature and humidity, or in polyethylene plastic bags in situ.

In Michigan, the most feasible method of sludge application to forest land appears to be spraying liquid sludge. As a result, sludge will not be incorporated into the soil as in agricultural applications. Instead, much of the solid portion will be caught in the forest floor with only the supernatant reaching the soil.

In this study, differences in nitrogen transformations resulting from surface application of liquid sludge to four forest types were examined by incubating intact cores containing both forest floor and soil. Both laboratory and field incubations were used. Incubation results were compared with nitrate-N concentrations occurring in soil leachate and groundwater following sludge application to the four types to determine if any relationship existed.

MATERIALS AND METHODS

Study Sites

Study sites are located in Montmorency County, Michigan, in the stands of the forest fertilization demonstration project being conducted by Michigan State University. They include four forest types: aspen, northern hardwoods, oak, and pine. The aspen site is in a stand of 12- to 13-year-old aspen regeneration. Bigtooth aspen (*Populus grandidentata* Michx.) predominates, with scattered oak (*Quercus* spp. L.), cherry (*Prunus* spp. L.), and other species occurring. The northern hardwoods study site is in a 50-year-old stand of red maple (*Acer rubrum* L.) and sugar maple (*Acer saccharum* Marsh.) containing remnants of American beech (*Fagus grandifolia* Ehrh.) and birch (*Betula* spp. L.). The stand also contains red oak (*Quercus rubra* L.), American basswood (*Tilia americana* L.), eastern hemlock (*Tsuga canadensis* [L.] Carr.), white ash (*Fraxinus americana* L.), and associated species. The oak study site is in a 70-year-old oak stand containing red oak (*Quercus rubra* L.), white oak (*Q. alba* L.), and red maple (*Acer rubrum* L.) with scattered pines (*Pinus* spp. L.) and aspen (*Populus* spp. L.). The pine study site is in a plantation of jack pine (*P. banksiana* Lamb.) and red pine (*P. resinosa* Ait.) which also includes minor amounts of northern pin oak (*Q. ellipsoidalis* E. J. Hill).

A soil survey of the study areas was conducted in October 1982. The soils at the aspen site are a mixture of the Rubicon series, a sandy, mixed, frigid Entic Haplorthod (Soil Conservation Service 1979a), and the Montcalm series, a coarse-loamy, mixed, frigid Eutric Glossoboralf (Soil Conservation Service 1984). The Rubicon series are deep, excessively drained soils formed in sandy, glacial drift deposits. The Montcalm series are deep, well-drained soils formed in sandy and loamy glacial drift deposits. The northern hardwoods study location occurs on soils of the Mancelona series, a sandy, mixed, frigid Alfic Haplorthod (Soil Conservation Service 1982a). They are deep, somewhat excessively drained soils formed in sandy and gravelly glacial drift deposits. The soils at the oak study site belong to the Graycalm series, a mixed, frigid Alfic Udipsamment (Soil Conservation Service 1979b). These are deep, somewhat excessively drained soils formed in sandy, glacial drift deposits. Soils at the pine site belong to the Grayling series, a mixed, frigid Typic Udipsamment (Soil Conservation Service 1982b) characterized as deep, excessively drained soils formed in deep, sandy deposits on outwash and lake plains.

Site Characterization

Samples taken for site characterization included five forest floor samples collected using a 0.09 square meter (1 square foot) metal frame, and samples of the upper 10 cm (3.94 inches) of soil under each forest floor sample. The forest floor material was divided into samples designated as 01 (litter, 0i), intact organic material that is recognizable and not discolored by decompositional processes; and 02 (fermentation and humus layer, 0a and 0e), finely divided decomposed organic materials.

Soil samples were oven dried at 105°C. Bulk density samples were taken at each site. Additional samples were collected, sieved, and subsampled for analysis of pH, total N, and organic carbon. Soil pH was measured in a 1:1 soil-water mixture. Organic carbon was determined using a Walkley-Black titration method (Black 1965). Total N was determined using a micro-Kjeldahl procedure with analysis on a Technicon Autoanalyzer II system (Technicon 1977a).

Forest floor samples were oven dried at 65°C, weighed, ground, and subsampled for analysis of total N and organic carbon. Forest floor weights and soil bulk densities were used to calculate contents of N forms on an areal basis.

Site characteristics are shown in Table 1. Differences discussed are significant unless otherwise noted. The main difference among the forest floors of the four types was

TABLE 1. Mean characteristics of the forest floor and surface 10 cm of mineral soil.

Component	Forest Type			
	Aspen	N. Hardwood	Oak	Pine
Forest floor (kg/ha)	28,600a	72,000c	59,200c	36,700b
01 total N (%)	1.72a	1.55b	1.54b	1.10c
02 total N (%)	1.18a	0.92b	0.71b	0.91b
01 organic C (%)	43.3	45.2	46.2	49.3
02 organic C (%)	24.3	22.8	21.0	29.4
01 C:N ratio	25.2a	29.4a	30.1a	45.1b
02 C:N ratio	20.6a	20.7ab	30.3bc	34.7c
Soil pH	4.73b	5.08c	4.43ab	4.20a
Soil bulk density	1.35	1.37	1.34	1.36
Soil total N (%)	0.083a	0.027b	0.026b	0.043b
Soil organic C (%)	1.48a	0.70b	0.61b	1.08ab
Soil C:N ratio	18.2	23.8	23.4	23.1

Row means followed by a different letter are significantly different at an alpha = 0.05 level.

weight. Northern hardwood and oak forest floors were heavier than those at aspen and pine. The pine forest floor was also heavier than that at aspen. The aspen forest floor had higher total N concentrations in both the 01 and 02 than the forest floors of the other types. Pine 01 had a lower total N content than was found in 01 at the oak and northern hardwood stands. Organic carbon percentages showed no differences among the four forest types in either the 01 or 02 layers. The pine 01 had a higher C:N ratio than the other three types. In the 02 layer, C:N ratios measured in the aspen and northern hardwood stands were less than those in the pine stand, with the aspen C:N ratio also being less than that in the oak. Aspen and northern hardwood C:N ratios in both the 01 and 02 were in a range where neither mineralization nor immobilization should predominate. Oak 01 and 02 layers had C:N levels that suggested immobilization may exceed mineralization, and pine 01 and 02 had C:N ratios that indicated that net immobilization would occur.

Soil at the aspen site had higher levels of both total N and organic carbon than the other three sites. However, there were no significant differences in soil C:N ratios at the four sites. The C:N ratio in the soil at all four sites was in the range in which neither mineralization nor immobilization is reported to dominate. The soil under pine had the most acidic reaction and was significantly more acidic than soil at the aspen or hardwood sites. The northern hardwoods site had less acidic soil than the other sites.

Sludge Analysis

Anaerobically digested municipal sludge used in the experiment was obtained from Alpena, Michigan. Samples were taken, dried at 65°C, weighed to determine percentage of solids, and analyzed for total N and organic carbon. Duplicate liquid sludge samples were extracted with 2N KCl and analyzed for ammonium-N and nitrate-N on a Technicon Autoanalyzer II system (Technicon 1971, 1977b). Sludge characteristics are shown in Table 2, with loading rates shown in Table 3. It was desired to load the cores at a rate of 10 Mg/ha of sludge solids, but because of limited water capacity in the cores and the low

TABLE 2. Liquid sludge composition.

Component	Sludge Composition
Solids (%)	2.1
Organic N (mg/l)	395.0
Organic C (mg/l)	4,305.0
C:N ratio of solids	10.9
pH	7.5
Ammonium-N (mg/l)	817.0
Nitrate-N (mg/l)	6.2

TABLE 3. Sludge loading rates.

Component	Loading Rate (kg/ha)
Solids	4,200.0
Organic N	79.0
Ammonium-N	163.4
Nitrate-N	1.2

solids content of the sludge, a loading rate of 4.2 Mg/ha was used to avoid core saturation.

Incubation and Sampling Procedures

At each site, twenty pairs of intact cores containing the forest floor and the upper 10 cm (3.94 inches) of mineral soil were collected in 3.81 cm (1.5 inches) inside-diameter PVC pipe for use in the laboratory incubation portion of the experiment. An additional twenty pairs of cores were taken at the oak and pine sites for field incubation.

Sludge was applied to the surface of one core of each pair. The other core served as a control and received deionized water in an amount equivalent to water added to treated cores during sludge application. The amount of moisture added to the cores placed them at approximately field capacity. Cores from all four sites were incubated in the laboratory at 80% relative humidity and 25°C, with five randomly selected pairs of cores from each site destructively sampled at 0, 2, 4, and 8 weeks. At the oak and pine sites, an additional set of treated and control cores was enclosed in polyethylene plastic bags and incubated in situ. Randomly selected pairs of these cores were sampled for analysis on the same schedule as used for laboratory incubated cores.

Sample Analysis

Sample analyses were performed at the Michigan State University Department of Forestry laboratory. Core samples were separated into forest floor and soil fractions, each of which was thoroughly mixed. Subsamples of 5 grams for forest floor and 20 grams for soil were shaken in 100 ml of $2N$ KCl for one hour. The extracts were filtered and analyzed for ammonium-N and nitrate-N on a Technicon Autoanalyzer II system (Technicon 1971, 1977b). Duplicate 5 and 20 gram samples were oven dried at 65°C to determine the dry weight of the extracted samples. The dried soil samples were then passed through a 2 mm sieve and the dried forest floor samples were ground, after which total N was determined using a micro-Kjeldahl procedure with analysis on a Technicon Autoanalyzer II (Technicon 1977a).

Statistical Analysis

Statistical analysis was performed using SPSS statistical programs (Nie et al. 1975, Hull and Nie 1981) on the Control Data Corporation main computer system at Michigan State University. For each time period data from the four sites were statistically analyzed using an analysis of variance. Means were separated using Duncan's multiple range test.

RESULTS AND DISCUSSION

Results discussed are statistically significant unless otherwise noted.

Laboratory Incubation

Table 4 shows ammonium-N and nitrate-N contents resulting from laboratory incubation of sludge treated and control cores from the four sites. Control cores did not nitrify during the eight-week incubation period even though final ammonium-N levels exceeded half those resulting from sludge application. This suggests: (1) the organisms responsible for nitrification were introduced with the sludge, or (2) sludge application created pH and nutrient conditions that promoted activity of existing populations of ni-

TABLE 4. Mean ammonium-N and nitrate-N contents for sludge and control treatments for the four forest types.

Component	Time (weeks)	Forest Type (kg/ha)			
		Aspen	N. Hardwood	Oak	Pine
Control Ammonium-N	0	8.4ab	13.8c	11.7bc	6.2a
	2	5.4	16.6	17.2	14.7
	4	30.4a	45.9ab	56.3b	31.1a
	8	62.4	50.2	69.8	73.9
Control Nitrate-N	0	3.37b	5.75c	3.16b	1.46a
	2	0.82	0.65	0.50	0.74
	4	2.02	1.64	1.59	3.01
	8	2.00	1.99	1.26	1.19
Sludge Ammonium-N	0	88.8	108.0	122.0	108.1
	2	62.2a	111.0b	101.8b	144.0c
	4	110.0a	112.4a	109.9a	157.8b
	8	89.7a	82.0a	144.2b	153.4b
Sludge Nitrate-N	0	4.8b	5.7b	3.2a	3.5a
	2	7.2a	12.5b	13.7b	5.2a
	4	38.1	41.1	34.1	24.0
	8	47.3b	53.1b	48.5b	20.1a

Row means followed by a different letter are significantly different at an alpha = 0.05 level.

trifiers. Although not presented in this paper, an additional experiment was performed using a liquid sludge similar in pH and nutrient content that was prepared from a sludge that had been freeze dried, destroying any nitrifying bacteria contained. The lack of nitrification in cores receiving this sterilized sludge supports the first explanation.

Nitrification in sludge-treated cores from the pine site was less than at the other three sites. This is illustrated in Figure 1, which shows differences between nitrate-N contents in treated and control cores over time. Nitrate-N content in cores from the pine site tended to be about half that of the other three sites. After two weeks, pine nitrate-N was less than that of the oak and northern hardwoods types. At four weeks, pine nitrate-N was lower, but not significantly different from the other types; and at eight weeks, pine nitrate-N levels were less than those of the other three types. Ammonium-N levels in sludge-treated pine cores were correspondingly higher than those of the other three types. Pine ammonium-N content was higher than that of all other types at two and four weeks, and higher than aspen and northern hardwood ammonium-N content at eight weeks. Ammonium-N levels for the aspen, northern hardwood, and oak types followed similar patterns, with aspen levels being lower at two weeks and oak levels higher at eight weeks.

The lower nitrification for the pine type is attributed to the more acidic conditions at the pine site. Tables 5 and 6 show the ammonium-N and nitrate-N concentrations in the forest floor and soil of the sludge-treated and control cores respectively. As can be seen in Table 5, pine forest floor nitrate-N concentrations following sludge treatment were similar or less than those for the other sites, and pine soil nitrate-N concentrations were much less at four and eight weeks than those in the aspen soil that received similar ammonium-N loading. The reduced soil nitrate-N concentrations at the oak and northern hardwoods sites through four weeks were the result of much of the initial sludge being retained by the thicker forest floors of these types. The increase in nitrate-N levels in the

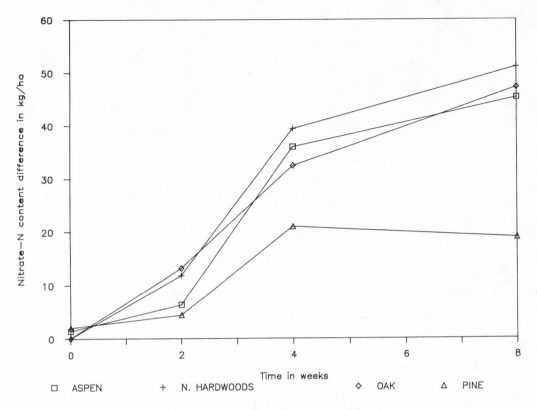

Figure 1. Sludge-control nitrate-N content differences.

soils of these types at eight weeks is attributed to increases in ammonium-N concentration over time as soil organic N mineralized, and the growth of populations of nitrifiers as more ammonium-N was made available.

The concentration of ammonium-N in the forest floors of sludge-treated cores for all four types decreased over time. Part of this is attributed to nitrification. Additional ammonium-N losses due to volatilization in high pH zones in the surface sludge layer and/ or immobilization had evidently occurred. The soils of treated cores of all four types were sites of net mineralization. Higher ammonium-N levels in the forest floor of treated cores from the pine site were probably due to its higher acidity, which would lessen volatilization losses. Also, it appears that the sludge penetrated the pine litter layer more deeply than for the other types, further reducing volatilization. Finally, even though pine C:N ratios were more favorable for immobilization, it is possible that immobilization was less because of lower biological activity caused by the more acidic conditions at the site. The higher ammonium-N concentrations in the pine soil following sludge treatment were probably the result of lower nitrification rates and more sludge reaching the soil than in the oak and hardwood types where much more was retained in the thicker forest floors.

Because of the high amounts of mineralization taking place in control cores and the ammonium-N losses from sludge-treated cores, ammonium-N differences between treated and control cores showed a decreasing trend over time, as can be seen in Figure 2. As a result, total inorganic N (ammonium-N plus nitrate-N) showed a decreasing trend over time, making it unusable as an estimate of mineralization differences among

TABLE 5. Mean ammonium-N and nitrate-N concentrations in the forest floor and soil of sludge-treated cores.

Component	Time (weeks)	Forest Type (mg/kg)			
		Aspen	N. Hardwood	Oak	Pine
Forest floor Ammonium-N	0	1,220	965	1,570	1,730
	2	756a	894a	899a	1,560b
	4	495a	952b	758ab	1,420c
	8	444a	458a	547a	1,510b
Soil Ammonium-N	0	33.8	18.2	13.1	27.9
	2	30.0a	34.3a	36.2a	63.2b
	4	70.9b	32.3a	48.5a	77.3b
	8	57.0b	36.1a	83.3c	71.5bc
Forest floor Nitrate-N	0	35.5	31.2	46.1	42.3
	2	168.0ab	155.0ab	204.0b	100.0a
	4	438.0	528.0	442.0	416.0
	8	530.0b	490.0ab	336.0a	324.0a
Soil Nitrate-N	0	2.8c	2.5bc	0.3a	1.4ab
	2	1.8	1.0	1.2	1.1
	4	18.9c	2.3a	5.9b	6.4b
	8	23.8b	13.1ab	21.3b	6.0a

Row means followed by a different letter are significantly different at an alpha = 0.05 level.

TABLE 6. Mean ammonium-N and nitrate-N concentrations in the forest floor and soil of control cores.

Component	Time (weeks)	Forest Type (mg/kg)			
		Aspen	N. Hardwood	Oak	Pine
Forest floor Ammonium-N	0	76	104	100	81
	2	139	70	132	223
	4	240a	328a	631b	254a
	8	503ab	344a	592ab	762b
Soil Ammonium-N	0	4.6	4.4	4.3	2.4
	2	2.4	8.2	7.0	4.7
	4	17.4	16.2	14.1	15.9
	8	35.5	18.5	25.9	33.5
Forest floor Nitrate-N	0	20.6ab	30.5b	25.3ab	14.2a
	2	10.0	8.2	6.1	7.8
	4	13.8	8.8	20.8	15.2
	8	19.8	12.7	17.3	13.7
Soil Nitrate-N	0	2.06bc	2.62c	1.24ab	0.69a
	2	0.41	0.03	0.11	0.34
	4	1.21bc	0.72ab	0.29a	1.79c
	8	1.07	0.78	0.17	0.50

Row means followed by a different letter are significantly different at an alpha = 0.05 level.

the four types. Also, because organic N added to the sites in sludge application was small compared with organic N native to the forest floors and soils of the four types, changes in total N could not be used to make mineralization estimates. Because of the high ammonium-N and low organic N contents of the sludge used in this study, min-

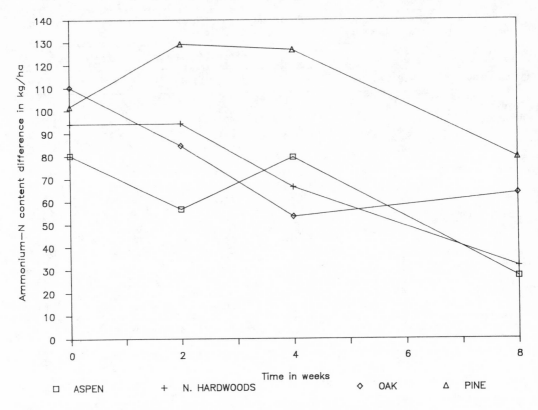

Figure 2. Sludge-control ammonium-N content differences.

eralization of sludge organic N was apparently of minor importance in controlling nitrate-N and ammonium-N levels in the cores.

Field Incubation

A comparison of the results of the field and laboratory incubations for the oak and pine forest types is presented in Table 7. In the field incubation, nitrate-N levels in sludge-treated pine cores were less than for oak, as was the case in the laboratory. Both oak and pine field nitrate-N levels were reduced below laboratory levels. No significant nitrification appears to have occurred in the pine field samples. Ammonium-N concentrations were similar in both the field and laboratory sludge incubations. For the control cores, mineralization in field incubations was lower for both types, as evidenced by the lower ammonium-N levels in the control cores at four and eight weeks.

The reduced nitrate-N levels in sludge-treated cores and lowered ammonium-N content in control cores in field incubations were probably a result of reduced rates of mineralization and nitrification caused by the lower average temperature encountered by field cores, especially in the later weeks of the experiment. This study was carried out in August and September 1984. A preliminary experiment was performed in the first portion of July 1984, when daily temperatures at the study site were higher. No difference in nitrate-N or ammonium-N contents was found between laboratory and field cores incubated for ten days at that time.

TABLE 7. Comparison of mean nitrate-N and ammonium-N contents of field and laboratory incubations.

Forest Type and Treatment	Time (weeks)	Nitrate-N (kg/ha)		Ammonium-N (kg/ha)	
		Field	Laboratory	Field	Laboratory
Pine-Sludge	0	3.5	3.5	102	102
	2	2.7	5.2*	101	144*
	4	2.5	24.0*	152	158
	8	3.4	20.1*	153	153
Pine-Control	0	1.5	1.5	6.2	6.2
	2	0.7	0.7	10.2	14.7
	4	2.0	3.0	13.2	31.1*
	8	1.0	1.2	26.4	73.8*
Oak-Sludge	0	3.2	3.2	110	110
	2	4.2	13.7*	103	102
	4	17.2	34.1*	113	110
	8	33.7	48.4	123	112
Oak-Control	0	3.2	3.2	11.7	11.7
	2	1.4	0.5	19.6	17.2
	4	1.4	1.6	122.8	56.3*
	8	1.9	1.3	33.2	69.8*

*Indicates a significant difference between laboratory and field incubation means at an alpha = 0.05 level.

Soil and Groundwater Quality

Lysimeters and wells were installed to monitor the quality of soil leachate and groundwater following sludge application to the four forest types in the more extensive forest fertilization demonstration project initiated by Michigan State University in 1981. In that study, average peak nitrate-N concentrations for suction lysimeters installed at a depth of 120 cm (47.2 inches) in sludge-treated plots were 21.1 mg/l at the aspen site, 2.4 mg/l at the northern hardwoods site, 3.7 mg/l for the oak site, and 15.9 mg/l at the pine site. Peak nitrate-N concentrations in groundwater monitoring wells located on sludge-treated plots, as of December 1984, were 11.1 mg/l at the aspen site, 0.6 mg/l at the northern hardwoods site, and 3.0 mg/l at the pine site (Urie et al. 1984).

These results cannot be predicted by incubation experiment nitrate-N contents for the four types alone, indicating that factors other than short-term nitrification rates must be taken into account in determining the potential for nitrate-N leaching following sludge application to such sites. Possible explanations for the elevated nitrate-N levels in groundwater and soil leachate at the pine and aspen sites compared with oak and hardwood include: (1) differential abilities of the four ecosystems to utilize nitrate-N as a result of differences in plant species, age, and stocking; (2) retention of a larger proportion of added sludge in the thicker forest floors of the oak and hardwood sites, where increased losses through ammonia-N volatilization would reduce the amount of ammonium-N available for nitrifiers; and (3) increased competition for ammonium-N among plants, nitrifiers, and other microbes at the oak and hardwood sites.

SUMMARY AND CONCLUSIONS

Sludge addition promoted nitrification in the forest floors and surface soils of four forest types that showed no nitrification without sludge addition. This suggests that the organisms responsible for a majority of the nitrification were added with the sludge. It is also felt that sludge addition enhanced nitrification by creating microsites with favorable pH and ammonium-N availability.

Nitrification in the pine type during laboratory incubation was less than for the other types, as was ammonium-N loss. The acidic conditions existing at this site are believed to be at least partly responsible. Nitrate-N in soil leachate and groundwater following sludge application at the four sites could not be predicted by incubation results, indicating factors other than nitrification need to be taken into account in estimating nitrate-N leaching losses following sludge application.

Rates of nitrification in sludge-treated cores and mineralization in control cores were less for field incubations than for laboratory incubations, presumably owing to lower temperatures in the field. Field sludge-treated plots showed evidence of large ammonium-N losses, evidently a result of ammonia-N volatilization.

As a result of ammonium-N losses in sludge-treated, laboratory-incubated cores, total inorganic N increased less over time than it did in control cores, making mineralization of sludge organic N impossible to estimate. It is felt that mineralization of sludge organics provided a minor amount of ammonium-N compared with that added in sludge treatment. In surface application of liquid sludge, it is believed that sludge ammonium-N is primarily responsible for nitrate-N leaching losses, making it desirable to use sludges low in ammonium-N. This would allow a higher loading rate for a given risk of nitrate-N leaching. Also, most of the applied N would be in organic form, which would be slowly released instead of lost to the atmosphere through ammonia-N volatilization.

ACKNOWLEDGMENTS

Although the information in this document has been funded in part by the U.S. Environmental Protection Agency under assistance agreement No. S005551-01 to the Michigan Department of Natural Resources and Michigan State University, it has not been subjected to the Agency's publication review process and therefore may not necessarily reflect the views of the Agency, and no official endorsement should be inferred. Mention of trade names or commercial products does not constitute endorsement or recommendation for use.

Special appreciation is extended to Connie Bobrovsky and Theresa Clark for their assistance in sample preparation and analysis, and to Dr. Phu V. Nguyen for his assistance in data analysis.

Michigan Agricultural Experiment Station journal article number 11863.

REFERENCES

Black, C. A. (ed.) 1965. Methods of soil analysis: Part 2. American Society of Agronomy, Madison, Wisconsin.

Bremmer, J. M. 1965. Nitrogen availability indexes. In C. A. Black (ed.) Methods of soil analysis: Part 2. American Society of Agronomy, Madison, Wisconsin.

Cassman, K. G., and D. N. Munns. 1980. Nitrogen mineralization as affected by soil moisture, temperature, and depth. Soil Sci. Soc. Am. J. 44:1233–1237.

Donovan, W. C., and T. J. Logan. 1983. Factors affecting ammonia volatilization from sewage applied to soil in a laboratory study. J. Environ. Qual. 12:584–590.

Epstein, D., D. B. Keane, J. J. Meisinger, and J. O.

Legg. 1978. Mineralization of nitrogen from sewage sludge and sludge compost. J. Environ. Qual. 7:217–221.

Focht, D. D., and W. Verstraete. 1977. Biochemical ecology of nitrification and denitrification. Adv. Microbiol. Ecol. 1:135–211.

Foth, H. D. 1978. Fundamentals of soil science. 6th ed. John Wiley and Sons, New York.

Gilmour, J. T. 1984. The effects of soil properties on nitrification and nitrification inhibition. Soil Sci. Soc. Am. J. 48:1262–1266.

Hull, C. H., and N. H. Nie. 1981. SPSS update 7-9 new procedures and facilities for releases. McGraw-Hill, New York. 402 p.

Lindemann, W. C., and M. Cardenas. 1984. Nitrogen mineralization potential and nitrogen transformations of sludge-amended soil. Soil Sci. Soc. Am. J. 48:1072–1077.

Magdoff, F. R., and F. W. Chromec. 1977. Nitrogen mineralization from sewage sludge. J. Environ. Qual. 9:451–455.

Nie, N. H., C. H. Hull, J. G. Jenkins, K. Steinbrenner, and D. H. Bent. 1975. SPSS statistical package for the social sciences. 2nd ed. McGraw-Hill, New York. 675 p.

Parker, C. F., and L. E. Sommers. 1983. Mineralization of nitrogen in sewage sludges. J. Environ. Qual. 12:150–156.

Sabey, B. R., N. N. Agbim, and D. C. Markstrom. 1975. Land application of sewage sludge: III. Nitrate accumulation and wheat growth resulting from addition of sewage sludge and wood wastes to soils. J. Environ. Qual. 4:388-393.

Soil Conservation Service. 1979a. Rubicon series. Nat. Coop. Soil survey. USDA.

————. 1979b. Graycalm series. Nat. Coop. Soil survey. USDA.

————. 1982a. Mancelona series. Nat. Coop. Soil survey. USDA.

————. 1982b. Grayling series. Nat. Coop. Soil survey. USDA.

————. 1984. Montcalm series. Nat. Coop. Soil survey. USDA.

Sommers, L. E. 1977. Chemical composition of sewage sludges and analysis of their potential use as fertilizers. J. Environ. Qual. 6:225–232.

Sommers. L. E., and D. W. Nelson. 1981. Nitrogen as a limiting factor in land application of sewage sludges. p. 425–488. In Fourth Annual Madison Conference of Applied Research and Practice on Municipal and Industrial Waste. Department of Engineering and Applied Science, University of Wisconsin, Madison.

Stanford, G., and S. J. Smith. 1972. Nitrogen mineralization potentials of soils. Soil Sci. Soc. Am. J. 36:465–472.

Technicon Industrial Method. 1971. Method no. 108-71W. Technicon Industrial Systems, Tarrytown, New York.

————. 1977a. Individual/simultaneous determination of nitrogen and/or phosphorus in BD acid digests. Method no. 334-74W/B. Technicon Industrial Systems, Tarrytown, New York.

————. 1977b. Nitrate and nitrite in water and wastewater. Method no. 102-70W/C. Technicon Industrial Systems, Tarrytown, New York.

Terry, R. E., D. W. Nelson, and L. E. Sommers. 1981. Nitrogen transformations in sewage sludge-amended soils as affected by soil environmental factors. Soil Sci. Soc. Am. J. 45:506–513.

Urie, D. H., J. B. Hart, P. V. Nguyen, and A. J. Burton. 1984. Hydrologic and water quality effects from sludge application to forests in northern lower Michigan. Annual progress report. Department of Forestry, Michigan State University, East Lansing. 97 p.

Chapter 15

Effect of Sewage Sludge from Two Sources
on Element Flux in Soil Solution
of Loblolly Pine Plantations

C. G. WELLS, C. E. MURPHY, C. DAVIS, D. M. STONE,
and G. J. HOLLOD

ABSTRACT   Sludge was applied to plantations at establishment and ages 3, 9, and 28 years to provide 400 and 800 kg/ha (360 and 720 1b/acre) of nitrogen from a liquid anaerobic source containing approximately 7% N (oven dry) and 630 kg N/ha (560 lb/acre) from a solid aerobic source which was about 1.3% N (oven dry). Ammonium and quickly mineralized organic N in the liquid sludge caused nitrate-N in soil water at 1 m (3.3 ft) depths to exceed 60 mg/l four months after application at the 800 kg N/ha rate for treatments applied at time of establishment and in 28-year-old stands. Nitrate-N then fluctuated from 10 to 20 mg/l for about eighteen months after application, when it declined to near the nontreated level. Estimated eighteen-month leaching of nitrogen past the 1 m depth was 7, 18, and 22% of the 800 kg N/ha applied as liquid sludge to the 3-, 9-, and 28-year-old stands, respectively. Except for application on the establishment age stand, nitrate-N was less than 5 mg/l for 400 kg N/ha of liquid sludge and 630 kg N/ha of solid sludge. Leaching of calcium and magnesium was related to application rates and nitrate leaching. Zinc, copper, lead, and cadmium concentrations in the soil solution were not significantly increased at the 0.5 and 1 m soil depths.

Sewage sludge is of high value as fertilizer, and its use in forests can improve productivity and provide an effective waste management alternative. It is generally assumed that forests can tolerate greater amounts of potentially toxic materials than agronomic systems; however, information on the movement of nutrients and heavy metals in the soil solution of forest systems is scarce where application rates approximate fertilization rates.

In studies of Douglas-fir (*Pseudotsuga menziesii* [Mirb.] Franco) forests and in clearcut areas, large amounts of nitrate leached from the soil into groundwater in the first seven months after application of 5,000 or 10,000 kg N/ha in 10 and 25 cm of dewatered sludge (Riekerk 1981, Breuer et al. 1979). The studies were near Seattle, Washington, on glacial outwash soils; therefore, conditions for leaching nitrate from such large applications were favorable.

Application of sewage sludge to forests in Michigan indicated substantial movement of nitrate-N into groundwater below the rooting zone after application of 23 Mg/ha or more municipal sludge containing 1,380 kg N/ha or more total N (Brockway and Urie 1983). Leaching of cations, especially calcium, has been associated with nitrate movement through the profile of forest soils (Van Miegroet and Cole 1984, Riekerk and Zasoski 1979).

The objective of this investigation was to determine the movement of elements in soil solution following the application of sewage sludge to loblolly pine stands. The estimates of leaching indicate the fertility benefits to the soil-tree system and the potential for water contamination.

MATERIALS AND METHODS

Sewage sludge studies were established in loblolly pine (*Pinus taeda* L.) plantations at planting time and ages 3, 9, and 28 years. The studies were on the Savannah River Plant, Aiken, South Carolina (Wells et al. 1984). The soils, ranging from clayey to sandy, were upland, well-drained, moderately permeable, acid, and low in CEC and fertility (Table 1). The trees treated at planting time, and one study area for the 3-, 9-, and 28-year-old plantations, were on loamy sand with the duplicate study on a soil with a more sandy B horizon (Table 2). The establishment age study was not duplicated by soil. Orangeburg

TABLE 1. Soil properties of the study areas.

Soil	Organic Matter	Nitrogen	Phosphorus	Potassium	Calcium	Magnesium	pH
			(kilograms/hectare)				
Forest floor							
Age 9	6,400	95	4.3	5.0	33	5.5	4.7
Age 28	24,100	294	15.2	12.8	68	15.0	5.4

Mineral soil	Organic Matter	Clay	pH			
	0.9-1.5%	8-10%	5.0-5.2			

0 to 10 cm (range for studies)	Total N (mg/kg)	Ext P (mg/kg)	CEC (c mol/kg)	Potassium	Calcium (c mol/kg)	Magnesium
	245-350	16-36	1-2	0.02-0.06	0.5-0.6	0.1-0.2

TABLE 2. Stand characteristics and treatment date by age of stand.

Age	Date Applied	Density (trees/ha)	Height (m)	Basal Area (m^2/ha)
1	3/81	1,200	--	--
3	10/81	900	1.7	--
9	9/81	963	5.6	6.2
28	7/81	484	18.4	22

and Wagram series represent the less sandy soils, whereas Dothan and Fuquay are series with more sandy B horizons. The soils are classified as fine-loamy, siliceous, thermic Typic and Arenic Paleudults. Soil properties of the study area and stand characteristics are given in Tables 1 and 2.

The study was designed to add specified quantities of nitrogen to the study plots (Table 3). Treatments consisted of liquid sludge as L 400 or L 800 kg N/ha (360 or 720 lb/

TABLE 3. Average concentrations and application rates for liquid and solid sludge.

Sludge Property	Average Concentrations		Application Rates		
	Liquid	Solid	Liquid		Solid
			Low N (400)	High N (800)	
	(%)	(%)	(kg/ha, oven-dry)		
Sludge, oven-dry	2.48	55.72	5,555	11,110	49,925
Ash, oven-dry	46.02	36.11	2,556	5,113	18,027
Carbon, oven-dry	23.2	25.2	1,300	2,600	12,500
	(mg/kg, oven-dry)		(kilograms/hectare)		
Kjeldahl-N	72,374	12,676	402	804	632
Ammonium-N	30,361	240	169	338	12
Nitrate-N	149	0	0.82	1.64	0
Phosphorus	16,209	7,546	90	180	377
Potassium	2,661	305	15	30	15
Calcium	14,556	20,898	81	162	1,043
Magnesium	2,460	3,486	14	28	174
Sodium	20,926	2,939	116	232	147
Manganese	137	300	0.76	1.52	15
Zinc	1,330	2,158	7.39	14.78	108
Copper	318	1,249	1.77	3.54	62
Lead	238	176	1.32	2.64	8.79
Nickel	44	43	0.24	0.48	2.15
Cadmium	44	6	0.24	0.48	0.29
Sulfur	5,967	3,644	33	66	181
Iron	10,808	42,449	565	1,130	2,119
Boron	254	268	1.41	2.82	13
Chromium	173	1,386	0.96	1.92	69

acre), or solid sludge as S 630 kg N/ha (560 lb/acre), diammonium phosphate (DAP), surface-applied or disked (D), and with herbicide (H) or without. Control received no fertilizer.

The liquid was an anaerobic processed sludge mainly of municipal origin, whereas the aerobic processed solid sludge contained significant amounts of waste from textile mills. Since the two sludges differed in nitrogen as well as other element concentrations, so did the quantities of these materials applied (Table 3). The pH averaged 7.3 in the liquid sludge and 7.4 in the dry sludge. The carbon to nitrogen ratio was strikingly different in the two materials—3.2 for the liquid and 19.8 for the dry.

Solid sludge was applied with conventional farm equipment, including a 70 hp farm tractor and a manure spreader. Liquid sludge was applied with pressure sprays from tankers. Treatment plots covered approximately 0.2 ha (0.5 acre), but measurements were confined to interior areas of approximately 0.1 ha (0.25 acre).

Fritted glass suction lysimeter tubes (Long 1978) were installed at 0.5 and 1 m depths in the soil. These lysimeters were constructed in such a way that solution contacted only glass and polyethylene tubing. In the 28-year-old stands, two lysimeters were installed at 0.5 m and two at 1 m depths in each plot. One lysimeter was installed at each depth only in the soil with the most clay for the age 3 and age 9 stands. Some lysimeters failed during operation, but data pooled across soils were sufficient for statistical analyses.

When soils were sufficiently moist, water samples were collected monthly from the time of treatment in 1981 until June 1983, and then from February through May 1984 in all except the youngest stand. To extract soil solution, the lysimeters were pumped empty and 0.05 MPa (0.5 bar) vacuum was applied to the system approximately 24 hours before the solutions were collected. From 10 to 50 ml of solution were collected. The samples were split: one portion was shipped with cold packs to the Forestry Sciences Laboratory at Research Triangle Park, North Carolina, for pH, nitrate, ammonium, phosphorus, and sulfate analyses, and the other portion was preserved with nitric acid and shipped to the University of Georgia Institute of Ecology, Athens, Georgia, for nutrient cation and trace metal analyses. Inorganic N and sulfate ions were determined by autoanalyzer, the other elements by atomic absorption and ICAP methods.

Sludge samples were dried at 105°C and ashed at 475°C in a muffle furnace. The ash was dissolved in concentrated nitric acid, evaporated at near the boiling point, and the residue dissolved in 0.3 M nitric acid. Metals were analyzed by flame atomic absorption and phosphorus by colorimetric methods. Kjeldahl-N was determined colorimetrically after block digestion.

Statistics

Data from soil solution analyses were pooled by three-month periods beginning in April 1981. Treatment effects by age and lysimeter depth were tested by analysis of variance and Duncan's multiple range tests. Comparisons for a plantation age could not be made, because sludge application dates were March, July, September, and November, respectively, for the plantations of age 1, 28, 9, and 3.

Water Drainage and Element Leaching

Water balance estimates were used to calculate water flux (drainage) at the 0.5 and 1 m depths. Nutrient leaching was estimated as the product of water flux times the element concentration in the soil solution at 0.5 and 1 m.

In the forested, permeable soils of this area, overland flow of water and lateral drainage at depth above 1 m can usually be ignored in water balance calculations. The change in water content of the soil column above a lysimeter is the difference between the precipitation and the sum of the evaporation, interception of precipitation, and the vertical drainage from the column.

The close proximity of a detailed evaporation study allowed evaporation to be estimated with great confidence for the forested sludge plots (Murphy et al. 1981, Murphy 1985). Interception and rainfall relationships have been determined for loblolly pine plantations of similar age and density (Rogerson 1966). Vertical drainage was estimated with the model proposed by Davidson et al. (1969). The results of these calculations were judged to be adequate to estimate drainage for periods of a week to a year on the basis of the agreement with measured soil water contents taken on one intensively measured stand.

RESULTS

Nitrates in Soil Solution

The concentration of nitrate-N in soil solution was the greatest for the stand treated at planting (age 1) followed by ages 28, 9, and 3 (Figure 1). Except for L 800 treatment, concentration was 10 mg/l or less for all three-month sample periods for ages 3, 9, and 28. The effect of low nitrogen and high carbon concentration is evident in the low nitrate-N levels for the solid sludge in all age classes.

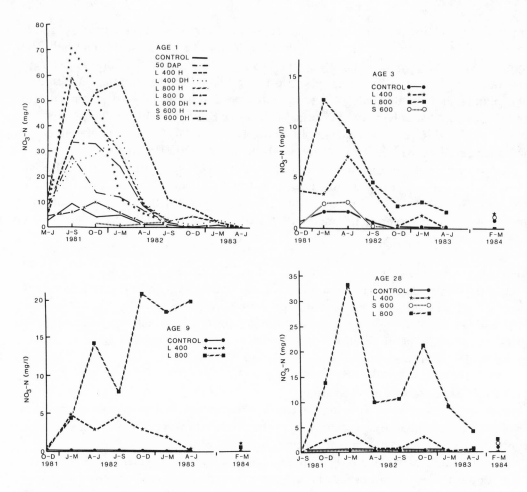

Figure 1. Nitrate-N concentration in soil solution at 1 m following application of two sources of sludge to 1-, 3-, 9-, and 28-year-old loblolly pine plantations.

In the age 1 stand, nitrification and leaching of nitrate occurred soon after application. The first samples, collected early in May after treatments in March, showed nitrate-N levels in the 50 and 70 mg/l range at 0.5 m and from 5 to 20 mg/l at 1 m. After more rain, nitrates leached to the 1 m depth in concentrations exceeding 70 mg/l for the high rate of liquid sludge that had been disked into soil.

At the L 800 treatment, disking significantly increased leaching of nitrate-N at the 0.5 m but not at the 1 m depth. At the L 400 rate, disking tended to decrease nitrate concentration, and the effect was significant in the spring and summer one year after application. All the treatment plots with lysimeters, except one at L 800, were also treated with herbicide. Where the weedy vegetation was allowed to grow, nitrate-N in the soil solution at 1 m in the July to September period averaged 34 mg/l compared with 72 mg/l where weeds were controlled. This was the only three-month period that the nonherbicide treatment was significantly lower in nitrate-N than where weeds were controlled. Competition and tree mortality during establishment were not severe without weed control on this site.

With the exception of L 400 without disking treatment, nitrate-N at 1 m had declined to below 100 mg/l twelve months after application. At 0.5 m, nitrate-N did not exceed 10 mg/l after December 1981. The delayed peak and decline for the L 400 without disking could not be attributed to unusual replicate data or soil variability. Delayed nitrification may be expected from surface application; however, large within-treatment variation is the only explanation for the lack of similarity in the high- and low-rate liquid treatments applied to the surface. The surface-applied and disked treatments were similar in nitrate-N at the 0.5 m depth, indicating that variation in soil drainage could have caused the sustained high nitrate-N at 1 m for the nondisked application.

For the 28-year-old stand, nitrate-N in the soil solution at 1 m was ten times greater for the L 800 treatment than for the L 400. The highest level of nitrate-N at the 0.5 m depth was 253 mg/l in October to December for the high rate treatment and 26 kg/ha in July to September for the L 400 treatment. The system's nitrogen loading capacity was exceeded by the 800 kg N/ha rate, but the 400 kg N/ha failed to approach the loading capacity. Solid sludge at 630 kg N/ha released no nitrate-N at the 0.5 or 1 m depth. Nitrate-N levels were less than 1 mg/l for all treatments in samples collected from January to June 1984, thirty months after application.

Tree roots apparently took up a large part of the nitrate-N before it leached to the 1 m depth. In the age 28 stand where nitrate-N was a maximum of 253 mg/l at 0.5 m, the maximum at 1 m was 33 mg/l; in the age 1 stand, nitrate-N decreased from 147 to 72 mg/l between the 0.5 and 1 m depths.

In the age 3 stand, nitrate-N averaged 12 mg/l at 1 m for the L 800 treatment in the period January to March after application in October. Lysimeters were installed in only the higher clay soil of the age 3 stands. With large variation between lysimeters in the same treatments, the April to June period of 1982 was the only time that treatments were statistically different. The results show a large capacity for the system to utilize or immobilize the applied N.

As in age 3 stands, lysimeters were installed in only the higher clay soil at age 9, and variation was large between lysimeters. A more sustained nitrate-N level was found in the soil solution at 1 m for the age 9 stands than for those younger. The delay in nitrate-N release probably was associated with slow nitrogen mineralization of the sludge applied to the litter layer, which had only begun to accumulate in the age 9 stands. All samples were less than 1 mg/l in the January to June period of 1984, twenty-eight months after application.

Nutrient Element Concentrations in Years 1 and 2 After Treatment

The average ion concentrations of nitrate-N, ammonium-N, potassium, calcium, magnesium, and sulfate were calculated for the periods from application to June 30, 1982, and July 1, 1982 to June 30, 1983, and designated year 1 and year 2, respectively. Average nitrate-N at 0.5 m was less than 2 mg/l in year 2 for all treatments in ages 1 and 28 stands except for L 800 in the 28-year-old stand (Table 4). The higher concentration in the older stand can be attributed to later application, July versus March, or possibly to less leaching in the stand with establishment trees. A concentration of 8 mg ammonium-N per liter at the 0.5 m depth indicates a different leaching pattern for the L 800 treatment in the 28-year-old stand.

At the 1 m depth, nitrate-N averaged 10 to 15 mg/l for the L 800 treatment in the 9- and 28-year-old stands and less than 6 mg/l for the highest treatment in ages 1 and 3 stands (Table 5). The age 3 stand, with sludge applied in October 1981, was also treated with

TABLE 4. Concentration of components in soil solution at 0.5 m during year 1 and year 2* following sludge application.

Stand Age	Treatment	Nitrate-N		Ammonium-N		Potassium		Calcium		Magnesium		Sulfate-S	
		1	2	1	2	1	2	1	2	1	2	1	2
						(milligrams/liter)							
1	Control H	10.1c	0.2cd	0.2c	0.02a	1.2b	0.58ab	20bc	2.2d	5.7bc	0.6bc	2.8b	2.3b
	DAP 50 NH	23.2c	0.1d	0.1c	0.02a	1.9b	0.36b	23bc	2.4d	7.4bc	0.5c	1.5b	3.3b
	L 400 H	29.0bc	1.0abcd	0.2c	0.01a	1.9b	0.60ab	23bc	4.0d	5.4bc	0.9bc	1.7b	1.9b
	L 400 DH	22.5c	0.6bcd	0.2bc	0.08a	0.9b	0.12b	28bc	2.9d	8.8bc	0.7bc	1.8b	2.9b
	L 800 H	47.6b	1.8a	1.2a	0.20a	6.6a	0.96a	42b	5.5bc	10.9b	1.2b	4.4b	2.3b
	L 800 D	22.7c	0.2cd	0.2bc	0.18a	0.7b	0.46b	27bc	4.0cd	6.8bc	0.8bc	2.7b	2.9b
	L 800 DH	74.2a	1.2abc	0.7ab	0.18a	4.0ab	0.43b	74a	3.9cd	20.8a	1.0bc	1.7b	2.3b
	S 630 H	14.8c	0.7bcd	0.1c	0.12a	1.2b	0.11b	18bc	8.5a	3.5c	2.1a	9.4a	8.2a
	S 630 DH	10.2c	1.3ab	0.1c	0.03a	0.9b	0.40b	13c	6.1b	3.7c	1.7a	9.0a	3.7b
28	Control	0.2b	0.0b	0.08b	0.04a	--	0.24abc	4.0b	2.9c	0.8b	0.6c	3.4b	3.4b
	L 400	6.5b	0.7b	0.15b	0.03a	--	0.18c	8.0b	4.6c	2.6b	1.1c	10.1ab	5.4b
	L 800	114.1a	9.5a	8.74a	0.04a	--	0.67a	61.7a	17.1a	18.1a	4.2a	10.3ab	10.1a
	S 630	0.4b	0.1b	0.19b	0.06a	--	0.48ab	4.3b	11.0b	0.9b	2.6b	18.5a	6.2b

Within stand age, values in a column followed by a common letter are not significantly different at the 0.05 level.

*Year 1 = First sample May 1981 to June 1982. Year 2 = July 1982 to June 1983.

TABLE 5. Concentration of components in soil solution at 1 m during year 1 and year 2* following sludge application.

Stand Age	Treatment	Nitrate-N		Ammonium-N		Potassium		Calcium		Magnesium		Sulfate-S	
		1	2	1	2	1	2	1	2	1	2	1	2
		(milligrams/liter)											
1	Control H	4.4b	0.5b	0.07ab	0.03b	0.4b	0.0b	14.5bcd	3.1bc	3.0bcd	0.6b	2.2a	2.0b
	DAP 56N H	12.9ab	0.7b	0.05ab	0.01b	0.4b	0.2ab	14.0bcd	3.6bc	3.2bcd	0.9b	1.4ab	1.7b
	L 400 H	25.8a	5.7a	0.11ab	0.05b	0.3b	0.1b	32.6abc	8.2a	9.5a	2.4a	1.8ab	1.4b
	L 400 DH	16.7ab	2.3b	0.09ab	0.62a	0.2b	0.3ab	22.7bcd	4.5b	7.8ab	2.2a	1.8ab	1.5b
	L 800 H	26.8a	3.8b	0.12a	0.02b	1.4a	0.2ab	51.5a	6.1ab	8.4a	1.0b	1.2b	1.7b
	L 800 D	26.2a	0.5b	0.09ab	0.11b	--	0.2ab	19.8bcd	3.8b	6.0abc	0.8b	1.7ab	3.1ab
	L 800 H	27.6a	1.4b	0.08ab	0.05b	1.9a	0.4a	37.5ab	4.2bc	7.9ab	0.8b	1.5ab	2.6ab
	S 630 DH	3.3b	1.8b	0.04b	0.05b	0.2b	0.1b	4.2d	3.9b	1.0c	1.0b	2.3a	4.2a
	S 630 DH	5.5b	1.1b	0.07ab	0.06b	0.3b	0.1b	8.6cd	2.5b	2.4cd	0.8b	2.1a	2.6ab
3	Control H	1.3b	0.3b	0.06b	0.04b	--	0.05a	2.6a	1.8b	1.1a	0.6a	1.7ab	1.4a
	L 400 H	4.2b	0.3b	0.42a	0.30a	--	0.14a	11.3a	1.6b	4.6a	0.9a	1.1b	1.2a
	L 800 H	9.6a	2.7a	0.07b	0.02b	--	0.05a	10.2a	6.7a	2.4a	1.7a	0.6b	1.2a
	S 630 H	2.0b	0.1b	0.09b	0.11b	--	0.13a	15.1a	3.9ab	4.6a	1.1a	3.2a	3.7a
9	Control	0.1a	0.0b	0.05a	0.01a	--	0.06a	1.1	1.0b	0.2b	0.2b	0.7a	1.2a
	L 400	4.1a	2.8b	0.13a	0.04a	--	0.08ab	4.0a	3.1b	1.3ab	0.9b	0.4a	1.0ab
	L 800	5.6a	15.6a	0.11a	0.02a	--	0.14a	15.2a	8.7a	4.8a	6.0a	0.5a	0.3b
28	Control	0.2b	0.0b	0.12b	0.04b	--	0.08a	2.2b	2.2b	0.9b	0.9b	2.2b	3.7ab
	L 400	2.8b	0.3b	0.28ab	0.22a	--	0.06a	4.3b	2.5b	1.3b	0.9b	1.7b	3.6ab
	L 800	22.9a	9.7a	0.54a	0.05b	--	0.06a	25.6a	6.9a	9.6a	3.0a	1.3b	2.2b
	S 630	0.2b	0.1b	0.09b	0.02b	--	0.15a	4.4b	4.8ab	1.5b	1.5b	5.1a	8.1a

Within stand age, values in a column followed by a common letter are not significantly different at the 0.05 level.

*Year 1 = First sample May 1981 to June 1982. Year 2 = July 1982 to June 1983.

herbicide, but sufficient understory vegetation survived to utilize large quantities of nitrogen. In the age 1 treatment without herbicide (L 800 D), nitrate-N was about one-third that of the same sludge treatment with herbicide for the 0.5 m depth in year 1; however, at the 1 m depth, there was no difference in nitrate-N levels as a result of herbicide treatment.

The concentrations of calcium in soil solution for the age 1 control treatment at 0.5 m and 1 m depths were 20 and 14 mg/l, respectively. The only significantly greater concentration was for the L 800 D treatment. The S 630 D treatment with lower nitrate-N had consistently lower calcium and magnesium levels than the control and other sludge treatments. Potassium, although at lower concentrations than calcium and magnesium, followed the same trends with treatments in the age 1 stand.

Base cation concentrations were lower in the control of the ages 3, 9, and 28 stands than in age 1, with an average of 1.7 and 0.6 mg/l, respectively, for calcium and magnesium for the three stands over the two years. Concentrations of 1.0 and 3.6 mg Ca/l and 0.8 and 1.5 mg K/l were found in soil solution at 1.2 m for nonfertilized and fertilized 14-year-old loblolly pine plots in the Piedmont of North Carolina (Wells et al. 1975). Nitrate-N in soil solution at 1.2 m averaged 2.1 mg/l the first year after fertilization, with 226 kg N/ha as ammonium nitrate. Excluding the high sludge rate, the results are similar for the fertilization and sludge studies on all except the age 1 stand.

Sulfate-S was significantly increased in soil solution at 1 m by the S 630 treatment in the 28-year-old stand but not in the younger stands. The concentration at 1 m was less than that at 0.5 m for those treatments where sulfate-S was increased; however a major part of the added sulfate was immobilized in the soil above 0.5 m.

Trace Elements and Phosporus in Soil Solution

Zinc, copper, cadmium, and chromium concentrations in soil solutions at 0.5 and 1 m were not influenced by sludge treatments (Table 6) when data were pooled for three-month periods or for the entire sampling period. The significant treatment effect for lead at age 1 was the result of 119 µg/l concentration in treatment S 630 D, which was significantly greater than for any other treatment. This higher concentration of lead was attributed to the influence of one plot; however, sludge applied to that plot was not higher in lead than in other plots receiving the same treatment. This unusually high lead concentration was probably caused by contamination of the lysimeter before or during installation.

In the age 28 stand, nickel, iron, and aluminum were significantly greater, and cobalt lesser for the L 800 treatment at the 0.5 cm depth, but concentrations were not different at the 1 m depth.

The significantly higher manganese concentrations are a result of an increase from the L 800 treatment and a decrease from the S 630 treatment in the age 1 stand. In the age 28 stand, concentrations of about 2,000 µg/l were found at 0.5 m depth for L 800 in the more sandy soil the first year after treatment. Considering that only 1.5 kg Mn/ha were applied in the sludge, the elevated levels must be a result of soil reaction.

The four age classes varied in average concentrations of trace elements at the 1 m depth. This was probably a soil and not a vegetation effect, because there are no strong trends relating concentrations or treatment effects with stand age. As expected, the indicated manganese increase from treatment on soil solution at 1 m occurred in the age 1

TABLE 6. Means for some chemical elements in soil solutions at 0.5 and 1.0 m over sample period.

Age	Date	Depth	Zn	Cu	Cd	Pb	Ni	Cr	Co	Fe	Al	Mn	P†
		(m)					(micrograms/liter)						
1	7/81-6/83	0.5	152	6.0	2.1	33	21	31.4	5.5	14.8	116	360**	8.6
		1.0	114	5.4	2.2	44*	24	10.0	5.7	8.7	28	30**	6.2
3	10/81-6/83	1.0	113	25.7	15.9	103	67	10.1	16.4	20.1	67	19	10.1
9	10/81-6/83	1.0	62	7.6	4.7	96	52	24.8	9.7	13.8	138	13	4.9
28	7/81-6/83	0.5	84	4.5	3.8	66	39**	15.2	6.5**	16.2**	130**	529**	8.6
		1.0	71	2.3	3.5	67	34	14.0	5.7	23.1	39	12	15.7

*, **Significant treatment effect at 0.05 and 0.01 level for treatment by depth.
†Phosphorus analyses from September 1982 to June 1983.

TABLE 7. Leaching at 0.5 m depth for year 1 and year 2* following sludge application.

Stand Age	Treatment	Nitrate-N		Ammonium-N		Potassium		Calcium		Magnesium		Sulfate-S		Phosphorus	
		1	2	1	2	1	2	1	2	1	2	1	2	1	2
								(kilograms/hectare)							
1	Control D	65	3	1.5	0.3	8	8	54	25	9	8	15	29	0.2	0.1
	DAP 56N H	97	2	1.1	0.2	12	6	104	30	53	6	16	43	0.2	0.1
	L 400 H	144	12	1.1	0.1	10	9	81	42	18	9	17	28	0.2	0.1
	L 400 DH	95	8	1.9	0.7	4	6	88	33	22	9	20	49	0.2	0.1
	L 800 H	225	24	2.8	1.7	48	14	87	61	21	14	44	33	0.3	0.1
	L 800 D	78	3	3.3	1.3	8	6	103	43	26	9	21	40	0.1	0.1
	L 800 DH	400	15	4.3	2.0	21	6	258	41	68	11	14	31	0.2	0.1
	S 630 H	77	8	0.4	1.0	6	5	159	102	20	26	100	117	0.2	0.0
	S 630 DH	50	21	0.2	0.4	7	8	96	82	21	24	89	56	0.2	0.0
28	Control	1	0	0.4	0.4	8	5	25	34	5	6	12	43	0.1	0.0
	L 400	15	9	0.5	0.2	4	4	36	53	12	13	47	63	0.1	0.0
	L 800	430	101	40.0	0.5	24	17	293	209	86	51	53	120	0.1	0.1
	S 630	2	1	1.3	0.8	4	7	24	125	6	29	106	96	0.2	0.1

*Year 1 = First sample May 1981 to June 1982. Year 2 = July 1982 to June 1983.

stand. The greater concentrations for the manganese in the treated plots were still near that for a number of untreated plots of the older age classes.

Amounts of phosphorus in soil solution at 0.5 and 1 m were in the range of 10 to 40 μg/l, and there were no indications of leaching at the 0.5 or 1 m depths. Investigations being conducted will provide additional information on changes in elemental concentrations in the soil profile. The results from soil solution data are in agreement with other reports that indicate low mobility of the trace elements investigated in this study (Brockway 1983, Riekerk and Zasoski 1979).

Nutrient Leaching at 0.5 and 1 m Depths

Precipitation was 33% greater and water percolation was 100% greater through the 1 m depth in year 2 (July 1982 to June 1983) than in year 1 (July 1981 to June 1982). This higher percolation rate in year 2 caused greater than normal nutrient leaching rates for control and treated plots. Twenty-year average annual precipitation for the area is 1,196 mm (47.10 inches).

Estimates indicate as much as 400 kg N/ha leached during year 1 to the 0.5 depth of the age 1 and age 28 stands where 800 kg N/ha were applied as liquid sludge (Table 7). For the same extreme treatments, about 100 kg N/ha reached the 1 m depth (Table 8). A summary of data for L 800 treatments in age 1 and age 28 shows respectively that 63 and 34% nitrate-N at 0.5 m depth leached to the 1 m depth. The same percolation values were assumed for the age 1 stand as calculated for the age 28 stand, an assumption that probably underestimates percolation and leaching in the age 1 stand.

Comparison of the L 800 D and the L 800 DH treatments shows a large difference in leaching at 0.5 m but no differences at the 1 m depth. Denitrification, which may tend to equalize concentrations, offers an explanation for the lack of treatment differences in leaching at the 1 m depth when there were large differences at 0.5 m. Sludge treatments had relatively small effects on leaching of ammonium-N, potassium, and sulfate, and no effect on phosphorus.

Leached calcium and magnesium were related to nitrate, with the highest rates in the age 1 stand. These soils, naturally low in bases, could be adversely affected by leaching as a result of greatly increased inorganic anion mobility. Eighty and 160 kg Ca/ha were added at L 400 and L 800 rates, and increased leaching at 1 m was in excess of or near the calcium added for several liquid sludge treatments in the age 1 stand. In the established stands, leaching losses were less than application. Similar estimations indicate induced magnesium leaching in excess of application in all stands. Preliminary soils data also show trends toward accelerated leaching of bases within the profile for L 400 and L 800 treatments.

The leaching of sulfate was interesting in that leaching was greater in year 2 than year 1 for all treatments as a result of more percolation and greater concentrations at the 1 m depth. Annual phosphorus leaching was from 0.1 to 0.2 kg/ha with no effects of stand age, soil, or treatment.

Leaching rates declined sharply in year 2 compared with year 1. The increased nitrate leaching for the L 800 treatment in age 9 is a notable exception. Sustained large nitrate-N concentrations for the treatment and doubled percolation for the second year caused the extreme leaching.

Soil solution was sampled in the 3-, 9-, and 28-year-old plantations in February, March, April, and May 1984, nearly three years after sludge application, and in the age

TABLE 8. Leaching at 1.0 m depth for year 1 and year 2* following sludge application.

Stand Age	Treatment	Nitrate-N 1	Nitrate-N 2	Ammonium-N 1	Ammonium-N 2	Potassium 1	Potassium 2	Calcium 1	Calcium 2	Magnesium 1	Magnesium 2	Sulfate-S 1	Sulfate-S 2	Phosphorus 1	Phosphorus 2
								(kilograms/hectare)							
1	Control DH	18	1	0.2	0.2	1	1	40	22	6	4	8	15	0.1	0.0
	DAP 50N DH	42	4	0.2	0.1	1	3	44	26	11	6	4	13	0.1	0.0
	L 400 H	190	35	0.3	0.5	1	2	140	56	41	16	5	10	0.1	0.0
	L 400 DH	121	17	0.3	4.7	1	4	83	27	33	17	6	12	0.1	0.1
	L 800 H	138	20	0.5	0.1	3	3	193	41	30	6	5	13	0.1	0.0
	L 800 D	94	5	0.2	0.6	3	3	33	28	18	6	7	26	0.1	0.0
	L 800 DH	97	12	0.4	0.4	17	5	128	33	24	7	6	21	0.1	0.0
	S 630 H	7	4	0.2	0.4	2	2	15	33	3	8	5	36	0.1	0.0
	S 630 DH	30	7	0.4	0.3	3	1	27	17	8	6	12	20	0.1	0.0
3	Control	4	1	0.2	0.3	1	2	8	10	4	4	6	11	0.0	0.0
	L 400 H	11	4	1.6	1.4	0	2	53	13	20	10	3	7	0.1	0.0
	L 800 H	34	18	0.2	0.2	1	4	28	49	7	12	2	8	0.1	0.1
	S 630 H	5	0	0.3	0.7	0	6	10	27	3	7	10	38	0.0	0.1
9	Control	1	0	0.2	0.1	2	3	4	11	1	2	2	8	0.0	0.1
	L 400	14	18	0.4	0.2	0	2	14	21	5	6	1	7	0.0	0.0
	L 800	18	126	0.4	0.2	13	2	64	64	17	44	2	2	0.0	0.0
28	Control	1	0	0.3	0.2	4	2	7	17	3	7	7	8	0.1	0.1
	L 400	12	4	0.6	1.2	9	2	17	21	4	8	6	24	0.1	0.1
	L 800	97	83	1.0	0.3	6	2	86	50	33	22	5	20	0.1	0.1
	S 630	2	3	0.6	0.3	8	5	29	71	11	22	39	139	0.1	0.1

*Year 1 = First sample May 1981 to June 1982. Year 2 = July 1982 to June 1983.

28 stand in February 1985. In 1984, significant treatment effects at the 1 m depth were found for potassium in the age 3 stand and magnesium in the age 28 stand. No treatment effects were detectable in the February 1985 samples. Therefore, except for potassium and magnesium, where concentrations in the high treatments were 0.3 and 1.1 mg/l, leaching had declined to the nontreatment levels two years after application of sludge to established loblolly stands.

SUMMARY AND CONCLUSIONS

Nitrate-N at the 1 m depth of soil exceeded 60 mg/l four months after application of 800 kg N/ha in sludge high in nitrogen to a stand at planting and a 28-year-old stand, but reached only 20 mg/l in 3- and 9-year-old stands. Application of 400 kg N/ha in the same sludge failed to increase nitrate-N levels above 10 mg/kg in 3-, 9-, and 28-year-old stands. Application of 630 kg N/ha in a sludge low in nitrogen increased nitrate-N in soil solution only slightly. Nitrate-N levels for all treatments had declined in eighteen months to near 10 mg/l at the 1 m depth, and by three years treatments were near the control levels.

Leaching of nitrogen past the 1 m depth over an eighteen-month period was 7, 18, and 22% of the 800 kg/ha applied in the L 800 treatment to the 3-, 9-, and 28-year-old stands. In the same stands, leaching of nitrogen from the L 400 treatment was less than 8%.

Leaching of potassium, calcium, and magnesium was related to nitrate levels in the soil solution. For the liquid sludge with a high nitrogen content, increased cation leaching was greater than the application rate, thus indicating a possible risk of cation depletion of the upper soil horizons.

Zinc, copper, lead, cadmium, chromium, nickel, aluminum, iron, and cobalt increased at the 1 m depth; however, there were increased concentrations of nickel, iron, aluminum, and manganese at the 0.5 m depth. These increased trace metal levels at 0.5 m are probably related to changes in soil reaction as a result of sludge application.

Results of these studies indicate that this loblolly pine well-drained sandy soil system can be fertilized with sludge without extreme risk of excessive element concentrations in the soil solution. Soils higher in clay may utilize greater quantities of sludge than these soils did. In all applications of sludge to forests vegetation, soil and sludge properties must be carefully considered in the evaluation of potential risks to the water resource.

ACKNOWLEDGMENTS

This work was supported in part by contract DE-A109-80SR10711 with the U.S. Department of Energy.

REFERENCES

Breuer, D. W., D. W. Cole, and P. Schiess. 1979. Nitrogen transformation and leaching associated with wastewater irrigation in Douglas fir, poplar, grass, and unvegetated systems. p. 19–33. In W. E. Sopper and S. N. Kerr (eds.) Utilization of municipal sewage effluent and sludge on forest and disturbed land. Pennsylvania State University Press, University Park.

Brockway. D. G. 1983. Forest floor, soil, and vegetation responses to sludge fertilization in red and white pine plantations. Soil Sci. Soc. Am. J. 47:776–784.

Brockway, D. G., and D. H. Urie. 1983. Determining sludge fertilization rates for forests from nitrate-N in leachate and groundwater. J. Environ. Qual. 12:487–492.

Davidson, J. M., L. R. Stone, D. R. Nielsen, and M. E. Larue. 1969. Field measurement and use of soil-water properties. Water Resour. Res. 5:1312–1321.

Long, F. L. 1978. A glass filter soil solution sampler. Soil Sci. Soc. Am. J. 834–835.

Murphy, C. E., Jr. 1985. Carbon dioxide exchange and growth of a pine plantation. For. Ecol. and Manage. 11:203–224.

Murphy, C. E., Jr., J. F. Schubert, and A. H. Dexter. 1981. The energy and mass exchange characteristics of a loblolly pine forest. J. Appl. Ecol. 18:271–281.

Riekerk, H. 1981. Effects of sludge disposal on drainage solutions of two forest soils. For. Sci. 27:792–800.

Riekert, H., and R. J. Zasoski. 1979. Effects of dewatered sludge applications to a Douglas fir forest soil on the soil, leachate, and groundwater composition. p. 35–45. In W. E. Sopper and S. N. Kerr (eds.) Utilization of municipal sewage effluent and sludge on forest and disturbed land. Pennsylvania State University Press, University Park.

Rogerson, T. L. 1966. Throughfall in pole sized pine as affected by stand density. p. 187–190. In W. E. Sopper and H. W. Lull (eds.) Forest hydrology. Pergamon Press, New York.

Van Miegroet, H., and D. W. Cole. 1984. The impact of nitrification on soil acidification and cation leaching in a red alder ecosystem. J. Environ. Qual. 13:586–590.

Wells, C. G., A. K. Nicholas, and S. W. Buol. 1975. Some effects of fertilization on mineral cycling in loblolly pine. p. 754–764. In F. G. Howell, J. B. Gentry, and M. H. Smith (eds.) Mineral cycling in southeastern ecosystems. ERDA Symposium Series (CONF–740513), Augusta, Georgia.

Wells, C. G., K. W. McLeod, C. E. Murphy, J. R. Jensen, J. C. Corey, W. H. McKee, and E. J. Christensen. 1984. Response of loblolly pine plantations to two sources of sewage sludge. p. 85–94. In 1984 TAPPI Research and Development Conference, Appleton, Wisconsin. Technical Association of the Pulp and Paper Industry, Technology Park, Atlanta, Georgia.

Chapter 16

Heavy Metal Storage in Soils of an Aspen Forest Fertilized with Municipal Sludge

A. RAY HARRIS and DEAN H. URIE

ABSTRACT   Municipal sewage was surface applied to aspen (*Populus grandidentata*) on a simulated clearcut, a six-year-old sapling stand, and a standard operational clearcut. The soil was a Montcalm loamy sand (Alfic Haplorthod) with low fertility and water holding capacity. Single applications of sludge were applied at the rate of 11.5, 23, and 46 Mg/ha in the spring of 1976, 1977, and 1978. Some plots received a second application at the 11.5 and 23 Mg/ha rate the year following initial application. The aspen humus layer effectively immobilized a high proportion of the metals present in the sludge. Over a five-year period, most metal movement seemed to be limited to the upper 5 cm soil layer. The vegetation recycled heavy metals back to the humus layer. Repeated applications in the following year appear to increase leaching to the soil. Pretreatment history of the aspen determines to some extent the overall immobilization of the heavy metals. More metals were lost from humus layers that had been recently disturbed after years of accumulation. Humus on plots receiving heavier dosage rates of sludge seemed more efficient in the storage of heavy metals. The mineral soil system does not, over a three to five year period, appear to play a major part in heavy metal storage or movement at the treatment rates used.

Aspen has the capacity to utilize many of the elements in sewage sludge. Rapid growth characteristics and short rotation time make aspen forests likely recycling sites for municipal sludge (Cooley 1979). Because forests minimize the transfer of harmful metals in sludge to the human food chain, they may be suitable for the land treatment of sewage sludges that contain metal concentrations above the suggested limits for agricultural uses (Urie 1979).

Because sludge must be surface applied in forested areas, the immediate stabilization of the waste material largely depends on weather factors (temperature, moisture, and air movement) and the organic layer at the soil surface (Kirkham 1977). Assimilation of the sludge-borne materials into the ecosystem does not reduce the presence of heavy metals, some of which have been found to accumulate in soils (Heinrichs and Mayer 1977, 1980, Chang et al. 1981). This is especially true in forests where even harvest removes only minimal amounts of the metals (Heinrichs and Mayer 1980).

Heavy metal buildup in forest soils is not as serious as in agricultural soils, because of the reduced food-chain hazard, although wildlife could be subject to increased metal levels in forage (West et al. 1981). Sludge application opportunities are dependent on clearcutting cycles and short postharvest periods, so overloading or high accumulations of heavy metals on well-managed sites should not be a problem, especially with low metal sludges.

This study investigated the movement and accumulation of sludge-borne heavy metals in an aspen forest and the persistence of these metals in the soil and organic matter.

MATERIALS AND METHODS

The study site was in the Pine River Experimental Forest, part of the Manistee National Forest, in Michigan's Lower Penninsula (latitude 44°15′N, longitude 83°W). The 30 hectare aspen (*Populus grandidentata* Michx.) stand was located on Montcalm loamy sand (Alfic Haplorthod), a well-drained soil low in fertility and water holding capacity. About 15 hectares of the stand were old growth, and about 15 had been clearcut in 1970, five years before the study. In 1975, 8 hectares of the five-year-old saplings on the old clear-cut area were roller chopped to simulate clearcut stage. In the winter of 1977-78 the old growth was also clearcut. Circular plots 15 m in diameter and 15 m apart were established in each stand.

Simulated clearcut, operational clearcut, and sapling stages of aspen growth were treated over a period of several years with surface-applied liquid sludge. Simulated clearcut (roller chopped) plots were treated in the fall of 1975 and spring of 1976 and 1977. Saplings were treated in the spring of 1977. The 1977-78 operational clearcut plots were treated in the spring of 1978 and 1979.

Treatments included one-time applications at three rates of municipal sludge with four replications. Half of the low and medium rate plots were retreated a following year to determine the effect of yearly retreatment in the simulated and operational clearcut areas. The following describes the field trials, treatment rates, and the time of application: (1) 7.7, 15.4, and 23.1 Mg/ha applied to simulated clearcut 0-age sprouts in the fall of 1975, (2) 11.5, 23, and 46 Mg/ha applied to simulated clearcut 0-age sprouts in the spring of 1976, (3) 11.5, 23, and 46 Mg/ha applied to 6-year-old saplings in the spring of 1977 and retreatments of half of the 11.5 and 23 Mg/ha original simulated clearcut 0-age sprouts in the spring 1976 plots, and (4) 11.5 and 23 Mg/ha applied to operational clear-cut 0-age sprouts in the spring of 1978, followed by retreatments at the same rates in the spring of 1979.

Anaerobically digested sludge was trucked from Cadillac, Michigan, pumped to the plots, and surface spread using a hand-held fire hose. Solids content of the sludge was determined from samples withdrawn from each tanker load, while composite samples of individual tanker load samples were analyzed for chemical characteristics for each year's sludge applications. Table 1 shows the average concentration and loading rate of heavy metals for each year applied. Poor control of plating waste in a local plating plant resulted in large variations in heavy metals content in the sludge, particularly for cadmium and chromium. Solids content ranged from 4.5 to 6.4%; but with random application to plots, average sludge loadings calculated at 5.5% appeared to be representative.

Humus, 0-5, and 5-10 cm soil samples were collected using a standard bulk density sampler. Soil sample depth intervals 10-15, 15-30, 30-60, and 60-90 cm were collected with a tube sampler. Samples were collected each year in October starting in 1976 and ending in 1981. Standard drying and storing procedures were used in sample preparation. Total metal content was analyzed on all humus samples from 1976 through 1981. Total metal contents were measured on soils starting in 1977 through 1981.

Heavy metal analyses were performed in 1976 and 1977 at the University of Minnesota

TABLE 1. Average metal concentrations and calculated loading rates derived from municipal sludge applications at the 11.5 Mg/ha total dry solids loading rate.

Year	Zinc	Cadmium	Chromium	Copper	Lead	Nickel	Iron	Manganese
				(milligrams/kilogram)				
1975	2,300	55	565	700	445	85	87,300	500
1976	3,400	240	705	845	600	100	98,900	730
1977	4,140	520	1,620	660	715	110	81,500	740
1978	3,330	250	895	800	670	100	91,000	770
1979	1,940	155	350	725	500	75	67,900	630
				(kilograms/hectare)				
1975	26	0.6	6.5	8.0	5.1	1.0	1,000	5.8
1976	39	2.8	8.1	9.7	6.9	1.2	1,140	8.4
1977	48	6.0	19.0	7.6	8.2	1.3	1,940	8.5
1978	38	2.9	10.0	9.2	7.7	1.2	1,050	8.8
1979	22	1.8	4.0	8.3	5.8	0.8	780	7.2

Metal loadings at 23.0 and 46.0 Mg/ha dry solids loading rates are 2X and 4X table values, respectively.

Analytical Laboratory using dry ashing with concentration determination on an ICP emission spectrometer (Dahlquist and Knoll 1978). After 1978, analyses were performed at the USDA Forest Service and Michigan State University Forest Soil and Water Co-op laboratory using nitric-perchloric digestion with concentration determined on a DCP emission spectrometer (McHard et al. 1979).

RESULTS AND DISCUSSION

Because of the high variability of the heavy metal concentrations in the sludge, loading rates of the metals studied varied widely over the several years of application (Table 1). Plots treated in different years at the same application rate received different metal loadings. For instance, at the 46 Mg/ha dry solids loading rate, the loading for chromium in 1976 was 32 kg/ha and in 1977 was 76 kg/ha. Cadmium and several other metals also showed similar variances.

Humus-Retained Metal

Generally, applied sludge could not be distinguished from the humus layer of the forest floor within a few months after application. In microdepressions in the forest floor, sludge pooled, dried, and hardened. Clumps of hard sludge were found in these locations four to five years after treatment. These fairly insoluble sludge clumps may have contributed to the high heavy metal retention ratios.

Concentrations of the heavy metals in the humus layer in 1977 for the simulated clearcut and the sapling stands of aspen treated for the three sludge rates are shown in Table 2. At the time these samples were collected, sludge had been on the plots for two years in the simulated clearcut, and one year in the saplings.

Table 3 lists the concentrations in the humus in all three of the aspen stands in the fall of 1980. These data represent concentrations five years after treatment in the simulated clearcut, four years later in the sapling stand, and three years after application in the operational clearcut. The humus in the operational clearcut had the highest overall loss

TABLE 2. Average pH and heavy metal concentration in the humus layer with respect to sludge application rate on two growth stages of aspen sprouts, fall 1977.

Treatment	pH	Zn	Cd	Cr	Cu	Pb	Ni	Fe	Mn
(Mg/ha)				(milligrams/kilogram)					
Simulated Clearcut									
Control	5.6	131	3	4	6	57	4	1,850	1,630
11.5	6.0	610	33	126	61	186	15	10,990	1,260
23.0	6.1	1,350	54	116	151	268	23	14,100	1,340
46.0	6.3	1,580	92	201	252	394	38	22,400	1,220
11.5 + 11.5	6.1	847	70	136	118	244	24	12,700	1,110
23.0 + 23.0	6.3	1,810	173	370	280	484	50	23,560	886
Sapling									
Control	5.6	92	6	10	7	59	5	3,158	1,240
11.5	6.5	1,380	160	350	155	358	35	17,159	1,240
23.0	6.5	1,970	237	536	252	511	51	23,100	1,120
46.0	6.6	2,450	299	652	318	636	68	27,000	1,080

even though it had received the most recent sludge applications. Concentrations of total zinc, copper, lead, nickel, iron, and manganese were decreased significantly from application levels at all dosage rates compared with the simulated clearcut and sapling treatments. Higher relative metal concentrations in the humus under the 46 Mg/ha sludge rate suggests that the dried cake remnants found in depressions isolated sludge metals over the long term in the litter-humus complex.

The highest level of metal retention in the humus layer occurred in the simulated clearcut plots. Because of the well-established sampling stand, litter additions to the humus layer may have diluted concentrations of the metals (Harris et al. 1984).

Manganese loss was highest in the operational clearcut, and manganese was consistently lost in all other treatments. The loss of added manganese plus native manganese may be attributed to the addition of other heavy metals as found by Korcak and Fanning (1981).

Concentrations of the heavy metals other than manganese in the humus were significantly higher after sludge was added. Concentrations were also significantly different between loading rates. Variations between treatments, however, were unique to each aspen site.

Both time and type of pretreatment seemed to affect the persistence of the heavy metals in the humus. At the 23 Mg/ha rate the simulated clearcut treatments retained more of the heavy metals in the organic matter than the other treatments at the same level. Except for cadmium and chromium, the metal concentration was lower in the operational clearcut three years after application than in the other treatments four and five years after application. Added manganese was only about 8% of the total soil manganese, and for the lower loading rates there was a net loss of manganese from the humus layer (Table 3).

Metal loss seems to be related to the amount of the more readily soluble organic matter present in the forest floor before sludge was applied (Harris 1979). Complexed metals can move with the more readily soluble organic matter that would be present in the disturbed organic matter associated with recently clearcut old-growth aspen stands (Hodgson et al. 1965). Very little of the less resistant organic matter would be present in

TABLE 3. Average pH and heavy metal concentration in the humus layer with respect to sludge application rate on three growth stages of aspen sprouts, fall 1980.

Treatment	pH	Zn	Cd	Cr	Cu	Pb	Ni	Fe	Mn
(Mg/ha)				(milligrams/kilogram)					
Simulated Clearcut									
Control	5.5	110	3	5	10	55	5	5,250	1,200
11.5	5.6	390	25	70	95	125	15	16,800	990
23.0	6.0	940	55	100	220	200	20	24,500	1,140
46.0	6.1	1,420	85	160	305	315	35	31,600	1,050
11.5 + 11.5	6.3	720	60	130	180	230	20	23,600	1,150
23.0 + 23.0	6.3	1,720	150	330	280	350	45	45,000	1,240
Sapling									
Control	5.6	85	3	7	7	45	5	5,650	1,120
11.5	6.4	805	90	285	120	195	20	20,800	1,210
23.0	6.4	1,510	165	535	220	330	40	33,800	1,250
46.0	6.5	2,010	250	650	375	425	65	51,000	1,400
Operational Clearcut									
Control	5.6	85	3	8	10	50	5	5,420	1,350
11.5	5.6	325	50	105	55	85	10	9,800	750
23.0	6.0	590	90	275	105	135	15	15,400	840
23.0 + 23.0	6.1	1,000	120	405	215	185	30	30,800	900

the sapling stand or the simulated clearcut that had been clearcut five years previous to treatment. Even though the simulated clearcut was roller chopped, little of the more readily soluble organic matter would have reaccumulated in the humus layer over such a short time.

Plots that received sludge in two equal annual concurrent treatments usually had concentrations and retentions of metals between those for the single loading rate and the next higher single loading rate. Although the loading was doubled on these plots, resulting concentrations were not. However, mineralization of the sludge was higher on retreated plots, as evidenced by an increased nitrate-N leaching rate to the soil water (Urie et al. 1978).

The average retention percentage was directly related to the dosage rates (Figure 1). The heavy metals were retained more efficiently at the higher dosage rates. Increased vegetative recycling and dried sludge accumulations in small depressions may have tied up metals that would otherwise have been mobilized. Accumulation of sludge solids in these small depressions was obviously greater at the higher sludge dosage rates.

The effect of adding a calcium-rich sludge is still evident after five years, as reflected by the pH (Table 3). In the sapling stand, the effect of dosage rate on soil pH was more evident than in the two clearcuts. Presumably, this was another reflection of the retention of sludge in the surface organic layer because of the protective crown cover, advanced sprout growth, and undisturbed litter.

The patterns of pH change in the humus layer suggest that pH initially increases for the first two years after clearcutting and then gradually decreases during the third year until it reaches a level slightly below that before harvest (Figure 2). This effect on humus pH could help explain the differences in retention between the three aspen sites under similar sludge loading rates. Also, pH may have affected the metal solubilities, and thus

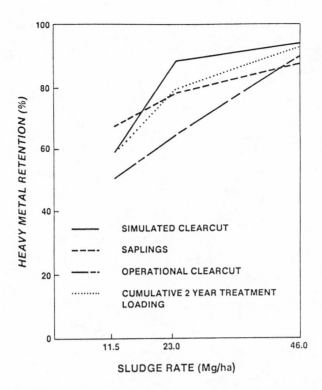

Figure 1. Mean percentage retention of all elements in the humus layer in 1980 with respect to sludge loading rate on three growth stages of aspen sprouts.

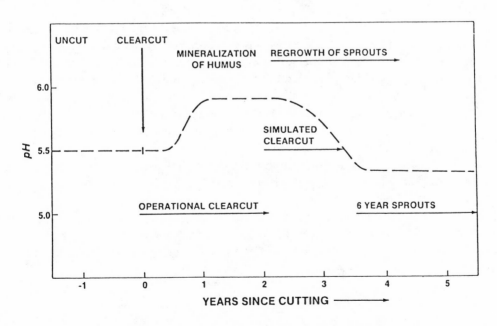

Figure 2. pH changes in humus layer over a clearcutting cycle.

TABLE 4. Average pH and heavy metal concentration in the 0-5 cm soil layer with respect to sludge application rate on three growth stages of aspen sprouts, fall 1980.

Treatment	pH	Zn	Cd	Cr	Cu	Pb	Ni	Fe	Mn
(Mg/ha)				(milligrams/kilogram)					
Simulated Clearcut									
Control	4.9	26	1	6	4	17	2	6,600	590
11.5	5.0	33	6	9	7	17	3	8,890	805
23.0	5.4	39	6	10	9	22	3	7,070	700
46.0	5.5	79	6	17	18	31	5	9,920	760
11.5 + 11.5	5.6	68	6	20	15	36	5	10,910	650
23.0 + 23.0	5.6	132	11	41	32	54	7	10,720	585
Sapling									
Control	5.0	28	1	5	3	16	2	5,000	685
11.5	5.4	208	22	65	33	74	8	10,465	735
23.0	5.5	50	6	15	9	25	3	5,210	650
46.0	5.7	109	10	31	18	44	5	6,932	760
Operational Clearcut									
Control	4.6	35	2	7	5	15	2	6,000	1,100
11.5	4.6	24	3	9	5	15	2	6,000	400
23.0	5.2	67	9	24	13	22	3	6,250	600
23.0 + 23.0	5.1	57	6	16	12	24	3	6,850	650

altered uptake in leaching under the higher pH of the heavier loading (Beveridge and Pickering 1980).

Soil-Retained Metal

Heavy metals in the sludge, to some extent, leached into the mineral soil surface layer. Breakthroughs into the 0-5 cm mineral soil depth occurred irrespective of treatment or loading rate (Table 4). The movement of zinc in the 11.5 and 46 Mg/ha loading rate in the sapling stand is very evident. No significant increase in metal concentration was found to occur beneath the 0-5 cm soil depth. Movement of metals did not seem to be correlated to time of application, which may be partly owing to a very droughty period from 1976 through 1979. Sidle and Kardos (1977) also found very little movement of heavy metals through the mineral soil in a mixed hardwood forest, although some loss was attributed partly to channelization through the soil to lower soil depths. No evidence was found of channelization in this study, based on the absence of any elevated concentrations of metals at lower soil depths.

Retreatment the following year seems to increase the susceptibility of metal loss from the humus to the soil. This is because the bacterial environment is primed by the first year's sludge application to attack the sludge immediately when applied the second year.

SUMMARY AND CONCLUSIONS

A large percentage of the heavy metals in sewage sludge applied to aspen sites was immobilized by the humus layer. Significant heavy metal leaching occurred only within the upper 5 cm soil layer after three to five years following sludge additions. Vegetative uptake of some of the heavy metals (e.g., zinc) would have been a major route for loss of heavy metals from the humus layer, and such metals would be recycled in subsequent years. This would help explain the increased retention of heavy metals at the higher sludge dosage rates. Effects from two concurrent annual loadings were similar to those on plots receiving one application equal to the sum of the two loadings; however, leaching to the upper soil layer from the humus layer was greater on plots receiving two concurrent loadings.

The pretreatment history of the humus layer appears important in the overall immobilization of the heavy metals. More metals were lost from the humus layer that had recently been disturbed after years of accumulation than were lost from the other two treatment sites that had been initially clearcut five to six years before sludge application. These metals probably moved as complexes with solubilized organic matter.

Humus receiving heavier dosage rates of sludge seemed more efficient in retaining the heavy metals. Unincorporated dried sludge, higher pH, and increased vegetative assimilation and recycling may have contributed to causing this effect.

The heavy metals are stored in the humus layer and may accumulate as they are recycled through vegetation. In these sandy soils, the mineral soil system does not, over a three to five year period, appear to play a major role in heavy metal movement or storage from applied sludge at the rates used.

REFERENCES

Beveridge, A., and W. F. Pickering. 1980. Influence of humate-solute interactions on aqueous heavy metal ion levels. Water, Air, and Soil Poll. 14:171–185.

Chang, A. C., A. L. Page, and F. T. Bingham. 1981. Re-utilization of municipal wastewater sludges—metals and nitrate. J. Water Poll. Control Fed. 53(2):237–245.

Cooley, J. H. 1979. Effects of irrigation with oxidation pond effluent on tree establishment and growth on sand soils. p. 145–153. In W. E. Sopper and S. N. Kerr (eds.) Utilization of municipal sewage effluent and sludge on forest and disturbed land. Pennsylvania State University Press, University Park.

Dahlquist, R. L., and J. W. Knoll. 1978. Inductively coupled plasma-atomic emission spectrometry: Analysis of biological materials and soils for major, trace, and ultra-trace elements. Applied Spectroscopy 32(1):1–29.

Harris, A. R. 1979. Physical and chemical changes in forested Michigan sand soils fertilized with effluent and sludge. p. 155–161. In W.E. Sopper and S. N. Kerr (eds.) Utilization of municipal sewage effluent and sludge on forest and disturbed land. Pennsylvania State University Press, University Park.

Harris A. R., D. H. Urie, and J. H. Cooley. 1984. Sludge fertilization of pine and aspen forests on sand soils in Michigan. p. 193–206. In E. L. Stone (ed.) Forest soils and treatment impacts, Proceedings of the Sixth North American Forest Soils Conference, University of Tennessee, Knoxville.

Heinrichs, H., and R. Mayer. 1977. Distribution and cycling of major and trace elements in two central European forest ecosystems. J. Environ. Qual. 6:402–407.

———. 1980. The role of forest vegetation in the biogeochemical cycle of heavy metals. J. Environ. Qual. 9:111–118.

Hodgson, J. F., H. R. Geering, and W. A. Norvell. 1965. Micronutrient cation complexes in soil solution: Partition between complexed and uncomplexed forms by solvent extraction. Soil Sci. Soc. Am. Proc. 29:665–669.

Kirkham, M. B. 1977. Organic matter and heavy metal uptake. Compost Sci. 18(1):18–21.

Korcak, R. F., and D. S. Fanning. 1981. Interaction between high levels of applied heavy metals and indigenous soil manganese. J. Environ. Qual. 10:69–72.

McHard, J. A., S. J. Foulk, S. Nikdel, A. H. Ullman, B. D. Pollard, and J. D. Winefordnes. 1979. Comparison study of four atomic spectrometric methods for the determination of metallic constituents in orange juice. Analytical Chemistry 51(11):1613–1616.

Sidle, R. C., and L. T. Kardos. 1977. Transport of heavy metals in a sludge-treated forested area. J. Environ. Qual. 6:431–437.

Urie, D. H. 1979. Nutrient recycling under forests treated with sewage effluents and sludge in Michigan. p. 7–17. In W. E. Sopper and S. N. Kerr (eds.) Utilization of municipal sewage effluent and sludge on forest and disturbed land. Pennsylvania State University Press, University Park.

Urie, D. H., A. R. Harris, and J. H. Cooley. 1978. Municipal and industrial sludge fertilization of forests and wildlife openings. p. 467–480. *In* First Annual Madison Conference of Applied Research and Practice on Municipal and Industrial Waste, Madison, Wisconsin.

West, S. D., R. D. Taber, and D. A. Anderson. 1981. Wildlife in sludge-treated plantations. p. 115–122. *In* C. S. Bledsoe (ed.) Municipal sludge application to Pacific Northwest forest lands. Institute of Forest Resources Contribution 41. College of Forest Resources, University of Washington, Seattle.

Chapter 17

Influence of Municipal Sludge on Forest Soil Mesofauna

GARY S. MacCONNELL, CAROL G. WELLS, and LOUIS J. METZ

ABSTRACT   The influence of municipal sludge on forest soil mesofauna was studied at the Savannah River Project, South Carolina. Loblolly pine (*Pinus taeda* L.) plantations established in 1953 and 1973 were treated with two different application rates of sludge, from two sources, beginning in July 1980. Soil mesofauna were collected from the sites in the spring or summer of 1982, the fall of 1982, and the spring of 1983. Mesofauna were then classified into three groups. It was determined that the physical properties and nitrogen concentrations of the applied sludge were the most important variables. A well-decomposed sludge with a low solids content resulted in the reduction of the total number of mesofauna, while a less decomposed sludge with a higher solids content produced an increase in the total numbers of mesofauna when compared with an untreated area. Other variables discussed include seasonal fluctuations, age class of site, soil type, application rate, and sludge source.

The need to dispose of municipal sludge in an environmentally safe manner has become a major concern of many municipalities. Present trends favor an increase in the application of municipal sludge to forest land as a method for tertiary sewage treatment (Pratt et al. 1977), forest fertilization (Smith et al. 1979), and municipal sludge disposal (Koch 1982). Soils make an ideal chemical filter because they possess the properties necessary for (1) ion exchange, (2) adsorption or precipitation, and (3) chemical alterations (Ellis 1973). In addition to being an excellent filter, soils can usually assimilate large quantities of wastes and prevent many harmful by-products from contaminating adjacent land or water systems (Hall 1978).

Municipal sludges typically contain substantial amounts of nitrogen, phosphorus, and potassium, the primary plant nutrients, along with the exchangeable bases calcium, magnesium, and sodium (Smith et al. 1979). The heavy metals manganese, cadmium, chromium, lead, zinc, copper, and nickel are often present in appreciable quantities, as are sulfur, iron, and boron (Riekerk and Zasoski 1979). Anthropogenic chemicals such as pesticides, polychlorinated biphenyls (PCBs), and polycyclic aromatic hydrocarbons which are dumped into waste treatment systems may be present in some sludges (Hollod 1981). Many of these chemicals and nutrients are bound in the organic matter of the sludge and are released during decomposition.

Land application of municipal sludge has been shown to be less expensive than incineration or landfilling, two traditional methods of sludge disposal (Pritchett 1979). Although sludge has been used successfully in various areas such as sod production, nurseries (Sopper 1973), and strip mine reclamation (Lejcher and Kunkle 1973), agricultural applications are more widely used. Sludge applications to forest and agricultural lands offer the greatest potential for the long-term disposal of sludge, in that both areas are often found in multiacre units.

Forest land disposal provides many advantages over agricultural land disposal. Forests are able to assimilate a higher rate of sludge application (Riekerk 1978), are often located near waste treatment facilities, and are relatively inexpensive (Nutter 1978). In addition, forest land is more acceptable aesthetically than agricultural land for waste disposal, and forests are better suited to withstand land erosion. The utilization of forests for sludge disposal also reduces the chance that heavy metals and toxics will enter the human food chain, requires minimum management, and presents few interruptions of treatment schedules necessary with seasonal crops (Urie 1979). Municipal sludge is capable of increasing the growth rate of many tree species (Smith et al. 1979) and helps to produce trees of higher quality wood with respect to pulp production (Murphy et al. 1973).

Municipal sludge, when applied to land, changes many of the physical and chemical properties of the soil. The organic matter in the sludge improves the aggregation of soil and reduces surface runoff and erosion, while increasing nutrient loading and infiltration rates (Khaleel et al. 1981). Nutrients and heavy metals are concentrated in the foliage of trees and are recycled when the leaves drop, thereby increasing the concentration of these substances in the organic ped (David and Struchtemeyer 1980). Nitrogen in the form of nitrates is the only nutrient or heavy metal typically leached in large quantities, and the concentration of nitrogen decreases with soil depth (Riekerk and Zasoski 1979).

While the application of municipal sludges to forests may provide many benefits to the forests, negative consequences may also occur. One area of concern is the effect of municipal sludges on soil fauna, which are organisms residing in the forest floor and mineral soil. These invertebrates are thought to be responsible either directly or indirectly for (1) organic matter decomposition and incorporation within the mineral soil, (2) production of the humus complex, (3) enhancement of soil aggregation, (4) increasing water holding capacity and soil porosity, and (5) the recycling of nutrients and chemical compounds within the terrestrial system (Dindal et al. 1979).

Soil fauna may be grouped in a number of ways, with size being a common method of distinguishing various soil invertebrates (Metz and Farrier 1973). Microfauna are organisms less than 100 mm in size and include the protozoans and nematodes. Mesofauna are animals that range in size from 100 mm to 1 cm, with mites, Collembola, "other insects," and other small invertebrates belonging to this group. Macrofauna are larger than 1 cm and include such animals as large insects, earthworms, and snails.

For purposes of this discussion, the mesofauna are divided into three distinct groups. The mites (class Arachnida, subclass Acari), which make up the largest number of soil mesofauna, include the oribatid or beetle mites (order Cryptostigmata), which account for the largest portion of mites, and the orders Prostigmata (also known as Trombidiformes) and Mesostigmata (Metz and Farrier 1973). Collembola or springtails (class Insecta, subclass Apterygota, order Collembola) are another group of mesofauna. All other soil animals that do not fall into one of these three groups are listed as "other fauna," and include ants, spiders, pseudoscorpions, and other invertebrates.

In forest soils, the majority of these soil invertebrates live in the surface organic layers, or forest floor (Metz and Dindal 1980). The forest floor is composed mostly of vegetative organic matter from the trees and is subdivided into three layers. The L (litter) layer is the surface layer and consists of recently fallen, not yet decayed organic matter. The F (fermentation) layer is below the L layer and consists of partly decomposed material. The H (humus) layer consists of decomposed organic material whose origin is not distinguishable. The L layer, which is often quite dry, offers the least favorable habitat for the organisms, and few are found there (Metz and Farrier 1973). Increasing numbers of animals are found in the F and H layers, with the H layer typically having the highest number, since it is a moister habitat and contains organic matter that is more decomposed. In the mineral soil, animal numbers usually decrease with depth.

Research has shown that some human activities influence soil invertebrate communities (Dindal 1977). Activities that have been studied include DDT spraying, urban street salting, prescribed burning (Metz and Farrier 1973), and fertilizer application (Hill et al. 1975). While much research has dealt with the influence of wastewater irrigation on soil mesofauna (Dindal et al. 1975, Dindal et al. 1979), there is little information on the influence of municipal sludge on soil mesofauna. The objective of this paper is to fill this void and discuss the effects of municipal sludge on forest soil mesofauna.

EXPERIMENTAL DESIGN AND METHOD

This experiment resulted from a program initiated in July 1980 at the Savannah River Plant (SRP), Aiken, South Carolina, to determine the response of a southern pine-soil system to two distinct sources of municipal sludge. One source was the Horse Creek Pollution Control Facility (Aiken), a 90 million liter per day (20 mgd) capacity treatment facility operated by the Aiken County Public Services Authority. Approximately 80% of the wastewater inflows are industrial (textile mills), with the remaining 20% domestic. The municipal sludge is aerobically digested, and the solids are treated for 20 minutes at 225°C under 17 atmospheres (250 psi) of pressure, with the treated sludge being dewatered to 30% solids. The treated sludge, which contained 1.3% nitrogen oven-dried weight, was brought to the SRP site in a 18 tonne (20 ton) dump truck and spread uniformly over the forested experimental plots using conventional manure spreading equipment. Quality control was checked by analyzing a subsample from each load.

The second source of sewage sludge was the Augusta, Georgia, Wastewater Treatment Plant (Augusta), which produces 190,000 liters per day (50,000 gpd) of anaerobically digested sewage sludge. The wastewater inflow at the Augusta plant is approximately 80% domestic and 20% industrial. The wastewater is processed by conventional aerobic treatment, and the secondary sludge is anaerobically digested, resulting in a 2% solids suspension. This sludge, with a nitrogen content of 7% oven-dried weight, was delivered to the SRP site in 21,000 liter (5,500 gal) tankers and transferred to an 8,000 liter (2,100 gal) tanker. The smaller tanker was pulled by a tractor and the sludge applied evenly over the experimental plots. A subsample was collected from each 21,000 liter (5,500 gal) load to monitor the quality of the sludge.

Four loblolly pine (*Pinus taeda* L.) plantations were utilized, with two of the sites established in 1953 and two in 1973. A sandy soil site and a sandy clay soil site were used for each of the two age classes. The sandy soil was of the Lucy series and the heavier sandy clay soil was of the Orangeburg series. Both soils are well to excessively drained, acidic, and infertile by agronomic standards. Each site had nine 0.2 ha (0.5 acre) plots in a randomized block design. The plot dimensions were 45 by 45 m (150 by 150 ft) with an interior test plot 30 by 30 m (100 by 100 ft) for intensive environmental and biological sampling.

The sludge applications were based on an equivalent loading rate for N of 0, 400 kg N/ha (360 lb N/acre), and 800 kg N/ha (720 lb N/acre). The Aiken sludge was applied at 630 kg N/ha (560 lb N/acre), which corresponds to approximately 36 wet Mg (40 wet tons) of sludge. The Augusta sludge was applied at both the 400 kg N/ha and 800 kg N/ha rates, which are equivalent to 38,000 l/ha (25,000 gal/acre) and 76,000 l/ha (50,000 gal/acre) of sludge, respectively.

For each of the two soil types, each 1953 plantation received Aiken sludge at a 630 kg N/ha dose on two plots each, Augusta sludge at 400 kg N/ha and 800 kg N/ha rates on two plots each, and no sludge on three plots (controls). The two 1973 sites received sludge from only the Augusta treatment plant. The Augusta sludge was applied at rates of 400 kg N/ha and 800 kg N/ha on three sites each for both soil types, with no sludge applied to the three remaining plots (controls) at each site.

Each site was sampled for soil mesofauna in the spring or summer of 1982 (sample 1), the fall of 1982 (sample 2), and the spring of 1983 (sample 3), with each soil type being sampled on the same date for each respective age class. The 1973 sites were sampled July 19, 1982, November 18, 1982, and June 9, 1983. The 1953 sites were sampled April 20, 1982, November 4, 1982, and May 18, 1983.

On each sampling date, five locations were selected at random on each of the nine plots, and for each of the two soil types. Samples of the forest floor and the first surface inch of mineral soil were collected by cutting around a 20 cm^2 area metal ring. The forest floor and mineral soil samples were stored in separate plastic containers, labeled, and taken to the laboratory. Thus there were ninety samples for each soil type in each age class for each sampling. The soil mesofauna were extracted for seven days into vials containing 85% ethanol, using modified Tullgren funnels (Murphy 1955). The specimens collected were grouped, using a steroscopic microscope, into mites, Collembola, and "other fauna." "Other fauna" were only included in total fauna because of their few numbers.

RESULTS AND DISCUSSION

The effects of sewage sludge or any other impact on a mesofaunal population may be twofold, with a change in total numbers being one factor and a change in the community structure being another. In this experiment there were five controlled variables: (1) date of sampling, (2) age class of stand, (3) soil type, (4) application rate, and (5) sludge type or source.

Figures 1 and 2 show the average total number of mesofauna for each age class on each sampling date, with the 1973 sites exhibited in Figure 1 and the 1953 sites in Figure 2. The 400 kg N/ha application of Augusta sludge reduced mesofauna for all samplings, except the third sampling on the 1953 sites, in which the control sites had the same number of animals (Figure 2). The 800 kg N/ha application of Augusta sludge reduced the mesofauna more than the 400 kg N/ha rate in all samplings, except for the first sampling on the 1973 sites (Figure 1). The 630 kg N/ha application of Aiken sludge, which was applied only to the 1953 sites, increased the mesofaunal populations for each sampling over the control and the Augusta sludge application levels. Results indicate that there

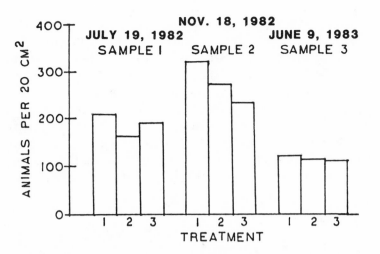

Figure 1. Average number of mesofauna for the 1973 sites. Treatment 1 = control; 2 = application of 400 kg N/ha Augusta sludge; 3 = application of 800 kg N/ha Augusta sludge.

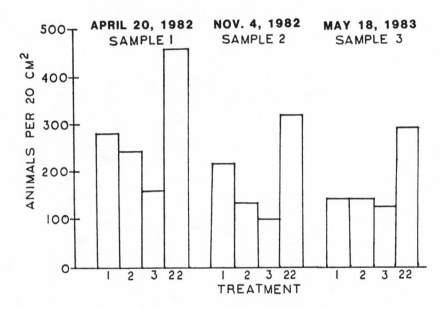

Figure 2. Average number of mesofauna for the 1953 sites. Treatment 1 = control; 2 = application of 400 kg N/ha Augusta sludge; 3 = application of 800 kg N/ha Augusta sludge; 22 = application of 630 kg N/ha Aiken sludge.

are changes in the total numbers of mesofauna at different times of the year; however, it is not possible to compare the numbers of 1973 sites with the 1953 sites, since samples were collected on different dates. The only conclusion that can be made from comparing total numbers of mesofauna for each age class at different sampling dates is that mesofaunal populations do exhibit seasonal fluctuations.

Seasonal fluctuations in total mesofaunal numbers may have several causes. Intrinsic factors such as the life cycle of each soil animal may be important in these variations. Soil temperatures and moisture content also play a role in faunal fluctuations. The compactness and pore space of the soil and variables related to specific tree stands, such as the type of cover and the percentage of crown cover, are also important.

While it was difficult to use the effects of stand age as a variable, since the stands were sampled on different dates, total numbers of mesofauna were compared for each soil type and for each sampling date. Total numbers of animals varied for each soil type on respective sampling dates for each age class. However, the Duncan range test showed that there were no statistical differences between soil types (95% level).

The effects of the sludge type and application rate are illustrated in Figures 3 through 8. In these figures, treatments indicated by bars that have the same letter(s) are statistically similar with respect to the Duncan range test (95% level). Since seasonal fluctua-

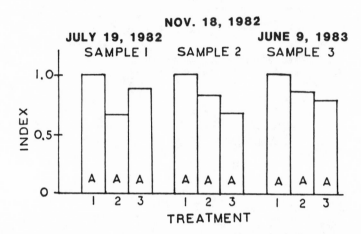

Figure 3. Mite indices for the 1973 sites. Treatment 1 = control; 2 = application of 400 kg N/ha Augusta sludge; 3 = application of 800 kg N/ha Augusta sludge. Treatments with like letters are statistically similar (95% level) for Duncan's range test.

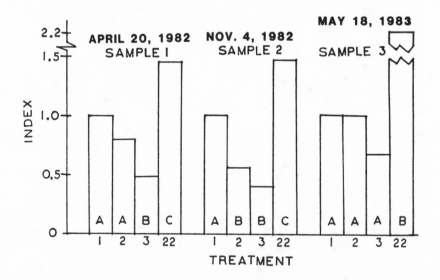

Figure 4. Mite indices for the 1953 sites. Treatment 1 = control; 2 = application of 400 kg N/ha Augusta sludge; 3 = application of 800 kg N/ha Augusta sludge; 22 = application of 630 kg N/ha Aiken sludge. Treatments with like letters are statistically similar (95% level) for Duncan's range test.

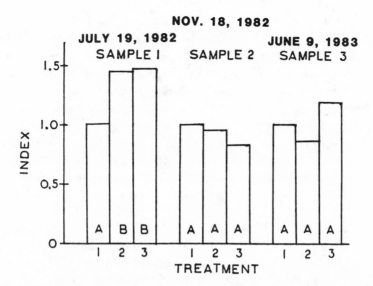

Figure 5. Total mesofauna indices for the 1973 sites. Treatment 1 = control; 2 = application of 400 kg N/ha Augusta sludge; 3 = application of 800 kg N/ha Augusta sludge. Treatments with like letters are statistically similar (95% level) for Duncan's range test.

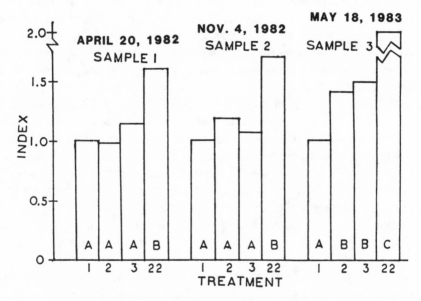

Figure 6. Total mesofauna indices for the 1953 sites. Treatment 1 = control; 2 = application of 400 kg N/ha Augusta sludge; 3 = application of 800 kg N/ha Augusta sludge; 22 = application of 630 kg N/ha Aiken sludge. Treatments with like letters are statistically similar (95% level) for Duncan's range test.

tions play a role in the total number of animals, an index is used in which the average number of animals for each treatment is divided by the average number of animals for the control, each treatment, and each data sampling.

Figures 3 and 4 illustrate the effects of the sludge type and application rate on the mite populations for the 1973 and 1953 age classes, respectively. The 1973 plots showed no statistical differences for the control, the 400 kg N/ha, and the 800 kg N/ha Augusta sludge treatments for all the samplings. For sample 1, the 800 kg N/ha Augusta sludge and the 630 kg N/ha Aiken sludge treatments were statistically different from each other, and from the 400 kg N/ha Augusta sludge treatment and the control for total numbers of mites. The latter two were not statistically different from each other for the 1953 sites. The 400 kg N/ha and 800 kg N/ha Augusta sludge treatments were not statistically different for the second sampling. The 630 kg N/ha Aiken sludge treatment was statistically different from all other treatments for the third sampling.

The statistical results for total fauna, which are shown in Figures 5 and 6, were the same as for the mite plots with one exception. The 400 kg N/ha Augusta sludge treatment was similar to both the control and the 800 kg N/ha Augusta sludge treatment. However, the control and the 800 kg N/ha Augusta sludge treatment were statistically different.

Only three treatments were statistically different for the Collembola populations. Figure 7 shows that for the 1973 sites, the 400 kg N/ha Augusta sludge treatment was statistically different from the control and the 800 kg N/ha Augusta sludge treatment for the first sampling. Figure 8 indicates that the 630 kg N/ha Aiken sludge treatment was statistically different from all other treatments for samples 1 and 2 on the 1953 sites.

The results, which indicate a general increase in fauna for the Aiken sludge treatments and a decrease in fauna for the Augusta treatments, can be accounted for when considering the sludge type. The Augusta sludge, as noted earlier, had approximately 2% solids in suspension. The Aiken sludge has undergone aerobic digestion and been dewatered to 30% solids. The fact that the Augusta sludge is primarily from domestic sources and the Aiken sludge primarily from industrial sources is not as important as the physical and chemical characteristics of the respective sludges. The Augusta sludge (7% nitrogen) was in a more dissolved state than the Aiken (1.3% nitrogen) sludge upon application (see Wells et al. in this volume, Chapter 15, for more information on chemical properties). The Augusta sludge coated vegetative matter and puddled on the forest floor, while the Aiken sludge formed aggregates of irregular shapes, ranging in size up to several centimeters in diameter. Much of the dissolved Augusta sludge was able to leach into the mineral portion of the soil, providing nutrients to plants. Since this sludge was in a well-digested state, it is believed that bacteria were able to increase in numbers and aid in decomposing the forest floor. A reduction of the forest floor depth with clogging of the interstitial space would then result in a decrease in the numbers of mesofauna. A high soluble nitrogen concentration, as reported for the Augusta treatment by Wells et al. (this volume), may decrease mesofauna populations (Hill et al. 1975). The Aiken sludge increased the forest floor, owing to its less decomposed state and greater solids content, but did not greatly increase nitrogen in the soil solution. It is believed that the Aiken sludge provided a better source of food for the mesofauna than the Augusta sludge, and the aggregates of sludge increased the surface area and mass of organic matter upon which the soil animals live, thus resulting in an increase in mesofaunal numbers. This theory can be further substantiated in that the Augusta plots exhibited darker crown covers and an increase in understory growth, which is characteristic of fertilizer applications, indicating that the Augusta sludge was well decomposed and its nutrients were available for uptake.

The effect of sewage sludge on the mesofaunal community structure is illustrated by comparing Figures 3 through 8, which show the indices of each mesofaunal group for both the 1973 and 1953 sites. The Augusta sludge generally reduced the oribatid mite populations, although the differences may not have been statistically significant (Figures 3 and 4). Collembola populations increased with the Augusta sludge treatment on the 1973 sites at sample 1, decreased with sample 2, and them remained constant (Figures 7 and 8). The Aiken sludge treatment resulted in a statistically significant increase in both the mite and Collembola populations (Figures 3, 4, 7, and 8). This information would

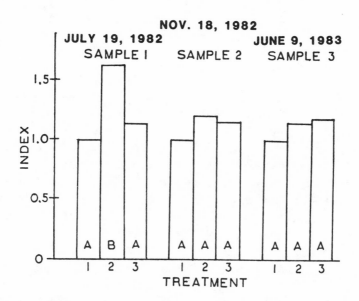

Figure 7. Collembola indices for the 1973 sites. Treatment 1 = control; 2 = application of 400 kg N/ha Augusta sludge; 3 = application of 800 kg N/ha Augusta sludge. Treatments with like letters are statistically similar (95% level) for Duncan's range test.

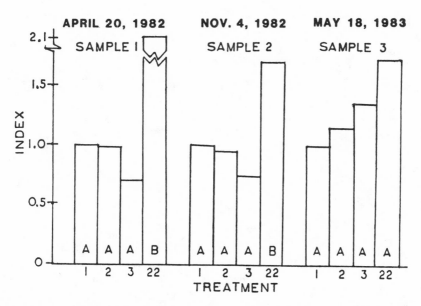

Figure 8. Collembola indices for the 1953 sites. Treatment 1 = control; 2 = application of 400 kg N/ha Augusta sludge; 3 = application of 800 kg N/ha Augusta sludge; 22 = application of 630 kg N/ha Aiken sludge. Treatments with like letters are statistically similar (95% level) for Duncan's range test.

indicate that sludge has an impact on the structure of the mesofaunal community. Indications are that a sludge with a high solids content such as the Aiken sludge would increase both mite and Collembola populations as a result of increased food and surface area. A well-digested sludge with a low solids content such as the Augusta sludge would reduce the mite population as a result of reduced surface area brought about by interstitial clogging. Collembola would not be affected, since they are more mobile than mites and may increase as a result of an increase in food.

CONCLUSION

When applied to forest soil, municipal sludge has an impact on the soil mesofauna population and structure. It was determined that mesofaunal populations do exhibit seasonal fluctuations, although the total numbers of mesofauna may be decreased or increased as a result of sludge applications. The sandy Lucy soil series showed no statistical difference from the sandy clay Orangeburg soil series with respect to mesofaunal populations. In general, a well-digested anaerobic sludge with a low solids content, such as the Augusta sludge, reduced the total number of mesofauna, with application rates serving as a function. An aerobically digested sludge with a high solids content, such as the Aiken sludge, increased the total number of mesofauna within the forest soil. More information is needed to determine the duration of the effects of sludge treatments on mesofauna.

Municipal sludge application may also result in a change of the mesofauna community structure. The Augusta sludge generally reduced the number of mites on both age classes, with an eventual increase in the number of Collembola for both the 1973 and 1953 sites. The Aiken sludge treatment increased both mites and Collembola but did not appear to change the community structure.

ACKNOWLEDGMENTS

This research was funded by the U.S. Department of Energy, contracts DE-AC09-76SR00819, DE-AC09-SR00001, and DE-A109-80SR10711. Research for this paper was conducted while Gary S. MacConnell was a graduate student at Duke University; it was presented while he worked for the Florida Department of Environmental Regulation, Bureau of Wastewater Management and Grants; he is currently employed by Camp Dresser and McKee Inc.

REFERENCES

David, M. B., and R. A. Struchtemeyer. 1980. Effects of spraying sewage effluent on forested land at Sugarloaf Mountain, Maine. Bulletin 773. Life Science and Agriculture Experiment Station, University of Maine, Orono.

Dindal, D. L. 1977. Influence of human activities on oribatid mite communities. p. 105–120. In Biology of oribatid mites. SUNY College of Environmental Science and Forestry, Syracuse, New York.

Dindal, D. L., L. T. Newell, and J.–P. Moreau. 1979. Municipal wastewater irrigation: Effects on community ecology of soil invertebrates. p. 197–205. In W. E. Sopper and S. N. Kerr (eds.) Utilization of municipal sewage effluent and sludge on forest and disturbed land. Pennsylvania State University Press, University Park.

Dindal, D. L., D. Schwert, and R. A. Norton. 1975. Effect of sewage effluent disposal on community structure of soil invertebrates. p. 419–427. In J. Vanck (ed.) Progress in soil zoology. Dr. W. Junk and Academia Publ., The Hague and Prague.

Ellis, B. G. 1973. The soil as a chemical filter. p. 45–70. In W. E. Sopper and L. T. Kardos (eds.) Recycling treated municipal wastewater and sludge through forest and cropland. Pennsylvania State University Press, University Park.

Hall, G. F. 1978. Site selection considerations for sludge and wastewater application on agricultural land: A planning and educational guide. U.S. Environmental Protection Agency, Washington, D.C.

Hill, S. B., L. J. Metz, and M. H. Farrier. 1975. Soil mesofauna and silvicultural practices. In B. Bernier and C. H. Winget (eds.) Forest soils and forest land management. Les Presses de l'Université Laval, Quebec.

Hollod, G. J. 1981. Environmental effects of biomass fuels program. Draft.

Khaleel, R., K. R. Reddy, and M. R. Overcash. 1981. Changes in soil physical properties due to organic waste applications: A review. J. Environ. Qual. 10:133–141.

Koch, P. 1982. Wastewater sludge: Will it help the forest? National Woodlands, Jan.-Feb. p. 13–16.

Lejcher, T. R., and S. H. Kunkle. 1973. Restoration of acid spoil banks with treated sewage sludge. p. 184–199. In W. E. Sopper and L. T. Kardos (eds.) Recycling treated municipal wastewater and sludge through forest and cropland. Pennsylvania State University Press, University Park.

Metz, L. J., and D. L. Dindal. 1980. Effects of fire on soil fauna in North America. p. 450–459. In Soil biology as related to land use practices. Proceedings of the VII International Colloquium of Soil Zoology.

Metz, L. J., and M. H. Farrier. 1973. Prescribed burning and populations of soil mesofauna. Environ. Entomology 2:433–440.

Murphey, W. K., R. L. Brisbin, W. J. Young, and B. E. Cutter. 1973. Anatomical and physical properties of red oak and red pine irrigated with municipal wastewater. p. 295–310. In W. E. Sopper and L. T. Kardos (eds.) Recycling treated municipal wastewater and sludge through forest and cropland. Pennsylvania State University Press, University Park.

Murphy, P. W. 1955. Note on processes used in sampling, extraction, and assessment of the mesofauna. p. 338–340. In D. K. McE. Kevan (ed.) Soil zoology. Academic Press, New York.

Nutter, W. D. 1978. Wastewater irrigation in southern forests. p. 138–147. In Proceedings, Soil Moisture … Site Productivity Symposium. USDA Forestry Service, Asheville, North Carolina.

Pratt, P. F., M. D. Thorn, and F. Wiersma. 1977. The future directions of waste utilization. p. 632–634. In Soils for management of organic wastes and wastewaters. ACA, USSA, SSSA, Madison, Wisconsin.

Pritchett, W. L. 1979. Properties and management of forest soils. John Wiley and Sons, New York.

Riekerk, H. 1978. The behavior of nutrient elements added to a forest soil with sewage sludge. Soil Sci. Soc. Am. J. 42:810–816.

Riekerk, H., and R. J. Zasoski. 1979. Effects of dewatered sludge applications to a Douglas fir forest soil on the soil, leachate, and groundwater composition. p. 35–45. In W. E. Sopper and S. N. Kerr (eds.) Utilization of municipal sewage effluent and sludge on forest and disturbed land. Pennsylvania State University Press, University Park.

Smith, W. H., D. M. Post, and F. W. Adrian. 1979. Waste management to maintain or enhance productivity. p. 304–320. In Proceedings, Impact of Intensive Harvesting on Forest Nutrient Cycling. State University of New York, Syracuse.

Sopper, W. E. 1973. Crop selection and management alternatives. p. 143–153. In Proceedings of the Joint Conference on Recycling Municipal Sludges and Effluents on Land. National Association of State Universities and Land Grant Colleges, Washington, D.C.

Urie, D. H. 1979. Nutrient recycling under forests treated with sewage effluents and sludges in Michigan. p. 7–17. In W. E. Sopper and S. N. Kerr (eds.) Utilization of municipal sewage effluent and sludge on forest and disturbed land. Pennsylvania State University Press, University Park.

Chapter 18

Deer and Elk Use of Forages Treated with Municipal Sewage Sludge

HENRY CAMPA III, DAVID K. WOODYARD,
and JONATHAN B. HAUFLER

ABSTRACT   The effects of browsing by white-tailed deer (*Odocoileus virginianus*) and elk (*Cervus elaphus canadensis*) on vegetation in sludge-treated plots in a 10-year-old aspen (*Populus* spp.) clearcut in Michigan were monitored. Browsing effects on vegetative composition and structure were determined by comparing areas open to browsing with areas in exclosures. Ungulates browsed sludged-treated forages significantly more than untreated vegetation. Browsing decreased vertical and horizontal cover and the densities of key browse species. Samples of selected forage species were analyzed for nutritive quality, and sludge-treated vegetation was found to contain higher levels of protein. In addition, organs from three deer harvested on the study area were analyzed for concentrations of selected elements.

The application of sewage sludge to appropriate forest lands could increase the quantity and quality of wildlife forages. Increases in plant productivity following sludge application have been reported by Anderson and Barrett (1982) and Woodyard (1982). Increases in plant nutritive quality after sludge application have been reported by Campa (1982) and Anderson (1983), with increases in crude protein levels being the major difference.

As a result of the increase in quantity and quality of vegetation treated with municipal sludge, herbivore browsing pressure may increase on treated areas. While this could represent an improvement in habitat quality for herbivores, it could also lead to two problems. First, large herbivores, such as deer and elk, can have a significant effect on vegetation if grazing or browsing pressures are great enough. Intense browsing can change vegetative structure and composition, and may also mask increases in plant productivity produced by sludge application. If the effect of ungulate foraging is not identified and monitored, all other vegetation studies on sludged areas may become meaningless. Second, deer and elk are important game species that are prized for their meat. Sludge applied to the study area may contain a number of potentially toxic metals which may accumulate in the tissues of deer or elk and thus enter the human food chain. For this reason, it is important to determine the degree of dietary exposure of ungulates to potentially toxic metals, and to determine if accumulation of these metals in the tissues of organs consumed by humans is high enough to be of concern.

This study investigated the extent of ungulate—white-tailed deer (*Odocoileus virginianus*) and elk (*Cervus elaphus canadensis*)—browsing on sludge-treated and untreated areas of an aspen clearcut, and determined the influence that the level of browsing had on vegetative composition and structure. In addition, the levels of selected metals in organs of deer were investigated to evaluate potential concerns of human consumption.

STUDY AREA

The study area was a 14 hectare, 10-year-old aspen (*Populus* spp.) clearcut, located in the northern half of Section 24, 45°N, 84°10'W Montmorency County in Michigan's northern lower peninsula. The climate of the area is typical of northern lower Michigan, with long, severe winters, short cool summers, and short growing seasons. Average annual precipitation is 765 mm (30 inches).

Soils on the study area are of Rubicon (Entic Haplorthod), Montcalm (Eutric Glossoboralf), and Grayling (Typic Udipsamment) series (Nguyen and Hart 1984). All soils from these series are sandy and well drained. The study area in the past was repeatedly roller chopped and burned in an unsuccessful attempt to maintain it as a wildlife opening. These treatments resulted in a mosaic pattern of aspen clumps and grassy openings. Overstory vegetation consisted of bigtooth aspen (*Populus grandidentata*), quaking aspen (*P. tremuloides*), and pin cherry (*Prunus pennsylvanica*). The major understory species were panic grass (*Panicum viragatum*), brambles (*Rubus* spp.), sweet fern (*Myrica asplenifolia*), sedge (*Carex* spp.), and bracken fern (*Pteridium aquilinum*).

METHODS AND MATERIALS

The study area was divided into nine plots, each approximately 1.5 ha and separated by buffer zones (Figure 1). Trails 5 m wide for sludge application were cleared 15 m apart on six of the plots. Because trail construction was expected to have a treatment effect, the

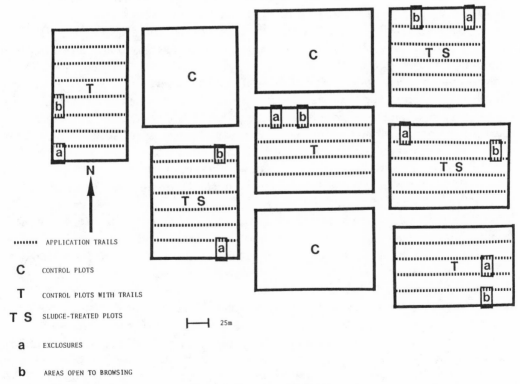

Figure 1. Experimental design.

study was designed with two treatments and a control. The nine study plots were randomly assigned into three plots with application trails and sludge treatment, three plots with application trails only, and three control plots. During October 1981, anaerobically digested, nonindustrial municipal sewage sludge from Alpena, Michigan was applied to treatment areas at a loading rate of approximately 9,980 kg/ha (8,904 lb/acre).

Browsing estimates were made on all plots of the study area. Estimates were conducted using randomly located belt transects. Browse utilization was determined as the percentage of stems browsed of the first 100 stems counted <2 m in height (Gysel and Lyon 1980). A height limit of 2 m was used because it was assumed that primary production higher than that was unavailable for deer and elk consumption.

Estimates of the influence of deer and elk on the composition and structure of sludge-treated and untreated vegetation were determined by conducting various vegetation measures in exclosures and areas open to ungulate browsing. Exclosures were constructed on each of the sludge-treated plots and on the plots with application trails only. The exclosures were 25 by 15 by 2.1 m and encompassed portions of two application trails and an interior area. On each plot with exclosures, areas open to browsing were delineated that had similar vegetation composition and structure as within the exclosures at the start of the investigation. The line intercept method (Canfield 1941) was used to measure the vertical cover in three height strata (0-0.5, 0.5-2.0, and >2.0 m) above randomly located line transects. Horizontal cover estimates were made using randomly located profile boards as described by Gysel and Lyon (1980). Stem densities of woody vegetation were determined in three height classes (0-0.5, 0.5-2.0, and >2.0 m) by counting all stems of each dominant woody species in both the exclosures and areas open to browsing. Frequency of occurrence for grasses and forbs was determined as the presence of each species in randomly located 1 by 1 m quadrats.

Selected plant species were collected from all plots during the spring, summer, and winter. These included a variety of herbaceous and woody species that were sufficiently abundant to be collected on all plots and that were important as wildlife forages. Three samples of each species were collected from randomly located belt transects in each plot. Sample material from approximately 200 individual plants was collected to eliminate individual plant variability. Enough sample material was collected to supply approximately 100 grams of dried, ground sample. For herbaceous species, the aboveground portion of the plants was clipped. For spring and summer samples of woody species, only leaves were removed. Winter samples of woody species consisted of twigs.

Samples were dried at 60°C until weight was constant. After drying, all samples were ground in a Wiley mill to pass a 1 mm sieve.

Vegetation samples were analyzed for ash percentage by methods stated in AOAC (1980). Total nitrogen and phosphorus were determined with a Kjeldahl digestion method using a Tecator Block Digestor, Model DS-40 (Tecator, Inc., Boulder, Colorado). Once samples were digested, nitrogen and phosphorus values were obtained using a Technicon Autoanalyzer II (Technicon Industrial Systems, Tarrytown, New York). Percentage of crude protein was calculated as described by AOAC (1980), using total nitrogen values. Fiber analyses were conducted as described by Goering and Van Soest (1970). In vitro dry matter digestibility was determined using the Tilley and Terry (1963) method modified as described by Campa et al. (1984). Inoculum for IVDMD trials was obtained from a fistulated nonlactating Holstein cow. Percentage of dry matter and ether extract were determined as described by Campa (1982).

Potential accumulation of elements in selected organs of white-tailed deer feeding on sludge-treated forages was investigated by collecting three does on the study area in November 1982. The organs analyzed included livers, kidneys, hearts, and skeletal muscles. Tissues were analyzed with a wet digestion procedure using nitric and perchloric acid. Element levels were determined using a DC-argon plasma atomic emission spectrometer (Spectrametrics, Inc., Andover, Massachusetts) (Dahlquist and Knoll 1978, De-Bolt 1980).

All data sets were tested for homogeneity of variance, and transformations were conducted if necessary to achieve homogeneity. A one-way analysis of variance was performed on the nutrition and browse utilization data. Duncan's new multiple range test was used to test the significant differences among treatments. Measures of vegetative composition and structure in the application trails and interior areas were compared between the exclosures and areas open to browsing by the use of paired t-tests (Steel and Torrie 1980).

RESULTS

Browse utilization data from 1984 and 1985 showed a trend from greatest to least utilization on sludge-treated plots, plots with application trails only, and control plots respectively (Table 1). Bigtooth and quaking aspen were not sampled for browse utilization in 1985, owing to the lack of current annual growth <2.0 m for sampling.

Mean stem densities on the study area tended to be greater in exclosures than in areas open to ungulate browsing (Table 2). In 1983, 56% of the species in all height strata within the trails and interiors had greater stem densities in the exclosures than in areas open to browsing. This percentage increased to 78% of all species in 1984.

Vertical vegetative cover had few significant differences between exclosures and areas open to browsing (Table 3). In the application trails, vertical cover in the upper height stratum was significantly greater in the exclosures than in browsed areas. Cover in the interior areas was significantly greater in the exclosures in only the second height stratum.

TABLE 1. Percentage of browsing ($\pm$ SE) of woody species on all plots in the study site (1984 and 1985).

Species		Sludge treated	Trails only	Controls without trails	Probability
			Treatments		
Quaking aspen	1984	36.1 $\pm$ 6.7a	22.9 $\pm$ 6.2b	18.0 $\pm$ 2.1b	p < 0.10
	1985	NAB*	NAB	NAB	--
Bigtooth aspen	1984	60.3 $\pm$ 9.8a	29.1 $\pm$ 2.5b	23.0 $\pm$ 4.4b	p < 0.10
	1985	NAB	NAB	NAB	--
Pin cherry	1984	46.3 $\pm$ 1.9a	37.8 $\pm$ 2.8b	46.0 $\pm$ 3.0a	p < 0.10
	1985	17.0 $\pm$ 1.0a	13.7 $\pm$ 1.5b	12.8 $\pm$ 1.0b	p < 0.10

Any two means within a row with different letters indicate a significant difference.
*No available browse <2.0 m.

Horizontal cover measurements, and both absolute and relative frequency estimates, showed few significant differences in vegetative structure and composition respectively between the exclosures and browsed areas. Overall, a trend was observed of a greater percentage of cover in the exclosures than in browsed areas. Relative and absolute frequencies of some herbaceous species tended to be lower in the exclosure than in the browsed areas. One year after sludge application, 25% of the herbaceous species in the

TABLE 2. Mean stem densities per hectare (± SE) of trees (0-0.5, 0.5-2.0, and >2.0 m in height) in exclosures and browsed areas (1983 and 1984).

Species	Stratum (m)	Exclosure		Browsed	
		1983	1984	1983	1984
Trails					
Quaking aspen	0-0.5	567 + 136	167 + 58a	520 + 220	745 + 333
	0.5-2.0	573 ∓ 253a	645 ∓ 247	167 ∓ 128	289 ∓ 162
	>2.0	0 ∓ 0	33 ∓ 22a	0 ∓ 0	0 ∓ 0
Bigtooth aspen	0-0.5	1,400 + 340	622 + 146	1,466 + 720	378 + 190
	0.5-2.0	1,666 ∓ 420a	1,811 ∓ 380	920 ∓ 406	900 ∓ 526
	>2.0	0 ∓ 0	33 ∓ 14a	0 ∓ 0	0 ∓ 0
Pin cherry	0-0.5	678 + 309	800 + 313	353 + 101	367 + 128
	0.5-2.0	867 ∓ 33a	911 ∓ 342a	166 ∓ 72	244 ∓ 117
	>2.0	0 ∓ 0	0 ∓ 0a	0 ∓ 0	33 ∓ 14
Interior Areas					
Quaking aspen	0-0.5	0 + 0a	22 + 15	33 + 15	8 + 7
	0.5-2.0	400 ∓ 112	385 ∓ 114	400 ∓ 179	370 ∓ 160
	>2.0	2,622 ∓ 666	2,563 ∓ 657	1,857 ∓ 893	2,200 ∓ 1,164
Bigtooth aspen	0-0.5	102 + 80a	30 + 14	0 + 0	52 + 29
	0.5-2.0	118 ∓ 48	215 ∓ 68	163 ∓ 88	111 ∓ 35
	>2.0	1,276 ∓ 271	948 ∓ 281	786 ∓ 518	1,081 ∓ 530
Pin cherry	0.0.5	355 + 172	1,733 + 785	377 + 67	852 + 339
	0.5-20	2,244 ∓ 867a	2,578 ∓ 965	875 ∓ 145	1,096 ∓ 355
	>2.0	788 ∓ 305a	1,170 ∓ 578a	320 ∓ 106	407 ∓ 152

a = Significantly different (p < 0.10) from browsed within years.

TABLE 3. Mean percentage of vegetative cover (± SE) for height strata in exclosures and browsed areas (1983 and 1984).

Stratum (m)	Exclosure		Browsed	
	1983	1984	1983	1984
Trails				
0-0.5	91.1 + 4.8	66.2 + 3.3	87.5 + 4.6	60.8 + 2.8
0.5-2.0	0 ∓ 0a	19.2 ∓ 3.0	11.8 ∓ 2.7	20.1 ∓ 2.9
>2.0	17.9 ∓ 4.1	18.2 ∓ 5.0b	5.3 ∓ 1.9	0.2 ∓ 0.2
Interior Areas				
0-0.5	57.0 + 2.8	79.1 + 2.1	53.8 + 4.2	76.3 + 1.8
0.5-2.0	21.1 ∓ 4.5	44.5 ∓ 3.3a	19.0 ∓ 2.9	36.6 ∓ 2.7
>2.0	21.8 ∓ 4.4	47.9 ∓ 3.4	28.4 ∓ 3.4	47.5 ∓ 3.3

a = Significantly different (p < 0.10) from browsed within years.
b = Significantly different (p < 0.01) from browsed within years.

interiors had frequencies greater in browsed areas than in the exclosures. In the application trails, 33% of the herbaceous species exhibited this type of response. In 1984, two years after treatment, the percentage of herbaceous species that had greater frequencies in the browsed areas increased to 50% in both the interiors and application trails.

Application of municipal sewage sludge significantly increased crude protein concentrations in most of the important wildlife forages for the first spring and summer seasons following sludge treatment (Tables 4 and 5). Phosphorus concentrations were also significantly greater in all species on sludge-treated plots during the spring, but only for herbaceous species in the summer. No other nutrient demonstrated an obvious trend with respect to sludge treatment or trail construction during the first growing season following sludge treatment. After the first growing season, the greater crude protein and phosphorus levels declined to control levels (Table 6).

Deer kidneys and livers were the tissues in which most of the selected metals were concentrated (Table 7). Kidneys had the highest concentrations of zinc and cadmium, while liver was the tissue highest in copper and chromium. Nickel was in the highest concentration in skeletal muscle.

DISCUSSION

Results of the browsing estimates made in 1984 and 1985 on the study area indicate that ungulates browsed more heavily on sludge-treated vegetation than on untreated vegetation. They selectively chose vegetation that had been amended with additional nutrients. As documented by Bayoumi and Smith (1976), deer will selectively browse on forages with a higher protein content. As demonstrated, bigtooth and quaking aspen collected from sludge-treated plots had crude protein levels significantly greater than on plots of the two other treatments. Such increased browsing pressure on vegetation treated with sewage sludge has been documented previously (Woodyard 1982).

The trend of heavier browsing on untreated plots with trails than on control plots without trails may be attributed to the increased accessibility the trails provided. The application trails thus may serve as travel lanes for ungulates. The trails also increase the availability of forage species, possibly attracting more animals to these plots.

In 1983 the three woody species sampled had significantly greater stem densities for the middle height stratum in the trails of the exclosures than on the browsed areas. On the study area in 1984, both aspen species showed significantly greater stem densities in the upper height stratum in the trails of the exclosures than on the areas open to browsing. Decreased numbers of taller aspen stems in trails may be the result of ungulate browsing that inhibited the growth of aspen shoots into the upper height stratum. Similar results were observed for pin cherry in the middle height stratum.

An increase in the percentage of vertical cover from 1983 to 1984 in the interiors of exclosures may be attributed to protection of vegetation from browsing. Cover in the lowest height stratum within the application trails of the exclosures decreased, perhaps because of the development of vegetation in the upper height strata, thus shading the ground-level vegetation. This conclusion is supported by the increase in frequency of herbaceous species in the browsed areas.

Results from analyses of the trace metal content of deer tissues were quite similar to those of Munshower and Neuman (1979), in that most metals were selectively concentrated in the kidneys and livers. However, these tissues had mean metal burdens

TABLE 4. Comparison of percentage of ash, in vitro dry matter digestibility (IVDMD), ether extract (EE), phosphorus (P), crude protein (CP), cell soluble material (CSM), hemicellulose (HEMI), acid detergent lignin (ADL), and cellulose (CELL) for forages collected on the study area in the spring of 1982.

Species	Treatment	Ash	IVDMD	EE	P	CP	CSM	HEMI	ADL	CELL
Bigtooth aspen	Control**	7.5	40.8	7.5	0.27a*	15.6a*	51.6	9.0	27.0	12.0
	Trails only	7.1	40.4	5.8	0.24a	14.2a	48.4	8.5	29.4	14.5
	Sludge	7.9	37.9	6.1	0.38b	21.6b	42.8	10.4	31.6	16.3
Quaking aspen	Control	6.1	39.9	7.2	0.25ab	17.2a	57.6	10.5	21.9	11.5
	Trails only	6.7	35.8	7.7	0.20a	16.1a	56.1	9.9	23.2	9.9
	Sludge	6.4	41.6	6.5	0.32b	23.4b	53.8	9.5	23.7	13.2
Pin cherry	Control	4.7	42.8	4.5	0.22a	15.2a	65.0	16.4	5.1	13.6
	Trails only	5.5	46.3	4.1	0.25a	15.6a	71.1	12.5	6.3	10.1
	Sludge	5.1	50.8	3.6	0.35b	22.9b	65.8	17.8	6.3	12.2
Wild strawberry	Control	9.0	42.8	4.7	0.19ab	10.1a	75.7	1.7	3.7	18.9
	Trails only	8.1	47.9	4.0	0.13a	11.4a	79.0	2.9	3.8	14.3
	Sludge	7.1	45.7	5.9	0.23b	15.2b	76.7	4.8	3.1	15.4
Orange hawkweed	Control	13.5	53.4	5.8	0.10a	11.7a	58.4	3.3	5.1	33.2
	Trails only	10.9	65.2	4.2	0.12ab	9.3a	57.2	5.1	7.1	30.6
	Sludge	12.0	57.9	4.9	0.15	14.2b	64.5	8.1	6.3	21.1
Panic grass	Control	4.7	39.7	3.5	0.08a	9.5a	65.4	26.0	4.9	3.7
	Trails only	5.2	38.6	1.3	0.09a	9.1a	57.7	34.4	4.5	3.4
	Sludge	7.4	45.5	2.0	0.17b	17.0b	54.8	37.8	5.1	2.3

*x̄ values within a column for a species with different letters indicate a significant difference (p < 0.10).

**x̄ sum from three plots of each treatment.

TABLE 5. Comparison of percentage of ash, in vitro dry matter digestibility (IVDMD), ether extract (EE), phosphorus (P), crude protein (CP), cell soluble material (CSM), hemicellulose (HEMI), acid detergent lignin (ADL), and cellulose (CELL) for forages collected on the study area in the summer of 1982.

Species	Treatment	Ash	IVDMD	EE	P	CP	CSM	HEMI	ADL	CELL
Bigtooth aspen	Control**	5.8	34.9	5.4	0.22	12.6a*	42.1	11.9	32.0	14.0
	Trails only	4.8	37.2	3.8	0.21	14.8ab	40.8	13.1	30.5	15.6
	Sludge	5.7	30.0	4.3	0.25	17.4b	37.4	14.7	31.1	16.8
Quaking aspen	Control	5.7	34.4	7.5	0.13	11.2a	47.2	11.0	29.7	12.1
	Trails only	5.7	36.5	7.3	0.20	12.7a	44.2	13.3	27.8	14.7
	Sludge	6.3	30.5	5.4	0.29	16.5b	48.2	14.0	24.9	12.9
Pin cherry	Control	8.0	40.7	2.7	0.24	14.1a	58.3	15.1a	8.4	18.2
	Trails only	8.4	45.3	4.2	0.20	14.9a	57.3	20.6b	7.2	14.9
	Sludge	9.1	48.0	4.8	0.22	25.3b	61.0	16.8a	8.9	13.3
Wild strawberry	Control	6.7	43.9	2.3	0.14a	9.5a	69.8	6.6	3.6	20.0
	Trails only	7.9	40.0	2.1	0.13a	8.6a	62.8	8.4	2.7	26.1
	Sludge	7.6	47.2	1.1	0.31b	12.7b	64.4	8.5	4.9	22.2
Orange hawkweed	Control	13.0	39.8	2.0	0.04a	8.9a	57.3	5.2	7.2	30.3
	Trails only	10.4	36.6	2.6	0.13b	13.6b	55.4	4.7	8.9	31.0
	Sludge	19.4	37.2	2.9	0.26c	16.9c	55.1	4.9	5.2	34.8
Panic grass	Control	6.6	28.2a	1.3	0.06ab	10.7a	55.1a	34.1	5.6	5.2a
	Trails only	7.7	32.1b	2.1	0.09ab	12.5a	46.0b	37.8	6.8	9.4b
	Sludge	7.1	50.0b	1.2	0.16b	15.8b	56.7a	34.4	5.4	3.5a

*x̄ values within a column for a species with different letters indicate a significant difference (p < 0.10).

**x̄ sum from three plots of each treatment.

TABLE 6. Comparison of percentage of ash, in vitro dry matter digestibility (IVDMD), ether extract (EE), phosphorus (P), crude protein (CP), cell soluble material (CSM), hemicellulose (HEMI), acid detergent lignin (ADL), and cellulose (CELL) for forages collected on the study area in the winter of 1982.

Species	Treatment	Ash	IVDMD	EE	P	CP	CSM	HEMI	ADL	CELL
Bigtooth aspen	Trails only**	4.3	26.1	5.1	0.07	6.1a*	29.8	11.6	37.4	21.2
	Sludge	5.2	24.8	5.6	0.07	10.3b	31.2	8.8	35.6	24.4
Quaking aspen	Trails only	5.1	29.9	5.0	0.05	7.8	34.7	9.7	32.6	23.0
	Sludge	4.9	27.5	4.9	0.08	9.5	42.8	6.3	33.4	17.5
Pin cherry	Trails only	2.9	25.7	8.4	0.04	8.8	35.6	9.1	29.5	25.8
	Sludge	3.5	31.4	7.4	0.05	6.5	33.7	10.2	31.7	24.4

*x̄ values within a column for a species with different letters indicate a significant difference (p < 0.10).

**x̄ sum from three plots of each treatment.

TABLE 7. Mean trace metal content of skeletal muscle, heart, kidney, and liver of white-tailed deer harvested on the study area in November 1982.

Tissue	Mean Trace Metal Content (ug/g dry weight)				
	Zinc	Cadmium	Copper	Nickel	Chromium
Skeletal muscle	399.2 + 26.2*	1.08 + 0.29	9.39 + 0.89	1.70 + 0.17	1.26 + 0.65
Heart	397.3 + 62.8	0.81 + 0.32	19.57 + 3.53	0.60 + 0.44	0.43 + 0.14
Kidney	858.0 + 184.0	31.36 + 1.28	21.16 + 3.88	0.99 + 0.39	0.85 + 0.20
Liver	688.5 + 74.6	3.13 + 0.32	473.8 + 71.5	1.53 + 0.27	1.59 + 1.0

*$\bar{x}$ + SE from one deer per age class: 1½ years, 2½ years, 3½ years.

(cadmium and zinc) that were greater than those reported by Munshower and Newman (1979) for mule deer (*Odocoileus hemionus*) in Montana. Zinc concentrations, however, were within the limit considered to be nontoxic (1,000 µg/g), as stated by Underwood (1977) and Torrey (1979). In addition, other researchers have documented that there is little potential for zinc accumulation by herbivores (Roberts and Johnson 1978).

The cadmium concentrations in liver tissue were within the naturally occurring range of levels found in deer from Illinois (0.02 to 6.5 µg/g dry weight) (Woolf et al. 1982). The cadmium levels in kidneys, although higher than those reported by Woolf et al. (1982), were not believed to present toxicity problems to deer or to human consumers. Cadmium toxity to deer is unlikely, owing to (1) the low loading rate of cadmium (0.28 kg/ha) used in this study (Nguyen and Hart 1984), which was below the maximum annual loading rate of 0.5 kg/ha set by the U.S. Environmental Protection Agency (1979), and (2) the low cadmium concentrations in deer forages, which were well below levels found to produce chronic toxicities in laboratory animals (Woodyard et al., this volume). In addition, results may differ from those of Woolf et al. (1982) because of a difference in the chemical analysis methods used and the wide variation of metal concentrations often found among individual animals (Munshower and Neuman 1979).

CONCLUSIONS

From this study it may be concluded that application of nonindustrial municipal sewage sludge, prior to the growing season, may increase the quantity and quality of forages for wildlife use owing to the input of sludge-borne nutrients. Wildlife populations will respond to these increases with greater foraging activity on sludge-treated areas. Metal concentrations from tissues of deer collected from sludge-treated areas were not felt to represent a threat to deer or human health. Negative effects, however, may also occur. In areas with high ungulate populations, sludge treatment may increase browsing to the point of restricting the regeneration of palatable woody species. Heavy foraging may also mask or negate the changes in vegetation produced by sludge treatment. Therefore, if effects of ungulate foraging are not identified and monitored, studies of sludge treatment on plant communities may be misleading.

ACKNOWLEDGMENTS

Although the information in this document has been funded in part by the U.S. Environmental Protection Agency under assistance agreement No. S005551-01 to the Michigan Department of Natural Resources and Michigan State University, it has not been subjected to the Agency's publication review process and therefore may not necessarily reflect the views of the Agency, and no official endorsement should be inferred. Mention of trade names or commercial products does not constitute endorsement or recommendation for use.

Michigan Agricultural Experiment Station journal article number 11925.

REFERENCES

Anderson, D. A. 1983. Reproductive success of Columbian black-tailed deer in a sewage-fertilized forest in western Washington. J. Wildl. Manage. 47:243–247.

Anderson, T. J., and G. W. Barrett. 1982. Effects of dried sewage sludge on meadow vole (*Microtus pennsylvanicus*) populations in two grassland communities. J. Appl. Ecol. 19:759–772.

A.O.A.C. 1980. Official methods of analysis of the Association of Official Analytical Chemists. 13th ed. 1018 p.

Bayoumi, M. A., and A. S. Smith. 1976. Response of big game winter range vegetation to fertilization. J. Range Manage. 29:44–48.

Campa, H., III. 1982. Nutritional response of wildlife forages to municipal sludge application. M.S. thesis, Michigan State University, East Lansing. 88 p.

Campa, H., III, D. K. Woodyard, and J. B. Haufler. 1984. Reliability of captive deer and cow *in vitro* digestion values in predicting wild deer digestion levels. J. Range Manage. 37:468–470.

Canfield, R. H. 1941. Application of the line intercept method in sampling range vegetation. J. For. 39:388–394.

Dahlquist, R. L., and J. W. Knoll. 1978. Inductively coupled plasma-atomic emission spectrometry analysis of biological materials and soils for major trace and ultra-trace elements. Applied Spectroscopy 32:1–30.

DeBolt, D. C. 1980. Multielement emission spectroscopic analysis of plant tissue using DC argon plasma source. A.O.A.C. 63:802–805.

Freese, F. 1978. Elementary forest sampling. USDA For. Serv. Handbook 232. 91 p.

Goering, H. K., and P. J. Van Soest. 1970. Forage fiber analyses (apparatus, reagents, procedures, and some applications). USDA Handbook 379. 20 p.

Gysel, L. W., and L. J. Lyon. 1980. Habitat analysis and evaluation. p. 305–403. *In* S. D. Scheminitz (ed.) Wildlife management techniques manual. Wildlife Society, Washington, D.C. 686 p.

Munshower, F. F., and D. R. Neuman. 1979. Metals in soft tissues of mule deer and antelope. Bull. Environ. Contam. Toxicol. 22:827–832.

Nguyen, P. V., and J. B. Hart, Jr. 1984. Ecological monitoring of sludge fertilization on state forest lands in northern lower Michigan. Annual report. On file, U.S. EPA, Chicago. 192 p.

Roberts, R. D., and M. S. Johnson. 1978. Dispersal of heavy metals from abandoned mine workings and their transference to terrestrial food chains. Environ. Poll. 16:293–310.

Steel, R. G. D., and J. H. Torrie. 1980. Principles and procedures of statistics. McGraw-Hill, New York. 633 p.

Tilley, J. M. A., and R. A. Terry. 1963. A two-staged technique for the *in vitro* digestion of forage crops. J. Brit. Grassl. Soc. 18:104–111.

Torrey, S. 1979. Sludge disposal by landspreading techniques. Noyes Data Corporation, Park Ridge, New York. 372 p.

Underwood, E. J. 1977. Trace elements in human and animal nutrition. Academic Press, New York. 545 p.

U.S. Environmental Protection Agency. 1979. Criteria for classification of solid waste disposal facilities and practices. Federal Register 44:53438–53468.

Woodyard, D. K. 1982. Response of wildlife to land application of sewage sludge. M.S. thesis, Michigan State University, East Lansing. 64 p.

Woolf, A., J. R. Smith, and L. Small. 1982. Metals in levels of white-tailed deer in Illinois. Bull. Environ. Contam. Toxicol. 28:189–194.

Chapter 19

The Influence of Forest Application of Sewage Sludge on the Concentration of Metals in Vegetation and Small Mammals

DAVID K. WOODYARD, HENRY CAMPA III, and JONATHAN B. HAUFLER

ABSTRACT   Nonindustrial, municipal sewage sludge was applied to four forest types in Montmorency County, Michigan. Monitoring of selected wildlife forages and small mammal tissues for cadmium, copper, chromium, nickel, and zinc demonstrated no accumulation of these selected trace metals over two years of posttreatment.

Legislation enacted to halt the direct discharge of sewage waste into water systems has accelerated production of sewage sludges (U.S. EPA 1976), and resultant disposal problems have prompted many communities to evaluate land recycling of nutrient-rich sludges. Forests are of interest for land application because they are often on nutrient deficient sites unsuitable for agriculture. Especially appealing are the possible remunerative effects from recycling; double benefits may be achieved by lessening problems of water quality while enhancing forest production (Urie 1979).

To determine how to avoid nitrate degradation of groundwater quality, Brockway and Urie (1983) experimented with sludge application rates for some of Michigan's sand soils. Other environmental issues lead to additional research. Of particular concern were the fates of potentially toxic metals concentrated in sewage sludge during the waste treatment process (Chaney 1980). A critical question was whether metals would accumulate in the food base of sensitive, higher tropic level wildlife species, such as raptors.

As part of a research project undertaken to identify problems with sludge-borne metals entering wildlife food chains, this study's objective was to determine concentrations of selected metals in forages and small mammal tissues collected from both sludge and no-sludge sites. The hypothesis tested was that sludge application would significantly increase metal concentrations in forages and tissues of small mammals when applied at rates compatible with maintenance of groundwater quality.

MATERIALS AND METHODS

Study Areas

Sludge fertilization experiments were conducted on the Mackinac State Forest in northeastern lower Michigan (44°59′N, 84°10′W). Climate is typical of northern lower Michigan, with long, severe winters, short, cool summers, and abbreviated growing seasons. Average annual precipitation is 765 mm (30 inches). Four forest types were selected for sludge application in Montmorency County on the basis of their importance in this region, and included stands of 10-year-old aspen, 70-year-old oak, 50-year-old pine, and 50-year-old mixed hardwoods.

Past habitat manipulation of the aspen type included repeated roller chopping and burning in an unsuccessful effort to create a wildlife opening. The result was a mosaic pattern of clumps of bigtooth aspen (*Populus grandidentata*) and quaking aspen (*P. tremuloides*) interspersed with scattered pin cherry (*Prunus pennsylvanica*), red maple (*Acer rubrum*), and oaks (*Quercus* spp.). Major understory vegetation consisted of sweetfern (*Comptonia*), bracken fern (*Pteridium*), sedge (*Carex*), grasses (*Panicum*), and brambles (*Rubus*). Excessively and well-drained sands of the Rubicon (Entic Haplorthod) and Montcalm series (Eutric Glossoboralf) (Nguyen and Hart 1984) were the predominant soils of the area.

The upland oak stand supported an overstory mixture of red oak (*Q. rubra*), white oak (*Q. alba*), and red maple, and an understory dominated by bracken fern, wintergreen (*Gaultheria*), and sedges, with shrubs of witch hazel (*Hamamelis virginiana*) and serviceberry (*Amelanchier*). Excessively and somewhat excessively drained sands from the Grayling (Typic Udipsamment) and Rubicon series were predominant soils on the site.

The pine plantation consisted of jack pine (*Pinus banksiana*) interspersed with red pine (*P. resinosa*). A sparse midstory consisted of red oak and red maple. Dominant understory species were blueberry (*Vaccinium*), sweetfern, bracken fern, and sedges. Soils consisted of excessively drained sands of the Grayling series (Typic Udipsamment) and a well-drained coarse loam of the Montcalm series (Eutric Glossoboralf).

Sugar maple (*A. saccharum*), red maple, American beech (*Fagus grandifolia*), and American basswood (*Tilia americana*) were the important overstory species for the mixed hardwoods type. A well-developed midcanopy included individuals of the overstory species and white ash (*Fraxinus americana*), eastern hop hornbeam (*Ostrya virginiana*), and striped maple (*A. pennsylvanicum*). The understory was sparse but diverse and was described by Thomas (1983). Soils were sands of the Mancelona and Menominee series (Alfic Haplorthods) with smaller areas consisting of somewhat poorly drained soils of the Kawkawlin (Aquic Eutroboralfs) and Sims (Mollic Haplaquept) series.

Experimental Design

The experimental design of the study consisted of nine study plots, each 1.5 ha, delineated in each vegetation type. The basis for plot selection was first to minimize vegetative differences among plots in terms of composition and structure, and then to leave buffer zones between adjoining plots. Random selection was used to designate plots for treatment or control. Three plots were assigned as control plots with no planned manipulation. Since the sludge application method selected by Michigan's Department of Natural Resources required application trails, trail construction was considered a treatment. Application trails were cleared, but no sludge was applied on three plots. The remaining three plots had application trails cut and received sludge. Forage species collected for metal analysis from each of the four forest types are listed in Table 1.

In 1982 and 1983, small mammals were collected during late summer when populations were highest. Species that were collected infrequently were not included in statistical analyses. Adequately abundant species (Table 2) were obtained from all plots via Sherman live-traps baited with a mixture of oats, fat, and anise extract. Specimens were identified with a number and frozen for future analysis of metals. Livers, kidneys, hum-

TABLE 1. Forage species collected for metal analysis from each forest type.

Aspen	Oak	Pine	Hardwoods
Wild strawberry (Fragaria virginiana)	Red maple (Acer rubrum)	Red maple (Acer rubrum)	Sugar maple (Acer saccharum)
Orange hawkweed (Hieracium aurantiacum)	Bracken fern (Pteridium aquilinum)	Sedge (Carex spp.)	American beech (Fagus grandifolia)
Panic grass (Panicum viragatum)	White oak (Quercus alba)	Bracken fern (Pteridium aquilinum)	White ash (Fraxinus americana)
Bigtooth aspen (Populus grandidentata)	Red oak (Quercus rubra)	Red oak (Quercus rubra)	Hop hornbeam (Ostrya virginiana)
Quaking aspen (Populus tremuloides)			
Pin cherry (Prunus pennsylvanica)			
Bracken fern (Pteridium aquilinum)			

TABLE 2. Small mammal species collected for trace element monitoring on each forest type (August 1982 and 1983).

Forest Type	Species	Treatment	Number of Samples or Composite Samples per Treatment	
			1982	1983
Aspen	Thirteen-lined ground squirrel (Citellus tridecemlineatus)	Control	6	-
		Trails only	6	-
		Sludge	6	-
	Eastern meadow vole (Microtus pennsylvanicus)	Control	6	-
		Trails only	6	3
		Sludge	6	3
	Eastern chipmunk (Tamias striatus)	Control	4	-
		Trails only	4	-
		Sludge	4	-
	Woodland jumping mouse (Napaeozapus insignis)	Control	3	3
		Trails only	3	3
		Sludge	3	3
Oak	Deer (white-footed) mouse (Peromyscus spp.)	Control	6	6
		Trails only	6	6
		Sludge	6	6
Pine	Eastern chipmunk	Control	-	5
		Trails only	-	4
		Sludge	-	4
Hardwoods	Deer (white-footed) mouse	Control	4	4
		Trails only	6	6
		Sludge	6	6
	Eastern chipmunk	Control	6	6
		Trails only	6	0
		Sludge	6	6

Metals in Vegetation and Mammals 201

eri, and muscles of the hind legs were removed, dried at 60°C and analyzed separately. Composite samples of tissues from three individuals were required for the smaller species, *Peromyscus* and *Napaeozapus*.

For all samples, organic matter was broken down by wet digestion with nitric and perchloric acids. After digestion, samples were diluted with deionized water and analyzed with a DC-argon plasma atomic emission spectrometer (Spectrametrics, Inc., Boulder, Colorado) for cadmium, chromium, copper, nickel, and zinc.

The U.S. Forest Service and Michigan State University cooperative analytical laboratory (East Lansing, Michigan) conducted all sampling and analysis of the applied sludges. Their analytical scheme is presented elsewhere (Nguyen and Hart 1984). One-way analysis of variance was used to test for differences within each vegetation type between treatment means (Steel and Torrie 1980). The study design used a completely randomized model. All data were examined with Bartlett's test for homogeneity of variance and subjected to a log transformation when necessary.

RESULTS AND DISCUSSION

Concentrations of the selected trace metals in the sewage sludges used in this study were generally within limits considered appropriate for land application (Table 3). Exceptions were sludge-borne copper applied to the hardwoods type and the cadmium to

TABLE 3. Mean concentration of trace metals in municipal sewage sludge applied to forest types in Michigan (Nguyen and Hart 1984).

Element	Maximum Suggested Concentration (Logan and Chaney 1983)	Mean Concentration by Forest Type (ug/g wet weight, with dry weight in parentheses)			
		Aspen	Oak	Pine	Hardwoods
Cadmium	50	0.88 (27)	1.86 (54)	1.57 (60)	0.42 (9)
Chromium	1,000	5.82 (181)	3.65 (106)	2.77 (106)	3.22 (64)
Copper	1,000	18.3 (570)	26.6 (775)	13.5 (515)	59.7 (1,182)
Nickel	200	1.37 (43)	1.35 (39)	1.13 (43)	1.17 (23)
Zinc	2,000	22.6 (705)	39.4 (1,150)	24.4 (931)	47.6 (942)

zinc ratio of sludges applied to the other three forest types. Loading rates for cadmium ranged from 0.08 kg/ha on the hardwood site to 0.42 kg/ha on the oak site. All sites received less than the maximum annual cadmium application rate of 0.5 kg/ha established by the U.S. Environmental Protection Agency (1979) for human food-chain crops. Analyses were completed for eleven additional elements (Nguyen and Hart 1984), none of which were found in high enough concentrations to be of concern.

Gross changes in trace metal concentrations in plant and small mammal tissues due to sludge application were not evident. Metal concentrations in plant tissues were well below (sometimes by magnitudes) doses that elicit chronic toxicities in laboratory animals (Underwood 1977). Significantly different metal concentrations between plants collected from sludge and no-sludge plots were no greater than expected by chance and demonstrated no apparent trend (Table 4). Values from both control plots and plots with only trails were not significantly different, consequently control values are not presented.

TABLE 4. Forage species demonstrating a significant difference in metal concentration.

Forest	Year	Season	Species	Metal	No Sludge (µg/g dry weight)	Sludge
Aspen	1982	Spring	Quaking aspen	Zn	38.8	97.5**
			Bracken fern	Cd	2.28	3.40*
				Cu	2.23	3.71**
			Panic grass	Zn	124	156*
				Cd	0.26*	0.12
	1983	Spring	Panic grass	Cd	1.31*	0.95
			Bracken fern	Cu	2.93*	2.36
		Summer	Bracken fern	Cd	1.88	2.56*
	1984	Winter	Pin cherry	Zn	1.47*	69.7
Oak	1982	Spring	Red oak	Cu	1.89	2.70
				Cr	0.70	1.29*
		Summer	Red oak	Cr	0.98**	0.28
	1983	Spring	Red oak	Cu	1.49	2.35*
	1984	Winter	Red maple	Cu	6.11*	4.28
Pine	1982	Summer	Sedge	Cu	3.80*	2.93
	1983	Spring	Sedge	Cd	0.89	1.64**
				Cu	3.41*	2.76
			Bracken fern	Cd	1.35	1.61*
Hardwoods	1982	Summer	Hop hornbeam	Cd	0.47*	0.16
	1983	Spring	White ash	Cu	8.96	12.3*
		Summer	White ash	Cu	6.62	10.8*

Level of significance: * = 0.10.   ** = 0.01.

Similar results were obtained from analyses of small mammal tissues. As expected, metal concentrations in small mammal tissues (Table 5) demonstrated differences in organ affinities for trace elements. Generally, kidney metal concentrations were highest, followed by liver, bone, and muscle, regardless of the species' feeding strategies (e.g., herbivore, omnivore). As observed for the vegetation, small mammal tissues did not accumulate metals from exposure to sludge-amended habitat.

Previous research has reported small herbivores (i.e., meadow voles) to accumulate sludge-borne cadmium (Williams et al. 1978, Anderson et al. 1982). However, in this study plant species did not accumulate any of the observed metals, possibly because of the soil-plant barrier (Chaney 1980), so lack of accumulation in the tissues of herbivorous small mammals was expected. For the omnivorous small mammals, additional factors may have prevented accumulation. At the selected loading rates, their life spans may have been too short to allow time for significant accumulation of sludge-borne metals, and their diet may include a variety of sources at least some of which may be unaffected by the additional metals. Thus exposure may be less than expected.

Results from this study suggest that at the application rates used, sludge does not present a metal toxicity problem to wildlife consuming vegetation or to higher trophic groups consuming the small mammal species studied. However, the insectivorous small mammal, detritivorous soil macrofauna food chain was not a major component of the communities studied. Sludge-borne metals have been reported to concentrate in components of such predator-prey relationships (Beyer et al. 1982, Wade et al. 1982), and cadmium has been shown to concentrate through this relationship (Hunter and Johnson 1982). Consequently, forest soils supporting habitat more suitable for wildlife that con-

TABLE 5. Small mammals demonstrating a significant difference in metal concentration.

Forest	Year	Species	Tissue	Metal	No Sludge (ug/g dry weight)	Sludge (ug/g dry weight)
Aspen	1982	Thirteen-lined ground squirrel	Kidney	Zn	95.9	147**
		Meadow vole	Bone	Cu	2.33	4.69*
		Eastern chipmunk	Kidney	Cu	8.19*	3.96
		Meadow jumping mouse	Liver	Cd	0.37	0.64*
			Kidney	Cd	0.22	0.83*
	1983	Eastern chipmunk	Kidney	Cu	6.34*	4.69
			Liver	Cd	0.64*	0.47
Oak	1983	Deer (white-footed) mouse	Kidney	Cu	22.8*	8.6
Pine	1983	Deer (white-footed) mouse	Kidney	Cu	13.8*	10.9
Hardwoods	1982	Deer (white-footed) mouse	Kidney	Cr	0.69	8.64*
		Eastern chipmunk	Liver	Cu	18.4*	7.39
	1983	Eastern chipmunk	Kidney	Cu	13.9*	6.71

Level of significance: * = 0.10.   ** = 0.01.

sume invertebrate detritivores (e.g., earthworms) require further study to determine safe application rates. In addition, industrial sewage sludges, with the potential for higher concentrations of metals such as cadmium, need additional evaluation to assess human and wildlife food-chain hazards.

ACKNOWLEDGMENTS

Although the information in this document has been funded in part by the U.S. Environmental Protection Agency under assistance agreement No. S005551-01 to the Michigan Department of Natural Resources and Michigan State University, it has not been subjected to the Agency's publication review process and therefore may not necessarily reflect the views of the Agency, and no official endorsement should be inferred. Mention of trade names or commercial products does not constitute endorsement or recommendation for use.

Michigan Agricultural Experiment Station journal article number 11806.

REFERENCES

Anderson, T. J., G. W. Barrett, C. S. Clark, V. J. Elia, and V. A. Majeti. 1982. Metal concentrations in tissues of meadow voles from sewage sludge-treated fields. J. Environ. Qual. 11:272–277.

Beyer, W. N., R. L. Chaney, and B. M. Mulhein. 1982. Heavy metal concentrations in earthworms from soil amended with sewage sludge. J. Environ. Qual. 11:381–385.

Brockway, D. G., and D. H. Urie. 1983. Determining sludge fertilization rates for forests from nitrate-N in leachate and groundwater. J. Environ. Qual. 12:487–492.

Chaney, R. L. 1980. Health risks associated with toxic metals in municipal sludge. p. 59–83. In G. Bitton, B. L. Damron, G. T. Edds, and J. M. Davidson (eds.) Sludge: Health risks of land application. Ann Arbor Science Publishers, Ann Arbor, Michigan. 367 p.

Hunter, B. A., and M. S. Johnson. 1982. Food chain relationships of copper and cadmium in contaminated grassland ecosystem. Oikos. 38:108–117.

Logan, J. T., and R. L. Chaney. 1983. Utilization of municipal wastewater and sludge on land—metals. p. 235–323. In A. L. Page, T. L. Gleason III, J. E. Smith, Jr., I. K. Iskandar, and L. E. Sommers (eds.) Proceedings of the 1983 Workshop on Utilization of Municipal Wastewater and Sludge on Land. University of California, Riverside.

Nguyen, P. V., and J. H. Hart. 1984. Ecological monitoring of sludge fertilization on state forest lands in northern lower Michigan. Annual report. On file, U.S. EPA, Chicago. 192 p.

Steel, R. G. D., and J. H. Torrie. 1980. Principles and procedures of statistics. McGraw-Hill, New York. 633 p.

Thomas, A. H. 1983. First-year responses of wildlife and wildlife habitat to sewage sludge application in a northern hardwoods forest. M.S. thesis, Michigan State University, East Lansing. 81 p.

Underwood, E. J. 1977. Trace elements in human and animal nutrition. Academic Press, New York. 545 p.

Urie, D. H. 1979. Nutrient recycling under forests treated with sewage effluents and sludges in Michigan. p. 7–17. In W. E. Sopper and S. N. Kerr (eds.) Utilization of municipal sewage effluent and sludge on forest and disturbed land. Pennsylvania State University Press, University Park.

U.S. Environmental Protection Agency. 1976. Municipal sludge management: EPA construction grants program. EPA 430/9-76-009. 64 p.

————. 1979. Criteria for classification of solid waste disposal facilities and practices. Federal Register 44:53438–53464.

Wade, S. E., C. A. Bache, and D. J. Lisk. 1982. Cadmium accumulation by earthworms inhabiting municipal sludge-amended soil. Bull. Environ. Contam. Toxicol. 28:557–560.

Williams, P. H., J. S. Shenk, and D. E. Baker. 1978. Cadmium accumulation by meadow voles from crops grown on sludge-treated soil. J. Environ. Qual. 7:450–454.

Forest Response

Chapter 20

Growth Response of Forest Trees
to Wastewater and Sludge Application

ROBERTA CHAPMAN–KING, THOMAS M. HINCKLEY,
and CHARLES C. GRIER

ABSTRACT In many forested areas productivity is moderately to severely limited by lack of nutrients or water. Amending forest soils with sludge or wastewater can potentially increase the nutrient base and moisture regime of the site. A variety of responses have been documented for forest trees receiving applications of wastewater and sludge. In general, the potential for increased growth can be expected for most species, but the nature, extent, and duration of the response will vary according to many factors, including species, age, site, and rates and composition of the material applied. This paper categorizes the response of evergreen and deciduous species, various site classes, the effect of tree age on the response as well as the management constraints associated with different age material, and the reaction of the individual tree compared with the stand. From these generalizations, growth response patterns of trees and stands should be predictable. This paper addresses both the physiological and site principles underlying a positive growth response as well as the soil and plant characteristics resulting in either a neutral or negative response to excess water or nutrients.

That trees generally respond in a positive fashion to the application of wastewater or dewatered sludge has been well documented (Berry 1982, Brockway 1982, Cole et al. 1984, Gouin et al. 1978, Sopper and Kardos 1973, Zasoski et al. 1983) (Figure 1). Species and hybrids of *Populus*, especially, have shown significant increases in growth when treated with wastewater or sludge in studies in Washington, Pennsylvania, and Michigan (Cole et al. 1984, Schiess and Cole 1981, Zasoski et al. 1983, Kerr and Sopper 1982, Cooley 1982, Brockway 1982). Growth of conifers has also increased significantly in response to applications, including Douglas-fir (*Pseudotsuga menziesii* [Mirb.] Franco) (Cole et al. 1984, Zasoski et al. 1983), loblolly (*Pinus taeda* L.), Scotch (*P. sylvestris* L.), and white pine (*P. strobus* L.) in the United States (Wells et al. 1984, Berry 1982, Brockway et al. 1979, Brockway 1983), and Monterey pine (*P. radiata* D. Don) in Australia (Cromer 1980). But not all species respond in the same manner to treatment nor do all species respond similarly at all stages of their life cycles.

While water and nitrogen are limiting on most of the forest land in the temperate regions of the world, certain sites, particularly eroded or reclaimed strip mines, are especially unproductive. Use of wastewater and sludge as site amendments has greatly improved establishment of seedlings and their subsequent growth (Kerr and Sopper 1982, Berry 1977, 1982). However, survival of outplanted tree seedlings treated with sludge or wastewater has often been reported as a problem, because of competition from herbaceous vegetation (Kerr and Sopper 1982, Berry 1977, Cole 1982, Brockway 1982). Where

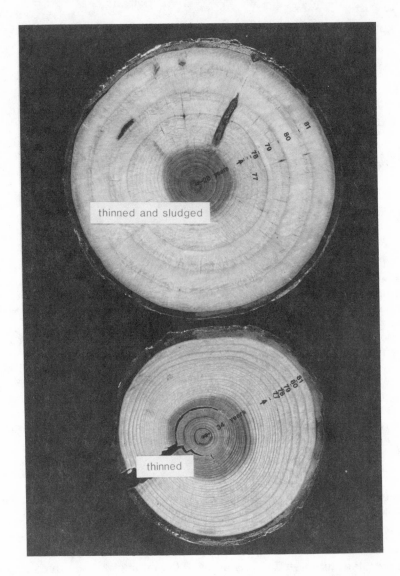

Figure 1. Disks removed at breast height (1.37 m) from a Douglas-fir tree treated with sludge (1978) and thinned (1976) (top), and a tree thinned in 1976 (bottom). The average radial increment between the pith and 1976 was 1.15 and 1.24 mm/yr in the top and bottom disks, respectively. Radial increments for 1976 and 1977 (first two years of thinning treatment) averaged 2.60 and 2.00 mm/yr, respectively, while increments for the period 1978-81 averaged 10.63 and 3.43 mm/yr, respectively.

herbicides or mechanical cultivation is used, as in tree nurseries or Christmas tree plantations (Zasoski et al. 1983, Gouin et al. 1978, Berry 1980, Cooley 1982), problems associated with weed competition are overcome and survival and growth of seedlings have been enhanced by sludge or wastewater applications. In addition to competition from weeds, there are several other problems associated with treatment, including ice breakage, deer browse, rodent damage, windthrow, and the appearance of nutrient deficiency symptoms (Cromer 1980, Kerr and Sopper 1982, Sopper and Kardos 1973). Although irrigation with wastewater may be beneficial for tree growth by enhancing the availability of water on a site, it may also result in shallow root systems which could lead to problems with windthrow or decreased drought tolerance (Cooley 1982, Sopper and Kardos 1973).

Because of the lack of a uniform response to treatment with either dewatered sludge or wastewater, the authors have chosen to develop a conceptual model describing how trees respond to growth amendments rather than to catalog the various responses. First, growth will be defined, followed by a discussion of the various factors that can limit growth, then the model will be developed, and finally the model will be discussed in terms of wastewater or dewatered sludge treatment.

GROWTH

Growth consists of cell division, cell elongation, cell differentiation, and dry weight accumulation. Water and nutrients are important in all four steps; however, water is most critical for cell elongation. For the purposes of this paper, growth will be defined as any increase in dry weight.

FACTORS LIMITING GROWTH

Maximum rates of growth depend on the presence of adequate substrates as well as an optimal environment. Since all factors are rarely at optimum levels, growth is typically limited by one or more factors. The simplest model to describe how one factor might limit growth was originally formulated by Liebig (1843) as the law of the minimum. Factors affecting growth can be viewed as staves in a barrel with each stave having a height in proportion to its abundance. Therefore, the lowest or shortest stave limits growth. In order to increase growth, the abundance or concentration of this limiting factor must be increased. Any increase in this factor will only affect the rate of growth as long as some other factor is not limiting.

Although Liebig's model nicely conveys the concept of limiting factors, it has several weaknesses. First, it assumes that a limiting factor can never be in excess. Theoretically all the staves in the barrel have unlimited height. However, too much water, for example, can have as severe an impact on trees as too little. Second, this model fails to deal with differences between organisms, even of the same species. Growth is not simply a function of the presence of a given substrate, such as sugar or nitrogen, but is the result of a process that takes substrates and, depending on the environmental conditions and the genetic capacity, converts them into dry matter. The genetic makeup of an individual may well set the upper limit, and not any environmental factor or substrate. Third, it is assumed that two factors do not interact in either a positive or negative fashion. In reality, there are interactions; for example, increased precipitation will not only supply more water but also increase the rate of decomposition and the potential availability of mineral nutrients. The addition of nitrogen will increase growth directly by supplying an important element, and indirectly by increasing the ability of a root to absorb water. Fourth, it is assumed that the effect of a factor at any stage remains the same for any other stage. For example, exposure to drought will change the reaction of a tree to subsequent droughts.

A biologically sound, but still weak, concept of limiting factors is presented in Figure 2. Growth is initiated when a threshold level of a factor, such as water or nitrogen (shown on the x-axis), is reached. Additional increases in the factor will increase the

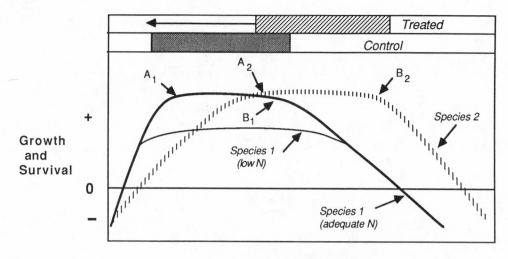

Figure 2. The effect of changing one environmental factor (e.g., water) on the growth and survival of two species having different tolerances to low and high concentrations of the factor under study. Control represents the range of concentrations observed for the environmental factor under study, and treated represents the situation for which a new range is created (e.g., addition of wastewater) or found (e.g., change of habitat or site). The arrow illustrates the change in concentration associated with irrigation system failure. A = saturation point; B = threshold for decline.

quantity of growth until a saturation point (A_1 or A_2) is reached. Growth will then remain constant until a higher threshold concentration (B_1 or B_2) of the factors is reached. At this point, productivity is impaired and growth losses occur. At even higher concentrations, the tree may die. This diagram has many of the weaknesses of Liebig's law of the minimum, including the inability to show complex interactions between other factors and the factor under consideration. It does, however, illustrate the simple interaction between two factors (e.g., water and nitrogen). For example, if the x-axis denotes the concentration or supply of water, initially increasing the concentration of water should increase growth until either the saturation concentration (species 1, adequate N) is reached or nitrogen becomes limiting (species 1, low N); however, increases in water supply may result in increased decomposition and/or the flux of nitrogen to the root, thus affecting the point where saturation is found.

An additional weakness associated with the model presented in Figure 2 is the dimensionless nature of the y- and x-axes. Four broad factors can be identified that will affect the nature of the dimensions to be assigned to the y- or x-axis: (1) species, (2) time of year, (3) age, and (4) site quality. The following discussion will deal only with defining the dimensions of the x-axis (Figure 2). Species 1, a tree species, such as Douglas-fir, shortleaf pine (*Pinus echinata*), or white oak (*Quercus alba*), has excellent drought resistance characteristics, but does not tolerate high concentration of water, while species 2, such as black cottonwood (*Populus trichocarpa*) or eastern cottonwood (*P. deltoides*), bald cypress (*Taxodium distichum*), or swamp white oak (*Q. bicolor*), has the opposite characteristics (Chapman et al. 1982). If the site to be treated with wastewater had the range of moisture shown in Figure 2 and the site was occupied by Douglas-fir, then the likelihood of losses in productivity and growth would increase with treatment. Therefore, one might choose to use species 2. The use of species 2 would demand greater management care, since any loss of ability to maintain irrigation schedules would readily lead to damaging or lethal water deficits.

By presenting these two simple models of how factors limit tree growth, and their associated weaknesses, one has a better understanding and appreciation of how difficult it is to describe growth in two dimensions. The following model is an effort to go beyond these simple, conceptual models.

MODEL

Two assumptions have been made in developing the model shown in Figure 3 (Grier and Hinckley 1986). First, water and nutrients are the principal limiting factors to forest tree growth and productivity. Second, tree age affects mostly site occupancy rather than altering a species tolerance or need for a factor. The model has inputs from nutrients and precipitation where either dewatered sludge or wastewater is considered to influence or modify these inputs. Outputs include evapotranspiration, runoff, drainage, and above- and below-ground productivity. The first three outputs are important in terms of site management and are mostly a function of the soil characteristics of the site and the climate of the region. It should be noted that changing the quantity of foliage on a site can dramatically affect both interception and transpirational losses (McNaughton and Jarvis 1983). The last two outputs are of concern in terms of realizing positive changes in productivity based on changes in nutrient and (or) water inputs (i.e., through changes in site quality or treatment).

Soil water storage is a function of the input of water, the outputs of evapotranspiration, runoff, and deep drainage, and the character of the soil as it affects storage capacity (depth, texture, and percentage organic matter). Soil water storage then defines the water available for absorption by roots and microorganisms.

Available Water and Leaf Area

The growth of a tree is closely related to the quantity of water available during the growing season. Water deficits affect growth directly through their impact on cell elongation and indirectly through their influence on stomatal aperture. Since stomata are key in the exchange of gases by the tree, then the quantity of foliage should be linked with the transpirational and photosynthetic potential of the tree. Site quality is in a sense an integrative measure of the availability of water, nutrients, and adequate growing temperatures. Unfortunately, site quality has only been indirectly quantified through measures of tree growth rather than through measures of these controlling factors. Therefore, this study's approach has been to focus on the controlling factors.

The relation between available water and leaf area has been described by Grier and Running (1977) for mature stands of conifers from the coast to the interior of Oregon (Figure 4). Grier and Running selected stands along a dramatic precipitation gradient; however, stands were fully stocked and were chosen on sites of high quality. It is important to note that as available water increases, stand leaf area increases mostly as more shade tolerant species replace intolerant species—for example, from western juniper (*Juniperus occidentalis* Hook) to ponderosa pine (*Pinus ponderosa* Dougl.) to Douglas-fir and western hemlock (*Tsuga heterophylla* [Raf.] Sarg.) to western hemlock and Douglas-fir to Sitka spruce (*Picea sitchensis* [Bong.] Carr.). Obviously, total leaf area would in-

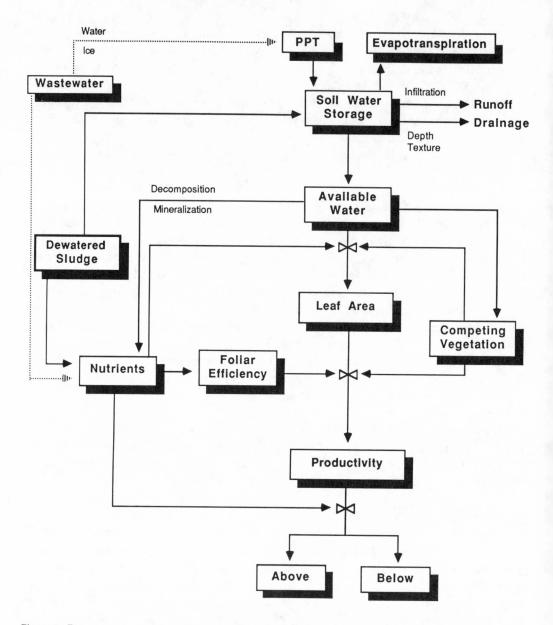

Figure 3. Diagrammatic model depicting the effect of nutrients and water on forest stand productivity (after Grier et al. 1985). Potential effects of wastewater and dewatered sludge have been included.

crease for stands of ponderosa pine as available water increases; however, the rate of increase in leaf area would be slower than that depicted in Figure 4 because of the inability of ponderosa pine to carry as much foliage as a more shade tolerant species. In summary, available water defines potential total leaf area of forested stands. From a management standpoint, available water can be effectively modified by changing the quantity of water present through irrigation.

As shown in Figures 3 and 5, the potential leaf area for a given level of available water can be modified by competing vegetation and nutrient supply. Competing vegetation can also utilize the supply of available water and reduce the quantity of light and the

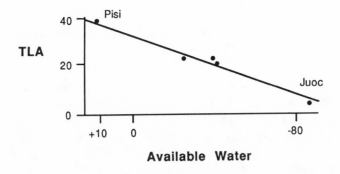

Available Water

Figure 4. The relation between available water and total leaf area (m²/m²) for five different conifer stands spanning from *Picea sitchensis* on the Oregon coast to *Juniperus occidentalis* in the interior of Oregon (after Grier and Running 1977).

concentration of nutrients. Clearly the negative impacts of competing vegetation can be ameliorated or eliminated through management. Generally the quantity of competing vegetation increases as site quality increases. Irrigation and fertilization will increase site quality.

Once stand closure has occurred, competing vegetation becomes an insignificant component of the system's leaf area. At this point, nutrient supply clearly controls the ability of a stand to realize its potential leaf area (a given level of available water is assumed). As site quality improves (Figure 5), a codominant to dominant tree of a given diameter will carry more foliage. The addition of nutrients will increase the quantity of foliage carried on a given site; however, the increase will be larger on poor sites compared with good sites (Brix and Mitchell 1983, Grier et al. 1984). This is supported by the growth response of 40- to 50-year-old Douglas-fir to sludge application, where the greatest response was noted on the poor compared with the high quality site (Zasoski et al. 1983). It would be an interesting experiment to return to the sites used by Grier and Running (1977) and

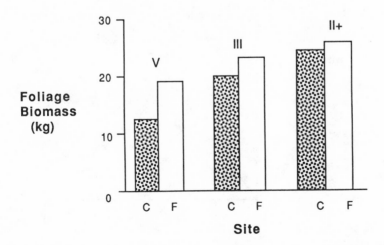

Site

Figure 5. The interaction between site (V-low, II+-high) and fertilization as they affect foliage biomass on a Douglas-fir tree with 250 cm² of sapwood basal areas at breast height, growing in a fully stocked stand (after Grier et al. 1984). C = control; F = fertilizer added.

fertilize half of their study stands. If the line shown in Figure 4 is truly a potential, then fertilization will have little effect on it. This should be true if all stands are from high sites (see Figure 5).

Leaf Area and Productivity

As discussed above, the presence of stomata, hence the quantity of foliage, is important in determining the transpirational and photosynthetic potential of foliage. Gholz (1982) has observed a close relationship between total leaf area and aboveground net primary productivity (Figure 6) while Schroeder et al. (1982) have shown a close agreement between stand leaf area and stemwood production in mixed conifer forests in eastern Washington. Indeed, Gholz (1982) used the same stands as Grier and Running (1977) as well as others. The authors of this paper assume that leaf area would also be related to total productivity. Others have noted a close relation between leaf area and total productivity (Isebrands and Michael 1985, Van Volkenburgh et al. 1985). However, both nutrients and competing vegetation can affect this relationship. Competing vegetation, through shading, reduces foliar photosynthetic capabilities. Foliar nutrient levels affect the potential rate of net photosynthesis, with an optimum rate at 1.78%; rates of net photosynthesis decrease as foliar nitrogen either increases or decreases from this point (Figure 7). Too much nitrogen can impair physiological efficiency (see Figure 2). In addition, there is evidence that superoptimal levels of foliar nitrogen, from treatments using dewatered sludge, can reduce the ability of foliage of Douglas-fir to resist drought stress (Chapman 1983). Chapman (1983) observed that stomata were more open in high nitrogen seedlings; and, with exposure to drought, high nitrogen seedlings showed a reduced ability to acclimate and resist the drought.

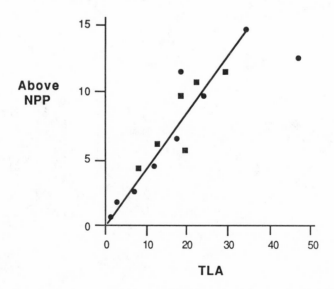

Figure 6. The relation between total leaf area (m²/m²) and aboveground net primary productivity (Mg/ha·yr) of fifteen conifer stands along a transect from the coast to the interior of Oregon (after Gholz 1982).

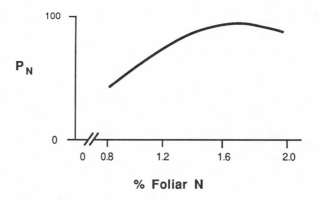

Figure 7. The relation between percentage foliar nitrogen and net photosynthesis (% of maximum) for current foliage of Douglas-fir (after Brix 1981).

Productivity: Above- and Below-ground Components

Leaf area is related to photosynthetic potential and, once respirational demands have been accounted for, net primary productivity. The larger the leaf area, the greater the net primary productivity. However, stands with large leaf areas do not necessarily have large stemwood volumes.

Net primary production represents the total net carbon gain for a tree or a stand in a year. This carbon gain can be viewed as a pool that can be allocated to above- and below-ground processes, and this allocation process appears to be dependent on site quality (Figure 8; Keyes and Grier 1981). Although the total net primary productivity does not vary appreciably between a high (18.3) and a low site Douglas-fir stand (15.7 Mg/ha·yr), the amount of carbon allocated to aboveground growth (foliage, branches, bark, and stemwood) versus below-ground growth (coarse root stemwood and bark and fine root) is vastly different. It is clear that a considerable proportion of the total production of a tree can be allocated to below-ground processes, particularly to the growth and turnover of fine roots. The quantity of nitrogen available in the forest floor appears to be a major

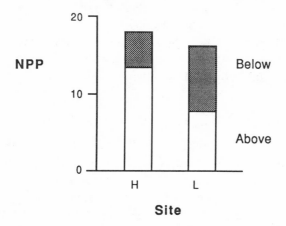

Figure 8. The effect of site quality on total net primary productivity (NPP) and the allocation of net primary productivity to above- and below-ground processes in 40 plus-year-old Douglas-fir stands from a high and low site (after Keyes and Grier 1981).

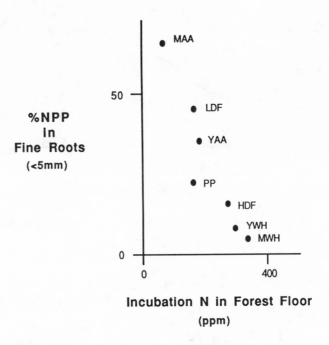

Figure 9. The relation between incubation N in the forest floor (ppm) and percentage of net primary production in fine roots (after Grier et al. 1985).

environmental factor controlling the percentage of net primary productivity allocated to fine roots (Figure 9; Grier et al. 1985). Any addition of nitrogen should increase the amount of aboveground growth because of its impact on fine root growth. Sites with high levels of nitrogen should respond to nitrogen amendments only in a minimal way.

Figure 3 illustrates the potential effects that changing available water or nutrients can have on forest stand productivity and the relative proportion of above- to below-ground productivity. The addition of either wastewater or sludge can affect both of these inputs. For example, results of several studies indicate that sludge application may enhance growth more than chemical fertilizers, such as urea nitrogen (Berry 1980, 1982, Cole et al. 1983). The duration of this response may also be greater from sludge application than from fertilization (Cole et al. 1984, Zasoski et al. 1983). The increased response and duration due to sludge application may be attributed to several factors—greater nutrient loading, increased organic matter in the soil, long-term nitrogen mineralization, and the availability of nutrients other than nitrogen which may have been limiting on the site.

CONCLUSIONS

Thus it is the authors' contention that productivity of conifer forests, at least those of western United States, is regulated primarily by the physical environment interacting with factors influencing mineral nutrition. The physical environment sets upper bounds on the productivity of a site; mineral nutrition determines the amount of this potential actually realized. Clearly, sludge and wastewater applications can affect site potential.

Although the model shown in Figure 3 provides a useful means of organizing what is known about the response of stands of trees to water and nutrients, information to test this model in systems other than the Pacific Northwest is very limited. There are data from deciduous systems which would suggest that this model is reasonably universal (e.g., Yen et al. 1978).

Most of the literature on growth response to sludge and wastewater applications concentrates on the correlation between application rates and aboveground productivity, thereby not addressing the interrelationship of factors contained in this model. While different tree species have varied potentials for response to applications of either wastewater or sludge, understanding the principles that affect tree growth is useful when one is confronted by the many, sometimes contradictory, results presented in the literature. Design of forest land application must be site and species specific in order to realize the potential for increased tree growth.

REFERENCES

Berry, C. R. 1977. Initial response of pine seedlings and weeds to dried sewage sludge in rehabilitation of an eroded forest site. USDA For. Serv. Res. Note SE-249.

——. 1980. Sewage sludge affects soil properties and growth of slash pine seedlings in a Florida nursery. Proceedings, Southeast Area Forest Tree Nursery Conference, Lake Barkley, Kentucky.

——. 1982. Dried sewage sludge improves growth of pines in the Tennessee Copper Basin. Reclam. Reveg. Res. 1:195–201.

Brix, H. 1981. Effects of nitrogen fertilizer source and application rates on foliar nitrogen concentration, photosynthesis, and growth of Douglas-fir. Can. J. For. Res. 11:775–780.

Brix, H., and A. K. Mitchell. 1983. Thinning and nitrogen fertilization effects on sapwood development and relationships of foliage quantities to sapwood area and basal area in Douglas-fir. Can. J. For. Res. 13:384–389.

Brockway, D. G. 1982. Tree seedling responses to wastewater irrigation on a reforested old field in southern Michigan. p. 165–179. In F. M. D'Itri (ed.) Land treatment of municipal wastewater. Ann Arbor Science Publishers, Ann Arbor, Michigan.

——. 1983. Forest floor, soil, and vegetation responses to sludge fertilization in red and white pine plantations. Soil Sci. Soc. Am. J. 47:776–784.

Brockway, D. G., G. Schneider, and D. P. White. 1979. Municipal wastewater renovation, growth and nutrient uptake in an immature conifer-hardwood plantation. In C. T. Youngberg (ed.) Forest soils and land use. Proceedings, Fifth North American Forest Soils Conference, Colorado State University, Fort Collins.

Chapman, R. J. 1983. Growth, nitrogen content and water relations of sludge-treated Douglas-fir seedlings. M.S. thesis, University of Washington, Seattle. 85 p.

Chapman, R. J., T. M. Hinckley, L. C. Lee, and R. O. Teskey. 1982. Impact of water level changes in woody riparian and wetland communities. Volume 10. Index and Addendum to Volumes 1–8. FWS/OBS-82/23. 111 p.

Cole, D. W. 1982. Response of forest ecosystems to sludge and wastewater applications: A case study in western Washington. p. 274–271. In W. E. Sopper, E. M. Seaker, and R. K. Bastian (eds.) Land reclamation and biomass production with municipal wastewater and sludge. Pennsylvania State University Press, University Park.

Cole, D. W., C. L. Henry, P. Schiess, and R. J. Zasoski. 1983. The role of forests in sludge and wastewater utilization programs. p. 125–143. In A. L. Page, T. L. Gleason III, J. E. Smith, Jr., I. K. Iskandar, and L. E. Sommers (eds.) Proceedings of the 1983 Workshop on Utilization of Municipal Wastewater and Sludge on Land. University of California, Riverside.

Cole, D. W., M. L. Rinehart, D. G. Briggs, C. L. Henry, and F. Mecifi. 1984. Response of Douglas-fir to sludge application: Volume growth and specific gravity. p. 77–84. In 1984 TAPPI Research and Development Conference, Appleton, Wisconsin. Technical Association of the Pulp and Paper Industry, Technology Park, Atlanta, Georgia.

Cooley, J. H. 1982. Growing trees on effluent irrigation sites with sand soils in the upper Midwest. p. 155–174. In F. M. D'Itri (ed.) Land treatment of municipal wastewater. Ann Arbor Science Publishers, Ann Arbor, Michigan.

Cromer, R. N. 1980. Irrigation of radiata pine with waste water: A review of the potential for tree growth and water renovation. Aust. For. 43:87–100.

Gholz, H. L. 1982. Environmental limits on aboveground net primary production, leaf area and biomass in vegetation zones of the Pacific Northwest. Ecology 63:469–481.

Gouin, F. R., C. B. Link, and J. F. Kundt. 1978. Forest tree seedlings thrive on composted sludge. Compost Sci./Land Util. 19(4):28–30.

Grier, C. C., and T. M. Hinckley. 1986. Productivity in Douglas-fir: Its relation to moisture and nutrition. In Douglas-fir: Stand management for the future. College of Forest Resources, University of Washington (in press).

Grier, C. C., and S. W. Running. 1977. Leaf area of mature Northwestern coniferous forests: Relation to site water balance. Ecology 58:893–899.

Grier, C. C., K. M. Lee, and R. M. Archibald. 1984. Effect of urea fertilization on allometric relations in young Douglas-fir trees. Can J. For. Res. 14:900–904.

Grier, C. C., K. A. Vogt, K. M. Lee, and R. O. Teskey. 1985. Factors affecting root production in subalpine forests of the Northwestern United States. In H. Turner and W. Tranquillini (eds.) Establishment and tending of subalpine forest: Research and management. Eing. Anst. Forstl. Versuchwes. 270:143–149.

Isebrands, J. G., and D. A. Michael. 1985. Effects of leaf morphology and orientation on light interception and photosynthesis in *Populus*. In Proceedings, IUFRO Workshop, Tokyo, Japan (in press).

Kerr, S. N., and W. E. Sopper. 1982. Utilization of municipal wastewater and sludge for forest biomass production on marginal and disturbed land. In W. E. Sopper, E. M. Seaker, and R. K. Bastian (eds.) Land reclamation and biomass production with municipal wastewater and sludge. Pennsylvania State University Press, University Park.

Keyes, M. R., and C. C. Grier. 1981. Above- and below-ground net production in 40-year-old Douglas-fir stands on low and high productivity sites. Can. J. For. Res. 11:599–605.

Liebig, J. 1843. Chemistry and its application to agriculture and physiology. 3d ed. Peterson, Philadelphia.

McNaughton, K. G., and P. G. Jarvis. 1983. Predicting effects of vegetation changes on transpiration and evaporation. p. 1–47. In T. T. Kozlowski (ed.) Water deficits and plant growth. Vol. 8. Academic Press, New York and London.

Schiess, P., and D. W. Cole. 1981. Renovation of wastewater by forest stands. p. 131–147. In C. S. Bledsoe (ed.) Municipal sludge application to Pacific Northwest forest lands. Institute of Forest Resources Contribution 41. College of Forest Resources, University of Washington, Seattle.

Schroeder, P. E., B. McCandlish, R. H. Waring, and D. A. Perry. 1982. The relationship of maximum canopy leaf area to forest growth in eastern Washington. Northwest Sci. 56:121–130.

Sopper, W. E., and L. T. Kardos (eds.) 1973. Recycling treated municipal wastewater and sludge through forest and cropland. Pennsylvania State University Press, University Park.

Sopper, W. E., and S. N. Kerr (eds.) 1979. Utilization of municipal sewage effluent and sludge on forest and disturbed land. Pennsylvania State University Press, University Park.

Van Volkenburgh, E., C. Ridge, and T. M. Hinckley. 1985. Limits to poplar leaf growth. Plant Physiol. (Suppl.) 77:136.

Wells, C. G., K. W. McLeod, C. E. Murphy, J. R. Jensen, J. C. Corey, W. H. McKee, and E. J. Christensen. 1984. Response of loblolly pine plantations to two sources of sewage sludge. p. 85–94. In 1984 TAPPI Research and Development Conference, Appleton, Wisconsin. Technical Association of the Pulp and Paper Industry, Technology Park, Atlanta, Georgia.

Yen, C. P., C. H. Pham, G. S. Cox, and H. E. Garrett. 1978. Soil depth and root development patterns of Missouri black walnut and certain Taiwan hardwoods. p. 36–43. In E. V. Eerdon and J. M. Kinghorn (eds.) Proceedings, Root Form of Planted Trees Symposium. B.C. Ministry of Forests/Canadian Forestry Service, Joint Report 8.

Zasoski, R. J., D. W. Cole, and C. S. Bledsoe. 1983. Municipal sewage sludge use in forests of the Pacific Northwest, U.S.A.: Growth responses. Waste Management and Research 1:103–114.

Chapter 21

Wastewater and Sludge Nutrient Utilization in Forest Ecosystems

DALE G. BROCKWAY, DEAN H. URIE, PHU V. NGUYEN, and JAMES B. HART

ABSTRACT   Although forest ecosystems have evolved efficient mechanisms to assimilate and retain modest levels of annual geochemical input, their productivity is frequently limited by low levels of available nutrients. A review of research studies conducted in the major U.S. forest regions indicates that the nutrients and organic matter in wastewater and sludge represent a resource of substantial potential benefit to augment site nutrient capital and ameliorate the environment for plant growth. Wastewater irrigation provides phosphorus that is strongly held in upper layers of mineral soil and cations (potassium, calcium, sodium) that accumulate but are subject to loss with leaching anions (sulfate, nitrate, chloride) during periods of groundwater recharge. Applied nitrogen that is not lost to the atmosphere by volatilization or denitrification accelerates forest floor decomposition, accumulates in soil in association with organic matter, is taken up by plants, or, following nitrification, is leached as nitrate to groundwater. Nitrogen utilization is greatest in young poplar forests growing in association with understory vegetation. Sludge applications provide phosphorus, potassium, and calcium that largely remain in the forest floor and upper layers of mineral soil. Calcium and potassium are subject to loss by leaching with anions (principally nitrate) during recharge periods. Applied nitrogen that is not lost through volatilization or denitrification is initially stored in the forest floor, where the resulting decrease in C:N ratio accelerates the decomposition of organic matter, and eventually the nitrogen is assimilated by vegetation or leached to groundwater as nitrate. Nitrogen uptake is highest on sites occupied by young poplar (200 to 400 kg/ha·yr) and somewhat less in middle-aged stands of Douglas-fir (90 kg/ha·yr) and loblolly pine (105 kg/ha·yr). Nitrogen application rates of 400 to 500 kg/ha have been associated with tree growth increases of up to 40% without producing soil leachate concentrations of nitrate that exceed the 10 mg/l U.S. EPA standard. Repeated sludge applications should provide a cumulative positive effect on forest site quality that could lead to permanent increases in productivity.

In the United States, disposal of waste effluent and sludge has become an increasing problem in recent decades because of expanding industrialization, population growth in urban and suburban areas, and legislation requiring a higher standard for wastewater treatment. Walsh (1976) estimated 1970 sludge production at 3.6 million Mg (4 million tons) and projected a doubling by 1985. Total 1975 discharge of domestic sewage was 90.5 billion liters (24 billion gallons), containing approximately 733 million kg (1.6 billion lb) of nitrogen, 674 million kg of phosphorus, and 428 million kg of potassium (Freshman 1977). The value of these nutrients amounted to $561 million.

With adoption of the Federal Water Pollution Control Act Amendments (PL 92-500) in 1972, land application of waste effluent and sludge was cited as a major alternative for eliminating nutrient-rich discharges into navigable waters (Morris and Jewell 1977). Preliminary research and experience have shown land application to be an innovative, cost-effective technology for environmentally sound waste treatment (Forster et al. 1977). Although agricultural land has received more study in this regard, forest land offers several unique advantages in terms of site characteristics, ecological structure, and mode of nutrient cycling (Smith and Evans 1977).

In application of wastewater and sludge to forest land, as with other lands, the main goal is to deliver nutrients to the site at a rate that does not exceed the assimilation capacity of the ecosystem. In so doing, deleterious side effects such as contamination of groundwater, impairment of biological production, and degradation of environmental aesthetics are avoided. However, forest ecosystems differ in their manner of and capacity for nutrient cycling.

Diverse research endeavors have been under way in the major U.S. forest regions in the hope of more precisely calibrating rates of waste-borne nutrient assimilation, loss, and cycling in the forest types of commercial importance (Sopper and Kerr 1979, Bledsoe 1981, Brockway 1983, Henry and Cole 1983, Urie et al. 1984, Nutter and Red 1984, Wells et al. 1984). This work is important for improved understanding of how wastewater and sludge application to forest land can become operationally useful in completing the nutrient cycle. Before examining these studies in detail, a brief review of nutrient cycling in natural, undisturbed forests will provide an ecosystem prospective for further discussion of the utilization of nutrient additions.

NUTRIENT CYCLING IN FOREST ECOSYSTEMS

Nutrient utilization in forest ecosystems is the extent to which and how nutrients are used (Winburne 1962) on the forest site. Although thousands of physical, chemical, and biological processes are involved in forest nutrient cycles, an examination of the dynamic interaction among ecosystem components within the "geochemical nutrient cycle" and "biological nutrient cycle" provides a concise overview.

Forests are systems bounded by atmosphere, geologic strata, and adjacent environments (Figure 1). Water, energy, and nutrients may cross their boundaries through various input and output mechanisms and accumulate in numerous ecosystem components: overstory trees, understory vegetation, forest floor, mineral soil, and saturated groundwater zone or geologic strata within the plant rooting zone. The magnitude of nutrient inputs from external environments and the efficiency of nutrient storage in ecosystem components largely determine the degree of nutrient loss from a forest site. Over time, the levels of site nutrients may increase, decrease, or remain unchanged, depending on the relative rates of nutrient input and output.

Within the forest ecosystem, nutrients are cycled on a seasonal basis among the numerous components present (Figure 2). Nutrients available in the soil and forest floor are taken up by plants (uptake) and used in growth and metabolism. Some of these nutrients are used in production of plant structures (retention) and others are cycled back either through crown wash or in litterfall to the forest floor (return). Decomposition and incorporation of organic matter in the forest floor and soil make these returned nutrients available for another turn of the annual nutrient cycle.

The relative magnitude of nutrient flux in temperate forest ecosystems is shown in Table 1, constructed from data compiled by Pritchett (1979). While by no means exhaustive, the information provides a range of values measured by numerous studies in recent

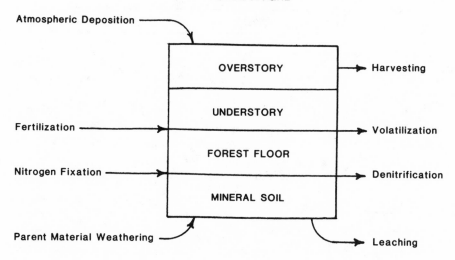

Figure 1. The geochemical nutrient cycle.

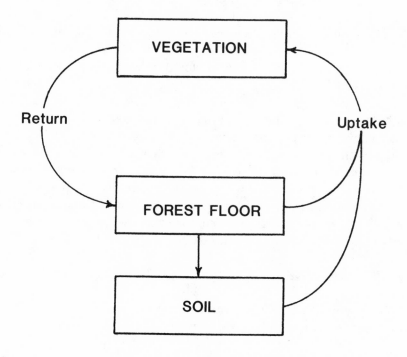

Figure 2. The biological nutrient cycle.

TABLE 1. Nutrient dynamics of temperate forest ecosystems.

	Nitrogen	Phosphorus	Potassium	Calcium	Magnesium
			(kg/ha·yr)		
Geochemical Inputs	13-26	0.1-0.5	8-20	19-40	5-11
Atmospheric	3-13	0.1-0.5	0.8-4.6	2-15	0.5-3.0
Dinitrogen fixation					
Nonsymbiotic	0.3-3.0				
Symbiotic	9-10				
Geologic weathering	--	--	7-15	17-25	4-8
Geochemical Outputs	1-5	0.1-0.5	1-4	4-11	3-12
Leaching and runoff	0.4-4.8	0.1-0.5	1-4	4.5-11.4	3.2-12.4
Volatilization	Negligible without fire				
Exports	Negligible without harvesting				
Net Geochemical Change	12-21	0	7-16	15-29	0-2
Ecosystem Internal Cycling					
Uptake	34-61	4-12	6-29	--	--
Retention	12-24	1-10	2-14	--	--
Return	22-40	3-10	4-10	--	--

decades. These data reflect rates of nutrient input to and output from ecosystems and rates of nutrient cycling within ecosystems.

In examining the net geochemical change (inputs minus outputs) of the forest ecosystem, more nutrients are input than output each season. As this balance is maintained year after year, large nutrient amounts accumulate until an event such as wildfire or timber harvest occurs, resulting in a massive nutrient loss. Mortality of individual or groups of trees from natural aging, insects, disease, or atmospheric causes and their subsequent decomposition and release of stored nutrients may also result in high outputs of nutrients.

Annual geochemical nutrient gains are largely assimilated by vegetation. Although lower nutrient quantities in the forest ecosystem are retained each year by plants than are returned to the forest floor, nutrients accumulated in plant biomass can reach substantial proportions over forty to fifty years. Much smaller nutrient quantities are accumulated in the forest floor and soil over the long term. While younger forest ecosystems may accumulate nutrients in the forest floor, such accretions are not present in middle-aged and mature forests, where the rate of litterfall addition approximates that of decomposition. Nutrient accumulation in soils is even less, being largely associated with living roots, organisms, and soil organic matter.

FERTILIZER ADDITIONS

Forest fertilization is a carefully controlled geochemical input to a forest ecosystem which is typically conducted when tree growth is limited by a deficiency of one or two plant nutrients. In the Pacific Northwest, where nitrogen is often limiting, nitrogen fertilizer applications of 100 kg/ha (89 lb/acre) every four years or 150 kg/ha every five years are commonly used (Miller and Webster 1979). In the Southeast, where phosphorus may be limiting, phosphorus fertilizer additions of 40 kg/ha are applied every twenty years (Pritchett and Smith 1975). These rates, equivalent to 25 to 30 kg N/ha and 2 kg P/ha per year, are very similar to natural nitrogen and phosphorus inputs in temperate forest ecosystems.

The addition of fertilizer increases the level of available nutrients in the forest floor and soil and increases nutrient uptake, nutrient retention, and tree growth. Fertilizer nitrogen is rapidly incorporated into the forest floor and soil, from where it is taken up by plants and retained in association with increased biomass production or returned to the forest floor by leaching and litterfall. After four to six years, the rate of nitrogen cycling declines to pretreatment levels, indicating the need for another application. Fertilizer phosphorus also enters the nutrient cycle to enhance growth, but most of it forms insoluble iron and aluminum compounds in the soil. Slow resolution of these compounds over an extended period provides small annual inputs of available phosphorus, similar to those from geologic weathering, to the forest ecosystem. Because of the economic considerations in conducting a forest fertilization program, fertilizer application rates are controlled to levels that promote tree growth but do not result in leaching losses great enough to degrade groundwater quality.

WASTEWATER ADDITIONS

Wastewater application to forest land provides the ecosystem with nutrients that improve soil fertility and enhance vegetation growth (Nutter and Red 1984). Of these, nitrogen and phosphorus are of greatest concern, because they are most limiting to production in aquatic ecosystems. Characteristic fertilizer nutrient concentrations found in various waste effluents are listed in Table 2. As is obvious, the chemical form of each element is quite variable and yet has a profound influence on its overall fate in each plant-soil system.

Nitrogen, for example, may be present as organically bound or in one or more of its mineralized forms, ammonia or nitrate. Biological wastewater treatment plants often discharge effluent that is higher in the mineralized forms of nitrogen, while sewage lagoon systems may be managed to convert most effluent nitrogen to organic forms. If sewage is stored in lagoons prior to land application, nitrate may become an important effluent constituent, because ammonia-N is typically lost as a volatilized gas. This process of "ammonia stripping" may be useful in lowering effluent nitrogen content so as to enable use of higher hydraulic application rates. The average concentrations for the various forms of nitrogen reported in effluents used in several forest application studies are

TABLE 2. Nitrogen and phosphorus concentrations in domestic wastewater (U.S. EPA 1981).

Component	Wastewater Nutrient Level		
	High (mg/1)	Medium (mg/1)	Low (mg/1)
Nitrogen	85	40	20
Organic	35	15	8
Ammonia	50	25	12
Nitrate	0	0	0
Phosphorus	15	8	4
Organic	5	3	1
Inorganic	10	5	3

shown in Table 3. The high degree of variation directly influences nutrient loading rates and highlights the need to evaluate study results in terms of the nutrient dynamics unique to each locale.

TABLE 3. Concentrations of nitrogen in wastewater used in forest irrigation studies (Hook and Kardos 1978, Brockway et al. 1979, Schiess and Cole 1981, Harris and Urie 1983, Nutter and Red 1984, Urie et al. 1984).

Study Site	Total N (mg/1)	Mineralized N (mg/1)	Organic N (mg/1)
Unicoi State Park, Georgia	18.0	7.6	10.4
Pennsylvania State University	27	21	6
Pack Forest, Washington	18.6	17.1	1.5
Harbor Springs, Michigan	5.4	2.1	3.3
Middleville, Michigan	12.6	7.6	5.0
East Lansing, Michigan	13.8	10.6	3.2

Renovation Capacity

The ability of a forest ecosystem to renovate wastewater is dependent on processes such as nutrient uptake by plants and nutrient accumulation in the soil. Renovation efficiency is commonly expressed as the ratio of the nutrient amount applied in irrigation water and the amount leaving in water drained from the site. Renovation capacities for nitrogen are listed in Table 4 for numerous forest ecosystems. In general, renovation capacities in juvenile forests exceed those of established stands because of the higher growth rate and nutrient uptake of young trees and associated herbaceous cover (Urie et al. 1978).

Over time, the ability of a forest site to renovate wastewater becomes progressively diminished. The vegetation can be managed so that site longevity is only limited by the assimilative capacity of the soil (McKim et al. 1982), but the ability of the soil to adsorb nutrients in solution is dependent on the amount of organic matter and clay it contains. Approximately 4 to 6% organic matter and clay has been suggested as a minimum to maintain soil renovation capacity for wastewater (Murrman and Koutz 1972). Forest ecosystem longevity is, as yet, not fully predictable (McKim et al. 1982). Longevity estimates for phosphorus removal through chemical precipitation and adsorption exceeding 100 years have been reported for wastewater sites in Washington, Michigan, and Pennsylvania. One may, nonetheless, expect disparity between empirical data and such predictions based on theory.

Phosphorus. Phosphorus applied to forest sites in irrigated wastewater is predominantly orthophosphate accompanied by lesser amounts of organically bound forms. These organic forms become available, also as orthophosphate, as soil organic matter decomposition proceeds. Phosphates entering the soil are adsorbed onto the surfaces of clay minerals and precipitated with iron and aluminum hydrous oxides. At typical slow-rate wastewater applications to forest ecosystems, nearly complete retention of phosphorus can be anticipated, leading to high renovation efficiency (McKim et al. 1982).

Wastewater-applied phosphorus is largely retained in the surface mineral soil (Harris 1979). As much as 34% of that applied has been reported to have been taken up by poplar plantations in the Pacific Northwest (Cole and Schiess 1978) and Great Lakes regions (Urie 1979). Soil phosphorus tends to become less soluble over time. This process, in addition to plant uptake, tends to increase the period during which a forest site may function as an effective renovator for phosphorus.

Potassium. Potassium applied to forest soils during wastewater irrigation is adsorbed by clay minerals and can be leached from the profile under an expanded presence of sodium and ammonium cations (McKim et al. 1982). The magnitude of leaching loss may be related to irrigation rate. In Pennsylvania, exchangeable potassium levels decreased in the upper meter of soil beneath an old field and a mixed hardwood forest that were irrigated for eight growing seasons with 5 cm (2 inches) of waste effluent per week. Beneath a nearby red pine stand and mixed hardwood forest irrigated with 2.5 cm of waste effluent per week, soil exchangeable potassium levels either increased or remained unchanged from controls (Sopper and Kardos 1973). In northern Michigan, exchangeable potassium levels increased in the sandy soil where a northern hardwoods forest was irrigated for five growing seasons with 7.5 cm of stabilization pond effluent per week (Harris and Urie 1983). In this study, the effluent contained an equivalent ratio of total N to total K, and the overall nutrient loading rate was relatively low. However, soil leachate samples collected at a depth of 120 cm (4 ft) contained elevated potassium concentrations.

Potassium depletion on sites irrigated with wastewater may be a problem where plant uptake of potassium increases or when effluent nitrogen concentrations greatly exceed those of potassium. Forage grass irrigated with wastewater has been shown to assimilate

TABLE 4. Nitrogen renovation capacities for numerous forest ecosystems.

Vegetation	Soil	State	Effluent Treatment	Irrigation Rate (cm/wk)	Nitrogen Loaded (kg/ha·yr)	Leached (kg/ha·yr)	Renovation (%)
Established Forests							
Douglas-fir	grav. sand	WA	secondary	5	327	75	77
Northern hardwoods	sand	MI	lagoon	7	82	15	82
Red pine	sand	MI	lagoon	8	131	40	70
Maple	loam	MI	secondary	5	217	200	8
Maple	loam	MI	secondary and lagoon	5	40	40	0
Red pine	loam	PA	secondary	5	728	250	66
Mixed hardwoods	loam	PA	secondary	5	726	100	86
Pine and hardwoods	loam	GA	lagoon	7.6	703	233	67
Juvenile Forests							
Douglas-fir	grav. sand	WA	secondary	5	484	20	96
Poplar	grav. sand	WA	secondary	5	484	20	96
Mixed conifer	loamy sand	MI	lagoon	7	105	12	89
Hardwood-conifer	loam	MI	secondary	5	117	30	74

greater amounts of potassium than that supplied in the applied effluent (Palazzo and Jenkins 1979). Analysis of tree foliage indicates that leaves typically accumulate less than half as much potassium as nitrogen, suggesting that potassium deficiencies induced by imbalanced wastewater nutrient composition are quite unlikely.

Cations. Sodium, calcium, and magnesium delivered to forest sites with irrigated waste effluent will tend to exchange with cations in the soil exchange complex. The wastewater levels of those cations will influence establishment of a new equilibrium of soil cations, the solution leached from soil being adjusted accordingly. Renovation efficiencies of 85% and 24%, respectively, for calcium and sodium reflect the tenacity with which each is held in the soil (McKim et al. 1982). Like potassium, calcium and magnesium are increasingly leached from the profile under the expanded presence of sodium and ammonium ions. Total dissolved salts remain relatively unchanged in the soil solution, normally at concentrations that do not adversely affect plants or soils in the humid climates of forested regions.

Anions. Sulfate in irrigation-applied wastewater is weakly adsorbed by hydrous iron oxides present in forest soils, but the amount of sulfate thus retained in relation to that applied is generally negligible. Chloride and nitrate are not adsorbed in mineral soils, generally passing rapidly through the profile during periods of recharge to groundwater. Levels of chloride, nitrate, and sulfate found in soil leachate may be expected to approximate those in applied effluent minus adjustments for plant uptake and denitrification of nitrate.

Nitrogen. Under high rates of nitrogen application with irrigated waste effluent, the biological decomposition rates of litter and humus have been increased in eastern mixed hardwood forests (Richenderfer and Sopper 1979). Nitrogen accumulation in wastewater-irrigated forest soils has also been reported. At weekly irrigation rates up to 88 mm which supplied as much as 550 kg N/ha over five seasons, the upper 10 cm of loamy sand beneath a red pine stand accumulated approximately 600 kg N/ha (Harris 1979). This increase was related to increases in soil organic matter ranging from 50 to 100%, presumably including decomposed forest floor materials. Ammonia adsorption onto clay minerals may also contribute to increases in soil nitrogen.

Wastewater-supplied ammonia may be subject to volatilization loss directly to the atmosphere but is largely converted to nitrate under well-aerated, acidic soil conditions. Nitrate is then susceptible to denitrification or leaching during periods of excess soil moisture. Nitrate leaching and nitrogen gas diffusion have been suggested as the primary means of nitrogen assimilation in forest land treatment systems in the Southeast (Nutter and Red 1984). However, studies in the Pacific Northwest have assumed denitrification loss to be of little consequence (Schiess and Cole 1981). In most studies, only applied nitrogen and leached nitrogen have been directly measured.

Wastewater irrigation of forests has, with minor exceptions, universally stimulated increased production of vegetation and a concomitant increase in nitrogen uptake. Significant increases in nitrogen levels in bark, twigs, and leaves were reported for effluent-irrigated conifers and hardwoods in Georgia (Brister and Schultz 1981). Increased crown biomass and nitrogen assimilation of red pine in Michigan were linearly related to rate of wastewater irrigation (White et al. 1975). Increases in foliar nitrogen concentrations for numerous effluent-irrigated forests are shown in Table 5. Values are similar to those associated with maximum growth rates in mineral fertilizer studies.

Nitrogen uptake by waste effluent-irrigated forests may reach substantial proportions

TABLE 5. Foliar nitrogen levels in forests irrigated with wastewater.

		Total Kjeldahl Nitrogen (%)	
Vegetation	State	Control	Irrigated
Conifer seedlings	MI	2.4	2.9
Hardwood seedlings	MI	2.1	2.2
Hybrid poplar cuttings	MI	1.9	2.3
Northern hardwoods, mature	MI	1.6	2.2
Northern hardwoods, understory	MI	2.0	2.6
Mixed hardwoods	PA	2.2	3.0
Red pine	PA	1.3	2.2
White spruce	PA	1.5	2.2
Mixed hardwoods	GA	1.6	2.0
Mixed pine	GA	1.1	1.6
Hardwood-pine, understory	GA	1.2	1.6

TABLE 6. Nitrogen uptake by vegetation irrigated with wastewater.

Vegetation	State	Irrigation Period (years)	Applied (kg N/ha)	Assimilated (kg N/ha)	Efficiency (%)
Grass	WA	5	2,215	627	28
Douglas-fir seedlings	WA	5	1,811	893	49
Poplar	WA	5	2,171	1,247	57
Hybrid poplar	MI	4	500	400	80
Old field	MI	1	150	128	85
Old field	PA	9	208	195	94
Pine and hardwoods	GA	6	703	470	67
Hardwood seedlings with grass	GA	1	521	152	29

(Table 6). Although the highest nitrogen uptake efficiencies appear in systems dominated by herbaceous cover, the greatest total uptake is observed on sites with young, rapidly growing stands. Douglas-fir and poplar seedlings in the Pacific Northwest can reportedly assimilate as much as 893 and 1,247 kg N/ha, respectively, over a five-year period of irrigation that supplied about 2,000 kg N/ha (Schiess and Cole 1981). Nitrogen uptake by hybrid poplar in Michigan was 400 kg/ha over a four-year irrigation period during which approximately 500 kg N/ha were loaded on the site (Cooley 1979). A mature mixed hardwood and pine forest in Georgia assimilated 470 kg N/ha, some of which was no doubt denitrification, during one irrigation year in which over 700 kg N/ha were applied (Nutter and Red 1984). A pole-sized red pine plantation in Michigan took up as much as 70% of the nitrogen applied to the site during the first three years of effluent irrigation (White et al. 1975). However, nitrogen removal rates of only 20% were measured in an effluent-irrigated mature mixed hardwood stand in the same locale (Burton and Hook 1979).

Harvesting trees is a useful means of removing accumulated nitrogen from an effluent-irrigated forest site. The storage of nitrogen in aboveground vegetation can be substantial, ranging from 112 to 224 kg/ha per year in established eastern forests to rates approaching 300 kg/ha per year in rapidly growing juvenile stands (McKim et al. 1982). Nitrogen is known to accumulate differentially within tree tissues, concentrating most in foliage and least in xylem and phloem of branches and stems (Ralston and Prince 1963). Therefore, the type as well as the timing of harvest may be a major management concern. If only stemwood is harvested in a young hybrid poplar stand, one-third as much nitrogen would be removed from the site as when whole trees containing foliage are taken. As the proportion of stemwood increases with age, the importance of short rotations to maximize nitrogen removal becomes evident. Whole tree harvest at five to eight year intervals could remove over 80% of the nitrogen stored in the aboveground biomass of an irrigated poplar plantation (Cooley 1978).

In juvenile stands, understory vegetation constitutes an important sink for nitrogen. The grass component in young Douglas-fir and poplar stands accounted for 57 and 37% of the nitrogen uptake, respectively, on effluent-irrigated sites in the Pacific Northwest (Schiess and Cole 1981). With crown closure the importance of understory vegetation declines. Ground cover has also been established as a component instrumental in abating nitrate leaching in young plantations.

Models of Nitrogen Utilization

Study results from the major forest regions have been compiled to assess ecosystem nitrogen utilization trends (Table 7). Direct measurements have been made of nitrogen applied, taken up, and leached. Values for soil storage and volatilization loss were estimates inferred from other measures.

Nitrogen application rates have typically been moderate, ranging from 40 to over 500 kg/ha annually, and produced nitrate discharges to groundwater generally less than the 10 mg/l U.S. EPA standard. With minor exception, this standard was not exceeded until applied inorganic nitrogen approached 200 kg/ha per year. High rates of effluent irrigation and nitrogen application resulted in higher concentrations of nitrate in leachate and increased rates of overall loss of nitrate-N from the site.

Generally, younger forests capable of substantial rates of nitrogen uptake have shown the best on-site retention of this nutrient. Older forests that have shifted their mode of nutrient cycling, from rapid assimilation of exogenous nutrients to conservation and recycling of endogenous nutrients, were less efficient sites for wastewater renovation. Although forest soils offered some increased storage for nitrogen, greater storage generally resulted from increases in site organic matter following irrigation. Volatilization and denitrification losses, while discounted by some studies, are believed important in others.

The fragmentary nature of these data does not allow for an integrated overview of each of the regions. Proper system selection and management will remain an exercise of balancing nitrogen application rates with volatilization, denitrification and leaching losses, plant uptake, and soil storage. Further studies will no doubt be required before a comprehensive construct of nitrogen utilization in various wastewater-irrigated forest ecosystems becomes available.

SLUDGE ADDITIONS

As previous research has indicated, the major benefit of nutrient additions in forest ecosystems is to remediate natural nutrient deficiencies (Cunningham 1976), increase foliar efficiency (Brix and Mitchell 1980), enhance foliage production, accelerate crown closure, and shorten the overall rotation period (Miller 1981). However, when large

TABLE 7. Nitrogen transformation and utilization in waste effluent-irrigated forests.

Vegetation	Application Rates Wastewater (mm/wk)	Organic N (kg/ha·yr)	Inorganic N (kg/ha·yr)	Volatilized N (kg/ha·yr)	Soil Storage (kg/ha·yr)	Plant Uptake (kg/ha·yr)	Leached Nitrate-N (kg/ha·yr)	Leachate Nitrate-N (mg/l)
Pacific Northwest Region								
Poplar	50	100	200	------134------		200-300	50	10
Douglas-fir, young	50	100	200	------96------		150-200	87	14
Douglas-fir, mature	50	100	200	--		175	125	8
Great Lakes Region								
Poplar	35	25	30	--	--	100	25	12
Poplar	70	50	60	--	--	100	50	25
Mixed hardwoods	50	48	32	------30------		--	51	2
Mixed hardwoods	50	48	147	------50------		--	146	5
Mixed hardwoods	72	90	20	------20------		40	20	2
Red pine	25	35	5	--	--	--	5	1
Red pine	50	120	20	--	--	60	50	5
Northeastern Region								
White spruce	50	100	180	------30------		200	50	5
Mixed hardwoods	50	100	180	------10------		84	200	15
Mixed hardwoods	50	------548------		--	--	--	--	25
Mixed hardwoods	25	------241------		--	--	--	--	10
Southeastern Region								
Mixed hardwoods	76	150	150	------50------		50	200	9

quantities of nutrients are added to a site, as in the application of wastewater sludge, substantial improvement of site quality can be effected (Miller 1981) through a change in moisture relations and the cycling of nutrients, and this may lead to permanent increases in productivity (Zasoski et al. 1983). Organic matter additions with sludge may also be instrumental in promoting site aggradation; the importance of the forest humus in nutrient storage and supply is well documented (Foster and Morrison 1983). Growth responses to sludge additions in Pacific Northwest Douglas-fir forests are reportedly of greater magnitude and longer duration than the average 23% response obtained from applications of commercial fertilizer (Zasoski et al. 1983).

The potential nutritional and growth benefits of sludge applications in forests appear to be intuitively clear, but environmental concerns persist about the possible enrichment of groundwater by sludge-borne nutrients that leach from the rooting zone (Koterba et al. 1979, Sidle and Kardos 1979, Riekerk 1981, Urie et al. 1984). While nitrogen loss to surface and groundwaters is generally less than 3% of the total applied as commercial fertilizer (Groman 1972), several researchers have measured higher rates of nitrate-N leaching below the rooting zone following applications of wastewater sludge (Riekerk and Zasoski 1979, Sidle and Kardos 1979, Vogt et al. 1981, Brockway and Urie 1983). This large flush of nitrate anions has been associated with a reduction of exchangeable cations in lower soil horizons (Riekerk 1978, Wells et al. 1984). Adjustment of sludge nutrient application rates to those suited for the particular combination of sludge, site, and vegetation characteristics has been shown to abate nutrient leaching losses (Riekerk 1982, Brockway and Urie 1983).

Fate of Applied Nutrients

The nutrients of major concern in forest ecosystems (nitrogen, phosphorus, potassium, and calcium) are, with the exception of potassium, found in good supply in wastewater sludges. Nitrogen may typically range from 2 to 7%, phosphorus from 1 to 4%, potassium from 0.2 to 1%, and calcium from 0.4 to 2%. The nutrient composition of three sludges applied to forests in the Pacific Northwest, Great Lakes, and Southeast is presented in Table 8. If a 20 Mg/ha (9 tons/acre) sludge application rate were assumed, individual nutrient application rates would result as shown in Table 9. As is obvious, sludges are highly variable in nutrient content.

Nutrients reaching the forest floor through sludge application are subject to numerous processes of transformation and translocation. Nitrogen is perhaps the most studied of the major nutrients, having been identified as an element that most often limits vegetation growth (Cunningham 1976, Edmonds and Mayer 1981). It may be found in sludge primarily in organic forms, but is also present in soluble forms as ammonia and, at lower concentrations, as nitrate. Nitrogen added to a forest site characteristically undergoes volatilization, mineralization, nitrification, denitrification, uptake by plants, leaching as nitrate, and storage in the forest floor and soil (Cole et al. 1983). Other nutrients—phosphorus, potassium, and calcium—are somewhat less dynamic and have received less attention in sludge application studies.

Phosphorus. Sludge applications have typically resulted in significant increases of phosphorus levels in the forest floor (Brockway 1983, Wells et al. 1984). Although a portion may be present as orthophosphate, most of the phosphorus is organically bound and slowly released as the humus and sludge slowly degrade (Edmonds and Mayer 1981). Phosphorus entering the soil was largely retained in the upper profile (Stednick

TABLE 8. Nutrient composition of three wastewater sludges applied to forest land, (Zasoski 1981, Cole et al. 1983, Brockway 1983, Wells et al. 1984).

	Seattle, WA (%)	Cadillac, MI (%)	Augusta, GA (%)
Total nitrogen	4.3	6.0	7.2
Ammonia-N	0.9	1.7	3.0
Nitrate-N	--	0.002	0.01
Phosphorus	1.5	7.8	1.6
Potassium	0.16	1.54	0.27
Calcium	0.4	1.4	1.5

TABLE 9. Nutrient application rates of three wastewater sludges applied to forest land, assuming a 20 Mg/ha sludge application.

	Seattle, WA (kg/ha)	Cadillac, MI (kg/ha)	Augusta, GA (kg/ha)
Total nitrogen	860	1,200	1,400
Ammonia-N	180	340	600
Nitrate-N	--	0.4	2
Phosphorus	300	1,560	320
Potassium	32	308	54
Calcium	80	280	300

and Wooldridge 1979, Richter et al. 1982, Urie et al. 1984), where fixation with free iron oxides caused increases (from 0.06 to 1.5%), up to 13-fold (Riekerk 1978), which were capable of substantially altering soil chemical properties (Zasoski 1981). Phosphorus concentrations in understory vegetation have reportedly doubled following sludge application (Richter et al. 1982, Brockway 1983), indicating increased levels of plant uptake. Phosphorus increases in foliage of loblolly pine have been less, though significant (Wells et al. 1984). Leaching losses of this nutrient have been consistently reported as minimal (Urie et al. 1978, Riekerk and Zasoski 1797, Hornbeck et al. 1979, Richter et al. 1982). Many forest ecosystems appear to accommodate sludge-borne phosphorus additions and store this nutrient in a manner of benefit to short- and long-term site quality.

Potassium. Potassium additions with sludge application are, because of lower concentrations, generally of lesser magnitude than those of other major nutrients (Bledsoe and Zasoski 1979). These smaller additions reportedly show minor changes in forest floor (Brockway 1983) and soil (Hornbeck et al. 1979) potassium levels. A very soluble nutrient, potassium is observed to move readily from sludge into the upper soil profile (Stednick and Wooldridge 1979), where it is presumably retained through exchange. Increased plant uptake of potassium following sludge application has been infrequently reported (Brockway 1979). However, potassium losses through leaching have been noted (Riekerk 1978, Urie et al. 1984, Wells et al. 1984) to accompany major intervals of nitrate leaching which cause cation stripping in the soil profile (Riekerk 1981, 1982). This interaction between leachable anions and potassium may result in induced nutrient deficiencies in certain potassium limited soils (Bledsoe and Zasoski 1979). Following sludge application, this stripping phenomenon is observed in the soils of several forest ecosystems.

Calcium. Additions of calcium at rates as high as 1,600 kg/ha have produced watershed level changes, which were relatively small and of short duration (Hornbeck et al. 1979). Calcium from sludge applications is largely held in the forest floor (Brockway 1979) and moves into the upper soil profile (Urie et al. 1984), where it is retained (Stednick and Wooldridge 1979) by cation exchange. Although increases in plant uptake of calcium following sludge application are not reported, several studies have identified calcium leaching losses in forest ecosystems (Riekerk 1978, Koterba et al. 1979, Riekerk and Zasoski 1979, Wells et al. 1984). These losses are temporarily associated with anion leaching occurring during periods of groundwater recharge.

Nitrogen Transformation and Translocation

Aside from potential toxicants, nitrogen is generally the most important element that limits land application rates of sludge (Sommers and Nelson 1981). This is true largely because nitrogen levels in sludges are high relative to other nutrients present, and nitrogen may be transformed into soluble forms, such as nitrate, that would represent a potential risk to environmental quality and public health if they were to enter groundwater in sufficient quantity. In addition, nitrogen is the element that appears as most frequently limiting in the nutrition and growth of tree species in a number of forest ecosystems (Edmonds and Mayer 1981). Nitrogen applied in forests with applications of wastewater sludge may undergo several different biological or chemical changes once it is delivered to the surface of the forest floor.

Volatilization. Surface applications of liquid sludge are normally subject to drying conditions wherein that portion of the supernatant that does not immediately infiltrate the soil begins to evaporate. During this period, much of the nitrogen present in the sludge as soluble ammonia is transformed to ammonia gas, which readily volatilizes into the atmosphere when sludge conditions exceed pH 7, as is commonly the case (Cole and Henry 1983). Ammonia loss by volatilization may range from 20 to 60% of that present (Sommers and Nelson 1981) depending on temperature, relative humidity, infiltration rate, and pH of sludge and soil. Typically, about 50% of the available nitrogen in a surface-applied sludge is volatilized as ammonia gas (Cole et al. 1984), most of it lost in the first three days following application (Sommers and Nelson 1981).

Mineralization. Nitrogen not lost through volatilization consists of mostly organic nitrogen and lesser amounts of more soluble forms. Through various biological decomposition processes, organic nitrogen is mineralized to form ammonia-N. In assessing the nitrogen mineralization rate for sludge, various studies have assumed a rate of 20% of the organic nitrogen as becoming available during the first year following application (Cole and Henry 1983, Urie et al. 1984). However, studies (Sommers and Nelson 1981, Parker and Sommers 1983) have shown that the nitrogen mineralization rate varies from 3 to 42% as a function of the biological process used in sludge treatment (Table 10). Also, experiences with incorporated (Richter et al. 1982) and surface-applied (Wells et al. 1984) dewatered sludge cake indicate that mineralization rates for these are much lower than those of liquid sludges. It appears then that mineralization of sludge-borne nitrogen is dependent on the nature of the sludge as well as the physical, chemical, and biological characteristics of the forest environment.

Nitrification and Denitrification. These two processes are of major importance in considering sludge nutrient utilization, because they both result in nitrogen losses from the application site. Nitrification occurs in nearly all adequately drained forest soils and is

TABLE 10. Nitrogen mineralization rates in the first year following sludge application (Sommers and Nelson 1981, Parker and Sommers 1983).

Sludge Type	Nitrogen Mineralization Rate (%)
Waste activated	42
Primary	29
Primary plus CaO	28
Aerobic digested	25
Anaerobic digested	15
Composted	9
Primary, wet-air oxidized	3

responsible for converting ammonia-N to nitrite and then nitrate-N. Nitrification is a biochemical process that occurs in a substrate dependent manner where, other factors remaining constant, rates of nitrate production will increase as ammonia availability increases, until the on-site microbial populations have become saturated. This largely accounts for the progressive abundance of nitrate produced under increasing application rates of nitrogen-rich sludge (Sidle and Kardos 1979, Cole et al. 1983, Brockway and Urie 1983). Nitrate enrichment of surface and groundwater is a frequently noted environmental and public health concern.

Nitrogen losses through denitrification may reach 20% of that available (Cole et al. 1984) when nitrate-N is abundant under reducing conditions in the soil. Such conditions were reported when perched water tables temporarily formed above lenses of fine sand in a coarse-textured outwash soil (Brockway and Urie 1983). Denitrification losses of nitrogen gases may, under certain circumstances, be an added measure of protection for phreatic aquifers which may become threatened by nitrate leaching from the rooting zone.

Forest Floor and Soil Storage. A portion of the nitrogen added to the forest floor during sludge application soon becomes indistinguishable from native nitrogen. Much of this supplemental nitrogen rapidly enters the surface soil, resulting in significant increases of this nutrient (Zasoski 1981, Brockway 1983). Soil nitrogen increases as high as threefold have been reported (Riekerk 1978), from approximately 700 to 2,000 kg N/ha in the A horizon and from 2,000 to 5,800 kg N/ha in the B horizon of a soil receiving nearly 5,600 kg N/ha with an anaerobically digested sludge application in the Pacific Northwest. Smaller increases in soil nitrogen were noted in soils treated with approximately half the above nitrogen application rate in the Great Lakes (Urie et al. 1978).

Most of the nitrogen contained in applied sludge is initially retained in the forest floor, where it is slowly released as degradation of organic matter ensues (Brockway 1983, Wells et al. 1984, Urie et al. 1984). Studies in the Pacific Northwest have shown that approximately 80% of the total nitrogen initially contained in sludge applied to Douglas-fir forests remains in the sludge after one year (Mayer 1980) and 71% is still present at the end of two years (Edmonds and Mayer 1981). The impact of this nitrogen on the forest floor is a measurable narrowing of the carbon to nitrogen (C:N) ratio, from values of 60:1 and 80:1 to ratios of 23:1 and 42:1 (Brockway 1979, Stednick and Wooldridge 1979). This has typically resulted in an accelerated rate of humus decomposition (Harris 1979, Stednick and Wooldridge 1979, Edmonds and Mayer 1981, Brockway 1983) and an enhanced availability of on-site nutrients. Sludge applications to loblolly pine plantations have resulted in an increased mass of litterfall needles that had higher nitrogen (and phosphorus) concentrations, and this has produced a doubling in nitrogen transfer rates in that ecosystem (Wells et al. 1984). Little evidence of microbial immobilization of nitrogen on sites receiving sludge applications has been reported (Edmonds and Mayer 1981).

Plant Uptake. Mineralized nitrogen not lost from the application site is available for uptake as ammonia or nitrate. Indeed, a prominent result of sludge applications in forests has been a dramatic increase in foliar nitrogen levels in understory and overstory plant species (Wells et al. 1984, Cole et al. 1984, Urie et al. 1984). Nitrogen additions with sludge application to Douglas-fir seedlings grown in native soils of western Washington have resulted in foliar nitrogen increases from background levels near 1.4 to 2.4% to concentrations ranging from 2.5 to 3.5% (Chapman 1983). Apparently 2.5% foliar nitrogen is optimal for growth of Douglas-fir, and increases beyond this level were of questionable benefit to tree growth. In northern Michigan, sludge applications increased foliar nitrogen levels from values near 1.1% to levels approaching 1.7% in red and white pines (Brockway 1983). The increased nitrogen levels facilitated optimization of the N:P ratio in the pines from 5:1 to 10:1 (van den Driessche 1974), thereby promoting an overall increase in canopy weight and photosynthetic capacity in the plantations (Brockway 1979). Similar increases have been reported for sludge-treated loblolly pines in South Carolina, where foliar nitrogen levels rose from 1.26 to 1.45% following applications as low as 400 kg N/ha (Wells et al. 1984). These increases in foliar nitrogen were associated with volume increases of approximately 40% in 9- and 28-year-old stands.

The uptake of nitrogen by forests can often be as great as that of agricultural crops (Cole et al. 1984) and, as can be seen in Table 11, is highly variable depending on factors such as climate, species, stand age, and rate and form of nitrogen addition. The greatest assimilators of nitrogen appear to be various poplars. Their rates of nitrogen assimilation

TABLE 11. Nitrogen uptake rates of various forests receiving waste-effluent or sludge applications.

Species	Age (years)	Uptake Rate (kg/ha·yr)	Reference
Pacific Northwest			
Poplar	seedlings	300 to 400*	Schiess and Cole 1981
Douglas-fir	seedlings	150 to 250*	Schiess and Cole 1981
Douglas-fir	young stands	up to 225	Cole and Henry 1983
Douglas-fir	55	90	Cole and Henry 1983
Douglas-fir	conceptual	112 to 168	Cole and Henry 1983
Great Lakes			
Hybrid poplar	young stands	up to 200	Cooley 1979
Aspen	less than 10	60 to 166	Urie et al. 1978
Mixed hardwoods	50	100*	McKim et al. 1982
Northeast			
Mixed hardwoods	--	95 to 224*	McKim et al. 1982
Red pine	--	112*	McKim et al. 1982
Southeast			
Mixed hardwoods	40 to 60	140 to 220*	McKim et al. 1982
Loblolly pine	20	220 to 330*	McKim et al. 1982
Loblolly pine	28	105	Wells et al. 1984

*Uptake rate with waste-effluent irrigation.

range from near 200 to 400 kg/ha per year in seedlings and young stands (Cooley 1979, Schiess and Cole 1981). Middle-aged mixed hardwood and loblolly pine forests also approach this range when nitrogen is supplied with waste effluent irrigation (McKim et al. 1982). Sludge-treated Douglas-fir of the Pacific Northwest is thought to assimilate up to 225 kg N/ha per year in young stands and has been reported to take up 90 kg N/ha per year in 55-year-old stands (Cole and Henry 1983). Aspen and mixed hardwoods in the Great Lakes region approximate this latter value, ranging from 60 to 166 kg N/ha per year (Urie et al. 1978). Nitrogen uptake of sludge-amended loblolly pine in the Southeast, estimated at 105 kg/ha annually, is similar (Wells et al. 1984). Because of the large biomass accumulation on forest sites during stand rotation, a substantial potential exists to store nitrogen in overstory trees.

Nitrogen uptake by understory vegetation has been reported to increase nitrogen levels from background near 1% to values approaching 3% (Brockway 1983). Although such increases are beneficial in understory growth, leading to biomass increments ranging from 50 to 100%, total assimilation is limited to less than 1% of the nitrogen applied during sludge application. Because many understory species annually return their nutrients to the forest floor, they can be useful in nitrogen cycling and leaching abatement but may not be effective long-term accumulators.

Leaching Losses. Nitrate losses from the rooting zone to groundwater following sludge applications in forests have been a major environmental and health concern. Available nitrogen that is not taken up by plants or immobilized by microbes is readily converted to the nitrate form, thus susceptible to leaching (Cole et al. 1984). Groundwater nitrate enrichment may be the most prominent factor limiting nitrogen additions with sludge applications to forest sites (Zasoski et al. 1984), as water yielded by forested watersheds is normally expected to be of high quality (Brockway and Urie 1983).

In the mild climate of the Pacific Northwest the opportunity for year-round nitrogen assimilation poses a reduced risk of nitrate leaching beneath evergreen forests (Zasoski et al. 1984). Rates of ammonia (McKane 1982) and nitrate (Vogt et al. 1981) leaching are nonetheless closely related to sludge application rates, suggesting that caution be used in prescribing high rate applications. Sludge applied at 1,080 Mg/ha (4,800 kg N/ha) in a 45-year-old Douglas-fir stand in western Washington resulted in nitrate concentrations peaking near 120 mg/l in groundwater and remaining elevated for several years following a single application (Riekerk 1981). Nitrogen losses from a cleared Douglas-fir forest soil amended with 247 Mg of anaerobically digested sludge per ha (5,750 kg N/ha) totaled 8% from nitrate leaching (Riekerk 1978). Though a small proportion of the total applied nitrogen, this value of 460 kg N/ha lost through leaching did acutely affect groundwater quality, producing nitrate peaks near 30 mg/l.

Excessive nitrate leaching need not always result from sludge application. Liquid sludge additions at 5.5 dry Mg/ha (402 kg N/ha) and dewatered sludge applications of 50 Mg/ha (632 kg N/ha) and 275 Mg/ha (1,500 kg N/ha) in loblolly pine plantations produced leachate nitrate levels that did not exceed the 10 mg/l drinking water standard (Richter et al. 1982, Wells et al. 1984). On northern hardwoods sites in New Hampshire, sludge additions up to 25 Mg/ha (477 kg N/ha) resulted in soil water nitrate increases of only 3 mg/l (Hornbeck et al. 1979). However, the more typical result of sludge-borne nitrogen applications is a rapid production of nitrate in well-aerated forest soils and subsequent leaching during periods of groundwater recharge. In groundwater 3 m below a 40-year-old red pine plantation growing on a coarse-textured sandy outwash soil in Michigan, nitrate peaks of 49 mg/l were recorded nearly two years after a single sludge application of 32 Mg/ha containing 2,260 kg N/ha (Brockway and Urie 1983). In Pennsylvania, sludge applications of 27 Mg/ha (3,034 kg N/ha) to mixed hardwood stands produced peak nitrate flushes as high as 290 mg/l in soil percolate (Sidle and Kardos 1979). Peak nitrate levels near 60 mg/l were measured beneath loblolly pine plantations that received sludge applications of 11 Mg/ha containing 804 kg N/ha (Wells et al. 1984). Sludge applications of 512 Mg/ha which added 20,750 kg N/ha (3,250 kg available N/ha) to a 42-year-old Douglas-fir stand growing on gravelly outwash resulted in peak nitrate concentrations of 420 mg/l in leachate moving immediately below the soil surface (Zasoski et al. 1984). When three parts sawdust were added to one part sludge, the C:N ratio of the amendment was increased from 10:1 to over 19:1, thus encouraging nitrogen immobilization by microbial action and a resulting decrease in nitrate leaching losses (Vogt et al. 1979, 1981). It should be noted that soil percolate is yet in the biologically active zone of the ecosystem and that high percolate nitrate values do not necessarily translate into high groundwater nitrate levels.

Soil structure and texture of the sludge application site may also have a major effect on the amount of nitrate leached to groundwater. High nitrate levels found percolating below a mixed hardwood stand growing on a clay loam soil were largely a result of solute channeling through macropores present in the soil (Sidle and Kardos 1979). On pine plantations in Michigan, higher nitrate levels were recorded in leachate collected from sands without textural bands than from soil containing lenses of finer sand and clay (Brockway and Urie 1983). These finer-textured bands served as denitrification sites during seasonal periods of moisture saturation. In western Washington greater nitrate losses from the rooting zone were observed in a coarse-textured, gravelly outwash than from a finer-textured, loamy residual soil (Riekerk and Zasoski 1979).

The temporary soil storage capacity for nitrogen is quite variable, ranging from 450 to 1,350 kg/ha (Cole and Henry 1983), depending on numerous characteristics of the particular forest site. Leaching of excess nitrate should theoretically not occur until the storage limits are exceeded, but nitrate leaching is observable once nitrogen reaches 50% of these maximums. Studies in Michigan have related sludge applications to the magnitude of nitrate leaching, in an effort to establish rates of nitrogen addition that were compatible with maintaining soil leachate and groundwater nitrate levels below the 10 mg/l potable water standard (Brockway and Urie 1983). While various site characteristics such as soil and vegetation type and age were of importance in determining the nitrogen storage capacity of each ecosystem, the chemical characteristics and mineralization potential of each sludge type were also germane.

If nitrogen additions with sludge application in forests are limited to 400 to 500 kg/ha per year, significant leaching of nitrate below the rooting zone can be avoided (Riekerk 1981, 1982). Other studies in the Pacific Northwest support this conclusion, finding very little nitrate leaching resulting from sludge applications of 21 Mg/ha, which supplied 400 kg N/ha to forest sites. In addition, sludge applications at rates as high as 40 to 47 Mg/ha have also been proposed as minimizing nitrate loss to groundwater (Stednick and Wooldridge 1979, Cole et al. 1983, 1984). Concern has been noted that sludge applications of no more than 40 Mg/ha may be of little benefit to increasing Douglas-fir growth (Cole et al. 1983). However, sludge nitrogen applications exceeding 400 kg/ha appear to be of no added benefit to growth of loblolly pine (Wells et al. 1984). In the Great Lakes region, sludge application rates ranging from 10 to 19 Mg/ha (670 to 1,140 kg N/ha) have been identified as maximums for pine and aspen forests, beyond which nitrate leaching to groundwater will become excessive (Brockway and Urie 1983). In this region, however, sludge application rates greater than 10 Mg/ha have proved of no added benefit to aspen growth (Urie et al. 1978).

Regional Conceptual Models of Nitrogen Utilization

In an effort to summarize the nutrient utilization characteristics of various forest ecosystems, conceptual models of nitrogen dynamics following sludge application have been constructed for three major forest regions (Table 12). Sludges used in each model are typical of those applied in each locale and, as such, are somewhat variable in their composition. Although all sludges are anaerobically digested, that applied to Douglas-fir forests in the Pacific Northwest has been dewatered and is therefore lower in ammonia-N than those applied to Great Lakes aspen and southeastern loblolly pine, which were applied as liquid slurries. The general correspondence between sludge application rates and overall nitrogen additions is generally similar.

Mineralization rates of organic nitrogen are, from the literature, assumed to be 15% in the first year (Sommers and Nelson 1981) in the aspen and pine forests; however, a rate of 25% was reportedly assumed for Douglas-fir stands (Cole et al. 1983). Although this value may be somewhat overestimated for a dewatered anaerobically digested sludge, the abundant moisture and moderate climatic conditions of the Pacific Northwest may justify such an assumption. Plant uptake rates, though variable, appear to fall into a range that does not differ radically. Nitrogen uptake in aspen and pine was estimated to vary with corresponding changes in sludge-supplied available nitrogen.

In examining these models, the most apparent regional difference is the variation in the threshold sludge application rate that produces leachate nitrate concentrations that exceed the 10 mg/l U.S. EPA standard. The threshold rate is near 40 Mg/ha (715 kg available N/ha) in the Pacific Northwest, 19 Mg/ha (466 kg available N/ha) in the Great Lakes region, and 9 Mg/ha (334 kg available N/ha) in the Southeast. From these calculations one might conclude that Pacific Northwest forest ecosystems possess inherently greater nitrogen assimilation capacities than those in the Great Lakes and Southeast. However, although these models may be useful starting points for comparative discussion, they are based on assumptions and data that must, at best, be considered only preliminary. Further studies will undoubtedly be required before a complete construct of sludge nitrogen utilization in various forest ecosystems becomes available.

SUMMARY OF MANAGEMENT CONSIDERATIONS

Forest ecosystems have evolved efficient mechanisms to assimilate and retain modest levels of annual geochemical input. As inputs exceed outputs over time, a slow, but progressive, process of nutrient accumulation leads to a natural aggradation of forest site quality. Nutrients accumulated on site are largely stored in association with an expanding plant biomass and to a lesser degree with forest floor and soil.

Fertilizer additions are carefully controlled geochemical inputs that are intended to alleviate deficiencies on sites where one or two nutrients limit plant growth. Increases in available nutrient levels in the forest floor and soil, nutrient uptake and retention, and tree growth are typical results of fertilization. Because of economic considerations, fer-

TABLE 12. Nitrogen transformation and utilization during first year following sludge application (conceptual models by region).

Application Rate (Mg/ha)	Organic N (kg/ha)	Ammonia-N (kg/ha)	Mineralized N (kg/ha)	Volatilized N (kg/ha)	Forest Floor Storage (kg/ha)	Soil Storage (kg/ha)	Plant Uptake (kg/ha)	Leached Nitrate-N (kg/ha)	Leachate Nitrate-N (mg/l)
Douglas-fir in Pacific Northwest Region									
23	784	224	196	168	588	140	112	0	0
47	1,568	448	392	420	1,176	224	112	84	15
93	3,136	896	784	907	2,352	448	112	213	42
140	4,704	1,344	1,176	1,445	3,528	560	112	403	79
187	6,272	1,792	1,568	1,982	4,704	672	112	594	117
280	9,408	2,688	2,352	3,058	7,056	896	112	974	192
467	15,680	4,480	3,920	5,208	11,760	1,344	112	1,736	341
Aspen in Great Lakes Region									
5	150	100	23	62	127	1	60	0	0
10	300	200	45	123	255	13	89	20	4
20	600	400	90	245	510	78	111	56	11
40	1,200	800	180	490	1,020	171	166	153	30
Loblolly Pine in Southeastern Region									
6	233	169	35	102	198	10	80	12	3
11	466	338	70	204	396	19	105	80	20

Assumptions:

1. Anaerobically digested sludges (analyses in Table 1) applied as 18.4%, 5.5%, and 2.5% solids in PNW, GL, and SE, respectively.
2. Organic nitrogen mineralized in first year is 25% in PNW and 15% in GL and SE (Sommers and Nelson 1981).
3. Volatilization loss is 50% of available nitrogen (Sommers and Nelson 1981).
4. Forest floor storage is unmineralized organic nitrogen remaining in sludge and O horizons.
5. Soil storage is the residual of available nitrogen inputs and losses in the short term. Long-term changes are unknown.
6. Plant uptake is estimated from field studies (Cole et al. 1983, Urie et al. 1978, Wells et al. 1984, Powers 1976).
7. Average nitrogen leached is based on regional precipitation and evapotranspiration: 114, 96, and 144 cm of precipitation and 64, 46, and 104 cm of evapotranspiration for the PNW, GL, and SE, respectively.

tilizers are applied at rates that simulate natural geochemical inputs and enhance tree growth but do not result in leaching losses that would degrade groundwater quality.

Wastewater additions provide forest sites with rates of soluble nutrients that generally exceed rates by natural input. Phosphorus is retained well by mineral soil, and cations (potassium, sodium, calcium) may also accumulate to some extent. However, anions (sulfate, nitrate, chloride) are quite subject to leaching from the soil and are accompanied by cations during periods of groundwater recharge. The chemical composition of waste effluent may influence the relative balance of cations and anions in the soil, often leading to a new equilibrium in the soil solution. Irrigation-applied nitrogen accelerates forest floor decomposition and may accumulate in soil in association with increases in organic matter. Volatilization of ammonia and denitrification of nitrate may be important avenues of nitrogen loss from sites irrigated with wastewater.

Two universal results of waste effluent irrigation are increased nitrogen uptake and retention by vegetation and increased rates of nitrate-N leaching below the rooting zone. Nitrogen utilization is greatest in young forests of poplar that are rapidly growing in association with understory vegetation. Mature forests are less efficient in assimilating and retaining wastewater-applied nitrogen. Substantial leaching of nitrate from waste effluent-irrigated forests has been reported where rates of soluble nitrogen application exceeded site assimilation capacities. Generally, rates approaching 200 kg of available N/ha per year lead to nitrate-N concentrations in leachate that exceed the 10 mg/l U.S. EPA standard.

Although not fully predictable, forest site renovation capacity for wastewater appears to diminish over time. Given that wastewater provides a high proportion of soluble nutrients and little organic matter, effluent irrigation remains limited in promoting forest site aggradation. Wastewater irrigation can be an effective means of recycling nutrients back into the ecosystem. However, nutrient application rates (especially nitrogen) should be adjusted to the assimilation capacity of the forest site, if unacceptable leaching losses and groundwater degradation are to be avoided.

Sludge applications provide forest sites with rates of nutrient addition that exceed those of wastewater, fertilizer, or natural inputs. Applied phosphorus, potassium, and calcium are largely retained in the forest floor and upper layers of mineral soil. While leaching of phosphorus is nil, potassium and calcium losses are reported to accompany leaching of anions (primarily nitrate) during periods of groundwater recharge. Sludge nitrogen is initially stored in the forest floor, where it becomes slowly available as organic matter degrades. A lowering of the C:N ratio in the forest floor results in an accelerated rate of decomposition. Volatilization of ammonia following surface application of liquid sludge and denitrification during periods of excess soil moisture represent two potential pathways of nitrogen loss from a treated forest site.

Increased nitrogen uptake and rates of nitrate leaching are two widespread results of sludge application to forest land. Nitrogen assimilation is greatest on sites occupied by young, rapidly growing poplar (200 to 400 kg/ha·yr) and somewhat less in middle-aged stands of Douglas-fir (90 kg/ha·yr) and loblolly pine (105 kg/ha·yr). While foliar nitrogen increases in both overstory and understory vegetation following sludge application, the nitrogen storage capacity is substantially greater in the larger biomass of overstory trees. Increases in tree growth rates up to 40% have been associated with sludge nitrogen applications of 400 kg/ha. Excessive nitrate-N leaching losses have been related to high rates of sludge application. However, when sludge nitrogen additions were limited to

400 or 500 kg/ha, nitrate concentrations in soil leachate did not exceed the 10 mg/l U.S. EPA water quality standard.

The addition of nutrients and organic matter with wastewater sludge applications may produce a substantial beneficial effect on overall site quality. Organic matter, in addition to nutrients, may prove a significant modifier of forest floor and soil characteristics, changing moisture relations and the manner in which nutrients are stored and cycled on the site. Short-term transfer rates for nitrogen and phosphorus are known to double within one year of sludge applications on pine sites. Abundant evidence has been noted in regard to the enhanced site nutrient relations and vegetation growth resulting soon after application of sludge to forest ecosystems. In the long term, repeated sludge applications should provide a cumulative positive effect on site quality. Although the prospect is quite speculative at this juncture, one would anticipate a continued buildup of site nutrient capital and development of a rich forest humus, resulting in a greatly ameliorated environment for plant growth. This process of accelerated site aggradation could lead to permanent increases in forest production.

Michigan Agricultural Experiment Station journal article number 11865.

REFERENCES

Bledsoe, C. S. (ed.) 1981. Municipal sludge application to Pacific Northwest forest lands. Institute of Forest Resources Contribution 41. College of Forest Resources, University of Washington, Seattle. 155 p.

Bledsoe, C. S., and R. J. Zasoski. 1979. Growth and nutrition of forest tree seedlings grown in sludge amended media. p. 75–79. In R. L. Edmonds and D. W. Cole (eds.) Use of dewatered sludge as an amendment for forest growth: Management and biological assessments. Vol. 3. Center of Ecosystem Studies, College of Forest Resources, University of Washington, Seattle.

Brister, G. H., and R. C. Schultz. 1981. The response of a southern Appalachian forest to waste water irrigation. J. Environ. Qual. 10:148–153.

Brix, H., and A. K. Mitchell. 1980. Effects of thinning and nitrogen fertilization on xylem development in Douglas-fir. Can. J. For. Res. 10:121–128.

Brockway, D. G. 1979. Evaluation of northern pine plantations as disposal sites for municipal and industrial sludge. Ph.D. diss., Michigan State University, East Lansing. University Microfilms, Ann Arbor, Mich. (Diss. Abstr. 40–2919B).

———. 1983. Forest floor, soil, and vegetation responses to sludge fertilization in red and white pine plantations. Soil Sci. Soc. Am. J. 47:776–784.

Brockway, D. G., G. Schneider, and D. P. White. 1979. Dynamics of municipal wastewater renovation in a young conifer-hardwood plantation in Michigan. p. 87–101. In W. E. Sopper and S. N. Kerr (eds.) Utilization of municipal sewage effluent and sludge on forest and disturbed land. Pennsylvania State University Press, University Park.

Brockway, D. G., and D. H. Urie. 1983. Determining sludge fertilization rates for forests from nitrate-N in leachate and groundwater. J. Environ. Qual. 12:487–492.

Burton, T. M., and J. E. Hook. 1979. A mass balance study of application of municipal waste water to forests in Michigan. J. Environ. Qual. 8:589–596.

Chapman, R. J. 1983. Growth, nitrogen content and water relations of sludge-treated Douglas-fir seedlings. M.S. thesis, University of Washington, Seattle. 85 p.

Cole, D. W., and C. L. Henry. 1983. Leaching and uptake of nitrogen applied as dewatered sludge. p. 57–66. In C. L. Henry and D. W. Cole (eds.) Use of dewatered sludge as an amendment for forest growth. Vol. 4. Institute of Forest Resources, University of Washington, Seattle.

Cole, D. W., and P. Schiess. 1978. Renovation of wastewater and response of forest ecosystems: The Pack Forest study. p. 323–332. In Land Treatment of Wastewater, International Symposium, Hanover, New Hampshire.

Cole, D. W., C. L. Henry, P. Schiess, and R. J. Zasoski. 1983. The role of forests in sludge and wastewater utilization programs. p. 125–143. In A. L. Page, T. L. Gleason III, J. E. Smith, Jr., I. K. Iskandar, and L. E. Sommers (eds.) Proceedings of the 1983 Workshop on Utilization of Municipal Wastewater and Sludge on Land. University of California, Riverside.

Cole, D. W., M. L. Rinehart, D. G. Briggs, C. L. Henry, and F. Mecifi. 1984. Response of Douglas-fir to sludge application: Volume growth and specific gravity. p. 77–84. In 1984 TAPPI Research and Development Conference, Appleton, Wisconsin.

Technical Association of the Pulp and Paper Industry, Technology Park, Atlanta, Georgia.

Cooley, J. H. 1978. Nutrient assimilation in trees irrigated with sewage oxidation pond effluent. p. 328–340. In Proceedings of the Central Hardwoods Forest Conference, Purdue University, West Lafayette, Indiana.

——. 1979. Fertilization of populus with municipal and industrial waste. p. 101–108.In Proceedings of the Poplar Council, Crystal Mountain, Michigan.

Cunningham, H. 1976. Nutrient management update. USDA Forest Service In-house paper. Timber Management Staff Conference, Eastern Region, Milwaukee, Wisconsin. 30 p.

Edmonds, R. L., and K. P. Mayer. 1981. Biological changes in soil properties associated with dewatered sludge application. p. 49–57. In C. S. Bledsoe (ed.) Municipal sludge application to Pacific Northwest forest lands. College of Forest Resources, University of Washington, Seattle.

Forster, D. L., L. J. Logan, R. H. Miller, and R. K. White. 1977. State of the art in municipal sewage sludge landspreading. p. 603–618. In R. C. Loehr (ed.) Land as a waste management alternative. Ann Arbor Science Publishers, Ann Arbor, Michigan.

Foster, N. W., and I. K. Morrison. 1983. Soil fertility, fertilization and growth of Canadian forests. Great Lakes Forest Research Centre Information Report O-X-353. Department of Environment, Canadian Forestry Service. 21 p.

Freshman, J. D. 1977. A perspective on land as a waste management alternative. p. 3–8. In R. C. Loehr (ed.) Land as a waste management alternative. Ann Arbor Science Publishers, Ann Arbor, Michigan.

Groman, W. A. 1972. Forest fertilization. Environmental Protection Technology Series. EPA R2-72-016. U.S. Environmental Protection Agency, Pacific Northwest Water Laboratory, Corvallis, Oregon. 57 p.

Harris, A. R. 1979. Physical and chemical changes in forested Michigan sand soils fertilized with effluent and sludge. p. 155–161. In W. E. Sopper and S. N. Kerr (eds.) Utilization of municipal sewage effluent and sludge on forest and disturbed land. Pennsylvania State University Press, University Park.

Harris, A. R., and D. H. Urie. 1983. Changes in sandy forest soils under northern hardwoods after five years of sewage effluent irrigation. Soil Sci. Soc. Am. J. 47:800–805.

Henry, C. L., and D. W. Cole (eds.) 1983. Use of dewatered sludge as an amendment for forest growth. Vol. 4. Institute of Forest Resources, University of Washington, Seattle. 110 p.

Hook, J. E., and L. T. Kardos. 1978. Nitrate leaching during long-term spray irrigation for treatment of secondary sewage effluent on woodland sites. J. Environ. Qual. 7:30–34.

Hornbeck, J. W., M. T. Koterba, and R. S. Pierce.

1979. Sludge application to a northern hardwood forest in New Hampshire: Potential for dual benefits? p. 137–143. In W. E. Sopper and S. N. Kerr (eds.) Utilization of municipal sewage effluent and sludge on forest and disturbed land. Pennsylvania State University Press, University Park.

Koterba, M. T., J. W. Hornbeck, and R. S. Pierce. 1979. Effects of sludge applications on soil water solution and vegetation. J. Environ. Qual. 8:72–78.

Mayer, K. P. 1980. Decomposition of dewatered sewage sludge applied to a forest soil. M.S. thesis, University of Washington, Seattle. 176 p.

McKane, R. B. 1982. Nutrient and metal losses by leaching from municipal sewage sludge. M.S. thesis, University of Washington, Seattle. 58 p.

McKim, H. L., W. E. Sopper, D. W. Cole, W. L. Nutter, D. H. Urie, P. Schiess, S. N. Kerr, and H. Farquhar. 1982. Wastewater applications in forest ecosystems. Cold Regions Research and Engineering Laboratory Report 82-19. Hanover, New Hampshire.

Miller, H. G. 1981. Forest fertilization: Some guiding concepts. Forestry 54:157–167.

Miller, R. E., and S. R. Webster. 1979. Fertilizer response in mature stands of Douglas-fir. p. 126–132. In S. P. Gessel, R. M. Kenady, and W. A. Atkinson (eds.) Proceedings, Forest Fertilization Conference. Institute of Forest Resources Contribution 40. College of Forest Resources, University of Washington, Seattle.

Morris, C. E., and W. J. Jewell. 1977. Regulations and guidelines for land application of wastes: A 50 state overview. p. 9–28. In R. C. Loehr (ed.) Land as a waste management alternative. Ann Arbor Science Publishers, Ann Arbor, Michigan.

Murrman, R. P., and F. R. Koutz. 1972. Role of soil chemical processes in reclamation of wastewater applied to land. p. 48–76. In S. C. Reed (ed.) Wastewater management by disposal on the land. Cold Regions Research and Engineering Laboratory, Hanover, New Hampshire.

Nutter, W. L., and J. T. Red. 1984. Treatment of wastewater by application to forest land. p. 95–100. In 1984 TAPPI Research and Development Conference, Appleton, Wisconsin. Technical Association of the Pulp and Paper Industry, Technology Park, Atlanta, Georgia.

Nutter, W. L., R. C. Schultz, and G. H. Brister. 1978. Land treatment of municipal wastewater on steep forest slopes in the humid southeastern United States. p. 265–274. In H. L. McKim (ed.) Land treatment of wastewater. Vol. 1. Cold Regions Research and Engineering Laboratory, Hanover, New Hampshire.

Palazzo, A. J., and T. F. Jenkins. 1979. Land application of waste water: Effect on soil and plant potassium. J. Environ. Qual. 8:309–312.

Parker, C. F., and L. E. Sommers. 1983. Mineralization of nitrogen in sewage sludges. J. Environ. Qual. 12:150–156.

Powers, R. F. 1976. Principles and concepts of forest soil fertility. *In* First Annual Earth Science Symposium, Fresno, California.

Pritchett, W. L. 1979. Properties and management of forest soils. John Wiley and Sons, New York. 500 p.

Pritchett, W. L., and W. H. Smith. 1975. Forest fertilization in the U.S. Southeast. p. 467–476. *In* B. Bernier and C. H. Winget (eds.) Forest soils and forest land management. Laval University Press, Quebec.

Ralston, C. W., and A. B. Prince. 1963. Accumulation of dry matter and nutrients by pine and hardwood forests in the lower piedmont of North Carolina. p. 77–94. *In* C. T. Youngberg (ed.) Forest-soil relationships in North America. Second North American Forest Soils Conference. Oregon State University Press, Corvallis.

Richenderfer, J. L., and W. E. Sopper. 1979. Effect of spray irrigation of treated municipal sewage effluent on the accumulation and decomposition of the forest floor. p. 163–177. *In* W. E. Sopper and S. N. Kerr (eds.) Utilization of municipal sewage effluent and sludge on forest and disturbed land. Pennsylvania State University Press, University Park.

Richter, D. D., D. W. Johnson, and D. M. Ingram. 1982. Effects of municipal sewage sludge-cake on nitrogen and phosphorus distributions in a pine plantation. p. 532–546. *In* Fifth Annual Madison Conference of Applied Research and Practice on Municipal and Industrial Waste. Department of Engineering and Applied Science, University of Wisconsin, Madison.

Riekerk, H. 1978. The behavior of nutrient elements added to a forest soil with sewage sludge. Soil Sci. Soc. Am. J. 42:810–816.

——. 1981. Effects of sludge disposal on drainage solutions of two forest soils. For. Sci. 27:792–800.

——. 1982. How much sewage nitrogen on forest soils? A case history. Biocycle 23:53–57.

Riekerk, H., and R. J. Zasoski. 1979. Effects of dewatered sludge applications to a Douglas fir forest soil on the soil, leachate, and groundwater composition. p. 35–45. *In* W. E. Sopper and S. N. Kerr (eds.) Utilization of municipal sewage effluent and sludge on forest and disturbed land. Pennsylvania State University Press, University Park.

Schiess, P., and D. W. Cole. 1981. Renovation of wastewater by forest stands. p. 131–147. *In* C. S. Bledsoe (ed.) Municipal sludge application to Pacific Northwest forest lands. Institute of Forest Resources Contribution 41. College of Forest Resources, University of Washington, Seattle.

Sidle, R. C., and L. T. Kardos. 1979. Nitrate leaching in a sludge-treated forest soil. Soil Sci. Soc. Am. J. 43:278–282.

Smith, W. H., and J. O. Evans. 1977. Special opportunities and problems in using forest soils for organic waste application. p. 429–454. *In* L. F. Elliott and F. J. Stevenson (eds.) Soils for management of organic wastes and waste waters. American Society of Agronomy, Madison, Wisconsin.

Sommers, L. E., and D. W. Nelson. 1981. Nitrogen as a limiting factor in land application of sewage sludges. p. 425–448. *In* Fourth Annual Madison Conference, of Applied Research and Practice on Municipal and Industrial Waste. Department of Engineering and Applied Science, University of Wisconsin, Madison.

Sopper, W. E., and L. T. Kardos. 1973. Vegetation responses to irrigation with treated municipal wastewater. p. 271–294. *In* W. E. Sopper and L. T. Kardos (eds.) Recycling treated municipal wastewater and sludge through forest and cropland. Pennsylvania State University Press, University Park.

Sopper, W. E., and S. N. Kerr (eds.) 1979. Utilization of municipal sewage effluent and sludge on forest and disturbed land. Pennsylvania State University Press, University Park. 537 p.

Stednick, J. D., and D. D. Wooldridge. 1979. Effects of liquid digested sludge irrigation on the soil of a Douglas fir forest. p. 47–60. *In* W. E. Sopper and S. N. Kerr (eds.) Utilization of municipal sewage effluent and sludge on forest and disturbed land. Pennsylvania State University Press, University Park.

Urie, D. H. 1979. Nutrient recycling under forests treated with sewage effluents and sludge in Michigan. p. 7–17. *In* W. E. Sopper and S. N. Kerr (eds.) Utilization of municipal sewage effluent and sludge on forest and disturbed land. Pennsylvania State University Press, University Park.

Urie, D. H., A. R. Harris, and J. H. Cooley. 1978. Municipal and industrial sludge fertilization of forests and wildlife openings. p. 467–480. *In* First Annual Madison Conference of Applied Research and Practice on Municipal and Industrial Waste. Department of Engineering and Applied Science, University of Wisconsin, Madison.

Urie, D. H., A. R. Harris, and J. H. Cooley. 1984. Forest land treatment of sewage wastewater and sludge in the Lake States. p. 101–110. *In* 1984 TAPPI Research and Development Conference, Appleton, Wisconsin. Technical Association of the Pulp and Paper Industry, Technology Park, Atlanta, Georgia.

U.S. Environmental Protection Agency. 1981. Process design manual: Land treatment of municipal wastewater. CERI, Cincinnati, Ohio.

van den Driessche, R. 1974. Prediction of mineral nutrient status of trees by foliar analysis. Botanical Review 40:347–394.

Vogt, K.A., R. L. Edmonds, and D. J. Vogt. 1979. Regulation of nitrate levels in sludge, soil and groundwater. p. 53–65. *In* R. L. Edmonds and D. W. Cole (eds.) Use of dewatered sludge as an amendment for forest growth: Management and biological assessments. Vol. 3. Center for Ecosystem Studies, University of Washington, Seattle.

——. 1981. Nitrate leaching in soils after sludge application. p. 59–66. *In* C. S. Bledsoe (ed.) Municipal sludge application to Pacific Northwest forest lands. Institute of Forest Resources Contribution 41. College of Forest Resources, University of Washington, Seattle.

Walsh, L. M. 1976. Application of sewage sludge to cropland: Appraisal of potential hazards of heavy metal to plants and animals. Council for Agricultural Science Technical Report 64. Iowa State University, Ames.

Wells, C. G., K. W. McLeod, C. E. Murphy, J. R. Jensen, J. C. Corey, W. H. McKee, and E. J. Christensen. 1984. Response of loblolly pine plantations to two sources of sewage sludge. p. 85–94. *In* 1984 TAPPI Research and Development Conference, Appleton, Wisconsin. Technical Association of the Pulp and Paper Industry, Technology Park, Atlanta, Georgia.

White, D. P., G. Schneider, E. A. Erickson, and D. H. Urie. 1975. Changes in vegetation and surface soil properties following irrigation of woodlands with municipal wastewater. Project completion report. Institute of Water Research, Michigan State University, East Lansing.

Winburne, J. H. (ed.) 1962. A dictionary of agricultural and allied terminology. Michigan State University Press, East Lansing. 903 p.

Zasoski, R. J. 1981. Effects of sludge on soil chemical properties. p. 45–48. *In* C. S. Bledsoe (ed.) Municipal sludge application to Pacific Northwest forest lands. Institute of Forest Resources Contribution 41. College of Forest Resources, University of Washington, Seattle.

Zasoski, R. J., D. W. Cole, and C. S. Bledsoe. 1983. Municipal sewage sludge use in forests of the Pacific Northwest, U.S.A.: Growth responses. Waste Management and Research 1:103–104.

Zasoski, R. J., R. L. Edmonds, C. S. Bledsoe, C. L. Henry, D. J. Vogt, K. A. Vogt, and D. W. Cole. 1984. Municipal sewage sludge use in forests of the Pacific Northwest, U.S.A.: Environmental concerns. Waste Management and Research 2:227–246.

Chapter 22

Effect of Sludge on Wood Properties: A Conceptual Review with Results from a Sixty-year-old Douglas-fir Stand

D. G. BRIGGS, F. MECIFI, and W. R. SMITH

ABSTRACT Expected changes in wood properties due to forest cultural practices are reviewed. The hormone theory and results reported in the literature for thinnings and fertilization provide a basis for hypotheses on how wood properties are affected by sludge treatments to forest land. The effect of municipal sludge applied on a low site, 60-year-old Douglas-fir stand is reported for several wood properties—changes in specific gravity, tracheid characteristics, and strength properties. Implications of these changes are discussed.

A difficult problem when evaluating the effect of a cultural practice on wood quality is determining what is meant by the term "wood quality." Although the precise wording varies from author to author, wood scientists generally agree that wood quality is the appropriateness or suitability of wood for a particular end use. There are three major elements to this definition. The first is that different end uses have different performance requirements; for example, a piece of paper and a roof truss have very different requirements. Research has improved our understanding of what happens to a product in service, therefore the performance requirements have undergone substantial change and will continue to do so. Part of this change is due to a changing resource. The second major element is that the substantial natural variation in most wood properties and modern forest cultural practices can influence many of the properties considerably. The third element is the grading rules, building codes, and product standards that were developed as commercial mechanisms to facilitate matching the other two elements. An engineer or architect would probably think of wood quality primarily from the perspective of end-use performance, a product manufacturer would probably focus on grading rules and product standards, while a wood technologist or biologist would be most interested in the properties of wood itself. It should also be noted that change can occur in any of these three elements.

Thus wood quality is not something that can be measured; one can specify those properties of major importance with respect to an end use, measure the changes in those properties that a practice brings about, and finally make a judgment, preferably in economic terms, as to whether the overall effect is good or bad.

The literature is not very helpful in providing quantitative estimates useful for financial evaluation of wood quality issues. There are several reasons for this, including species differences, inadequate or inconsistent documentation of stand and treatment conditions, differences in sampling methods and data presentation, inconsistencies in use of different growth measures, and confounding the effect of growth with the effect of age on wood properties. These problems with the literature have been discussed in more detail elsewhere (Briggs and Smith 1986), but the problems with growth measurement and confounding growth with age effect will be repeated here.

Growth can be measured in terms of either basal area increment or radial increment. Unfortunately, it is common for basal area to be increasing while radial increment is decreasing, as shown in Figure 1. Rather than increase basal area by increasing ring width, a cultural treatment could cause basal area increment to increase by simply slowing the rate at which ring width had been decreasing, as was the case in the tree illustrated. Of course, both radial and basal area increments may increase simultaneously. Thus it would be possible for a property to show a positive or negative relation with growth rate depending on the measure used.

Any growth rate relation is secondary to the strong influence that age (ring count from the pith) has on a large number of wood properties (Figure 2). Generally, the region of rapid change near the pith, commonly called juvenile wood, has undesirable properties for virtually all products. It is formed when this region of the tree has a young vigorous crown and is under the strong influence of growth regulators produced by the crown. Thus, a half-inch wide ring produced at age 4 in a tree will have very different properties from a half-inch ring produced at age 40. The confusion of age with growth rate occurs because the rings produced near the pith under natural conditions are usually wider than those produced later. Many authors have compared wood from a young, fast growing tree with that of a slower growing, older tree (or outer part of the same tree) and mistakenly concluded that growth rate was the major factor involved. However, when wood of the same age is compared, many authors have concluded that growth rate differences have small effects on properties (Bendtsen 1978, Larson 1957, Pearson and Gilmore 1980, Turnbull and Plessis 1946). The growth rate effect, if it exists, may be noticeable only at extremely slow or fast rates.

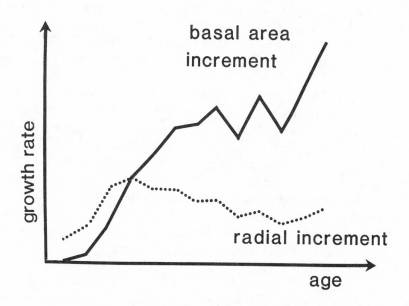

Figure 1. Different measures of growth may produce opposing trends.

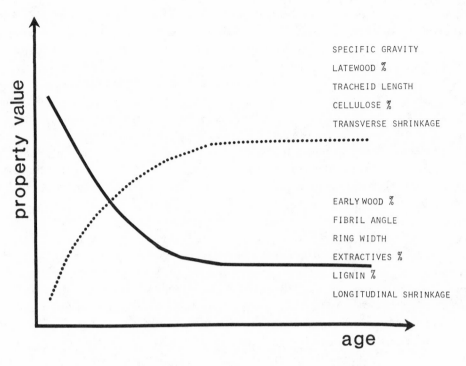

Figure 2. Effect of age on the value of various clear wood properties.

PHYSIOLOGICAL BASIS FOR HYPOTHESES

Age effects on wood formation occur by influence of growth regulators produced in the crown. These affect the process of wood formation, the rate of cambial division, and subsequent tracheid differentiation (Larson 1969, Denne and Dodd 1981). The size and arrangement of cells in the ring depend on these regulatory mechanisms, which in turn are affected by changes in the growth environment that affect the size and vigor of the crown. Cultural practices primarily affect wood properties by changing the growth environment.

In tracheid structure development, length is controlled by the length of the cambial initial from which it was derived. In early season, the cambium is stimulated to divide more rapidly than later, and this leads to cambial initials that are relatively short. The tracheids that derive from these initials are also relatively short. Microfibril angle, the orientation of cellulose molecular strands to the tracheid longitudinal axis, is greater in short cells than in long ones. Increased fibril angle reduces tensile strength of the tracheid and increases longitudinal shrinkage. Tracheid diameter has been demonstrated to be controlled by hormones (auxin); under high concentrations, larger diameters are produced. This is the basis of earlywood formation. As these hormone concentrations change, smaller diameters form, starting at the stem base since it is farther from the hormone source in the crown. As the season advances, all diameters become smaller, progressing upward from the base. Finally, wall thickness is modified by growth regulators, and substrate changes as well, leading to thicker walls as the season progresses. Since these growth regulators occur as gradients from the tip to base of the tree and continuously change, a pattern of wood formation occurs that we tend to arbitrarily oversimplify as earlywood and latewood. The word "arbitrarily" is used here since there may be varying amounts of intermediate or transition wood formed depending on conditions.

Thus wood formation is dependent on distance from the tip of the tree (internode number, age) as well as time of year. The wood formed within the vigorous upper part of the live crown under very strong hormonal influence is juvenile wood. In the southern pines, juvenile wood is produced for seven to ten years before the age curve flattens out into mature wood. In Douglas-fir, juvenile wood appears to form for at least fifteen years, and it is common to see the age curve not flatten out as in the southern pine until it is much older. Juvenile wood, therefore, forms a central core of wood in the tree from base to top.

KEY WOOD PROPERTIES

A few of the wood features and properties discussed here may need further explanation. Density is the relative mass of cell substance per unit volume. It reflects the overall amount of matter produced by wood formation, hence is sensitive to changes in wood formation. Specific gravity is the density of wood relative to the density of water. Specific gravity has been found to be well correlated with many other properties, and since it is easy to measure, it is a popular indicator used in wood quality research. Compression wood is an abnormal type of wood typically formed on the lower side of leaning trees and branches. It can also be formed in straight trees and has characteristics leading to poor strength and poor dimensional stability in products.

Modulus of rupture (MOR) and modulus of elasticity (MOE) are two mechanical properties of interest. MOR has to do with strength and MOE with stiffness. Wood scientists usually measure these values on small clear samples, in which case they are well correlated with specific gravity. In dealing with lumber, other defects, such as knots and associated grain distortion, greatly complicate the relation with specific gravity and become the principal determinants of strength.

The natural variation in wood properties within a species is quite large. This variation is high whether in geographic range, between trees on the same plot, within a tree, or even within a ring. For example, in Douglas-fir, specific gravity varies from 0.33 to 0.59 over its geographic range (USDA Forest Service 1965). We have already seen (Figure 2) that whole-ring specific gravity is lowest in juvenile wood, climbs with age, and eventually levels out. The change involved is on the order of 20 to 30% or more. Also, there are two factors that work to reduce specific gravity with height in a tree. The average specific gravity of disks taken from various heights will decrease because of progressively greater proportions of juvenile growth, but this is only part of the story. Also, within a ring, specific gravity tends to be lowest at the top of the tree, gradually increasing to a maximum point somewhere along the bole, and slowly declining to the stem base. The location of this culmination point along the bole varies with age and growth conditions. Finally, across a ring a three- to fourfold change in specific gravity from earlywood to latewood is common (Megraw 1985).

Similar patterns of variation can be described for other properties, such as tracheid length, fibril angle, and chemical composition, although magnitudes vary from property to property. In juvenile wood, tracheid length may be less than half that produced later in mature wood. Within a ring, earlywood tracheids may be 10 to 20% shorter than late-wood tracheids. The general patterns described in the preceding discussion resemble those found in conifers such as Douglas-fir or southern pines with an abrupt change from earlywood to latewood. These patterns are not universal among all species (Panshin and de Zeeuw 1980).

Some may take comfort in this wide degree of natural variation by arguing that any observed changes caused by applying intensive cultural practices fall well within this pattern of natural variation. This misses the point that product standards and grading rules are designed to screen out extremes, and, furthermore, widespread adoption of cultural practices will cause a general shift in population characteristics upon which the various rules and standards are based. Anyone familiar with the techniques for deriving design stress values for lumber based on population estimates will recognize that shifts in properties leading to lower design values could have serious consequences. Another possibility is to regard material grown under intensive culture to be a new subspecies requiring derivation of a new set of parameters.

EFFECTS OF CULTURAL PRACTICES

Growth stimulating treatments such as wider initial spacing, thinning, and fertilization (sludge included) tend to promote more vigorous crowns and slow the rate of crown recession. Pruning, in contrast, may be passive in just removing dead or dying branches; if carried into the vigorous live crown, it may accelerate crown recession, with effects opposite those of stocking control and fertilization. These changes in the crown affect many features and properties of wood. In considering the possible effects of sludge treatment on wood properties, one should be cautioned that both direction and magnitude vary considerably depending on age of tree when treated, degree of treatment, and various other site factors that may interact with the treatment. To date, no studies have been of the scope necessary to provide understanding of these interactions.

General Tree Features

Table 1 presents a list of the more important tree and clear wood properties that can be affected by cultural treatments. Beside each property is indicated the type of change that various cultural practices may cause. Although often ignored in studies of cultural effects on wood quality, changes in the general tree features may completely override effects on the properties of clear wood that are usually studied in detail. These features are often elements of log grading and scaling practices or are elements of product grading, or create other general utilization problems.

Knots. Knots are a serious defect affecting the aesthetics of a piece of wood as well as the mechanical properties of sawn and peeled products. Knotwood is also undesirable in pulping, since it is denser and has a large amount of compression wood. Except for pruning, which removes knots, other practices intended to accelerate growth also accelerate the growth of live branches, thus increasing knot sizes. Since these practices promote healthier crowns and reduce crown recession, knot development can be a critical problem. These changes are especially notable since the area of knot, with the distorted grain surrounding it, increases in proportion to the square of its diameter. It is the knot (with associated area of distorted grain) that most directly affects mechanical properties.

The importance of knot size depends on when treatments are practiced during stand

TABLE 1. Effects that various cultural practices may have on wood characteristics.

Wood Characteristic	Wider Spacing	Precommercial Thinning	Commercial Thinning	Pruning	Fertilization	Sludge Application
Knots	Larger, sound	Larger, sound	Larger, sound	Removed	Larger, sound	Larger, sound
Rings	Wide, uniform	Wide, abrupt change	Wide, abrupt change	Narrowed	Wide, abrupt change	Wide, abrupt change
Juvenile wood	Enlarged all logs	Enlarged most logs	Enlarged upper logs	Reduced	Enlarged	Enlarged
Compression wood	Increase	Increase	Increase	?	Increase	Increase
Taper	Increase	Increase	Increase	Decrease	Increase or unchanged	Increase or unchanged
Spiral grain	Increase	Increase	Increase	?	Increase	Increase
Crookedness	More prevalent	Reduced	Reduced	No effect	No effect? (snow, ice)	No effect? (snow, ice)
Specific gravity	Little change in young trees, since age effect dominates (may be an effect with extreme treatment)		± 5 to 10%	Increase	Decrease 5 to 15% (except severely deficient sites)	Decrease
Fibril angle	Same as above	Same as above	Increase	Decrease	Increase	Increase
Cell dimensions						
Length	Same as above	Same as above	Small decrease	Small increase	Small decrease	Decrease
Diameter	Same as above	Same as above	Increase	Decrease	Increase	Increase
Wall thickness	Same as above	Same as above	Decrease	Increase	Decrease	Decrease

development. Stands under intensive regimes from the time of planting can develop large knot problems throughout the bole, greatly reducing product value from all logs. If stands are treated closer to maturity, these effects are limited to the present and future live crown region, generally affecting smaller, less valuable upper logs in the stem.

Ring Structure. In lumber, wide-ringed stock or a change from narrow to wide rings can cause problems in wood machining (Fielding 1967), can create warpage problems, and may be aesthetically undesirable in some markets. Visual lumber grading rules have limitations on ring width, often in combination with latewood percentage. These limitations attempt to sort out unusual pieces that would have low mechanical properties. These grading rule features were designed primarily to screen out naturally fast grown juvenile wood portions of trees, but there are practical problems in attempting to separate fast grown juvenile and mature wood. More widespread use of machine grading may allow more of the faster grown wood to be used than is presently allowed by visual grading.

There are also problems with wide rings in veneer and plywood manufacture. A change from narrow to wide rings within a log often creates serious problems with veneer roughness leading to degrade of veneer and increased glue consumption. Although lathes can be set to produce smooth veneer from fast grown logs, it is not feasible to make changes from log to log, let alone make adjustments when growth rate abruptly changes within a log. Thus the key factor is to have fairly large batches of logs with similar and fairly uniform ring structures. Wide rings can create another problem, primarily in developing veneer sheets that are almost entirely of either earlywood or latewood. In assembling sheets to make plywood panels, it is possible to get a combination in which the panel is entirely earlywood and quite weak. Also, adjacent sheets, if entirely latewood, present gluing difficulties (Koch 1972).

In pulp and paper production, as well as in manufacturing reconstituted panels from chips, wide-ring material creates serious problems of chip quality and pulping variability (Hatton 1979).

Juvenile Wood. Juvenile wood will be increased by all cultural practices that increase growth rate in a young stem. This is because of the increased width of rings and the enhanced crown vigor that may cause the crown region to produce juvenile wood for more years than normal. The critical question is where in the tree does this juvenile core enlargement take place. If the growth enhancing cultural practice begins when the tree is still young, then the juvenile core will be quite large throughout the tree. If the stimulation occurs when the tree is much older, the juvenile core can only be enlarged within the current or future crown, and this may be such that it will only influence the eventual top log or two.

A critical problem with juvenile wood is that it has properties that are undesirable for all products, and the usual purpose of intensive culture is to produce young trees so as to minimize rotation length. This tends to produce trees with very high percentages of juvenile wood and consequently low product value. The same management philosophy will also produce large knots.

Compression Wood. Compression wood is an abnormal form usually thought of in terms of leaning trees or the underside of branches. It creates serious problems of strength, dimensional stability, and aesthetics in sawn and panel products and is not desired for pulping. Intensive practices often stimulate formation of compression wood even in straight trees (Smith 1968, Cown 1974, Johnston 1962). Compression wood in the bole has been associated with large knots (Zobel and Haught 1962). Cultural practices that open stands up and stimulate a heavier, more vigorous crown apparently lead to greater wind stresses, which create the formation of compression wood.

Taper. Growth stimulating practices frequently increase the rate of taper in trees. When this happens, the percentage recovery of high value sawn or peeled products decreases rapidly, and low value residues, principally chips, increase (Williston 1981). It has also been reported that sawmilling costs rise with greater log taper because of lower recovery (Williston 1981).

Spiral Grain. Although the evidence is not strong, there have been reports that accelerated growth may lead to the development of spiral grain (Johnston 1962). In lumber and veneer, this would produce cross grain—a serious weakening defect.

Crookedness. Except for thinning, which can select against crooked trees, other cultural practices have little influence on crookedness, which can increase logging and milling costs, as a result of difficult handling, and reduce yield of sawn and peeled products. It is possible, however, that practices that stimulate heavier crowns may, in stands in some areas, lead to increased wind or ice breakage that could increase crookedness. Wide initial spacing may also favor crookedness by reducing the opportunity to eliminate such trees during thinnings.

The impact of sludge treatment on the features discussed so far takes the same direction as stocking controls or fertilization. The only issue is degree of change, and unfortunately little quantitative research has been done in this area for any of these practices. A hypothesis might be expressed that the impact of a growth stimulating practice on these properties would be proportional to the basal area response of the bole.

Clear Wood Properties

Table 1 also serves as a checklist of cultural practices and the direction of change and, in some cases, general magnitude of change for specific gravity, fibril angle, and tracheid dimensions. The effects of cultural practices such as wide spacing and precommercial thinning are not clear in young trees; perhaps the rapid changes in the juvenile region due to age override these effects. Also, research on young stands is sparser. There are indications that extreme conditions may lead to serious problems.

Commercial thinning has been shown to cause specific gravity to increase or decrease. The increases usually result on sites where dry summers truncate normal latewood production. Thinning increases available moisture to fewer trees, so more latewood is produced. Average fibril angle and tracheid dimension changes similarly depend on these site influences.

With fertilization, the typical response is a decrease in specific gravity, an increase in fibril angle, and a tendency toward a transition wood type of tracheid. However, on sites extremely deficient in nutrients, with trees producing abnormally thin-walled cells, the opposite effects have been reported. Sludge should fit into the same pattern of effects as these traditional stocking and fertilization measures, in that sludge also will promote crown vigor and slow crown recession. A major possible exception with sludge is that it may have an effect of longer duration.

Pruning, if extended far enough into the live crown, can cause most of the effects to be reversed by artificially accelerating crown recession. A more detailed discussion of the effects of these practices is given elsewhere (see Briggs and Smith 1986).

For clear wood properties we have results from a 60-year-old, low site, sludge-treated

TABLE 2. Four-year average percentage of change in wood properties of 60-year-old Douglas-fir due to sludge application.

Treatment	Ring Width	% Latewood	Specific Gravity	Fiber Length	Fibril Angle
Control	-18	+19	+ 2*	0*	-3*
Thinned	+ 8*	+15	+ 4	-4	-2*
Sludge applied	+54	- 6	-15	-8	+9
Thinned and sludge	+72	- 3*	-13	-5	+9*

*Not significant at 95% level.

Douglas-fir stand (Tables 2 and 3). Details of the treatment and sampling are discussed elsewhere (Cole et al. 1984). These results are based on a small sample of only two trees per treatment and represent an average of four sample positions along the bole. The changes reported represent comparison of the four-year average change since treatment to an equal amount of pretreatment growth in each tree. These results should be regarded as a case study and not as having broad applicability to stands of other ages, stocking levels, or sites.

Column 1 of Table 2 indicates that while ring width declined in the controls, it increased with the treatments. The thinning by itself did not produce a statistically significant ring width change in the study trees. However, since the thinning substantially altered the natural pattern of ring width change from declining width in the controls to no decline with thinning, the thinned trees must have had a significant basal area and volume response, which indeed was reported (Cole et al. 1984).

Specific Gravity. The hormonal theory suggests that practices that increase crown size and vigor would lead to lowered specific gravity. Thus, wide spacing, thinning, fertilization, and sludge application should reduce specific gravity. On the other hand, pruning a substantial portion of the live crown should increase specific gravity.

In general, research reported in the literature supports this theory, but extreme care is needed in interpreting results, because other factors of the site may be involved. This is why some have reported that specific gravity increased or did not change after application of a growth stimulating treatment. Thinning may cause specific gravity to increase on sites where dry summers truncate normal latewood production. Fertilization may increase specific gravity on sites that are extremely deficient in nutrients, where trees are producing abnormally thin-walled cells. Another complicating factor is that the degree of specific gravity response appears to change depending on age at time of treatment. There appears to be less effect in young trees than in older trees. Perhaps this is because

TABLE 3. Four-year average percentage change in shrinkage and mechanical properties of 60-year-old Douglas-fir due to sludge application.

Treatment	Shrinkage				Modulus of Elasticity	Modulus of Rupture
	Longitudinal	Tangential	Radial	Volumetric		
Control	+ 5*	- 4*	+ 2*	- 3*	0*	+ 2*
Thinned	+ 5*	- 2*	- 2*	- 1*	+ 1*	+ 1*
Sludge applied	+50	-10	-11	-13	-16	-19
Thinned and sludge	- 9*	-16	-11	-14	-13	-14

*Not significant at 95% level.

the age influence is so strong in young trees that we usually do not observe change until this age effect diminishes. This, combined with varying degrees of treatment, means that this literature base is very diverse and superficially appears to be contradictory. Unfortunately, few comprehensive studies have been carried out.

Research with thinning shows changes in specific gravity of plus or minus 5 to 10%. Except as noted previously, fertilization also typically shows a decrease in specific gravity of 5 to 10%, and combinations of thinning and fertilization show changes of up to 20% or more. Examination of the 60-year-old stand (Table 2) showed no change in the control, a small increase in the thinned trees (this is a dry site in late summer), and decreases of about 15% with sludge application. The latewood percentage values, while indicating direction of change, fail to show a similar magnitude. This is probably because of the arbitrariness of any definition of latewood, as noted earlier. There may be more substantial changes in stands with more spectacular growth response. The lack of additional change when thinning and sludge application are combined is probably due to the lack of ring width response of the sample trees to the thinning and the slight positive change in specific gravity with thinning alone.

Tracheid Dimensions. Tracheid dimensions are important properties, particularly in pulp and paper manufacture. The changes in dimensions caused by sludge (Table 2) showed a tendency for treatments to reduce tracheid length by 4 to 8%. The average tracheid length measured was 3.9 to 4.3 mm. These tracheids may have a somewhat larger diameter and thinner wall. With these changes, they will be more than adequate for papermaking and may be more conformable for improved paper strength. Here we are discussing tracheids of mature wood, which in both earlywood and latewood are much longer than those formed in juvenile wood.

Fibril Angle. A low fibril angle leads to high fiber tensile strength and low longitudinal shrinkage. A high angle means reduced tensile strength and greater longitudinal shrinkage and associated problems with dimensional stability of products. It is a useful indicator property that has been very tedious to measure. New data collection devices simplify the process, showing that sludge caused a 9% increase (Table 2). Fibril angle is inversely correlated with tracheid length—larger angles being associated with shorter tracheids, which are found in early wood and in juvenile wood.

The changes noted for fibril angle and specific gravity suggest that the treatment with sludge should influence shrinkage and mechanical properties. Table 3 indicates that sludge increased longitudinal shrinkage and decreased tangential, radial, and volumetric shrinkage. This suggests some potential problems with dimensional stability, drying, and drying defects of sludge-treated wood. Special sorting and drying methods may need to be devised to prevent unacceptable levels of degrade.

Table 3 also indicates that the stiffness (MOE) and strength (MOR) of clear sludge-treated wood are substantially lowered. When these reductions in clear wood values are combined with the effects of knots and other defects, there may be some serious problems with sawn and peeled products. The best way to define this issue precisely is to run a mill recovery study and examine the net effect on value by both visual and machine grading methods.

A recent report on a mill study in the South may provide some insight on the effects on product value of intensive culture combined with short rotations (Burkart et al. 1984). This study compared mill recovery, visual grade yield, and ability of graded lumber to meet grade design criteria. The more intensively managed stand was 20 years old, 36.3

cm (14.3 in.) dbh, 22.0 m (72 ft) tall, and had been managed to a stocking of 100 trees/ha (250 trees/acre). The less intensively managed stand was 50 years old, 37.3 cm (15.1 in.) dbh, 32.0 m (105 ft) tall, and managed to essentially the same level of stocking.

The intensive stand averaged two logs per tree with a lumber recovery factor (LRF) of 5.0 to 5.8, depending on log position (LRF measures the mill tally of finished lumber per cubic foot of log input). The less intensive stand averaged four logs per tree with an LRF of 6.3 to 7.1. The lower LRF in the intensive stand was attributed to greater taper and more crookedness.

Visually graded lumber from the stands showed a greater amount of higher grade lumber from the less intensive stand. More important, the lumber from the butt and top logs was tested and it was found that while all the lumber from the less intensive stand met the design criteria of the grades into which it was classified, only 8 to 19% of the intensive stand lumber met the appropriate visual grade stiffness (MOE) criteria and 77 to 100% met the appropriate strength (MOR) criteria.

One reason for these differences is that the intensive stand had 55% juvenile wood as opposed to 16% in the less intensive stand. However, it seems that both stands were old enough to show differences other than juvenile wood alone, which normally occupies only the first seven to ten rings. The faster growth effects of wide rings, lower specific gravity, greater fibril angle, and other factors must also account for some of this difference. The visual grading took into account differences in knots and other defects.

CONCLUSIONS

1. The hormone theory, combined with knowledge of age of trees at time of treatment and insight into other factors that may be limiting on a given site, allows us to predict the direction of change likely to result from a treatment. The effect of sludge application on wood properties will generally be along the lines of any accelerated growth regime or practice. A more critical question is the magnitude of change and its duration. Fertilization alone rarely produces changes exceeding 10 to 15%, and these changes usually recover three to five years after treatment. Sludge research has not been conducted long enough to address these important issues, but results from the 60-year-old stand raise some serious questions, particularly with respect to stands that have produced much more spectacular growth responses.

2. The results of combining sludge with other treatments are not clear. Fertilization and thinning when combined often produce effects on specific gravity that are at least additive. However, the combination of sludge treatment with thinning in the 60-year-old stand had little effect beyond sludge treatment alone. This may simply reflect the lack of overall response by the sample trees to the thinning that was conducted. In some cases the combined effect was less than that of sludge alone, perhaps a result of the tendency of thinning alone on this dry site to increase specific gravity, compensating the sludge effect. It seems that combinations will often occur that yield spectacular growth with concomitant effects on wood properties. Further research is needed here.

3. Perhaps the most important aspect of sludge treatment is the timing of its application during the rotation. Early application affects wood quality throughout more of the tree by producing more juvenile wood and larger knots in all logs. If this is combined with short rotations, the harvested trees may have very large diameters for their age but may have so much juvenile wood as to make them unsuitable for many of the higher

valued end uses. Late application affects only the wood grown in the live crown—thus it affects only upper stem, low volume logs. It also produces growth around the periphery of lower logs. In sawn and peeled products much of this material may end up as chip residue because of the nature of these manufacturing processes. A key problem in such products may be the abrupt change from slow to fast growth.

4. We have a lot to learn about effects of sludge treatments on wood properties. In fact, we have a lot to learn about the effects of all intensive cultural practices. We must keep in mind the requirements and desires of the various markets and work with them to remove any artificial barriers to faster grown wood. We must also remember the important role of age and not be lured into the trap of producing large diameter trees in minimum time. Such trees with high juvenile wood fractions and other problems will have much lower values than the assumed values included in financial calculations to justify such regimes.

REFERENCES

Bendtsen, B. A. 1978. Properties of wood from improved and intensively managed trees. For. Prod. J. 28(10):61–72.

Briggs, D. G., and W. R. Smith. 1986. Effects of silvicultural practices on wood properties of conifers: A review. In Douglas-fir: Stand management for the future. College of Forest Resources, University of Washington (in press).

Burkhart, L. F., M. D. MacPeak, and D. Weldon. 1984. Quality and yield of lumber produced from fast growth, short rotation slash pine: A mill study. Unpublished report. On file at Stephen F. Austin University, Nacodoches, Texas.

Cole, D. W., M. L. Rinehart, D. G. Briggs, C. H. Henry, and F. Mecifi. 1984. Response of Douglas-fir to sludge application: Volume growth and specific gravity. p. 77–84. In 1984 TAPPI Research and Development Conference, Appleton, Wisconsin. Technical Association of the Pulp and Paper Industry, Technology Park, Atlanta, Georgia.

Cown, D. J. 1974. Comparison of the effects of two thinning regimes on some wood properties of radiata pine. New Zealand J. For. Sci. 4(3):540–551.

Denne, M. P., and R. S. Dodd. 1981. The environmental control of xylem differentiation. In J. R. Barnett (ed.) Xylem cell development. Castle House Publications, Ltd.

Fielding, J. M. 1967. The influence of silvicultural practices on wood properties. p. 95–126. Int. Rev. For. Res. Academic Press, New York and London.

Hatton, J. V. 1979. Chip quality monograph. Pulp and Paper Technology Series 5. TAPPI Press, Atlanta, Georgia.

Johnston, D. R. 1962. Growing conifers in South Africa. Emp. For. Rev. 41(1):37–43.

Koch, P. 1972. The three-ring-per-inch dense southern pine: Should it be developed? In Proceedings,

Symposium on the Effect of Growth Acceleration on the Properties of Wood. USDA Forest Service, Forest Products Laboratory, Madison, Wisconsin.

Larson, P. R. 1957. Effect of environment on the percentage of summer wood and specific gravity of slash pine. Bulletin 63. School of Forestry, Yale University.

———. 1969. Wood formation and the concept of wood quality. Bulletin 74. School of Forestry, Yale University.

Megraw, Robert. 1985. Wood quality factors in loblolly pine: Influence of tree age, position in tree and cultural practices on wood specific gravity, fiber length and fibril angle. TAPPI Press, Atlanta, Georgia.

Panshin, A. J., and C. de Zeeuw. 1980. Textbook of wood technology. 4th ed. McGraw-Hill, New York.

Pearson, R. G., and R. C. Gilmore. 1980. Effect of fast growth rate on the mechanical properties of loblolly pine. For. Prod. J. 30(5):47–53.

Smith, D. M. 1968. Wood quality of loblolly pine after thinning. USDA For. Serv. Res. Paper FPL-89.

Turnbull, J. M., and C. P. du Plessis. 1946. Some sidelights on the rate of growth bogey. J. South African Forestry Assoc. 14:29–36.

U.S. Forest Service. 1965. Western wood density survey. Report no. 1. USDA For. Serv. Res. Paper FPL-27.

Williston, E. M. 1981. Small log sawmills: Profitable product selection, process design and operation. Miller Freeman Publications, San Francisco.

Zobel, B. J., and A. E. Haught. 1962. Effect of bole straightness on compression wood of loblolly pine. Technical Report 15. North Carolina University School of Forestry, Raleigh.

Chapter 23

Growth Response, Mortality, and Foliar Nitrogen Concentrations of Four Tree Species Treated with Pulp and Paper and Municipal Sludges

CHARLES L. HENRY

ABSTRACT   The purpose of this project is to demonstrate how particular tree species respond to different treatments of primary and secondary pulp and paper sludge and municipal sludge. The project is located at the University of Washington Pack Forest. Four nursery beds had eight treatments applied to Douglas-fir, noble fir, white pine, and hybrid cottonwood seedlings. Each sludge and the sand were analyzed for total solids, total carbon, total nitrogen, ammonia-N, total phosphorus, and total potassium. Material for each treatment was surface applied and mixed into the top 15 cm of sand. Height and diameter measurements were taken following planting in April 1984 and in February 1985. Foliage was sampled during the winter and analyzed for nitrogen. Addition of pulp and paper sludge alone and in combination with municipal sludge provided predictable first-year growth responses when compared with the C:N ratio of each treatment. Where the C:N ratio of the planting mixture was altered to be more favorable than the control, average response was positive. Conversely, at the very high C:N ratio caused by the addition of primary P&P sludge only, average growth response was negative. Results of the C:N ratio were also evident in foliar nitrogen concentrations. Although the higher level nutrient treatments showed increased growth, mortality also increased. Grass and weed establishment was also much greater with these treatments.

Land application of municipal sludge has been successfully practiced for many years and has proved to be an effective fertilizer and soil conditioner when applied to forest lands or tree species (Bledsoe 1981, Henry and Cole 1983). Like municipalities, the pulp and paper industry is faced with increasing quantities of residuals from primary and secondary treatment of its wastewaters. At present, however, only minor proportions of pulp and paper (P&P) sludges are applied to land, even though these sludges have soil conditioning and fertilizing properties. Also, combining cellulose primary P&P sludge with nitrogenous municipal sludge results in more favorable C:N ratios for land application. It follows, then, that application of these sludges with or without municipal sludge could improve soil productivity, and provide P&P sludge management with a viable alternative to disposal.

The purpose of this research project was to demonstrate how particular tree species respond to different treatments of primary and secondary P&P and municipal sludges. This paper reports first-year findings.

LITERATURE REVIEW

The estimated wastewater sludge production from the U.S. pulp and paper industry is over 3 million dry tonnes (Mg) per year (Rock and Alexander 1983), compared with annual municipal sludge production of about 6 million dry tonnes per year (Metcalf and Eddy 1979). Although land application of municipal sludge is the most popular management alternative in the United States, 86% of the P&P sludge was used as landfill in 1979 and another 11% was incinerated (Thacker 1984).

It is somewhat ironic that land application has played such a minor role for sludge management within the industry. Concentrations of heavy metals are typically much lower (particularly cadmium), and pathogens are essentially nonexistent, compared with municipal sludges (Hermann 1983). Burial of P&P wastes, however, was one of the earliest practices developed. Treatment of wastewaters in the past required only sedimentation, yielding primary sludges consisting mainly of clay and fiber. This material was easy to dewater and stable when placed in a landfill. With the advent of biological treatment, sludge characteristics changed drastically. The secondary sludge is harder to dewater and usually has a higher moisture content than primary sludge, and contains much higher nutrient concentrations, particularly of nitrogen and phosphorous. Microorganisms are able to decompose this material much more rapidly, resulting in release of excess liquid as leachate and instability of the sludge mass (Wardwell and Charlie 1981).

The very conditions that create problems in using secondary or combined P&P sludges for landfill are obviously advantageous for land application: high nutrient concentrations and decomposition. The benefits of land application identified by Harkin (1983) include (1) a slow release of available nutrients from decomposition rather than the immediate burst characteristic of commerical fertilizers, (2) an improved soil texture due to the fiber structure of the sludge, which can improve the nutrient and moisture holding characteristics of the sludge, (3) a source of trace nutrients, (4) adjustment of soil pH, particularly with lime-containing sludges, and (5) more rapid pesticide degradation. Einspahr et al. (1984) found increased cation exchange capacity and available soil moisture with three different pulp and paper sludges, concluding that less frequent irrigation was required and fertilizer could be more efficiently utilized.

Soils in the Pacific Northwest are typically deficient in nitrogen, thus one of the major benefits from sludge applications is the nitrogen in sludge. Because secondary P&P sludge is relatively high in nitrogen, as is the case with municipal sludge, a significant growth response can be expected. Conversely, primary P&P sludge containing little nitrogen can immobilize soil nitrogen and adversely affect tree growth.

Although municipal sludge application by itself has been shown to be an excellent amendment (Bledsoe 1981, Henry and Cole 1983, Brockway 1983), in combination with primary P&P sludge it has the following advantages: (1) Municipal sludge can provide available nitrogen for a more favorable carbon to nitrogen (C:N) ratio than primary P&P sludge has alone. (2) Conversely, primary P&P sludge can immobilize excess available nitrogen from municipal sludge. (3) Primary and secondary P&P sludges, generally having less heavy metals, can dilute the concentration of heavy metals in municipal sludge.

Land application of sludge has been shown by many research projects to enhance the productivity of soils when properly managed. Studies have been conducted showing growth response in crops grown in soil amended with P&P sludge applications. Many of the results from these studies reflect the type of sludge used (primary or secondary). Hermann (1983) reported sweet corn yields 20% higher using secondary paper sludge compared with commercial fertilizers. Thiel (1984) reported that the sweet corn yield from combined primary and secondary sludge was equal to control with commercial fertilizers added. At high applications of primary sludge, Dolar et al. (1972) found growth in oats diminished, even though nitrogen, phosphorus, and potassium fertilizers were added. Simpson et al. (1983) reported higher rye grass production from secondary sludge additions over control and fertilized plots, and comparable yields of corn and sorghum from primary sludge compared with fertilized plots. Secondary papermill sludge caused foliar nitrogen content and biomass to increase significantly in both over- and understory vegetation in a red pine plantation as reported by Brockway (1983).

METHODS AND MATERIALS

The project site is located at the University of Washington Charles Lathrop Pack Demonstration Forest. Four 24 by 1.5 m by 0.6 m deep nursery beds were filled with subsoil taken from a sandy glacial outwash soil (Indianola). Each bed was divided into eight plots with thirty-two seedlings or cuttings planted in each plot. Treatments were: (1) control (soil only), (2) 98 dry Mg/ha municipal sludge from Tacoma, (3) 98 dry Mg/ha primary P&P sludge from Crown Zellerbach's Wauna, Oregon mill, (4) 98 dry Mg/ha primary sludge from Boise Cascade's St. Helens, Oregon mill, (5) 98 dry Mg/ha secondary sludge from Crown Zellerbach's Wauna, Oregon mill, (6) 49 dry Mg/ha municipal sludge plus 49 dry Mg/ha primary sludge from Crown Zellerbach, (7) 49 dry Mg/ha municipal sludge plus 49 dry Mg/ha primary sludge from Boise Cascade, and (8) 49 dry Mg/ha municipal sludge plus 49 dry Mg/ha secondary sludge from Crown Zellerbach.

Each sludge and the soil were analyzed for percentage of oven-dry solids, total carbon using a Leco induction furnace, total nitrogen, ammonia-N, total phosphorus, and total potassium by a wet oxidation procedure (Parkinson and Allen 1975). Material for each treatment was surface applied and mixed into the top 15 cm of soil. Plots were separated by vertical plywood dividers. Bed 1 was planted with 2-0 Douglas-fir (*Pseudotsuga menziesii*) bare root seedlings, bed 2 with noble fir (*Abies procera*) plug seedlings, bed 3 with 2-0 white pine (*Pinus monticola*) bare root seedlings, and bed 4 with hybrid cottonwood cuttings (*Populus deltoides* x *Populus trichocarpa*) graded for uniformity. All seedlings were planted at a spacing of 30 by 30 cm.

Seedlings were watered at regular intervals throughout the dry part of the year. This eliminated the effect of differences in moisture holding capacity for the treatments. Grass and weeds were clipped on all plots when their height approached that of the seedlings, to keep light competition from affecting growth. No other weed and grass control was used, so that problems with management of vegetation could be observed.

Height and diameter measurements were taken following planting in April 1984 and again in February 1985. Diameter measurements were taken 15 cm above the soil on the cottonwood, and 5 cm on the other species. Cottonwood foliage was sampled in October 1984. Nitrogen analysis was made on a composite of leaves from five trees from each treatment. Other foliage was sampled in February 1985. A composite sample for each treatment was made by collecting needles from each tree. Foliage was analyzed for nitrogen content.

RESULTS

Sludge Analysis

Results of the nutrient analysis of the soil and different sludges are presented in Table 1, and calculated nutrient concentrations of the sludge and soil mixtures are shown in Table 2. As can be seen in Table 2, the concentration of potassium in the mixtures varies little. Phosphorus concentration does vary up to a maximum increase of 77% higher than the control, but the big difference is in the nitrogen concentrations. Increases of nitrogen range from 13% for the Boise Cascade (BC) primary sludge treatment to 1,982% in the Crown Zellerbach (CZ) mixture. This has a tremendous effect on the C:N ratio, also shown, ranging from 9:1 to 116:1, which in turn has a major effect on available nitrogen.

TABLE 1. Comparison of characteristics of sludges and control.

Characteristic	Control Soil	Tacoma Municipal Sludge	CZ Primary Sludge	BC Primary Sludge	CZ Secondary Sludge
Total solids (%)	85.8	42.5	18.1	37.8	13.8
Total carbon (%)	1.12	17.2	37.4	18.8	41.5
Total nitrogen (%)	0.02	2.36	0.30	0.04	7.77
C:N ratio	75:1	7:1	126:1	482:1	5:1
Ammonia-N (mg/kg)	2	1,521	5	0	1,828
Total phosphorus (mg/kg)	499	11,194	501	494	9,299
Total potassium (mg/kg)	3,075	2,324	340	706	3,454
pH	5.4	7.4	8.0	8.4	5.8

TABLE 2. Calculated nutrients in top 15 cm of sand with addition of each treatment (kg/ha).

Nutrient	Control Soil	Tacoma Muni Sludge	CZ Pri Sludge	BC Pri Sludge	CZ Sec Sludge	Muni & CZ Pri Sludge	Muni & BC Pri Sludge	Muni & CZ Sec Sludge
				Treatments				
Total nitrogen	308	2,162	542	349	6,413	1,352	1,250	4,288
C:N ratio	75:1	18:1	102:1	116:1	9:1	33:1	38:1	11:1
Total phosphorus	1,147	2,026	1,186	1,185	1,878	1,607	1,606	1,952
Total potassium	7,070	7,253	7,097	7,125	7,341	7,174	7,189	7,297

Growth Response

The first-year diameter and height growth responses compared with controls for the four species grown in the seven sludge treatments are presented in Tables 3 and 4. In general, response is as expected. A grouping of average response of the four species for each treatment into high, fair, and low to negative matches exactly a grouping of C:N ratios for the treatments. A high average response resulted from treatments of secondary P&P, municipal, and municipal plus secondary sludges (they had C:N ratios of 9:1 to 18:1). Both primary P&P plus municipal sludge treatments resulted in fair average responses (they had C:N ratios of 33:1 to 38:1). When primary sludges were used alone

TABLE 3. First-year diameter response to sludge amendments (percentage of growth compared with control).

Species	Tacoma Muni Sludge	CZ Pri Sludge	BC Pri Sludge	CZ Sec Sludge	Muni & CZ Pri Sludge	Muni & BC Pri Sludge	Muni & CZ Sec Sludge
					Treatments		
Hybrid cottonwood	17	-17	3	87*	31	24	23
Douglas-fir	116*	-11	6	300*	141*	73*	168*
Noble fir	66*	-54*	-35*	91*	-24	-43*	80*
White pine	5	-40*	-16	26	-9	-20	19
Average	51	-31	-11	126	35	9	73

*Statistically significant (p < 0.05) difference compared with control.

TABLE 4. First-year height response to sludge amendments (percentage of growth compared with control).

Species	Muni Sludge	CZ Pri Sludge	BC Pri Sludge	CZ Sec Sludge	Muni & CZ Pri Sludge	Muni & BC Pri Sludge	Muni & CZ Sec Sludge
					Treatments		
Hybrid cottonwood	14	-18*	-2	45*	30*	26	0
Douglas-fir	42	14	16	100*	55*	16	80*
Noble fir	77*	-9	15	58*	-3	-2	114*
White pine	86*	14	-21	59	30	23	83*
Average	55	0	2	66	28	16	69

*Statistically significant (p < 0.05) difference compared with control.

(with C:N ratios of 102:1 to 116:1), low or negative responses occurred. A Newman-Keuls multiple range test for analysis of variance (Zar 1974) was performed for the treatments of each of the species. Statistically significant differences are shown in Tables 3 and 4.

Noble fir appeared to be the most sensitive to the C:N ratio. Height and growth responses were very good in all treatments that did not include primary P&P sludges, and low to negative in all treatments that included primary P&P sludges. This was the result even where the C:N ratio for the primary P&P sludges was more favorable than the control because of the addition of municipal sludge, probably owing to the available nitrogen from the municipal sludge being immobilized by the primary P&P sludges. It is possible that this result will reverse itself next year as more nitrogen becomes available from the municipal sludge and the primary sludges continue to decompose.

White pine generally showed the least effect from the treatments in terms of diameter response: only the CZ primary P&P sludge showed a statistically significant difference—a negative response. This species did, however, show a good height response where only secondary P&P and municipal sludges were used.

Douglas-fir showed the greatest positive and least negative diameter responses to the treatments compared with any of the other species. Height response was greatest for Douglas-fir in four of the treatments and average in the other three.

Foliar Nitrogen Concentration

Results of foliar nitrogen analysis are shown in Table 5 and percentages of increase or decrease over control are presented in Table 6. Again, it is not surprising that the greatest average increase in foliar nitrogen concentration occurred where the C:N ratio was the lowest—in the secondary P&P and the secondary P&P plus municipal sludge treatments. The foliar nitrogen concentrations in the primary P&P plus municipal sludge treatments, however, were greater than might be expected relating them to the growth response. This may have been because sampling was done after the growing season, at which time more nitrogen may have been available than was available during the majority of the growing season. One other notable difference was that the BC primary P&P sludge treatment caused an average increase in foliar nitrogen concentration compared with the control, while the CZ primary P&P sludge treatment caused a large average decrease compared with the control.

TABLE 5. First-year foliar nitrogen concentration on dry weight basis.

Species	Control (Sand)	Tacoma Muni Sludge	CZ Pri Sludge	BC Pri Sludge	CZ Sec Sludge	Muni & CZ Pri Sludge	Muni & BC Pri Sludge	Muni & CZ Sec Sludge
Hybrid cottonwood	1.4	2.2	1.9	2.0	3.6	2.2	2.0	3.5
Douglas-fir	1.4	1.8	0.5	1.6	3.2	2.8	2.7	2.7
Noble fir	1.8	2.1	0.8	1.2	2.3	1.7	1.3	2.8
White pine	1.4	1.8	1.2	2.1	2.5	2.2	2.7	2.9
Average	1.5	2.0	1.1	1.7	2.9	2.2	2.2	3.0

TABLE 6. First-year foliar nitrogen concentration on dry weight basis (percentage increase or decrease compared with control).

Species	Muni Sludge	CZ Pri Sludge	BC Pri Sludge	CZ Sec Sludge	Muni & CZ Pri Sludge	Muni & BC Pri Sludge	Muni & CZ Sec Sludge
Hybrid cottonwood	54	36	42	154	52	77	146
Douglas-fir	25	-65	13	126	99	87	92
Noble fir	16	-58	-34	29	-6	-25	58
White pine	27	-15	48	81	55	96	112
Average	31	-26	17	98	50	59	102

Mortality

Table 7 presents the percentage of mortality for the four different species in each treatment. In general, high mortality was observed in treatments that also showed the greatest growth response—the treatments where only secondary P&P and municipal sludges were used. Since watering occurred at regular intervals, mortality was probably not due to moisture stress. It is possible that, because of the high, readily degradable organic

TABLE 7. Mortality of four species of trees treated with sludges (%).

Species	Control (Sand)	Tacoma Muni Sludge	CZ Pri Sludge	BC Pri Sludge	CZ Sec Sludge	Muni & CZ Pri Sludge	Muni & BC Pri Sludge	Muni & CZ Sec Sludge
							Treatments	
Hybrid cottonwood	6	16	0	3	34	16	9	25
Douglas-fir	3	9	3	0	16	3	9	6
Noble fir	25	16	3	3	28	13	9	16
White pine	3	9	0	0	19	3	0	13

loading in some of the treatments, anaerobic conditions existed for periods of time following irrigation that may have resulted in the higher mortality. Since most of the nutrients were applied in an organic form, mortality was probably not due to too high a concentration of available nutrients, such as ammonia.

Another factor may have been the vigorous growth of grasses and weeds in the high level nutrient treatments. In contrast to these, very little competition established with the primary P&P sludge treatments only, and it wasn't until late in the season that any established in the primary P&P plus municipal sludge treatments. Since weeds and grasses were kept clipped, light competition should not have been a factor.

SUMMARY

This project demonstrates the potential for use of pulp and paper sludge with and without municipal sludge as an amendment to soils for tree growth. Although most P&P sludge is currently used for landfill, increasing pressure is forcing evaluation of alternatives such as land application. Soils in the Pacific Northwest are ideal for accepting both the nutrients and the soil conditioning aspects of P&P sludges. In many cases, primary P&P sludges are produced at a much greater rate than secondary P&P sludges; since they have a C:N ratio that can cause a reduction in growth, they could be mixed with municipal sludge for a more favorable amendment.

Addition of pulp and paper sludge alone and in combination with municipal sludge provided predictable first-year growth responses when compared with the C:N ratio of each treatment. Where the C:N ratio of the planting mixture was altered to be more favorable than the control, average response was positive. Conversely, at the very high C:N ratio caused by the addition of primary P&P sludge only, average growth response was negative. Results of the C:N ratio were also evident in foliar nitrogen concentrations.

Whereas the higher level nutrient treatments showed increased growth, mortality also increased. Grass and weed establishment was also much greater with these treatments.

REFERENCES

Bledsoe, C. S. (ed.) 1981. Municipal sludge application to Pacific Northwest forest lands. Institute of Forest Resources Contribution 41. College of Forest Resources, University of Washington, Seattle. 155 p.

Brockway, D. G. 1983. Forest floor, soil, and vegetation responses to sludge fertilization in red and white pine plantations. Soil Sci. Soc. Am. J. 47:776–784.

Dolar, S. G., J. R. Boyle, and D. R. Keeney. 1972. Paper mill sludge disposal on soils: Effects on the yield and mineral nutrition of oats (*Avena sativa* L.). J. Environ. Qual. 1(4):405–409.

Einspahr, D., M. A. Fiscus, and K. Gargan. 1984. Paper mill sludge as a soil amendment. p. 253–257. *In* TAPPI Proceedings, 1984 Environmental Conference. TAPPI Press, Atlanta, Georgia.

Harkin, J. M. 1983. Wise use of Wisconsin's papermill sludge. p. 65–78. *In* C. A. Rock and J. A. Alexander (eds.) Proceedings of the Symposium on Long Range Disposal Alternatives for Pulp and Paper Sludges, University of Maine, Orono.

Henry, C. L., and D. W. Cole (eds.) 1983. Use of dewatered sludge as an amendment for forest growth. Vol. 4. Institute of Forest Resources, University of Washington, Seattle. 110 p.

Hermann, D. J. 1983. Considerations for using wastewater sludge as an agricultural and silvicultural soil amendment. p. 79–94. *In* C. A. Rock and J. A. Alexander (eds.) Proceedings of the Symposium on Long Range Disposal Alternatives for Pulp and Paper Sludges, University of Maine, Orono.

McKeown, J. J. 1983. The management of paper industry residuals in Pennsylvania and the U.S. p. 1–12. *In* C. A. Rock and J. A. Alexander (eds.) Proceedings of the Symposium on Long Range Disposal Alternatives for Pulp and Paper Sludges, University of Maine, Orono.

Metcalf and Eddy, Inc. 1979. Wastewater engineering. McGraw-Hill, New York.

Parkinson, J. A., and S. E. Allen. 1975. A wet oxidation procedure suitable for the determination of nitrogen and mineral nutrients in biological material. Commun. Soil Sci. and Plant Analysis 6(1):1–11.

Rock, C. A., and J. A. Alexander. 1983. Pulp and paper sludge disposal: The problem, current practice and future directions. Land and Water Resources Center, University of Maine, Orono.

Simpson, G. G., L. D. King, B. L. Carlile, and P. S. Blickensderfer. 1983. Paper mill sludges, coal fly ash, and surplus lime mud as soil amendments in crop production. TAPPI J. 66(7):71–74.

Thacker, W. E. 1984. The land application and related utilization of pulp and paper mill sludges. NCASI Tech. Bull. 439. National Council of the Paper Industry for Air and Stream Improvement, New York.

Thiel, D. A. 1984. Sweet corn grown on land treated with combined primary/secondary sludge. p. 93–102. *In* TAPPI Proceedings, 1984 Environmental Conference. TAPPI Press, Atlanta, Georgia.

Wardwell, R. E., and W. A. Charlie. 1981. Effects of fiber decomposition on the compressibility and leachate generation at a combined sludge landfill area. p. 223–238. *In* TAPPI Proceedings, 1981 Environmental Conference. TAPPI Press, Atlanta, Georgia.

Zar, J. H. 1974. Biostatistical analysis. Prentice-Hall, Englewood Cliffs, New Jersey.

Chapter 24

Aspen Mortality Following Sludge Application in Michigan

JOHN H. HART, JAMES B. HART, and PHU V. NGUYEN

ABSTRACT   Sludge fertilization of a ten-year-old aspen stand in Michigan increased mortality from 4% in control to 27% in treated plots in 1982. Part of the mortality resulted from sunscald injury to stems along the north side of the east-west application trails. During June and July, elk frequently broke aspen stems 1 to 2 m above ground to reach the foliage. In July 1983, *Cystospora chrysosperma*, a pathogenic fungus, was observed on all of 132 recently killed stems of bigtooth aspen (*Populus grandidentata*). Of 149 bigtooth aspen stems broken by elk in 1982 or 1983 and alive in July 1983, 46% were dead by June 1984. Mortality of 82 elk-broken quaking aspen (*Populus tremuloides*) was 24%. Twenty bigtooth and twenty quaking aspen were broken in July 1983 to simulate elk injury. By June 1984, three bigtooth aspen had died and cankers (caused by *C. chrysosperma*) averaged 38 cm in length; none of the quaking aspen had died and cankers averaged 14 cm. These results indicate that sludge application led to an increase in elk damage, thereby predisposing aspen clones to pathogenic fungi. Quaking aspen was more resistant than bigtooth aspen to these stresses. The differential susceptibility of the two aspen species to elk browsing and *C. chrysosperma* infection following sludge application appears to be an example of plant diversity being mediated by herbivory and susceptibility to a plant pathogen.

The application of nutrient-rich sewage sludge to forest ecosystems could increase the quantity and quality of wildlife forages. This might result in increased herbivore browsing on treated areas, producing changes in community structure. The primary objective of this study was to determine the interaction between sludge application, vertebrate injury, the presence of pathogenic fungi, and the decline of aspen.

METHODS

The study site was located in Montmorency County, Michigan. The vegetation type was predominantly bigtooth aspen (*Populus grandidentata*), with some scattered quaking aspen (*Populus tremuloides*), oak (*Quercus* spp.), and black cherry (*Prunus serotina*) (Campa et al., this volume). The experimental design (Campa et al., this volume) and sludge loading levels have previously been published (Hart et al. 1983). Application trails were constructed and sewage sludge from Alpena, Michigan, was applied to the site in October 1981.

Because of large openings throughout the site where aspen regeneration was not present, twelve pairs of measurement subplots were installed based on aspen densities. One of each matched pair received sludge and the other served as a control. During the summer of 1982, mortality of all aspen (*P. grandidentata*) was determined within the subplots. Four pairs of treatment-control plots, in a unique topographic position within the study area, appeared to have unusually high aspen mortality (due to sunscald). These four plots were deleted and another paired t-test run to test the hypothesis that aspen mortality was equal in both treatments.

To study the use of sludge-treated vegetation by deer (*Odocoileus virginianus*) and elk (*Cervus elaphas*), browsing estimates were conducted on all plots during the summers of 1983 and 1984. For the purposes of this study the data for the two years were combined; more detailed information is available (Campa et al., this volume).

A disease survey of bigtooth aspen in the study area was conducted during the summer of 1983. At least 25 recently killed stems per plot for plots 5(TS), 6(T), 7(TS), and 8(C) were dissected and the probable cause of mortality determined (see Figure 1, Campa et al., this volume). In addition to browsing, elk commonly break over sapling aspen in early summer when the foliage is highly palatable (Figure 1). Approximately an equal number of broken and unbroken stems were dissected.

On July 19, 1983, 69 quaking aspen and 143 bigtooth aspen injured by elk during the summer of 1982 or during June 1983 were marked and monitored for their reaction to elk injury. The extent of injury and the amount of canker present were recorded. During

Figure 1. Bigtooth aspen damaged by elk in June 1984 (note stripping of leaves).

September 1983 and June and September of 1984 all trees were examined and mortality and canker development tallied.

To provide better comparative information on the difference in mortality and canker development between quaking and bigtooth aspen following elk injury, 20 stems of each species were broken on July 19, 1983, in a manner to simulate elk injury. Stems were broken 1 to 2 m above ground, retained at least one live branch below the break, and were generally along the margins of clones. Subsequent mortality and canker development were recorded in September 1983, June 1984, and September 1984.

RESULTS

A paired t-test was done on aspen mortality on the twelve pairs of sludged and control plots. The results showed a highly significant difference in aspen mortality (t = 3.288, P(t) = 0.0072) between sludged (27% average mortality) and control (4% average mortality) plots. The null hypothesis of equal mortality in aspen, independent of treatment, was thus rejected. The second t-test based on eight pairs of plots also was significant (t = 3.48, P(t) = 0.01), with average mortalities of 15% for sludged plots and 2% for control plots.

Many of the bigtooth aspen stems in the four plots with unusually high aspen mortality had basal lesions on the south or southwest side of the stems. These injuries were sunscald lesions which resulted from the east-west application trail orientation and subsequent exposure of the stems to increased solar radiation. Lesions were initiated by July 1982 and were restricted to the trails or trails plus sludge plots. Observations made in June 1983 revealed some bigtooth aspen clones were just beginning to leaf out while other clones were in nearly full leaf. The incidence of sunscald varied markedly from area to area, being most conspicuous on a brown-barked clone. All trembling aspen clones were in full leaf.

Woody vegetation in sludge-treated plots was more heavily browsed during 1983-84 than similar vegetation in plots with application trails only or in control plots (Campa et al., this volume). Both bigtooth and quaking aspen were more heavily browsed on sludge-treated plots than they were on control plots (Table 1, Campa et al. 1985). Bigtooth was preferred compared with quaking whether they occurred on sludge-treated or control plots (Table 1).

TABLE 1.   Percentage of browsing of quaking and bigtooth aspen, 1983-84 (Campa et al. 1985).

Treatment	Species	
	Quaking	Bigtooth
Control	16	27
Trails plus sludge	30	47

Cytospora chrysosperma, a pathogenic, canker-causing fungus, was observed on all 132 of the dead stems examined during the 1983 disease survey (Table 2). The fungus appeared to enter the stem at an injury and gradually moved downward to the ground

TABLE 2. Results of the disease survey of bigtooth aspen during the summer of 1983.

Broken by Elk	Cytospora Present	Armillaria Present	Number of Trees Infected (plot no.)				
			5 (TS)	7 (TS)	6 (T)	8 (C)	Total
No	Yes	No	8	9	3	2	22
No	Yes	Yes	14	10	9	7	40
Yes	Yes	No	8	9	2	26	45
Yes	Yes	Yes	5	3	13	4	25
			35	31	27	39	132

TS = trails plus sludge. T = trails only. C = control. (Campa et al. 1985).

line. *Armillaria mellea*, a fungus that causes a root rot, was observed infecting root systems of 65 of the 132 dead stems. *Armillaria* rhizomorphs were present on the outside of most root systems even when no infection was observed. Since both *C. chrysosperma* and *A. mellea* have readily recognizable signs, their presence could be diagnosed from field observation with a high degree of accuracy. The presence of *A. mellea* was not influenced by sludge treatment; 32 of the infected stems occurred in sludge-treated plots and 33 in nonsludged plots (Table 3). Since *Armillaria* caused root death on only one-half the stems, it was considered to be a secondary pathogen, while *Cytospora*, which was always present, was considered to be the primary pathogen.

Mortality of bigtooth aspen following elk injury was approximately twice that for

TABLE 3. Presence of *Armillaria mellea* on dying bigtooth aspen in 1983.

Armillaria Present	Number of Trees	
	Sludge-treated	No sludge
No	34	33
Yes	32	33

quaking aspen (Table 4). The death rate of both species far exceeded the mortality of 2% for control stems. One difficulty was locating aspen of both species wounded at the same time for valid comparative observations. The trees wounded in mid-June 1983 were probably wounded at the same or nearly the same time, hence providing the best

TABLE 4. Subsequent mortality of aspen injured by elk.

Species	Date of Elk injury	Number of Trees	Mortality (%)		
			Sept. 1983	June 1984	Sept. 1984
Quaking aspen	Summer 1982	8	13	13	38
Quaking aspen	Mid-June 1983	61	8	23	36
Bigtooth aspen	Summer 1982	96	17	46	51
Bigtooth aspen	Mid-June 1983	47	4	53	60

data for comparative purposes. Mortality of elk-injured bigtooth aspen was independent of sludge treatment: 49% mortality in untreated plots versus 42% in sludge-fertilized areas as of June 1984.

Similar mortality and canker development to that observed following elk injury was observed following simulated elk injury. Bigtooth aspen developed larger cankers—t = 3.26, P(t) = 0.005—and had greater mortality than quaking aspen (Table 5).

TABLE 5. Mortality and canker development of twenty quaking and twenty bigtooth aspens which received simulated elk injury during July 1983.

Species	Mortality (no. of trees)			Average Canker Length (cm)		
	Sept. 1983	June 1984	Sept. 1984	Sept. 1983	June 1984	Sept. 1984
Quaking	0	0	2	4	14	26
Bigtooth	1	3	5	6	38	44

DISCUSSION

These results indicate that construction of application trails increased sunscald injury and that sludge application increased elk damage, thereby predisposing aspen stems, and perhaps clones, to pathogenic fungi. C. chrysosperma occurred in the stand as a primary pathogen and was the direct cause of most aspen mortality. It was an especially significant cause of mortality of bigtooth aspen injured by elk or by sunscald. A. mellea occurred in the stand primarily as a secondary pathogen.

Changes that expose large areas of previously shaded bark may cause sunscald. The resulting lesions serve as infection points for pathogenic fungi. Serious sunscald has been reported in the Lake States in recently disturbed aspen stands (Graham et al. 1963). Sunscald followed the construction of roads and campsites in Colorado aspen stands (Hinds 1976). The immediate cause of mortality was canker fungi, but sunscald was the initiating cause.

Herbivores may cause injuries on trees, providing ideal conditions for the spread of plant pathogens. When cervids are concentrated in relatively small areas, they cause more severe injury to a stand than would be expected from an evenly distributed herd. In areas of moderate to light herbivore pressure the damage caused by the animals would not be severe enough to cause stand deterioration, except for the secondary action of pathogens and insects. The most common reason for concentrating cervids is harsh winter weather.

In the present study the concentrating factor appears to have been the increase during early summer in relative palatability of the sludge-treated aspen compared with adjacent untreated aspen. Interactions between cervids, aspen, and pathogenic fungi have been studied previously (Packard 1942, Graham et al. 1963, Krebill 1972, Mielke 1943, Walters et al. 1982). Tree scars caused by elk barking provided ideal conditions for the spread of C. chrysosperma through the stands, resulting in significant aspen mortality.

A conceptual model of some components of the aspen ecosystem following sludge applications is presented in Figure 2. The predatory organisms are in three levels: those exerting a controlling influence (elk and deer); those (C. chrysosperma) that are dependent

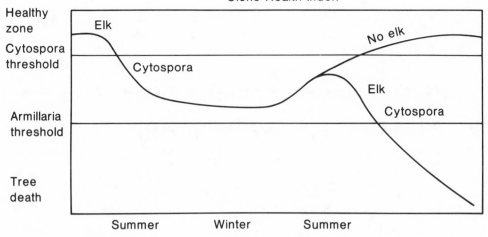

Figure 2. Model of the interaction of aspen, elk, and pathogenic fungi following sewage sludge application. Vertical axis is clone vigor. With low or moderate level of elk injury (no sludge) aspen clones survive. With repeated elk injury (sludge applied) aspen clones become susceptible to *Armillaria mellea*, which is potentially lethal.

for their success on conditions created by the first; and those (*A. mellea*) that follow the predisposing effects of the first two levels.

In summary, these results indicate that sludge application led to an increase in sunscald injury and elk damage, thereby predisposing aspen clones to pathogenic but nonlethal fungi. Quaking aspen was more resistant than bigtooth aspen to these stresses. Aspen clones can recover provided growing conditions are favorable and provided the clones have not been lethally invaded by opportunistic organisms (*A. mellea*) that can attack the clones in their weakened state. The differential susceptibility of the two aspen species to the elk and *C. chrysosperma* complex appears to be an example of predators altering the species composition of plant communities following sludge application.

REFERENCES

Crouch, G. L., and M. A. Radwan. 1981. Effects of nitrogen and phosphorus fertilizers on deer browsing and growth of young Douglas fir. USDA For. Serv. Res. Note PNW–368.

Graham, S. A., R. P. Harrison, Jr., and C. E. Westell, Jr. 1963. Aspens: Phoenix trees of the Great Lakes Region. University of Michigan Press, Ann Arbor. 272 p.

Hart, J. B., P. V. Nguyen, J. H. Hart, and C. W. Ramm, D. M. Merkel, and C. Thomas. 1983. Ecological monitoring of sludge fertilization on state forest lands in northern lower Michigan. Annual progress report. Department of Forestry, Michigan State University, East Lansing. 101 p.

Hinds, T. E. 1976. Aspen mortality in Rocky Mountain campgrounds. USDA For. Serv. Res. Paper RM–164.

Krebill, R. G. 1972. Mortality of aspen on the Gros Ventre elk winter range. USDA For. Serv. Res. Paper INT–129.

Mielke, J. L. 1943. Elk and dying of aspen. U.S. Bur. Plant Industry, Albuquerque, N.M. Typewritten report.

Packard, F. M. 1942. Wildlife and aspen in Rocky Mountain National Park, Colorado. Ecology 23:478–482.

Walters, J. W., T. E. Hinds, D. W. Johnson, and J. Beatty. 1982. Effects of partial cutting on diseases, mortality, and regeneration of Rocky Mountain aspen stands. USDA For. Serv. Res. Paper RM–240.

Chapter 25

Growth Response of Loblolly Pine to Municipal and Industrial Sewage Sludge Applied at Four Ages on Upper Coastal Plain Sites

W. H. McKEE, JR., K. W. McLEOD, C. E. DAVIS,
M. R. McKEVLIN, and H. A. THOMAS

ABSTRACT   Sewage sludge as either a liquid anaerobic sludge at 0, 402, or 804 kg N/ha or a solid aerobic material at 632 kg N/ha was applied to 1-, 3-, 8-, and 28-year-old loblolly pine stands on sandy and clayey upper coastal plain soils in South Carolina. Over four years, applications of liquid sludge at the low rate of N resulted in increased diameter and basal area growth of the 8- and 28-year-old trees. Liquid sludge applied at the high level did not result in further increases in growth for either of these groups. Application of solid sludge did not improve growth of the 28-year-old trees. Solid or liquid sludge applied before planting resulted in modest growth increases after four years, but growth was dependent on tip moth treatment, control of competing vegetation, and incorporation in the soil of the solid sludge. Sludge applied to 3-year-old trees did not increase growth, primarily because of weed competition. Results indicate that the growth of loblolly pine can increase with sludge treatment over a range of upper coastal plain sites, but the amount of increased growth is highly correlated with age.

The use of sludge has proved effective for land reclamation and for improving tree growth (Edmonds and Cole 1980, Sopper et al. 1982). However, little information is available for southern pines concerning optimal rates of sludge application, stand age when sludge application is beneficial, and site characteristics that will yield the highest growth response to sludge. The need for other silvicultural practices such as control of competing vegetation and stand density is not fully understood in relation to fertilization, and must be determined for the effective use of sludge on southern pine stands. Background and justification for this study are presented by Corey et al. in this volume. This paper presents growth responses of loblolly pine (*Pinus taeda* L.) in four age classes, growing on heavy- and light-textured soils, following the application of liquid and solid sewage sludge.

METHODS AND MATERIALS

The study was conducted on the Savannah River Plant (SRP) at Aiken, South Carolina, on upper coastal plain terraces with long agricultural histories. The areas were reforested about thirty years ago. Orangeburg loamy sand (Typic Paleudults, fine-loamy, siliceous, thermic) is typical of soil for many of the study plantations. Other stands are on a Fuquay soil (Arenic Plinthic Paleudults, loamy, siliceous, thermic), with a sandier B horizon, and Lucy loamy sand (Arenic Paleudults, loamy, siliceous, thermic), a soil similar to Orangeburg but with a lighter textured surface soil, and a Wagram loamy sand (Arenic Paleudults, loamy, siliceous, thermic), a soil similar to Lucy but with a lighter subsoil. These series are extensive in the upper coastal plain and represent a cross section of forested sites in the region. Soils are well to slightly excessively drained, have a pH of 5 to 5.5, contain 200 to 500 mg/kg total N in the surface 10 cm, and have a CEC of 0.01 to 0.02 moles/kg throughout the profile.

Four age classes of loblolly pine were studied: (1) 1-year-old trees on Orangeburg soil, (2) 3-year-old trees on Orangeburg and Lucy soils, (3) 8-year-old trees on Wagram and Fuquay soils, and (4) 28-year-old trees on Fuquay and Orangeburg soils.

Sewage sludge from two sources was used to add specific quantities of nitrogen to the study plots. A liquid anaerobic sludge, primarily of municipal origin, was applied at either a low or high N level equal to 402 and 804 kg/ha. A solid, aerobically digested sludge which contained significant amounts of waste from textile mills was applied at 632 kg/ha N. Amounts of nitrogen and other materials applied per plot are shown in Table 1. Further discussion of the sludge properties is given by Wells et al. (1984). Liquid sludge was sprayed uniformly over plots from a tank spreader, and solid material was spread uniformly from a manure spreader. Treatment plots were arranged by either ran-

TABLE 1. Chemical composition and average amounts of liquid and solid sludge applied to the loblolly pine stands.

| Sludge Property | Liquid Sludge | | Solid Sludge |
	Low N	High N	
	(kilograms/hectare, oven-dry)		
Sludge, oven-dry	5,555	11,110	49,925
Ash, oven-dry	2,556	5,113	18,027
Carbon, oven-dry	1,300	2,600	12,500
Kjeldahl-N	402	804	632
Ammonium-N	169	338	12
Nitrate-N	0.82	1.64	0
Phosphorus	90	180	377
Potassium	15	30	15
Calcium	81	162	1,043
Magnesium	14	28	174
Sodium	116	232	147
Manganese	0.76	1.52	15
Zinc	7.39	14.78	108
Copper	1.77	3.54	62
Lead	1.32	2.64	8.79
Nickel	0.24	0.48	2.15
Cadmium	0.24	0.48	0.29
Sulfur	33	66	181
Iron	565	1,130	2,119
Boron	1.41	2.82	13
Chromium	0.96	1.92	69
Antimony	1.04	0.08	0.09
Selenium	0.45	0.90	3.79
Arsenic	0.58	1.16	1.74
Tin	0.56	1.12	2.00
Cobalt	0.04	0.08	0.39
Mercury	0.02	0.04	1.55

dom or randomized complete block designs with three replications on each soil (except for the 1-year-old trees, which used one soil with three replications).

One-year-old Stand

Sludge was applied before planting in January 1981. The site had supported a pine plantation that was harvested in early 1980. The site was sheared, root raked, and disked before application of sludge materials. The following eleven treatments were applied before planting: (1) control, (2) Furadan®, (3) inorganic fertilizer (25-65-0) plus Furadan, (4) inorganic fertilizer (50-130-0) plus Furadan, (5) low N liquid plus Furadan, (6) low N liquid (disked in) plus Furadan, (7) high N liquid plus Furadan, (8) high N liquid (disked in), (9) high N liquid (disked in) plus Furadan, (10) solid plus Furadan, and (11) solid (disked in) plus Furadan.

Tip moth treatment consisted of 80 grams of granular Furadan® per seedling, and was applied in late April after planting. In mid-April each of the 0.202 ha treatment plots was split into three square subplots with treatments for control of competing vegetation: (1) no herbicide, (2) Velpar® at 0.7 kg/ha, or (3) Goal® at 4.5 kg/ha sprayed over the subplot. In early March 1981 the site was planted with nursery-run 1-0 seedlings at a 1.80 by 3.05 m spacing. Annual growth measurements consisted of height and ground-line diameter until 90% of the trees were 1.37 m high, after which diameters were taken at 1.37 m. The center four rows of each gross plot of eight rows were used for a measurement plot with 40 planting spaces.

Three-year-old Stands

Sludge was applied to 3-year-old planted loblolly pine stands in a randomized complete block design in October 1981. The sites had previously supported 30-year-old pine stands that were harvested in 1978. The sites were sheared and then machine planted in early 1979 with a 1.80 by 3.05 m spacing. Survival averaged about 55% after two years. The stands were fully covered with hardwood sprouts, most of which were less than 1 m tall at the time of sludge application. The following treatments were applied: (1) control, (2) low N liquid sludge, (3) high N liquid sludge, and (4) solid sludge. Each of the square 0.202 ha plots was split into three subplots to which a granular Velpar® was applied in 2 foot strips between rows at either 0.67 or 1.34 kg/ha in April 1981. Tree heights and diameters were measured annually beginning in early 1981. Diameters were taken at ground line until 90% of the trees exceeded 1.37 m, and then diameters were taken at this height.

Eight-year-old Stands

Sludge treatments were applied to the 8-year-old loblolly pine in a randomized complete block design. Treatments consisted of (1) control, (2) low N liquid sludge, and (3) high N liquid sludge. The site had previously supported a pine stand, which had been harvested and machine planted in early 1974. Trees were nursery-run stock and were planted at a 1.80 by 3.05 m spacing. Survival of the 8-year-old plantation was about 60%, and hardwood competition was sparse on both sites. Sludge was applied to square 0.202 ha plots in September 1981. Canopy closure was nearly complete at this time, but live limbs were present on most trees to within 0.5 m of the ground. Internal measurement plots of approximately 0.07 ha were established inside gross treatment plots. Heights and diameters at 1.37 m were taken annually for four years.

Twenty-eight-year-old Stands

The 28-year-old pine stands were treated with sludge in July 1981 in a random design. The sites were planted in 1953 with nursery-run 1-0 seedlings in a 1.80 by 3.05 m spacing on old agricultural fields. The stands were thinned at age 20, leaving trees at one-third to one-fourth the original stocking. Hardwood competition at time of sludge application was sparse (see McLeod et al., this volume).

Four treatments were used: (1) control, (2) low N liquid sludge, (3) high N liquid sludge, and (4) solid sludge. Internal measurement plots were established inside gross treatment plots, which were 0.202 ha square. Measurement plot boundaries were about halfway between tree rows, with a border of about 8 to 10 m between measurement- and treatment-plot boundaries. Measurement plots were 0.06 to 0.08 ha and contained 16 to 35 trees. Tree heights were measured biennially beginning in early 1981, and diameters were measured annually at 1.37 m above the ground.

RESULTS

Application to One-year-old Stand

Sludge treatments have increased diameter growth and to a lesser extend height growth of trees four years after planting. However, seedlings were also influenced by treatment of tip moth infestations, control of competing vegetation, and incorporation of the sludge (Table 2). Height and diameter were measured; basal area and initial volume

TABLE 2. Response of one-year-old trees to sludge treatments, Furadan[®], and disking four years after application.

Treatment	Tree Height (m)	Dbh (cm)	Basal Area (m^2/ha)	Tree Volume (m^3/ha)
Control	2.91b	3.35c	1.17a	1.14a
Furadan[®] alone	3.67a	4.64ab	2.29a	2.18a
Inorganic fertilizer	3.26ab	3.86bc	1.54a	1.70a
Low N liquid sludge	3.62a	4.71ab	2.17a	2.62a
High N liquid sludge	3.43ab	4.46abc	1.79a	2.09a
Solid on surface	3.10ab	3.53bc	1.40a	1.45a
Solid incorporated	3.47ab	5.11a	2.69a	3.11a

Values in the same column followed by the same letter are not significantly different at the 0.05 level.

All sludge and fertilizer treatments shown received Furadan[®].

(computed as one-third height × cross-sectional area at breast height, by tree) were obtained. Survival averaged about 60% and was not affected by treatments. Mortality of the Velpar®-treated seedlings was higher, and these areas were replanted. Thus, this treatment was excluded from the results. Application of Furadan® increased tree heights by 26% and diameters by 39% over control.

Application of low N liquid sludge plus tip moth treatment increased heights by 25% and diameters by 45% over control. Other treatments or treatment combinations did not significantly improve average height growth. The use of Goal® herbicide did not interact with sludge treatments but significantly increased average height across sludge treatments from 3.24 to 3.34 m, or 6%, and increased diameters from 4.0 to 4.4 cm, or 10%.

Plot basal area and stem volume were not altered by treatments, because the stand was two to five years from closing and trees were growing as individuals. Significant differences become considerably more difficult to detect with basal area and stem volume growth where variation in tree heights and diameters are compounded with survival. Basal area and stem volume give some indication of growth obtained in four years to compare treatment response of different age classes of pine.

An example of a treatment interaction with sludge application is shown to indicate how other practices influenced response of this planting to sludge application. The high liquid sludge or Furadan® applied alone did not increase diameters, but when the sludge and Furadan treatments were combined, diameters were increased by 1.79 cm. Heights were increased 0.76 m by Furadan alone. Sludge treatment combined with Furadan did not result in a further significant increase in height. Furadan improved growth by 26% without sludge and by 20% with sludge. The application of liquid sludge with high N alone did not significantly increase height growth, but this treatment plus Furadan improved diameter growth over untreated trees and those that received sludge only, by 53 and 35% respectively. Thus the treatment for tip moth had an equal or greater effect on early stand growth than nutrition did.

Incorporation of solid sludge into the soil was found to increase growth response over incorporated soil sludge where tree diameters were increased 1.42 cm by disking the solid sludge into the soil. Incorporation of liquid sludge did not affect diameter growth. Sludge type or incorporation did not affect seedling heights after four years.

Application to Three-year-old Stands

Sludge treatment did not significantly affect growth of 3-year-old trees (Table 3). Initially, trees were about 0.5 m tall and heavily infested with tip moth. The site contained a high density of hardwood sprouts. Herbicide treatments were effective for only one season, and a heavy growth of tall (2 m) herbaceous weeds was present, especially the second year after herbicide treatment where sludge was applied. Herbicide treatment did not significantly affect heights or diameters, which averaged 3.45, 3.63, and 3.51 m tall and 4.68, 5.37, and 5.36 cm dbh for the no herbicide, 0.67 kg Velpar®, and 1.37 kg Velpar/ha.

TABLE 3. Response after four years in height, diameter, basal area, and stem volume of loblolly pine treated with sludge at age 3.

Sludge Treatment	Height (m)	Dbh (cm)	Basal Area (m^2/ha)	Stem Volume (m^3/ha)
Control	3.74	5.36	1.80	3.01
Low N liquid	3.67	4.80	1.63	2.70
High N liquid	3.70	5.46	1.80	2.77
Solid	3.86	5.36	2.11	3.36

Treatments had no significant effects.

Application to Eight-year-old Stands

Soil type did not alter the effect of sludge treatment on trees at age 8. Hence results are averaged across soil types (Table 4). Initially, the two sites supported stands with 771 to 801 trees/ha, representing a stocking of 50% of the trees planted, which were 5.61 to 5.76 m tall and 8.6 to 9.1 cm dbh. The stands had basal areas of 5.74 to 6.22 m²/ha, with stand volumes of 11.62 to 13.22 m³/ha. Volume was computed as one-third the height times the area of the stem at 1.37 m.

Sludge treatments had no effect on height, which averaged 9.63 m after four years, but increased diameters by 11%. Basal area was 27% higher on plots with liquid sludge with high N than on nontreated plots. Low N sludge plots had basal areas that were not different from other treatments. Sludge appeared to have an effect on stem volume, but the differences were not significant, probably because of stocking differences.

Increase in tree and stand increment four years after treatment indicated no change in height growth but significant changes in diameter growth. The low N sludge treatment increased diameters by 25%, and the high N sludge treatment increased diameters by 38%. Basal area growth was increased by 46% with the application of sludge. Although not significant, sludge increased tree volume by about one-third.

TABLE 4. Stand characteristics and growth of loblolly pine treated at age 8 with liquid sludge.

Sludge Treatment	Stand Density (trees/ha)	Tree Height (m)	Dbh (cm)	Basal Area (m²/ha)	Stem Volume (m³/ha)
		Initial Characteristics			
Control	801	5.73*	9.1	6.22	12.04
Low N	773	5.61	9.1	5.74	11.62
High N	771	5.76	8.6	5.83	13.22
		Characteristics Four Years After Fertilization			
Control	--	9.63a	15.2b	13.68b	45.90a
Low N	--	9.60a	16.8a	15.98ab	53.18a
High N	--	9.66a	17.0a	17.31a	58.50a
		Growth Increment After Four Years			
Control	--	3.90	6.1c	7.46b	33.86a
Low N	--	3.99	7.7b	10.24a	41.56a
High N	--	3.90	8.4a	11.48a	45.28a

Values in the same column in a data set followed by the same letter are not significantly different at the 0.05 level.

*Initial values did not differ significantly.

Application to Twenty-eight-year-old Stands

Results indicated there was no interaction between sludge treatment and soil type, so results are combined (Table 5). Initially, stands had 418 to 524 stems/ha, and trees averaged 18.42 m tall and 23.75 cm in diameter. Basal area averaged 20.84 to 24.2 m²/ha. Stem volumes were computed by using the equation of Bailey and Clutter (1970), which resulted in initial values of 206.8 to 249.2 m³/ha.

TABLE 5. Stand characteristics and growth of loblolly pine treated at age 28 with liquid and solid sludge.

Sludge Treatment	Stand Density (trees/ha)	Tree Height (m)	Dbh (cm)	Basal Area (m²/ha)	Stem Volume (m³/ha)
		Initial Characteristics			
Control	524	18.26	22.9a	22.43a	220.8ab
Low N liquid	519	18.65a	24.1a	24.20a	249.2a
High N liquid	492	18.44	23.6a	23.23a	231.0ab
Solid	418	18.35a	24.4a	20.84a	206.8b
		Characteristics Four Years After Fertilization			
Control	--	20.21a	24.6a	25.80ab	250.5ab
Low N liquid	--	20.97a	27.2a	27.89a	282.4a
High N liquid	--	20.42a	26.4a	26.61ab	260.8ab
Solid	--	20.51a	26.9a	24.10b	235.9b
		Growth Increment After Four Years			
Control	--	1.95a	1.7b	3.37a	29.7a
Low N liquid	--	2.32a	3.1a	3.69a	33.2a
High N liquid	--	1.98a	2.8a	3.38a	29.8a
Solid	--	2.16a	2.5ab	3.26a	29.1a

Values in the same column in a data set followed by the same letter are not significantly different at the 0.05 level.

After four growing seasons, trees averaged 20.21 to 20.97 m tall and 24.6 to 27.2 cm in diameter. Mortality was less than 5% and not related to any treatment in the four years studied. Trees receiving low N liquid sludge had 3.76 m²/ha greater basal area and 46.5 m³/ha greater volume than trees on solid sludge plots. There were no differences between the low N liquid or solid sludge and the control.

Height increase over four years was not affected by treatments and averaged 1.95 to 2.23 m. Application of liquid sludge increased average stem diameters by 56 to 67% compared with untreated trees. Stem diameters from solid sludge treatments were not different from those of other treatments. Basal area growth averaged 3.44 m²/ha, and volume growth averaged 30.5 m³/ha. Treatments had no significant effect on these measurements. Use of initial basal area as a covariant did not appreciably change basal area and volume values or change the statistical inference.

DISCUSSION AND IMPLICATIONS

Soil type had only a small effect on growth. These results can probably be extrapolated to sites having similar soils with deep profiles, low pH, low fertility, and good drainage. Diameter growth response to sludge as a nutrient source at different ages was inconsistent. The authors found growth of newly planted seedlings to respond to tip moth treatment, vegetative competition control, and incorporation of the solid sludge. In contrast, solid sludge did not increase diameter growth of 28-year-old trees. One factor may be the 3.2 to 1 C:N ratio for liquid sludge compared with a 20 to 1 C:N ratio for solid sludge. Results suggest that older trees responded better to sludge with a narrow C:N ratio and higher N availability. A wide C:N ratio sludge may be most effective on younger stands where competition is a problem; however, the narrow C:N ratio is equally effective, especially if competing vegetation is controlled.

A more pressing question is: At what stand age should sludge be used to obtain the largest volume increase of wood per amount of sludge applied per hectare over a four-year period. Figure 1 illustrates the volume of wood produced at different ages above that produced with no sludge treatment. As noted in Tables 2 to 5, values for stem volumes were not statistically significant for treatments by age groups, but these volumes of wood produced do reflect relative differences between ages for treatments applied on these sites. Understandably, this does not solve all the problems nor provide a complete justification for timing of sludge treatments. Application at plantation establishment may, by promoting early growth, shorten rotations that may produce volume increases in stem wood not apparent in Figure 1, if competition is controlled. Specifically, depending on management objectives, increases in volume for small sawlog trees in the 28-year-old stand may have a much higher value than volume growth increases on 8-year-old

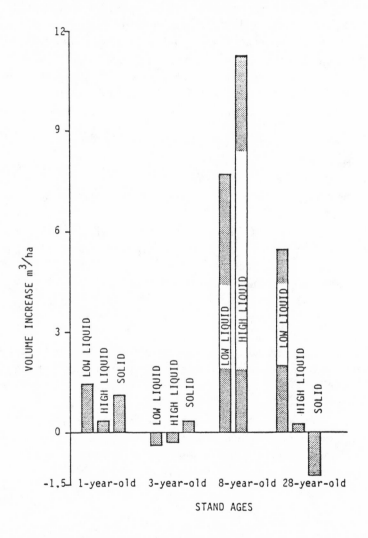

Figure 1. Four-year increase in stem volume (outside bark) of treated loblolly pine over unfertilized controls, for four age classes on upper coastal plain sites.

trees. Diameter growth of 1 to 2 cm on a 23 cm dbh tree has considerably more value than this growth on a 9 cm dbh tree. To assess the benefits of the sludge over an entire rotation is beyond the scope of this study. Other factors such as groundwater contamination must be considered in selecting stand ages and intensity of treatment (see Wells, Murphy, Davis, Stone, and Hollod, this volume). While this study has not attempted to report changes in nutrient content of foliage or physiology of trees, the growth response to sludge treatments above 400 kg/ha was marginal with physiological problems evident. Trees were more susceptible to wind damage because of heavy crowns, and premature needle drop was observed, especially with the older trees the first year after treatment. Measurement and analysis of data from this study are continuing.

SUMMARY

Measurements of loblolly pine in four age classes and growing on several upper coastal plain sites indicated that growth was increased by sewage sludge treatments. The findings suggest that control of competition and tip moth treatment is needed in younger stands when sludge is applied. Tip moth treatment alone can improve growth. The nature of the sludge, its nutrient content, and C:N ratio should be considered in the selection of stands for treatment. Growth of older stands, especially diameter, was increased more by a high N, narrow C:N ratio material, while younger stands appeared to be equally responsive to a low N, wide C:N ratio sludge, especially if incorporated in the soil.

ACKNOWLEDGMENTS

This work was supported in part by contract DE-A109-80SR10711 with the U.S. Department of Energy.

The use of trade, firm, or corporation names in this publication is for the information and convenience of the reader, and such use does not constitute an official endorsement or approval by the U.S. Department of Agriculture or the Forest Service of any product or service to the exclusion of others that may be suitable.

CAUTION

This publication reports research involving pesticides. It does not contain recommendations for their use, nor does it imply that the uses discussed have been registered. All uses of pesticides must be registered by appropriate state or federal agencies before they can be recommended.

Pesticides can be injurious to humans, domestic animals, desirable plants, and fish or other wildlife if they are not handled or applied properly. All pesticides should be used selectively and carefully, and recommended practices for the disposal of surplus pesticides and pesticide containers should be followed.

REFERENCES

Bailey, R. L., and J. L. Clutter. 1970. Volume tables for old-field loblolly pine plantations in the Georgia Piedmont. *In* Georgia Forestry Research Council Report 22, Series 2.

Edmonds, R. L., and D. W. Cole (eds.) 1980. Use of dewatered sludge as an amendment for forest growth. Vol. 3. Center for Ecosystem Studies, College of Forest Resources, University of Washington. 120 p.

Sopper, W. E., E. M. Seaker, and R. K. Bastian. 1982. Land reclamation and biomass production with municipal wastewater and sludge. Pennsylvania State University Press, University Park. 524 p.

Wells, C. G., K. W. McLeod, C. E. Murphy, J. R. Jensen, J. C. Corey, W. H. McKee, and E. J. Christensen. 1984. Response of loblolly pine plantations to two sources of sewage sludge. p. 85–94. *In* 1984 TAPPI Research and Development Conference, Appleton, Wisconsin. Technical Association of the Pulp and Paper Industry, Technology Park, Atlanta, Georgia.

Chapter 26

Municipal Sludge Fertilization on Oak Forests in Michigan: Short-term Nutrient Changes and Growth Responses

PHU V. NGUYEN, JAMES B. HART, JR., and DENNIS M. MERKEL

ABSTRACT   A research and demonstration project was initiated in 1981 in northern Michigan to explore the potential use of municipal waste on forest lands. Oak was one of four forest cover types studied. The forest stand was an upland mixed oak type with red oak (*Quercus rubra* L.), white oak (*Q. alba* L.), and red maple (*Acer rubrum* L.). Sludge was applied on three 1.5 ha plots at the rate of 8 Mg dry solids/ha to provide 400 kg/ha of nitrogen from a liquid anaerobic source. Early results indicate the following responses: (1) There has been no tree mortality due to sludge application. (2) Tree growth was increased by sludge over a three-year period: a significant increase of 63% in diameter growth over the control was observed for all species combined; basal area growth of all species combined exhibited a significant 44% increase over the control; a differential species response also occurred for diameter growth. (3) There were no effects on sapling diameter and basal area growth. (4) A large portion of the nonsoluble nutrients applied remained in the forest floor two years after application. (5) There were no differences in chemical properties of surface and subsurface soils two years after sludge application. (6) Nitrate-N concentration in soil leachate increased five months after treatment, but the peak concentration was less than 3 mg/l and decreased to background level within one year after sludge application.

Sludge application to forest lands appears to be a favorable practice since it can serve both waste disposal and forest fertilization. In developing guidelines for sludge application that can promote and maintain forest productivity without endangering environmental quality and polluting the groundwater, forest responses must be examined using ecosystem approaches, allowing evaluation of both short-term tree growth and long-term stand growth. Such approaches should include considerations of native site nutrient resources, sludge nutrient additions, nutrient balances, and cycling. Short-term nutrient growth changes of an upland oak stand in relation to sludge application are examined; long-term effects on forest growth are presented in another report (see Merkel et al., this volume).

METHODS AND MATERIALS

Study Area

Located in Montmorency County, Michigan, the study area was occupied by a 70-year-old oak stand that was a mixture of red oak (*Quercus rubra* L.), white oak (*Q. alba* L.), and red maple (*Acer rubrum* L.) with scattered pines (*Pinus* spp. L.) and aspen (*Pop-*

ulus spp. L.). The stand averaged 388 trees/ha with an average combined basal area of more than 21.5 m²/ha (Table 1). Three species, red oak, white oak, and red maple, make up 95% of the total number of trees. Basal area distribution by percentage among species at the oak site reflects the dominance of red oak (44%), white oak (28%), and red maple (22%) as the three major commercial tree species.

Sapling species composition was dominated by red maple (72%), followed by iron-wood (*Carpinus caroliniana* W.) (11%), white oak (9%), and red oak (5%). Red maple, white oak, and red oak make up 86% of the total saplings and 95% of the sapling basal area, reflecting the dominance of these three commercial species in the overstory.

The soils were quite uniform and classified as Graycalm series, a mixed, frigid Alfic Udipsamment; with a few small inclusions of the Rubicon series, a mixed, frigid Entic Haplorthod (Hart et al. 1984). Both series are distinguished by being deep, excessively drained and formed in sand on till and outwashed plains and moraines (Soil Conservation Service 1976). On the oak site the Graycalm series consisted of a weakly developed sandy outwash cap overlying a calcareous, indurated moraine, features not unlike other soils in the immediate vicinity (Farrand 1982). Depth to groundwater was in excess of 25 meters.

TABLE 1. Pretreatment tree diameters and basal areas of areas to receive treatment.

Species	Control	Trail	Sludge	Stand Average
		Treatment		
	Diameter at Breast Height (cm)			
Red maple	11.6	10.7	10.5	11.0
White oak	15.9	11.2	15.8	14.3
Red oak	17.6	19.6	19.6	18.9
All species	17.1	17.3	16.9	17.1
	Basal Area (m²/ha)*			
Red maple	5.05	5.00	4.00	4.68
White oak	7.02	3.90	6.72	5.88
Red oak	8.14	11.93	7.79	9.29
All species	22.0	22.5	20.0	21.5

*Basal area was calculated from tree dbh on an areal basis.

Experimental Design

The experiment is a completely random design involving three treatments: C, T, and TS (control, application trails only, and application trails with sludge application). All measurements and samplings were carried out adjacent to or within five measurement subplots (15 by 20 m) within each treatment plot (1.5 ha); measurement subplots were divided into three zones (one center zone and two edge zones).

Anaerobically digested liquid municipal sludges from treatment facilities in Rogers City and Alpena, Michigan were applied using an all-terrain vehicle, equipped with a modified three-nozzle spray system. Characteristics and loading levels of nutrients and metals are given in Table 2. The study plots cover 13.6 ha, of which 4.5 ha were treated.

Foliage specimens of 24 red oak and 24 white oak trees, 8 from each treatment, were sampled in 1981 and in 1984, before and three years after sludge application, respec-

TABLE 2. Average chemical concentrations of applied sludge and nutrient application rates.

Component	Average Concentration	Loading
	(%)	(kg/ha, oven-dry)
Sludge solids	3.43	8,019
	(mg/kg, wet basis)	(kg/ha)
Kjeldahl-N	1,700	401
Phosphorus	1,200	272
Calcium	2,641	619
Potassium	92	21
Magnesium	219	51
Aluminum	625	146
Iron	2,110	492
Zinc	40	9
Manganese	27	6
Copper	27	6
Cadmium	2.0	0.4
Nickel	1.4	0.3
Chromium	3.7	0.9

tively. Samples were collected from the south side of the sunlit upper crown for foliar nutrient diagnosis.

Forest floor and soils were sampled in the fall before and two years after sludge application. Forest floor samples, collected from 90 randomly located points, 30 per treatment, using a 30 by 30 cm sampling frame, were systematically separated into 01 and 02 fractions. The 01 is equivalent to the L layer, while the 02 is the F and H layers. Soils directly beneath the forest floor to the depth of 45 cm were sampled and separated into surface soils (mineral soil to bottom of the E horizon) and subsurface soils (top of B horizon to 45 cm). Percolating soil water at 1.2 m beneath treatment plots was collected from a network of 27 pressure-vacuum lysimeters (PVL).

Woody plants taller than 1.80 meters (6 feet) and with a diameter at breast height (dbh) of less than 10 cm (4.0 inches) were considered saplings. Saplings on measurement plots were identified by species and measured for dbh prior to sludge application and remeasured three years later in 1984.

In 1981, before sludge application, all trees of sawtimber and pole sizes (dbh > 10 cm) within the measurement subplots were individually tagged, had species identified, and dbh measured to the nearest mm. In addition, vigor rating and condition coding, as well as crown class, were also added. Remeasurements one year later ensured the accuracy of the initial measurements. In 1984 these trees were rechecked for condition coding and vigor status, and their dbh was remeasured. Basal area by treatment and zone was calculated from dbh based on zone areas.

Chemical Analysis

Plant tissue and samples were oven dried at 70°C and prepared for chemical analysis. Soil samples were air dried and sifted through a 2 mm sieve before being subsampled for analysis. Sludge samples were digested with HNO_3 and $HClO_4$ while plant tissues were ashed in a muffle furnace for metal analysis (Blanchar et al. 1965). Soil samples were extracted with $1N$ NH_4OAc and with $0.1N$ HCl for exchangeable cations and micronutrients, respectively (Black 1965, Ellis et al. 1976). Soil organic matter was determined by the Walkley-Black method (Black 1965). Metals and cations were determined using DCP atomic emission spectroscopy. Kjeldahl-N, phosphorus (Technicon 1977a), and soil water nitrate-N were determined colorimetrically (American Public Health Association 1975, Technicon 1977b, U.S. EPA 1979).

Statistical Analysis

The study was analyzed by the analysis of variance technique on a randomized split-plot experiment with three treatments as whole-plot factors, and the zones within the measurement units and the treatment X zone interactions as split-plot factors (Steel and Torrie 1980).

RESULTS AND DISCUSSIONS

Generally, statistical analyses indicated no zone effects or treatment X zone interactions. In a few instances interactions were significant at a lower statistical level than main effects and only main effects were reported.

Growth Responses

Pretreatment data analyses indicated there were no significant differences in tree composition, diameters, or basal areas on areas to receive different treatment or zone effects.

Tree Mortality. Sludge application has not caused tree mortality, as only 4 of 1,219 trees within the monitoring subplots died during the 1981–84 period.

Diameter Growth. Red maple trees showed a nonsignificant increasing trend in diameter growth with sludge treatment (Table 3). White oak diameter growth exhibited significant sludge treatment effects for the three-year period; the diameter growth of 0.66 cm was 61% and 78% greater than the growth in control and trail treatment, respectively. There were no zone, or treatment X zone, interaction effects.

There appears to be a differential species response to treatments, since red oak showed a similar response to sludge application as white oak but at a lower magnitude. A 39% diameter growth increase over control was observed on the sludge areas. Differ-

TABLE 3. Sludge and trail treatment effects on three-year tree diameter and basal area growth.

	Treatment		
Species	Control	Trails	Sludge
	Diameter at Breast Height (cm)		
Red maple	0.46a	0.48a	0.86a
White oak	0.41a	0.37a	0.66b
Red oak	0.56a	0.66ab	0.78b
All species	0.54a	0.62ab	0.88b
	Basal Area (m^2/ha)		
Red maple	0.38a	0.46a	0.63a
White oak	0.37a	0.28a	0.55a
Red oak	0.53a	0.80a	0.65a
All species	1.41a	1.66ab	2.03b

Treatment means in a row followed by the same letter are not significantly different at the 0.10 level by Duncan's multiple range test.

ences between control and trail treatments, and between trail and sludge treatments, were not statistically significant.

The diameter growth for all trees receiving sludge treatments (0.88 cm) was 63% greater than for trees in the control areas (0.54 cm). However, the 42% growth increase of sludge over trail treatments, and the 15% increase from control to trail treatments, were not significant (Table 3).

Basal Area Growth. Responses of total basal area growth for all species with sludge application were similar to those exhibited by diameter growth (Table 3). Basal area growth of 2.03 m²/ha for trees of all species on sludge areas represented a 44% increase over control. The trail treatment, compared with the control, showed a 18% increase in basal area over the three-year period. Although not statistically significant, any increase would help compensate for the 20 to 25% basal area cut during trail preparation. A basal area growth difference of 0.37 m²/ha, a 22% increase of sludge over trail treatments, suggests a result of sludge treatment alone. However, this difference is not statistically significant.

In contrast to diameter growth, basal area growth showed no significant treatment effects for the major individual species. However, there were trends of increasing basal area growth with either sludge or trail treatments and they will be evaluated over longer periods.

Tree Foliage Nutrient Levels

There was no significant effect on diagnostic foliar nitrogen and phosphorus concentrations three years after sludge application; foliar nitrogen concentration showed consistent increases in all treatments for both red oak (from 2.0% to 2.3%) and white oak (from 2.1% to 2.3%) over the three years following sludge treatment. The foliar phosphorus concentration remained at 0.2% for both species.

Sapling Growth Responses

Application effects, both treatment and zone, on number of saplings, diameter, and basal area per zone were not significant over the 1981–84 period. This may be a result of no growth increase or growth increases combined with increased ungulate browsing activities in the area.

Ground Vegetation

In a study to determine effects of sludge application on annual wildlife food plants less than 2 meters tall, it was found that both trail and sludge treatments have affected annual primary production (Haufler and Woodyard 1984). Sludge-treated areas exhibited the greatest primary production during the three years after sludge application, with treatment response most pronounced from the herbaceous group. Also, nitrogen and phosphorus concentrations increased in current-year growth of red maple, the most dominant species in the understory. The increases over the control, in nitrogen (60% in spring, 23% in summer) and phosphorus (43% in spring, 60% in summer), existed only the first year after sludge application.

Forest Floor

From 1981 to 1983, sludge treatment significantly increased forest floor weights: weights of the sludge treatment displayed a nearly twofold increase (Table 4). The in-

TABLE 4. Changes in forest floor weight for the two-year period, 1981-83.

Forest Floor Component	Treatment		
	Control	Trails	Sludge
	(kilograms/hectare)		
01	-2,338a	-2,419a	-1,506a
02	16,432a	17,422a	28,624b
FF (01+02)	14,094a	15,003a	27,122b

Row mean differences followed by the same letter are not significantly different at the 0.10 level by Duncan's multiple range test.

crease was more pronounced in the decomposed 02 layer, which comprised up to 90% of the total forest floor weight. Litter (01) weights tended to decrease from 1981 to 1983 for all treatments, apparently owing to yearly differences in mineralization or other processes. Within sludge-treated areas, the 01 component exhibited less change in weight, perhaps as a result of increased canopy production from sludge application.

The magnitude of elemental contents in the 01 horizon, compared with the 02 horizon (Table 5), suggests that the 02 was the major ecosystem component retaining the applied sludge elements. Sludge addition resulted in significant increases in total contents of most of the macronutrients (nitrogen, phosphorus, magnesium) in the 02 horizon compared with the control and trail treatments. Calcium and potassium showed noticeable, although not statistically significant, increases with sludge treatment. The sludge material was low in potassium relative to nitrogen and phosphorus and would be considered an unbalanced fertilizer.

Sludge treatment also significantly increased total micronutrient contents in the 02 layer, with the exception of manganese and aluminum. Of the trace elements, nickel

TABLE 5. Elemental contents of the forest floor by components for 1983 (two years after trail cutting and sludge application).

Component	01 Layer (litter)			02 Layer (F and H)		
	Control	Trails	Sludge	Control	Trails	Sludge
	(kilograms/hectare)			(kilograms/hectare)		
Nitrogen	65.7a	63.8a	79.0a	391.6a	420.3a	584.1b
Phosphorus	9.2	8.5	19.8	19.5a	23.2a	70.7b
Potassium	4.3a	4.1a	5.4a	23.2a	22.6a	33.2a
Calcium	57.5a	58.1a	70.9a	169.5a	220.3a	422.9b
Magnesium	7.2	7.4	9.5	37.9a	41.0a	68.9a
Aluminum	1.7	1.6	5.5	64.3a	67.6a	118.4a
Iron	3.9	2.6	23.2	96.8a	105.4a	438.5b
Manganese	8.8a	8.9a	9.6a	97.8a	123.0a	121.4a
Copper	0.04	0.03	0.21	0.3a	0.3a	3.2b
Zinc	0.27	0.22	0.62	1.8a	2.3a	7.4b
Cadmium	0.004a	0.005a	0.007a	0.04a	0.04a	0.08b
Nickel	0.009a	0.008a	0.021b	0.15a	0.17a	0.35a
Chromium	0.010	0.007	0.044	0.14a	0.16a	0.63b

Contents based on 30 points per treatment sampled in 1983.
Row means within a component followed by the same letter are not significantly different at the 0.10 level.
Row means without letters indicate interaction effect at the 0.10 level.

Oak Forests: Short Term 287

TABLE 6. Changes in elemental contents over the two-year period (1981-83) for the 02 forest floor layer.

Component	Treatment		
	Control	Trails	Sludge
	(kilograms/hectare)		
Nitrogen	162.1a	157.5a	336.7*/a
Phosphorus	3.2a	4.2a	53.5b
Potassium	9.6a	7.3a	17.7*/a
Magnesium	11.6a	6.0a	40.0*/a
Calcium	55.2a	72.0a	238.3*/a
Aluminum	37.3a	31.7a	89.0*/a
Iron	53.1a	45.7a	391.6*/b
Manganese	41.9a	25.1a	44.9a
Copper	0.09a	0.07a	2.98*/b
Zinc	0.52a	0.49a	5.96*/b
Cadmium	0.02a	0.01a	0.05*/a
Nickel	0.06a	0.02a	0.25*/a
Chromium	0.07a	0.08a	0.56*/a

Changes based on differences between predetermined adjacent points sampled in 1981 and 1983.
Row mean differences followed by the same letter are not significantly different at the 0.10 level by Duncan's multiple range test.
*Indicates the mean difference is significantly different from zero at the 0.10 level.

displayed a nonsignificant increase, while cadmium and chromium were significantly increased with sludge treatment, but at very low magnitudes (Table 2).

Changes in elemental contents of the 02 layer over time (Table 6) reflect the effects of sludge treatment on weights of this component (Table 4). The difference in 02 mean weights of 11,202 kg/ha between sludge and trail treatments in the third year represents 140% of the sludge solids applied. This might suggest that the sludge was not readily degraded and released nutrients bound within the sludge matrix to soils and vegetation at a slow rate. The increase was also partly associated with increased growth, litter production, and humification to 02 materials.

Two years after sludge application, 45% of the applied nitrogen, 22% of phosphorus, 56% of potassium, 61% of magnesium, and 35% of calcium was found in the forest floor. Of the micronutrients, 40% of aluminum, 73% of iron, 50% of copper, and 59% of zinc applied as sludge was retained in the forest floor. Nickel and chromium displayed high retention, 64% and 60%. The stability of portions of the sludge, and the role of the 02 horizon as a repository for the applied elements, are major responses to sludge application.

Surface and Subsurface Soils

Surface and subsurface soil pH were not changed by sludge application from 1981 to 1983 (Table 7). Concentrations of most macronutrients and micronutrients in 1983, and changes over the two-year period following sludge treatment, for both surface and subsurface soils, were not significantly different for treatments or zones (Table 7). The E horizons which dominated the surface soils would not be expected to accumulate nutrients. The B horizons were also sandy with low cation exchange capacity and nutrient retention capacity. Nutrient accumulations in the soil might also be precluded by plant uptake rates exceeding mineralization rates.

TABLE 7. Chemical properties of surface and subsurface soils two years after trail cutting and sludge application.

Component	Surface Soils (S1)			Subsurface Soils (S2)		
	Control	Sludge	Trails	Control	Sludge	Trails
pH	4.15	4.22	4.21	4.72	4.61	4.79
Organic matter (%)	1.43	1.36	1.40	0.41	0.40	0.39
	(micrograms/gram)			(micrograms/gram)		
Nitrogen (TKN)	602	614	599	212	201	213
Phosphorus (TKN digest)	79	102	91	128	122	130
Exchangeable*						
Magnesium	19.62	24.07	23.86	14.96	9.35	15.32
Potassium	27.32	28.66	26.51	14.86	15.83	15.56
Calcium	62.44	105.16	94.81	34.76	24.68	35.93
Sodium	4.96	4.94	7.54	4.31	3.00	3.50
Extractable**						
Zinc	3.69	3.34	4.21	3.69	3.34	4.21
Manganese	67.49	60.47	79.41	13.56	12.91	15.53
Iron	50.88	56.41	55.72	44.05	43.58	47.96
Copper	0.75	1.03	1.24	0.59	0.43	0.64

*$1\underline{N}$ NH_4OAc at pH 7.   **$0.1\underline{N}$ HCl.

Nitrate-N Concentration in Soil Leachate

Sludge application resulted in measurable changes in nitrate-N in soil leachate at the 1.20 m depth (Figure 1). Increases in nitrate-N were detected after snowmelt in 1982, the first year after sludge application. A nitrate-N peak occurred in August 1982, eight months after sludge application, at the time excess sludge supernatant ammonia would be nitrified and leached down (Urie et al. 1984). This nitrate-N increase exceeded effects of soil disturbance caused by lysimeter installation which occurred in both control and sludge-treated plots. The peak level of nitrate-N, however, is much lower than 10 mg/l, the U.S. EPA standard for drinking water. The nitrate-N level decreased rapidly and receded to the background level within one year after sludge application.

SUMMARY AND CONCLUSIONS

Sludge application at the rate of 401 kg of nitrogen and 272 kg of phosphorus and other associated nutrients per hectare increased growth compared with the control. Over three years, diameter growth increased 63%, stand basal area growth increased 44%, and ground vegetation production also increased. These responses are comparable to those reported elsewhere for sludge fertilization. Increased volume growth of approximately 40% has been reported for 9- and 28-year-old loblolly pine plantations (Wells et al. 1984). Similar growth responses also occurred in established Douglas-fir stands of the Pacific Northwest, where 40 to 60% volume increases from sludge exceeded growth response of 23% from chemical fertilization (Zasoski et al. 1983).

Two years after treatment, large portions of the applied nutrients (22 to 60%) and trace elements (40 to 70%) from the applied sludge were found in the forest floor, primarily the 02 horizon. Similar changes did not occur in the soil. In a series of the U.S. Forest

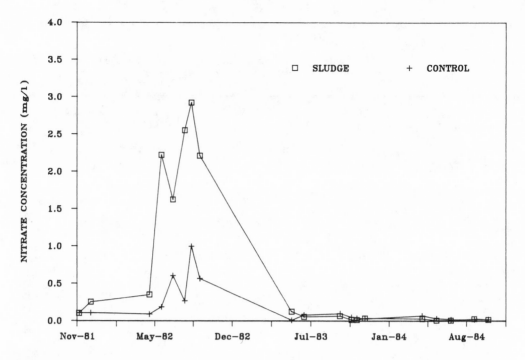

Figure 1. Nitrate-N concentrations in leachate from 120 cm soil depth in control and sludge-treated plots. Sludge was applied in November 1981.

Service studies in the Manistee National Forest of the northern Lower Peninsula, the forest floor had immobilized and stored most of the nutrients and metals either directly or after cycling through vegetation (Harris et al. 1984).

The peak of soil solution nitrate-N did not exceed the 10 mg/l level. The increases were very close to predicted concentrations for 10 Mg sludge applications to pine plantations on sandy soils in northern Michigan (Brockway and Urie 1983). They were also compara-ble to results for loblolly pine plantations treated with equivalent nitrogen loading rates (Wells et al. 1984).

Additional studies are required to determine the specific fate of added nutrients in forest ecosystems and their effects on forest long-term productivity. However, results to date indicate that most nutrients added from sludge application either enter the nutrient cycle or are retained in components of the oak forest ecosystem in a safe manner. The authors interpret the results as follows: nutrients were added to the forest ecosystems; some nutrients were readily available, or rapidly mineralized to available forms, and taken up by the trees, understory, and ground vegetation; biomass production and biomass nutrient contents were increased; and diameter and basal area of the stand were significantly increased. Greater litterfall probably occurred and the litter mineralized more rapidly, producing accelerated biological nutrient cycling. Significant amounts of the macronutrients and micronutrients were retained in the 02 horizon in unavailable and undecomposed forms after two years. Since nutrients moving from the 02 to the mineral soil would be in available forms, they would be rapidly taken up from the sandy soils by plants. Apparently the sludge, and associated nutrient loadings, were applied at a sufficiently low rate that nitrate-N production did not exceed plant uptake sufficiently to cause excessive nitrate-N leaching in the soil solution.

ACKNOWLEDGMENTS

Results from this research are part of the Ecological Monitoring of Sludge Fertilization on State Forest Lands in Northern Lower Michigan project.

Although the information in this document has been funded in part by the U.S. Environmental Protection Agency under assistance agreement No. S005551-01 to the Michigan Department of Natural Resources and Michigan State University, it has not been subjected to the Agency's publication review process and therefore may not necessarily reflect the views of the Agency, and no official endorsement should be inferred.

The authors are grateful to the above agencies for their support and to the many individuals who assisted with field and laboratory activities.

Michigan Agricultural Experiment Station journal article number 11862.

REFERENCES

American Public Health Association. 1975. Standard methods for the examination of water and wastewater. APHA, New York. 1193 p.

Black, C. A. (ed.) 1965. Methods of soil analysis: Part 2. American Society of Agronomy, Madison, Wisconsin.

Blanchar, R. W., G. Rehm, and A. C. Caldwell. 1965. Sulfur in plant materials by digestion with nitric and perchloric acid. Soil Sci. Soc. Am. J. 29:71–72.

Brockway, D. G., and D. H. Urie. 1983. Determining sludge fertilization rates for forests from nitrate-N in leachate and groundwater. J. Environ. Qual. 12:487–492.

Ellis, R., Jr., J. J. Hanway, G. Holmgren, D. R. Keeney, Subcommittee of NC 118. 1976. Sampling and analysis of soils, plants, wastewaters, and sludge: Suggested standardization and methodology. North Central Regional Publication 230.

Farrand, W. R. 1982. Quaternary geology of Michigan (map). State of Michigan Department of Natural Resources Geologic Survey.

Harris, A. R., D. H. Urie, and J. H. Cooley. 1984. Sludge fertilization of pine and aspen forests on sand soils in Michigan. p. 193–206. In E. L. Stone (ed.) Forest soils and treatment impacts. Proceedings of the Sixth North American Forest Soils Conference, June 1983, University of Tennessee, Knoxville.

Hart, J. B., P. V. Nguyen, C. W. Ramm, J. H. Hart, and D. M. Merkel. 1984. Ecological monitoring of sludge fertilization on state forest lands of northern lower Michigan. Annual progress report. Department of Forestry, Michigan State University, East Lansing. 192 p.

Haufler, J. B., and D. K. Woodyard. 1984. Influences on wildlife populations of the application of sewage sludge to upland forest types. Annual progress report. Department of Fisheries and Wildlife, Michigan State University, East Lansing.

Soil Conservation Service. 1976. Rubicon series. Nat. Coop. Soil survey. USDA.

———. 1976. Graycalm series. Nat. Coop. Soil survey. USDA.

Steel, R. G. D., and J. H. Torrie. 1980. Principles and procedures of statistics: A biometric approach. McGraw-Hill, New York. 633 p.

Technicon Industrial Method. 1977a. Nitrate and nitrite in water and wastewater. Method no. 102–70W/C. Technicon Industrial Systems, Tarrytown, New York.

———. 1977b. Individual/simultaneous determination of nitrogen and/or phosphorus in BD acid digests. Method no. 334–74W/B. Technicon Industrial Systems, Tarrytown, New York.

Urie, D. H., J. B. Hart, P. V. Nguyen, and A. J. Burton. 1984. Hydrologic and water quality effects from sludge application to forests in northern lower Michigan. Annual progress report. Department of Forestry, Michigan State University, East Lansing. 97 p.

U.S. Environmental Protection Agency. 1979. Methods for chemical analysis of water and wastes. Environmental Monitoring and Support Laboratory, Office of Research and Development, U.S. EPA, Cincinnati, Ohio.

Wells, C. G., K. W. McLeod, C. E. Murphy, J. F. Jensen, J. C. Corey, W. H. McKee, and E. J. Christensen. 1984. Responses of loblolly pine plantations to two sources of sewage sludge. In 1984 TAPPI Research and Development Conference, Appleton, Wisconsin. Technical Association of the Pulp and Paper Industry, Technology Park, Atlanta, Georgia.

Zasoski, R. J., D. W. Cole, and C. S. Bledsoe. 1983. Municipal sewage sludge use in the forests of the Pacific Northwest, U.S.A.: Growth responses. Waste Management and Research 1:103–114.

Chapter 27

Municipal Sludge Fertilization on Oak Forests in Michigan: Estimations of Long-term Growth Responses

DENNIS M. MERKEL, J. B. HART, JR., PHU V. NGUYEN, and CARL W. RAMM

ABSTRACT   Relationships between mean annual increment (MAI) and total nitrogen and phosphorus contents in forest floor and 0 to 45 cm soil samples were studied using regression methods. Twenty-nine forest stands were used with growths ranging from 1.19 to 4.26 m³/ha per year. Nitrogen and phosphorus contents in the surface soil (A and E horizons), and nitrogen content in the fermentation and humus horizon (02), accounted for nearly 70% of the variation in MAI. Nutrient contents in surface soil and 02 layers from control and sludge fertilized plots were inserted in the regression equation. A 29% increase in growth (MAI) was predicted based on two-year sludge effects on forest floor and soil nutrient contents. A three-year, 44% increase in basal area growth was found using conventional fertilizer trial techniques on the same area. Long-term responses of this magnitude would require that soil nutrient changes persist or can be maintained through retreatment. Use of site nutrient contents is a useful approach to the assessment of long-term growth.

Forest growth in regions of similar climate, soils, and stand histories is dependent on an adequate and balanced supply of nutrients. Nutrient contents (kg/ha) to a given depth represent a site resource based not only on nutrient concentrations but also on soil or horizon depths. Forest ecosystems are understood to cycle nutrients between biotic and abiotic components. Available nutrients are taken up from the forest floor and soil by forest vegetation and utilized for growth. Therefore, nutrient increases in forest floor and surface soils should be related to increases in overall stand growth. The interpretation of forest growth correlations with site nutrient contents has received little attention in the literature but has yielded significant associations when applied. Site index of mature ponderosa pine trees in California showed a positive correlation with nitrogen content in the top 1.22 m of soil (Zinke 1960), and total height of 30-year-old red pine trees in New York had a significant correlation with extractable potassium contents in the top 5 feet of soil (White and Leaf 1964).

Trees growing on sandy outwash sands in northern lower Michigan rely on mineralization of soil organic matter for nitrogen and phosphorus requirements. In soils of similar origin and texture in northern New York it was found that organic amendments increase soil nutrient reserves, site productivity, and therefore site quality (Heiberg and Leaf 1961). The addition of organic matter in the form of municipal sewage sludge to sandy soils has been shown to increase nitrogen and phosphorus levels significantly in the forest floor of red and white pine stands in northern lower Michigan (Brockway 1983) and can be expected to increase water holding capacity of surface soil horizons.

Few sludge studies have been concerned with long-term growth changes in mature forest stands. Intuitively it is expected that sludge additions will result in increased forest growth as do fertilizer applications. However, studies of basal area and radial dbh growth from detailed stem analysis at two years (Koterba et al. 1979) and fourteen months (Brockway 1983) after sludge applications found no significant increases in growth of mixed northern hardwoods or red pine, respectively. An inherent difficulty is the poor resolution obtained when measuring forest growth differences over a short period. Growth lag periods, from time of fertilization until tree response, such as the two- to five-year period reported for fertilized red pine in northern New York (Leaf et al. 1970), may further obscure short-term examinations.

The growth effects of sludge application are commonly measured for relatively short periods, while effects of sludge application on forested systems may take fifty years or more to assess completely. One method of evaluation is by establishing long-term sludge fertilization study sites. Such monitoring efforts were installed at the University of Washington in 1974, and at Michigan State University in 1981. Results from long-term measurements will not be available for some time. Until these studies reach fruition, other methods of assessing long-term growth effects are needed.

The objectives of this study are (1) to determine if a correlation exists between site nutrient resources and stand growth and identify important nutrient componets, and (2) to use the nutrient resource approach to estimate changes in forest stand growth with sludge application.

MATERIALS AND METHODS

The forest stand was chosen as the unit for measurement of nutrient contents and stand growth. Mean annual increment (MAI) was used as a measure of long-term growth and may be thought of as average growth over the age of a stand which integrates site factors and climate. In this study, MAIs of stands averaging 72 years old were used. Because stand growth is a complex response to many soil and environmental factors, no single site characteristic or measurement of growth can completely quantify site quality. This study uses MAI as a growth measurement while also recognizing the inherent limitations arising from variations in stand age and species composition.

Because of small nutrient fluctuations in soil at depths below 15 cm (6 inches) in sludge-amended stands on similar soils in northern Michigan (Brockway 1983), and on this study area (Hart et al. 1984), only forest floor and surface soil contents were used. The Kjeldahl total contents for nitrogen and phosphorus are reported here. Acid extractable calcium, potassium, and magnesium should also be related to MAI and are being evaluated. No heavy metals were examined, owing to the extremely low metal loadings with sludge application (Nguyen et al., this volume).

Estimation of long-term growth changes encompassing a range of nutrient resources was accomplished by sampling 29 stands in the region. The stands were located across an upland productivity gradient ranging from low productivity scrub oak to high productivity sugar maple stands. Regression equations were developed using data from the 29 regional stands for the nutrient contents of soil and forest floor components exhibiting the strongest correlations to stand growth.

Site nutrient resources for sludge and control plots on the sludge study area two years after treatment were inserted into the equation to predict changes in long-term growth

(MAI) from sludge applications. The oak ecosystem in the sludge study has stand growth and composition that fall within the range sampled in the regional study; therefore, implications for changes in stand growth should be applicable.

Sludge Study Area

Located in Montmorency County, Michigan, the sludge study area was on a 70-year-old oak stand composed of red oak (*Quercus rubra* L.), white oak (*Q. alba* L.), and red maple (*Acer rubrum* L.) with scattered pines (*Pinus* spp. L.) and aspen (*Populus* spp. L.). The experimental design, sludge loading rates, and further details of the site and methodology are presented elsewhere (Hart et al. 1984; Nguyen et al., this volume). The soils at this site belonged to the Graycalm series, a mixed, frigid Alfic Udipsamment; with small inclusions of the Rubicon soil series, a mixed, frigid Entic Haplorthod (Soil Conservation Service 1979, 1976). Both series are distinguished by being deep, excessively drained soils formed in sand on glacial drift materials (Soil Conservation Service 1979). On the oak site the Graycalm series consisted of a weakly developed sandy outwash cap overlying a calcareous, indurated till, with textural bands of varying thickness occurring at fluctuating depths—features not unlike other soils in the immediate vicinity (Farrand 1982).

Regional Study Areas

The 29 stands were located in the Manistee National Forest, 80 to 90 miles southwest of the sludge study plots. Figure 1 shows the general locations of the two study areas. Soils on the two areas are similar, predominantly sandy outwash materials overlying glacial drift, with analogous soil series occurring in both areas. Stands were sampled by strata to give equal representation across a range of productivities and stand compositions. Specific stands were randomly selected from a list of stands provided by the Huron-Manistee National Forest. Stands occupied by pioneer species, or showing signs of disturbance in the past forty years, were excluded from sampling. The minimum basal area of stands sampled was 18.24 m^2/ha. Other stand selection criteria were as follows: (1) the overstory must be at least fifty-five years old, (2) the stand must be normally stocked (i.e., the canopy should be closed as far as site conditions will permit), (3) stocking must be uniform throughout the stand with no extensive open areas, (4) the topography must be representative of upland conditions, (5) the soils must be well drained, and (6) no more than 30% of the dominant overstory may have multiple stems. The average stand age was 68 to 76 years.

SAMPLE COLLECTION

Regional Stands

Samples were collected for the regional nutrient resource study in the summer of 1983. A soil pit 1.5 m deep was located near the center of a homogeneous one hectare portion of the stand. The soils were then characterized and classified to the series level. Three additional sample points were randomly located, and soil, forest floor, and understory samples were collected. Forest floor samples were collected from six sampling points (three at the pit and one at each of the three additional points) with a 30 by 30 cm metal frame and systematically separated into fractions designated 01 and 02. Litter (01) samples included recognizable and nearly entire leaf material not discolored (blackened) by

Figure 1. Locations of sludge and off-site study areas in Michigan.

decompositional processes, and woody material. Fermentation and humus layer (02) samples consisted of finely divided decomposed organic materials extending to a distinct "salt and pepper" layer which marked the upper boundary of the mineral soil. Soil samples collected directly beneath the forest floor samples were divided into surface (S1, A, and E horizons) and subsurface (S2, upper B horizon to a depth of 45 cm) layers. Bulk density samples were collected to allow conversion of nutrient concentration data from the laboratory to a content by depth basis.

Forest growth was measured at each sampling point using a 10 basal area factor (BAF) prism. All trees at each prism point were measured for dbh, total height, and merchantable height to a 10.2 cm top. Selected trees were measured for age. Stand growth was calculated by averaging measurements of the four subplots.

Sludge Study Area

Site nutrient resources on the oak sludge study area were calculated from forest floor and soil samples collected in 1983, two years after application. Thirty points each on sludge and control treatments were sampled.

CHEMICAL AND STATISTICAL ANALYSIS

Forest floor samples were oven dried at 70°C, weighed, ground, and subsampled prior to analysis. Soil samples were air dried, sieved, and subsampled prior to analysis. All chemical analyses were performed in the Michigan State University Forestry Department laboratory using 10% sample replication to ensure precision, and bulk sample analysis with each sample set to ensure accuracy. The quality assurance procedures confirmed with a probability of 0.95 the determination of mean nitrogen and phosphorus concentrations with a 10% confidence interval.

Total nitrogen and phosphorus were determined by a micro-Kjeldahl digestion procedure and analyzed on a Technicon Autoanalyzer II system (Technicon 1977). The data were used with forest floor (01 and 02) weights and areas, and soil (S1 and S2) depths and soil bulk densities, to calculate nitrogen and phosphorus contents.

Scatter plots between MAI and stand nutrient contents of the eight forest floor and soil components revealed several nonlinear relationships. More linear scatter plots were obtained with log (base 10) transformations of the variables.

A multiple regression equation was formulated using a stepwise procedure available in the MICROSTAT statistical package (Ecosoft 1984). The equation was of the form:

$$Y = B_0 + B_1X_1 + B_2X_2 \ldots + B_nX_n$$

where
Y = growth (MAI, m³/ha·yr)
n = number of variables in equation
B_n = regression coefficients
X_n = nutrient contents in horizon (kg/ha)

The normal matrix solution for the vector of regression coefficients (B) is:

$$(X'X)^{-1}X'Y \qquad \text{(Draper and Smith 1981).}$$

Confidence intervals for the predictions on sludge and control plots in the sludge study were calculated using:

$$Y \pm t(v, 1\text{-}1/2\alpha)^*s \sqrt{1/g + X_0'CX_0}$$

where

Y = predicted value
v = sample size of regional study − number of parameters in regression equation including B_0, $(29-4)$
s = standard error of estimate
g = number of observations in X_0
X_0 = a (nx1) data matrix
C = a (nxn) variance-covariance matrix $(X'X)^{-1}$

(Draper and Smith 1981).

RESULTS AND DISCUSSION

The regression equation developed from the 29 regional stands is:

MEAN ANNUAL INCREMENT = 1.1883
(m³/ha · yr)
+ (−5.5847 * log nitrogen content in S1)
+ (5.7873 * log phosphorus content in S1)
+ (2.6515 * log nitrogen content in 02)

The standard error of the estimate is 0.5241 m³/ha·yr. The contribution of individual variables to the equation can be evaluated through their partial correlation coefficients (calculated with the effects of all other predictor variables removed). The partial correlation coefficients were 0.357 for nitrogen in the surface soil, 0.437 for phosphorus in the surface soil, and 0.519 for nitrogen in the 02 horizon. The three partial correlation coefficients had similar magnitudes, indicating that their influences on stand growth were similar. Log of nitrogen in the 02 horizon was the strongest contributor to long-term growth. If the 02 horizon is thought of as a nutrient repository for sludge-applied nutrients, then mineralization should result in greater nutrient availability in the 02 horizon and greater growth. The nitrogen concentrations are total determinations, and do not reflect the availability or immobilization of sludge nitrogen applied to the forest floor.

The residuals were examined and no deviations from normality were noted. Nearly 70% of the variability in MAI was explained (adjusted R square of 0.691). The results of the regression analysis confirm that a site's nutrient resources have correlations with stand growth as measured by MAI.

Table 1 presents mean nitrogen and phosphorus contents for forest floor and soil components on sludge and control plots. Increases from control existed with sludge treatment for all variables in the regression equation for MAI. Insertion of control and sludge

TABLE 1. Means of N and P contents for forest floor and soil components two years after treatment on the sludge fertilization study.

Nutrient Component	Treatment	
	Control n = 30	Sludge n = 30
	(kilograms/hectare)	
Nitrogen in S1	833.8	895.9
Phosphorus in S1	110.2	149.3
Nitrogen in 02	391.6	584.1

TABLE 2. Predicted mean annual increments over two years for sludge and control treatments.

Treatment	Predicted MAI (m³/ha·yr)	10% Confidence Interval
Control	3.57a	± 0.3403
Sludge	4.62b	± 0.4998

Predictions followed by the same letter are not significantly different at an alpha = 0.10 level.

treatment means into the regression equation yielded the results presented in Table 2. The predicted sludge MAI of 4.62 m³/ha per year is greater than the control MAI of 3.57 m³/ha per year by 29%. This suggests that sludge application has resulted in nutrient changes in soils and forest floor that may have long-term growth effects on this site, assuming that the two-year changes will persist or can be maintained by retreatment. A significant 44% increase in three-year basal area growth and a significant 63% increase in three-year diameter growth were found using conventional fertilizer trial techniques (Nguyen et al., this volume).

The difference between treatments exceeded the confidence intervals for the sludge and control predictions and was therefore statistically significant. The growth predicted for the sludge treatment was 8.5% higher than the maximum MAI from the regional study (4.62 compared with 4.26 m³/ha per year). This is an extrapolation of the data, and results should be interpreted cautiously.

It should be pointed out that although the predicted sludge MAI was beyond the range of the 29 regional stands, it was not greater than measured growth in the region. Stands in the region that had MAI measured in the summer of 1983 had MAIs up to 5.11 m³/ha per year. The small magnitude of the overrange, high statistical significance, and results of short-term observations tend to reinforce the interpretations reached here: there was a significant difference between sludge and control treatments equivalent to 1.05 m³/ha per year.

The use of MAI-nutrient resource relationships to assess potential changes in long-term growth from sludge is a new approach. Development of this approach and regression model are still being refined, and verification of the technique must be completed.

The MAI-nutrient resource relationship was also used to calculate changes in potential long-term growth with hypothetical retreatments of the oak site. This requires three major assumptions. First, the differences between sludge and control nutrient contents of components (Table 1) were assumed to represent two-year retentions of sludge-added nutrients on the oak site. Second, subsequent reapplications were assumed to have loading rates and retention rates equivalent to the initial loading and two-year retention. Third, nitrogen and phosphorus contents from previous sludge additions have different retention rates than nutrients and organic matter from newly applied sludge.

Figure 2 presents MAI predictions for initial application and three retreatments over eight years using three different retention rates of sludge more than two years old. Curve A assumes a retention rate of 100% for sludge nutrients after the second year. The MAI predicted in the eighth year using this assumption was 70% greater than the highest MAI observed in the 29 regional stands and represents an extrapolation that must be interpreted with caution. The 100% retention rate would result in maximum nutrient accumulations with no further degradation of applied organic matter and nutrient release. This may not be a reasonable assumption for a forest system. Curve B presents growth when sludge nutrients were assumed to have nutrient retention rates equivalent to those of the first two years after application. The MAI in curve B in the eighth year was 23% above the maximum sampled MAI. However, retention rates for the first two years after application may be unreasonably low for "older" (and probably more resistant) organic nutrient pools. For this study area, Hart et al. (1984) report that significant increases in nutrients and organic matter were found in the forest floor following sludge application. With this in mind, use of a 75% retention rate, which falls between the two extremes, seems a reasonable stand response as represented by curve C.

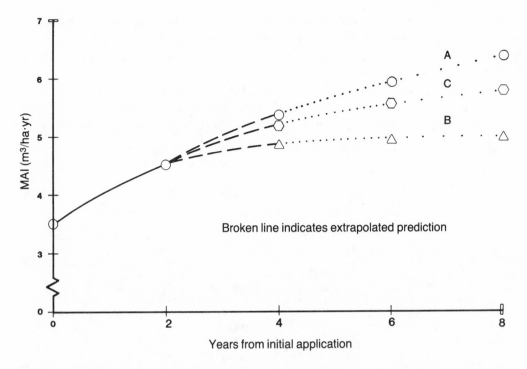

Figure 2. Hypothetical MAI curve: Changes over time with three different nutrient retention rates (A = 100% retention of sludge nutrients older than two years; B = retention of sludge nutrients older than two years equals initial retention rate; C = 75% retention of sludge nutrients older than two years).

From the first application onward, the MAIs exceed the range of MAIs in the 29-stand regional sample from which the regression was developed, and therefore constitute extrapolations that may limit interpretations. From the point of view of regional stand growth, curve B is the only one that does not exceed MAIs measured in the field (5.11 m³/ha·yr), although it does exceed the maximum growth rate of the regional study stands (4.26 m³/ha·yr). The figure should be interpreted based on the general form of the curve from a biological standpoint rather than strictly accepting the accuracy of the MAI predictions. The shape of the curve is similar to many growth curves with nutrient additions, and the model appears to be sensible from a biological point of view.

If no retreatments are made to the stand, one would predict that nutrient contents might decrease over time until they attain levels close to control levels. Stand growth should also decrease over time until it reaches a rate similar to the control. The length of this response period is not known.

In summary, site nutrient resources explained almost 70% of the variation in stand growth for 29 regional stands, and use of the nutrient resource approach with sludge and control plots predicted significant increases in growth with sludge application.

ACKNOWLEDGMENTS

Although the information in this document has been funded in part by the U.S. Environmental Protection Agency under assistance agreement No. S005551-01 to the Michigan Department of Natural Resources and Michigan State University, it has not been subjected to the Agency's publication review process and therefore may not necessarily reflect the views of the Agency, and no official endorsement should be inferred. Mention of trade names of commercial products does not constitute endorsement or recommendation for use.

Special appreciation is extended to Connie Bobrovsky, Barbara Kinnunen, Leslie Loeffler, Laurie Zwick, Michael and David Richmond, Robert Morell, and Brenda Ellens for their invaluable assistance in sample preparation, laboratory analysis, and data entry.

Michigan Agricultural Experiment Station journal article number 11864.

REFERENCES

Brockway, D. G. 1983. Forest floor, soil, and vegetation responses to sludge fertilization in red and white pine plantations. Soil Sci. Soc. Am. J. 47:776–784.

Draper, N. R., and H. Smith. 1981. Applied regression analysis. John Wiley and Sons, New York.

Ecosoft, Inc. 1984. Microstat, an interactive general-purpose statistics package. Release 4.0. Ecosoft, Inc., Indianapolis, Indiana.

Farrand, W. R. 1982. Quaternary geology of Michigan (map). State of Michigan Department of Natural Resources Geologic Survey.

Hart, J. B., P. V. Nguyen, C. W. Ramm, J. H. Hart, and D. M. Merkel. 1984. Ecological monitoring of sludge fertilization on state forest lands of northern lower Michigan. Annual progress report. Department of Forestry, Michigan State University, East Lansing.

Heiberg, S. O., and A. L. Leaf. 1961. Effect of forest debris on the amelioration of sandy soils. Rec. Adv. in Bot. 1622–1627.

Koterba, M. T., J. W. Hornbeck, and R. S. Pierce. 1979. Sludge application in a northern hardwood site. p. 252–268. In S. Torrey (ed.) Sludge disposal by landspreading techniques. Noyes Data Corp. 372 p.

Leaf, A. L., R. E. Leonard, J. V. Berglund, A. R. Eschner, P. H. Cochran, J. B. Hart, G. M. Marion, and R. A. Cunningham. 1970. Growth and development of Pinus resinosa plantations subjected to irrigation-fertilization treatments. p. 97-118. In Tree growth and forest soils. Proceedings of the Third North American Forest Soils Conference, August 1968, North Carolina State University, Raleigh. Oregon State University Press, Corvallis.

Soil Conservation Service. 1976. Rubicon series. Nat. Coop. Soil survey. USDA.

———. 1979. Graycalm series. Nat. Coop. Soil survey. USDA.

Technicon Industrial Method. 1977. Individual/simultaneous determination of nitrogen and/or phosphorus in BD acid digests. Method no. 334–74W/B. Technicon Industrial Systems, Tarrytown, New York.

White, E. H., and A. L. Leaf. 1964. Soil and tree potassium contents related to tree growth I: HNO_3-extractable potassium. Soil Sci. 98:395–402.

Zinke, P. J. 1960. Forest site quality as related to soil nitrogen content. p. 411–418. In Seventh International Congress of Soil Science, Madison, Wisconsin.

Chapter 28

Response of Loblolly Pine to Sewage Sludge Application: Water Relations

GAIL L. RIDGEWAY, LISA A. DONOVAN, and KENNETH W. McLEOD

ABSTRACT   Seasonal and diurnal measuring of xylem pressure potential (XPP) was done to determine if increased amounts of foliage on fertilized loblolly pine trees resulted in greater water stress, which might account for an observed altered needlefall pattern in fertilized plots. Experimental variables included two soil series and three levels of sewage sludge application. Xylem pressure potential of individual needles was determined using the pressure chamber technique. Seasonal and diurnal data indicated that the XPP of loblolly pine trees on fertilized plots on either Wagrum or Fuquay soils did not differ significantly from that of nonfertilized trees. Fertilization did not induce greater water stress, even though the trees supported a greater amount of foliage.

Soils of the coastal plain of the southeastern United States are usually very sandy with low cation exchange and water holding capacities. Because of this poor site quality, nutrient amendments are frequently needed to support timber production. In several field trials throughout the Southeast, inorganic fertilizer applications of nitrogen and phosphorous, and nitrogen and lime, resulted in increased growth in semimature loblolly pines (Fisher and Garbett 1980, Van Lear 1980). Organic fertilizers, such as sewage sludge, are an alternate form of nutrients being applied to some forested lands (Bengtson and Cornette 1973, Brockway 1983, Riekerk 1978, 1981, Sopper and Kardos 1973, Sopper and Kerr 1979).

In the fall of 1981, sewage sludge was applied to several loblolly pine (*Pinus taeda* L.) plantations on the Savannah River Plant (SRP) near Aiken, South Carolina. In the three years following sludge application, annual mass of needlefall in the fertilized plots was found to be greater, with more occurring in the summer and less in the autumn, than in adjacent unfertilized plots (Wells et al. 1984). Increases in tree height, diameter, and amount of foliage, and in competing understory biomass, were also observed in fertilized plots. Since all plots received equal precipitation, soil moisture levels were assumed to be similar or greater in unfertilized plots because of potentially greater water use by the increased quantity of foliage resulting from fertilization. On similar xeric coastal plain soils, Carter et al. (1984) found that removal of competing herbaceous and woody vegetation within proximity of loblolly pines decreased water stress, by increasing water and mineral nutrient availability.

Premature shedding of leaves, which reduces evaporative surface, has been suggested as an adaptive response to water stress (Levitt 1972). Therefore, it was hypothesized that the increased amount of foliage of the fertilized trees was being subjected to greater water stress, causing the premature drop of needles in the summer. To investigate this hypothesis, needle xylem pressure potential (XPP) was measured on a seasonal and diurnal basis in trees from these sewage sludge fertilized plots.

MATERIALS AND METHODS

Two plantations of loblolly pine located on the SRP were selected for study. Each plantation was growing on a different soil series (Wagrum or Fuquay). These soil series are typical of forested sites in the upper coastal plain and are well to slightly excessively drained, have a pH of 5 to 5.5, contain 200 to 500 mg/kg total N in the surface 10 cm and have a CEC of 0.01 to 0.02 moles/kg (McKee et al., this volume). Within each plantation, nine 50 m^2 plots were delineated. Sewage sludge was applied at either 0 (control), 400 (LAG), or 800 kg N/ha (HAG) to triplicate plots. These 10-year-old pines were in the second growing season since the liquid (2.5% solids) sludge was applied to the soil. A complete description of the fertilizer material and general soil description can be found in Wells et al. (1984) and McKee et al. (this volume).

Needle samples from each soil series and fertilizer treatment were collected between the hours of 1100 and 1200 on a weekly schedule from late June to early August 1983. In the seasonal study, all nine plots on a given soil series were sampled. For the diurnal study, one plot from each treatment was sampled at 1.5 hour intervals over a 15-hour period from dawn to dusk, alternating soil series from week to week. For each sampling, from each of three trees on each plot, three needles were taken of the previous year's growth approximately 2 m from the ground. A third aspect of the study described the pattern of xylem pressure potential at three crown positions (top, middle, and lower) within the tree. To minimize alterations in the pressure potentials due to transpiration, the needles were placed in polyethylene bags, rolled tightly, and stored in a cooled ice chest. A storage experiment showed that needles could be stored for at least four hours without statistically significant ($p < 0.05$) XPP change. Measurements were made in the field for the diurnal study. For seasonal data, the needles were returned to the laboratory and XPP measured. The pressure chamber method (Ritchie and Hinckley 1975) was used for XPP measurements.

As the study progressed, additional measurements of soil moisture and solar radiation were incorporated. Gravimetric soil moisture was determined for samples (0 to 30 cm) taken near each sample tree on the eighteen plots. Solar radiation levels incident to the forest floor were measured at 0.5 m height within the plantation with four pyrheliometers, two on each soil series, one each in a control and a heavily fertilized plot. Rainfall data were collected from the Savannah River Forest Station by C. E. Davis.

Data were analyzed using one- and two-way analysis of variance procedures in the Statistical Analysis System (SAS 1982). A Tukey's test was used to examine differences between treatment means (Kirk 1968).

RESULTS AND DISCUSSION

Since the soil-plant water continuum is already well documented (Hinckley et al. 1978, Kramer 1983), seasonal fluctuations in XPP reflect the influence of the local precipitation pattern on soil moisture content. Xylem pressure potential of trees in all treatments and soil series decreased from mid-June through early July in response to low precipitation (Figure 1). Several small (<1 cm) rain events caused a minor recovery period in XPP of

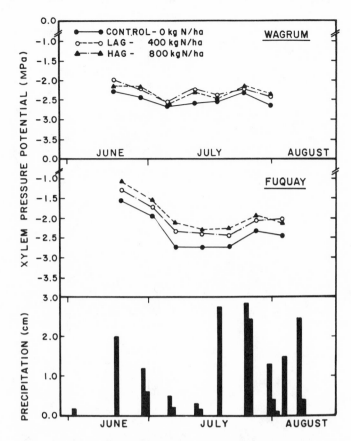

Figure 1. Xylem pressure potential (-MPa) of loblolly pine needles from plantations fertilized at 0, 400, and 800 kg N/ha (control, LAG, and HAG, respectively), and the seasonal rainfall pattern.

trees on the Wagrum soil but not at the Fuquay site. Only after several events of approximately 3 cm/event did the XPP of trees on the Fuquay site begin to increase. This response agrees with the seasonal XPP pattern found for longleaf pine (Ginter et al. 1979) and loblolly pine (Carter et al. 1984) on similar soil series. In the study, much lower seasonal fluctuations (-2.0 to -2.7 MPa) were observed in XPP of trees grown on the Wagrum site than on the Fuquay site. However, for both soil series, XPP of trees on fertilized plots was not statistically different from that of the unfertilized trees ($p > 0.05$).

The dawn-to-dusk XPP pattern (Figure 2) agrees with the pattern often reported for pine (Ginter et al. 1979, Cline and Campbell 1976, Hellkvist and Parsby 1976, Carter et al. 1984) and other species (Hinckley et al. 1978). For days where there was no precipitation (Figure 2, 7/6, 7/13, 7/21, and 7/27), XPP began a substantial decline after sunrise, with maximum water stress occurring shortly after solar noon. By sunset (≈ 2000 hr), XPP was returning to predawn XPP levels. On days where precipitation occurred (Figure 2, 6/22 and 6/29), the XPP recovered to values equal to or less than initial predawn values.

The diurnal studies also indicate no statistical differences ($p > 0.05$) in XPP due to treatment. All treatments followed the same general trend (Figure 2). Minimal XPP of trees on the two different soil series throughout the summer was -2.5 to -3.0 MPa. Response of the trees on the two different soils cannot be compared, since diurnal curves were not determined on the two soils on a common day.

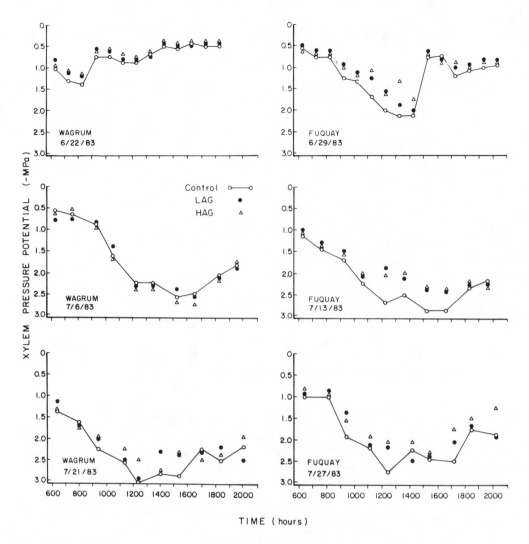

Figure 2. Diurnal xylem pressure potential (-MPa) of loblolly pine needles from fertilized plantations. See Figure 1 for abbreviations.

Fertilization treatments did not alter the relative differences in XPP due to crown position, with XPP tending to be slightly less negative as height increased. This is in contrast to the pattern of more negative XPP occurring with increasing height in the crown of longleaf pine (Ginter et al. 1979). The differences in pattern may result from different crown morphology of loblolly pine or the effort to measure only branches exposed to full sun. In this study, an open canopy still existed in this 8-year-old plantation, whereas Ginter et al. (1979) worked in the closed canopy of a 25-year-old plantation. Hinckley et al. (1978) attribute many crown position differences to differences in incident radiation and hence leaf temperature.

Base XPP, which is correlated with soil moisture and indicates the overnight water recharge (Ritchie and Hinckley 1975), did not differ significantly between treatments (Figure 3). The minimum number of observations and the relative time since rainfall preclude observation of the seasonal trend in base XPP. In agreement with base XPP measurements (Figure 3), there were no significant differences ($p > 0.05$) in soil moisture levels between treatments. The mechanism by which the trees maintain similar XPP with equal soil water content, while continuing to maintain different leaf areas, probably involves different stomatal activities. Fertilized trees must restrict stomatal opening, which reduces transpirational loss but also reduces photosynthetic capacity. Thus fertilized trees with their greater leaf area and production capacity are limited by their response to restrict water loss. It appears then that the fertilized trees are not experiencing greater water stress (as reflected by XPP) even though the fertilized trees support greater amounts of needles on the same soil water content as unfertilized trees.

A greater degree of shading in the fertilized plots was the only difference observed between treatments (Figure 4). Thus self-shading was probably responsible for the premature needlefall, which occurred during the three years following fertilization, including this study period. Regardless of what may have initiated the premature needlefall, the leaf area of the fertilized trees was reduced, but these trees still retained larger leaf areas than unfertilized trees.

In summary, loblolly pine trees on fertilized plots are not more water stressed than control trees; soil moisture was correlated with general trends in base XPP of loblolly

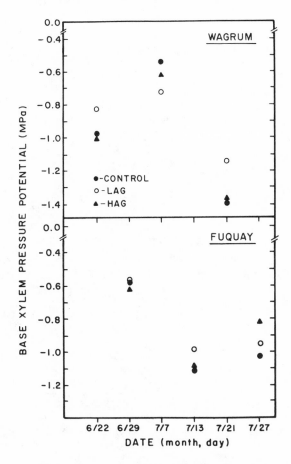

Figure 3. Base xylem pressure potential (-MPa) of loblolly pine needles from fertilized plantations. See Figure 1 for abbreviations.

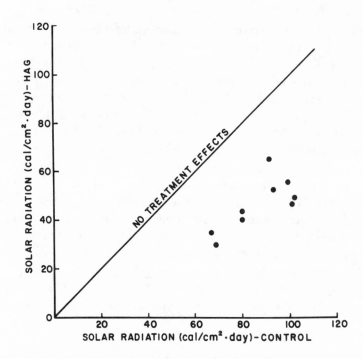

Figure 4. Solar radiation (cal/cm2·day) incident on the forest floor of control and fertilized (HAG) loblolly pine plantations. Each point represents the total daily radiation received at the two pyrheliometers. Points below the "no treatment effects" line indicate radiation received in control plot is greater than in HAG plot. Points above the line indicate greater solar radiation received in HAG than control plot.

pines; and factors other than water stress are responsible for early litterfall in the study plots. Sewage sludge fertilization alleviated a nutrient limitation, leading to greater productivity, but a second (water) limitation occurred before the full production capacity, of the greater leaf area caused by fertilization, could be realized.

ACKNOWLEDGMENTS

The authors gratefully acknowledge support by contract number DE-AC09-76SR00819 between the U.S. Department of Energy and the University of Georgia's Institute of Ecology. Field and laboratory assistance provided by T. G. Ciravolo, C. E. Davis, and K. C. Sherrod was greatly appreciated. A very constructive review by Dr. L. C. Lee is also acknowledged. G. L. Ridgeway was supported as a participant in the SREL Undergraduate Research Participation Program.

REFERENCES

Bengtson, G. W., and J. J. Cornette. 1973. Disposal of composted municipal waste in a plantation of young slash pine: Effects on soil and trees. J. Environ. Qual. 2(4):441–444.

Brockway, D. G. 1983. Forest floor, soil, and vegetation responses to sludge fertilization in red and white pine plantations. Soil Sci. Soc. Am. J. 47:776–784.

Carter, G. A., J. H. Miller, D. E. Davis, and R. M. Patterson. 1984. Effect of vegetative competition on the moisture and nutrient status of loblolly pine. Can. J. For. Res. 14:1–9.

Cline, R. G., and G. S. Campbell. 1976. Seasonal and diurnal water relations of selected forest species. Ecology 57:367–373.

Fisher, R. F., and W. S. Garbett. 1980. Response of semimature slash and loblolly pine plantations to fertilization with nitrogen and phosphorus. Soil Sci. Soc. Am. J. 4:850–854.

Ginter, D. L., K. W. McLeod, and C. Sherrod, Jr. 1979. Water stress in longleaf pine induced by litter removal. For. Ecol. and Manage. 2:13–20.

Hellkvist, J., and J. Parsby. 1976. Water relations of *Pinus sylvestris*. Physiol. Plant. 38:61–68.

Hinckley, T. M., J. P. Lassoie, and S. W. Running. 1978. Temporal and spatial variations in the water status of forest trees. For. Sci., Monograph 20.

Kirk, R. E. 1968. Experimental design: Procedures for the behavioral sciences. Brooks/Cole Publ. Co., Belmont, California. 577 p.

Kramer, P. J. 1983. Water relations of plants. Academic Press, New York. 489 p.

Levitt, J. 1972. Responses of plants to environmental stresses. Academic Press, New York. 697 p.

Riekerk, H. 1978. The behavior of nutrient elements added to a forest soil with sewage sludge. Soil Sci. Soc. Am. J. 42:810–816.

———. 1981. Effects of sludge disposal on drainage solutions of two forest soils. For. Sci. 27:792–800.

Ritchie, G. A., and T. M. Hinckley. 1975. The pressure chamber as an instrument for ecological research. Adv. Ecol. Res. 9:165–254.

SAS Institute Inc. 1982. SAS user's guide: Basics. SAS Institute, Cary, North Carolina. 923 p.

Sopper, W. E., and L. T. Kardos (eds.) 1973. Recycling treated municipal wastewater and sludge through forest and cropland. Pennsylvania State University Press, University Park. 479 p.

Sopper, W. E., and S. N. Kerr (eds.) 1979. Utilization of municipal sewage effluent and sludge on forest and disturbed land. Pennsylvania State University Press, University Park. 537 p.

Van Lear, D. H. 1980. Effects of nitrogen, phosphorus, and lime on the forest floor and growth of pole-size loblolly pine. Soil Sci. Soc. Am. J. 44:383–841.

Wells, C. G., K. W. McLeod, C. E. Murphy, J. R. Jensen, J. C. Corey, W. H. McKee, and E. J. Christensen. 1984. Response of loblolly pine plantations to two sources of sewage sludge. p. 85–94. *In* 1984 TAPPI Research and Development Conference, Appleton, Wisconsin. Technical Association of the Pulp and Paper Industry, Technology Park, Atlanta, Georgia.

Chapter 29

Understory Response to Sewage Sludge Fertilization of Loblolly Pine Plantations

K. W. McLEOD, C. E. DAVIS, K. C. SHERROD, and C. G. WELLS

ABSTRACT Loblolly pine (*Pinus taeda* L.) plantations were fertilized with sewage sludge to determine the growth response, and environmental effects, of sewage sludge additions to coastal plain soils of the southeastern United States. Four ages of loblolly pine plantations (establishment, 3, 8, and 27 years old) growing on a range of soils were treated with approximately 0, 400, or 800 kg N/ha of a liquid municipal sludge, or 630 kg N/ha of solid industrial sludge. Understory vegetation was sampled and divided into four components: grass, herbaceous, shrub and vine, and woody. The understory response is important owing to (1) competition with pine and management needed to control this competition, and (2) potential uptake and bioaccumulation of heavy metals. Understory biomass production was increased by sewage sludge fertilization in all age classes. The herbaceous component showed the greatest response, but increases in the shrub-vine and woody components indicate a potential long-term effect. Although an herbicide treatment was used in split plots of the establishment and 3-year-old plantations, understory production was still increased by sludge addition. Understory response in the establishment, 3-, and 8-year-old plantations could pose management problems owing to increased competition and fire fuel loading. In addition, heavy metal concentrations (primarily cadmium) of the understory increased in the sludge-fertilized plots. It is suggested that application rates not exceed 400 kg N/ha to minimize heavy metal concentrations and maximize nutrient utilization. A more effective understory control will be necessary if sludge is applied to plantations of less than 8 years of age.

Sewage sludge is a resource with both positive and negative attributes, containing both large quantities of nutrients needed for plant growth and potentially harmful pathogens and toxic materials (Bastian 1977, Weaver et al. 1977). To use sewage sludge wisely, we must maximize its potential benefit while minimizing its negative impact. The presence of pathogens and toxic materials has generally restricted the use of sewage sludge on agricultural lands (Weaver et al. 1977). However, the forest industry uses large quantities of fertilizer, and forest products do not directly enter the human food chain. Hence forests are potentially an ideal recipient for sewage sludge. The obvious benefit of the sewage sludge as a fertilizer, at a cost usually lower than inorganic fertilizers, must be weighed against the possible negative factors. Benefits of sewage sludge disposal in forested lands, other than fertilization, include (1) renovation of the sludge by immobilizing nutrients and heavy metals at exchange sites in the forest floor and soil, or incorporation into plant tissue, (2) increase of wildlife cover and browse, and (3) odor control through rapid incorporation into the forest floor and soil system (Loehr et al. 1979). Negative effects of sludge addition include (1) nutrients and heavy metals applied at excessive rates which degrade soil or water quality, (2) excess growth response of understory, and (3) possible phytotoxicity due to heavy metal incorporation. Therefore, application rates must be such that the forest system can assimilate or fix the nutrients (especially nitrates) and harmful heavy metals, thus preventing their entry into groundwater (Breuer et al. 1979). Following fertilization the release of understory vegetation frequently occurs (Brockway 1983). This not only increases competition for water and nutrients but also provides a sink for heavy metals and nitrates, potentially reducing leaching problems (Loehr et al. 1979). Especially in younger plantations, increased competition can be severe, and can affect plantation tree growth. In addition, the greater understory biomass increases the fire hazard.

The benefit of fertilization increases as the inherent fertility of the soil decreases. In the southeastern coastal plain, soils are characteristically acidic, sandy in texture, with low water holding and cation exchange capacities. Sewage sludge fertilization on forest lands of the southeastern coastal plain should be particularly advantageous, owing to the low soil fertility; however, the assimilation rate may dictate a low application rate to avoid groundwater contamination.

To address the feasibility of sewage sludge application to coastal plain soils, a large-scale forest fertilization program was initiated on the U.S. Department of Energy's Savannah River Plant, near Aiken, South Carolina. This paper discusses the response of the understory vegetation in loblolly pine plantations to fertilizer additions, gauged by biomass production and elemental content, and taking into account the potential problems associated with understory release, including competition, fire hazard, and uptake of heavy metals.

MATERIALS AND METHODS

The overall experimental design involved several soil series, four ages of loblolly pine (*Pinus taeda* L.) plantations, two sources of sewage sludge, an inorganic fertilizer, and different rates of fertilization. Soil series, treatments, and treatment acronyms for the four plantation ages used (establishment, 3, 8, and 27 years old at treatment), are given in Table 1. Four soil series (Fuquay, Lucy, Orangeburg, and Wagrum) were selected to represent a range of soils typical of the southeastern coastal plain. These soil series are light to medium in texture and moderate to excessively well drained. Within each plantation age, except at establishment, two soil series were used which represented differing characteristics. Augusta sludge was an anaerobically digested liquid (2.5% solids), primarily of domestic waste. Aiken sludge was aerobically digested, dewatered to about 55% solids, and consisted of both domestic and textile mill wastes. More complete descriptions of both soils and sludges are found in Wells et al. (1984) and McKee et al. (this volume). All sludges were surface applied with some specific treatments in the establishment plots incorporating the sludge by disking. Split plots in the establishment and 3-year-old plantations tested the effect of herbicide (Velpar®) control of understory growth. The herbicide (0.8 kg/ha) was applied as a liquid to the establishment plots, while in the 3-year-old plantations it was applied in granular form at 6.7 kg/ha. All treatments were repeated on two soil series except for the establishment plots, where only the Orangeburg soil was used.

Within the 0.07 ha split plots of the establishment and 3-year-old plantations, three randomly selected subplots per split plot were harvested to determine understory biomass during September of 1982 and 1983. Understory vegetation, defined as all vege-

TABLE 1. Soils and treatments used in this study.

			Plantation Age		
Soils and Treatments	Acronym	Establishment	3 years old	8 years old	27 years old
Soil Series					
Wagrum		-	-	X	-
Orangeburg		X	X	-	X
Lucy		-	X	-	-
Fuquay		-	-	X	X
Treatment*					
Control	CONT	3**	3	3	3
Sludge:					
Augusta at about 400 kg N/ha	LAG	3	2	3	2
plus disk incorporation	LAGD	3	-	-	-
Augusta at about 800 kg N/ha	HAG	3	2	3	2
plus disk incorporation	HAGD	3	-	-	-
Aiken at about 630 kg N/ha	LAK	3	2	-	2
plus disk incorporation	LAKD	3	-	-	-
Inorganic:					
Diammonium phosphate at 25 kg N/ha	LDAP	3	-	-	-
Diammonium phosphate at 50 kg N/ha	HDAP	3	-	-	-

*Treatments used for each soil type within that age class.

**Value is number of replicate plots used per treatment. Establishment, 3-, 8-, and 27-year-old plantations had 27, 9, 9, and 9 total plots per soil type, respectively.

tation less than 2.5 cm dbh other than the planted pines, was divided into grass, herbaceous, shrub and vine, and woody categories. The woody category included any plants that could eventually have a dbh of 2.5 cm or greater. Seedlings of volunteer pine and hardwoods were included in this category. In the establishment and 3-year-old plantations, no vegetation exceeded this limit. Subplots varied in size: herbaceous, shrub and vine, and woody samples were collected from 0.85 m² subplots, while the grass subplot was 0.10 m² nested in the upper left corner of the larger subplot. Vegetation was clipped at ground level, sorted, and oven-dry weight (65°C until constant weight) determined. The methods were identical in the 8- and 27-year-old plantations, except that six randomly selected subplots were sampled per 0.20 ha main treatment plot. Sampling was conducted on plots of all soil series in 1982. Only plots of the heavier soil series were sampled in 1983.

It became obvious that natural pine regeneration under the canopy of the 27-year-old plantations was being affected. Therefore, in the summer of 1985, pine seedlings were counted in fifteen 1 m² quadrats per main treatment plot to document these differences.

For elemental analyses, samples were composited by vegetation type and plot or split plot, if split plots existed, then ground in a stainless steel Wiley mill until passed through a 40-mesh stainless steel sieve. Kjeldahl-N was determined colorimetrically following block digestion with sulfuric acid. For the other elements, the homogenized samples were dried, ashed at 475°C, and dissolved in concentrated nitric acid. Phosphorus concentration was determined colorimetrically while metals were analyzed by flame atomic absorption. Biomass and elemental concentrations were tested for statistical significance ($p < 0.05$) using SAS one and two-way analysis of variance and Tukey's multiple range test (SAS 1982). Analysis of variance within each age class of plantation was conducted using herbicide treatment, year of sampling, vegetation type, fertilization treatment, and soil series as main effects (when appropriate), plus the appropriate interaction terms. If significant interaction terms existed ($p < 0.05$), then treatment effects were examined with one-way analysis of variance.

RESULTS

Biomass: Establishment Plantation

Total understory biomass production in the establishment plantation with no herbicide during the first year following fertilization was greatest in the HAGD treatment at 717.5 g/m² (260% of CONT, Figure 1). Biomass production decreased in the order of

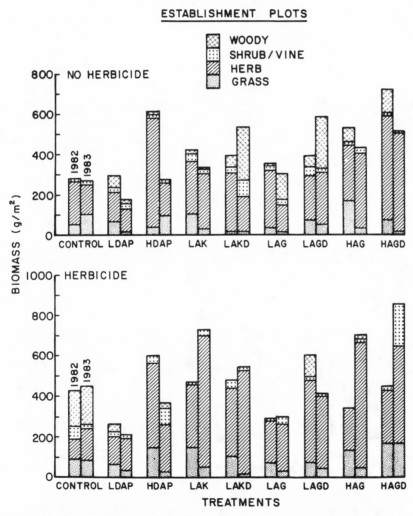

Figure 1. Understory biomass (g/m²) in treated establishment plantations. See Table 1 for treatment acronyms. The left bar within each treatment is biomass for 1982; the right bar indicates biomass for 1983.

HAGD > HDAP > HAG > LAK > LAKD > LAGD > LAG > LDAP > CONT. In the herbicide-treated split plots, maximum biomass in the first year was 601.6 g/m² (LAGD), 140% greater than CONT. The least understory biomass production was generally in the CONT, LDAP, and LAG treatments. Increasing the rate of Augusta sludge or inorganic fertilizer application usually resulted in greater understory biomass. The only statistically significant effects of the fertilizer treatments were higher herbaceous biomass in 1983 in the HAG than CONT plots and higher total understory biomass in the HAGD than CONT. There was no effect of herbicide on total biomass or biomass of any separate vegetation type except for an increase in herbaceous biomass in 1983 in the split plot treated with herbicide in 1982. There were no statistical differences in either the separate components or total biomass between 1982 and 1983.

The bulk of the understory biomass was herbaceous, with grasses secondary. Shrub-vine and woody components were generally of lesser importance but were locally abundant in some treatment-with-herbicide combinations. For instance, the woody biomass (~150 g/m²) in the CONT with herbicide plots was about 40% of the total biomass, and this persisted between years. The shrub-vine and woody production accounted for the entire difference in the CONT plots between treatments with and without herbicide.

Three-year-old Plantations

Overall biomass production (g/m²) was similar in the establishment and 3-year-old plantations. Shrub-vine and woody components accounted for more of the biomass production in the 3-year-old plots, especially in the treatments without herbicide (Figure 2). Greatest biomass production was due to the HAG treatment within the plots without

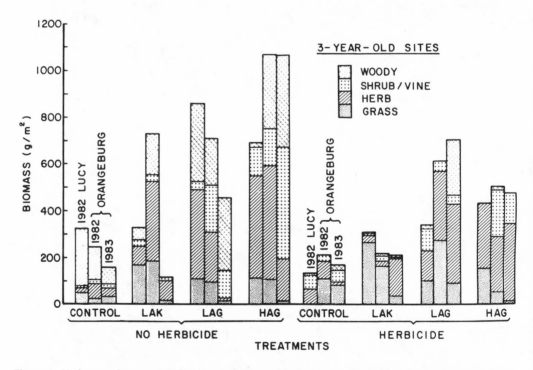

Figure 2. Understory biomass (g/m²) in treated 3-year-old plantations. See Table 1 for treatment acronyms. Within each treatment, the bars left to right indicate biomass at the Lucy site in 1982, at the Orangeburg site in 1982, and at the Orangeburg site in 1983.

herbicide and the LAG treatment in the herbicide plots. The LAK treatment showed little difference from CONT. There were significant effects on total biomass as a result of herbicide treatment in 1982 but not 1983, and as a result of fertilizer treatment in both years. The herbicide treatment reduced the total biomass by 44% (583 versus 325 g/m²) during the first year. In both 1982 and 1983, HAG had statistically greater total biomass than the CONT treatment. The LAG treatment was also statistically greater than the CONT but only in 1982. There were no significant differences in total biomass between the two soil series.

Analysis of the individual vegetation components revealed no statistical differences caused by soil, treatment, or herbicide on grass biomass. In the herbaceous component on the Orangeburg soil, the HAG treatment had statistically greater biomass than CONT, with no difference due to herbicide treatment. On the Lucy soil, statistical increases in herbaceous biomass occurred both with no herbicide and with the HAG fertilization. Both the LAG and HAG treatments had statistically greater shrub-vine biomass on the Orangeburg soil, but no differences were observed on the Lucy soil. The only overall difference in woody biomass was a reduction due to herbicide treatment.

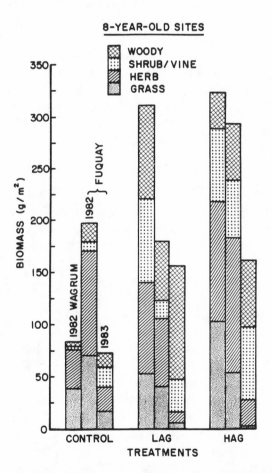

Figure 3. Understory biomass (g/m²) in treated 8-year-old plantations. See Table 1 for treatment acronyms. Within each treatment, the bars left to right indicate biomass at the Wagrum site in 1982, at the Fuquay site in 1982, and at the Fuquay site in 1983.

Eight-year-old Plantations

Total understory biomass production in the 8-year-old plantations was about half that of the younger plantations (Figure 3). Both sludge fertilization treatments used in the 8-year-old plantations on Wagrum soil statistically increased understory biomass; however, the response on the Fuquay soil was not significant. Without fertilization, the Wagrum soil supported about 50% of the total understory biomass of that found on the Fuquay soil. Following fertilization, the Wagrum soil supported greater understory biomass than the Fuquay soil did. Statistically significant decreases in biomass on the Fuquay site were observed from 1982 to 1983, with losses occurring primarily in the grass and herbaceous components, while shrub-vine and woody components increased. The herbaceous, short-lived species were being replaced by long-lived, woody components. Within the component vegetation types, the only statistical differences were an increase in the grass biomass in 1983 and a decrease in the shrub-vine biomass in 1982, both associated with the HAG treatment. There were no differences associated with soil type in either total biomass or biomass of individual vegetation types.

Twenty-seven-year-old Plantations

Total understory biomass production decreased as the age of the plantation increased. The 27-year-old plantations were the only ones in this study where canopy closure had occurred. In the CONT plots, the vast majority of the understory was made up of the shrub-vine and woody components (Figure 4). No difference in total understory biomass production was found that could be attributed to soil type, but the HAG treatment did produce significantly greater total biomass. The LAK treatment did not increase total biomass production. Shifts between proportions of vegetation types were evident, with herbaceous biomass production significantly increased in 1982 in the HAG treatments. By 1983, herbaceous biomass was similar to CONT levels.

Pine Seedlings

Trends in pine seedling numbers were similar for both soil types, with LAK having the greatest number, followed by CONT, LAG, and HAG. Apparently the solid Aiken sludge provided a good substrate for pine germination, being statistically different at the Orangeburg site from the other three treatments (LAK = 14.2, CONT = 1.9, LAG = 1.3, HAG = 0.2 seedlings/m^2) and at the Fuquay site from LAG and HAG (LAK = 1.6, CONT = 0.9, LAG = 0.7, HAG = 0.2 seedlings/m^2). No statistical differences in seedling numbers were observed between CONT, LAG, and HAG on either soil type. More seedlings were generally found in the Orangeburg site than in the Fuquay site. The lower stem density in the Orangeburg site—428 versus 542 stems/ha (W. H. McKee, pers. comm.)—allowed greater light penetration and increased forest flow light levels and temperatures. This, combined with the solid sludge substrate, created a favorable environment for pine seed germination and establishment.

Elemental Concentrations. Elemental analyses of the 1983 understory samples have not been completed. Therefore, only the results from the 1982 sampling can be discussed. Since there were no differences in elemental concentrations between herbicide treatments, the data were combined within each fertilization treatment. The establishment plantation data showed no statistically significant interaction terms, thus the effects of vegetation type and treatment were tested directly. Significant differences in elemental concentrations existed among vegetation types for all age classes of plantations, but will

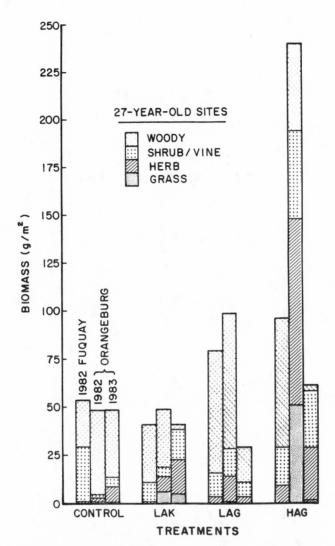

Figure 4. Understory biomass (g/m²) in treated 27-year-old plantations. See Table 1 for treatment acronyms. Within each treatment, the bars left to right indicate biomass at the Fuquay site in 1982, at the Orangeburg site in 1982, and at the Orangeburg site in 1983.

not be discussed since the primary concern of this research is the effect of fertilization treatments. Statistically significant ($p < 0.05$) differences existed in phosphorus, potassium, magnesium, copper, nickel, and zinc concentrations among treatments (Table 2) in the establishment plantation. All these differences were increased concentrations associated with fertilization except for potassium, which decreased in the LAKD treatment. Other elements, such as nitrogen, calcium, chromium, and lead, tended to have higher concentrations associated with sludge addition but were not statistically different, owing to large variability in data.

No significant interactions in the analysis of variance of the elemental concentrations from the 3-year-old plantations were found, allowing direct testing of treatment effects. Elevated concentrations of nitrogen, phosphorus, magnesium, and cadmium were associated with the HAG treatment, while increased concentrations of phosphorus, mag-

TABLE 2. Mean elemental concentrations of all understory vegetation types from the establishment plantation.

Element*					Treatment				
	CONT	LDAP	HDAP	LAK	LAKD	LAG	LAGD	HAG	HAGD
Nitrogen	0.96a	1.30a	1.01a	1.04a	1.11a	1.31a	1.27a	1.30a	1.11a
Phosphorus	1,311cd	1,759abcd	2,072a	1,235d	1,316bcd	2,435a	1,871abcd	2,062ab	2,019abc
Potassium	11,300ab	9,575abc	9,806abc	8,431bc	8,234c	10,503abc	11,416a	10,398abc	11,695a
Calcium	7,873a	8,491a	9,407a	7,239a	6,634a	7,945a	8,951a	7,771a	8,296a
Magnesium	1,837b	2,157b	2,471ab	1,968b	2,125b	2,493ab	3,443a	2,757ab	2,648ab
Manganese	355ab	228b	188b	303ab	255ab	338ab	374ab	348ab	433a
Sodium	166a	174a	160a	155a	128a	204a	157a	245a	210a
Cadmium	0.32ab	0.16b	0.25ab	0.27ab	0.23b	0.56a	0.30ab	0.46ab	0.55a
Chromium	0.57a	2.17a	2.03a	1.59a	0.95a	1.16a	1.51a	1.03a	0.88a
Copper	5.69b	6.51ab	7.71a	5.56b	4.88b	6.22ab	6.23ab	6.25ab	5.96ab
Lead	1.35a	1.17a	1.58a	1.10a	2.53a	1.42a	2.09a	1.78a	2.30a
Nickel	1.48b	1.42b	1.80ab	1.53b	1.65ab	1.90ab	2.24ab	1.93ab	2.58a
Zinc	28.6cd	56.3abcd	94.51a	22.9d	32.7bcd	85.0ab	54.48abcd	73.6abcd	80.1abc

Unlike letters within an element row indicate significant differences at the 5% level.

*Nitrogen concentration is %; all others are ppm.

TABLE 3. Mean elemental concentrations of all understory vegetation types from both soils of the 3-year-old plantations.

Element*	Treatment			
	CONT	LAK	LAG	HAG
Nitrogen	0.94c	1.09bc	1.26b	1.55a
Phosphorus	920b	1,433a	1,330a	1,393a
Potassium	7,467a	7,662a	7,598a	8,250a
Calcium	4,925ab	6,751a	4,395b	5,693ab
Magnesium	1,641b	2,399a	2,158ab	2,481a
Manganese	330a	374a	318a	444a
Sodium	121b	207ab	247ab	294a
Cadmium	0.37b	0.25b	1.06a	1.57a
Chromium	0.67a	0.79a	0.69a	0.59a
Copper	5.45b	7.27a	5.79ab	6.20ab
Lead	1.12a	1.11a	0.83a	0.82a
Nickel	0.92a	0.86a	0.83a	1.20a
Zinc	26.8a	47.4a	32.7a	44.8a

Unlike letters within an element row indicate significant differences at the 5% level.
*Nitrogen concentration is %; all others are ppm.

nesium, and copper were associated with the LAK treatment (Table 3). Other elements, again including zinc and sodium, show increased concentrations associated with treatment but were not statistically different.

Concentrations of sodium, cadmium, and copper in the understory vegetation in 8-year-old plantations showed significant statistical interactions with vegetation type, soil, or treatment. For these elements, direct treatment effects were examined by one-way analysis of variance of elemental concentrations of different vegetation types and soils. Of those elements without significant interaction terms, only nitrogen concentration in the HAG treatment was significantly greater than CONT (Table 4). No significant understory differences in sodium, cadmium, or copper concentrations due to treatment were found in the different vegetation types from the Wagrum site. Understory samples of grass from the Fuquay site exhibited statistically higher concentrations of cadmium in LAG and HAG treatments than in CONT. In the herbaceous vegetation type on the Fuquay soil, sodium and cadmium concentrations were significantly increased by the LAG and HAG treatments, while copper concentrations were significantly decreased. Sodium concentrations of the shrub-vine component were statistically greater in LAG and HAG treatments than in CONT, but copper concentrations in only the HAG treatment were greater than CONT. No significant differences among treatments were found in the woody component from either soil.

Of the elements analyzed from the understory samples of the 27-year-old plantations, phosphorus, sodium, cadmium, nickel, and zinc all showed significant interaction effects. These elemental concentrations were therefore statistically analyzed in the same manner as the data from the 8-year-old plantations. The results showed that cadmium concentrations of herbaceous and shrub-vine components from the Orangeburg site were elevated by the LAG and HAG treatments, but not by the LAK treatment. In the woody component from the Fuquay site, LAK, LAG, and HAG all had significantly greater phosphorus concentrations, while only LAG and HAG treatments had greater cadmium concentrations.

TABLE 4. Mean elemental concentrations of all understory vegetation types from both soils of the 8-year-old plantations.

	Treatment		
Element*	CONT	LAG	HAG
Nitrogen	1.24b	1.33ab	1.60a
Phosphorus	1,419a	1,419a	1,402a
Potassium	7,957a	8,010a	7,493a
Calcium	7,285a	7,102a	6,028a
Magnesium	1,713a	1,915a	1,942a
Manganese	272a	318a	299a
Sodium**	128.9	201.9	204.6
Chromium	0.48a	0.68a	1.00a
Cadmium**	0.31	1.01	1.25
Copper**	5.79	6.83	6.45
Lead	1.38a	1.62a	1.50a
Nickel	0.90a	1.14a	1.24a
Zinc	37.8a	47.8a	45.8a

Unlike letters within an element row indicate significant differences at the 5% level.

*Nitrogen concentration is %; all others are ppm.

**Main effects of treatment cannot be tested directly owing to significant interaction effects.

Aiken sludge addition increased only the copper concentration of those elements with no significant interaction terms (Table 5). The HAG treatment statistically increased nitrogen, magnesium, manganese, and copper concentrations, while the LAG treatment increased nitrogen, magnesium, and copper concentrations.

TABLE 5. Mean elemental concentrations of all understory vegetation types from both soils of the 27-year-old plantations.

	Treatment			
Element*	CONT	LAK	LAG	HAG
Nitrogen	0.98b	1.17b	1.58a	1.56a
Phosphorus**	921	1,418	3,054	2,470
Potassium	5,285a	6,567a	8,722a	8,280a
Calcium	6,483a	6,822a	8,080a	5,518a
Magnesium	1,307c	1,877bc	2,592ab	2,917a
Manganese	162b	198b	360ab	494a
Sodium**	116	204	160	180
Cadmium**	0.55	0.48	1.87	3.77
Chromium	0.80a	1.52a	1.03a	0.95a
Copper	4.22b	9.23a	8.13a	7.64a
Lead	0.97a	1.23a	1.00a	1.20a
Nickel**	1.17	1.17	1.57	1.59
Zinc**	44.5	57.4	104.0	137.4

Unlike letters within an element row indicate significant differences at the 5% level.

*Nitrogen concentration is %; all others are ppm.

DISCUSSION

With the increase in elemental concentrations and understory biomass associated with sludge fertilization, the standing crop inventory of elements contained in the forest understory increased. Increased concentrations of plant nutrients and heavy metals in the understory associated with fertilization treatments should be intuitive, and examples of this type of uptake are common (see symposia proceedings edited by Sopper and Kardos 1973, Sopper and Kerr 1979, Sopper et al. 1982, and Page et al. 1983). Elemental inventories as affected by increased elemental conncentrations and biomass are given for the HAGD and CONT treatments in the establishment plantation (Table 6). Elemental inventories in the HAGD treatment range from 1.70 (potassium) to 4.60 (zinc) times greater than CONT treatment. Thus both greater absolute and relative amounts of plant nutrients and heavy metals are found in the forest understory.

Based on nitrogen concentrations found in the grass and herbaceous components of the HAG treatments, nitrogen effects from the sludge additions increased forage quantity and quality. Phosphorus fertilization (Kinard 1977) and wastewater irrigation (Wood et al. 1973, Anthony and Wood 1979) have also shown increases in quantity and quality of wildlife forage.

Since not all plant species are equal in the crude protein to nitrogen concentration ratio, palatability, crude fiber content, or other factors that affect browse quality, and since preferred browse species change seasonally (Harlow and Hooper 1971), it is not known whether the fertilization actually improved the wildlife forage or possibly degraded it by excessive heavy metal amounts. Although this study was not designed to find an answer to that question, it can be said that even though the potential exists for forage degradation by heavy metal contamination, a major portion of those elements is not readily available for wildlife either because of storage in tissue not normally consumed or storage in nonpreferred browse species.

It has been suggested that sludge addition to forests could adversely affect forest wild-

TABLE 6. Standing crop (g/m^2) of biomass and elemental inventory* for CONT and HAGD treatments in the establishment plantation. HAGD/CONT ratio indicates proportion increased standing crop due to HAGD treatment.

Biomass and Elements (g/m^2)	CONT	HAGD	HAGD/CONT
Biomass	352.76	580.04	1.64
Nitrogen	3.39	6.44	1.90
Phosphorus	0.46	1.17	2.54
Potassium	3.99	6.78	1.70
Calcium	2.77	4.81	1.74
Magnesium	0.65	1.54	2.37
Manganese	0.13	0.25	1.92
Sodium	0.06	0.12	2.00
Cadmium	0.0001	0.0003	3.00
Chromium	0.0002	0.0005	2.50
Copper	0.0020	0.0034	1.70
Lead	0.0005	0.0013	2.60
Nickel	0.0005	0.0015	3.00
Zinc	0.010	0.046	4.60

*Computed as the summation of the weighted averages of all vegetation types.

life through heavy metal incorporation in the food chain (West et al. 1981). Brockway (1979) reported levels of 10.7 to 22.7 ppm for understory cadmium concentrations and suggested a possible food-chain transfer hazard. This study found that there were significantly elevated levels of cadmium in the 3-, 8-, and 27-year-old plantations. In the 27-year-old plantation, mean cadmium concentration in HAG treatment with the four vegetation types and both soils composited was 3.77 ppm. Control mean values were 0.55 ppm. For the herbaceous and grass vegetation types in these stands, cadmium concentrations ranged from 0.14 to 0.72 and 3.13 to 13.29 ppm, respectively. Cadmium concentrations in excess of 1 ppm have been indicated to be excessive for wildlife consumption and could constitute a potential hazard to the wildlife food chain (Allaway 1968, Logan and Chaney 1983). Therefore, a substantial quantity of cadmium was added to our plantations, potentially reducing the quality of the browse. The analyses of the 1983 understory samples will determine the persistence of the high cadmium levels from this one application. More research is needed on the food-chain transfer of heavy metals to wildlife.

Concentrations of heavy metals (cadmium, chromium, copper, manganese, nickel, lead, and zinc) in grasses from this study were similar or lower than concentrations reported by Horton et al. (1977) from the Savannah River Plant except for manganese and cadmium, which were slightly higher. Horton et al. (1977) were investigating the effect of distance from a coal-fired power plant and the associated fly ash deposition on plant heavy metal concentrations. Heavy metal concentrations from control plots also compared similarly with those reported by Kabata-Pendias and Pendias (1984) for grasses of the United States and other global locations. According to the toxicity levels listed by Allaway (1968), Logan and Chaney (1983), and Kabata-Pendias and Pendias (1984), manganese and zinc concentrations were close to the toxic range depending on the plant species. It is interesting to note that plant and animal cadmium toxicities differ such that cadmium appears to be more of a concern for animal consumption than for plant toxicity (Logan and Chaney 1983). We should also consider the fact that these soils, characteristic of the southeastern coastal plain, were loaded with up to 0.48, 69, 62, 15, 2.15, 8.79, and 108 kg/ha of cadmium, chromium, copper, manganese, nickel, lead, and zinc, respectively, yet plant toxicity levels were apparently not exceeded. This was a single application, and the cumulative effect of multiple applications cannot be assessed from this study. The relative uptake of the heavy metals shows that cadmium, nickel, and manganese were taken up proportionately greater than the other heavy metals relative to the quantity added. Zinc and particularly chromium were not found in proportion to other metals relative to the quantity added in the sludge. Thus, zinc and chromium seem to be less available to the plants relative to the other metals, either owing to soil sorption or root discrimination.

Greater competition for water and nutrients will result from the greater understory biomass. Carter et al. (1984) showed greater water and nutrient availability to loblolly trees when the competing vegetation was removed. Ridgeway et al. (this volume) showed no significant differences in xylem pressure potential due to sludge fertilization in the 8-year-old plantations of this study. Nutrient limitations might occur owing to increased competition by the understory vegetation, but, overall, nutrient capital has been increased. The degree of competition, as appraised by the amount of biomass, differed among ages of plantations, being most severe in the establishment and 3-year-old plantations, and less so in the 8- and 27-year-old plantations. The closed canopy of the

27-year-old plantations and the impending closure of the 8-year-old plantations will limit understory growth, but it is questionable whether the establishment and 3-year-old plantations can successfully compete with the increased understory vegetation. This is especially true in the 3-year-old plantations where many hardwood sprouts occurred prior to treatment.

A wildfire in any of the three youngest stands used in this study would destroy the entire plantation. The increased understory response due to fertilization would compound the wildfire hazard with its additional fuel loading. Additional fire prevention methods should be considered in the fertilized stands. From the fire management perspective, only the 27-year-old plantations do not have significantly increased fire management problems.

Optimal rate of sewage sludge application must be determined by the growth response and nutrient content of the plantation trees and understory, plus the elemental content of soil solution below the rooting zone. Groundwater monitoring will indicate if excess elements were leaching from the forest soil system, while the growth response associated with specific fertilization rates will indicate efficient nutrient use. Both leaching and efficient nutrient use will depend on the root mass of the forest. A large root mass will contribute to the soil's ability to fix nutrients and heavy metals. Thus a leachate collected below the rooting depth will show reduced elemental content owing to both soil fixation and root uptake. In younger plantations, root mass was low—made up primarily of understory roots. This would dictate lower application rates to avoid groundwater contamination. As the root mass of understory or plantation trees increased, application rates could be increased. Wells et al. (1984) found least nitrate leaching and hence more efficient nitrate assimilation at rates of 400 kg N/ha.

This study shows that understory biomass was greatest for the higher application rate but never proportionately greater than the low application rate. In addition, at the higher application rate, cadmium concentrations in the understory may be excessive, creating a hazard for wildlife consumption. From this perspective, application rates should not exceed 400 kg N/ha for the liquid Augusta sludge if fertilizer efficiency is the main criterion. Higher rates could be used if disposal is the main goal, depending on soil water or groundwater quality.

SUMMARY

The understory response to fertilization, whether inorganic or organic, was increased biomass and increased concentrations of certain elements. The herbicide did not significantly control the understory in the establishment plantation, but was more effective in the 3-year-old plantations. Peak understory biomass in all plantations occurred in the first year following fertilization, with lower biomass production during the second year. Increases in growth of perennials might lead to long-term persistent understory. The increased nitrogen content and biomass produced a greater quantity and quality of wildlife food; however, occasional increases in nickel, zinc, copper, and cadmium concentrations associated with sewage sludge treatments could reduce the food quality, depending on browse species involved or the storage of metals in the various plant tissues. Greater understory biomass also increases potential competition for water and nutrients, and increases the wildfire hazard. All of these variables need to be considered, along with tree-growth response and groundwater quality information, in making management decisions regarding sewage sludge fertilization of forest lands of the southeastern coastal plain that will maximize the beneficial use of sewage sludge.

ACKNOWLEDGMENTS

This research was supported under contract no. DE-AC09-76SR00819 between the U.S. Department of Energy and the University of Georgia's Institute of Ecology. Field and laboratory assistance by J. Batson, T. Ciravolo, J. Earnest, T. Perkins, K. Risher, A. Shumpert, and others is greatly appreciated. Reviews of this manuscript by T. G. Ciravolo, R. Wolf, and Dr. L. C. Lee contributed to its present form. Exceptional support by the drafting (J. Coleman and L. Orebaugh) and word processing (M. Stapleton, S. Hemmer, and C. Turnipseed) staffs through power outages and "short deadlines" was tremendously appreciated.

The use of trade, firm, or corporation names in this publication is for the information and convenience of the reader. Such use does not constitute an official endorsement or approval of any product or service to the exclusion of others that may be suitable.

REFERENCES

Allaway, W. H. 1968. Agronomic controls over the environmental cycling of trace elements. Adv. Agron. 20:235–274.

Anthony, R. G., and G. W. Wood. 1979. Effects of municipal wastewater irrigation on wildlife and wildlife habitat. p. 213–223. In W. E. Sopper and S. N. Kerr (eds.) Utilization of municipal sewage effluent and sludge on forest and disturbed land. Pennsylvania State University Press, University Park.

Bastian, R. K. 1977. Municipal sludge management. p. 673–692. In R. C. Loehr (ed.) Land as a waste management alternative. Ann Arbor Science Publishers, Ann Arbor, Michigan.

Breuer, D. W., D. W. Cole, and P. Schiess. 1979. Nitrogen transformation and leaching associated with wastewater irrigation in Douglas fir, poplar, grass, and unvegetated systems. p. 19–33. In W. E. Sopper and S. N. Kerr (eds.) Utilization of municipal sewage effluent and sludge on forest and disturbed land. Pennsylvania State University Press, University Park.

Brockway, D. G. 1979. Evaluation of northern pine plantations as disposal sites for municipal and industrial sludge. Ph.D. diss., Michigan State University, East Lansing.

———. 1983. Forest floor, soil, and vegetation responses to sludge fertilization in red and white pine plantations. Soil Sci. Soc. Am. J. 47:776–784.

Carter, G. A., J. H. Miller, D. E. Davis, and R. M. Patterson. 1984. Effect of vegetative competition on the moisture and nutrient status of loblolly pine. Can. J. For. Res. 14:1–9.

Harlow, R. F., and R. G. Hooper. 1971. Forages eaten by deer in the Southeast. Proceedings, Annual Conference of the Southeastern Association of Fish and Wildlife Agencies 25:18–46.

Horton, J. H., R. S. Dorsett, and R. E. Cooper. 1977. Trace elements in the terrestrial environment of a coal-fired powerhouse. DP-1475. DuPont de Nemours and Company, Savannah River Laboratory, Aiken, South Carolina. 49 p.

Kabata-Pendias, A., and H. Pendias. 1984. Trace elements in soils and plants. CRC Press, Boca Raton, Florida. 315 p.

Kinard, F. W., Jr. 1977. Phosphorus fertilization and nutrient composition of forage. Proceedings, Annual Conference of the Southeastern Association of Fish and Wildlife Agencies 31:24–28.

Loehr, R. C., W. J. Jewell, J. D. Novak, W. W. Clarkson, and G. S. Friedman. 1979. Role of vegetation cover. p. 97–108. In Land application of wastes. Vol. 1. Van Nostrand Reinhold, New York.

Logan, T. J., and R. L. Chaney. 1983. Utilization of municipal wastewater and sludge on land—metals. p. 235–323. In A. L. Page, T. L. Gleason III, J. E. Smith, Jr., I. K. Iskandar, and L. E. Sommers (eds.) Proceedings of the 1983 Workshop on Utilization of Municipal Wastewater and Sludge on Land. University of California, Riverside.

Page, A. L., T. L. Gleason III, J. E. Smith, Jr., I. K. Iskandar, and L. E. Sommers (eds.) 1983. Proceedings of the 1983 Workshop on Utilization of Municipal Wastewater and Sludge on Land. University of California, Riverside. 480 p.

SAS Institute Inc. 1982. SAS user's guide: Basics. Cary, North Carolina.

Sopper, W. E., and L. T. Kardos (eds.) 1973. Recycling treated municipal wastewater and sludge through forest and cropland. Pennsylvania State University Press, University Park. 470 p.

Sopper, W. E., and S. N. Kerr (eds.) 1979. Utilization

of municipal sewage effluent and sludge on forest and disturbed land. Pennsylvania State University Press, University Park. 537 p.

Sopper, W. E., E. M. Seaker, and R. K. Bastian (eds.) 1982. Land reclamation and biomass production with municipal wastewater and sludge. Pennsylvania State University Press, University Park. 524 p.

Svoboda, D., G. Smout, G. T. Weaver, and P. L. Roth. 1979. Accumulation of heavy metals in selected woody plant species on sludge-treated strip mine spoils at the Palzo site, Shawnee National Forest. p. 395–405. *In* W. E. Sopper and S. N. Kerr (eds.) Utilization of municipal sewage effluent and sludge on forest and disturbed land. Pennsylvania State University Press, University Park.

Weaver, D. E., J. L. Mang, W. A. Galke, and G. I. Love. 1977. Potential for adverse health effects associated with the application of wastewater or sludges to agricultural lands. p. 363–370. *In* R. C. Loehr (ed.) Land as a waste management alternative. Ann Arbor Science Publishers, Ann Arbor, Michigan.

Wells, C. G., K. W. McLeod, C. E. Murphy, J. R. Jensen, J. C. Corey, W. H. McKee, and E. J. Christensen. 1984. Response of loblolly pine plantations to two sources of sewage sludge. p. 85–94. *In* 1984 TAPPI Research and Development Conference, Appleton, Wisconsin. Technical Association of the Pulp and Paper Industry, Technology Park, Atlanta, Georgia.

West, S. D., R. D. Taber, and D. A. Anderson. 1981. Wildlife in sludge-treated plantations. p. 115–122. *In* C. S. Bledsoe (ed.) Municipal sludge application to Pacific Northwest forest lands. Institute of Forest Resources Contribution 41. College of Forest Resources, University of Washington, Seattle.

Wood, G. W., D. W. Simpson, and R. L. Dressler. 1973. Effects of spray irrigation of forests with chlorinated sewage effluent on deer and rabbits. p. 311–323. *In* W. E. Sopper and L. T. Kardos (eds.) Recycling treated municipal wastewater and sludge through forest and cropland. Pennsylvania State University Press, University Park.

Chapter 30

Remote Sensing Forest Biomass for Loblolly Pine Using High Resolution Airborne Remote Sensor Data

JOHN R. JENSEN, MICHAEL E. HODGSON, and HALKARD E. MACKEY, JR.

ABSTRACT   The amount and spatial distribution of biomass are important parameters useful in agriculture and forestry research. Biomass data may be collected in situ by harvesting a sample of the vegetation and determining the amount of biological matter. This approach provides estimates that are expensive to collect and difficult to extend through space in order to map the geographic distribution of biomass. It is possible to obtain estimates of biomass by analyzing remotely sensed data. This paper describes the analysis of high resolution, remotely sensed data of loblolly pine (*Pinus taeda*) in a controlled sludge application experiment in South Carolina. Pine plots on both sandy and clay soils were treated with sewage sludge to provide 0, 445, or 890 kg (approximately 0, 990, or 1,980 lb) of nitrogen per hectare. Indices of biomass from remote sensing data were significantly correlated with in situ biomass measurements in each plot. A ratio of infrared (0.9 to 1.1 μm) and red (0.65 to 0.70 μm) channels yielded the best correlation between the remote sensing data and the in situ biomass measurements. Also, remote sensing provided a quantitative map of the spatial distribution of the biomass in the forest plots.

The amount and spatial distribution of vegetation are important parameters in agriculture, forestry, and wetland research. Biomass data may be collected in situ by harvesting a sample of the vegetation and determining the amount of dry, wet, alive, or dead biological matter. This provides point estimates that are expensive to collect and difficult to extend through space to map the geographic distribution of biomass.

It is possible to obtain estimates of biomass from remotely sensed data. This paper describes the analysis of high resolution, remotely sensed data of loblolly pine (*Pinus taeda*) in a controlled sludge experiment in South Carolina. Loblolly pine represents over 70% of the commercial forest in the Southeast. The concept that remotely sensed data might provide accurate biomass information is not new (Curran 1980, Myers 1983). The studies supporting this concept are based on the use of low resolution, satellite data and a few field biomass measurements. For example, some remote sensing biomass studies used Landsat satellite multispectral scanner (MSS) data or NOAA Advanced High Resolution Radiometer (AVHRR) data and evaluated the results using only a few measurements (Harlan et al. 1979, Tucker et al. 1984). The satellite MSS data had a spatial resolution of 80 by 80 meters in four bands of the electromagnetic spectrum; the AVHRR data had a spatial resolution of 1.1 by 1.1 kilometers and measured reflectance in five spectral bands. Such low spatial resolution precluded any type of per tree analysis to document quantitatively the ability of remote sensing to perform accurate biomass estimation.

High resolution data were acquired for this study and represent a rigorous mechanism for biomass estimation.

METHODOLOGY

To evaluate the ability of high resolution, remote sensor data to measure biomass accurately it was necessary to (1) have access to unbiased in situ biomass information, (2) collect remote sensor imagery with sufficient spectral and spatial resolution, (3) model the remote sensor spectral information using a variety of biomass vegetation transformations, (4) determine how well the remote sensing biomass information correlated with the in situ data, and (5) produce a map of the spatial distribution of biomass.

In Situ Data Collection Experimental Design

In August 1980 an 8-year-old loblolly pine plantation in Aiken, South Carolina, was selected for this study. The site was divided into sandy and clay soil areas which were then subdivided into three treatment blocks (Figure 1). Sewage sludge was applied to the plots at the rate of 0 kg of nitrogen per hectare or 445 kg (990 lb) of nitrogen per hectare.

An in situ biomass measurement program was initiated in August 1982 to determine the biomass of the individual trees in the plots. The diameter at breast height (dbh) in centimeters and height of each tree (h) in meters were measured. In addition, sample trees in each plot were completely harvested, dried, and the total dry weight obtained as a measurement of per tree biomass (McKee et al. 1982). Such data allowed the following equation to be derived for the remainder of the trees in each plot:

$$\text{Biomass (grams)} = 3{,}626.25 \text{ grams} + \{[(22.47)(\text{dbh})]^2 * (\text{h})\}.$$

Per tree biomass values became the standard against which the remote sensing biomass estimates were compared.

Remote Sensing Imagery

To evaluate the ability of remote sensor data to provide biomass information, high spectral and spatial resolution imagery data were collected. Spectral resolution refers to the size and number of bands or channels to which the instrument is sensitive. Spatial resolution is the size of the instantaneous-field-of-view (IFOV) of the sensor—that is, the pixel size.

High resolution imagery was obtained using a Daedalus multispectral scanner operated for the Department of Energy by EG&G, Inc. Data were acquired at 609 meters (2,000 feet) above ground level (AGL). The upwelling radiant energy from each IFOV was measured in ten channels of the electromagnetic spectrum (Table 1). Eight channels were used in these analyses. These eight channels provided greater spectral resolution than aerial photography (Rennie and Cress 1974) or Landsat MSS data (Kan and Dillman 1975).

At this altitude the spatial resolution was 1.4 by 1.4 m per pixel. The average diameter of the loblolly pine tree crown was 3 m. Therefore, it was possible to identify at least one homogeneous pixel within each tree crown. Sensor systems such as Landsat MSS with 80 by 80 m IFOVs cannot provide this type of detail.

EXPERIMENTAL DESIGN

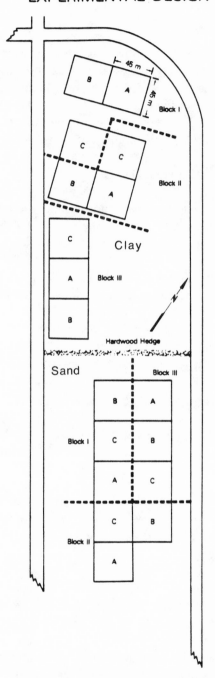

Figure 1. Map of the study area (not to scale) showing the location of the clay and sandy soil types, the blocks (I, II, and III separated by dashed lines) of the randomized block design, and the nitrogen treatment plots (A = no nitrogen applied; B = 160 kg of nitrogen per plot; C = 320 kg of nitrogen per plot). Each plot was 45 meters on a side. The area is near Aiken, South Carolina.

TABLE 1. Original spectral variables remotely sensed by the multispectral scanner and additional linear combinations of variables used to estimate biomass.

Variable	Original Spectral Variables	Description
a*	0.38 - 0.42 um	near ultraviolet
b*	0.42 - 0.45 um	blue
Chan1	0.45 - 0.50 um	blue
Chan2	0.50 - 0.55 um	green
Chan3	0.55 - 0.60 um	green/yellow
Chan4	0.60 - 0.65 um	orange/red
Chan5	0.65 - 0.70 um	red
Chan6	0.80 - 0.89 um	near-infrared
Chan7	0.90 - 1.10 um	near-infrared
Chan8	8.00 - 14.0 um	thermal IR
	Linear Combination Variables	
9	chan6/chan4	
10	chan6/chan5	
11	chan7/chan4	
12	chan7/chan5	
13	SQRT[(chan6-chan4)/(chan6+chan4)+.5]	
14	SQRT[(chan6-chan5)/(chan6+chan5)+.5]	
15	SQRT[(chan7-chan4)/(chan7+chan4)+.5]	
16	SQRT[(chan7-chan5)/(chan7+chan5)+.5]	
17	chan6-chan4	
18	chan6-chan5	
19	chan7-chan4	
20	chan7-chan5	
21	[2(chan7-chan6)-(chan4-chan2)]+40	
22	[2(chan7-chan6)-(chan5-chan2)]+40	
23	[2(chan7-chan6)-(chan4-chan2)]+40	
24	[2(chan7-chan6)-(chan5-chan3)]+40	
25	(chan1+chan4)/(chan6+chan7)	
26	(chan1+chan5)/(chan6+chan7)	
27	chan1/chan6	
28	chan1/chan7	
29	chan1/chan2	
30	chan1/chan3	
31	Greenness	

*Data obtained in channels a and b by the Daedalus DS-1260 multispectral scanner (MSS) were not used because of low signal-to-noise ratios caused by reduced bandwidths and atmospheric attenuation.

Black and white infrared images (channel 7 = 0.9 to 1.1 μm) of the clay and sandy soil plots are shown in Figures 2A and 2B. The sludge treatment plots are scribed in white on the images.

Modeling Remote Sensor Data Using Biomass Transformations

It is useful to review the fundamental logic of remote sensor biomass estimation. There is an inverse relation between the amount of biomass and the spectral reflectance from the plant in the blue and red portions of the spectrum (Figure 3). In healthy green pine needles, chlorophyll a and b absorb blue and red light for photosynthetic purposes; chlorophyll a at wavelengths of 0.43 and 0.60 μm and chlorophyll b at 0.45 and 0.65 μm (Curran 1980, Kondratyev et al. 1984, Leeman 1971, Weber and Polycyn 1972). The greater the amount of photosynthetic matter, the lower the reflectance in these chlorophyll absorption bands. Less absorption in the wavelengths between the two chlorophyll bands produces a peak at approximately 0.54 μm in the green portion of the electromagnetic spectrum (Gates et al. 1965). This causes the vegetation to appear green to human observers.

Reflected near-infrared energy in the spectral region from 0.74 to 1.1 μm is believed to be positively correlated with plant biomass (Tucker 1978). The amount of infrared reflectance in this region is controlled primarily by the spongy mesophyll cells in typical green leaves (Jensen 1983, Leeman 1971). Healthy green vegetation is characterized by high reflectance (45 to 50%), high transmittance (45 to 50%), and little absorptance (Sinclair et al. 1971). The increase in reflectance beginning at about 0.70 μm (Figure 3) is the result of scattering at the interfaces of the spongy mesophyll cell walls. As the biomass increases and the canopy becomes more dense, it has been suggested that the relation between near-infrared reflectance and biomass can be considered linear (Curran 1980).

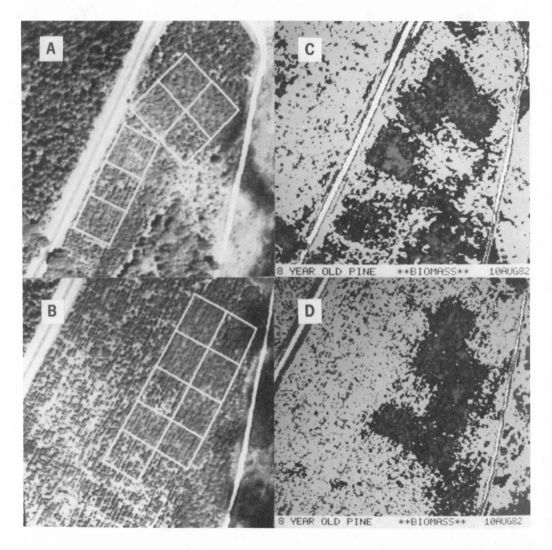

Figure 2. Black and white infrared images (channel 7, 0.9 to 1.1 μm) of the clay (2A) and sandy soil study areas (2B). The nitrogen treatment plots are scribed in white on the imagery. Images 2C and 2D are black and white biomass images of the clay and sandy soil study areas produced using data derived from variable 12 (an infrared to red ratio).

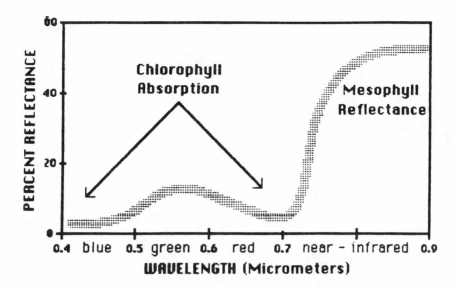

Figure 3. General spectral reflectance characteristics of healthy green leaves or pine needles for the wavelength interval 0.4 to 0.9 μm (modified after Deering et al. 1975, Jensen 1983, Kondratyev et al. 1984, Leeman 1971, Weber and Polycyn 1972).

Based on these relationships, numerous remote sensing biomass transformations (indices) have been developed that use measurements in the visible and near-infrared region. The more promising include the Normalized Difference Index (NDI), the Transformed Vegetation Index (TVI), the Perpendicular Vegetation Index (PVI), and the Green Vegetation Index (GVI) (Deering et al. 1975, Harlan et al. 1979, Holben et al. 1980, Jackson 1983, Kauth and Thomas 1976, Norwine and Greegor 1983, Richardson and Wiegand 1977). Most indices use measurements from at least one band in the near-infrared region from 0.7 to 1.1 μm and one band in the red region from 0.6 to 0.7 μm (Perry and Lautenschlager 1984, Tucker 1979). The result may be a linear combination that is more correlated with biomass than either red or reflective infrared measurement alone (Bartlett and Klemas 1980, Curran 1982 and 1983, Hardisky et al. 1983).

In this study, the in situ biomass measurements were correlated with (1) the raw spectral reflectance or emittance brightness values, and (2) the results of applying biomass transformations to the raw spectral data (Table 1). The remote sensing measurements were made by selecting random coordinates within each plot until they fell on the crown of an individual tree. At least four were selected from each plot. Then the raw brightness value or the scaled biomass transformation value for each pixel was extracted. These values derived from remote sensing were then compared with the in situ biomass measurements for the same trees. Thus 28 trees were sampled in the clay soil area and 36 were sampled in the sandy soil area.

Statistical Analysis

The following hypotheses were evaluated statistically with the field measurement of biomass as the dependent variable and the remote sensing biomass estimate as the independent variable: (1) There is no statistically significant correlation between field biomass estimates and estimates derived from remote sensing. (2) The type of soil (clay and sandy) has no significant effect on the remotely sensed biomass estimates. (3) The application of various sludge treatments resulting in the addition of nitrogen (0, 445, or 890 kg per hectare) has no significant effect on the remotely sensed biomass estimates. If the first hypothesis is rejected, a least squares equation could be derived to compute the biomass present for each pixel in the remotely sensed imagery.

RESULTS

Hypothesis 1

Numerous biomass transformations were significantly correlated with the field biomass measurements (Table 2). The most highly correlated was variable 12, which was computed as [infrared (0.9 to 1.1 μm)/red (0.65 to 0.7 μm)], i.e., channel 7/channel 5. For the clay soil this resulted in an r-squared of 0.67 or a correlation coefficient of 0.82. For the sandy soil, an r-squared of 0.63 or a correlation coefficient of 0.79 was realized. For both soils combined, an r-squared of 0.57 with a correlation coefficient of 0.76 was obtained. Jackson et al. (1982) concluded that an infrared/red ratio was a good indicator of vegetation quantity when agricultural ground cover was greater than 50%.

TABLE 2. Correlation of remote sensing biomass indices and in situ biomass measurements.

	Correlation Coefficients*		
Variable	Both Soils	Clayey	Sandy
Chan1	-0.489	-0.608	-0.521
Chan2	-0.580	-0.804	-0.582
Chan3	-0.575	-0.818(2)	-0.595
Chan4	-0.569	-0.790	-0.623
Chan5	-0.569	-0.791	-0.585
Chan6	0.345	0.444	0.403
Chan7	0.568	0.662	0.561
Chan8	0.528	0.633	0.533
9	0.598	0.706	0.712
10	0.621	0.728	0.706
11	0.720	0.803	0.766
12	0.760(1)	0.820(1)	0.790(3)
13	0.622	0.721	0.742
14	0.640	0.743	0.732
15	0.717	0.802	0.778
16	0.753(2)	0.817(3)	0.798(1)
17	0.551	0.629	0.667
18	0.576	0.663	0.672
19	0.691	0.763	0.721
20	0.720	0.782	0.745
21	0.480	0.514	0.468
22	0.499	0.531	0.483
23	0.492	0.517	0.479
24	0.512	0.535	0.494
25	-0.705	-0.739	-0.796(2)
26	-0.724(3)	-0.757	-0.706
27	-0.561	-0.573	-0.682
28	-0.708	-0.715	-0.763
29	0.505	0.690	0.516
30	0.538	0.725	0.627
31	0.706	0.754	0.763

*All coefficients were significant at the 0.05 level.

Several of the other indices were correlated with variable 12, suggesting that they might provide acceptable biomass estimation results. This is not surprising given the results of Perry and Lautenschlager (1984), who found equivalence in some indices. Therefore, because variable 12 (the infrared/red ratio) accounted for the most variance in predicting biomass and because it was correlated with several of the other indices, it was used to accept or reject hypotheses 2 and 3.

Hypothesis 2

In this study there was no significant difference between the field biomass measurements and the estimates derived from remote sensing for the sandy and clay soil types (Table 3). Thus all 16 plots (7 in clay and 9 in sandy soil) were treated as a single, homogeneous soil type for subsequent analyses. This suggests that for southeastern loblolly pine forests, the difference between clay and sandy soil is not a contributing factor in conducting remote sensing derived biomass estimates, as long as the canopy is sufficiently developed to obscure much of the underlying ground.

TABLE 3. Two-way analysis of variance to determine if soil type or sludge application had a significant effect on the remote sensing biomass estimates.

Source	Sums of Squares	Degrees of Freedom	Mean Standard Error	F
Soil type treatment	0.38	1	0.38	0.39*
Sludge treatment	12.13	1	12.13	12.24**
Interaction	0.09	1	0.09	0.09*
Error	59.52	60	0.99	--
Total	72.12	63	--	--

*Not significant at the 0.05 level.
**Significant at the 0.05 level.

Hypothesis 3

There was a significant difference between the remote sensing measurements obtained on treated and untreated pine plots (Tables 3 and 4). However, there was no significant difference between the results obtained for the treated plots (445 and 890 kg N/ha, Table 4). This suggests that once a certain level of sludge has been added, the effect of adding additional fertilizer to the area does not have a significant effect on either the creation of biomass or its spectral reflectance characteristics.

Computer Graphic Display

To prepare a thematic map depicting the spatial distribution of biomass for the study area, it was necessary to scale the remotely sensed data. First, because the best biomass transformation was derived from a ratio of brightness values, channel 7/channel 5, it was necessary to scale the data to a range from 0 to 255 compatible with the digital image processing systems. This involved the use of the following equations:

$$\text{If Old } 7/5 < 1 \text{ then New } 7/5 = [(\text{Old } 7/5 * 127) + 1], \text{ or}$$
$$\text{If Old } 7/5 > 1 \text{ then New } 7/5 = \{255 - [(1/\text{Old } 7/5) * (127)]\}.$$

TABLE 4. Analysis of variance to determine what levels of sewage sludge treatment had a significant effect on remote sensing biomass estimates.

Source	Sums of Squares	Degrees of Freedom	Mean Standard Error	F
[A]				
0 versus 445 kg N/ha	7.05	1	7.05	11.68**
Error	22.95	38	0.60	--
Total	30.00	39	--	--
[B]				
0 versus 890 kg N/ha	11.01	1	11.01	11.68**
Error	39.60	42	0.94	--
Total	50.61	43	--	--
[C]				
445 versus 890 kg N/ha	0.29	1	0.29	0.22*
Error	57.37	42	1.36	--
Total	57.66	43	--	--

*Not significant at the 0.05 level.
**Significant at the 0.05 level.

With the data scaled, it was possible to use the following least squares equation derived from an analysis of all 16 plots to relate the output of variable 12 to the in situ biomass measurement:

$$y = 0.0176 \times - 0.477.$$

This resulted in the creation of biomass images where each pixel (1.4 by 1.4 m) could take on a value from 0 to 4 kilograms of biomass. The biomass images (Figures 2C and 2D) were created using the class interval logic given in Table 5. Note the increase of biomass in the 445 and 890 kg N/ha in both the clay and sandy soil areas. The road and much of the region adjacent to the study areas exhibit very low biomass.

TABLE 5. Class intervals used in the creation of the 8-year-old loblolly pine biomass images.

Kilograms per Pixel	Gray Scale Code
0 - 1.90	Black [e.g., roads]
2.0 - 2.79	Darkest gray
2.8 - 2.89	Dark gray
2.9 - 2.99	Gray
3.0 - 3.09	Lighter gray
3.1 - 3.19	Lightest gray
3.2 - 4.00	White

SUMMARY

Biomass and its spatial distribution throughout a geographic region continue to be important information useful in many biogeographical problems. Unfortunately, such information is often difficult to acquire using field measurement techniques. This research has demonstrated that accurate biomass information can be extracted from aircraft remote sensor data of southeastern loblolly pine. In particular, an infrared/red ratio was highly correlated with the amount of biomass present. It has been shown that the remotely sensed biomass estimates are sensitive to sewage sludge application and that the type of soil (clay or sandy) did not influence the measurements. These results have significance for rapid, accurate measurement of loblolly pine biomass in the southeastern United States.

ACKNOWLEDGMENTS

The calibrated data tapes of the forest stands were obtained by EG&G, Inc., Las Vegas, Nevada, using a Daedalus multispectral scanner operated for the U.S. Department of Energy.

The information contained in this article was developed during the course of work under contract No. DE-AC09-76SR00001 with the U.S. Department of Energy.

REFERENCES

Bartlett, D. S., and V. Klemas. 1980. Quantitative assessment of tidal wetlands using remote sensing. Environ. Manage. 4:337–345.

Curran, P. J. 1980. Multispectral remote sensing of vegetation amount. Progress in Physical Geography 4:315–341.

——. 1982. Multispectral photographic remote sensing of green vegetation biomass and productivity. Photogrammetric Engineering and Remote Sensing 48:243–250.

——. 1983. Estimating green LAI from multispectral aerial photography. Photogrammetric Engineering and Remote Sensing 49:1709–1720.

Deering, D., J. W. Rouse, R. H. Haas, and J. A. Schell. 1975. Forage production of grazing units from Landsat MSS data. p. 1169–1178. In Proceedings, Tenth International Symposium on Remote Sensing of Environment. Environmental Research Institute of Michigan, Ann Arbor.

Gates, D. M., H. J. Keegan, J. C. Schletes, and V. R. Weidner. 1965. Spectral properties of plants. Applied Optics 4:11–12.

Hardisky, M. A., R. M. Smart, and V. Klemas. 1983. Seasonal spectral characteristics and above ground biomass of the tidal marsh plant, Spartina alterniflora. Photogrammetric Engineering and Remote Sensing 49:85–92.

Harlan, J. C., D. Deering, R. H. Haas, and W. E. Boyd. 1979. Determination of range biomass using Landsat. p. 101–115. In Proceedings, Thirteenth International Symposium on Remote Sensing of Environment. Environmental Research Institute of Michigan, Ann Arbor.

Holben, B., C. Tucker, and C. Fan. 1980. Spectral assessment of soybean leaf area and leaf biomass. Photogrammetric Engineering and Remote Sensing 46:651–656.

Jackson, R. D. 1983. Spectral indices in n-space. Remote Sensing of Environment 13:409–421.

Jackson, R. D., P. N. Slater, and P. J. Pinter. 1982. Discrimination of growth and water stress in wheat by various vegetation indices through clear and turbid atmospheres. Remote Sensing of Environment 12:200–230.

Jensen, J. R. 1983. Biophysical remote sensing. Annals of the Association of American Geographers 73:111–132.

Kan, E. P., and R. D. Dillman. 1975. Timber type separability in southeastern United States on Landsat-1 MSS data. p. 135–158. In Proceedings, NASA Earth Resource Symposium, Houston. NASA 1-A.

Kauth, R. J., and G. S. Thomas. 1976. The tassled cap: A graphic description of the spectral temporal development of agricultural crops as seen by Landsat. p. 41–51. In Proceedings, Symposium on Machine Processing of Remotely Sensed Data.

Kondratyev, K. Y., L. A. Grinenko, L. A. Fedchenko, and V. V. Kozoderov. 1984. Spectrophotometry of the chlorophyll content in leaves of plants. p. 53–56. In Proceedings, Symposium on Machine Processing of Remotely Sensed Data.

Leeman, V. 1971. The NASA Earth Resources Spectral Information System. National Aeronautics and Space Administration, Washington, D.C.

McKee, W. H., G. J. Hollod, and C. G. Wells. 1982. Influences of nutrient amendments on the growth and nutrient cycling of 8 year old loblolly pine for well drained sandy clayey soils. Report 4110. USDA Forest Service, Charleston. 25 p.

Myers, V. I. 1983. Remote sensing applications in agriculture. 2:2150–2151. In R. N. Colwell (ed.) The manual of remote sensing. American Society of Photogrammetry, Falls Church, Virginia.

Norwine, J., and D. H. Greegor. 1983. Vegetation classification based on advanced very high resolution radiometer (AVHRR) satellite imagery. Remote Sensing of Environment 13:69–87.

Perry, C. R., and L. F. Lautenschlager. 1984. Functional equivalence of spectral vegetation indices. Remote Sensing of Environment 14:169–182.

Rennie, J. C., and D. H. Cress. 1974. Multispectral imagery for detection of nutrient deficiencies in pine plantations. p. 345–354. *In* Proceedings, Fourth Annual Remote Sensing of Earth Resources Conference. University of Tennessee Space Institute, Tullahoma.

Richardson, A. J., and C. L. Wiegand. 1977. Distinguishing vegetation from soil background information. Photogrammetric Engineering and Remote Sensing 13:1541–1552.

Sinclair, T. R., R. M. Hoffer, and M. M. Schreiber. 1971. Reflectance and internal structure of leaves from several crops during a growing season. Agronomy J. 633:864–868.

Tucker, C. 1978. A comparison of satellite sensor bands for vegetation monitoring. Photogrammetric Engineering and Remote Sensing 44:1369–1380.

——. 1979. Red and photographic infrared linear combinations for monitoring vegetation. Remote Sensing of Environment 8:127–150.

Tucker, C., J. A. Gatlin, and S. R. Schneider. 1984. Monitoring vegetation in the Nile Delta with NOAA-6 and NOAA-7 AVHRR imagery. Photogrammetric Engineering and Remote Sensing 50:53–61.

Weber, F. P., and F. C. Polycyn. 1972. Remote sensing to detect stress in forests. Photogrammetric Engineering 38:163–175.

Program Implementation

Policies and Guidelines: The Development of Technical Sludge Regulations

ELLIOT D. LOMNITZ

ABSTRACT   Section 405 of the Clean Water Act requires the U.S. Environmental Protection Agency to develop and issue regulations that (1) identify uses for sludge including disposal, (2) specify factors to be taken into account in determining the measures and practices applicable for each use or disposal (including costs), and (3) identify concentrations of pollutants that interfere with each such use or disposal. In 1982 EPA established a Sludge Task Force to assess the magnitude of the task and the management approaches needed for municipal sludge reuse and disposal nationwide, and to evaluate the strengths and weaknesses of past regulatory activities. Based on the conclusions and recommendations of the Task Force, EPA is proceeding with a regulatory program to develop technical regulations for five major reuse and disposal options: distribution and marketing, land application to food-chain and nonfood-chain crops, landfilling, ocean dumping, and incineration. This paper discusses (1) the work plan and steps taken by EPA to develop such regulations, (2) the recommendations of the Sludge Task Force, and (3) data gaps and informational needs related to determining the health and environmental impacts of reusing or disposing of municipal sludge. This paper emphasizes the reuse option of land application.

The need for policies, regulations, and guidelines related to reuse or disposal of municipal sewage sludge has increased significantly in this decade as the management of sludge has become more complex. The complexity of sludge management is directly related to the larger volumes of sludge being generated because of (1) the installation of more sophisticated wastewater treatment systems and (2) population growth. Increasing public awareness of environmental issues has also created pressure for the development of regulations to ensure protection of human health and the environment. The purpose of this paper is to describe the U.S. Environmental Protection Agency's (EPA) current effort for issuing new sludge regulations. This paper concentrates on the development of the technical regulations and the associated scientific evaluations and briefly describes the proposed state management regulation related to municipal sludge programs.

EPA SLUDGE TASK FORCE RECOMMENDATIONS

In 1982 an EPA Sludge Task Force was formed and included representatives from all major EPA programs and functions. The objectives of this Task Force were to (1) evaluate the Agency's past regulatory framework for regulating municipal sludge, including an assessment of strengths and weaknesses, (2) assess the magnitude of the task and management approaches to municipal sludge reuse or disposal nationwide, and (3) delineate the data and research needs for future regulatory actions. The Task Force was also mandated to generate several reports, including (1) a policy statement that would establish a general EPA philosophy or position on the key issues of sludge management, (2) a general guidance document that would consolidate existing information on sludge management issues, and (3) an overall Agency work plan for improving its regulatory program and identifying the responsible entities.

In 1983 the Task Force presented its recommendations for improving the regulatory program related to municipal sludge and issued an Agency work plan. The Task Force recommended that two sets of sludge regulations be issued under Section 405 of the Clean Water Act. Section 405 requires EPA to develop and issue regulations that (1) identify uses for sludge including disposal, (2) specify factors to be taken into account in determining the measures and practices applicable for each use or disposal, and (3) identify concentrations of pollutants that interfere with each such use or disposal. Using this authority, the Task Force recommended that a set of technical regulations be generated by the Office of Water Regulations and Standards (OWRS) and that state management regulations should be issued by the Office of Municipal Pollution Control (OMPC). The Task Force envisioned that these regulations would provide a comprehensive regulatory framework and set of guidelines needed by state decision makers to make informed decisions on sludge reuse or disposal options.

OWRS PROGRAM FOR DEVELOPING TECHNICAL REGULATIONS

Based on the recommendations of the Task Force, OWRS is currently proceeding with a regulatory program to develop technical regulations for five major reuse or disposal options: distribution and marketing, land application (including agriculture, silviculture, reclamation, and dedicated sites), landfilling, ocean dumping, and incineration. Several key questions need to be considered while developing these regulations: (1) Which pollutants in municipal sludge can pose a human health or environmental concern for each of the reuse or disposal options? (2) What are the levels at which such pollutants pose risk, and should numeric criteria or limits be established? (3) Can management practices be established that would diminish risks? (4) Are there site-specific factors related to fate and transport that need to be assessed? (5) What are the relative risks between the various reuse and disposal options? In order to answer these key questions, OWRS developed an overall work plan for developing the five regulations. Figure 1 delineates the major steps being pursued to generate these regulations.

The initial step in this regulatory development process is to identify pollutants that may interfere with each reuse or disposal option because of environmental or health considerations (Work Element 1). This step was accomplished by (1) convening expert committees on each of the options and (2) developing environmental profiles containing data compilations for a specific pollutant as well as a set of hazard indices. The output from this step is a list of pollutants of potential concern for each reuse and disposal option.

Upon completion of Work Element 1, the development of risk assessment methodologies commenced (Work Element 2). The development of methods and procedures for determining maximum permissible contaminant levels and identifying management practices is a key step in the process. Such methods and procedures are complex, owing to the number of environmental pathways as well as the need for extensive modeling.

Once the risk assessment methodologies are completed, the derivation of criteria, ap-

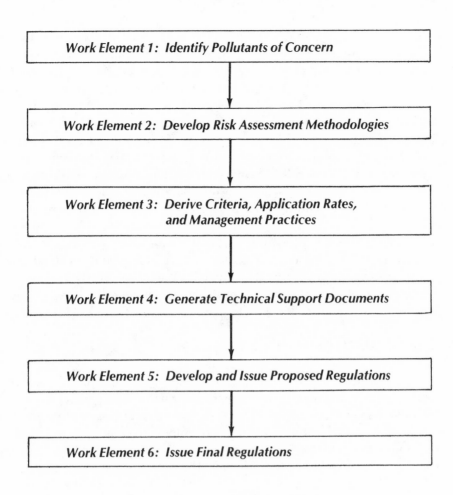

Figure 1. Steps for developing sludge regulations.

plication rates, and management practices can be accomplished (Work Element 3). The pollutants of concern will then be analyzed for degrees of risk for specific application rates or input rates by varying the inputs into the models and methodologies. The output of this step may be either maximum numeric criteria (e.g., sludges containing X μg/kg may not be land applied) or management practices (e.g., incinerators should be operated in such a manner).

Work Element 4 consists of assembling all the data and analyses conducted in the first three work elements as well as compiling data on (1) the fate and transport of pollutants, (2) site-specific factors and variability, (3) economic evaluations and analyses of benefits, and (4) compliance monitoring requirements. All the data and analyses will be incorporated into a technical support document, one for each reuse or disposal option. This document will serve as the background document for each of the regulations and will be subject to public comment with the proposed regulation. Upon completion of the technical support documents, the actual regulation writing can commence, ultimately leading to proposal in the Federal Register (Work Element 5).

In developing the silvicultural portion of the land application regulation, at least eleven major environmental pathways need to be considered: (1) effects on soil biota, (2) effects on soil biota predators, (3) phytotoxicity, (4) animal toxicity from incidental sludge ingestion, (5) animal toxicity resulting from plant consumption, (6) human toxicity from consuming plants in forest where sludge has been applied, (7) human toxicity from consumption of animals resident to forests, (8) human toxicity from consumption of animals that incidentally ingested sludge, (9) air particulates and volatiles, (10) surface water runoff, and (11) groundwater contamination. Each of these pathways will be considered during the regulatory development process for various pollutants of potential concern. The prevention of deleterious effects from sludge application in a silvicultural setting may be controlled by establishing maximum permissible concentrations of sludge constituents as well as establishing best management practices. The use of best management practices may easily control potential hazards for several of the environmental pathways.

The magnitude of this regulatory effort is as great as the number of scientific and policy questions is large. The development of these technical regulations will require consideration of many scientific factors and parameters in order to develop regulations that will be protective of human health and the environment while maintaining and upholding EPA's policy for the beneficial reuse of municipal sludge.

STATE SLUDGE MANAGEMENT REGULATIONS

As previously mentioned, EPA is also developing a state sludge management regulation. The purpose of this regulation is to provide information and general requirements applicable to the management of sewage sludge and to require states to develop programs to ensure that municipalities and other generators of sewage sludge comply with federal minimum requirements. These regulations lay out the procedural requirements for receiving EPA's approval of a state program. This rule establishes the goals for state programs while providing states with the flexibility to choose the means of achieving them. Some states may choose to operate sludge programs through existing solid waste programs; others may choose water pollution control programs to manage sludge. States have this flexibility as long as they are able to demonstrate to EPA that such programs will be effective and will provide compliance with federal requirements, specifically the technical regulations issued under Section 405 of the Clean Water Act.

In conclusion, the need for sludge regulations is obvious, and a vigorous effort is being made to fulfill this need in terms of scientific and technical guidance as well as management. The goal of this effort is clear—to provide a comprehensive set of regulations that will ensure the safe reuse and disposal of sewage sludge.

Chapter 32

Planning for the Public Dimension in Forest Sludge and Wastewater Application Projects

R. BEN PEYTON and LARRY M. GIGLIOTTI

ABSTRACT   Public participation in a project can provide the greatest benefit and present the fewest hazards when it is properly implemented. To facilitate this process a summary is presented of some key findings of public dimension studies of wastewater and sludge projects, including a Michigan study of public acceptance of forest application. The considerations are then applied to recommendations for increasing public approval of appropriate forest application projects. Recommendations include: focus on selection from alternatives rather than acceptance of a single proposal; use "local" sludge sources; keep projects small and publicize results highly while the public develops a concept of sludge as a resource.

Scientists, technicians, and others who work with the treatment of wastewater and its products are primarily concerned with technical questions of effectiveness, public and environmental safety, and cost. In the past twenty years, they have also been forced increasingly to consider public acceptance of waste treatment projects. The need for public acceptance has placed a greater emphasis on public participation.

Public participation can provide the greatest benefit when it is properly implemented. A good basis for implementing public involvement processes in forest application issues may be found in the literature related to public acceptance of wastewater reuse and agricultural land application of sludge. This paper briefly summarizes some trends reported in studies of the public dimension of wastewater and sludge projects, including a Michigan study of public acceptance of forest application. These trends are used as a basis to make recommendations for increasing public approval of appropriate forest application projects.

PUBLIC ATTITUDES RELATED TO WASTE TREATMENT

The public dimension of issues involving wastewater and sludge application to agricultural lands has received considerable attention in the literature (Bruvold 1972, Christensen 1982, D'Itri and D'Itri 1977, Donnermeyer 1977, Dotson 1982, Ellis and Disinger 1981, Musselman et al. 1980, Olson and Bruvold 1982, Stitzlein 1980, Walker 1979). An extensive review of many of these studies has been reported by Forster and Southgate (1983). In addition, a survey of sixteen successful or attempted projects was analyzed by Deese et al. (1981) to determine public factors relating to the success or failure of the projects. This section will be based on these two comprehensive reports.

Deese et al. (1981) surveyed managers of several projects using sewage sludge for land reclamation and biomass production to determine what may have contributed to the final public acceptance or rejection. Some of the factors that played an important role are summarized here: (1) Public attitudes that sludge is malodorous, disease causing, and repulsive were problems in all cases. The public was usually not familiar with the proposed use of sludge and tended to react in an emotional rather than rational way. (2) Projects that required transporting sludge into other communities tended to have greater problems because of more complex institutional barriers or latent antagonisms between sludge generating communities and receiving communities. (3) Public opposition varied directly with the nature of the adjacent land uses (residential, forest land, etc.). (4) A public relations program was recommended as a necessary but not always sufficient prerequisite to acceptance for projects likely to receive public opposition. Findings supported employing trained experts to conduct public relations programs. (5) Successful public relations were initiated early, before negative public attitudes were formed without benefit of accurate beliefs about the project.

In their review of the literature, Forster and Southgate (1983) offered a similar list of constraints pertaining to several forms of sludge use on land. Those factors relating to public attitudes are briefly described as follows: (1) Prevailing concerns for health risks (pathogens, parasites, metals), environmental effects, and nuisance factors (odor, traffic, spills) are predictable barriers for most land application proposals. (2) The public often mentioned the negative effects on land values as a reason for opposing land application. (3) Farmers have strong principles of individualism, conservation, or agrarianism that foster an attitude that municipal waste is an urban problem and should be solved in the city. (4) Historic precedents have shaped attitudes toward land treatment. During the last fifty years, a trend away from land treatment toward technologies that only partly treat wastes has resulted in little public experience with land treatment concepts. (5) Attitudes toward land application were found to be influenced by demographic factors. Women usually had less positive views on the subject than men. Age was found to be negatively correlated with acceptance. Formal education and experiences with the concept are positively correlated with acceptance. (6) Municipal officers responsible for waste treatment tended to be opposed to land application. In addition, they often overestimated the extent of opposition among the general public.

Both reports recommended that the following institutional changes should be made to facilitate better acceptance of effective land application technologies. Public involvement in decision making should not only be increased but changed from an ineffective public hearing format to more interactive advisory groups, workshops, and seminars. Public participation should be planned for and initiated early in the planning process. Incentives (benefits) should be clarified continually. Odor, health, and environmental issues should be addressed not only through technical and supervisory improvements but in communications and educational messages to the public as well. The technology of land application should be sufficiently developed to be recognized as a tested method with reliable and adequate safeguards, monitoring, record keeping, and other processes of implementation.

THE PUBLIC DIMENSION IN FOREST APPLICATION

Two studies by Michigan State University provide further insights into public acceptance of sludge application on state forest lands. The Michigan Department of Natural

Resources obtained EPA grants to conduct feasibility studies of forest application. Sites for the pilot study were initially selected in Kalkaska County. These were rejected by public opposition and sites in neighboring Montmorency County were used. The authors designed a study to determine why the pilot sites were rejected by one county and accepted by the second (Lagerstrom 1983). Another survey was implemented in seven other northwestern Michigan counties to assess general public perception and acceptance of forest application in that region (Gigliotti 1983). The findings of these two studies are only briefly reported here to illustrate major concepts. Details of the latter study are given in a second paper included in this volume (Gigliotti and Peyton).

Both studies used a mail questionnaire and sampled general public as well as a subsample of county officials. The surveys were designed to reflect the important factors associated with attitude formation and behavior discussed earlier. The questionnaire measured not only attitudes toward forest sludge application, but relative acceptance of alternative technologies, accuracy and extent of belief systems about sludge application in forests, and value priorities used in evaluating the forest application concept. In addition, researchers attempted to anticipate and measure intervening attitudes that might have an impact on both public acceptance and involvement in forest application.

Although public apathy was expected, the general survey indicated considerable interest in the problem. The final response rate (60%), obtained with three mailings, indicated high public interest. Many respondents (67%) further indicated their interest by agreeing with a statement that the sludge disposal problem was an important one for many municipal areas in Michigan, and over 85% stated an interest in learning more about the topic.

Incineration was the disposal method most preferred by respondents, but forest application was the second choice, followed by landfilling and agricultural application methods. In response to whether they would support a forest application proposal in their area, 45% were undecided, 30% would oppose the proposal, and 25% would favor it. However, it is important to know the basis (values and beliefs) for those preferences.

Most respondents agreed that the principal concern was public health, followed by environmental, economic, and aesthetic concerns. Responses to items that measured understanding of benefits, risks, and technical aspects of municipal sludge application in forests indicated that knowledge levels of most respondents were very low. Many "no opinion" responses indicated a lack of information rather than inaccurate beliefs. Thus the respondents agreed on the values to be protected but had inadequate belief systems with which to evaluate consequences of forest application. Conflicting views would be primarily conflicts of beliefs (and perceived consequences) rather than differences in value priorities.

Attitude on forest application was related to use of forests for recreation. This introduces an additional value concern for that subgroup utilizing forests for recreation. A small proportion (7.6%) expressed moralistic or preservationist values by strongly agreeing that it was wrong to put sludge in forests, regardless of benefits or risks.

The survey also measured the intention of respondents to become involved if an effort was initiated to apply sludge to a forest near them. Almost 32% intended to become involved; 23% would not. More important, 71% of respondents who were opposed to the proposal would become involved in the issue, but only 42% of those in favor intended to become involved.

Local public officials who responded were more in favor (46%) of forest application than the general public. Only 30% were undecided. Local public officials also ranked the alternatives differently, preferring, in order, forest application, agricultural use, incineration, and landfilling. The officials agreed that human health concerns should receive priority, but considered economic concerns of nearly equal importance as environmental concerns.

Two intervening attitudes were found to influence responses. A large majority of the respondents to the general survey agreed that sludge should not be brought into their forest area from other parts of Michigan. The initial pilot proposal in Kalkaska County was to use sludge from a city in southern Michigan. Lagerstrom (1983) found this to be a contributing factor in the public rejection of that proposal. The Montmorency County project used sludge from a neighboring county.

The second intervening attitude dealt with the credibility of the Department of Natural Resources (MDNR). In the general survey, opposition to the forest application concept was strongly related to the degree of skepticism respondents had about the MDNR and its accuracy as an information source. Respondents were also more likely to intend to become actively opposed to a forest sludge application project if they were skeptical of the MDNR's management capability and accuracy. In the Kalkaska study, this was even more clear. In fact, public opposition to the pilot program was led by an informally organized group of citizens who had opposed the MDNR on an earlier, successful attempt to bury PBB-contaminated cattle in Kalkaska County. The group had formed a strong mistrust of the MDNR's concern for that county's welfare.

The Kalkaska County proposal was rejected not on the merits of the experimental plan, but because of a combination of three factors: (1) The plan proposed using sludge from another area. (2) Residents were mistrustful of the managing agency. (3) A "watch dog" citizen group was mobilized and prepared to oppose any MDNR plan they perceived as potentially harmful.

IMPLICATIONS FOR INCREASED PUBLIC ACCEPTANCE

With a few notable exceptions, what is known about the public acceptance of other wastewater municipal sludge recycling technologies is applicable to silvicultural use of municipal sludge. However, forest application offers some unique features that make it potentially more acceptable. Forest sites can more readily be selected in isolated areas, thus reducing nuisance concerns and some of the health concerns associated with more populated areas. Because sludge is not applied to a crop in the human food chain, risk of contamination by organic and inorganic chemicals will be substantially lessened. In addition, many states have suitable forest application sites in public ownership. Managers of public lands are more receptive to land application than private owners are (Deese et al. 1981).

Disadvantages are also associated with forest application. Application costs may be higher, and selecting isolated sites may require increased transportation costs. In addition, lowered concerns by the general public may be replaced by concerns of special interest groups such as enviornmental preservationists or those using forests for recreation.

Managers who wish to increase the acceptable use of municipal sludge as a resource for silvicultural and wildlife habitat management must work to familiarize the public with the technology. Success stories must be publicized. Procedures and regulations must be standardized so that citizens view these methods as reliable, regulated, and productive techniques rather than risky innovations.

The natural response of the public is to reject a proposed change on the basis of any suspected risk when they do not have the opportunity to consider risks associated with alternatives. It is imperative that citizens be confronted with a sludge disposal problem to be solved rather than a single proposal to apply sludge to forest lands. Approaching the overall problem in this manner allows alternatives (landfill, incineration, etc.) to be identified and the relative benefits and risks, both known and unknown, to be compared. The emphasis thus becomes a selection from among alternatives rather than an acceptance or rejection of any given proposal.

Because municipal sludge is a resource that may be used to increase forest productivity or enhance wildlife habitat on nutrient-poor soils, forest and wildlife managers may also need to gain local public acceptance of forest sludge application. Their situation differs from that of the waste treatment manager, who should get the public to select from among several disposal alternatives. The forest or wildlife manager needs to get public acceptance for forest sludge application where other disposal alternatives are not a part of the issue. The manager in this situation must stress predictable use benefits which can be compared with potential risks and known impacts. The focus in communicating to the public should be the enhancement of forestry and wildlife values. Discretion is required when deciding whether to emphasize, in the public relations program, the need to solve another region's municipal waste problem.

Obviously, in early forest application programs, sources of municipal sludge should be identifiable by area decision makers and citizens as being "local." It is reasonable to expect that continued successful experience with forest application in a region could increase public acceptance of sludge from distant sources as a welcome resource. Well-managed demonstration plots with continued publicity and on-site interpretation (e.g., signs) are an essential part of this educational program.

Public Involvement

The utility of repeating a public perception study as a strategy to gain public input in a region considering forest application has been shown by Michigan's experience. The survey results have helped agency managers better understand the concerns and informational levels of public segments and have suggested strategies for further public participation and education programs. To be effective, such surveys should be designed to measure components (value priorities and beliefs) of broad public perceptions and to detect both anticipated and unsuspected intervening factors. The survey should be implemented early enough to provide input to planning and to avoid disruptive stages of a resource issue where public groups have become polarized and emotionally defend their positions rather than rationally seek alternatives. Another paper in this volume (Gigliotti and Peyton) has provided more details on survey design. However, it is likely that Deese et al. (1981) are correct in their assessment that expert assistance should be obtained for this and other public relations efforts.

In most cases, the data collection (e.g., survey) phase will suggest additional efforts at public involvement and education. Northern Michigan respondents stated a preference that resource managers should first assess public concerns and ideas pertaining to resource problems, then develop alternative management plans that the public could select from. When implementing these efforts, every attempt should be made to ensure representative public participation. Forest application proposals in northern Michigan (and other areas) are likely to stimulate active involvement of preservation-oriented groups and people who use those forest areas for recreation. In addition, a common experience of resource managers and one predicted by this study is that most citizens who actively respond to a proposed project by attending public meetings, writing letters, and so forth, are opposed to the project. To get representative input and improve decision making, the manager must solicit views and involvement from all affected public segments.

Beginning the public involvement process early will help encourage a representative and rational response to the issue, especially if all affected public segments are identified and represented in the planning process. The activities of special interest groups can also be made more productive if these groups interact with other segments with a common problem-solving goal. This study indicated that formal organizations (hunter groups, Audubon Society members, etc.) could not be relied on to represent all the many public interests that should be considered.

The impact of special interest groups will also be modified by focusing attention on the need to solve municipal sludge problems. This allows more time to educate participants on the comparative risks of all alternatives, before the polarized responses occur (e.g., NIMBY—Not In My Back Yard).

Public Education

Recommendations have been made for educational programs to heighten public awareness of municipal sludge problems and alternatives, prompt the public to consider sludge as a potential resource, and to increase public understanding of technical aspects of municipal sludge treatment (Deese et al. 1981, Gigliotti 1983). The Michigan study indicated that the public is interested in such information and will most likely seek information from news media even though they do not consider these sources to be accurate. There was a tendency for the more involved citizens to seek information from other sources (brochures, workshops, managing agencies, etc.). Some public segments should be targeted during early planning phases for special educational emphasis using brochures, films, and guest speakers. A more technical information bulletin is being developed by the authors for use in Michigan by decision makers (e.g., political officials, advisory groups, opinion leaders) in local areas where a municipal sludge problem may be considered.

Perceived accuracy was found to vary considerably for various information sources in northern Michigan. Educational campaigns should use university information sources, such as extension programs and researchers, where possible. State agencies were also perceived as accurate by the general public, but local government officials and industrial sources of information were considered even less accurate than news media. Given the reliance on the news media for information by most of the public, even though they do not trust the accuracy of these sources, managers should work closely with news facilities to improve this situation. Not only must accurate information be communicated, but the integrity of its source (such as university or state agency) should be assured. Including media reporters as participants in interactive educational and involvement activities (seminars, workshops) may help to increase use of media in the communication program.

The Michigan study suggested a number of messages to include in a communication and education program. For example, education programs should include information on all values involved in the issue (health, environmental, economic, etc.). Messages should be accurate in comparing known consequences and assessing possible risks associated with sludge treatment alternatives. Because most Michigan respondents appeared to have well-defined value priorities but weak information, educational messages should assist citizens in evaluating forest application and other alternatives based on their values. In addition, the need for changes (i.e., nature of the problem) and the benefits of alternatives must be communicated early in the process (Zaltman and Duncan 1977).

SUMMARY

There are environmental and public health risks associated with forest application procedures. However, when properly implemented, this sludge recycling technique offers more benefits and fewer risks than many sludge disposal alternatives. Studies of public perception of wastewater treatment products as resources, including the Michigan study on forest application of sludge, are encouraging. Many citizen groups are interested in the concept of recycling sludge in forests. Technological procedures can be demonstrated and information communicated which will allow most public segments to evaluate forest application as presenting the least threat to health concerns and providing environmental and economic benefits in many geographical areas. At the same time, care must be taken to avoid having forest application projects rejected because of issues centering on intervening attitudes. Appropriate strategies include focusing on selection from alternative technologies, using "local" sludge sources, and keeping projects small and successes highly publicized while technology improves and the public gain experience with the concept of sludge as a resource. In addition, the process must develop from an innovative technology stage to a regulated and accepted procedure, and portrayed to the public as such.

ACKNOWLEDGMENTS

The authors wish to thank Dr. Frank D'Itri, Institute of Water Research, Michigan State University, for his technical and editorial suggestions, and Sue Plesko, Department of Fisheries and Wildlife, Michigan State University, for accurately preparing the manuscript copy. The project was supported by U.S. Environmental Protection Agency grant No. S005551-01 and Michigan Agricultural Experiment Station project No. 3248.

Although the information in this document has been funded in part by the U.S. Environmental Protection Agency under assistance agreement No. S005551-01 to the Michigan Department of Natural Resources and Michigan State University, it has not been subjected to the Agency's publication review process and therefore may not necessarily reflect the views of the Agency, and no official endorsement should be inferred.

Michigan Agricultural Experiment Station journal article number 11655.

REFERENCES

Bruvold, W. H. 1972. Public attitudes toward reuse of reclaimed water. Contribution 137. University of California, Water Resources Center, Berkeley. 54 p.

Christensen, L. A. 1982. Irrigating with municipal effluent: A socioeconomic study of community experiences. ERS-672. U.S. Department of Agriculture, Washington, D.C. 49 p.

Deese, P. L., J. R. Miyares, and S. F. Fogel. 1981. Institutional constraints and public acceptance barriers to utilization of municipal wastewater and sludge for land reclamation and biomass production. Office of Water Program Operations, U.S. EPA, Washington, D.C. EPA 430/9-81-013.

D'Itri, F. M., and P. A. D'Itri. 1977. Public attitudes toward the renovation and reuse of municipal wastewater. p. 603–642. In F. M. D'Itri (ed.) Wastewater renovation and reuse. Marcel Dekker, Inc., New York.

Donnermeyer, J. 1977. Socio-cultural factors associated with the utilization of municipal waste on farmland for agricultural purposes. p. 154. In C. E. Young and D. J. Epp (eds.) Wastewater management in rural communities: A socio-economic perspective. Institute for Research on Land Water Resources, Pennsylvania State University.

Dotson, K. 1982. Public acceptance of wastewater sludge on land. U.S. Environmental Protection Agency, MERL, Cincinnati, Ohio.

Ellis, R. A., and J. F. Disinger. 1981. Project outcomes correlate with public participation variables. J. Poll. Control Fed. 53:1564–1567.

Forster, D. L., and D. D. Southgate. 1983. Institutions constraining the utilization of municipal wastewater and sludge on land. p. 19–49. In A. L. Page, T. L. Gleason III, J. E. Smith, Jr., I. K. Iskandar, and L. E. Sommers (eds.) Proceedings of the 1983 Workshop on Utilization of Municipal Wastewater and Sludge on Land. University of California, Riverside.

Gigliotti, L. M. 1983. A public assessment of concerns and beliefs about forest application of sludge. M.S. thesis, Michigan State University, East Lansing.

Heberlein, T. A. 1976. Some observations on alternative mechanisms for public involvement. Nat. Res. J. 16:197–212.

Lagerstrom, T. R. 1983. Comparison of citizen reaction to a proposed sludge demonstration project in two Michigan counties. M.S. thesis, Michigan State University, East Lansing.

Langenau, E. E., Jr., and R. B. Peyton. 1983. Policy implications of human dimensions research for wildlife information and education programs. Transactions, Northeastern Fish and Wildlife Conference 39:119–135.

Musselman, N. M., L. G. Welling, S. C. Newman, and O. A. Sharp. 1980. Information programs affect attitudes toward sewage sludge use in agriculture. EPA 600/2-80-103. U.S. Environmental Protection Agency, Cincinnati, Ohio. 51 p.

Olson, B. H., and W. Bruvold. 1982. Influence of social factors on public acceptance of renovated wastewater. p. 55–73. In E. J. Middlebrooks (ed.) Water reuse. Ann Arbor Science Publishers, Ann Arbor, Michigan.

Peyton, R. B., and B. A. Miller. 1980. Developing an internal locus of control as a prerequisite to environmental action taking. p. 173–192. In A. B. Sacks, L. L. Burrus-Bammel, C. B. Davis, L. A. Iozzi (eds.) Current issues IV: The yearbook of environmental education and environmental studies. ERIC Center for Science, Mathematics and Environmental Education, Ohio State University, Columbus.

Rotter, J. B. 1966. Generalized expectancies for internal versus external control of reinforcement. Psychological Monographs 80.

Shafer, R. B., and J. L. Tait. 1981. A guide for understanding attitudes and attitude change. North Central Regional Extension Publication 138. Cooperative Extension Service, Iowa State University, Ames.

Stitzlein, J. N. 1980. Public acceptance of land application of sewage sludge. In Utilization of wastes on land: Emphasis on municipal sewage. U.S. Department of Agriculture, Washington, D.C.

Walker, J. M. 1979. Overview: Costs, benefits, and problems of utilization of sludges. p. 167–174. In Proceedings of Eighth National Conference on Municipal Sludge Management, Miami Beach, Florida.

Zaltman, G., and R. Duncan. 1977. Strategies for planned change. John Wiley and Sons, New York. 404 p.

Chapter 33

Technology and Costs of Wastewater Application to Forest Systems

RONALD W. CRITES and SHERWOOD C. REED

ABSTRACT   Land treatment of municipal wastewater on forest land has been practiced experimentally for over twenty years and on a full-scale basis for over ten. The technology of land application consists of sprinkler irrigation using solid-set (fixed) sprinklers. Most sprinkler systems have been installed in existing forests using either buried or aboveground laterals. Design guidance for sprinkler spacing and operating pressures for solid-set systems in forests is presented. Costs of installed forest land application systems are also given. Costs and design factors are reviewed for systems at Snoqualmie Pass, Washington; Wolfeboro, New Hampshire; Lake of the Pines, California; Clayton County, Georgia; and State College, Pennsylvania. Operation and maintenance costs are provided for systems at Clayton County, Georgia; West Dover, Vermont; and Kennett Square, Pennsylvania. Reduction of the cost of future systems can be accomplished by minimizing the amount of effluent storage provided. Most forest systems can operate with thirty days storage or less. New technology and new plantations can allow reductions in the cost of wastewater application. Potential revenue from tree harvest can also reduce overall costs.

Forest land treatment has been practiced successfully for many years (Reed, et al. 1971, Sepp 1965). The Seabrook Farms system in Seabrook, New Jersey, has been operating since 1950 (Pound and Crites 1973). In 1963, research was initiated at the Pennsylvania State University (Sopper and Seaker 1984) on forest land treatment; in 1973, research on wastewater application to forest systems began at the University of Georgia (Nutter et al. 1978) and at the University of Washington (Cole et al. 1983). By 1984 the technology was well established and a number of full-scale forest land treatment systems were in operation (Crites 1984). In addition, forest land treatment is now being recognized in undergraduate environmental engineering texts (Tchobanoglous and Schroeder 1985). A summary of operating forest land treatment systems is presented in Table 1.

APPLICATION TECHNOLOGY

Forest wastewater application systems usually consist of solid-set arrangements of laterals with rotating impact sprinklers. Laterals may either be buried or aboveground. Aboveground laterals are less expensive, easier to install and maintain, and their installation has potential for less erosion. Buried laterals are less susceptible to vandalism, are more pleasing aesthetically, and can facilitate harvesting of the trees without damaging or moving the pipes.

Spacing of laterals typically ranges from 12 to 24 m (40 to 80 ft). Although wider spac-

TABLE 1. Operation and forest land treatment systems in the United States receiving municipal wastewater.

Location	Year Started	Flow (m^3/d)	Forest Type
State College, PA	1963	15,385	Mixed hardwood, red pine plantation, spruce plantation
Mount Sunapee State Park, Newbury, NH	1971	26	Mixed hardwood and conifers
Unicoi State Park, GA	1973	76	Mixed hardwood and pine
Kennett Square, PA	1973	189	Mixed hardwood
Mackinaw City, MI	1976	760	Aspen, white pine, birch
West Dover, VT	1976	2,080	Hardwood, balsam, hemlock, and spruce
Wolfeboro, NH	1976	1,140	Mixed hardwood and pine
Lake of the Pines, CA	1978	946	Mixed hardwood and pine
Clayton County, GA (E. L. Huie System)	1981	73,800	Loblolly pine plantation and hardwood
Clayton County, GA (Shoal Creek System)	1982	4,160	Mixed pine and hardwood
Kings Bay Submarine Support Base, St. Marys, GA	1981	1,250	Slash pine plantation
Snoqualmie Pass, WA	1983	4,000	Douglas-fir
Dalton, GA (~ 6,000 acres)	1986	151,400	Mixed hardwood and pine
Covington, GA (~ 1,000 acres)	1985	26,500	Mixed hardwood and pine

ings between laterals results in a lower cost of installation, closer spacings provide improved distribution, less wind drift, and allow lower operating pressures. A 24 m (80 ft) spacing between laterals and an 18 m (60 ft) spacing between sprinklers along the lateral are often recommended (McKim et al. 1982).

Operating pressures at the sprinkler nozzle should not exceed 0.38 MPa (55 psi) to avoid potential damage to the trees. The use of traveling gun sprinkler systems in forests is limited because of this pressure restriction and the need for wide, 10 m (30 ft), lanes. Some thick-barked hardwood species may tolerate pressures up to 0.59 MPa (70 psi).

Sprinkler risers are usually 2.5 cm (1 inch) galvanized pipes that extend 0.3 to 1.5 m (1 to 5 ft) above the ground. Risers are usually connected directly to the lateral, although for buried laterals a swing joint is sometimes used in an attempt to dampen the vibration caused by the impact sprinkler. At Clayton County some of the swing joints became clogged with solids, and the swing joint prevented cleaning out the solids. As a result, the swing joints are being replaced when clogging occurs.

Considerable differences in elevation often occur over forest land application sites. These differences make it difficult to maintain uniform distribution, since pressure variations will usually occur. To compensate, each lateral or each riser may be fitted with a pressure reducing valve or a flow control valve. Alternatively, lower pressure zones can be used at higher elevations, or the sprinklers and laterals may be placed closer together.

Wintertime forest application requires special design considerations where subfreezing conditions occur. Drain valves should be placed at all low points in the system, so that water can escape before it freezes in the line. At State College, Pennsylvania, the forest land application operates year-round without storage. Generally, storage is provided for extreme cold or wet weather.

At West Dover, Vermont, nozzles that spray downward are installed at low points in each line (Cassell et al. 1979). These specially modified nozzles have allowed wastewater applications with ambient air temperatures as low as -18°C (0°F).

COSTS OF EXISTING SYSTEMS

Construction costs of five existing systems are presented along with key design features that affect costs. In addition, operation and maintenance (O&M) costs are presented for three systems.

Snoqualmie Pass, Washington

Snoqualmie Pass is a popular ski area at the summit of the Cascade Mountains 80 km (50 miles) east of Seattle. Wastewater is treated in a 1,400 m³/day (0.37 mgd) capacity aerated lagoon and sprinkler irrigation system. The land application site consists of 19.2 ha (47 acres) of Douglas-fir and brush (Card 1983).

The aerated lagoon system includes thirty days of emergency storage capacity, although the land treatment system was designed to operate year-round at a loading rate of 5 cm/wk (2 in./wk). Costs in 1982 for total system construction were $1,851,946 or $1,323/m³ per day ($5.03/gpd). Items included in the cost are the pumping, aerated lagoon, distribution piping and nozzles, telemetry system, laboratory and office facilities, and engineering services. Although the unit cost is relatively high, a nearby ski area with an advanced wastewater treatment system is paying six times the cost of the Snoqualmie Pass facility (Card 1983).

Wolfeboro, New Hampshire

Secondary effluent is applied to 32 ha (80 acres) of an existing forest of mixed hardwood and pine. Constructed in 1976 with a design flow of 1,140 m³/day (0.3 mgd), the system contains a 138 day storage lagoon. The distribution system is aboveground with aluminum laterals hung on metal posts, and impact sprinklers on about 15 m (50 ft) centers. Construction costs for the Wolfeboro system are presented in Table 2. As shown in

TABLE 2. Construction costs of Wolfeboro, New Hampshire forest irrigation system (1976).

Item	Cost
Effluent pump station	$ 76,800
Storage lagoon	670,000
Distribution pumping	215,000
Distribution piping	120,600
Total	$1,082,400
Unit cost	$ 949/m³ per day

Table 2, the 138 day storage lagoon accounts for 62% of the project construction costs. The cost of the distribution system (pipe, risers, nozzles, and pipe supports) was 11% of the total or $3,800/ha ($1,500/acre).

Lake of the Pines, California

Lake of the Pines is one of the several sites in California where effluent is used to sprinkler irrigate forest land. In 1926 a hillside spray irrigation system was constructed for the Montezuma School near Los Gatos, and thirty other hillside spray systems in California were surveyed in 1963–64 (Sepp 1965).

Lake of the Pines is in Nevada County between Auburn and Nevada City. The sprinkler irrigation system consists of 13 ha (32 acres) of mixed hardwood and pine. The site is grazed by beef cattle during the summer to keep down the herbaceous growth. Secondary treatment is achieved through a series of aerated ponds and stabilization ponds. The current flow is 946 m³/day (0.25 mgd).

The distribution system consists of buried mains and laterals with laterals spaced 30 m (100 ft) apart. The construction cost in 1978 was $368,000 for floating aerators, pumping, distribution piping, fencing, and controls.

Clayton County, Georgia

The Clayton County land treatment system is the largest forest irrigation system in the United States using secondary effluent. The irrigated area of 1,020 ha (2,520 acres) consists of native hardwood stands and loblolly pine. Cost and design factors are presented in Table 3.

The land treatment system began operation in 1981, applying secondary effluent through a buried solid-set sprinkler system. The laterals are 20 m (60 ft) apart, with sprinklers 24 m (80 ft) apart on the lateral. The risers are 1.8 cm (0.75 inch) galvanized steel about 1 m (3 ft) high. The sprinkler nozzles operate at 0.34 MPa (50 psi).

TABLE 3. Cost and design factors at Clayton County, Georgia.

Factor	Value
Flow (m³/day)	73,800
Treatment area (ha)	1,020
Storage time (days)	12
Loading rate (cm/wk)	6.25
Capital cost	$18,000,000
Unit capital cost ($/m³ per day)	$244
Operation and maintenance cost ($/yr)	$712,265
Unit O&M cost ($/m³)	$0.047

State College, Pennsylvania

After twenty years of operation as a research facility, the State College facility was expanded to 200 ha (500 acres) to accommodate the full wastewater flow of 15,385 m³/day (4 mgd). The expansion cost about $10,000/ha ($4,000/acre), including pumping, aboveground piping, drains, and monitoring wells. There is no storage in the system, so application is continuous year-round. Lateral lines are 25 m (80 ft) apart with sprinklers spaced 25 m (80 ft) apart. Sprinklers are on risers ranging from 30 cm (1 ft) to 90 cm (3 ft) in height.

West Dover, Vermont

West Dover operates a forest land treatment system with thirty-three days of storage. Storage is required by the state of Vermont during snowmelt periods. Secondary effluent is applied to 14 ha (35 acres) at an annual loading rate of 1.7 m/yr (5.6 ft/yr). Operating costs for 1980 are presented in Table 4 (U.S. EPA 1982). The land treatment costs are 33% of the total treatment costs.

TABLE 4. Operation and maintenance costs for sprinkler irrigation at West Dover, Vermont.

Item	Cost ($/yr)
Labor	$10,550
Power	2,510
Materials and administration	6,065
Total	$19,125
Unit cost ($/m^3)	$ 0.149
Unit cost ($/ha)	$ 1,388

Kennett Square, Pennsylvania

The Kennett Square land treatment serves two retirement communities 48.3 km (30 miles) southwest of Philadelphia. Aerated lagoon effluent is spray irrigated on a wooded 2.3 ha (8 acre) site. The annual loading is 2.1 m/hr (6.9 ft/hr). Operating costs for 1980 are presented in Table 5. The land treatment cost of this 189 m^3/day (0.05 mgd) facility is 27% of the total treatment system O&M cost.

TABLE 5. Operation and maintenance costs for sprinkler irrigation at Kennett Square, Pennsylvania.

Item	Cost ($/yr)
Labor	$1,070
Power	2,200
Materials and administration	800
Total	$4,070
Unit cost ($/m^3)	$ 0.06
Unit cost ($/ha)	$1,255

COSTS OF FUTURE SYSTEMS

As shown, the construction cost of existing systems varies from $3,800 to $10,000/ha ($1,500 to $4,000/acre). The O&M costs vary from $700 to $1,400/ha ($280 to $560/acre). The cost of future application systems can be reduced by minimizing storage, through the use of alternative technology, and through the use of new plantations. Overall operational costs can be reduced by revenue from tree harvest.

Reduced Storage

For Wolfeboro, New Hampshire, the 138 days of winter storage accounted for 62% of the total construction cost. Reduced storage to thirty-three days (as for West Dover, Vermont) would have a significant effect on construction costs. The State College and Snoqualmie Pass systems operate with no winter storage (Snoqualmie Pass has emergency storage available for thirty days). As regulatory agencies accept the ability of forest land treatment systems to operate year-round, more minimal storage systems can be expected.

Alternative Technology

Solid-set sprinkler systems are the most expensive type of application for land treatment (Reed and Crites 1984). Alternatives such as center pivot or traveling gun sprinkling could reduce construction costs where they are applicable. Christmas trees can be irrigated with center pivot sprinklers and can yield an attractive economic return (Shelton and Parnell 1984).

Traveling gun sprinkling systems can cost as little as 22% as much as solid-set sprinkling (Reed and Crites 1984). The units require relatively even terrain for their traveling lanes, however, and currently require 0.67 MPa (80 psi) nozzle pressure. This high nozzle pressure will cause harmful impacts on most trees. If the needed nozzle pressure can be reduced in future equipment designs, traveling guns may have a wider potential for forest systems, particularly for new plantations.

New plastic nozzles that are as much as $10 less than conventional nozzles can offer some reduction in costs of future systems. The savings may range from $220 to $660/ha ($90 to $270/acre), depending on the spacings.

New Plantations

The current forest land application systems have been installed in existing forests. As a result, the sprinkler systems had to be fitted to existing tree density, or selective site clearing was required. The clearing of trees and vegetation to install laterals can be quite expensive.

The effluent application system installed in a new plantation can be less expensive than in an existing forest. If the site is relatively flat, it can be leveled to allow surface irrigation. The cost of surface irrigation can be 20 to 30% of the cost of sprinkler irrigation. Traveling gun sprinkling systems can also be considered in new plantations, resulting in a low-cost sprinkler system. If subsurface drainage is required, the drains can be planned for and installed prior to planting the trees, thus minimizing costs.

Revenue Potential

Work conducted in California by the Chapman Forestry Foundation has shown that several varieties of eucalyptus (*E. grandis*, *E. globulus*, and *E. camaldulensis*) have good potential for biomass production (Chapman, pers. comm. 1985). For spacings of seedlings 3 by 3 m (10 by 10 ft) the tree density would be 1,067/ha (432/acre). Five years of growth of these eucalyptus varieties can yield 16.9 m^3/ha (25 cords/acre) of firewood with a value in the field of $1,235/ha ($500/acre), a return not uncommon for many commercial operations. This type of return makes fast-growing trees an economically attractive crop for slow-rate land treatment systems. Similar plantation work is under way in Pennsylvania, with sludge and wastewater for nutrients, growing hybrid poplars, and in Georgia with wastewater growing five different native hardwood species.

REFERENCES

Card, B. 1983. Economical year-around land treatment at an Alpine ski area. Presented at the WPCF Annual Conference, Atlanta, Georgia.

Cassall, E. A., D. W. Meals, and J. R. Bouzoun. 1979. Spray application of wastewater effluent in West Dover, Vermont. Special Report 79-6. Cold Regions Research and Engineering Laboratory (CRREL), Hanover, New Hampshire.

Cole, D. W., C. L. Henry, P. Schiess, and R. J. Zasoski. 1983. The role of forests in sludge and wastewater utilization programs. p. 125–143. *In* A. L. Page, T. L. Gleason III, J. E. Smith, Jr., I. K. Iskandar, and L. E. Sommers (eds.) Proceedings of the 1983 Workshop on Utilization of Municipal Wastewater and Sludge on Land. University of California, Riverside.

Crites, R. W. 1984. Land use of wastewater and sludge. Environ. Sci. and Tech. 18(5):140A–147A.

McKim, H. L., W. E. Sopper, D. W. Cole, W. L. Nutter, D. H. Urie, P. Schiess, S. N. Kerr, and H. Farquhar. 1982. Wastewater applications in forest ecosystems. CRREL Report 82-19. CRREL, Hanover, New Hampshire.

Nutter, W. L., R. C. Schultz, and G. H. Brister. 1978. Land treatment of municipal wastewater on steep forest slopes in the humid southeastern United States. p. 265–274. *In* H. L. McKim (ed.) Land treatment of wastewater. Vol. 1. CRREL, Hanover, New Hampshire.

Pound, C. E., and R. W. Crites. 1973. Wastewater treatment and reuse by land application. EPA 600/2-73-006b.

Reed, S. C., and R. W. Crites. 1984. Handbook of land treatment systems for industrial and municipal wastes. Noyes Publications, Park Ridge, New Jersey.

Reed, S. C., et al. 1972. Wastewater management by disposal on the land. CRREL SR 171. CRREL, Hanover, New Hampshire.

Sepp, E. 1965. Survey of sewage disposal by hillside sprays. Bureau of Sanitary Engineering, California Department of Health, Berkeley.

Shelton, S. P., and K. G. Parnell. 1984. Cultivation of Christmas trees using reclaimed municipal wastewater. p. 428–451. *In* Proceedings of the Water Reuse Symposium III. Vol. 1. AWWA Research Foundation, Denver, Colorado.

Sopper, W. E., and E. M. Seaker. 1984. Fate of pollutants on a municipal wastewater land treatment system: A twenty year report. Presented at the WPCF Conference, New Orleans, Louisiana.

Tchobanoglous, G., and E. D. Schroeder. 1985. Water quality. Addison-Wesley Publishing Company, Reading, Massachusetts.

U.S. Environmental Protection Agency. 1982. Operation and maintenance considerations for land treatment systems. EPA 600/2-82-039. MERL, Cincinnati, Ohio.

Chapter 34

Technology and Costs of Forest Sludge Applications

CHARLES L. HENRY, CHARLES G. NICHOLS, and TERRILL J. CHANG

ABSTRACT   Forest environments create unique problems in sludge application methods compared with agriculture. Because forest stands are continuous crops, access may be limited to specific trails, thus preventing use of conventional agricultural methods. Sludge must therefore be propelled into stands to get even and complete coverage of the area. Equipment in many instances must be able to negotiate uneven terrain. This paper discusses the current alternatives for forest applications of municipal and industrial sludges. Three basic site types dictate equipment requirements: (1) recently cleared forest land, (2) young plantations, and (3) established stands. A major factor in equipment design is the character of the sludge as it leaves the treatment plant, specifically the percentage of solids. This affects choice of technique and also how transfer of sludge can be made to application equipment from trucks arriving from the treatment plant. On-site storage of sludge is also a major factor in equipment facility needs. Application techniques evaluated include set irrigation systems, traveling guns, application vehicles with mounted cannon, sludge spreading and incorporation methods, and use of a manure spreader.

Application of sludge to forest lands in the Pacific Northwest has been experimentally practiced for well over a decade (Bledsoe 1981, Henry and Cole 1983). This research has led to development of sludge handling and application practices, and in turn to state and federal guidelines (WDOE 1982, U.S. EPA 1983, Henry et al. 1984). With increased regulatory and public acceptance, and the possibility of increased revenue from accelerated tree growth, a number of municipalities and landowners are becoming interested in this alternative. In Washington, full-scale operations are now being or have been practiced by both large and small municipalities, such as Seattle, Bremerton, and Tacoma.

Compared with agricultural uses, full-scale operational use of sludge in the forest is a fairly recent alternative, thus creating unique challenges in sludge application methods. Since forest stands are long-term crops and tree spacings in older stands are often uneven, forest access may be limited to existing or specially constructed trails. Also, because sludge adhering to tree needles or leaves may adversely affect photosynthesis, under-the-canopy applications in mature forests or properly timed over-the-canopy applications in young plantations are required. Consequently, the conventional agricultural method—preplanting application—may not be suited to forests. As a result, forest application must rely on some spray or broadcasting method in order to obtain complete and even coverage of the area. Factors different from agricultural applications must be given special consideration, such as soil erosion from equipment usage on steep slopes, reduced forest productivity from soil compaction, and land taken out of productivity for use as trails and storage areas. Sludge characteristics also have major affects on the choice and design of sludge handling, storage, and application systems in forest sites, of which the most significant is the percentage of solids in the material coming from the treatment plant.

APPLICATION SITES

Three types of sites can be considered for sludge additions: (1) recent clearcuts, (2) young forest plantations (over the canopy), and (3) older forest stands (under the canopy). Each has different operational advantages and restraints that require that consideration be given to rates of application, timing of applications, and application system.

Clearcuts

Clearcuts offer the most flexible and most economical kind of site for sludge application. Depending on the site, sludge can easily be directly spread regardless of moisture content. Ease of delivery will depend on the amount of site preparation (stump removal, residual debris burning, etc.), slopes, soil conditions, and weather. Other options available are temporary spray irrigation systems, injectors and splash plates for more liquid material, or manure spreaders for more solid material. Again, site preparation and characteristics are major factors in application technique.

While clearcuts offer many application alternatives, they also have major drawbacks. Because a site amended with sludge suddenly has an abundance of nutrients, grass and weed growth can be vigorous. Plantation establishment becomes much harder because of competition from this unwanted vegetation, and high mortality will occur if grasses are not held in check by extensive maintenance with herbicides or disking. This vegetation also offers excellent cover for rodents such as voles, which can proliferate and add to seedling survival problems. Because of these problems, clearcuts are generally not recommended for sludge application prior to substantial plantation establishment in the Pacific Northwest, unless it is economically feasible to provide a high level of vegetation control.

Young Plantations

Applications of sludge to existing stands virtually eliminate problems with competition and rodents. Sludge is typically applied by a tanker-sprayer system, which can apply sludge with 14% solids over the canopy 45 m into the plantation. This method requires application trails at a maximum of 90 m (300 ft) intervals if total site coverage is desired. Timing of applications is important with over-the-canopy applications for two reasons: (1) In order to wash sludge from the foliage, spraying should take place during the rainy season. (2) Since sludge sticking to new foliage could retard growth, sludge should be applied during the dormant season. Liquid sludge has been successfully applied via an irrigation system, but clogging of nozzles has been a major drawback. Manure spreaders are useful for applying dewatered sludge too thick to spray, but application trails must be at close intervals, no more than twice the range of the spreader depending on obstructions such as tree limbs.

Older Stands

Applications to older stands have the advantage of year-round application potential in appropriate soils. Since spraying will take place under the canopy, no foliage will be affected. Application methods are similar to those described for young plantations. But since stands are not typically in rows, some of the alternatives available for young plantations may be eliminated.

APPLICATION RATE AND TIMING

Two application rate and timing philosophies exist: (1) *Annual applications designed to meet only the annual nutrient uptake requirements of the trees.* In most cases the nitrogen balance will dictate application rates, considering volatilization and denitrification losses, and decomposition from current and prior years. (2) *A heavier application rate one year, followed by a number of years when no applications are made.* This schedule depends on the soil storage to temporarily tie up excess nitrogen, which will become available in later years.

Application costs are lower for the second technique because of less frequent entry into a site, and the public can use the site for recreation in the nonapplication years. In addition, the excess available nitrogen present during the first year is no longer at elevated levels in following years, and leaching loss of nitrate in these years usually will not occur.

SLUDGE CHARACTERISTICS

One sludge characteristic, the percentage of solids, is vitally important to the technology used in forest sludge applications. The percentage of solids, calculated by weighing the solids and water in a sludge sample, dictates the kinds of machinery required, the application procedures, and application timing. Of secondary importance is the degree of sludge stability as measured by the volatile solids content and the nature and amount of any chemicals used to dewater the sludge.

The solids content of sludge will vary from a dark liquid at 2 to 3% solids to a solid-state, moist, cakelike material of up to 40% solids. Typical ranges that have been applied to forests include liquid sludge at 2 to 3% or 6 to 8% solids, which can easily be pumped; semisolid sludge at 8 to 18% solids, which can be pumped, but less efficiently than liquids; and solid sludge cake at 20 to 40% solids, which may be flung from a manure-type spreader.

The degree of stabilization, or digestion, of sludge is important when attempting to pump semisolid sludges at the high end of the solids content range. A relation has been observed that indicates that "fresh" sludge, which has a higher volatile solids content than "aged" sludge, has a greater viscosity. With currently available pump systems, the maximum solids content at which pumping appears successful at needed pressures and flow rates is 18% for aged sludge and about 14% for fresh sludge. These values are approximate observations and apply to Seattle Metro's combined primary and secondary activated digested, centrifuged sludge.

The importance of sludge stabilization in system design relates to sludge dewatering, transportation, storage, and means of application. For example, if sufficient capacity were available in a large basin to allow for six to nine months of sludge storage before application, then dewatered sludge at 18 to 20% could possibly be spray applied. If only daily storage were available, the fresh sludge could be transported at 20% solids, but would need to be diluted to an average of 13 to 14% solids in order to spray apply. It is likely that for substantial quantities of sludge production, such as 20 dry Mg/day or more, it would be less costly to dewater to 20% solids, truck the sludge, and rewet rather than store the sludge in a basin for many months. Of course, each circumstance is different and must be evaluated in light of local conditions.

SLUDGE STORAGE

Since sludge generally cannot be applied with the same regularity as it is provided from the plant, some temporary storage space is required. Options for storage depend on land availability and constraints of the treatment plant and application site, the nearness to and ownership of the sites, and operational and management objectives. The two options available are (1) no on-site storage (storage provided at the treatment plant) and (2) on-site storage.

No Storage

If storage is provided at the treatment plant, sludge can remain there until needed for application. This is an attractive option for the small landowner, since he can avoid the cost of building and operating a storage system, does not have to take the needed land out of productivity for a storage basin, and does not have the liability associated with a basin. This option also creates maximum flexibility for application. Use of multiple sites is not limited by the expense of multiple basin construction. Also, within a large site, transportation from the basin to the application area is eliminated. This option can be accomplished with either (1) multipurpose application vehicles that can travel both highway and forest roads or (2) transfer trailers to supply sludge to an off-road application vehicle.

If liquid sludge is to be hauled from the treatment plant and applied by a multipurpose vehicle, the treatment plant would provide off-site storage capacity. Operational costs would be relatively high because of the frequent trips necessitated by the relatively small capacity of the applicator vehicles, but this approach may be desirable for small treatment works. For larger systems, transfer trailers can provide greater capacity at higher speeds to move sludge over the highway to the site. Depending on distances, one tractor may work with two trailers by leaving a full one at the site to nurse the application vehicle and taking the empty one back to the plant. If the distance is great enough, a tractor and trailer combination can remain permanently connected, and a group of tractors and trailers may be required. This system might be the best option for small ownerships.

On-site Storage

There are several alternatives for on-site storage. The first and most commonly used is in-ground facilities, such as earthen basins. Basins can have substantial capacity, but may require extensive construction at remote sites. Regulatory scrutiny over the use of earthen basins has increased because of the potential for groundwater contamination from leachates; thus basins frequently require costly liners. Furthermore, some conveniently located sites may have geological disadvantages (e.g., surface bedrock) or topographical disadvantages (e.g., steep slopes), which will increase the costs. An alternative to earthen basins is single or multiple in-ground tanks of aluminum, steel, concrete, or fiberglass, strategically located to minimize distance to application areas. The major advantage of in-ground facilities is that they allow gravity unloading of long-haul sludge vehicles. The major disadvantage is the significant earthwork required for installation, and removal when no longer needed.

A third alternative uses aboveground, field-assembled tanks. Precast concrete sections with keyed and gasketed joints can be assembled to form a tank, but these are difficult to handle in the field, as well as heavy to transport. Open-top fiberglass tanks (similar to aboveground swimming pools) are lightweight and chemically compatible with sludge. Because of their one-piece construction, they cannot be disassembled for transport. A tank narrow enough to be transported easily has to be fairly tall to have adequate capacity. Furthermore, fiberglass is less durable than metal. Bladderlike, collapsible rubber storage tanks are used for water and wastewater, but sludge drawoff may be difficult. They also are subject to damage by sharp objects. Vertical metal tanks may have open tops or roofs. They may be assembled by welding or bolting; bolting allows them to be disassembled and moved to other sites. A large bolted tank, when disassembled, is likely to be easier to transport than a smaller welded vertical tank. Horizontal metal tanks generally have welded construction. All of these methods require sludge to be pumped from the tank, limiting their usefulness for liquid sludge.

A fourth alternative involves portable tanks. Horizontal trailer-mounted tanks are used to haul and store a variety of commodities including milk, chemicals, petroleum products, and cement. They are usually of welded metal such as aluminum, steel, or stainless steel. Advantages include maneuverability on forest roads and the ability to be pressurized or withstand vacuum (if designed for that service). Their major disadvantages are high capital cost and long lead time for delivery.

SLUDGE TRANSFER EQUIPMENT

In all alternatives discussed above, except for the multipurpose highway and application vehicle, sludge needs to be moved from one vessel to another. The two basic transfer mechanisms are direct dumping and pumping. The two most important design factors are the percentage of solids and the type of storage: the percentage of solids dictates whether transfer operations (i.e., pumping) will be effective; position and configuration of storage vessel determines if gravity is sufficient for facilitating transfer.

Direct Dumping

There are several ways of unloading the sludge brought to the site by trailer from the treatment plant. If the trailer has dumping capability, then gravity unloading is simple and relatively rapid: the tailgate or rear valve is opened and the bed is elevated. Obviously, gravity dumping requires either an in-ground facility or driving the truck onto a ramp above the storage vessel. The flow rate of the sludge depends somewhat on the angle of the truck bed, as well as the sludge characteristics; there is usually an initial burst of sludge out of the trailer when the gate is opened. A slide gate on the tailgate can provide some additional control over the rate at which sludge is dumped. Trucks without dumping beds can be unloaded by a hydraulic ram, which pushes sludge toward an outlet pipe on the tailgate. The trailer, however, must be specially designed to withstand pressures created by the ram. Another possibility is the use of an auger to pump sludge to an outlet pipe, although experience indicates that augers are not suitable for pumping high concentration sludge.

Pumping

Depending on the percentage of solids, sludge can also be pumped out of trailers. Below 15% solids, most sludges will flow as a semisolid and be able to feed a pump if enough positive head is present or the angle of the trailer bed is great enough. At higher percentage solids, bridging of the sludge commonly occurs, restricting or preventing flow of sludge to the pump. Positive displacement pumps include progressive cavity pumps, concrete pumps, and slurry pumps. Progressive cavity pumps employ a screw-type rotor and can handle sludges exceeding 18% solids, but are susceptible to rotor damage if allowed to run dry and are capable of low flow rates. Their capital cost is moderate. Plunger-type concrete pumps are capable of relatively low suction lifts and can handle up to 13% solid sludges. They require a hopper feed on the suction side of the pump, an arrangement difficult to achieve at remote forest sites. Furthermore, their capital cost is very high. One manufacturer has a plunger pump designed for very viscous high solid sludges and slurries. The design includes a hopper and mechanism for breaking up the sludge. While it appears capable of handling 18% solid sludge, its cost is about ten times that of the progressive cavity pump of equal capacity. The delivery rates are also low.

Centrifugal pumps can be used only on diluted sludge. Chopper-type centrifugal pumps are made by several firms; they employ some combination of cutter bar or cutter impeller and suction inducer in order to pump up to 16% solid sludge. Several manufacturers have more conventional centrifugal pumps designed especially for sludge, capable of pumping about 13% solids. Both chopper and sludge centrifugal pumps require a flooded suction to operate properly. A chopper pump would be considered low cost, and a sludge centrifugal pump, moderate. These pumps are capable of moderate flow rates and pressures and are suited to spray nozzles.

Sludge can also be conveyed using compressed air, vacuum, or a combination of both. This requires a storage tank and trailer designed as a pressure vessel, with attendant increased costs. Experience has shown that about 1.0 kg/cm^2 of air pressure or 38 cm Hg (5.2 m water) vacuum can be used to carry 16% solid sludge.

Transfer of Dewatered Sludge (more than 15% solids)

Usually the most costly portion of land application of sludge is transportation from the treatment plant to the site. For travel distances greater than 15 km, it is economically advisable to thicken or dewater the sludge. For 80 km or more, it is desirable to dewater to 18% or higher. At this percentage, pumps are generally ineffective at transfer. As a result, dumping has been the greatly favored alternative. This has made it difficult to eliminate expensive below-ground facilities—that is, the effect has been to limit the use of aboveground facilities. One system that has just recently been developed and successfully put into full-scale operation involves rewatering a 20 to 25% solid sludge to 12 to 14%, and pumping into a portable 115 cubic meter storage vessel (Nichols, Henry, and Chang, unpublished data). A specially designed water injector is placed inside the tailgate of a long-haul trailer where water is pumped through nozzles at 25 liters per second. This provides enough mixing and rewatering for a centrifugal pump to transfer this slurry at approximately 60 liters per second. A 23 cubic meter trailer load requires about 20 minutes for complete emptying.

Conveyor belts can also handle high solid sludges. Open-type conveyors can operate only at shallow angles and thus require substantial space. A disadvantage is the need for a carefully regulated flow rate.

SLUDGE APPLICATION SYSTEMS

As the sludge solids content increases, application options decrease. Sludge of 40% solids can be applied only by dumping (in clearcuts) or with a manure spreader. If a forest site is many miles from the sludge source, it may be more economical to dewater the sludge at the treatment plant than to transport the extra water in a liquid sludge. Whereas more options are available for application of a dilute sludge, applying semisolid or solid sludges results in handling reduced amounts of material on site. A comparison of alternatives dealing with varying solids concentration is crucial to a thorough economic evaluation of appropriate systems.

There are four forest application systems. Three are effective for liquid sludge, two for semisolid sludge, and one for solid sludge, as shown in Table 1. The principles behind these four systems are simple and rely on the use of pumps, gravity, or mechanical flinging. As shown in Table 1, the effective range of these systems varies from 3 m to more than 60 m. In long-distance spraying, the stand density is an important factor if under-the-canopy applications are to be used.

TABLE 1.   Sludge application system.

System	Application Range (m)
Liquid sludge	
Spray irrigation: Set system	60
Traveling gun	60
Application vehicle with spray cannon	60
Direct spreading	3
Semisolid sludge	
Application vehicle with spray cannon	45
Direct spreading	3
Solid sludge	
Manure-type spreader	15

Liquid Sludge

The three systems used for liquid sludge applications are spray irrigation, application vehicle with spray cannon, and direct spreading. Each operates effectively under different circumstances and site conditions. Important considerations in choosing a system are type of terrain, road conditions, stand conditions, and size of the site.

Spray Irrigation. Sludge spray irrigation can be accomplished with a permanently installed system, a portable aluminum pipe system, or a traveling system. The irrigation system is patterned after agricultural systems that pump liquids through pipes to nozzles, which direct a stream of sludge over the forest area. This type of system is useful only for liquid sludges, preferably with 2 to 3% solids, but may be adapted for use with 6 to 8% solids. Control of sludge application is monitored by flow meters, which totalize the volume of material applied to any area. By monitoring the percentage of solids, the volume of sludge can be converted into a dry weight of sludge applied.

The set irrigation system is a series of standpipes with spray nozzles on top, which are connected by a sludge delivery manifold. The manifold is connected to a sludge pump which boosts the pressure and flow rate of the material to suit the needs of spray range and application rate. The sludge pump is connected to a source of sludge such as the treatment plant digester, a sludge delivery truck, an on-site sludge storage tank, or a storage basin. The delivery manifold and standpipes may be permanently installed if the site will be used continually, or it may be made portable with aluminum irrigation pipe. A portable system can be disassembled, moved to a different location, and reassembled for use.

The traveling irrigation system features a single spray nozzle assembled on some kind of mobile base. Wheeled or tracked vehicles provide ample stability on even, flat terrain. Sludge is delivered by pump through a long, flexible hose mounted on a reel. As sludge is applied through the 360 degree rotating gun, the vehicle moves slowly through the forest along a predetermined track or guide wire. Once the traveling nozzle has reached the limit of its extension, it may be recoiled and set for application from another trail. With this type of system it is crucial to have even terrain and a uniformly spaced plantation providing the needed trail access.

Application Vehicle. A second major application system suited to liquid sludge is a vehicle with a tank and spray nozzle mounted on the rear. Depending on the site needs, a specially designed all-terrain vehicle may be used, or a simple heavy-duty truck chassis with rear-mounted tank may be acceptable. Both systems had been effective in the forests of the Pacific Northwest (Edmonds and Cole 1977, Nichols 1983). Operation is relatively simple. A sludge source is available either at the treatment plant, through a delivery truck, or from on-site storage where sludge is transferred into the application vehicle. Once full, the vehicle moves into the forest over the roads or trails. The sludge is unloaded in uniform thin layers whether the vehicle is moving or stationary. When empty, the vehicle returns to the sludge source for a refill and repeats the cycle.

The vehicle-tank spray system is patterned on a combination of fire-fighting systems and log skidders in the case of the all-terrain vehicle. Key features of the vehicular system include (1) high ground clearance, (2) walking beam or oscillating suspension to increase tire contact with the ground, (3) articulated steering to reduce vehicle turning radius, (4) oversized steel-belted, heavy lug tires for low ground pressure, high flotation, high traction, and puncture resistance, and (5) automatic torque-converter transmission for ease of shifting. Key parts of the tank spray system include (1) as large a tank as possible, mounted low on the chassis for a low center of gravity to reduce rollover potential, (2) a pressure-vacuum system for sludge transfer, and (3) a sludge or solids pump supplying material to a remotely controlled spray nozzle.

Use of the simpler heavy-duty truck with a rear-mounted tank results in less flexibility for use in the forests compared with the all-terrain vehicle. This vehicle is limited to normally constructed forest roads or improved limited slope skidder trails. This sacrifice in vehicle mobility may be acceptable depending on local circumstances and may be offset by ease of operation if the existing road network is adequate.

Direct Spreading. The third system suited to liquid sludge applications in forests is direct spreading, which relies on gravity and may be assisted by a pressurized tank. This system has a minimal application range of about 3 m and is suited for intensively managed plantations where stand layout is controlled to match the vehicle needs. Numerous equipment technologies are available to perform direct spreading, including farm manure wagons, all-terrain vehicles with a rear tank, and trucks with a tank.

Semisolid Sludge

At present, only two systems are effective in applying semisolid sludge (8 to 18% solids) to forests: (1) use of a spray cannon mounted on a tank and vehicle, or (2) direct spreading. Both were described in the section on liquid sludge.

Application Vehicle. The use of a spray application vehicle for semisolid sludge is similar to liquid sludge application. However, pumping semisolid sludge ranges from difficult to nearly impossible. As noted in the section on sludge characteristics, the degree of sludge stabilization may dramatically affect the sludge viscosity and its ability to be pumped. It is impossible to obtain pumps suited for high viscosity sludges that can match the head and flow rate desired in these systems. Very careful attention must be given to system hydraulics plus hands-on experience with individual systems and sludges. From the authors' experience, even slight variations in sludge characteristics may be the difference between pumpable and unpumpable sludge.

Direct Spreading. The approach to direct spreading of semisolid sludge is similar to liquid sludge applications. With sludge of higher solids content, it may be necessary to pressurize the delivery tank to force sludge into the forest at an economical flow rate.

Solid Sludge

Solid sludge can be applied to forests by using a manure-type spreader. A variety of equipment is available to move solid sludge into the forests, including agricultural tractors or skidders towing spreaders, or all-terrain vehicles with rear-mounted spreaders in place of tanks. This system, as with direct spreading, requires even terrain and controlled stand layout to provide access through the stand at close intervals.

COMPARISON OF DIFFERENT STORAGE AND APPLICATION SYSTEMS

In the 1983 Workshop on Utilization of Municipal Wastewater and Sludge on Land (Page et al. 1983) comprehensive, up-to-date information on costs was identified as one of the major needs concerning land application of sludge. This is certainly the case for forest applications, since few municipalities have operational programs. In fact, in a summary of forest sludge projects in the United States (Henry et al. 1984), only a few were fully operational programs: Bremerton, Washington, a city of 36,000 whose program had been operating for five years, and some other small communities in Washington. A number of research projects exist, such as the one in Seattle, where Metro has been involved in research and demonstration projects with the University of Washington for over twelve years, but is just now entering full-scale operations.

Operation and maintenance (O&M) cost data are nearly nonexistent, and only the authors' subjective ranking of costs are presented in this paper. Tables 2 and 3 show comparisons of storage systems and application methods, respectively, in terms of relative costs, advantages, and disadvantages.

TABLE 2. Comparison of storage systems.

System	Relative Costs	Advantages	Disadvantages
Multipurpose vehicles	Low capital, moderate O&M	No transfer of sludge required No in-ground storage needed	Must apply load immediately
Transfer trailers	Low capital, moderate O&M	Can bring sludge directly to area No in-ground storage needed	Ties up trailer while applying (may need more trailers)
Earthen basins	Moderate-high capital, low O&M	Can end dump sludge	Liability, must rehabilitate Usually not close to all of site
Multiple buried tanks	Moderate capital, low O&M	Can end dump sludge Can locate near to all areas	Liability, must remove from site
Aboveground tanks	Low capital, moderate O&M	Can reuse No in-ground storage needed	Must pump sludge May be hard to transport
Portable vessels	Moderate capital, moderate O&M	Can reuse Easy to transport No in-ground storage needed	Must pump sludge

TABLE 3. Comparison of application systems.

System	Relative Costs	Advantages	Disadvantages
Sludge spreading and incorporation	Low capital and O&M	Simple to operate Any % solids	Need cleared site Difficult plantation establishment with some species
Set irrigation system	High capital, low O&M	Simple to operate	Frequent clogging Uses only low % solids Brush interferes
Traveling big gun	Moderate capital, low O&M	Simple to operate on appropriate sites	Frequent clogging Uses only low % solids Brush interferes Needs even terrain
Application vehicle with mounted cannon	Low-moderate capital, high O&M	Any terrain Sludge up to 18% solids	May need special trails
Manure-type spreader	Low capital and O&M	Only effective way to apply high % solids sludge	Limited to high % solids, trails close together

INFORMATION NEEDS

Full-scale operational use of sludge in the forest is a fairly recent alternative challenged by unique sludge application methods needed for forest environments. Although several technologies have been described in this paper, much additional information is needed for a full evaluation of a forest application alternative.

One question becoming increasingly important, as forest landowners are asked to join with municipalities in sludge reuse projects, is how management of a forest site for sludge use differs from normal forest management practices. A complete model of operational technology that integrates sludge management practices with forest management practices is needed. And because development of appropriate application equipment is new, refinements in equipment handling will be necessary—some of them site specific—for more efficient operation. The energy requirements for the special equipment used in these operations must be identified, so that the appropriate power system for each condition can be chosen. A fourth need is complete data on operating costs for full-scale programs, which, together with more information about potential benefits, will provide the basis for economic cost-benefit analyses.

REFERENCES

Bledsoe, C. S. (ed.) 1981. Municipal sludge application to Pacific Northwest forest lands. Institute of Forest Resources Contribution 41. College of Forest Resources, University of Washington, Seattle. 155 p.

Edmonds, R. L., and D. W. Cole (eds.) 1977. Use of dewatered sludge as an amendment for forest growth: Management and biological assessments. Vol. 2. Center for Ecosystem Studies, College of Forest Resources, University of Washington, Seattle.

Henry, C. L. et al. (eds.) 1984. Silvicycle: Information Network of Waste Utilization in Forestlands. Institute of Forest Resources, University of Washington, Seattle. 2(2):5–11. 2(4):1–4.

Henry, C. L., and D. W. Cole (eds.) 1983. Use of dewatered sludge as an amendment for forest growth. Vol. 4. Institute of Forest Resources, University of Washington, Seattle. 110 p.

Nichols, C. G. 1983. Seattle sludge and silviculture. J. Water/Engineering and Management. January, pp. 36–37.

Page, A. L., T. L. Gleason III, J. E. Smith, Jr., I. K. Iskandar, and L. E. Sommers (eds.) 1983. Proceedings of the 1983 Workshop on Utilization of Municipal Wastewater and Sludge on Land. University of California, Riverside. 480 p.

U.S. Environmental Protection Agency. 1983. Process design manual: Land application of municipal sludge. EPA 625/1-83-016. CERI, Cincinnati, Ohio.

Washington Department of Ecology. 1982. Best management practices for use of municipal sewage sludge. WDOE 82-12.

Chapter 35

Utility of a Public Acceptance Survey for Forest Application Planning: A Case Study

LARRY M. GIGLIOTTI and R. BEN PEYTON

ABSTRACT   An important obstacle to the success of forest application of sewage sludge is the failure of publics to accept such disposal methods. A survey instrument was developed to assess Michigan public perception of sewage disposal technologies, especially forest application. It also determined preferred information sources, perceived accuracy of sources, preferred level of involvement, forest recreational activities, and demographic characteristics of respondents. Some findings of the Michigan survey are presented to demonstrate the utility of a survey in providing a basis for presenting sludge treatment alternatives to the public. Nearly half of the general public were undecided about forest application. About 25% were in favor of it. Beliefs about the alternative sludge disposal methods provided the best predictor of attitudes toward forest application. Technical understanding was low because of a lack of knowledge rather than inaccurate perceptions. Citizens who were opposed reported more of an intention to be involved in the issue than those who favored forest application. This study shows the utility of designing surveys to measure a broad range of belief and value components of public attitudes and to anticipate and measure possible intervening factors that may influence public acceptance of forest application. The questionnaire should ask respondents to evaluate all alternatives rather than focusing on one.

An important obstacle to the success of even safe land application systems is the nonacceptance of such procedures by the public (Loehr et al. 1979). Nonacceptance is often accompanied by disruptive tactics that prevent rational consideration of all alternatives and consequences. Since it is more difficult to manage an issue in this disruptive stage, managers must attempt to involve and inform publics in early stages of the decision making. This will require that managers first have an understanding of the public's beliefs and concerns about forest application to identify the appropriate educational and public involvement strategies. Surveys are especially useful in gathering information from the public on emerging issues, since at this stage the public is not motivated to respond to other information gathering techniques. Another possible use of the survey is to stimulate interest in an issue and to educate the public about an emerging issue. This paper will serve to demonstrate the use of a survey to determine public perception of forest sludge application in Michigan.

METHODS

Counties were selected from two regions of the state (Upper and Lower Peninsula) on the basis of (1) proximity to urban areas that may have future sludge disposal problems and (2) the availability of state forest lands (Figure 1). A general public sample of those

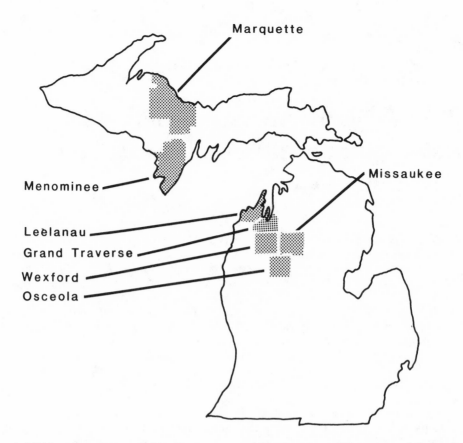

Figure 1. Michigan counties sampled in the survey on public perception and acceptance of sludge disposal alternatives.

counties (n = 2,789) was selected by computer from the population of licensed drivers (twenty years of age or older) by the Secretary of State's Office (Data Processing Division). A second group, political officeholders (n = 174) (township supervisors and county commissioners), was also sampled, since this group was expected to be more involved in resource issues and problems. The public officials were administered the same questionnaire received by the random public sample.

All questionnaires were mailed out at the same time (July 6, 1982) by bulk rate mailing. Prepaid return envelopes were enclosed with the questionnaire and cover letter. The first mailing was followed by a postcard reminder seven days later. A third mailing (second survey and new cover letter) followed the postcard reminder by seventeen days.

The objective of this research was to link current technical disposal research with the human aspects that influence opinion. To accomplish this, a survey instrument was developed to measure important factors that contribute to the public's attitudes and behavior. These included beliefs and knowledge about sludge disposal and its anticipated impacts, the values (concerns) used to evaluate sludge disposal methods, preferred informational sources, and the public's perception of the accuracy of informational sources. The public was also asked to indicate their preferred level of involvement in resource management, their past history of involvement, the extent to which they expect to be involved in future sludge disposal issues, and what they perceive to be their influence on planning and policy. In addition, demographic characteristics and state forest use data were collected.

Attitudes consist of at least two components—affect (the emotional component) and belief (the cognitive component) (Fishbein 1963). There is ample evidence that knowledge of a person's verbalized attitude will be inadequate for predicting behavior. This is partly due to the complex and dynamic nature of attitude formation. Any educational program intended to bring about change must be based more on the underlying beliefs and values that make up attitudes than the expressed opinions themselves.

The construction of the survey instrument reflected these principles of attitude theory (Figure 2). Behavior was hypothesized to be a function of several intervening variables including attitude toward forest application, preference for disposal methods, attitude toward the source of sludge, and perceived credibility of the managing agency. Key public beliefs and values were measured in addition to overall attitude statements. Since attitude statements were assessed at the end of the questionnaire, respondents had the opportunity to consider impinging knowledge and values before making an overall attitude commitment. This was intended to increase the reliability of the attitude responses. Relative value priorities were assessed in addition to merely describing absolute values. Finally, respondents were asked to evaluate not only sludge application for forest lands, but other sludge disposal methods, so that the evaluation could be kept in perspective.

Attitude toward forest application was developed from an item determining support or opposition to forest application and intention of becoming involved, and was used as the dependent variable. Response to this item was combined with the relative prefer-

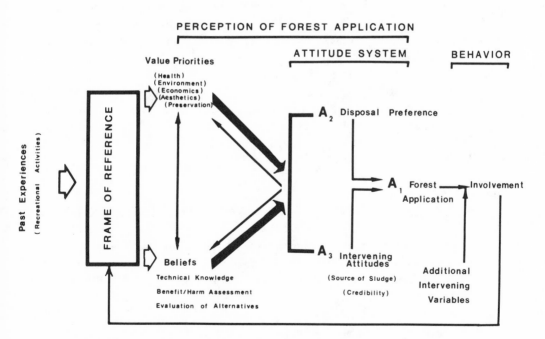

Figure 2. Proposed model of value, belief, attitude, and behavior relationships in a forest application issue. This model proposes that past experiences and frame of reference influence a person's value priorities and beliefs. These interact to form attitudes (e.g., A_1, A_2, A_3) which in turn constitute attitude systems. Attitudes and other intervening variables related to involvement determine a person's behavior in a forest application issue.

ence given forest application compared with three other alternatives to give a scale ranging from very favorable ($+8$) to very opposed ($+1$).

Respondents rated their perceived threat of forest application to health, environmental, and economic concerns in relation to the threats posed by alternative methods (incineration, landfilling, or agricultural application) to produce the relative forest impact variable. This belief scale ranged from low threat ($+3$) to high threat (-3). Two additional belief scales included (1) the technical knowledge variable, which was derived from twelve items scored as plus one for correct answers and zero for wrong or "don't know" responses, and (2) the benefits-or-harm assessment variable, which was a rating of positive or negative impacts forest application may have on nine factors (e.g., forest growth, land values, recreation, water quality).

Three variables were produced to measure the public's value system. The priorities variable was determined by the relative importance assigned to human health, economics, environmental quality, and aesthetics by the respondent. The environmental-or-economic index variable was a measure of the respondents' willingness to make trade-offs between economic and environmental values. The preservationist variable was determined by the response to "Regardless of whether or not any bad effects result, it is wrong for society to dump sludge in the forest."

Two intervening attitudes included in the model were: the source of sludge, which was an evaluation of the acceptance of nonlocal sludge for forest application; and the degree of credibility assigned to state and local government agencies by the respondents. The instrument was designed to measure two components of perceived credibility: trustworthiness and expertise (Petty and Cacioppo 1983). Respondents indicated whether they were skeptical of the Michigan Department of Natural Resources (MDNR), whether they trusted local and state agencies, and whether they considered MDNR an accurate source of information. Responses were used to calculate a credibility scale ranging from very low credibility (-4) to very high ($+4$).

In addition, data were collected on recreational activities (type of activities done on state lands in the past two years) and demographics. Demographics included sex, education, income, urban-rural residence, age, and membership in environmental-outdoor types of organizations.

SURVEY RESULTS AND DISCUSSION

Response Rate

A 60.5% return rate for the general public was obtained based on an adjusted sample size of 2,511 deliverable surveys (Gigliotti 1983). A 72.0% return rate was obtained for the public officials. An estimate of nonresponse bias was conducted by telephone survey. In general, there were few significant differences between respondents and nonrespondents. Nonrespondents were more undecided and less interested in the sludge disposal issue.

Perceived Significance of the Sludge Disposal Problem

Over two-thirds of the respondents agreed (45.0%) or strongly agreed (21.9%) that sludge disposal is a significant problem for many cities in Michigan. Less than 2% thought that sludge disposal was not a significant problem, and 31.3% did not have an opinion. Most (86.5%) of the public indicated an interest in learning more about sludge disposal. Thus, along with the high return rate and the perceived significance of the sludge disposal problem, this survey indicates that people are interested in sludge disposal problems and will be receptive to educational material.

Attitude Toward Forest Application

The general public was mainly undecided (45.1%) in response to a hypothetical proposal to apply sludge in a nearby state forest: 29.9% were opposed and 25.0% in favor of the proposal (Table 1). There was no significant relationship among the seven counties on attitude at the 0.05 significance level. However, when combined into Lower and Upper Peninsula samples, significantly larger groups in the Upper Peninsula were opposed and undecided and a smaller number were favorable.

These data indicate forest application to be an emerging issue in Michigan, since a large segment of the public has not formed an opinion. An educational program at this stage will be faced with a much smaller polarized audience that refuses to consider new information and additional values than would be the case for a more advanced, disruptive issue.

Other researchers have reported that age was negatively correlated with acceptance of land application (Musselman et al. 1980, Olson and Bruvold 1982). However, the Michigan study found that older people were more accepting of forest application of sewage sludge. The survey research did agree with the findings of Olson and Bruvold (1982) that formal education was positively correlated to acceptance of land application.

TABLE 1. Attitudes of selected Michigan publics by region (Upper and Lower Peninsula) toward a hypothetical forest application proposal ($\alpha = 0.05$).

	General Public*		General Public** Combined	Public Officials** Combined
Attitude	Upper Peninsula (n = 406)	Lower Peninsula (n = 1,040)	(n = 1,446)	(n = 115)
	(percent)			
Opposed	32.3	28.9	29.9	23.5
Favor	19.5	27.2	25.0	46.1
Undecided	48.3	43.8	45.1	30.4
	100.1	99.9	100.0	100.0

*Upper Peninsula vs. Lower Peninsula: $x^2 = 9.36$; df = 2; $p < 0.009$.
**General public vs. public officials: $x^2 = 24.44$; df = 2; $p < 0.001$.

Knowledge About Forest Application of Sludge

There was a significantly higher knowledge score for those in favor of a hypothetical proposal to apply sludge in a nearby state forest (5.9) compared with those opposed (3.1) and undecided (3.0) (F = 205.987; df = 2; $p < 0.001$). However, there is evidence that a person's overall attitude may have influenced responses to the knowledge items more than knowledge influenced attitude. First, the group favorable to forest application tended to miss items that when *correct* incriminated forest application. Second, neutral items were about equally missed by both groups. Third, the opposed group missed more of the items that when *incorrect* incriminated sludge. Since there were more items of this type, their overall scores were lower than those in favor of forest application.

A breakdown of the knowledge score shows that respondents were *inadequately* informed rather than *inaccurately* informed. Most (72.5%) of the incorrect responses in the knowledge score were due to "don't know" responses rather than to inaccurate responses. This condition is also preferable, since an educational program needs to form new belief structures rather than change existing belief structures, which may be protected by receivers.

Preferences for Sludge Disposal Methods

Incineration was the first choice of 32.5% of the public, followed by forest application (22.8%), burying in landfills (19.6%), and agricultural application (16.9%), with 8.3% having no opinion. There was a significant relationship between county of residence and first choice of disposal methods. Forest application was ranked first in Leelanau County and agricultural application was first in Grand Traverse County, while incineration was first for the other five counties.

As expected, there was a strong relation between a person's attitude toward a hypothetical proposal to apply sewage sludge in a nearby state forest and preference for an alternative sludge disposal method (Table 2). Those who preferred incineration were

TABLE 2. Relationship between the public's attitude toward a hypothetical forest application proposal and their preference (first choice) for other sewage sludge disposal methods ($\alpha = 0.05$).

| | | Attitude | | | |
Preference	N	Against (n = 432)	Favor (n = 362)	Undecided (n = 651)	Total
		(percent)			
Bury	285	36.8	12.6	50.5	99.9
Incineration	472	48.7	6.1	45.1	99.9
Agriculture	244	18.9	46.3	34.8	100.0
Forest	333	6.9	52.3	40.8	100.0
No opinion	111	25.2	9.0	65.8	100.0

x^2 = 388.33; df = 8; p < 0.001.

least in favor of forest application. There was a large percentage of respondents who perferred agricultural application yet were favorable toward forest application.

These and other results of the study indicate strongly that the citizens must be presented with a sludge disposal problem and asked to select one or more of the available technologies rather than confronted with the task of accepting or rejecting any specific technology in a single proposal. If an issue was proposed separately, special interest groups would focus on any real or perceived negative aspects associated with that technology. Also, evidence suggests that for most respondents, preference for a sludge disposal technology was not based on an informed choice. Preference for incineration or landfills was based on an intuitive "out-of-sight, out-of-mind, out-of-danger" perception.

Perceived Impact of Sludge Disposal Technologies

"Forest growth" was perceived as a beneficial impact of applying sewage sludge in the forest by the public (Table 3). "Long term environmental quality" was also perceived as a beneficial impact, while the seven remaining items (e.g., wildlife, health, water quality, recreation) were considered harmful impacts. "Don't know" responses ranged from 26.3% for "forest growth" to 43.6% for "long term environmental quality." Overall, this analysis showed that many respondents perceived forest application as having many negative impacts. Based on current technological information, it appears that much of their negative evaluation of forest application is unwarranted.

As would be expected from other belief findings, those opposed to a forest application proposal perceived a significantly more harmful impact than did those who favored the proposal or were undecided. These perceived impacts suggest several specific needs to be incorporated into an educational program.

Respondents' evaluation of the impacts of forest application in relation to other sludge disposal methods (the relative forest impact variable) was the best predictor of attitude toward forest application. This variable accounted for 35% of the variance in attitude in a linear multiple regression model (Table 4). Thus, an important component of attitude

TABLE 3. Perceived impact of forest application on nine separate items.

Item	Item Effects Score*			
	Mean	0.95 C.I.	SD	N**
Forest growth	2.18	+0.05	0.93	1,059
Long-term environmental quality	2.87	∓0.09	1.20	808
Wildlife habitat	3.42	∓0.07	1.03	952
Wildlife species	3.50	∓0.06	0.99	918
Public health	3.53	∓0.05	0.77	885
Groundwater quality	3.73	∓0.04	0.78	899
Recreation	3.75	∓0.05	0.81	985
Adjacent property value	3.85	∓0.05	0.82	1,051
Surface water quality	3.96	∓0.04	0.71	937

*Respondents scored 1 for very beneficial, 2 for beneficial, 3 for no impact, 4 for harmful, and 5 for very harmful.
**N excludes those who responded "don't know" or did not respond.

TABLE 4. Stepwise multiple regression: Prediction of attitude toward forest application of sewage sludge.

Variables	Standardized Beta	F Value	Significance of F	R Square	R Square Change	Overall F
Relative forest impact	0.699	352.99	0.000	0.350	0.350	352.99
Preservationist value	0.358	233.82	0.000	0.521	0.171	356.05
Technical knowledge	0.216	69.93	0.000	0.567	0.046	285.66
Benefits or harm assessment	-0.049	78.83	0.000	0.614	0.047	259.45
Degree of credibility	0.149	38.08	0.000	0.635	0.021	226.96
Source of sludge	0.199	15.26	0.000	0.643	0.008	195.81
Recreation activities	-0.037	7.25	0.007	0.647	0.004	170.49
Health priorities	-0.008	6.98	0.008	0.651	0.004	151.42

Listwise deletion of missing data was used; 658 cases were used in the equation.

toward forest application is composed of the beliefs that are held about the alternative technologies. This further supports the recommendation that educational and public involvement programs should include all sludge disposal technologies.

Credibility

Overall, 57.8% of the public perceived the responsible agency (Michigan Department of Natural Resources, MDNR) as credible (Table 5). The general public rated the MDNR high as an accurate source of information; however, a large proportion of respondents were skeptical of management programs proposed by the MDNR. This suggests that the problem lies in the area of implementing programs and may be due to past inadequate public involvement strategies.

Perhaps the most important aspect of this finding is that those opposed to forest application had ranked the agency as significantly less credible than those in favor or those undecided ($F = 97.963$; df $= 2$; $p < 0.001$). This suggests the importance of credibility as an intervening variable in determining public response to any forest application proposal. It is likely that many who would oppose this or any change proposed by a state agency would base their opposition more on a lack of trust in the agency than on the merits of the proposal.

TABLE 5. Frequency distribution of credibility scores for the general public.

Degree of Credibility*	Absolute Frequency	Relative Frequency	X̄	SD
-1	64	4.3	.730	2.081
-3	59	4.0		
-2	121	8.2		
-1	155	10.5		
0	225	15.3		
+1	227	15.4		
+2	237	16.1		
+3	342	23.2		
+4	45	3.1		
Total	1,475	100.1		

*-4 (very low credibility) to +4 (very high credibility). This credibility scale reflects a combination of (1) the public's skepticism of the MDNR, (2) the public's trust in local and state agencies, and (3) the public's rating of the MDNR as an information source.

Public Value Priorities

The instrument determined the relative priority given along a continuum between high economic and high environmental values by the respondents. Most respondents expressed priorities on the environmental side of the continuum (Figure 3). Those opposed to forest application had higher environmental values than those undecided and those in favor ($F = 15.039$; df $= 2$; $p > 0.001$). In addition, those preferring forest application had lower environmental values than respondents preferring other sludge disposal methods ($F = 3.973$; df $= 4$; $p > 0.003$).

Value priorities were determined by asking the respondents to divide 100 points among four categories to represent their relative concern for each (human health, eco-

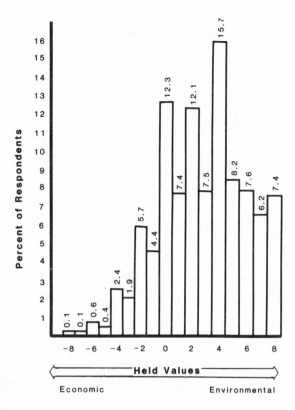

Figure 3. Frequency distribution of the economic-environmental score for the general public. This score was calculated from four questions that required respondents to choose between economic and environmental trade-offs that indicated their value priorities.

nomics or costs, environmental quality and wildlife, and aesthetics or beauty of the area). Health was by far the most important category (X = 50.9; SD = 19.1). Environmental quality ranked number two (X = 22.1; SD = 12.0) and economics (X = 13.5; SD = 10.5) and aesthetics (X = 13.4; SD = 9.1) tied for third.

Those in favor were significantly less concerned about human health impacts of sludge management decisions than those opposed. However, this difference is not large, and for both groups human health was the overriding concern (Table 6). Environmental quality and aesthetics were of equal concern to both groups, while those in favor were more concerned about economics than those opposed. Those in favor see no real threat to health, and perceive some economic benefit. Economic arguments will have little influence on the opposed group, and will not take precedent over health and environmental concerns of even those in favor of forest application.

Almost one-third of the respondents expressed some level of preservationist value by being opposed to forest application regardless of whether or not any bad effect occurs, although only 10.9% strongly expressed this value. In the multiple regression analysis, this was the second most important variable for predicting attitude toward forest application, accounting for an additional 17% of the variance (Table 4). While 10.8% of the sample stated that they strongly agreed that it was wrong to dump sludge in the forest, only 7.6% stated that they were also opposed to a hypothetical proposal to apply sludge in a nearby state forest.

TABLE 6. Number of points allotted to each of the values by respondents opposed to, in favor of, or undecided about forest application.

	Value Priorities		
	X̄	SD	N
*Human Health by Attitude (F = 9.720; df = 2; p < 0.000)			
Opposed	53.3	20.3	397
Favor	47.3	18.3	347
Undecided	51.3	18.4	615
Environment by Attitude (F = 0.446; df = 2; p < 0.640)			
Opposed	22.0	13.0	397
Favor	22.6	13.0	347
Undecided	21.9	10.6	615
*Economics by Attitude (F = 57.001; df = 2; p < 0.000)			
Opposed	10.9	9.8	397
Favor	17.0	11.3	347
Undecided	13.4	9.9	615
Aesthetic by Attitude (F = 0.484; df = 2; p < 0.616)			
Opposed	13.8	10.4	397
Favor	13.1	8.9	347
Undecided	13.4	8.3	615

*α = 0.05.

While findings indicated this small group was no more likely to become involved than the general sample, the potential does exist for this (or any other) minority viewpoint to be overrepresented in public involvement procedures. However, this problem and a number of other potential difficulties can be minimized by a proper introduction of the sludge disposal problem to a local area. A balanced approach identifying impacts of alternative sludge disposal solutions may generate a more supportive, rational response from this group.

Source of Sludge

A high percentage of respondents (72.2%) agreed that if sludge is going to be applied to nearby state forests, it should *not* be brought in from other parts of Michigan. Those who felt that sludge could be brought in from other areas were much more favorable toward forest application (Table 7). Lagerstrom (1983) found that the source of sludge was a major cause of Kalkaska County residents rejecting a proposed forest application demonstration study. The importance of sludge source as an intervening variable may change over time if the public has successful experience dealing with local sludge applied to forest lands.

Forest Recreation Activities

Forest use is expected to be an intervening variable that influences attitude toward forest application of sludge. The count of the number of types of forest recreation activities participated in during the past two years was used to indicate the importance of the state forests to the respondent. Those opposed to forest application reported significantly more forest activities (7.5) than those in favor (6.4) or those undecided (6.1) (F = 16.123; df = 2; $p < 0.001$).

Type of forest use was also explored by using the most important forest use as a variable. The five activities most frequently reported as the most important forest activity were scenic driving (16.6%), fishing (11.6%), firearm deer hunting (11.5%), camping (12.9%), and looking for wildlife (8.4%). There was no significant relationship between acceptance of a forest application proposal and these top five most important forest uses (Chi-square = 12.041; df = 4; $p < 0.149$).

These data were useful in predicting the role of forest use in determining acceptance of forest application, but since attitude cannot be associated with specific segments, groups of recreators cannot be identified and targeted. However, public education and involvement programs should address recreational interests and solicit their input.

Involvement

The general public overwhelmingly (98.0%) feels that citizens should be involved in natural resource decisions. The type of involvement most respondents preferred indicated resource agencies should first obtain the opinions of affected citizens and then provide alternatives from which citizens can select. However, those in favor tended to prefer a lower level of involvement than that preferred by those opposed to forest application (Table 8).

Respondents were also asked specifically whether they would be involved if a forest

TABLE 7. Relationship between respondents' feelings about the source of sludge used in nearby state forests and their attitude toward a hypothetical forest application proposal ($\pm$ = 0.05).

| Source of Sludge Should Be Local | N | Attitude | | | Total |
		Opposed (n = 424)	Favor (n = 351)	Undecided (n = 640)	
		(percent)			
Strongly agree	452	47.3	16.4	36.3	100.0
Agree	585	26.5	23.6	49.9	100.0
Don't know	159	17.6	21.4	61.0	100.0
Disagree	194	10.3	47.4	42.3	100.0
Strongly disagree	25	28.0	52.0	20.0	100.0

x^2 = 166.062; df = 8; p < 0.000.

TABLE 8. Preferred level of involvement on sludge disposal problems analyzed by attitude toward a hypothetical forest application proposal (α = 0.05).

| Attitude | N | Level of Involvement | | | | Total |
		Level A	Level B	Level C	Level D	
		(percent)				
Opposed	426	6.6	29.6	44.1	19.7	100.0
Favor	362	10.8	49.6	31.8	7.7	99.9
Undecided	648	7.9	38.0	46.0	8.2	100.0

x^2 = 72.808; df = 6; p < 0.000.

Level A: No citizen involvement.
Level B: Experts first obtain public views, then make final decisions.
Level C: Experts provide alternatives from which citizens can select.
Level D: Citizens in complete control of the process.

application proposal was made. About 32% of the general public reported that they would be involved, 23.3% would not, and 45.1% were undecided. Since this question was combined with attitude, "undecided" cannot be interpreted for attitude or predicted involvement separately. A majority of those who were opposed to forest application planned to get involved (71.1%). Yet only 42.0% of those in favor planned to become involved (Chi-square = 68.269; df = 1; $p < 0.001$).

It can thus be expected that those opposed to forest application are more likely to be represented in the public involvement process unless special efforts are taken to obtain equal representation. This finding is consistent with other studies and experiences in public participation. Heberlein (1976) reported this as a typical problem with public hearings, which tend to attract those people opposed and give the impression that the majority of the public is opposed. Citizens who favor the technology may find little at risk in the outcome of the proposal and would be less motivated to participate in the decision-making process. Those opposed to forest application would have more motivation to become involved when they perceive a threat to something they value. This was also supported by a strong positive correlation between the number of recreational uses made of forests and the types of actions intended if forest application were proposed (Figure 4).

Unless special efforts are made, public involvement opportunities will be dominated by citizens who are opposed to a specific change (e.g., forest application). Better representativeness may be achieved to some extent by educating those not intending to be involved as to how they may be effective through their participation. However, many

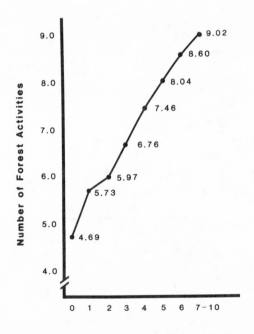

Number of Intended Citizen Actions

Figure 4. Relationship of respondent's number of intended citizen actions to a hypothetical forest application proposal and the number of forest activities the respondent had engaged in during the past two years (F = 17.634; df = 7; $p < 0.000$).

respondents prefer that experts do the planning and decision making in this issue, and it is likely that this group will not increase its participation. Surveys such as this one can help ensure that resource plans reflect the concerns of this group as well as those of the more vocal participants.

When asked what kind of citizen actions they would take to show support or opposition to a proposal to apply sludge to a nearby state forest, few persons reported that they planned to take high level actions such as starting a petition or organizing a meeting or action group. A large percentage reported that they would read material, talk about the issue, or simply sign petitions. A nearly equal group reported they would attend meetings, contact officials, or join a group.

Respondents planning to become involved perceived that they could have more influence on natural resource decisions than those not expecting to become involved. This perceived influence is a measure of locus of control (Peyton and Miller 1980), which may act as an intervening variable in behavioral response to forest application projects. A public involvement program should also teach citizens how to be effectively involved to increase the amount of influence citizens feel they have in natural resource management decisions. The data showed that all groups should be targeted with this type of effort. In addition, since the general public respondents perceived that they had more influence on local government agencies than on state or federal levels, public participation and decision-making processes should be local or regional.

Information Sources

Over one-third (38.9%) of the public reported that they had heard of the concept of applying sludge to land before receiving this survey. These respondents identified their information sources as newspapers (37.3%), television or radio (24.6%), magazine or journal (15.4%), another person (10.2%), pamphlet or brochure (1.2%), school or class (0.8%), and "other" (30.3%) (total is greater than 100%, since respondents frequently listed two sources). Respondents also identified future sources of information as newspapers, television or radio, and pamphlets, in that order (Table 9). Those opposed and those favorable to forest application were relatively similar on their preferred information sources.

TABLE 9. Likelihood that respondents will use various sources to seek more information on natural resource issues.

Information Sources	Rating of Use*			
	X̄	95% C.I.	SD	N
Newspapers	1.97	± 0.05	0.94	1,415
Television and radio	2.04	± 0.05	0.92	1,403
Pamphlets and brochures	2.08	± 0.05	0.85	1,396
Magazines and journals	2.13	± 0.05	0.91	1,391
Attending public hearings	2.52	± 0.06	1.04	1,393
Friend or relative	2.53	± 0.05	0.92	1,389
Library	2.56	± 0.06	1.07	1,396
Workshop, seminars, special lectures	2.76	± 0.06	1.05	1,394
Contacting a university or government agency	2.91	± 0.05	1.02	1,389
Adult education class	3.13	± 0.05	0.92	1,391

*1 = Definitely use.   2 = Probably use.   3 = Seldom use.   4 = Not likely to use.

TABLE 10. Mean rating of perceived accuracy of information sources by the general
public.

Accuracy of Information Sources	Rating*			
	X̄	95% C.I.	SD	N
University sources	2.26	+ 0.03	0.60	1,291
Michigan Department of Agriculture	2.42	∓ 0.04	0.69	1,293
Michigan Department of Public Health	2.44	∓ 0.04	0.74	1,322
Michigan DNR	2.65	∓ 0.05	0.83	1,347
Michigan United Conservation Clubs	2.66	∓ 0.05	0.75	1,011
Environmental organizations	2.70	∓ 0.06	0.84	1,187
Sporting organizations	2.73	∓ 0.05	0.82	1,145
U.S. Environmental Protection Agency	2.81	∓ 0.05	0.89	1,207
Magazines	3.02	∓ 0.04	0.73	1,244
Newspapers	3.14	∓ 0.04	0.78	1,280
Television and radio	3.15	∓ 0.04	0.79	1,251
Local government officials	3.28	∓ 0.05	0.82	1,201
Industrial sources	3.54	∓ 0.05	0.84	1,170

*Respondents rated each source: 1 = Always accurate. 2 = Usually accurate.
3 = Sometimes accurate. 4 = Seldom accurate. 5 = Never accurate.

Respondents rated the accuracy of thirteen information sources (Table 10). University sources had the highest perceived accuracy rating followed by the Michigan Department of Agriculture and the Michigan Department of Public Health. Industrial sources and local government officials were seen as the least accurate information sources. The accuracy of the three most popular sources (magazines, newspapers, and television and radio) ranked low.

These results suggest a number of recommendations. First, information should be linked with a source seen as credible. Petty and Cacioppo (1983) identified credibility of the source of a message as one of the most important features of successful persuasion. Second, information should be disseminated via news media, since respondents identified this as their preferred source of information. And, third, a major effort of the information and education program should be to provide workshops at appropriate stages to inform news media personnel. This is important because while the public stated that they would most likely use the news media as sources for information, they rated them relatively low in accuracy. Attempts should be made to improve the accuracy of news media information and to stress the original sources of news media information (university sources, unbiased research, etc.).

Public Officials

The separate sample of public officials were significantly more favorable to forest application than the general public (Table 1). If a forest application proposal is introduced to public officials, there is a greater chance of its being accepted; however, public officials may not represent the public's opinion. This reaction occurred in Kalkaska County, Michigan (Lagerstrom 1983). The MDNR initially received verbal approval from public officials to conduct a forest application demonstration study, but public protest later reversed the public officials' position. However, the findings of the present study are different from those of Jewell and Seabrook (1979), who stated that municipal officials were more opposed to land application.

Public officials were also significantly more economically orientated than the general public. This difference in value priorities could lead to problems when these values are in conflict. Public officials may opt for a cheaper solution to problems while the general public is willing to pay for increased safety. O'Riordan (1971) found that public officials would not consider a solution to solve a wastewater issue that involved increased taxes because they assumed that the public would be opposed. However, it was discovered that the public was more interested in solving the environmental aspect and was willing to endure higher taxes. This further emphasizes the need to assess opinion of the general public and of public officials and to provide for public participation.

SUMMARY AND RECOMMENDATIONS

Based on experience with this and other surveys, the authors have a number of recommendations for the use of surveys as a tool in sludge disposal management. To be effective, a survey must measure a wide range of related beliefs and the value priorities that might be affected. This will permit a better understanding of the underlying reasons for the public's acceptance or nonacceptance and will be important in guiding and focusing education and public participation programs. In this survey, the item measuring attitude was placed near the end of the questionnaire, thus encouraging respondents to consider the full range of their beliefs about the sludge disposal issue. It also provided an opportunity for respondents to think about their priorities and how they believe sludge disposal issues affect their values. This method tends to produce an attitude measurement that is more realistic and stable. The results of this survey give the manager an understanding of the dynamics of the public's attitudes toward forest application, providing a large number of implications for management. More important, these findings were identified before the issue reached the disruptive stage.

An important strategy in questionnaire design is to incorporate all sludge treatment alternatives for respondents to evaluate. This enables the planner to put attitudes toward forest application in perspective. Attempts to assess information levels and value priorities should apply to a broad range of alternatives as well.

This survey identified mass media as an important source of information for a sludge management issue. This information is probably applicable for most geographic regions in the United States. However, when a survey is used for a specific geographical area, it is necessary to measure credibility of specific offices and agencies that may be involved in that area. This and other studies also identified the source of sludge as an important factor. A survey designed for a specific region should explore this variable in more detail to determine whether the receiving residents view the sludge as a "local" problem and the extent to which that influences their acceptance.

One limitation in this research was in sampling design. This research failed to consider the nonresident landowners. This group in Michigan is likely to have a strong interest in any resource management issue that potentially affects forest recreational activities and should be included in the public education and involvement efforts.

ACKNOWLEDGMENTS

Research was supported by U.S. Environmental Protection Agency grant No. S005551-011 and Michigan Agricultural Experiment Station project No. 3248.

Michigan Agricultural Experiment Station journal article number 11656.

REFERENCES

Fishbein, M. 1963. An investigation of the relationship between beliefs about an object and attitude toward that object. Human Relations 16:233–239.

Gigliotti, L. M. 1983. A public assessment of concerns and beliefs about forest application of sludge. M.S. thesis, Michigan State University, East Lansing.

Heberlein, T. A. 1976. Some observations on alternative mechanisms for public involvement: The hearing, public opinion poll, workshop and quasi-experiment. Nat. Res. J. 16:197–212.

Jewell, W. J., and B. L. Seabrook. 1979. A history of land application as a treatment alternative. EPA 430/9-79-012. U.S. Environmental Protection Agency, Washington, D.C. 83 p.

Lagerstrom, T. R. 1983. Comparison of citizen reaction to a proposed sludge demonstration project in two Michigan counties. M.S. thesis, Michigan State University, East Lansing.

Loehr, R. C., W. J. Jewell, J. D. Novak, W. W. Clarkson, and G. S. Friedman. 1979. Land application of wastes. Vol. 1. Van Nostrand Reinhold Company, New York.

Musselman, N. M., L. G. Welling, S. C. Newman, and O. A. Sharp. 1980. Information programs affect attitudes toward sewage sludge use in agriculture. EPA 600/2-80-103. U.S. Environmental Protection Agency, Cincinnati, Ohio. 51 p.

Olson, B. H., and W. Bruvold. 1982. Influence of social factors on public acceptance of renovated wastewater. p. 55–73. In E. J. Middlebrooks (ed.) Water reuse. Ann Arbor Science Publishers, Ann Arbor, Michigan.

O'Riordan, T. 1971. Public opinion and environmental quality: A reappraisal. Environment and Behavior 3(2):191–214.

Petty, R. E., and J. R. Cacioppo. 1983. Attitudes and persuasion: Classic and contemporary approaches. Wm. C. Brown Company Publishers, Dubuque, Iowa.

Peyton, R. B., and B. A. Miller. 1980. Developing an internal locus of control as a prerequisite to environmental action taking. p. 173–192. In A. B. Sacks, L.L. Burrus-Bammel, C. B. Davis, L. A. Iozzi (eds.) Current issues IV: The yearbook of environmental education and environmental studies. ERIC Center for Science, Mathematics and Environmental Education, Ohio State University, Columbus.

Chapter 36

Relating Research Results to Sludge Guidelines for Michigan's Forests

DEAN H. URIE and DALE G. BROCKWAY

ABSTRACT   Guidelines for application of wastewater sludge to forest land in Michigan were developed by the Department of Natural Resources from research studies on small plots and large-scale demonstration sites. Growth response and groundwater quality data provided a basis for estimating appropriate application rates, selecting suitable application sites, and developing proper application procedures. Balancing nutrient additions with the assimilation capacity of each forest ecosystem was found to enhance site productivity while minimizing nutrient enrichment of groundwater. This principle, referred to as the periodic agronomic rate, is similar to the already widely accepted annual agronomic rate currently in place as policy for sludge application on cropland. The periodic agronomic rate, however, recognizes the unique nutrient cycling and long-term storage potential of aggrading forests. Sludge nutrient application rates and operational procedures reflect the need for a conservative approach in protecting the environment and public health when implementing regulatory programs that employ new technology.

In the process of developing regulatory direction for land application of wastewater sludge, it is important to draw upon the available research data and management experience that will aid in developing guidelines that provide adequate protection for environmental quality and the public health. In Michigan, a statewide program based on state-of-the-art knowledge regulates land application of sludge through a system of permits issued by the Department of Natural Resources (DNR). This paper describes how available research and development information has been used by DNR to formulate proposed guidelines for the forest land application option.

State Groundwater Rules specify that "quality degradation of any usable aquifer is prohibited" (MDNR 1980). Groundwater in the northern forested region of the state is widely used as a domestic water source with minimal treatment. Furthermore, the water yielded by forested watersheds is normally expected to be of high quality. Therefore, any surface land management activity is required by law to have minimal effect on a usable phreatic aquifer.

While some regulators have interpreted the legislative mandate to mean "zero discharge" to any usable aquifer, foresters and agronomists have, from their understanding of the normal rates of elemental flux in natural biogeochemical cycles, argued for a less restrictive policy of quality control for the groundwater. Plans for land application of sludge should consider rates of addition appropriate to each ecosystem's capacity for nutrient and trace element assimilation. Available research on forest land application indicates that excessive sludge applications, which result in available nitrogen at levels greater than site assimilation, storage, and denitrification capacities, can lead to ground-water enrichment by nitrate leaching. Balancing sludge application rates with rates of volatilization, mineralization, nitrification, denitrification, soil storage, and plant uptake is central to minimizing potentially adverse effects on groundwater.

EXISTING GUIDELINES

Existing guidelines for land application of sludge have been developed with primary emphasis on cropland (MDNR 1982). Since the mid-1970s, application to farmland has been the most frequently selected sludge recycling option. Under the requirements of the NPDES permit issued for operation of each municipal and industrial facility, a Program for Effective Residuals Management (PERM) must be drafted by the waste generator which outlines the proposed sludge management procedures.

The PERM must contain the following information: (1) a description of the treatment processes used and an estimate of the monthly production of sludge solids, (2) a list and description of all facility storage structures, including the capacity of each, (3) a description of the sludge transportation equipment available, (4) application site data, including maps, site dimensions, ownership, and significant site modifications, (5) a list of site management activities related to method and timing of sludge application, crops to be grown, and harvest frequency, (6) proposed rates of sludge application based on sludge chemical composition, soil fertility needs, and crop nutrient needs, (7) a comprehensive chemical analysis of the sludge, (8) a monitoring schedule to include frequency of analyses for sludge and soil (groundwater and plant tissue if more intensive monitoring is indicated), and (9) a list of contingency options. As long as an agronomic rate of sludge nutrient additions is used for land application, the state requirement for filing a hydro-geological study and a groundwater monitoring plan for each site is usually waived.

In conducting a local sludge management program under a state approved PERM, additional guidelines related to use on croplands are to be observed. Although primarily aimed at abating pollution of surface and groundwaters, these best management practices are also intended to minimize problems with sludge odors and citizen complaints.

These additional requirements include: (1) Sound farming practices are to be used. (2) Sludge use on soils with high available phosphorus levels (greater than 336 kg/ha or 300 lb/acre) is to be avoided. (3) Surface-applied liquid or solid sludge is to be incorporated into the soil within 48 hours of application. Liquid sludge applied to forage crops within 7 days of cutting is exempt from this requirement. (4) Grazing is prohibited for 30 days following sludge application. (5) For surface application of sludge, isolation distances of 152 m (500 ft) to homes and commercial buildings and 46 m (150 ft) to wells, surface waters, public roads, and property lines must be observed. (6) For subsurface application of sludge, isolation distances of 30 m (100 ft) to homes and commercial buildings, 46 m (150 ft) to wells, 15 m (50 ft) to surface waters, and 8 m (25 ft) to roads and property lines must be observed. (7) Maximum slope limitation for surface-applied sludge is 6%, and for subsurface-applied sludge 12%. (8) Use of sludge is to be avoided when the water table is less than 76 cm (30 inches) below the soil surface. (9) Sludge applications to frozen or snow-covered croplands are not generally approved.

The sludge management activities of each generator facility are monitored by the required submission of monthly operation reports, which provide detailed information for each period of sludge application to land. For each site receiving sludge a Waste Disposal Sheet (Figure 1) is completed, to document placement of the material and track the on-site accumulation of nutrients and trace elements.

Because of limited experience with silvicultural use of wastewater sludge, forest land application programs proposed by municipalities and industries have been evaluated on a case-by-case basis. However, recent research has provided sufficient data for state specialists to formulate guidelines for forest land applications. Although sludge application programs developed for forests must suit the individual aims of the generator and the land manager, meeting their objectives must be subordinated to sufficient protection for the environment and public health.

Research has shown that nitrate-N leaching is usually the factor limiting sludge application rates for most forest ecosystems (Brockway and Urie 1983). Rates of nitrogen volatilization, mineralization, nitrification, denitrification, and plant assimilation will all affect the allowable maximum application rate in each forest type. However, nitrogen assimilation rates in fertilized forests have received much less attention than those of natural, unamended ecosystems. First-year assimilation rates can be estimated from available research data for liquid sludge applications in several forest types (Urie et al. 1984).

The timing of repeated sludge applications will depend on the rate at which organic nitrogen in the sludge solids remaining from earlier treatments is mineralized. This rate is known to vary, depending on the treatment process used during sludge generation (Sommers and Nelson 1981). The nitrogen-enriched forest floor will undergo more rapid decomposition, and the native organic matter may also become mineralized more rapidly. Although available data are sketchy on these dynamics, nitrogen availability on sludge-treated forest sites and the need for sludge nitrogen inputs will be determined by the "new" nitrogen cycle.

Annual nitrogen and cumulative heavy metal loading rates are considered in sludge application guidelines for cropland. No criteria based on demonstrated toxicity currently exist for placing limits on heavy metal application rates in forests. State regulatory agencies generally discourage land application of sludges containing unusually high levels of toxicants. Although relatively high applications of heavy metals have been made to forests with no measurable adverse effects (Brockway 1983), heavy metal mobility is expected to be somewhat greater in the acid soils typically present on northern forest sites than in less acidic agricultural soils.

Groundwater monitoring for the first generation of operational forest land application programs, though expensive, may be required to develop public confidence in the new technology. This may be especially true when an added margin of safety is needed for projects operated in sensitive ecosystems. Monitoring will also serve to evaluate the rather conservative restrictions usually in effect during the early tentative stages of a new program. Using this additional body of experience, guidelines can be later refined to more closely match site assimilation capacity to sludge application rates.

PROPOSED GUIDELINES

A policy of sludge application for forest land is proposed which is consistent with the statutory directive of nondegradation of usable aquifers. Similar to the annual agronomic rate currently in use for cropland, this proposal is one of a periodic agronomic rate that supplies the nutrient needs of the forest crop for the interval (averaging five years)

______________________ 19 ____
(Month)

____
(Facility)

____
(Superintendent's Signature)

State of Michigan
Department of Natural Resources

WASTE DISPOSAL SHEET

Site ______________________
____
# of seasons utilized to date ______
Acres used this month ______
Total acres in site ______

DATE	WASTE APPLIED: Gallons or Cubic Yards	% Solids	% VS	Dry Tons per ac	WASTE ANALYSIS AND SOIL LOADING RATES: Nitrogen TKN %	Nitrogen NH₄ %	Nitrogen NO₃ %	Nitrogen AVAN lb/ac	Total Phosphorus TP %	Total Phosphorus TP lb/ac	Potassium K %	Potassium K lb/ac	Lead Pb mg/kg	Lead Pb lb/ac	Zinc Zn mg/kg	Zinc Zn lb/ac	Copper Cu mg/kg	Copper Cu lb/ac	Nickel Ni mg/kg	Nickel Ni lb/ac	Cadmium Cd mg/kg	Cadmium Cd lb/ac	Chromium Cr mg/kg	Chromium Cr lb/ac	mg/kg	lb/ac
Average																										
TOTALS This Month																										
TOTALS This Year																										
TOTALS Cumulative																										

CROP AND SOIL DATA

Crop to be fertilized ______

Subsequent Crop ______

CEC ______ meq/100g

pH ______ S.U.

Bray P1 ______ ppm

K ______ ppm

Crop Yield Goal ______

NITROGEN RECOMMENDED

N ______ lb/ac

Acceptable Metal Accumulations

	Total	Yearly	
Pb			lb/ac
Zn			lb/ac
Cu			lb/ac
Ni			lb/ac
Cd	4.5	0.45	lb/ac
			lb/ac
			lb/ac

Cd 0.45 lb/ac/yr (0.5 kg/ha/yr)

REMARKS:

Date of Waste Analysis

PR 5530
Rev 12/84

Figure 1. Waste disposal sheet.

between sludge applications. Nutrient additions in this manner will fully utilize the assimilation and storage capacities of the forest environment. Sludge nitrogen additions are limited by the rates of volatilization loss, mineralization, nitrification, denitrification loss, plant uptake, and soil storage so as to minimize leaching losses to groundwater.

Operational sludge application rates must be individually computed for each site, based on specific sludge nutrient and trace element concentrations, sludge stabilization method, and the nutrient uptake potential of the candidate forest type. Ammonia volatilization losses following land application are expected to vary from 20 to 60% (Sommers and Nelson 1981) depending on temperature and wind conditions. An average of 50% of the ammonia-N is estimated to be available in the first year. Sludge stabilization process provides an indication of the nitrogen mineralization rate (Table 1). Nitrogen uptake rate

TABLE 1. Nitrogen mineralization rates in the first year following sludge application (Sommers and Nelson 1981).

Sludge Type	Nitrogen Mineralization Rate (%)
Waste activated	42
Primary	29
Primary plus CaO	28
Aerobic digested	25
Anaerobic digested	15
Composted	9
Primary, wet air oxidized	3

is a prominent factor determining the periodic agronomic rate of sludge application. Those for several Michigan forests are listed from available literature (Table 2). Using these data, a sludge application rate that provides a substantial input of fertilizing nutrients and an adequate level of protection for the groundwater is computed.

Site selection criteria are the second major area of sludge management planning concern, since they determine the antecedent conditions within which program activities will be conducted. Selecting sites according to these guidelines will aid in management operation and minimize adverse effects on the forest environment.

The required information for and characteristics of a candidate forest site appropriate for land application of sludge are as follows: (1) Provide background data concerning site location (plat map), proximity to structures, roads, and drainage ways (aerial photo), soil types (soil survey map), soil fertility, site history, significant site modifications, stand composition, age, basal area, stocking (cover type map), and understory. (2) Soils should be no less than somewhat poorly drained, and the water table must be at least 76 cm (30 inches) below the soil surface at the time of sludge application. (3) Maximum slope limitation for surface-applied sludge is 6%. (4) For surface application of sludge, isolation distances of 152 m (500 ft) to homes and commercial buildings and 46 m (150 ft) to wells, surface waters, public roads, and property lines must be observed.

If sludge application to frozen or snow-covered soil in winter is proposed, these additional considerations must be met for each site: (1) Site access trails must not exceed 2% slope. (2) An isolation distance of 152 m (500 ft) to homes, commercial buildings, wells, surface waters, public roads, and property lines must be observed. (3) Soils should be no

TABLE 2. Recommended sludge available nitrogen (AVAN) application rates for forests in Michigan, over a five-year retreatment interval.

| Species | Age (years) | Sludge Application Rate | | Nitrate Leached (kg/ha·yr) |
		Dry solids (Mg/ha·5 yrs)	AVAN (kg/ha·yr)	
Aspen	0 to 5	9	56	17
Aspen	5 to 20	18	112	28
Aspen	over 20	9	56	17
Northern hardwoods	0 to 10	7	45	17
Northern hardwoods	10 to 30	14	90	24
Northern hardwoods	over 30	7	45	15
Oak and hickory	0 to 10	9	56	17
Oak and hickory	10 to 30	18	112	28
Oak and hickory	over 30	9	56	17
Scrub oak	0 to 20	5	34	12
Scrub oak	over 20	9	56	17
Red, white, jack pine	0 to 10	9	56	17
Red, white, jack pine	10 to 30	5	34	12
Red, white, jack pine	over 30	4	22	10
Spruce and fir	0 to 10	7	45	15
Spruce and fir	10 to 30	5	34	12
Spruce and fir	over 30	4	22	10
Northern white cedar	0 to 20	9	56	15
Northern white cedar	over 20	4	22	10

less than moderately well drained with no reasonable probability of surface runoff of sludge-applied solids. (4) The stand must be fully stocked with canopy cover of no less than 60%. (5) Liquid sludge application rates are limited to a maximum of 280,500 liters per hectare (10,000 gallons per acre). (6) There must be no established winter recreation uses on the site. (7) Each site must be selected by September 15 prior to the winter during which it is proposed for sludge application and clearly identified by signs that caution entry.

In addition to information on application rates and site selection, several points regarding management activities should be noted. Surface application of liquid, well-digested sludge is encouraged for use in Michigan forests. Research in other states (Richter et al. 1982, Wells et al. 1984) indicates that dewatered sludges have much lower rates of biological activity and provide little immediate benefit to forest nutrition, for they remain relatively inert while perched on the forest litter for long periods. Federal regulations (U.S. EPA 1979) require that land-applied sludge be properly stabilized by a process that significantly reduces pathogens to protect public health. This point is particularly pertinent where sludges are surface applied on publicly owned forest land, because well-stabilized sludges are also less likely to produce offensive odors.

The method of liquid application should effectively distribute a uniform cover of sludge on the forest floor. This will reduce the possibility of creating overloaded microsites, which may function as points of high nitrate discharge to groundwater. Equipment should be selected and operated to minimize damage to soil and standing trees. Such measures will diminish the risk of soil compaction, infection of trees through scars on stems or roots, and infestation by insects.

SUMMARY

Guidelines for application of wastewater sludge to Michigan's forest lands are proposed that include an empirical basis for estimating appropriate application rates, criteria for selecting suitable application sites, and general guidance for overall application program operation. The central approach of balancing nitrogen additions to the assimilation capacity of each forest type represents a rational way of stimulating biological productivity while ensuring adequate protection for the environment and the public health. The principle of balance, referred to as the periodic agronomic rate of application, is patterned closely after the already widely accepted annual agronomic rate currently in place as policy for sludge applications to cropland. The distinguishing feature between them lies in recognition of the unique nutrient cycling and long-term storage potential of aggrading forest ecosystems. Regional research data have been used to establish very conservative estimates of nitrogen assimilation for numerous forest types. These were established in recognition of the need to maintain continued yields of high quality water from forested watersheds. As knowledge grows with increasing experience in the area of forest land sludge applications, revised estimates of nutrient and trace element assimilation capacities for the numerous forest environments of Michigan will facilitate development of more finely tuned guidelines for maximizing forest site benefits and obtaining the lowest sludge nutrient recycling costs consistent with environmental protection.

REFERENCES

Brockway, D. G. 1983. Forest floor, soil, and vegetation responses to sludge fertilization in red and white pine plantations. Soil Sci. Soc. Am. J. 47:776–784.

Brockway, D. G., and D. H. Urie. 1983. Determining sludge fertilization rates for forests from nitrate-N in leachate and groundwater. J. Environ. Qual. 12:487–492.

Michigan Department of Natural Resources. 1980. Groundwater quality rules of Act. No. 245 of the Public Acts of 1929, as amended Part 22, Rule 323.2205.

————. 1982. The Michigan municipal wastewater sludge management program. Water Quality Division, Lansing. 14 p.

Richter, D. D., D. W. Johnson, and D. M. Ingram. 1982. Effects of municipal sewage sludge-cake on nitrogen and phosphorus distributions in a pine plantation. p. 532–546. In Fifth Annual Madison Conference of Applied Research and Practice on Municipal and Industrial Waste, Department of Engineering and Applied Science, University of Wisconsin, Madison.

Sommers, L. E., and D. W. Nelson. 1981. Nitrogen as a limiting factor in land application of sewage sludges. p. 425–448. In Fourth Annual Madison Conference of Applied Research and Practice on Municipal and Industrial Waste. Department of Engineering and Applied Science, University of Wisconsin, Madison.

Urie, D. H., A. R. Harris, and J. H. Cooley. 1984. Forest land treatment of sewage wastewater and sludge in the Lake States. p. 101–110. In 1984 TAPPI Research and Development Conference, Appleton, Wisconsin. Technical Association of the Pulp and Paper Industry, Technology Park, Atlanta, Georgia.

U.S. Environmental Protection Agency. 1979. Criteria for classification of solid waste disposal facilities and practices. Federal Register 44(179) Part IX, 40 CFR Part 257. September 13.

Wells, C. G., K. W. McLeod, C. E. Murphy, J. R. Jensen, J. C. Corey, W. H. McKee, and E. J. Christensen. 1984. Response of loblolly pine plantations to two sources of sewage sludge. p. 85–94. In 1984 TAPPI Research and Development Conference.

Case Studies: Municipal Wastewater

Forest Land Treatment of Wastewater in Clayton County, Georgia: A Case Study

WADE L. NUTTER

ABSTRACT   Clayton County, Georgia, within the metropolitan Atlanta area, was faced in the early 1970s with unique wastewater treatment and water supply problems. Situated along the eastern continental divide, streams in the county do not have sufficient flow to assimilate wastes or to meet the demands for water supply. After technical and cost-effective evaluations of a number of wastewater treatment plans, land treatment by spray irrigation to forests was selected as the cost-effective and environmentally sound alternative. Public information and education programs were essential to public acceptance of land treatment. Two forest land treatment systems are now in operation in Clayton County, 48 ha receiving 0.05 m³/sec flow and 1,460 ha receiving 0.85 m³/sec flow. The systems operate year-round with a hydraulic loading of 6.4 cm/week. Wastewater is applied at below freezing temperatures when necessary and on slopes as great as 30%. Forest biomass is harvested as an energy source to dry and pelletize the sludge. Since operation began in October 1982, there has been no change in quality of groundwater or surface water leaving the site. Annual streamflow has increased by 93%, primarily as baseflow. The 1,460 ha site is located within the water supply watershed and serves to recycle and increase the supply of drinking water.

Clayton County is the fourth smallest county in Georgia and is located in the southern portion of metropolitan Atlanta. Of the 159 counties in the state, Clayton County was the ninth most populated in 1980 and is expected to be fourth by the year 2000. The county's rapidly growing population (estimated at 200,000 in 1985), small land area, and geologic and topographic features create unique wastewater and water supply problems.

The county is located in the middle piedmont physiographic province and is characterized by deeply weathered soils with clay to clay loam subsoil textures, granitic bedrock, incised streams, relatively flat interfluves, and limited groundwater supplies. Straddling the eastern continental divide, streams in the eastern part of the county flow to the Atlantic Ocean and those in the west flow to the Gulf of Mexico. As a result, most streams originate within the county and have insufficient flow to provide adequate water supply or to assimilate the wastewaters generated. Clayton County's principal water supply source is located in an adjacent county on a stream in the Atlantic drainage which has its headwaters in Clayton County.

Wastewater generated at the two Clayton County treatment plants had historically been discharged to the Flint River in the Gulf of Mexico basin. The mean annual flow of the Flint River in the Clayton County watershed is approximately 1.1 m³/sec, and discharge of wastewater in 1974 to the river averaged 0.44 m³/sec. During many summer months, the wastewater discharge exceeded the background streamflow.

Water quality goals established in 1973 and 1974 by the Georgia Environmental Protection Division (EPD) for the Upper Flint River Basin were designed to protect the drinking water supply of the city of Griffin downstream from the Clayton County and other discharge points and to meet new state and federal water quality standards. The goals, expressed as effluent standards for all discharges above the Griffin water intake, were 10 mg/l five-day biological oxygen demand (BOD_5), 2 mg/l ammonia-N, 6 mg/l dissolved oxygen, 1 mg/l phosphorus, and disinfection.

LAND TREATMENT EVALUATION

The Clayton County Water Authority (CCWA) undertook in 1974 a study in accordance with Section 201 of Public Law 92-500, the Federal Water Pollution Control Act Amendments of 1972, to evaluate wastewater treatment alternatives to meet the stated water quality goals. Section 201 encourages the use of recycling technology and requires that land treatment be evaluated and compared with conventional treatment methods as part of the cost-effective analysis. At the time the study was initiated, the law required innovative and alternative technology (land treatment is specified as one such technology) to be selected if the cost did not exceed the cost-effective alternative by more than 15%.

Public Law 92-500 also requires areawide basin planning as part of the wastewater facilities evaluation. Thus the Section 201 Flint River Wastewater Facilities Plan study site was 570 square kilometers in area and included five counties and ten communities. All wastewater discharges within the basin were evaluated and wasteload allocations assigned by the Georgia EPD to each of the facilities. Clayton County's allocation was 0.59 m³/sec at the effluent standards previously described. Because the projected twenty-year wastewater flows were 0.95 m³/sec, the CCWA would have to treat the wastewater to yield a higher quality discharge (i.e., lower constituent concentrations) or limit growth to meet these allocations.

At the same time conventional wastewater treatment processes were evaluated as part of facilities planning, a spray irrigation land treatment system evaluation was also undertaken. The first task was a feasibility study using as source material published information such as USDA Soil Conservation Service soil surveys, topographic maps, climatic summaries, and engineering data such as projected wastewater flow volume and quality. Results indicated that general soil, topographic, and climatic conditions were favorable for land treatment, and a hydraulic loading between 5.0 and 6.4 cm/week was possible with a minimum wetted area of approximately 800 to 1,200 ha. Preliminary cost analyses at this point indicated that the land treatment cost was within an acceptable range of the conventional alternatives being considered and further study was justified.

The second phase of the land treatment evaluation process included identification of potential sites, evaluation of each site, selection of a preferred site, intensive study of the selected site, and development of cost estimates. Five sites with a total area of 4,800 ha were selected for study using tax maps and local real estate brokers' knowledge of land availability. Criteria used to select sites for evaluation included large contiguous areas, well-drained soils, forest as the dominant land use, and location within 12 km of the wastewater treatment plants. All but one of the study sites were within the Flint River basin planning area.

Detailed analyses of each site were conducted using published information which included soil surveys, aerial photographs, climatic summaries, and geologic, topographic, groundwater, and surface water maps. Overlay maps of soils, slope, land use (vegetation and structures), and water resources were constructed and used to estimate the land area within each site suitable for land treatment of wastewater. Evaluations such as winter temperatures, wind speed and direction, daily and monthly expected precipitation, and soil permeability and wetness limitations were conducted. Water balances based on five-year monthly return period precipitation, potential evapotranspiration, soil hydraulic properties, and season of the year were calculated. Projections of wastewater quality from the conventional secondary treatment plants were also developed. Based on suitable and available land area and other site factors and projected wastewater quality, the site outside the planning area was selected for further study and cost analysis. The site was in the headwaters of Big Cotton Indian Creek, the source of Clayton County's drinking water supply. Following selection of the site, a detailed geotechnical and soils investigation was conducted to obtain the additional site specific information necessary for development of land treatment design criteria.

The selected site was the largest contiguous site studied (1,850 ha), was 80% forested with slopes ranging from nearly level to 35%, had a well-defined drainage network, and would result in displacement of the fewest property owners. Total land area requirement (wetted and buffer areas) was initially set at 1,400 ha, the final requirement to be determined when property configurations and buffers were specified. Using these specifications, a detailed analysis of capital and operations and maintenance costs was completed for comparison with the other alternatives.

The principal alternative wastewater treatment strategies considered were (1) land treatment of the total flow by spray irrigation, (2) several methods of advanced wastewater treatment (AWT) of the total flow, and (3) AWT of the portion of flow that could be stream discharged in accordance with the wasteload allocations for the Flint River and land treatment of the remainder following secondary treatment. Also considered as part of each alternative was upgrading of the existing treatment plants or abandonment and the construction of new plants.

Land treatment of the total flow was the alternative selected based on cost-effective and environmental considerations. The projected cost (including capital and annual operating and maintenance for twenty years) was slightly less than the combined AWT and land treatment cost. Although abandonment of the existing plants and substitution of a multicell pond for treatment prior to land application was considered, it was less costly to upgrade the two existing facilities to the activated sludge process.

Because of suburban development downgradient from the two existing treatment plants, a third, small plant to handle approximately 6% of the projected wastewater flow (0.05 m³/sec) was to be built and the wastewater applied to a second land treatment site of 48 ha, because the flows could be collected and transported to the plant for the most part by gravity. This smaller system, known as the Shoal Creek system and developed using the same design criteria, is not included in the following discussion. The larger system was referred to as the Pates Creek system and was later dedicated as the E. L. Huie Land Treatment System.

The projected cost of the Pates Creek land treatment alternative is presented in Table 1. The actual capital cost of the land treatment system was about 1.5% less than projected. Annual operations and maintenance costs (based on 0.66 m³/sec flow) were esti-

TABLE 1. Projected capital cost of Clayton County Pates Creek land treatment system.

Item	Cost (in 1976 U.S.$ x 10^6)
Treatment plants	$ 8.4
Land treatment sites	
Transmission line	1.7
Storage reservoir	0.46
Pumping and distribution	5.3
Structures, roads, fencing, wells, etc.	0.64
Land	10.3
Total (not including plants)	$18.40

mated at $355,000 per year. On the basis of present worth calculations, which include a salvage value for the land, the present worth of the land treatment system was estimated at about $30 million (1978 base) (Roy F. Weston, Inc. 1983).

During the site evaluation phase of the 201 Wastewater Facilities Planning process, education and information sessions were held with local and state environmental groups, local elected and technical personnel, and state and federal regulatory officials. Once the land areas to be studied were identified, an education and information session was held for property owners within and near the study sites. Tours were also arranged for public officials and journalists to visit similar operating land treatment systems. At the end of the Facilities Planning Study, public hearings were held to review the planning process, costs of alternatives, and description of the selected alternative. The educational program was judged to be a principal factor in public acceptance of land treatment as the selected wastewater treatment plan.

The Wastewater Facilities Plan, including the treatment alternative evaluations and development of the land treatment design, is discussed in two engineering reports prepared for the Clayton County Water Authority by Robert and Company Associates (1976a and 1976b).

LAND TREATMENT SYSTEM DESCRIPTION

Operation of the Clayton County Pates Creek land treatment system began in October 1982. Present flows average 0.53 m³/sec and are expected to equal the design flow of 0.85 m³/sec by 1998. The total site area is 1,460 ha, of which 1,020 ha are irrigated with 6.4 cm/ week year-round. Wastewater is applied by a buried, solid-set irrigation system. Storage is provided on site for approximately twelve days' pretreated flow to permit curtailment of irrigation during inclement weather and equipment failure. The site is managed for woody biomass production as a source of energy for drying and pelletizing the sludge. Land use at the site is distributed as follows: irrigated area, 69%; buffer area, 13%; wastewater storage, 2%; and lakes, roads, and other areas, 16%. Buffer zones, to be maintained in forest cover, were established at 90 m from residences, 45 m along property boundaries, and 23 m along road rights of way.

The site for the most part is fully contained within the headwaters of Pates Creek and is bounded by the continental divide to the west and roads and residential development to the north and south. Several families maintain residences along a county road that traverses the interior of the site. A map of the site is presented in Figure 1. Topography is rolling to steep, with slopes on over a third of the site ranging from 11 to 25%. First-order streams are deeply incised, and there are many healed gullies remnant from past agricultural practices. The main stream valleys are filled with alluvium and contain many swampy areas.

Dominant soils at the site are the Pacolet and Cecil series (70% of the area) and are in the Typic Hapludult taxonomic group. The A horizon soil texture ranges from fine sandy loam to sandy clay loam, and the B horizon from sandy clay to clay. The A horizon is thin as a result of past erosion and rarely exceeds 15 cm in depth. The argillic B horizon is

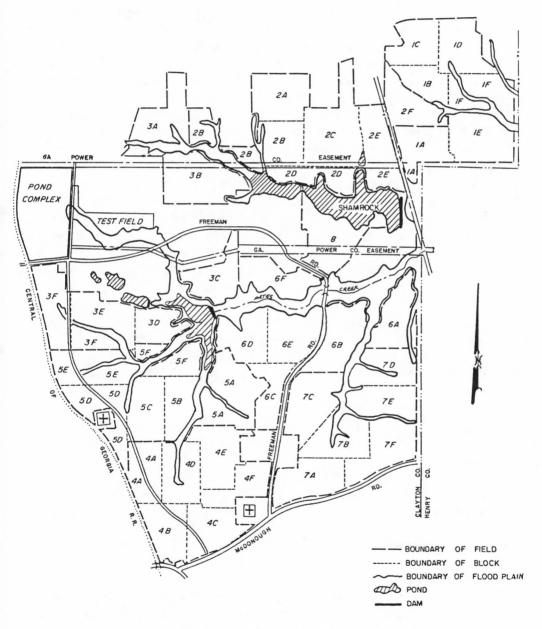

Figure 1. The Clayton County Pates Creek land treatment site.

thick, ranging from 0.4 m to over 1.0 m in thickness. The soils are underlain by saprolite (partially weathered rock) that extends to over 30 m in depth in many places. Below three meters, the weathered material grades from fine-grained silts to coarser-grained silty sands. With increasing depth, the material frequently becomes coarser, and slightly silty sands or relatively clean sands are encountered. These coarse-grained soils generally continue until the bedrock surface is encountered.

Saturated hydraulic conductivities measured in the most restrictive soil horizon between 0.3 and 1.2 m averaged 8.4×10^{-4} cm/sec. The range of the seventy-one laboratory falling head conductivity measurements was 1.4×10^{-6} to 6.6×10^{-3} cm/sec. The soils are well drained except in the alluvium along streams.

Geologic structure at the site is dominated by granitic gneiss with some fracturing and jointing. Groundwater occurs under water table conditions, and most of the recoverable water is above the bedrock at depths of 3 to 25 m below the surface. Permeabilities of the saprolite are low, averaging 5×10^{-4} cm/sec. Groundwater movement is generally in the direction of topographic slope, and the geotechnical investigation indicated that groundwater discharge points from the basin coincide with the stream courses.

Forests covered over 85% of the site at purchase, and all open land was subsequently planted to loblolly pine (*Pinus taeda*) during the construction period. About 80% of the site's forest was pine plantation, old-field pine, or mixed pine-hardwood stands. The original forest management objective for the site was to produce pine pulpwood on a twenty-year rotation to optimize nutrient uptake and subsequent removal by harvest. Pine pulpwood was chosen as the final product because markets existed and pine was easiest to regenerate. However, with escalating energy costs during project implementation, the forest management objective was changed to maximize woody biomass production for either sale or use internally in the sludge treatment process. This is currently achieved by promoting hardwood coppice growth where sufficient hardwood was present in the original stand and planting loblolly pine at the other locations.

Climatic conditions in the Georgia Piedmont favor year-round application of wastewater. Average precipitation at the site is 1,305 mm/year, with March generally the wettest month (154 mm) and October the driest (64 mm). Annual average temperature is 16.7°C, with no daily average minimum temperatures below freezing. About five days can be expected each winter in which temperatures do not rise above freezing, and one or two light snowfalls can be expected each year. Soil freezing rarely occurs. Six to nine days per month with 25 mm precipitation or more can be expected in the winter and summer months. Potential evapotranspiration is estimated to be 890 mm/year. Winds average 12 to 18 km/hour, with the strongest winds occurring in the winter when fronts advance from the northwest.

When land treatment design criteria were established during the evaluation phase, best estimates of wastewater quality were used, because the upgraded activated sludge treatment plants were not yet in service. Based on these estimates, nutrient and hydrologic budgets were calculated for the design flows, and at a 6.4 cm/week hydraulic loading, the site was considered to be both nitrogen and hydraulic loading limited. In other words, addition of greater quantities of either nitrogen or wastewater might result in surface and/or groundwater pollution as well as other problems related to site management. Current flows and concentrations of wastewater constituents are below design levels, but as flows steadily increase with population growth, the efficiency of the plants is decreasing and wastewater constituent concentrations are increasing. The design con-

TABLE 2. Comparison of design and current wastewater characterizations and loadings to the land treatment site.

Constituent	Design		1985	
	Concentration (mg/l)	Loading (kg/ha·yr)	Concentration (mg/l)	Loading (kg/ha·yr)
Total N	18	594	9.3	306
Organic-N	5		2.7	
Ammonia-N	12		3.4	
Nitrate-N	1		3.0	
Total P	10	329	4.9	162
Chlorine			37	1,221
Sodium	70	2,310	88	2,904
Magnesium	5	165	3.9	129
Calcium	20	660	17	561
Potassium	9	297	9.6	317
Iron	2	63	0.4	13
Boron	0.7	22		
Copper	0.4	12	<0.1	<3
Zinc	0.6	19	0.3	10
BOD_5	40		28	
TSS	40		63	
pH	6.5		6.6	

centrations and loadings are compared with present conditions in Table 2. To evaluate the design criteria, wastewater has been applied at the design loading (6.4 cm/week) to two fields (fields 2 and 7, Figure 1). Hydraulic loading over the entire site is increasing each year and is expected to reach design limits before the end of the planning period in 1998.

The irrigation system is solid-set and buried to a depth of at least 0.5 m for protection from heavy equipment traffic during harvesting and other activities. All pipe under 25 cm diameter is PVC, and large pipe is ductile iron. Steel risers 1.1 m high above ground level are topped with impact type sprinklers with a capacity of 0.7 l/sec at 380 kPa pressure. Flow control valves are installed where necessary to maintain the design sprinkler flow. Subsurface drains are placed at low points to eliminate standing water in the risers that could freeze. Over 17,000 sprinklers are in use at the site. For purposes of irrigation scheduling and forest harvesting, the site is divided into seven fields with six blocks each (Figure 1). The fields and blocks are sized such that a field of five blocks can be irrigated one day each week. The blocks average 24 ha. The sixth block in each field is reserved as a contingency block for use as storage reduction or while other blocks require maintenance, site work, and forest harvesting and regeneration. The contingency blocks add an opportunity for flexibility in the day-to-day operations. An eighth, smaller field was added later to make use of additional property acquisitions. The field and block layout is illustrated in Figure 1.

The twelve-day-flow storage facility consists of a central pumping pond surrounded by four additional ponds at slightly higher elevations so water can flow by gravity to the central pond. The combined water surface area of the ponds is 24 ha. The central pumping pond contains wastewater at all times and is used for flow equalization and short-term storage such as weekend flows. The wastewater is pumped 12 km from the two activated sludge treatment plants to the storage ponds.

LAND TREATMENT SYSTEM OPERATION

The spray irrigation system is designed to irrigate each day the expected flow of 0.85 m³/sec (73,800 m³/day) to five blocks within one of the seven fields (Figure 1). The application rate of 0.53 cm/hr for twelve hours to achieve the full hydraulic loading of 6.4 cm is repeated weekly. To maintain a relatively uniform discharge to the principal stream channels within the site, irrigation is staggered (two adjacent fields discharging to the same stream are not irrigated on successive days). Most irrigation takes place during daylight hours, but night irrigation is possible and is used occasionally for storage reduction.

Wastewater is irrigated year-round, but when air temperatures are below or near freezing, young pine stands are not irrigated, in order to avoid breakage. The operators observe the site during irrigation, and if the soils become saturated and runoff occurs or there is an intense rain (>2.5 cm/hr), irrigation ceases until conditions are favorable for irrigation to proceed. Constant observation of site conditions is necessary to ensure proper management.

Two blocks (approximately 48 ha), plus various buffer areas and floodplains not irrigated, are scheduled for harvest each year. When a block is harvested, it is taken out of service for a sufficient time for the soils to dry before equipment access. Risers are marked and left in place but sprinklers are removed prior to harvest. Whole-tree harvesting methods, consisting of a feller-buncher, skidder, on-site chipper, and vans, remove all woody material greater than 2.5 cm diameter, with as little soil disturbance as possible. The harvesting procedures provide adequate site preparation for planting improved loblolly pine seedlings (1-0 stock) in early spring or for hardwood coppice. Pine seedlings are not irrigated the first year after planting unless it is necessary to supplement low precipitation. Irrigation the second and third years after planting is at a reduced rate, about 20 and 50%, respectively, to prevent woody and herbaceous competition or saturated soil conditions due to reduced evapotranspiration. Resumption of the full 6.4 cm/week irrigation may occur in the third or fourth year following planting. With increased nutrient and water availability, control of competition during stand establishment is critical, and use of herbicide for control will be evaluated in the future.

Several harvested blocks that contained sufficient hardwood stocking prior to harvest have not been planted to pine but rather left to coppice regeneration to determine if biomass yields can be equal to or greater than the pine while avoiding the cost and problems associated with pine reproduction. In coppice stands, irrigation is resumed at about 1.3 cm/week soon after harvest. As the site becomes rehabilitated from logging damage and evapotranspiration increases, irrigation is increased incrementally to the full 6.4 cm/week.

Each block is scheduled to be harvested once every twenty years. More frequent harvesting will be possible only if irrigation can resume at greater rates than now possible following harvest. This may be possible with natural coppice hardwood stands and short-rotation hardwood stands; the latter possibility is currently being evaluated. Growth plots have been established in a variety of stands, and early indications are that growth of planted loblolly pine has been markedly increased. Trees five years old (three growing seasons with irrigation) average 9 cm diameter breast height (dbh) and 5.5 m in height. In the short-rotation hardwood experimental plots, eastern cottonwood appears to be the most responsive hardwood in terms of biomass production of the five species tested (yellow poplar, white ash, sweetgum, sycamore, and eastern cottonwood). After three growing seasons, each with irrigation, the cottonwoods average 7.5 cm dbh and 6 m in height.

Buffer zones are managed to maintain continual forest with a multistoried canopy for screening and reduction of wind velocities. The buffers are harvested by selection cutting at the same time the adjacent irrigation block is harvested to reduce frequency of equipment access.

The forest harvesting program has been cost effective in terms of providing an alternate source of energy for the sludge drying process. During one twelve-month period in 1984 and 1985, an average of 60 green Mg/ha of biomass were produced at a cost of $20/Mg. This provided a savings of $135,000 over the cost of natural gas during the same period. Sludge pellets produced by the drying process are sold as a soil amendment and fertilizer filler material, providing additional revenue. An added benefit of whole-tree harvesting over conventional harvesting methods is that more nutrients, especially nitrogen, are removed (summer is the preferred harvest time), thus improving the efficiency of the land treatment system.

To aid in scheduling irrigation, an interactive computer model is used that tracks daily information such as wastewater flow and storage pond levels, air temperatures, precipitation, active blocks and irrigation capacity, and irrigation history. Operators furnish specific information about each block in the form of status codes that indicate conditions such as time of last harvest, tendency for runoff during irrigation, soil erosion hazard due to disturbance such as maintenance activities, and so forth. A daily antecedent soil moisture index is calculated for both the growing and dormant seasons using weighted values of precipitation for the four days preceding the day of irrigation. The status codes and antecedent soil moisture index are compared with an operator-generated index based on experience. If the index is exceeded, the record is flagged and the operator reminded that several critical conditions must be evaluated before irrigation may proceed. The flagging may mean nothing more than more frequent observation during the irrigation cycle and/or a walkover of the site to observe actual site conditions before irrigation begins.

A harvest scheduling model was also developed that considers block location and access, stand age and biomass volume, projected growth rates, demand schedule for chips at the sludge drying facility, irrigation demands, and other factors. This model has not yet been implemented on a full-scale basis.

Storage for climatic reasons has not lasted for more than a week at any one time, and that has usually been in the winter and early spring, when a combination of precipitation, irrigation, and low evapotranspiration resulted in high soil water content levels.

A number of operational problems were faced in implementing the day-to-day operation of a system of this size. Except for the usual equipment start-up problems, most of the operational problems stemmed from the need to begin wastewater irrigation immediately following construction and before rehabilitation of construction scars had taken place—a restriction imposed by the regulatory directives Clayton County had to meet to cease discharge to the Flint River. During the construction phase, precipitation was below normal, especially during the growing season, and rehabilitation efforts on most of the disturbed areas failed. Most of the lanes cut through the forest for equipment access and placement of irrigation lateral lines (only 3.0 to 4.5 m wide) had considerable mineral soil exposure and compaction. The potential for erosion and low infiltration rates in these areas controlled the amount and timing of wastewater application; thus the capacity of the site to receive wastewater was controlled by not more than 10% of the area. From the experience gained at Clayton County, the importance of a rehabilitation program must be recognized. A successful rehabilitation program must be implemented and completed before irrigation begins. Irrigation may be phased in to aid in the rehabilitation process, but when the site cannot receive additional wastewater for fear of washing out the rehabilitation efforts, an alternate discharge method must be available.

SYSTEM PERFORMANCE

An extensive environmental monitoring program has been implemented at the Clayton County land treatment site which includes groundwater, surface water, soil, and vegetation. Twenty-two groundwater wells as well as several private water supply wells in and around the site are monitored. Monitoring is conducted monthly for critical indicator parameters at the wells located at the downgradient edge of the site and quarterly for other parameters and wells. A summary of the groundwater analyses for the four wells sampling the groundwater effluent from the site prior to irrigation and for 1985 (following over three years of irrigation) is presented in Table 3. It is evident from the data presented in Table 3 that there has been no change in groundwater quality discharged from the site except for a slight increase in chloride concentration, which is expected for this mobile anion. Nitrate-N, also a mobile anion, has not increased in concentration. Metals and organics concentrations in groundwater have not changed as a result of wastewater irrigation, although background levels of several metals exceed drinking water standards because of the local soil chemistry.

Pates Creek, the stream draining the irrigation site, is monitored for both streamflow and water quality. Water quality data for the period preceding irrigation and for 1985 are

TABLE 3. Mean groundwater quality prior to and following three years of irrigation for the four wells sampling groundwater effluent from the site.

Year	pH	Nitrate-N	Ammonia-N	Organic N	Total P	Chloride	Specific Conductance
			(milligrams/liter)				(umhos/cm)
1979-81	5.6	0.2	0.08	0.3	0.07	3.1	71
1985	6.0	<0.1	<0.05	0.1	0.02	4.4	54

TABLE 4. Mean quality of Pates Creek prior to irrigation and in 1985 following three years of irrigation.

Year*	pH	Nitrate-N	Ammonia-N	Organic N	Total P	Chloride	Specific Conductance
			(milligrams/liter)				(umhos/cm)
1978	6.7	<0.2	<0.06	0.4	<0.03	1.2	56
1985	6.6	0.5	<0.1	0.5	<0.25	13.5	135

*Water year from October through September.

presented in Table 4. As with groundwater, no changes in stream water quality are evident as a result of wastewater irrigation for over three years.

Results of soil analyses indicate essentially no detectable change in pH, nitrogen, phosphorus, or other parameters, including metals. Because of the natural variability of the occurrence of these parameters in the soils and the relative low concentrations in the wastewater, detectable accumulations and/or changes over background levels are not expected to be evident for several more years. Analyses of foliar tissue indicate there are increased concentrations of nitrogen and phosphorus in both pine and hardwood species.

Pates Creek streamflow has increased as a result of irrigation, and most of the increase appears as baseflow. Further increases are expected as wastewater flows increase. Approximately 33% of the gauged Pates Creek watershed (3,080 ha) is irrigated with wastewater. Total streamflow for the 1984 water year (October 1983 through September 1984) was 95.5 cm compared with the mean streamflow of 67.5 cm from the five nearby, or regional, gauged watersheds. The increase represents a return of approximately 45% of the wastewater applied. The influence of wastewater irrigation on baseflow during the summer is shown in Figure 2. Compared with the mean daily flow duration of the five regional watersheds, low flows in Pates Creek increased and there was little or no change in high flows (flows occurring less than 5% of the time). Prior to irrigation the Pates Creek low flow duration curve was equivalent to or slightly less than the regional

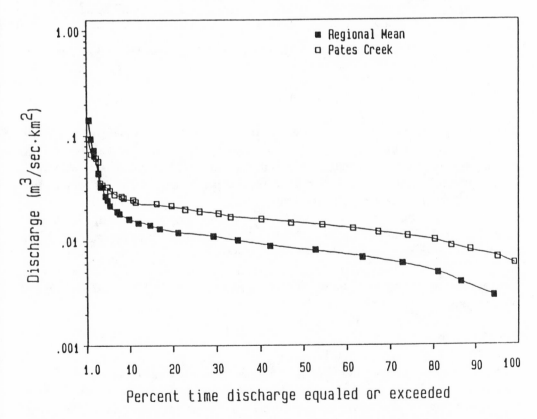

Figure 2. Duration of summer flow in Pates Creek during wastewater irrigation compared with summer flow from five regional watersheds.

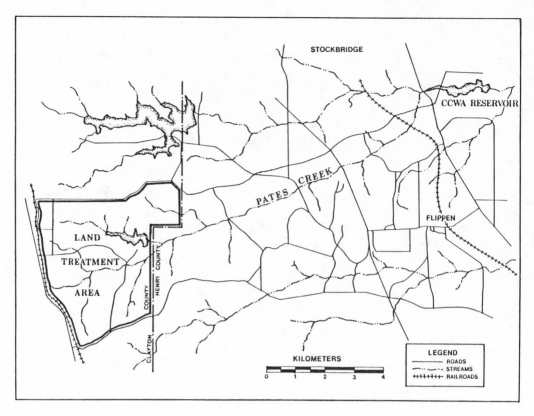

Figure 3. Map showing location of land treatment system within the Clayton County water supply watershed.

mean. Groundwater table elevations within the site also increased as the result of wastewater irrigation.

The land treatment system is located in the basin that is one source of Clayton County's drinking water supply. The irrigated area occupies 8% of the basin (Figure 3), and the irrigated wastewater return flow to the stream eventually makes its way to the water supply reservoir. Clayton County has traditionally been faced with a water shortage, particularly in the late summer months, and the additional streamflow resulting from wastewater irrigation as illustrated by the flow duration curve (Figure 2) will help to ease the problem (Snell et al. 1981). The reservoir water quality is also monitored, and in accordance with the Georgia EPD permit if the nitrate-N concentrations reach 5.0 mg/l (one-half the drinking water standard), Clayton County must develop alternate plans for either additional treatment of the water supply, modification of the land treatment system, or change to another water supply source.

CONCLUSIONS

A recent review commissioned by the U.S. Environmental Protection Agency (Roy F. Weston, Inc. 1983) concluded that the Clayton County land treatment system is performing as designed and represented an excellent demonstration of the application of land treatment technology.

The following conclusions may be drawn from the operational experience thus far:

1. The Clayton County land treatment system has successfully demonstrated that a large forest system and a system in close proximity to a large metropolitan area can be designed, constructed, and operated in a cost-effective and environmentally sound manner.

2. A hydraulic loading of 6.4 cm/week of secondary treated wastewater has resulted in no change in groundwater quality effluent from the site. No significant changes in soil or surface water quality of nutrients, metals, or organics have occurred.

3. Wastewater is applied year-round, including during below freezing temperatures.

4. Wastewater is irrigated on slopes as steep as 30%, and maintenance of good forest cover ensures infiltration of all the applied wastewater.

5. Forest productivity is increased by the application of wastewater.

6. Streamflow is increased, particularly during baseflow periods.

7. Recycling of drinking water has been successfully demonstrated at the Clayton County land treatment site.

8. Biomass produced on the land treatment site has been successfully and cost-effectively used as an energy source for drying and pelletizing the sludge, which is sold for its nutrient value.

9. Public information and education programs as part of the alternative selection and design process were essential to public acceptance of land treatment in Clayton County.

10. Provisions for a site rehabilitation period following construction and before full-scale wastewater application begins must be provided.

Monitoring of the performance of the Clayton County land treatment system will continue not only as required to meet regulatory standards but also to provide a better foundation for management of this and other forest land treatment systems.

ACKNOWLEDGMENTS

The author thanks the land management staff of the Clayton County Water Authority for making available their time and access to the site and data and for supporting cooperative research with the University of Georgia.

REFERENCES

Robert and Company Associates. 1976a. Upper Flint River Basin Wastewater Facilities 201 Plan. Report prepared for the Clayton County Water Authority (unpublished).

——. 1976b. Upper Flint River Basin Wastewater Facilities 201 Plan: Appendices. Vol. 1. Feasibility of land treatment. Report prepared for the Clayton County Water Authority (unpublished).

Snell, T. D., G. L. Taylor, N. L. Wellons, and W. L. Nutter. 1981. Clayton County: A case study on municipal water recycling. National Water Supply Improvement J. p. 5–13.

Roy F. Weston, Inc. 1983. I/A technology assessment post construction evaluation of Clayton County, Georgia land treatment-silviculture facility. Report prepared for U.S. Environmental Protection Agency, Municipal Environmental Research Laboratory, Cincinnati, Ohio (draft).

Chapter 38

Penn State's "Living Filter":
Twenty-three Years of Operation

WILLIAM E. SOPPER

ABSTRACT   One of the alternatives for treatment of municipal wastewater for groundwater recharge and reuse is land application. This method has been under investigation at The Pennsylvania State University since 1963. From these investigations the living filter concept evolved. The idea embodied in this concept is to apply wastewater on the land for direct recharge to the groundwater reservoir. This is accomplished by controlling application rates and by maintaining normal aerobic conditions within the soil. Under these conditions, organic and inorganic constituents in the wastewater are removed and degraded by microorganisms, chemical precipitation, ion exchange, biological transformations, and biological absorption through the root systems of the vegetative cover. The utilization of the vegetative cover as an integral part of the system to complement the soil is an essential component of the living filter concept and provides for maximum renovation capacity and durability of the system. Since 1963, treated municipal wastewater has been spray irrigated on cropland and several forest ecosystems at various application rates. Over a twenty-year period approximately 32 meters of wastewater were spray irrigated in the forest ecosystem that received the highest application rate. The irrigation rate was 5.0 cm per week from April through November of each year. The efficacy of the forest ecosystem to remove potential groundwater contaminants, particularly nitrate-N and trace metals, was monitored throughout the period. Results indicate that the wastewater, after passage through 120 cm of soil, was consistently renovated to potable water quality. In 1982, vegetation foliar samples were collected from the white spruce trees and dominant herbaceous vegetation, and soil samples were collected at the 0-5, 5-10, 10-15, and 15-30 cm depths and analyzed for copper, chromium, zinc, lead, cobalt, cadmium, and nickel. Results indicate that the wastewater applications did not significantly affect trace metal concentrations of either the vegetation or the soil. The original research and demonstration portable spray irrigation system has now been replaced with a permanent facility with the capacity to irrigate 15,140 cubic meters (4 mgd) of wastewater on 208 ha (514 acres) of cropland and forest land. The irrigation system is operated year around with no winter storage.

Research on land treatment of wastewater at The Pennsylvania State University began in 1962. A group of scientists were attempting to find solutions to two local water problems: (1) the pollution of Spring Creek, the area's only surface water, by the discharge of ever-increasing amounts of sewage effluent from the wastewater treatment plant that serves both the university and the town of State College; and (2) a dwindling supply of groundwater due to a seven-year drought. In those seven years there was a deficit of 127 cm (50 inches) of precipitation, which is almost equivalent to 1.3 years of normal precipitation. During the same period, millions of gallons of sewage effluent were discharged daily into Spring Creek and were not available for reuse as a water resource in the immediate area.

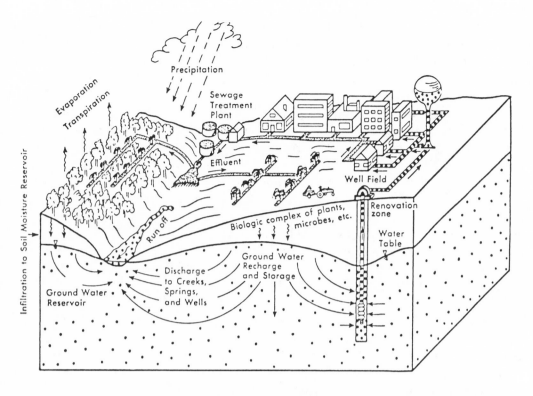

Figure 1. Diagram of the living filter concept.

The need to find solutions to these problems led to a full-scale investigation into the feasibility of applying large volumes of wastewater on the land. In 1963 a research facility was constructed to spray irrigate 1,890 cubic meters (0.5 mgd) of chlorinated secondary-treated wastewater on cropland and forest land. From these initial investigations evolved the living filter concept. The idea embodied in this concept is to apply treated wastewater on the land in a manner that utilizes the entire biosystem—soil, vegetation, and microorganisms—as a living filter to renovate the wastewater to potable water quality for direct recharge to the groundwater reservoir (Figure 1).

This can be accomplished by controlling the application rates and by maintaining normal aerobic conditions within the soil. Under these conditions organic and inorganic constituents in the wastewater are removed and degraded by microorganisms, chemical precipitation, ion exchange, biological transformation, and biological absorption through the root systems of the vegetative cover. The utilization of the vegetative cover as an integral part of the system to complement the microbiological and physiochemical systems in the soil is an essential component of the living filter concept and provides for maximum renovation capacity and durability of the system.

METHOD OF STUDY

The forest ecosystems used in this study consisted of two mixed hardwood forests, a red pine (*Pinus resinosa*) plantation, and a sparse white spruce (*Picea glauca*) plantation established on an abandoned old field. Detailed descriptions of these areas have been reported by Parizek et al. (1967). The two soils on the study sites are ultisols. The Hublersburg soil (Typic Hapludalf) has a silt loam topsoil with a silty clay or clay subsoil. It has moderate to moderately rapid permeability and is well drained. The Morrison soil (Ultic Hapludalf) is a sandy loam with a sandy clay loam subsoil, has a moderately rapid to rapid permeability, and is well drained.

Initial research was directed toward evaluating wastewater application rates and the wastewater renovating efficiencies of various forest ecosystems. Results of these studies were reported by Parizek et al. (1967), Pennypacker et al. (1967), and Sopper and Kardos (1973). Later investigations included effects of wastewater applications on tree growth, herbaceous vegetation and seedling reproduction, wood quality, physical, hydrological, and chemical properties of the soil, soil invertebrates, microclimate, and wildlife (rabbits, mice, deer, and songbirds). Results of these studies have been reported by Kardos and Sopper (1973a, 1973b), Sopper and Kardos (1973), Richenderfer et al. (1975), Sopper and Richenderfer (1979), Richenderfer and Sopper (1979), Murphey et al. (1973), Wood et al. (1973), Anthony and Wood (1979), Dindal et al. (1979), DeWalle (1979), Snetsinger et al. (1979), and Anthony and Kozlowski (1982).

Wastewater Composition and Application Rates

The wastewater receives secondary treatment at a standard trickling filter plant operated by the university. Although chemical composition has varied a little over the twenty-three years, typical concentrations of constituents in the wastewater are given in Table 1. Application rates ranged from 2.5 to 7.5 cm/wk (1 to 3 in./wk) over various

TABLE 1. Typical chemical composition of the wastewater.

Constituent	Average Concentration (mg/l)	Average Annual Application* (kg/ha)	Total Amount Applied** (kg/ha)
Nitrate-N	9.2	--	--
Ammonium-N	11.0	--	--
Organic N	5.2	--	--
Total N	25.4	310	6,200
Phosphorus	5.6	58	1,160
Calcium	47.8	482	9,640
Chlorine	47.1	484	9,680
Magnesium	11.9	110	2,200
Sodium	32.8	336	6,720
Iron	0.5	6	120
Potassium	10.3	109	2,180
Boron	169.0	1.8	36
Manganese	70.0	0.7	14
Copper	0.068	1.1	22
Zinc	0.197	3.2	64
Chromium	0.022	0.4	8
Lead	0.140	2.3	46
Cadmium	0.003	0.05	1
Cobalt	0.040	0.6	12
Nickel	0.050	0.8	16
Mercury	0.004	0.05	1
pH	7.5		

*Total amount in forest ecosystems that received wastewater applications at 5 cm/wk (2 in./wk).

**Applied over twenty years (1963 to 1982).

lengths of time ranging from the growing season (April to November) to the entire year. The normal weekly application rates were 2.5 and 5.0 cm/wk (0.63 cm/hr) throughout the study period except for 1972 and 1973. During these two years the application rates were increased by 50% to evaluate the effects of chronic applications of wastewater on ecosystem collapse and recovery. Composite wastewater samples were collected during each irrigation period for chemical analyses. Suction lysimeters were installed at the 120 cm (4 ft) depth to monitor soil percolate water quality and determine renovation efficiency. Monthly water samples were collected from the lysimeters and from a deep groundwater well located 300 m (984 ft) downgradient from the land treatment system. At the 5 cm/wk application rate, approximately 32 m (105 ft) of wastewater have been applied to the forest ecosystems over the study period.

Trace Metals

In August 1982, after twenty years of operation, a special study was conducted to determine the fate of trace metals in the white spruce, old field ecosystem on the Hublersburg soil, which had been irrigated every year at 5 cm/wk from April to November. Soil percolate samples were collected at the 120 cm depth monthly and analyzed for copper, zinc, chromium, lead, cobalt, cadmium, and nickel with an atomic absorption spectrophotometer. Soil samples were collected at the 0-5, 5-10, 10-15, and 15-30 cm depths at six locations and analyzed for extractable copper, zinc, chromium, lead, cobalt, cadmium, and nickel by the Dilute hydrochloric acid procedure and atomic absorption (Jackson 1958). In addition, soil samples collected at the 0-30 cm depth periodically during the twenty-year period and stored were also analyzed for the same trace metals, phosphorus, potassium, magnesium, calcium, and pH. Foliar samples were collected from the white spruce trees and predominant herbaceous species. Similar samples were collected in a control area.

RESULTS AND DISCUSSION

Nitrogen Management

Nitrogen is usually the most critical parameter that must be considered in the design and operation of a forest land treatment system. Most ecosystems have an initial high capacity to accept and renovate wastewater satisfactorily. However, continual application of large volumes of wastewater with high amounts of nutrients, particularly nitrogen, will ultimately affect ecosystem stability and all renovation processes. Most of these changes occur slowly and may not become obvious for a decade. Some of the more important changes that may influence wastewater renovation efficiency are vegetation species composition and density, foliar chemical composition, leaf litter and forest floor decomposition rates, humus mineralization rates, physical and chemical properties of soil, microbial and earthworm populations and activity, and microclimate.

Monitoring data collected on several of the forest ecosystems at the Penn State Facility provide some insight as to the long-term efficiencies of such ecosystems to renovate wastewater adequately.

Red Pine Ecosystem on Hublersburg Soil. This plantation was planted in 1939 on abandoned agricultural land. Trees were planted at a spacing of 2.4 by 2.4 m (8 by 8 ft). In 1963 the average tree diameter was 17.3 cm (6.8 inches) and the average height was 10.6 m (35 ft). Wastewater was irrigated in this plantation during each growing season at the

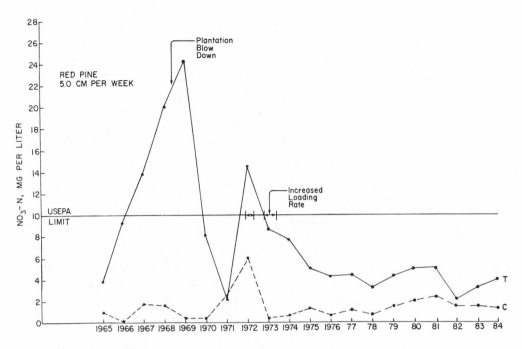

Figure 2. Average annual nitrate-N concentrations of soil water at the 120 cm soil depth in the red pine forest ecosystem irrigated with municipal wastewater at 5 cm/wk during the growing season, 1963-84.

rate of 5 cm/wk throughout the study period except for 1972 and 1973, when the application rate was increased by 50%. Average annual nitrate-N concentrations of soil percolate water at the 120 cm (4 ft) depth for both treated and control areas are shown in Figure 2. Results indicate that this forest ecosystem was able to renovate wastewater satisfactorily only during the first four years of operation (1963 to 1966).

Data for 1963 and 1964 are not plotted on the graph, since percolate samples were collected at shallow depths. However, mean annual nitrate-N concentrations were all less than 10 mg/l. Mean annual concentration of nitrate-N in soil water gradually increased from 3.9 to a peak of 24.2 mg/l in 1969. In November 1968 a snowstorm accompanied by high winds resulted in a complete blowdown of the plantation. In 1969 the area was clearcut and all trees were removed. Sewage effluent irrigation was continued, and immediately a dense cover of herbaceous and shrub vegetation developed. With the development and growth of this perennial herbaceous vegetative cover, the mean annual concentration of nitrate-N in the soil water decreased from 24.2 mg/l in 1969 to 8.3 mg/l in 1970 and to 2.9 mg/l in 1971. Increasing the application rate from 5.0 to 7.5 cm/wk in 1972 increased the nitrate-N concentration in the soil water to 14.5 mg/l. Even though this concentration exceeds 10 mg/l, it is obvious that the new developing ecosystem composed of invading pioneer species of herbaceous and tree vegetation was very efficient in renovating wastewater. In 1972 tropical storm Agnes increased the mean annual concentration of nitrate-N in the soil water on the control area from 2.6 to 6.0 mg/l. On the irrigated area, in 1973, with the increased application rate under normal climatic conditions, the nitrate-N concentration in soil water was only 8.7 mg/l. This shows how quickly this pioneer vegetation ecosystem had recovered from the chronic applications of wastewater. It is also obvious that the renovation efficiency of the new ecosystem is much greater than the original red pine ecosystem. Even though the red pine ecosystem was very inefficient in renovating the wastewater during the period 1967 to 1969, reaching a peak concentration of 24.2 mg/l of nitrate-N in soil water, the new developing pioneer vegetation ecosystem was extremely efficient, as indicated by the fact that mean annual nitrate-N concentration in the soil water continued to decrease, reaching a low value of 3.1 mg/l in 1978. The ecosystem has apparently stabilized, for nitrate-N concentration in soil water has not changed significantly since that time. The results obtained from this area dramatically illustrate the interrelation between the application rate, the type of vegetation, and the system of management.

White Spruce, Old Field Ecosystem on Hublersburg Soil. The white spruce plantation was planted about 1955 on an abandoned agricultural field. In 1963 the average height of the trees was 1.2 m (4 ft). Predominant groundcover was a poverty grass (*Danthonia spicata*), goldenrod (*Solidago* spp.), and dewberry (*Rubus flagellaris*) plant community.

Wastewater was irrigated in this ecosystem during each growing season at the application rate of 5 cm/wk throughout the study period except for 1972 and 1973, when the application rate was increased by 50%. Average annual nitrate-N concentrations of soil percolate water at the 120 cm depth for both treated and control areas are shown in Figure 3. This ecosystem has been somewhat exceptional in terms of nitrogen renovation compared with the other forest ecosystems. It received the highest application rate on the Hublersburg soil and yet has consistently maintained nitrate-N concentrations in soil water below 10 mg/l throughout the period 1963 to 1971. Increasing the application rate in 1972 resulted in a slight increase in concentrations of nitrate-N in the soil water. However, the ecosystem quickly recovered, as seen in Figure 3. At the start of the project in 1963 the area was primarily an open field with a few scattered white spruce saplings (1 to 2 m in height). Although the trees were more than 9 m in height in 1972, the spruce stand was still sparse, with fairly large open areas occupied by perennial herbaceous vegetation and shrubs. It appears that the annual and perennial vegetation occupying these open areas during the growing season (irrigation period) provides temporary storage for nitrogen and hence reduces the nitrate-N leaching losses. In the fall, vegetation growth ceases and nitrate-N is again available for leaching. But since irrigation has ceased by this time, the concentration of nitrate-N in soil water remains at an acceptable

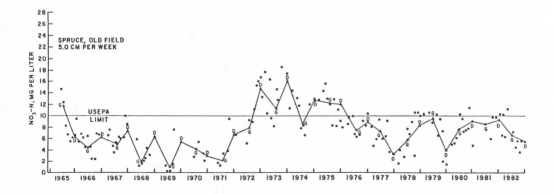

Figure 3. Average monthly and seasonal nitrate-N concentrations of soil water at the 120 cm soil depth in the white spruce, old field forest ecosystem irrigated with municipal wastewater at 5 cm/wk during the growing season, 1965-82. Growing season (G) and dormant season (D).

level. The desynchronization effect of nitrate application, vegetation utilization, and soil water leaching is illustrated in Figure 3. The mean seasonal concentrations of nitrate-N in soil water plotted in Figure 3 show that the concentrations for the dormant season are higher than those of the growing season, but are still well below U.S. EPA potable water standards. This same phenomenon explains the excellent renovation observed in the red pine area after the development of the pioneer vegetation ecosystem.

Another point illustrated dramatically in Figure 3 is what happens when one overloads a forest land treatment system with nitrogen. If a forest ecosystem is overloaded with nitrogen so that renovation is unsatisfactory, there are three management options available to correct the situation: (1) cease irrigation, (2) reduce application rate to a smaller amount, and (3) expand the irrigation system into new forest areas. Most overload situations occur when increasing volumes of wastewater are applied on a static facility because land area is not available for expansion. In such cases, the only feasible option is to reduce the application rate. This was the option selected for the Penn State forest ecosystem. In 1974 the application rate was reduced from 7.5 cm/wk back to the design rate of 5.0 cm/wk. Even with this action, recovery of the forest ecosystem in renovation efficiency was extremely slow. It took three years before the ecosystem was again capable of satisfactorily removing nitrogen from the wastewater.

Mixed Hardwood Forest Ecosystem on Morrison Soil. This hardwood stand consists primarily of a mixture of white oak (*Quercus alba*), black oak (*Q. velutina*), red oak (*Q. rubra*), and scarlet oak (*Q. coccinea*) in association with a few red maple (*Acer rubrum*) and hickory (*Carya tomentosa*). In 1965, average tree diameter was 35.5 cm (14 inches) and average height was 21.3 m (70 ft). Wastewater was irrigated in this ecosystem during the entire year at the rate of 5 cm/wk throughout the period 1965 to 1974. Average annual nitrate-N concentrations of soil percolate water at the 120 cm depth for both treated and control areas are shown in Figure 4. Results indicate that wastewater irrigation throughout the entire year of mature forest ecosystems on sandy soils is not feasible. Mean annual concentration of nitrate-N in soil water continually increased, reaching a peak of 42.8 mg/l within five years (1970). During the study period, unknown amounts of liquid-digested sludge were periodically injected into the sewage effluent and spray irrigated on the area. These sludge applications probably account for the unexplained fluctuations in the mean annual nitrate-N concentrations in the soil water during the period 1968 to 1971. The increase in nitrate-N concentration in soil water in 1972 was partly the result of tropical storm Agnes. At the end of the growing season in 1974, it was decided to cease wastewater irrigation in this forest ecosystem and to evaluate the rate of ecosystem recovery in terms of nitrate-N renovation efficiency. With complete cessation of wastewater application, ecosystem recovery is relatively rapid. Within one year the mean annual concentration of nitrate-N in soil water decreased from 14.3 to 9.0 mg/l. In 1976 during the growing season, 46.2 cm of wastewater was again irrigated in the ecosystem, but nitrate-N concentration in soil water remained at a low level (4.8 mg/l). During the remainder of the study period (1976 to 1984) this forest ecosystem has been irrigated during the growing season only at an application rate of 5 cm/wk, and nitrate-N concentrations in soil percolate water have remained consistently below 10 mg/l.

Trace Metals

Foliar Analyses. Wild strawberry (*Fragaria virginiana*) and goldenrod (*Solidago* spp.) were the most abundant ground vegetation common to both the wastewater-irrigated

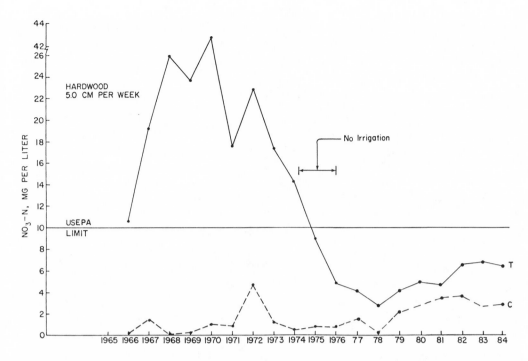

Figure 4. Average annual nitrate-N concentrations of soil water at the 120 cm soil depth in the mixed hardwood forest ecosystem irrigated with municipal wastewater at 5 cm/wk year around for the period 1966 to 1974 and during the growing season only for the period 1976 to 1984.

and control white spruce, old field plots. Results of foliar analyses for these two species as well as the white spruce trees are given in Table 2. Results indicate that the twenty years of wastewater irrigation had little effect on the trace metal uptake by the vegetation. For the irrigated white spruce, only the foliar concentration of cobalt was slightly

TABLE 2. Foliar concentrations of trace metals in selected vegetation species in the white spruce, old field plantation in 1982.

Plot	Copper	Zinc	Chromium	Lead	Cobalt	Cadmium	Nickel
			(micrograms/gram)				
			Goldenrod				
Control	19.7	100	0.33	7.3	0.25	0.313	2.1
Irrigated	13.3	33	0.17	8.3	0.33	0.060	0.4
			Wild Strawberry				
Control	6.0	42	0.75	8.7	1.50	0.184	3.2
Irrigated	8.7	29	2.08	9.3	2.67	0.097	2.6
			White Spruce				
Control	3.7	44	0.04	5.5	0.58	0.076	4.8
Irrigated	3.3	21	0.01	5.3	0.71	0.013	1.0
			Suggested Tolerance Level (Melsted 1973)				
	150	300	2	20	5	3	50

higher than the control trees. For the herbaceous species, concentrations of lead and cobalt were increased in goldenrod, and copper, chromium, lead, and cobalt were increased in wild strawberry by the wastewater irrigation. All increases were insignificant except for chromium in wild strawberry, where foliar concentrations of wastewater-irrigated plants were threefold higher than control plants. Concentrations of copper and zinc in all vegetation were within the normal range reported for agronomic crops. Concentrations of chromium, lead, cobalt, cadmium, and nickel exceeded the normal range in some of the vegetation samples, but all concentrations were below the suggested tolerance level for yield suppression (Melsted 1973). In many instances, foliar concentrations were higher in control vegetation. This may be the result of biological dilution. Previous studies have shown that the total biomass of the irrigated vegetation increased threefold (Sopper 1971).

Soils. Results of the soils analyses for extractable trace metals are given in Tables 3 and 4. Highest trace metal concentrations were found in the 0-5 cm depth in both the irrigated and control plots. After twenty years of wastewater irrigation, concentrations of copper and cobalt increased at all soil depths, zinc and cadmium concentrations increased at the 5-10 and 10-15 cm depths, and chromium concentrations increased at the 0-5 and 5-10 cm depths. There was no change in lead and nickel concentrations. In some instances, concentrations of trace metals were higher in the soil on the control plot. Lead and nickel concentrations were higher in the control soils at all depths, and zinc and cadmium concentrations were higher in the 0-5 cm depth.

Changes in trace metal concentrations in the 0-30 cm soil depth over the twenty years are shown in Table 4. Concentrations of copper, zinc, cadmium, and nickel show an increasing trend over the period in both the control and wastewater-irrigated plots. In 1982, only copper and zinc concentrations were significantly higher in the wastewater-irrigated soils.

Soil samples collected in 1982 were also analyzed for pH, phosphorus, potassium,

TABLE 3. Extractable trace metal concentrations in the soils of the white spruce, old field plantation on the Hublersburg soil in 1982.

Plot	Copper	Zinc	Chromium	Lead	Cobalt	Cadmium	Nickel
				(micrograms/gram)			
				0-5 cm Soil Depth			
Control	2.33	20.38	0.05	13.32	1.98	0.433	1.70
Irrigated	3.15	18.45	0.10	4.43	2.82	0.267	1.28
				5-10 cm Soil Depth			
Control	1.80	3.73	0.05	4.87	0.87	0.043	1.93
Irrigated	2.47	12.53	0.10	3.55	1.55	0.195	1.30
				10-15 cm Soil Depth			
Control	1.52	2.65	0.05	4.05	0.67	0.038	2.32
Irrigated	1.77	5.25	0.05	3.32	0.90	0.085	1.58
				Normal Range in Soils (Allaway 1968)			
	2-100	10-300	5-3,000	2-200	1-40	0.01-7.00	10-1,000

TABLE 4. Extractable trace metal concentrations in the 0-30 cm soil depth in the white spruce, old field plantation, 1963-82.

Plot	Copper	Zinc	Chromium	Lead	Cobalt	Cadmium	Nickel
				(micrograms/gram)			
Control							
1963	0.93ab	2.45a	0.07a	2.99a	1.23a	0.05a	0.31c
1965	0.66b	2.63a	0.08a	3.76a	2.12a	0.05a	0.28c
1967	1.16a	1.93a	0.08a	3.66a	1.81a	0.03a	0.30c
1971	0.92ab	3.91a	0.10a	3.69a	1.43a	0.06a	0.35c
1976	2.49c	2.85a	0.10a	3.75a	0.70a	0.07a	0.88b
1978	1.15a	4.44a	0.05a	4.48a	1.20a	0.09a	1.47b
1982	1.15a	4.98a	0.06a	5.45a	1.15a	0.09a	1.08a
Irrigated							
1963	0.65d	3.23b	2.23b	4.61a	1.80c	0.04b	0.56c
1965	0.95cd	3.78b	0.06a	4.21a	2.75b	0.04b	0.67bc
1967	1.43b	6.15ab	0.04a	4.45a	3.21ab	0.07b	0.89b
1971	1.23bc	6.01ab	0.07a	4.19a	3.73a	0.05b	0.54c
1976	1.92a	7.48a	0.01a	3.29a	1.87c	0.03b	0.73bc
1978	1.65ab	7.08a	0.06a	3.33a	0.92d	0.08a	1.18a
1982	1.63ab	6.84a	0.15a	3.63a	1.54c	0.09a	0.68bc

Means followed by same letter are not significantly different at the 0.05 level.

TABLE 5. Mean constituent concentrations in the soil of the white spruce, old field plantation in 1982.

Area	Depth (cm)	pH	Phosphorus (ug/g)	Potassium	Magnesium	Calcium
				(meq/100 grams)		
Control	0-30	5.1	9	0.19	0.2	1.3
	30-60	5.2	3	0.20	1.3	2.5
	60-90	5.3	5	0.20	2.2	2.0
	90-120	5.2	3	0.23	2.1	2.2
Irrigated	0-30	6.4	75	0.52	2.2	5.1
	30-60	5.9	6	0.41	1.8	2.9
	60-90	5.3	3	0.34	1.5	2.6
	90-120	5.1	4	0.30	1.3	2.1

TABLE 6. Base saturation in the soil of the white spruce, old field plantation in 1982.

Area	Depth (cm)	Saturation (%)		
		Potassium	Magnesium	Calcium
Control	0-15	1.8	2.0	9.9
	15-30	1.4	2.2	13.1
	30-60	1.5	9.2	20.3
	60-90	1.5	16.3	16.4
	90-120	1.8	16.6	16.0
Irrigated	0-15	4.1	21.5	55.2
	15-30	6.0	19.8	34.8
	30-60	4.6	19.9	30.6
	60-90	2.7	13.1	21.4
	90-120	2.2	10.2	16.1

magnesium, and calcium. Results of these analyses by soil depths are given in Table 5. It is obvious that wastewater irrigation has significantly increased the soil pH to a depth of 60 cm (2 ft). Concentrations of phosphorus, magnesium, and calcium in the 0-30 cm (1 ft) depth have all increased significantly. Concentrations of potassium were significantly increased at all depths. Base saturation percentages are given in Table 6. Potassium, magnesium, and calcium in the control soil are all in the deficiency range. However, it is obvious that the wastewater irrigation has significantly improved the nutrient status of the soil.

Results of the analyses for the 0-30 cm depth were compared with the analyses of other samples collected during the twenty-year period (Table 7). Concentrations of phosphorus, potassium, magnesium, and calcium in the 0-30 cm depth in the control plot did not significantly change over the twenty-year period. However, soil pH increased slightly from 4.8 to 5.1. In the wastewater-irrigated plot, pH increased significantly from 4.9 to 6.4. In addition, there were significant increases in the concentrations of phosphorus, magnesium, and calcium. There was essentially no change in potassium concentration. The amount of phosphorus in the 0-15 cm soil depth was significantly increased from 19 kg/ha (17 lb/acre) in 1963 to 246 kg/ha (220 lb/acre) in 1982.

Soil Percolate Water. Results of trace metal analyses of soil percolate water samples collected in 1982 are given in Table 8. Even after twenty years of wastewater irrigation, the forest ecosystem was still very efficient in renovating the wastewater to potable water quality. There was no significant difference in trace metal concentrations in soil percolate water at the 120 cm depth between the irrigated and control plots.

TABLE 7. Changes in mean constituent concentrations in the 0-30 cm soil depth in the white spruce, old field plantation, 1963-82.

Area	Year	pH	Phosphorus (ug/g)	Potassium	Magnesium	Calcium
				(meq/100 grams)		
Control	1963	4.8	8.7	0.40	0.27	1.43
	1967	4.8	13.5	0.03	0.20	1.20
	1971	5.0	10.9	0.10	0.23	2.00
	1982	5.1	9.0	0.19	0.20	1.30
Irrigated	1963	4.9	16.1	0.63	0.50	1.73
	1967	5.2	42.1	0.26	1.36	3.36
	1971	5.4	50.8	0.43	1.73	3.33
	1982	6.4	75.0	0.52	2.20	5.10

TABLE 8. Average annual trace metal concentrations in soil percolate water at 120 cm depth in the white spruce, old field plantation in 1982.

Source	Copper	Zinc	Chromium	Lead	Cobalt	Cadmium	Nickel
	(milligrams/liter)						
Wastewater	0.07	0.20	0.02	0.14	0.04	0.003	0.05
Irrigated plot	0.01	0.05	0.01	0.02	0.01	<0.001	0.02
Control plot	0.01	0.05	0.01	0.02	0.01	<0.001	0.01
U.S. EPA drinking water maximum	1.0	5.0	0.05	0.05	--	0.01	--

Groundwater

The amount of potable water recharged to the groundwater reservoir was estimated from data available on the total amount of wastewater applied annually, rainfall, and potential evapotranspiration. Annual recharge has ranged from 10,300 to 17,300 cubic meters per hectare (1.1 to 1.8 million gallons per acre), which represents approximately 90% of the wastewater applied. Wastewater irrigation has not had any significant effect on groundwater quality. Mean annual concentrations of nitrate-N in a deep groundwater monitoring well located 300 m (984 ft) downgradient from the land treatment system are given in Table 9. Depth to the water table is approximately 60 m (197 ft). Average annual concentrations of nitrate-N have ranged from 1.4 to 4.5 mg/l over the twenty-year period.

TABLE 9. Mean annual concentration of nitrate-N in a deep groundwater monitoring well adjacent to the wastewater irrigation areas.

Year	Nitrate-N (mg/l)	Year	Nitrate-N (mg/l)
1963	1.9	1975	3.5
1964	2.3	1976	3.5
1965	1.4	1977	4.1
1966	1.8	1978	4.5
1967	2.9	1979	4.3
1968	3.1	1980	3.3
1969	4.0	1982	2.5

NEW EXPANDED FACILITY

In the late 1970s plans were initiated to replace the 1,890 cubic meter (0.5 mgd) portable aluminum pipe irrigation system with a new permanent facility that would have the capacity to irrigate the total volume, 15,140 cubic meters (4 mgd) of wastewater from the university waste treatment plant. A new pumping station was constructed with a 1,893 cubic meter surge tank and two 0.26 MW turbine pumps. A buried 45 cm (18 inch) diameter ductile iron main supply line carries the wastewater 4.8 km (2.9 miles) to two irrigation sites. A solid-set irrigation system consisting of 96 km (60 miles) of galvanized pipe and over 3,000 sprinklers was installed to apply the wastewater on 208 hectares (514 acres) of cropland and forest land. Application rate is 5 cm per week applied at 0.42 cm/hr (0.17 in./hr). The week has been divided into fourteen irrigation periods, each of which is 12 hours in duration. At design capacity 175 l/sec (4 mgd), 15 ha (37 acres) would be irrigated during each 12 hour period. The 15 ha (37 acre) unit is further divided into 1.2 ha (3 acre) subplots to obtain hydraulic balance and nearly uniform pressure requirements. Typical layout of the 1.2 ha plots involves 18 sprinklers spaced on 26 by 26 m (85 by 85 ft) in the forest land and 22 by 31 m (72 by 100 ft) on cropland. Irrigation continues year around, and there is no storage in the system except for the surge tank at the pumping station. Forested areas are primarily used in winter. Laterals in the winter irrigation areas were installed with a 2 to 3% slope for rapid draining at the end of an irrigation period. All other lateral lines lay on the ground and follow the natural contour of the land. The performance of the system is monitored with a circle of groundwater wells that sample all groundwater leaving the site. Suction lysimeters have also been installed at the 1.8 m (6 ft) and 3.6 m (12 ft) depths to sample soil water. Samples are collected quarterly for chemical analyses. The new facility was completed and put into operation in April 1983.

CONCLUSION

Results of research conducted since 1963 at the Penn State living filter system indicate that forest ecosystems can be very efficient in removing potential groundwater contaminants from treated municipal wastewater. A properly designed and managed forest land treatment system should be able to renovate municipal wastewater satisfactorily for the design life of the physical facility.

ACKNOWLEDGMENTS

The research reported here is part of the program of the Waste Water Renovation and Conservation Project of the Institute for Research on Land and Water Resources and Hatch Project No. 2556 of the Agricultural Experiment Station, The Pennsylvania State University, University Park, Pennsylvania.

REFERENCES

Allaway, W. H. 1968. Agronomic controls over the environmental cycling of trace elements. Adv. Agron. 20:235–274.

Anthony, R. G., and R. Kozlowski. 1982. Heavy metals in tissues of small mammals inhabiting waste-water-irrigated habitats. J. Environ. Qual. 11(1):20–22.

Anthony, R. G., and G. W. Wood. 1979. Effects of municipal wastewater irrigation on wildlife and wildlife habitat. p. 213–223. In W. E. Sopper and S. N. Kerr (eds.) Utilization of municipal sewage effluent and sludge on forest and disturbed land. Pennsylvania State University Press, University Park.

DeWalle, D. R. 1979. Microclimate and wastewater spray irrigation in forests. p. 225–239. In W. E. Sopper and S. N. Kerr (eds.) Utilization of municipal sewage effluent and sludge on forest and disturbed land. Pennsylvania State University Press, University Park.

Dindal, D. L., L. T. Newell, and J.-P. Moreau. 1979. Municipal wastewater irrigation: Effects on community ecology of soil invertebrates. p. 197–205. In W. E. Sopper and S. N. Kerr (eds.) Utilization of municipal sewage effluent and sludge on forest and disturbed land. Pennsylvania State University Press, University Park.

Jackson, M. L. 1958. Soil chemical analysis. Prentice-Hall, Englewood Cliffs, New Jersey. 498 p.

Kardos, L. T., and W. E. Sopper. 1973a. Renovation of municipal wastewater through land disposal by spray irrigation. p. 148–163. In W. E. Sopper and

L. T. Kardos (eds.) Recycling treated municipal wastewater and sludge through forest and cropland. Pennsylvania State University Press, University Park.

———. 1973b. Effect of land disposal of wastewater on exchangeable cations and other chemical elements in the soil. p. 220–231. In W. E. Sopper and L. T. Kardos (eds.) Recycling treated municipal wastewater and sludge through forest and cropland. Pennsylvania State University Press, University Park.

Melsted, S. W. 1973. Soil-plant relationships. p. 121–128. In Proceedings of the Joint Conference on Recycling Municipal Sludges and Effluents on Land. National Association of State University and Land Grant Colleges, Washington, D.C.

Murphey, W. K., R. L. Brisbin, W. J. Young, and B. E. Cutter. 1973. Anatomical and physical properties of red oak and red pine irrigated with municipal wastewater. p. 295–310. In W. E. Sopper and L. T. Kardos (eds.) Recycling treated municipal wastewater and sludge through forest and cropland. Pennsylvania State University Press, University Park.

Parizek, R. R., L. T. Kardos, W. E. Sopper, E. A. Myers, D. E. Davis, M. A. Farrell, and J. B. Nesbitt. 1967. Waste water renovation and conservation. Penn State Studies 23. Pennsylvania State University, University Park. 71 p.

Pennypacker, S. P., W. E. Sopper, and L. T. Kardos. 1967. Renovation of wastewater effluent by irrigation of forestland. J. Water Poll. Control Fed.

39(2):185–296.

Richenderfer, J. L., W. E. Sopper, and L. T. Kardos. 1975. Spray irrigation of treated municipal sewage effluent and its effect on chemical properties of the soil. USDA For. Serv. General Technical Report NE-17. Northeastern Forest Experiment Station, Upper Darby, Pennsylvania. 24 p.

Richenderfer, J. L., and W. E. Sopper. 1979. Effect of spray irrigation of treated municipal sewage effluent on the accumulation and decomposition of the forest floor. p. 163–177. *In* W. E. Sopper and S. N. Kerr (eds.) Utilization of municipal sewage effluent and sludge on forest and disturbed land. Pennsylvania State University Press, University Park.

Snetsinger, R., D. L. Guthrie, J. Sprowls, M. Quinn, and W. Wills. 1979. Municipal wastewater irrigation and mosquito populations. p. 207–212. *In* W. E. Sopper and S. N. Kerr (eds.) Utilization of municipal sewage effluent and sludge on forest and disturbed land. Pennsylvania State University Press, University Park.

Sopper, W. E. 1971. Disposal of municipal waste water through forest irrigation. Environ. Poll. 1(4):263–284.

Sopper, W. E., and L. T. Kardos. 1973. Vegetation responses to irrigation with treated municipal wastewater. p. 271–294. *In* W. E. Sopper and L. T. Kardos (eds.) Recycling treated municipal wastewater and sludge through forest and cropland. Pennsylvania State University Press, University Park.

Sopper, W. E., and J. L. Richenderfer. 1979. Effect of municipal wastewater irrigation on the physical properties of the soil. p. 179–95. *In* W. E. Sopper and S. N. Kerr (eds.) Utilization of municipal sewage effluent and sludge on forest and disturbed land. Pennsylvania State University Press, University Park.

Wood, G. W., D. W. Simpson, and R. L. Dressler. 1973. Effects of spray irrigation of forests with chlorinated sewage effluent on deer and rabbits. p. 311–323. *In* W. E. Sopper and L. T. Kardos (eds.) Recycling treated municipal wastewater and sludge through forest and cropland. Pennsylvania State University Press, University Park.

Chapter 39

Forest Land Treatment with Municipal Wastewater in New England

SHERWOOD C. REED and RONALD W. CRITES

ABSTRACT   An overview of several case studies of forest land treatment with municipal wastewater in New England is presented. One of the earliest land treatment systems in this area in modern times was installed in 1971 by the state of New Hampshire at Sunapee State Park, in a mature forest of mixed hardwoods and conifers. The system is in excellent condition, and continued operation is planned for the foreseeable future. Municipal forest land treatment systems are also operating successfully at West Dover, Vermont; Wolfeboro, New Hampshire; and Greenville, Maine. Design and operating information is provided for all four systems. For West Dover the energy consumption is evaluated and the treatment performance is documented. West Dover operates throughout most winters with minimal storage. The improvements in water quality at several of these systems are also discussed, and a method for estimating phosphorus removal is described.

The potential for effective wastewater renovation via land treatment in forests has been well recognized for at least twenty years (Sepp 1965, Sopper 1971, Reed et al. 1972, Pound and Crites 1973). These experiences have long since been reduced to engineering criteria for successful planning, design, construction, and operation of forested systems. Such criteria have been presented in the U.S. EPA Process Design Manuals for Land Treatment commencing in 1977, and in numerous reports and textbooks (McKim et al. 1982, Reed and Crites 1984). However, the design procedure itself is anything but routine, since the concept, especially in forests, is strongly dependent on specific site conditions (soils, tree species, slopes, etc.). Thus the design details appropriate for one location will not automatically be acceptable elsewhere.

The rural character, topography, vegetation, climate, and hydrology of northern New England all favor the use of forests for land treatment with wastewater. The concept is the most common (and often the only) form of land treatment found in Maine, New Hampshire, and Vermont. A listing of municipal and private systems in these states is found in Table 1. It is the purpose of this paper to evaluate, on a case-study basis, several of these systems, and to identify strengths as well as concerns, so that the next generation of system designs may benefit.

CASE STUDIES

Extensive data and observations are available for Sunapee and Wolfeboro, New Hampshire, and West Dover, Vermont. Comparative data are also available for some of the resort operations in Vermont. Results observed are evaluated in relation to prevailing design procedures.

TABLE 1. Forested land treatment systems in New England.

Location	Operating Season (days/year)	Average Hydraulic Loading (m/yr)
Maine		
Greenville*	182	0.9
Sugarloaf Mountain	140	1.12
New Hampshire		
Sunapee**	56	0.5
Wolfeboro*	227	1.3
Vermont		
West Dover*	332	design 5.5
		actual 1.7
Resort areas	280-330	2 to 2.5
Stratton		
Sherburne		
Haystack		
Bromley		

*Municipal system.   **State owned and operated.

Sunapee, New Hampshire

One of the earliest land treatment systems in this area in modern times was installed by the state of New Hampshire at Sunapee State Park in 1971. It is located in a mature forest of mixed hardwoods and conifers. The soils are an acidic, loamy glacial till material common to the uplands of New England. Wastewater is collected in facultative lagoons from the park facilities. These lagoons are drawn down during late spring and summer and the undisinfected effluent pumped through aboveground aluminum pipe to the sprinkler nozzles (laterals are 8 cm aluminum pipe, with impact nozzles about 15 m apart on risers about 1 m above the ground). A visit was made to the site in May 1984 to observe conditions after twelve years of operation.

Both the site and the equipment seem to be in excellent condition. The original aluminum pipe is still in service, although some of the sprinklers have been replaced. Since there was no intention at this site to institute a harvest program, the original tree stand is still in place. The area receiving wastewater since 1971 appears essentially the same as the nearby undisturbed areas. The forest floor within the impact zone of the sprinklers seems to be essentially the same as adjacent undisturbed areas. Some of the larger trees in very close proximity to the nozzles have a lighter color (slightly bleached) zone on the trunk from the wastewater impact, but there appears to be no damage to the bark itself. Hydraulic loading on the site is about 0.5 m of wastewater per application season.

This hydraulic loading is quite low and does not impose any significant stress on the site. Very conservative criteria were used for design of this "first generation" system, which has an operating season of 56 days per year to apply the annual volume of wastewater at 5 cm per week. It was recognized in 1971 that a longer operating season is feasible. It is recognized today that something smaller than the 2 hectare treatment area would be sufficient. Although disinfection of the applied wastewater was discontinued some years ago, public access is not restricted. Since most of the park visitors are not aware the system exists, public contact is limited. Groundwater sampling has never indicated any problem. The wastewater applications have not stimulated tree growth (Earley and Hosker 1982). This is not surprising, since it was already a mature stand and the wastewater loading is too low to have a significant impact.

Wolfeboro, New Hampshire

This resort community, with about four thousand permanent residents, is on the eastern shore of Lake Winnipesaukee. Land treatment in a forested area was selected, after a thorough study, as the most cost-effective alternative for municipal wastewater treatment. Secondary treatment (activated sludge) is provided for the 1,140 m³/day design flow. Disinfection is not required by the state, but the operator elects to chlorinate the effluent prior to transmission to the 138 day storage lagoon. Water withdrawn from the lagoon can be pumped to one of five spray areas. The total area, including the 90 m buffer zone, is about 40 hectares. About 32 hectares of this are considered to be the treatment area, receiving water from the sprinkler nozzles. Sprinkling starts about mid-May and has continued until early November. The average design loading is 5.6 cm of wastewater, applied over a 14 to 19 hour period, every seventh day. The system has operated on this program since 1978.

The distribution network is similar to that at Sunapee, consisting of aboveground mains, aluminum laterals hung on metal posts, and impact sprinklers on about 15 m centers. This network can be drained, but freezing is not an operational concern. Wide lanes were not cleared during construction of the pipe network. Public access is not restricted in any way, and snow trails for cross country skiing are maintained on the site during the winter.

The treatment area is on a hillside with the historic local name Poor Farm Hill. The title is an apt description of the agronomic potential for the site. The soils are thin and irregular, with bedrock at 0.3 to 1 m, numerous outcrops, slopes at 8 to 15%, and shallow, seasonally high "perched" water in the low spots. The major surficial soils are well-drained sandy loams.

Vegetation is primarily mixed hardwoods of the beech, maple, and birch type with some areas of second-growth white pine. Earley and Hosker (1982) studied the tree responses during 1980–82, after the site had been operational for two years. Their results are summarized in Table 2.

Tree mortality was about 3.5 times greater on the spray plots compared with the controls (Earley and Hosker 1982). In both cases the affected trees were mostly small diameter (<5 cm). No pervasive insect or rodent damage was noted, probably because of the higher moisture in the soils on the spray plots. The growth rate of the remaining trees on the spray plots was significantly greater than on the control areas because of fewer trees,

TABLE 2. Tree responses at Wolfeboro, New Hampshire (Earley and Hosker 1982).

Species	1982 Density (% basal area)	Mortality, 1980-82 (%)	
		Spray area	Control
Hemlock	2	23	13
Sugar maple	4	20	5
Beech	14	20	5
Red maple	20	11	16
White pine	3	8	34
Red oak	38	3	14
Paper birch	11	--	--
White ash	2	--	--
White oak	2	--	--
Other	4	--	--

more light, and increased nutrients on the spray areas. Observations during a site visit in April 1985 indicated that the site seemed to be typical of an upland New England forest, with normal density and no evidence of persistent mortality in the spray areas.

A special effort was made to monitor water quality during 1978–82. Flanders (1983) has reported on the 1978–81 period, and the 1982 data are on file with the state of New Hampshire (D. Allen, New Hampshire Water Supply and Pollution Control Commission, pers. comm., 1985). Pan lysimeters were installed at 15 and 60 cm to monitor shallow percolate, wells were installed to or in bedrock to monitor groundwater, and samples were obtained from brooks draining the treatment area.

Nitrogen and phosphorus were the parameters of greatest concern because of the potential for eutrophication in nearby brooks and ultimately in the lake. Table 3 compares nitrogen and phosphorus results from the pan lysimeters for the 1982 operational season. Applied nitrogen in the wastewater was about 7.1 mg/l (60% ammonia, 39% organic) and the total phosphorus was 3.3 mg/l. The nitrogen content is lower than usual for secondary effluent, but there are probably significant losses in the storage lagoon (Reed 1984).

The mean phosphorus concentration for all of the lysimeters for the entire 1979–81 period was less than 0.05 mg/l. Table 4 compares the "background" phosphorus in 1978 with the 1982 values for the adjacent brooks and groundwater. It is clear that the operation has not had an impact on the phosphorus content of these receiving waters.

The irregular soils and nonuniform slopes on this site result in local runoff as well as temporary saturation in the low spots during wastewater application. Any such runoff does not leave the site but infiltrates down slope. These local conditions are gradually

TABLE 3. Lysimeter data, 1982 season, for Wolfeboro, New Hampshire.

Month	Total N (mg/l)		Total P (mg/l)	
	15 cm*	60 cm	15 cm	60 cm
May	1.4	1.4	0.02	0.03
June	1.7	1.3	0.01	0.01
July	1.1	1.2	0.02	0.04
August	1.2	2.0	0.04	0.02
September	2.4	1.6	0.02	0.02
October	2.3	2.1	0.01	0.01
Average	1.8	1.6	0.02	0.02

*Lysimeter depth.

TABLE 4. Phosphorus comparisons for Wolfeboro, New Hampshire.

Location	Total Phosphorus (mg/l)	
	Background, 1978	1982
Applied effluent	--	3.3
Percolate, at 60 cm	--	0.02
Brook A, above site	0.03	0.02
Brook B, below site	0.08	0.05
Groundwater	0.02	0.01

being corrected by the operator. The aboveground aluminum pipe has experienced local damage from large limbs or tree falls during high winds. A routine tree harvesting program is being considered for the operation.

West Dover, Vermont

The design of the West Dover system was based on "second generation" criteria after it was understood that very low application rates and short seasons were unduly conservative for a forested system. The nonoperational period (33 days) at West Dover is not due to winter conditions but to a state requirement that wastewater cannot be applied during the spring snowmelt and runoff period.

The West Dover land treatment site is on a wooded hillside immediately adjacent to the sewage treatment plant. The elevation difference to the top of the hill is about 30 m. Secondary treatment is provided in an oxidation ditch system. Three 1.3 m^3/min centrifugal pumps discharge to a manifold system which in turn connects to the twelve laterals on the hillsides. The main lines from the pump are buried plastic pipe. The laterals are above ground and are either 5 or 8 cm diameter galvanized steel pipe inside a PVC plastic jacket. The laterals are spaced about 23 m apart and are parallel to the general contours of the hillside. A lane not more than 6 m wide was cleared for installation of each lateral. No grading was performed, so the laterals, suspended from posts, follow the transverse undulations on the hillside, and the pipe varies from 1.5 to 4.5 m above the ground. These steel laterals have not suffered any damage from limb or tree falls.

The eastern hillside has slopes ranging from 8 to 15%, and the western side exceeds 25%. The dominant soils are well-drained, loamy glacial tills. About 40% of the area is covered with a mixture of maples, beeches, and birches with a dense understory of balsam, spruce, and hemlock. The remainder of the site is dominated by white spruce, spruce, and fir. The mean annual temperature at the site is about 6°C. Average annual precipitation is 1.4 m, with snowfall in excess of 2.5 m (as snow). On average the site is snow covered about 120 days per year and frost free for about 90 days (June-September).

Special measures were required to allow operation throughout the winter. Trial and error resulted in a specially modified, downward spraying nozzle. These were placed at all low spots in each lateral, so that all remaining liquid would rapidly drain at the end of the operational cycle. The water in the main lines drains back to the holding pond. Wintertime wastewater applications have been successful with ambient air temperatures as low as -18°C.

Another winter problem involved the accumulation of ice beneath the upward spraying nozzles. This accumulation can engulf both the pipe and the nozzle, imposing excess loads on the pipe and blocking flow from the nozzle. This was solved by adding a 1 meter long riser, at the former nozzle connection, inclined at about 20 degrees from the vertical, so that most of the spray would fall to one side rather than directly on the pipe. Problems still occur, and the operator continues to experiment with other approaches and nozzle types.

The energy use at this system was evaluated in 1980 (Martel et al. 1982). The total operation and maintenance cost for 1980 was $54,467, with a total flow of 91,000 cubic meters. Most of this flow occurs in the winter months, owing to the nearby ski resorts. These costs include the complete oxidation ditch and aerobic sludge digestion secondary treatment plant, as well as the land treatment portion of the system. About 25% of the annual costs are related to energy. Electrical costs account for about 80%, and the remainder was for heating oil. The calculated energy consumption is shown in Table 5. The 14% assigned in the table to the land treatment component represents about 88,000 kwh to pump effluent to the hillside for the year. The remaining energy was used to provide secondary treatment.

TABLE 5. Energy consumption at West Dover, Vermont.

Function	Equipment	% of Total Energy
Secondary treatment aeration	Ditch rotors, blowers	46
Other treatment components	Comminutor, pumps, sludge management	12
Heat	Oil furnace	27
Lighting, miscellaneous	--	1
Land treatment	Pumps	14

The West Dover design was based on the assumption that a fragipan layer (glacial induced hardpan) existed more or less uniformly on the treatment slopes at a depth of a meter or less. This barrier was expected to prevent deep percolation to the groundwater and to channel the percolate flow laterally away from the site. Fragipan is a common feature in the glaciated uplands of northern New England, and apparently its presence was taken for granted during design. A postconstruction evaluation (Bouzoun et al. 1982) determined that fragipan is essentially absent on the eastern slope, so deep percolation is only limited by the relatively shallow bedrock (1 to 3 m).

The secondary treatment plant and the holding pond were constructed in the narrow floodplain of Ellis Brook. This required construction of the holding pond relatively close to the toe of the eastern treatment hillside. It was noted during construction that natural runoff and subflow from the hillside were entering the holding pond in significant quantities. A cutoff ditch was constructed at the toe of the slope to correct this problem. This ditch drains to an evaporation pond constructed in sandy soil at the southern end of the site. The ditch does not collect all the subsurface flow from the hillside (Bouzoun et al. 1982). The western treatment slope drains directly to the Deerfield River.

The treatment areas contain numerous outcrops, irregular slopes, and low wet spots at random locations on the hillside. These conditions are not desirable for the "ideal" system and, where possible, are being corrected by the operator.

The special downward spraying nozzles are essential for winter operation, but they do induce significant local runoff. They do not open just for the drainage cycle but operate continuously during the entire application period. The large orifice needed to prevent freezing results in a flow rate of about 0.5 l/sec on an impact area less than 4 m in diameter. Saturation and runoff from this zone occur rapidly. But the runoff then infiltrates quickly down slope. This is not a problem during the summer. As soon as temperatures permit, the nozzle is rotated 180 degrees and sprays upward in the warm months.

The performance of this system with respect to water quality requirements has been consistently excellent. Water samples are obtained from six shallow wells on the treatment site (about 1 m deep) and from the cutoff ditch at the toe of the eastern slope. Special studies have also measured the characteristics in the adjacent Ellis Brook and the Deerfield River as well as the snow and ice adjacent to the sprinkler nozzles.

Table 6 compares the on-site water quality data from 1978–79 to 1984. Bouzoun et al.

reported the earlier results in 1982; the 1984 data were collected by the state of Vermont (T. Willard, State of Vermont Agency of Environmental Conservation, pers. comm., 1985). Table 7 is a similar comparison for the adjacent brook.

The removal of nitrogen and phosphorus has been excellent, and there was no significant difference between winter and summer performance as indicated by the 1978–79 data, and no significant differences after ten years of operation. Table 8 summarizes water quality data in the adjacent surface waters at West Dover and other sites in Vermont.

Since operations continue throughout the winter at West Dover, there was some concern regarding the impact of the applied wastewater on the adjacent snow, and then again on the receiving streams during the snowmelt and runoff period. Table 9 compares data collected during the 1978–79 winter on the treatment site and at a remote off-site location.

The ice mound immediately beneath the sprinkler still contained about 90% of the applied nitrogen; the snowpack within the impact zone contained 7.5 mg/l total nitrogen. The ice mounds were about 3 m in diameter at the base and about 1.5 m tall; the impacted snow zone was about 8 m in diameter and 0.7 m deep. Using these dimensions and the flow and water quality data reported, it is possible to show that about 95% of the applied nitrogen moved through the porous snowpack with the still unfrozen water droplets and then probably infiltrated into the still unfrozen soil. The ice mound is essentially composed of frozen wastewater. The 7.5 mg/l in the adjacent snowpack is for the most part due to entrapment of particulate matter.

The nitrogen and phosphorus that are contained in this ice and snow would have a negligible impact on receiving waters during snowmelt and runoff. The 600 sprinklers at West Dover affect only 3 hectares out of the entire 20 hectare site. The pristine snow covering the other 17 hectares would melt and run off at the same time. The calculated quality for the runoff would be: BOD, 0.09 mg/l; total N, 1.8 mg/l; and total P, 0.02 mg/l.

It seems clear that winter operations with large droplet sprinklers pose no environmental concern, since most of the applied water will infiltrate within the site. The situation is slightly different with snow-making operations where the intention is to freeze essentially all the applied water. In this case, the impacted zone may contain most of the applied nutrients. But in the typical case the impact zone will still be a small portion of the total watershed area, so the ultimate runoff impact should still be negligible. The concern over using wastewater for recreational snow making is not because of the nutrients it contains but because of the pathogens; appropriate levels of treatment, including filtration and disinfection, should be effective in eliminating them.

TABLE 6. On-site water quality at West Dover, Vermont.

Location	Total N (mg/l)	Total P (mg/l)
Applied effluent		
1978-79	15	4.0
1984	17	4.4
Cut-off ditch		
1978-79	6.3	0.2
1984	4.3	0.1
On-site wells, 1978-79		
East side	7.2	0.7
West side	5.0	0.1

TABLE 7. Water quality in Ellis Brook, West Dover, Vermont, 1978-79.

Parameter	Upstream	Downstream
Total N (mg/l)	0.3	0.3
Total P (mg/l)	0.07	0.07
Chlorine (mg/l)	1.0	2.4
Fecal coliforms (no./100 ml)	18	12

TABLE 8. Surface water quality at Vermont sites.

Location	Total N (mg/l)	Total P (mg/l)	Chlorine (mg/l)
Haystack			
Applied effluent	1.5	0.7	8.7
Brook: Upstream	0.6	0.03	2.8
Downstream	1.1	0.2	4.9
West Dover			
Applied effluent	15.0	4.0	57
Brook: Upstream	0.3	0.07	1.0
Downstream	0.3	0.07	2.4

TABLE 9. Quality of snow and ice at West Dover, Vermont.

Location	Total N (mg/l)	Total P (mg/l)	BOD (mg/l)
Background snow	0.97	0.0	0.0
Snow within sprinkler impact	7.5	0.0	2.7
Ice at sprinkler base	13.3	1.7	5.6
Applied effluent			
(annual average)	14.9	4.0	10.0

DESIGN LIMITS

The approach recommended by the U.S. EPA and other texts (Reed and Crites 1984) determines the wastewater constituent that is the limiting design parameter and then determines the treatment area required for that substance. This approach bases design directly on wastewater characteristics and site conditions rather than on other indirect secondary relationships. For typical domestic wastewaters, the limiting parameter will usually be either the volume of water itself (as constrained by the hydrological capabilities of the soils) or the nitrogen content—to protect the quality of groundwater, which may be a drinking water source.

If the potential for downgradient groundwater use exists, then the EPA recommends that percolate quality should be equivalent to drinking water standards at the project boundary. If extraction for drinking water is not possible, then the requirement does not prevail. At West Dover, for example, essentially all the percolate should emerge as sub-flow in the adjacent surface streams, with no intermediate extraction for drinking water.

The vegetation on the site is the major permanent pathway for nitrogen removal. Typical values for forests range from 100 to 300 kg/ha per year depending on the type of tree, age, harvesting practices, and other factors. Based on the data discussed previously, the calculated nitrogen removal at West Dover is about 145 kg/ha per year, but at Wolfeboro it is only 60 kg. These calculations are based on assumptions regarding the "actual" treatment area at both locations. It is possible that the actual wetted area at Wolfeboro is smaller than assumed, thereby accounting for the low calculated uptake. Taken together, the two values confirm the validity of the criteria in the design texts.

Phosphorus is an additional design concern because of the potential environmental impact on the receiving surface waters. A design based only on drinking water standards will not be concerned with phosphorus, so additional procedures are necessary. It is essential during the planning and design stages to assess the phosphorus impact and if removal is necessary to determine the site capabilities for that purpose. Some authorities specify a minimum percolate detention time in the soil to ensure that there is time (and space) for all essential reactions to occur prior to emergence in the surface water. A more accurate, and more expensive, procedure is to obtain and test soil samples from the potential flow path to determine their phosphorus retention capacity. A rapid method for preliminary site assessment is presented below. It is derived from the U.S. EPA design procedures for very high hydraulic loadings on coarse soils, so should provide a conservative estimate for most forest soils in New England. The basic equation is:

$$P_x = P_o \exp^{-kt}$$

where

P_x = total P at a distance x on the flow path, mg/l
P_o = total P in applied wastewater, mg/l
k = 0.048 at pH 7, d^{-1} (pH 7 gives the lowest value)
t = detention time, d = $(x)(W)/(K_x)(G)$
x = distance along flow path, m
W = saturated soil water content, use 0.4 m^3/m^3
K_x = hydraulic conductivity of soil in direction x, m/d.
   So: K_v = vertical, K_h = horizontal.
G = hydraulic gradient for flow system, G = 1 for vertical flow.

The equation is solved in two steps—first for the vertical flow component, from the soil surface to the subsurface flow barrier (if one exists), and then for the lateral flow to the adjacent surface water. The calculations are based on assumed saturated conditions, to produce the lowest possible detention time. Since the vertical flow in most cases will be unsaturated, the actual detention time in this zone will be much longer than calculated with this procedure. If the equation predicts acceptable removal, then there is some assurance that the site should perform reliably, and detailed tests should not be necessary for preliminary work. Detailed tests should be conducted for final design of large-scale projects.

CONCLUSIONS

1. Wastewater treatment by land application in forests is a reliable and environmentally compatible concept. It has become the most common form of land treatment in northern New England, and that use is likely to expand.

2. The glaciated nature of upland forested sites does not usually represent ideal conditions on a microscale, owing to irregular slopes, thin soils, shallow bedrock, and wet areas. But experience at operational systems indicates excellent overall performance even with these internal constraints. Many of these local conditions can be improved over time in an operational system. That approach is preferred to clearing, site regrading, and then reforestation.

3. The experience at West Dover clearly indicates that a thorough site investigation is essential for design of all large-scale systems. The potential subsurface flow path(s) of the applied wastewater must be defined and the flow velocities at least estimated.

4. Snowmelt and runoff from land treatment sites should not have a significant impact on receiving surface waters even if the site is operated in the winter months.

5. Both nitrogen and phosphorus can be effectively managed on forested sites at the design loadings currently in use. Sustained high level nitrogen removal will require a tree harvest program at some point in the operation. The phosphorus removal capability of most soils will exceed the "design life" of most engineering works. However, if that time is ever reached, there is no need to abandon the site, because phosphorus could still be removed by precipitation in a pretreatment step.

6. Wastewater disinfection should not be a routine process requirement, nor should secondary treatment, since the forest ecosystem will provide better treatment for the less oxidized forms of wastewater. Pathogen removal is essential for recreational operations where direct and immediate public access is expected.

ACKNOWLEDGMENTS

The assistance provided by Mr. Wally Bronson, chief operator at West Dover, Vermont, and Mr. Gordon Reade, chief operator at Wolfeboro, New Hampshire, was essential to this study. The assistance and information furnished by Mr. Dan Allen, NHWSPCC, and Mr. Tom Willard from the State of Vermont is gratefully acknowledged.

REFERENCES

Bouzoun, J. R., D. W. Meals, and E. A. Cassell. 1982. A case study of land treatment in a cold climate—West Dover, Vermont. Report 82-44. Cold Regions Research and Engineering Laboratory (CRREL), Hanover, New Hampshire.

Earley, D. J., and H. W. Hosker. 1982. Tree response to sewage effluent application in Wolfeboro, New Hampshire. Department of Forest Resources, University of New Hampshire, Durham.

Flanders, R. A. 1983. Evaluation of wastewater renovation by spray irrigation: Wolfeboro, New Hampshire, 1978–1981. Master's thesis, Civil Engineering Department, Northeastern University, Boston.

Martel, C. J., B. C. Sargent, W. A. Bronson. 1982. Energy conservation at the West Dover, Vermont water pollution control facility. CRREL SR 82-24. CRREL, Hanover, New Hampshire.

McKim, H. L., W. E. Sopper, D. W. Cole, W. L. Nutter, D. H. Urie, P. Schiess, S. N. Kerr, and H. Farquhar. 1982. Wastewater applications in forest ecosystems. CRREL Report 82-19. CRREL, Hanover, New Hampshire.

Pound, C. E., and R. W. Crites. 1973. Wastewater treatment and reuse by land application. Vols. 1 and 2. EPA 660/2-73-006.

Reed, S. C. 1984. Nitrogen removal in wastewater ponds. CRREL Report 84-13. CRREL, Hanover, New Hampshire.

Reed, S. C., et al. 1972. Wastewater management by disposal on the land. CRREL SR 171. CRREL, Hanover, New Hampshire.

Reed, S. C., and R. W. Crites. 1984. Handbook of land treatment systems for industrial and municipal wastes. Noyes Publications, Park Ridge, New Jersey.

Sepp, E. 1965. Survey of sewage disposal by hillside sprays. Bureau of Sanitary Engineering, California Department of Health, Berkeley.

Sopper, W. E. 1971. Effects of trees and forests in neutralizing wastes. Reprint Series 23. Pennsylvania State University, University Park.

U.S. Environmental Protection Agency. 1981. Process design manual: Land treatment of municipal wastewater. EPA 625/1-81-013.

Chapter 40

Irrigation of Tree Plantations with Recycled Water in Australia: Research Developments and Case Studies

H. T. L. STEWART, E. ALLENDER, P. SANDELL, and P. KUBE

ABSTRACT   Recent developments in Australia reflect a general trend in drier countries in that all potential sources of nonpotable water are being investigated and, where possible, used in preference to potable sources. Also, schemes that currently discharge, or propose to discharge, treated wastewater effluent to inland water systems are being critically reviewed. Thus, more attention is being focused on schemes for recycling effluents, especially those in which parks, recreation areas, and plantation forests are irrigated with recycled water. In Australia, research on irrigation of trees with effluents commenced in the 1970s. Initial trials tested the survival and early growth of more than sixty native and exotic species irrigated with municipal effluent. Later work has concentrated on species from the genera *Eucalyptus, Casuarina, Pinus,* and *Populus,* with the most detailed research being conducted with *Pinus radiata* D. Don (radiata pine). Aspects studied so far include the effects of effluent irrigation on tree production, wood quality, nutrition, photosynthesis, transpiration, and changes in soil properties. There are now several projects across Australia in which tree plantations are being irrigated with recycled water at an operational level. Described in this paper are three case studies: a 30 ha (74 acre) plantation of *Populus* sp. (poplars) established in 1980 in Victoria; a 20 ha (50 acre) plantation of *Eucalyptus* sp. (eucalypts) established in 1984 in South Australia; and a 25 ha (62 acre) plantation of *E. camaldulensis* Dehnh. (river red gum) for fuelwood production established in 1980 in the Northern Territory. The results of these projects to date, along with the results from similar projects, demonstrate that well planned and managed enterprises in which tree plantations are irrigated with recycled water can have important social, aesthetic, and environmental benefits.

Australia is the driest and flattest continent, yet the amount of sewage effluent that is recycled (or reused) is only about 4% of the total annual flow of sewage (Gutteridge, Haskins, and Davey 1983). This apparent anomaly is explained by such factors (Strom 1979) as: (1) most of Australia's population live in coastal cities where adequate water supplies have been available at reasonable cost; (2) there have been few health and environmental guidelines for recycling effluent; and (3) at the time when most major sewerage schemes were developed, recycling of effluent was not considered.

Although there have been lost opportunities for recycling effluents, an overview of reuse schemes in Australia (N.S.W. Govt. 1982) showed that recent developments reflect a general trend in drier countries, in that all potential sources of nonpotable water are being investigated and, where practical, used in preference to potable sources. Also, schemes that currently discharge, or propose to discharge, treated effluent to inland water systems are being critically reviewed. Most state governments in Australia have introduced guidelines for recycling effluents.

The most used method of recycling effluent in Australia has been flood irrigation of pastures that have been grazed by sheep and cattle. The main objective has generally been disposal of secondary-treated effluent. Many of the schemes in inland regions have developed because (1) it has not been necessary to augment water supplies by discharging effluent to rivers, unlike many rivers overseas where there is not sufficient water to allow for effluent disposal to land, and (2) ample and relatively cheap land has been available for disposing effluent by irrigation (Johnston 1984a).

Apart from irrigation of agricultural land, increasing amounts of effluent are being recycled by irrigating parks and recreation areas, and to a lesser extent plantation forests and land used to grow wine grapes. Irrigation of recreational areas increases their amenity value, especially if they are in dry inland regions. The use of recycled water for irrigating forests is an attractive proposition in that health considerations are not as demanding, because treatment of effluent need not include disinfection and can be limited to primary treatment by sedimentation (Gutteridge, Haskins, and Davey 1983). In addition, well-planned and well-managed projects in which trees are irrigated with recycled water can have important social, aesthetic, and environmental benefits.

This paper briefly reviews research in Australia on irrigation of trees with recycled water, and then describes three important projects in which effluents are being recycled successfully through irrigation of tree plantations.

RESEARCH DEVELOPMENTS

The stimulus for research in Australia was the studies in America on irrigation of forested land with municipal effluent, especially at Pennsylvania State University (e.g., Sopper 1971). The first experiments in Victoria were four screening trials established between 1973 and 1976 that tested the survival and early growth of over sixty native and exotic species irrigated with municipal effluent (Edgar and Stewart 1978, Murray Valley Dev. League 1979, Stewart et al. 1979). In 1973 a trial was established at Darwin in the Northern Territory to test the performance of six tropical eucalypts (Craciun 1978), but it was destroyed in the second year by a cyclone. Based on this early work, fourteen species (from the genera *Eucalyptus, Casuarina, Pinus,* and *Populus*) were selected for further evaluation in Victoria at four inland sites established between 1977 and 1979 (Stewart and Flinn 1984) and at a further site in 1980. Results to date in Victoria show that some eucalypts have high survival rates and height increments of around 3 m (10 ft) per year after four years, irrespective of whether effluent is applied by spray, sprinkler, or flood irrigation on either clay or loam soils.

Other related research with eucalypts includes field studies of herbicides suitable for spraying over planted seedlings (Flinn et al. 1979); diagnosis and correction of iron deficiency in trees planted on calcareous soils and irrigated with recycled water (Stewart et al. 1981); tests of basic density, strength properties, and preservative absorption on round posts from five-year-old trees irrigated with recycled water (McKimm 1984); and studies of the effects of a young plantation on water tables in an irrigation area (Heuperman et al. 1984). In progress are studies of root systems, water use of irrigated trees, and effects of irrigation with recycled water on the chemistry of soils.

The most detailed research has been conducted with *Pinus radiata* D. Don (radiata pine), commencing in 1972 with a four-year study of the growth of planted seedlings that were trickle irrigated with recycled water (Cromer and Turton 1979). Dramatic growth responses were recorded, and a similar result was obtained from a three-year experiment in which a fifteen-year-old stand growing on sand was sprinkler irrigated with recycled water. In this latter study, each increment of 500 mm of water—between the limits of 500 mm (19.7 inches) and 1,500 mm (59.1 inches)—produced an extra 11 m³/ ha (157 ft³/acre) per year of merchantable timber (Cromer et al. 1983). Other aspects studied have been intraspecific variation in the response of the species to saline water and effluent (Cromer et al. 1982); photosynthesis and transpiration (Attiwill and Cromer 1982) and litterfall (Cromer et al. 1984b) of a stand irrigated with recycled water; the effect on soil chemistry and groundwater composition of irrigation with recycled water (Cromer et al. 1984a); and the fertilizer value of a dried sewage sludge for established stands (de Vries 1981).

Research is continuing in several states in Australia. Nevertheless, from the work to date, a sufficient number of useful species suitable for irrigating with recycled water have been identified such that species can be recommended with some confidence for a variety of soils and climates. There are now some good examples across Australia of tree plantations being irrigated with recycled water at an operational level, and three of these are described next as case studies.

CASE STUDIES

The case studies described are at Wangaratta in Victoria, Loxton in South Australia, and Alice Springs in the Northern Territory (Figure 1). Details of the contrasting soils and climate of the sites are given in Table 1. At each site, summer temperatures often exceed 40°C (104°F) and severe frosts commonly occur during winter.

Irrigated Poplars at Wangaratta, Victoria

Until the late 1970s, the Wangaratta Sewerage Authority discharged all secondary-treated effluent to Reedy Creek, which flows close by the treatment lagoons. Summer flows in the creek are generally low, and during this period the Authority had difficulty in reaching the standards set for the quality of effluents discharged to watercourses. To overcome this problem, and to recycle a resource that was being wasted, the Authority decided in 1979 to purchase land and use it to recycle effluent by irrigating a tree plantation.

A 54 hectare (133 acre) site was purchased in 1979. The land had been used for graz-

TABLE 1. Features of three sites in Australia where tree plantations are irrigated with recycled water (soils data: Northcote et al. 1975, Northcote 1979).

Feature	Wangaratta	Loxton	Alice Springs
Soil	Red duplex soil (Dr 2.2)	Calcareous earth (Gc 1.1)	Red earthy sand (Uc 5.2)
Annual rainfall	650 mm	278 mm	298 mm
Annual class A pan evaporation	1,808 mm	2,100 mm	3,143 mm
Maximum mean monthly temperature	January 30.4°C	January 31.0°C	January 35.8°C
Minimum mean monthly temperature	July 3.4°C	July 4.0°C	July 4.1°C

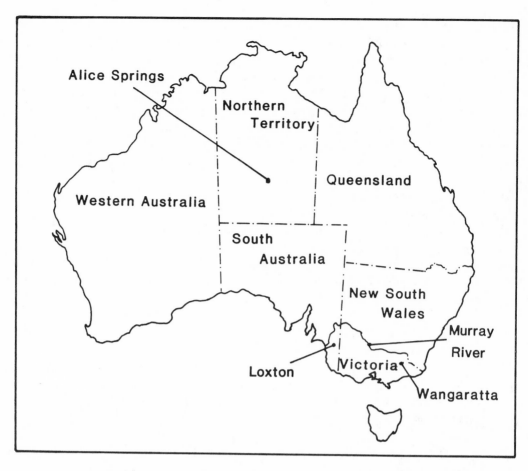

Figure 1. Locations of three sites in Australia where plantations of trees are irrigated with recycled water. Also shown is the Murray River, Australia's most important river.

ing, but had permeable soils suitable for irrigation. Over the period 1980 to 1983, 30 hectares (74 acres) were planted to *Populus* sp. (poplars), usually during the winter month of July. Three clones were planted: *P. deltoides* clone ANU 70/51 (poplar clone 70/51); *P. deltoides* × *P. nigra* semi-evergreen clone ANU 65/31 (poplar clone 65/31); and *P. deltoides* × *P. nigra* semi-evergreen clone ANU 65/1 (poplar clone 65/1). Rooted cuttings were planted at a spacing of 4.5 by 5.5 m (15 by 18 ft) in holes 30 cm (1 ft) in diameter and 60 cm (2 ft) deep, giving a stocking of 404 stems/ha (998/acre).

Effluent delivered to the treatment works is given primary treatment followed by retention of about twenty days through an aerated lagoon system. The treated effluent has, on average, the following characteristics: pH 7.1; EC < 780 μS/cm (<500 ppm); total phosphorus 8.0 ppm; and total nitrogen 17 ppm. The effluent is recycled through a fixed-sprinkler irrigation system. Sprinklers are 11 m (36 ft) apart on 0.5 m (1.6 ft) risers. The system is automated and designed to apply up to 70 mm (2.8 inches) per week, so that the average daily dry-weather flow of approximately 4 million liters (880,000 gallons) during the months November to April can be recycled by irrigating the plantation.

Growth statistics of two of the clones fifty-four months after establishment are given in Table 2. The semi-evergreen clone, with an estimated volume of 67 m³/ha (957 ft³/

acre), was considerably more productive than the deciduous clone. The rotation for irrigated poplars in northern Victoria is fifteen to twenty years, usually with two thinnings. On this site, a mean annual increment over that period of around 25 m³/ha (357 ft³/acre) could be expected.

TABLE 2. Growth fifty-four months after establishment of two clones of poplars irrigated with recycled water at Wangaratta, Victoria.

Growth Statistic	Clone 70/51	Clone 65/31
Trees per ha	404	404
Mean height (m)	12.5	16.0
Mean DBHOB* (cm)	17.8	20.0
Basal area (m²/ha)	10.1	12.5
1/3 BA x Ht (m³/ha)	42	67

*Diameter over bark at 1.3 m (4.3 ft) above ground level.

Pruning is essential to produce high quality logs. Trees at Wangaratta were pruned at ages two and four years, and a third and fourth pruning will probably be scheduled at six and eight years respectively, with the last pruning to a height of 8 m (26 ft). The timber has a variety of uses that include match splints, plywood (especially from the pruned sections of the stem), moldings, and pulpwood.

Weed control at Wangaratta has been done as required by mechanical slashing. Sheep that were introduced during the third year were quickly withdrawn, because they stripped bark from trees. The plantation is now probably at a stage where controlled grazing could commence with little risk of damage to the trees. Experience suggests that cattle are less likely to damage trees than sheep. A bonus from grazing a plantation of poplars is that the leaves have a high protein content and are palatable to animals.

Irrigated Eucalypts at Loxton, South Australia

The Murray River system (including the Darling and Murrumbidgee rivers and their tributaries) drains over one million square kilometers or one-seventh of the total area of Australia. The lower third of the Murray is in the state of South Australia, which is particularly dependent on the river for its water. Adelaide receives between 20 and 80% of its water—depending on climatic conditions—from the Murray; and almost half the state's domestic and industrial water, plus nearly all water for irrigation, is drawn from the Murray.

Although the quality of water from the Murray is good by world standards, it has significant problems that include salinity and elevated levels of nutrients. The salinity, owing in part to the hydrologic and geomorphologic characteristics of the natural river basin, has been aggravated by such activities as clearing native vegetation, irrigation, and river regulation; and increased nutrient levels are the result of runoff from agricultural land and discharge of sewage effluent to the river (Johnston 1984b).

In recognition of the need to maintain (and, where possible, improve) the quality of water in the Murray, state and federal authorities have taken measures to minimize inputs of saline waters and effluents. For example, extensive engineering works have been undertaken to relocate saline seepage waters into evaporation pans well away from the river, and in 1984, industries in the riverland region of South Australia were directed to stop discharging effluents into the Murray or associated evaporation pans. As a consequence, Moore Brothers Pty Ltd., which operates a large fruit processing factory at Loxton from which effluent has been discharged to the nearby Murray since the 1950s, was forced to rethink its method of effluent disposal. Therefore, it developed plans (in consultation with Land Energy Pty Ltd.) for recycling effluent by irrigating a plantation of eucalypts. A major factor that influenced the proposals is that fuelwood consumption in South Australia is now thought to exceed the yields that can be sustained from native vegetation within realistic haulage distances from the centers of population. Forty hectares (100 acres) were purchased adjacent to the sewage treatment works—some 5 kilometers (3 miles) from Loxton—so that the plantation could be irrigated with a mixture of municipal and industrial effluent. The site had been used for dryland agriculture, and wind erosion during dry years had left shallow soils in some areas.

The two species planted, *Eucalyptus grandis* W. Hill ex Maiden (flooded gum) and *E. camaldulensis* Dehnh. (river red gum, Lake Albacutya provenance), were selected largely on the basis of results from trials in Victoria (e.g., Stewart and Flinn 1984) and in the Northern Territory (see later). Characteristics common to both species that are important to the project are the ability to coppice and withstand periods of waterlogging and drought, and low to moderate susceptibility to frost and insects.

In the first stage of the project, 14 ha (35 acres) were planted to *E. grandis*, and 6 ha (15 acres) to *E. camaldulensis* at a spacing of 1.5 by 3.0 m (4.9 by 9.8 ft), this being equivalent to 2,200 trees/ha (890 trees/acre). The *E. grandis* is generally confined to the deeper soils. Seedlings were pit-planted in cultivated ground using a coring tool in December 1984, and watered on several occasions during planting to avoid mortality from hot, desiccating winds. Four months after planting, all seedlings have survived except in one small area where there was a temporary failure of the irrigation system. During this time, weeds have been controlled with chemicals only to the extent needed to provide the trees with a competitive advantage, because bare soil would result in wind damage and sand abrasion of the trees. The second (and final) establishment phase of the project is scheduled for late 1985, and will be preceded by two months of irrigation and then chemical control of most of the weed sward just prior to planting the seedlings.

For irrigation, effluent is pumped to three 100,000 liter (22,000 gal) concrete tanks adjacent to the plantation. The tanks are connected in series, and the first has a sloping base to allow periodic flushing of accumulated solids. A float valve at rim level in the third tank activates a high pressure, 20 hp pump that empties the tank. Water is pumped through three sand filters, each with a capacity of 2,700 liters (594 gal) per minute, to drippers (one per tree) that deliver 8 liters (1.8 gal) per hectare.

The plantation is divided into bays, each about 1 hectare (2.5 acres) in area. Each time the pump is activated, three bays are irrigated simultaneously with a volume predetermined on a programmed, solid-state controller in the pump house. In the event of a breakdown, the effluent is discharged via an overflow pipe to nearby evaporation ponds. The filters flush automatically and the solids are discharged also to the evaporation ponds.

Two-thirds of the effluent available for the project is from the fruit processing factory, and the remainder is primary-treated sewage from Loxton. Chemical characteristics are

TABLE 3. Chemical characteristics of effluent recycled by irrigating plantations of eucalypts at two sites in Australia.

Site and Source	pH	EC (uS/cm) (ppm)	Total P	Ca	Mg (g/m^3 or ppm)	K	Cl	Na
Loxton								
Sewage works	7.2	1,200 (768)	2	39	14	27	120	108
Factory	7.8	2,250 (1,440)	<1	80	46	16	291	370
Alice Springs								
Sewage works	7.8	1,640 (1,050)	13	38	29	26	143	179

given in Table 3. Provision has been made for either dilution or augmentation of this mixture with river water as required—for example, if salinity is a problem.

The total annual flow to the completed project is expected to exceed 400 million liters (88 million gallons), equivalent to about 10% of the effluent available for recycling in the locality. There is little variation in the daily flow of sewage, whereas flows of effluent from the factory are greatest during the summer months. When the plantation is fully developed, the annual irrigation should be at least 800 mm (32 inches).

It is expected that the plantation will be managed on a rotation of about eight years, and that at least one subsequent crop will be produced from coppice. Over an eight-year period, a mean annual increment of approximately 20 m³/ha (286 ft³/acre) is predicted, resulting in an estimated yield (air-dried) of 90 Mg/ha (36 tons/acre) per rotation. The plantation will be managed on a system of clearcutting with no pruning or thinning.

Moore Brothers may choose to sell the timber as firewood in Adelaide—where minimum prices are $70 (Aust.) per Mg (tonne)—or they may burn it to generate steam for their industrial process. They currently transport black coal some 960 km (600 miles) for generation of steam, and are investigating the possibility of substituting fuelwood for part of their annual consumption of coal.

Fuelwood Production from an Irrigated Plantation of Eucalypts at Alice Springs, Northern Territory

Two factors led to the project at Alice Springs: (1) there was a need to dispose of sewage effluent, and (2) there was an increasing demand for fuelwood. The first was the most pressing because overflow of effluent from the existing treatment lagoons had created a permanent swamp, which was an ideal breeding ground for mosquitoes and therefore a health hazard.

An increase in the popularity of wood for domestic heating had resulted in an estimated use of around 1,000 Mg (984 tons) of fuelwood per year by 1980. Traditionally, the demand had been met by harvesting slow-growing native vegetation, *Acacia aneura* F. Muell. ex. Benth. (mulga), which forms a tall shrubland. In order to avoid unnecessary denudation of land around the township (to minimize local dust problems and preserve the environment), a system of licensed cutting was introduced in the 1960s. Commercial suppliers now have to obtain fuelwood some 60 km (37 miles) from Alice Springs; thus costs of this fuelwood continue to rise as the haulage distance increases. The proposal to produce fuelwood from plantations irrigated with recycled water was therefore attractive, because as well as alleviating a health hazard, the project would reduce the intensity of harvesting of native vegetation that is part of a fragile ecosystem.

A 25 ha (62 acre) site adjacent to the sewage lagoon, with soils suitable for long-term irrigation, was selected. The major species selected was *E. camaldulensis* (Wiluna provenance), because it had grown successfully in earlier trials at Alice Springs and was known to tolerate limestone (hence alkaline) soils (Turnbull and Pryor 1978). The entire area was planted from September 1980 to April 1985. Initially, seedlings were pit-planted by hand; the later plantings were done by machine. Plant spacing has been 1.5 by 3.0 m (4.9 by 9.8 ft), giving a stocking of 2,200 trees/ha (890 trees/acre). All weed control has been mechanical, consisting of thorough cultivation of the soil before planting, and mowing between the tree rows as required with a tractor-drawn slasher during the first two years. Preemergent herbicides have not been effective, mainly because of the unstable, sandy surface soil. After two years, the tree canopy is generally dense enough to suppress weed growth to a level at which it causes no concern.

The system of irrigation is border-strip flooding of bays that are 12 m (39 ft) wide and 100 to 200 m (328 to 656 ft) long. Approximately 50 mm (2 inches) of wastewater is applied at each irrigation, every two weeks except during winter (June to August), when irrigation is applied about once per month. The annual application is therefore about 1,000 mm (39 inches) spread over an average of twenty events. Irrigation of the entire site takes one person seven days. Chemical characteristics of the primary-treated effluent used for irrigation are given in Table 3. Weeds growing in the earthen irrigation channels are mown before each watering using an angled mower mounted on the back of a tractor. Unless mown, the weeds virtually stop the flow of water; herbicides are not used because the channels quickly erode if not vegetated.

The growth results of the irrigated plantation (Table 4) show a mean annual increment of 19.4 m3/ha (277 ft3/acre) and a merchantable volume of 45.5 m3/ha (650 ft3/acre) on the best sites forty-five months after establishment. Growth has been much less on site quality 2 areas that have received considerably less irrigation because of their slightly elevated and uneven ground surface. On site quality 1 areas, the mean annual increment was 14.5 m3/ha (207 ft3/acre) at age 30 months, and is likely to exceed 20 m3/ha (286 ft3/acre) as the plantation matures. Assuming a mean annual increment of 20 m3/ha (286 ft3/acre) over the entire plantation, and given a measured basic density of wood of 510 kg/m3 (32 lb/ft3), a sustained yield (oven-dried) of at least 250 Mg (246 tons) per year should be produced by the plantation. The rotation length, though not set at this stage, is likely to be about seven years, and it is hoped that several successive crops will be harvested from coppice regrowth.

TABLE 4. Growth forty-five months after establishment of a plantation of Eucalyptus camaldulensis (river red gum) irrigated with recycled water at Alice Springs, Northern Territory.

Growth Statistic	Site Quality 1	Site Quality 2
Merchantable stems per ha	2,107	1,204
Mean height (m)	10.3	6.6
Basal area (m^2/ha)	13.6	5.5
Total overbark volume (ft^3/acre)	73	23
MAI (m^3/ha)	19.4	6.2
Merchantable underbark volume (m^3/ha)	46	8.5

Ten other eucalypts and seventeen other provenances of *E. camaldulensis* were evaluated in trials adjacent to the plantation. Results fifty-three months after planting show that *E. camaldulensis* is the best of the species tested—as measured by merchantable volume production. The Wiluna provenance was ranked thirteenth out of seventeen, using mean height sixteen months after planting. Seven other provenances had significantly better height growth; therefore, the production of the plantation to date (Table 4) could be increased somewhat by planting superior provenances.

Soil properties and groundwater have been monitored since irrigation commenced. As might be expected, soil pH and salinity have increased, and the infiltration rate and depth to groundwater have decreased, though none of these changes have been large enough to be of concern. Monitoring of these parameters, and tree vigor, will need to be long term.

DISCUSSION

Research in Australia, followed by application of the results in large-scale projects, demonstrates that treated municipal effluents can be successfully recycled by irrigating plantations of several tree species. The three case studies described are not isolated examples; in northwestern Victoria, Mildura Wines Ltd. irrigates a small woodlot with winery effluent; in South Australia, a project similar to that at Loxton is being planned; and in the Northern Territory, another project is under way at Yulara.

Species evaluation trials have led to the identification of a variety of native and exotic trees that grow satisfactorily across a range of climates and sites when irrigated with recycled water. Now it is possible to develop a whole farm plan of tree planting that includes species (whose main purpose is either wood production, shade and shelter, ornament, habitat, or suitability for saline or waterlogged sites) that are chosen depending on the soil conditions and location of the area in question on the sewage farm. On sites where soil or topography is not uniform, this is likely to result in a mosaic of irregular areas planted to different species. Some sites may be suited to trees planted at relatively wide spacings and integrated with animals grazing irrigated pastures.

The only species to be grown commercially in irrigated plantations for wood production in Australia are poplars. Thus, the management of irrigated plantations of eucalypts will be dictated to some extent by practices in those areas overseas where the species are extensively planted. Outside Australia, some eucalypts are regenerated three to five times by coppice following establishment of the first crop from seedlings, before high mortality and reduced growth rates make it necessary to replant. However, *E. grandis*, grown on short, coppice rotations overseas, will not coppice nearly as well in Australia, and may need to be regenerated by replanting seedlings rather than relying on coppice regrowth (Cremer et al. 1978). In either case, close attention will need to be paid to weed control during the establishment phase of successive crops.

Rotation length depends largely on the product sought. When eucalypts are grown primarily for fuelwood, the rotation length is a trade-off between a short rotation to produce a manageable piece size and a longer rotation to maximize the yield of the plantation. If a species does not coppice successfully, a longer rotation will be desirable to reduce the costs of production.

Many of the health problems associated with the recycling of effluents in agriculture do not apply where trees are the crop irrigated, since there is no risk of transmitting diseases to humans through a secondary animal host, nor is the crop consumed directly (Cromer 1980). As a result, it is permissible in Australia to recycle primary-treated and nondisinfected effluent by irrigating forests that have no public access. Aside from whether agriculture or forestry is the most suitable form of land use, Strom (1984) concluded that the main advantage of recycling effluents to land in northern Victoria was the elimination of potential health hazards that would be posed if secondary-treated effluent was discharged to a potable water system.

ACKNOWLEDGEMENTS

The authors gratefully acknowledge the information and support provided by sewerage authorities and industry, particularly the Wangaratta Sewerage Authority, the Loxton Sewerage Authority, the Transport and Works Department, Alice Springs, and Moore Brothers Pty Ltd.

REFERENCES

Attiwill, P. M., and R. N. Cromer. 1982. Photosynthesis and transpiration of *Pinus radiata* D. Don under plantation conditions in southern Australia. I. Response to irrigation with waste water. Aust. J. Plant Physiol. 9:749–760.

Craciun, G. C. J. 1978. *Eucalyptus* trials in the Northern Territory coastal region. Aust. For. Res. 8:153–161.

Cremer, K. W., R. N. Cromer, and R. G. Florence. 1978. Stand establishment. p. 81–135. *In* W. E. Hillis and A. G. Brown (eds.) Eucalypts for wood production. CSIRO, Canberra, A.C.T.

Cromer, R. N. 1980. Irrigation of radiata pine with waste water: A review of the potential for tree growth and water renovation. Aust. For. 43:86–100.

Cromer, R. N., K. G. Eldridge, D. Tompkins, and N. J. Barr. 1982. Intraspecific variation in the response of *Pinus radiata* to saline and waste water. Aust. For. Res. 12:203–215.

Cromer, R. N., D. Tompkins, and N. J. Barr. 1983. Irrigation of *Pinus radiata* with waste water: Tree growth in response to treatment. Aust. For. Res. 13:57–65.

Cromer, R. N., D. Tompkins, N. J. Barr, and P. Hopmans. 1984a. Irrigation of Monterey pine with wastewater: Effect on soil chemistry and groundwater composition. J. Environ. Qual. 13:539–542.

Cromer, R. N., D. Tompkins, N. J. Barr, E. R. Williams, and H. T. L. Stewart. 1984b. Litter-fall in a *Pinus radiata* forest: The effect of irrigation and fertilizer treatments. J. Appl. Ecol. 21:313–326.

Cromer, R. N., and A. G. Turton. 1979. CSIRO irrigates *Pinus radiata* with wastewater p. 21–25. *In* Wastewater renovation using trees. Proceedings of a Symposium, April 1979, Albury, N.S.W. Murray Valley Development League, Wodonga, Victoria.

Edgar, J. G., and H. T. L. Stewart. 1978. Wastewater disposal and reclamation using eucalyptus and other trees. *In* Developments in land methods of wastewater treatment and utilization. Proceedings of Conference, October 1978, International Association on Water Pollution Research, Melbourne, Victoria.

Flinn, D. W., H. T. L. Stewart, and P. J. O'Shaughnessy. 1979. Screening of weedicides for overspraying *Eucalyptus*, *Pinus* and *Casuarina* on clay soils irrigated with treated effluent. Aust. For. 42:215–225.

Gutteridge, Haskins, and Davey. 1983. Water technology, reuse and efficiency. I. Water reuse and new technology. *In* Water 2000. Department of Resources and Energy, Australian Government Publication Service, Canberra, A.C.T.

Heuperman, A. F., H. T. L. Stewart, and R. A. Wildes. 1984. The effect of eucalypts on water tables in an irrigation area of northern Victoria. Water Talk 52:4–8.

Johnson, K. E. 1984a. Introduction. p. 1–2. *In* Off-river disposal of treated sewage effluent to reduce nutrient levels in the River Murray. Proceedings of Seminar, November 1984, Wodonga, Victoria River Murray Commission, Canberra, A.C.T.

———. 1984b. Keeping the Murray clean: The role of the River Murray Commission. Water 11:22–24.

McKimm, R. J. 1984. Fence posts from young trees irrigated with sewage effluent. Aust. For. 47:172–78.

Murray Valley Development League. 1979. MMBW tree trials at Werribee. p. 18–20. *In* Wastewater renovation using trees. Proceedings of a Symposium, April 1979, Albury, N.S.W. Murray Valley Development League, Wodonga, Victoria.

N.S.W. Government. 1982. Report of the Task Force on the Use of Reclaimed Water. 198 p.

Northcote, K. H. 1979. A factual key for the recognition of Australian soils. 4th ed. Rellim Technical

Publications, Adelaide, S.A. 124 p.

Northcote, K. H., G. D. Hubble, R. F. Isbell, C. H. Thompson, and E. Bettenay. 1975. A description of Australian soils. CSIRO, Melbourne, Victoria. 170 p.

Sopper, W. E. 1971. Disposal of municipal waste water through forest irrigation. Environ. Poll. 1:263–284.

Stewart, H. T. L., F. G. Craig, and B. D. Dexter. 1979. Effluent treatment and reclamation in Victoria using tree plantations. p. 14–18. *In* Wastewater renovation using trees. Proceedings of a Symposium, April 1979, Albury, N.S.W. Murray Valley League, Wodonga, Victoria.

Stewart, H. T. L., and D. W. Flinn. 1984. Establishment and early growth of trees irrigated with wastewater at four sites in Victoria, Australia. For. Ecol. and Manage. 8:243–256.

Stewart, H. T. L., D. W. Flinn, P. J. Baldwin, and J. M. James. 1981. Diagnosis and correction of iron deficiency in planted eucalypts in north-west Victoria. Aust. For. Res. 11:185–190.

Strom, A. G. 1979. Use of sewage as a resource in Australia. Search 10:136–142.

————. 1984. Nutrient removal: High and low technology. p. 10–18. *In* Off-river disposal of treated sewage effluent to reduce nutrient levels in the River Murray. Proceedings of Seminar, November 1984, Wodonga, Victoria River Murray Commission, Canberra, A.C.T.

Turnbull, J. W., and L. D. Pryor. 1978. Choice of species and seed sources. p. 6–65. *In* W. E. Hillis and A. G. Brown (eds.) Eucalypts for wood production. CSIRO, Canberra, A.C.T.

Vries, M. P. C. de. 1981. Fertiliser value of a dried sewage sludge for *Pinus radiata* plantations. Aust. For. 44:190–193.

Chapter 41

Municipal Wastewater Renovation on a Coastal Plain, Slash Pine Land Treatment System

JANE T. RED and WADE L. NUTTER

ABSTRACT   In December 1981, a prescribed burn was performed on a 17-year-old slash pine (*Pinus elliottii* Morelet) plantation land treatment system for domestic wastewater in the Georgia coastal plain. Irrigation with secondary treated wastewater began in 1981, and approximately 2.54 cm per week were applied during 1982 and 1983. Satisfactory wastewater renovation was maintained throughout the study. Prescribed burning had no significant effect on wastewater renovation. Potassium, calcium, magnesium, and phosphorus storage in the forest floor was significantly increased as a result of irrigation. Burning did not affect forest floor nutrient storage. Wastewater irrigation resulted in increased concentrations of total and available phosphorus, calcium, magnesium, and potassium. Soil acidity and total Kjeldahl nitrogen concentrations were significantly reduced. Total phosphorus, calcium, magnesium, and potassium concentrations were further increased by burning. The major effect of prescribed burning was the alteration of species composition and density of the understory.

The Naval Submarine Base, Kings Bay, Georgia was selected as the location for the Trident Atlantic Coast strategic submarine base in October 1980. The 4,900 hectare (12,000 acre) facility is in the southeastern corner of the state, in Glynn County. When completed in 1990, the base will support one squadron of Trident submarines and will include facilities for crew training, weapons handling and storage, submarine maintenance, personnel support, and family housing. In 1981 the Department of the Navy installed a slow rate irrigation, land treatment system at Kings Bay. The system is designed to renovate the wastewater generated by the family housing units and operational facilities on the base.

The School of Forest Resources at the University of Georgia is evaluating the effectiveness of the land treatment system. The specific objectives are (1) to assess the effectiveness of the subsurface drainage system and the renovation capacity of the system by monitoring the soil percolate, (2) to evaluate the effects of prescribed burning on wastewater renovation and nutrient partitioning within the system, and (3) to devise a forest management plan that ensures wastewater renovation, as well as continued site productivity.

SYSTEM DESIGN

The raw sewage generated on the base is collected and pumped to the Kings Bay Sewage Treatment Plant. Pretreatment consists of an aerobic facultative lagoon system and chlorine contact chamber. The treated wastewater is transmitted via an effluent pump station to the spray irrigation fields. Table 1 summarizes the chemical composition of the wastewater prior to land application.

The land treatment area is divided into eight 12.15 hectare blocks. Each block is treated as a separate forest management area. Owing to the high water table of the irrigation blocks, a surface drainage system was installed to ensure a sufficiently aerated travel distance through the soil for wastewater renovation. An open-ditched drainage network of ditches 1.5 m deep, spaced 91.5 m apart, was designed to maintain a minimum aerated soil depth of 0.8 m.

The land application system was designed to accommodate the projected wastewater volume of 1,140,000 liters per day (300,000 gallons per day), with each irrigation block receiving 1.3 cm per week of wastewater at an application rate of 0.64 cm per hour. Upon reevaluation of system performance, the hydraulic loading was increased to 2.54 cm per week in March 1982. Block 8 is a contingency field to be used in emergency loading situations. The irrigation blocks are forested primarily with loblolly and slash pine plantations.

Impact type sprinklers are spaced 18.3 m apart on lateral PVC pipes that run parallel to the drainage ditches at a spacing of about 24.4 m. The lateral pipes are tapered to ensure uniform application rates at each sprinkler.

STUDY SITE DESCRIPTION

The site is on Irrigation Block 6, a slash pine plantation planted in 1968 on a 1.8 by 2.4 m spacing. The basal area in 1982 was 32 m^2/ha. The site is essentially level, with slopes less than 2%. The understory is dominated by saw palmetto (*Serenoa repens*) and gallberry (*Ilex glabra*). The acidic sandy soils belong to the Mandarin series, a Typic Haplohumod, which is characterized by a spodic horizon of fluctuating depth. The climate of the area is generally warm and humid, with an average annual temperature of 19.7°C. The long-term annual rainfall is 130.8 cm (SCS 1980).

EXPERIMENTAL DESIGN

In July 1981, twelve randomly located, 0.02 ha plots were established in two experimental areas each approximately 91 by 305 m (Figure 1). Forest floor and soil samples were collected in July 1981 (prior to treatment) and in December 1982 and 1983, to assess the variability of nutrients. An overstory biomass inventory was performed at the same times. The trees were inventoried by measuring the diameter at breast height (dbh) of each tree within each plot. Tree heights of five dominant trees per plot were also measured. Four forest floor samples were randomly collected on each plot by cutting around a plastic circle (0.073 m^2) and removing all organic matter down to the mineral soil surface. The soil was sampled at three depths (0 to 7.5 cm, 7.5 to 15.0 cm, and 30.0 to 60.0 cm) in five randomly located holes per plot.

In December 1981, half of block 6 was burned with strip fires. The burn, which was contained by the drainage ditches, eliminated most of the understory vegetation and forest floor, although the mineral soil was not exposed.

After burning, suction cup lysimeters and groundwater monitoring wells were installed on the burned and nonburned areas. Three lysimeters, at depths of 30, 60, and 90 cm, were centrally located on four plots in each treatment area. Two transects of two

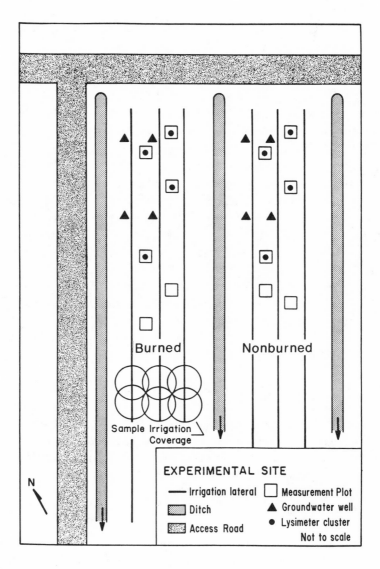

Figure 1. Experimental site.

groundwater wells each were installed on the burned and nonburned areas (Figure 1).

Lysimeter samples were collected monthly to evaluate wastewater renovation through the soil column. Sewage treatment pond samples were collected at the same time to ascertain the quality of applied wastewater. Groundwater recording sheets were changed monthly to monitor fluctuations in the water table.

In December 1982, eight trees on each treatment area, as well as a nearby nonirrigated plantation of similar characteristics, were destructively sampled following procedures outlined by Saucier (1979). Total height, total weight, diameter at 1.4 m (dbh), branch weight, foliage weight, and bark weight were determined for each tree. Subsamples of each component were analyzed for moisture content and nutrient composition.

A qualitative understory vegetation survey was also performed in December 1982. Two 1.0 m² subplots were randomly located adjacent to each permanent measurement plot. Within each subplot, all herbaceous material less than 0.6 cm in diameter 1.0 cm above the ground was destructively sampled and analyzed for nutrient composition. All larger material was recorded by species, diameter, and average height and subsampled for nutrient analysis.

SAMPLE PREPARATION AND ANALYSIS

Water samples were transported on ice and frozen if not analyzed within 24 hours of collection. Soil, forest floor, and vegetation samples were oven dried (40°C) to a constant weight, then ground, sieved, and subsampled prior to digestion. The samples were digested and analyzed on the Technicon Autoanalyzer II (EPA 1976, Black 1965).

STATISTICAL ANALYSIS

All data were summarized using the 1979 version of Statistical Analysis System (SAS) program PROC MEANS (Goodnight 1979). The general linear model procedure (PROC GLM) was used to test the relation between the dependent variable (nutrient content) and the independent variable (treatment) for each component (soils, forest floor, etc.). Tests for significant differences between sample means were conducted using an SAS version of Duncan's multiple range test. The level of significance was fixed at 5% for all tests.

RESULTS AND DISCUSSION

The mean quality of the wastewater applied to the site from January 1982 through December 1983 is presented in Table 1. Since the submarine base will not be fully operable until 1990, the base population and thus the volume and concentration of the applied wastewater were much lower than the design specifications.

A hydrologic budget was determined for the site, so that mass nutrient losses could be calculated for use in wastewater renovation determinations. The average annual potential evapotranspiration (PET) was estimated to be 102.7 cm by the Thornthwaite method

TABLE 1. Mean concentration and loading of wastewater constituents, January 1982 through December 1983.

Constituent	Concentration (mg/l)			Mean Annual Loading (kg/ha)
	Minimum	Maximum	Mean	
Calcium	3.5	56.3	32.6	352
Chlorine	9.0	63.5	43.0	465
Potassium	2.6	53.3	11.3	122
Magnesium	8.4	121.5	19.9	215
Total nitrogen	0.15	33.1	4.5	48.6
Ammonium-N	0	0.35	0.2	2.2
Nitrate-N	0.03	26.3	2.5	27.0
Organic N	0.12	6.4	1.8	19.4
Sodium	20.6	111.0	73.4	793
Total phosphorus	2.6	19.0	7.8	84.3

TABLE 2. Mean concentration of soil percolate by depth and treatment as measured by suction cup lysimeters, January 1982 through December 1983.

Treatment and Depth	Chlorine	Nitrate-N	TKN	Total Phosphorus	Sodium	Calcium	Potassium
				(milligrams/liter)			
Burned							
15 cm	41.2	0.43	2.67	0.89	55.6	4.9	3.0
30 cm	40.4	0.56	1.34	0.62	52.9	5.2	2.4
60 cm	41.7	0.30	1.40	0.59	58.4	3.7	2.7
Nonburned							
15 cm	37.7	0.63	1.87	0.80	47.6	5.4	1.8
30 cm	36.2	0.35	1.05	0.57	47.2	4.4	1.6
60 cm	37.0	0.14	1.06	0.90	50.7	4.3	1.8

(Thornthwaite and Hare 1965). Drainage from the site was calculated as the difference between total hydraulic loading and PET. Average annual values for the period of study for drainage, precipitation, and wastewater loading were 93.3 cm, 77.4 cm, and 114.6 cm respectively. The nutrient concentrations in the deepest lysimeter samples (Table 2) were used in conjunction with drainage to determine mass losses of nutrients from the site. These data are summarized in Table 3.

The system functioned well in wastewater renovation. The low renovation of chlorine and sodium (Tables 2 and 3) is due to chlorine's anion status and the ease of replaceability of sodium on exchange sites. Nitrogen renovation includes total Kjeldahl nitrogen (TKN) lost through leaching. Leachate nitrate-N levels remained well below the drinking water standard of 10 ppm throughout the study, as would be expected at such low application levels. If estimated nitrogen input from precipitation of 2.6 kg/ha annually (Haines 1976) is included in nitrogen input, renovation is decreased by less than 2%. No significant differences in wastewater renovation occurred as a result of burning.

Table 4 summarizes the changes observed in the forest floor. Decomposition rates were initially accelerated, as reflected by the significant reduction in dry weight of the forest floor in 1982. In addition, the forest floor dry weight on the burned area was significantly lower than that of the nonburned area in 1982, owing to the controlled burn performed in December 1981. By 1983, the forest floor dry weight on both treatments had returned to pretreatment levels.

TABLE 3. Mean nutrient loss in soil water percolate and percentage of renovation at 90 cm depth, January 1982 through December 1983.

Parameter	Nutrient Loss (kg/ha·yr)		Renovation (%)	
	Burned	Nonburned	Burned	Nonburned
Calcium	39.6	44.4	88.8	87.4
Chlorine	410	351	11.7	24.4
Potassium	21.6	14.4	82.3	88.2
Magnesium	57.6	39.6	73.2	81.6
Total nitrogen	16.8	13.2	65.4	72.8
Total phosphorus	7.2	9.6	91.5	88.6
Sodium	553	482	30.2	39.2

TABLE 4. Changes in forest floor resulting from wastewater irrigation and burning.

Parameter	Burned			Nonburned		
	1981	1982	1983	1981	1982	1983
	(kilograms/hectare)					
Potassium	9.8d (0.78)	10.3dc (1.02)	16.5a (1.38)	10.8dc (0.75)	13.2bc (0.83)	16.0ba (1.14)
Calcium	101.1c (6.8)	261.6b (23.4)	507.2a (37.2)	93.6c (5.4)	311.4b (25.6)	488.3a (42.6)
Magnesium	23.0b (1.8)	63.6a (5.4)	71.9a (6.8)	22.7b (1.3)	71.1a (5.4)	69.7a (5.7)
Total phosphorus	14.8d (1.1)	18.1dc (1.8)	27.5ab (2.6)	17.8dc (1.2)	22.3bc (1.4)	30.0a (2.4)
Total nitrogen	292.7b (24.7)	421.7a (42.1)	252.9b (25.8)	315.9b (18.5)	499.0a (40.6)	269.5b (22.5)
Dry weight	50,636ba (2,627)	33,972c (3,026)	54,849ab (5,591)	63,287a (3,171)	49,633b (3,408)	60,015ab (4,490)

Means with the same letter are not significantly different (alpha = 0.05).
Values in parentheses are the standard errors.

Wastewater irrigation resulted in significantly increased levels of potassium, calcium, magnesium, and phosphorus storage in the forest floor over time. Nitrogen storage increased significantly in 1982 but returned to pretreatment levels by 1983. No changes in nutrient storage of the forest floor occurred as a result of burning.

A summary of changes in soil nutrient concentration by treatment, year, and depth is presented in Table 5. (Unfortunately, because of the destruction of the control plot halfway through the study, the effects of drainage and wastewater irrigation are indistinguishable. Therefore, any changes observed on the nonburned area must be attributed to the combined effects of drainage and wastewater irrigation.) Total phosphorus levels were significantly increased at all depths, on both the burned and nonburned areas, from 1981 to 1983. In addition, depths 1 and 2 on the burned area were significantly higher than on the nonburned area in 1983. These results are consistent with those of Vlamis and Gowans (1961) and White et al. (1973), who documented increased phosphorus levels in soils following burning. Available phosphorus levels also increased significantly (from 2.5 to 17.3 ppm) in the surface horizon on both treatment areas.

Calcium and magnesium levels increased significantly in the surface 7.5 cm on both treatments, with 1983 concentrations of 275.5 ppm calcium and 151.4 ppm magnesium being approximately three times as high as pretreatment levels for calcium and five times as high for magnesium. The combined effect of burning, wastewater irrigation, and drainage resulted in significantly higher calcium and magnesium concentrations, 514.4 ppm and 267.7 ppm respectively, on the burned area surface soils in 1983. Wells (1971) also found increases in calcium and magnesium levels in the upper few inches of mineral soil following burning.

Potassium concentrations increased significantly in depth 1 of the nonburned area, and in depths 1 and 2 of the burned area. In addition, the potassium concentration of 78.9 ppm in depth 1 of the burned area was significantly higher than in the nonburned area, 37.6 ppm, in 1983. Burning has been shown to increase potassium concentrations of surface soils (Wells et al. 1979). Presumably the high mobility of this cation explains the increase in concentration in depth 2 of the burned area.

Total Kjeldahl nitrogen levels decreased significantly at all depths of the nonburned area and at depth 1 of the burned area from pretreatment levels. Although some TKN was found in the leachate from the site, reductions this great cannot be explained solely by leaching loss. One hypothesis is that the installation of the drainage network, which increased the aerobic depth of the soil, created conditions favorable to nitrification. The nitrate thus formed would be leached into the anaerobic zone, where conditions would be ideal for denitrification because of the presence of a carbon source in the wastewater. High levels of denitrification have been reported on land treatment sites as a result of the environmental conditions produced by wastewater application (Nutter et al. 1978).

Soil acidity was significantly reduced on both treatment sites. The average pH was increased over two units from pretreatment levels. Prescribed burning had no significant effect on soil reaction.

Results of the understory vegetation analysis are presented in Tables 6 and 7. No differences in nutrient concentrations (Table 6) were observed between treatments except for magnesium levels, which were significantly higher on the nonburned area. In the destructively sampled class (stem diameter < 0.6 cm) the burned area was dominated by gallberry and the nonburned area by gallberry and palmetto. The significantly larger mass of these two species on the burned area is due to the flush of new growth that occurred following the prescribed burn. In the nondestructively sampled class, the gallberry and palmetto on the burned area were small in diameter and height and less frequent than on the nonburned area.

TABLE 5. Changes in soil resulting from wastewater irrigation and burning.

Parameter and Depth	Burned				Nonburned			
	1981		1983		1981		1983	
	Mean	SE	Mean	SE	Mean	SE	Mean	SE
	(milligrams/kilogram)							
Total P								
0-7.5 cm	124.7a	7.6	325.4b	15.1	146.3a	9.9	237.1c	10.4
7.5-15 cm	115.7a	18.2	269.6b	10.5	117.8a	6.6	233.3c	12.0
30-60 cm	112.5a	7.8	270.4c	9.6	121.3a	4.2	240.4c	9.2
Calcium								
0-7.5 cm	169.0a	53.5	514.4b	62.0	73.8a	27.8	275.5c	39.0
7.5-15 cm	89.4a	28.8	159.2a	42.2	44.5a	21.2	72.4b	14.6
30-60 cm	13.2a	5.8	45.3a	18.2	9.9a	3.3	30.8b	9.7
TKN								
0-7.5 cm	1,353 b	197.2	960.7a	121.4	1,543 b	177.1	587.8a	32.0
7.5-15 cm	752.5a	121.9	511.9a	58.5	961.9c	108.2	461.4a	27.9
30-60 cm	537.9a	101.4	318.3a	26.4	698.7c	96.6	374.4a	27.4
pH								
0-7.5 cm	3.9a	0.1	6.5b	0.1	3.8a	0.1	6.8b	0.1
7.5-15 cm	4.1a	0.1	6.1b	0.1	4.0a	0.1	6.7b	0.1
30-60 cm	4.5a	0.1	6.1b	0.1	4.6a	0.1	6.8b	0.1

Means with the same letter are not significantly different (alpha = 0.05).
SE = standard error.

TABLE 6. Nutrient concentrations of the understory vegetation.

Constituent	Burned	Nonburned
Calcium (mg/kg)	7,302	8,216
Potassium (mg/kg)	2,009	1,965
Magnesium (mg/kg)	1,429	1,886
Phosphorus (mg/kg)	790	744
Nitrogen (%)	0.0101	0.0099

TABLE 7. Understory biomass distribution.

Sample	Burned	Nonburned
Destructively Sampled Understory		
(stem diameter < 0.6 cm)		
Major species	Gallberry	Saw palmetto and gallberry
Dry mass (g/m^2)	104	17.5
Nondestructively Sampled Understory		
(stem diameter > 0.6 cm)		
Major species	Saw palmetto and gallberry	Saw palmetto
Average diameter	0.95 cm	1.04 cm
Average height	0.42 m	1.02 m
Stems per hectare	30,000	82,000

Prescribed burning altered the species composition and density of the understory. The dominant postfire species was gallberry. Burning eliminates nutrient uptake and storage by the understory only temporarily. In fact, the flush of new growth following the fire may have removed larger quantities of nutrients than the prefire vegetation did. Further research is needed to determine if this is the case.

Table 8 summarizes the nutrient concentrations of overstory needles, branches, bark, and core wood by treatment. The calcium content of bark from the burned area was significantly higher than from the nonburned area. Phosphorus concentration in the branches was significantly higher on the nonburned area. When these concentration data were used in conjunction with the regression equations developed by the USDA

TABLE 8. Nutrient concentrations of the overstory vegetation.

Treatment and Constituent	Calcium (mg/kg)	Potassium (mg/kg)	Magnesium (mg/kg)	Phosphorus (mg/kg)	Nitrogen (%)
Burned					
Needles	3,364	2,195	1,285	949	0.006
Branches	2,687	1,528	669	541	0.003
Bark	2,055	827	495	295	0.006
Core wood	624	544	180	134	0.005
Nonburned					
Needles	3,138	2,147	1,206	870	0.008
Branches	2,888	1,643	787	711	0.003
Bark	1,780	873	455	305	0.006
Core wood	588	509	167	111	0.006

Forest Service to predict biomass based on dbh (Saucier 1979), no significant differences in mass nutrient storage were found.

No significant differences were found in overstory tree growth between the burned and nonburned areas. The overstory biomass will be sampled again in December 1986, at which time specific forest management guidelines will be developed from the five-year study.

CONCLUSIONS

Satisfactory wastewater renovation was maintained on a coastal, slash pine plantation land treatment system receiving 2.54 cm per week of domestic wastewater. Nitrate-N concentrations in groundwater effluent from the site were maintained below the drinking water standard of 10 ppm. Prescribed burning had no effect on wastewater renovation. A drainage system may be a necessary requirement for proper functioning of coastal land treatment systems.

Prescribed burning had no effect on forest floor nutrient storage. Potassium, calcium, magnesium, and phosphorus storage in the forest floor was significantly increased by wastewater irrigation.

Wastewater irrigation and drainage resulted in increased concentrations of total and available phosphorus, calcium, magnesium, and potassium in the soil. Soil acidity and TKN concentration were significantly reduced. Total phosphorus, calcium, magnesium, and potassium concentrations were further increased by the prescribed burning. Burning altered the species composition and density of the understory, with gallberry dominating as the postfire species. The flush of new growth following a prescribed burn may remove larger quantities of nutrients than the prefire vegetation does.

No significant differences in overstory growth or nutrient storage were observed as a result of treatment. The short period of study, small sample size, and semimature age of the stand are possible explanations of this observation.

The forest management plan will be completed in 1986, at the conclusion of the study. At this time, however, we have recommended that prescribed burning may be used on the land treatment system as a means of reducing wildfire hazard and understory competition. The management blocks may be burned at the forester's discretion, as frequently as every two years.

REFERENCES

Black, C. A. (ed.) 1965. Methods of soil analysis: Part 2. American Society of Agronomy, Madison, Wisconsin.

Goodnight, J. H. 1979. Statistical Analysis System (SAS) user's guide. p. 237–264, 303–306. SAS Institute, Cary, North Carolina.

Haines, E. B. 1976. Nitrogen content and acidity of rain on the Georgia coast. Water Res. Bull. 92(6):1223–1231.

Nutter, W. L., R. C. Schultz, and G. H. Brister. 1978. Land treatment of municipal wastewater on steep forest slopes in the humid southeastern United States. p. 265–274. *In* H. L. McKim (ed.) Land treatment of wastewater. Vol. 1. Cold Regions Research and Engineering Laboratory, Hanover, New Hampshire.

Saucier, J. R. 1979. Estimation of biomass production and removal. p. 172–190. *In* Proceedings, Impact of intensive harvesting on forest nutrient cycling. Syracuse, New York.

Soil Conservation Service. 1980. Soil survey of Camden and Glynn Counties, Georgia. USDA SCS.

Thornthwaite, C. W., and F. K. Hare. 1965. The loss of water to the air. Agricultural Meteorology, Meteorological Monographs. 6(28):163–180. American Meteorological Society.

U.S. Environmental Protection Agency. 1976. Methods for chemical analysis of water and wastes. EPA 600/4-84-017.

Vlamis, J., and K. D. Gowans. 1961. Availability of nitrogen, phosphorus, and sulfur after brush burning. J. Range Manage. 14(1):38–40.

Wells, C. G. 1971. Effects of prescribed burning on soil chemical properties and nutrient availability. p. 86–99. In Prescribed Burning Symposium Proceedings. USDA Forest Service, Southeastern Forest Experiment Station, Asheville, North Carolina.

Wells, C. G., R. E. Campbell, L. F. DeBano, C. E. Lewis, R. L. Fredriksen, E. C. Franklin, R. C. Froelich, and P. H. Dunn. 1979. Effects of fire on soil: A state of knowledge review. USDA For. Serv. General Technical Report WO-7.

White, E. M., W. W. Thompson, and F. R. Gartner. 1973. Heat effects on nutrient release from soils under ponderosa pine. J. Range Manage. 26(1):22–24.

Chapter 42

Fourteen Years of Wastewater Irrigation at Bennett Spring State Park

DON BARNETT and KEN ARNOLD

ABSTRACT   Bennett Spring State Park, in the Missouri Ozarks, has a natural spring that produces 379 million liters (100 million gallons) of clear, cool water a day. The spring teems with trout, drawing fishermen and vacationers from all over Missouri and the Midwest. To protect this resource, the Missouri Department of Natural Resources has adopted use of sewage for wastewater irrigation to keep sewage contamination out of Bennett Spring's waters. The degree of treatment is comparable to advanced treatment techniques. In the late 1960s and early 1970s, large quantities of algae and aquatic plants grew in the spring itself. When officials looked into the problem, they found the spring was getting doses of plant nutrients probably from nutrient-rich seepage from sewage disposal fields on adjoining properties. The irrigation system at Bennett Spring consists of twenty-seven sprinklers covering approximately 2 hectares (5 acres). The system was designed to apply 5 cm (2 inches) of wastewater per week for about 28 weeks a year. Use of the park has continued to expand, and the current application season approaches 36 to 40 weeks per year, with annual application estimated at over 178 cm (70 inches). In operation since April 1972, this land application system was the first of its kind in the state park system and one of the first in the state. As a result, it was the subject of a research project by the University of Missouri–Columbia during 1972. Although oak and pine did not significantly increase in rate of growth, the wastewater did prompt a significant response in ash, cottonwood, cypress, maple, sycamore, and walnut. In just one year, some trees grew more than 61 cm (2 ft) taller than similar trees not being irrigated.

Nestled in the Missouri Ozarks, Bennett Spring State Park is graced with a natural spring that produces 379 million liters (100 million gallons) of clear, cool water daily. For many years the Missouri Department of Conservation has used the spring water in a trout hatchery and rearing system. The stream receiving the spring flow is thus a trout fishing attraction that draws fishermen and vacationers from all over Missouri and the Midwest. Approximately 800,000 people visit the park each year.

In the late 1960s and early 1970s the spring and stream were plagued with unacceptable growths of algae and aquatic plants, greatly interfering with fishing in the spring water. An investigation revealed that the spring was enriched with plant nutrients probably from sewage seeping out of disposal fields in the area. Since the Missouri Ozarks are known for clear, swift, wild rivers, the cleanup and preservation of the spring and the Niangua River were considered essential. The Missouri State Park Board decided to employ the living filter concept of effluent disposal—a technique promoted, developed, and thoroughly tested at Pennsylvania State University by Sopper (1971) and others. This system would provide advanced (tertiary) treatment with no discharge to the receiving stream.

In 1970, a three-cell lagoon system was built to handle current and projected sewage flow demands. This system would also provide storage and primary treatment for an irrigation system. To guarantee that the wastewater would not find its way back to the spring, the lagoon system was located several thousand meters downstream of the spring.

The irrigation system was put in operation April 1972. This system consists of twenty-seven sprinklers ("rainbird" kicker type) covering about 2 hectares (5 acres) of mixed oak hillside and part of an old bottomland field along the Niangua River. The forested slopes are of the Lebanon, Nixa, and Clarksville soils association (Figure 1). Because of a characteristic shallow fragipan overlaid by a more permeable layer, these soils have a low water storage capacity. The slopes contribute both surface and subsurface flow to the bottomland section of the site. The bottom soils are of the Huntington series, which is a well-drained, highly permeable, silt loam. Huntington soils are alluvial with less than 5% slope. The vegetation is primarily mixed grass with several wild broadleaf species and some hardy multiflora roses. Concrete pads had to be added around the sprinklers to prevent the roses from interfering with sprinkler rotation.

Although not by design, this system operates as a combination overland flow and infiltration system. Because of the shallow fragipan in the hillside soils, some of the applied wastewater is treated by overland flow as it travels onto the more permeable bottomland soils. The higher permeability of the bottomland soils allows both the water applied by overland flow from the hillside and the water applied by sprinklers to infiltrate in the bottomland soil.

Bennett Spring State Park had the first land application system of its kind in a Missouri state park and one of the first in Missouri. As a result, the Bennett Spring treatment system was the subject of a research project by the School of Forestry at the University of Missouri–Columbia during 1972 (Turner 1973). The site was irrigated for twenty-eight weeks from April to November. During this period, 142 cm (56 inches) of effluent were applied through the sprinkler system. Two groundwater wells and several lysimeters at 30 and 91 cm depths (1 and 3 ft) were installed to measure the fluctuation of groundwater level and to collect samples for chemical analysis. All of these were located either on or near the irrigation site. As would be expected, the groundwater level did change with irrigation and with different rates of application (5 and 12 cm/wk; 2.0 and 4.6 in./wk).

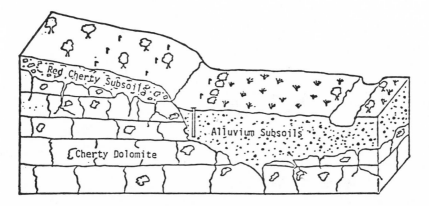

Figure 1. The Clarksville (red cherty subsoil) and Huntington (alluvium subsoil) association at the study site.

The soil water did not show any accumulation of nutrients during the irrigation season (Table 1). The wastewater effluent applied 20 kg/ha (18 lb/acre) of N, 26 kg/ha (23 lb/acre) of P_2O_5, and 52 kg/ha (46 lb/acre) of K_2O, without accumulation of nutrients in the soil water. It is safe to assume that there was very little leaching of nutrients and that most of the nutrients were taken up by the plants or held by the nutrient-deficient soil matrix.

Although the oak and pine did not significantly increase in rate of growth, the wastewater did prompt a significant response from ash, cottonwood, cypress, maple, sycamore, and walnut. During 1972 some trees grew more than 61 cm (2 ft) taller than similar trees not being irrigated. A 60% increase in general vegetation growth was estimated. However, in recent years many of the oaks and pines have died out in favor of the more water hardy species. The land application system of Bennett Spring State Park has continued to have a good performance record for the last fourteen years, protecting the Niangua River and utilizing the plant nutrients. Based on the degree of confidence established by the 1972 research and previous work by others in the wastewater field, additional research at this site has not been deemed necessary.

TABLE 1. Trends in ammonium-N, nitrate-N, and phosphate-P (ortho+meta) concentrations on irrigated and control plots at the Bennett Spring study area after 28 weeks of lysimeter sampling.

Depth		Initial (ppm)	Final (ppm)
	Ammonium-N		
30 cm (1 ft)	Irrigated	0.61	0.64
	Control	0.74	0.55
91 cm (3 ft)	Irrigated	0.60	0.31
	Control	0.66	0.34
	Nitrate-N		
30 cm (1 ft)	Irrigated	1.76	1.01
	Control	1.17	1.60
91 cm (3 ft)	Irrigated	5.68	0.68
	Control	4.58	2.62
	Phosphate-P		
30 cm (1 ft)	Irrigated	0.31	0.63
	Control	0.20	0.75
91 cm (3 ft)	Irrigated	0.39	0.96
	Control	0.21	0.82

No accumulation of ammonium-N, nitrate-N, or phosphate-P indicated.

This system has continued to operate with no discharge of sewage effluent to a receiving stream even though increased use of the park has increased the application rate to a range of 178 to 229 cm/yr (70 to 90 in./yr). The operating season begins about March 1, when trout season opens, and continues through the summer and into the early winter. Typical weekly operation is Monday, Wednesday, and Friday applying 15 to 18 mm (0.6 to 0.7 inch) per day, with additional application times being utilized to maintain the desired lagoon level. This system has also operated without nuisance odor complaints, an important feature considering that it is located adjacent to a public access area to the Niangua River and across the road from the Sand Spring Resort (Figures 2 and 3).

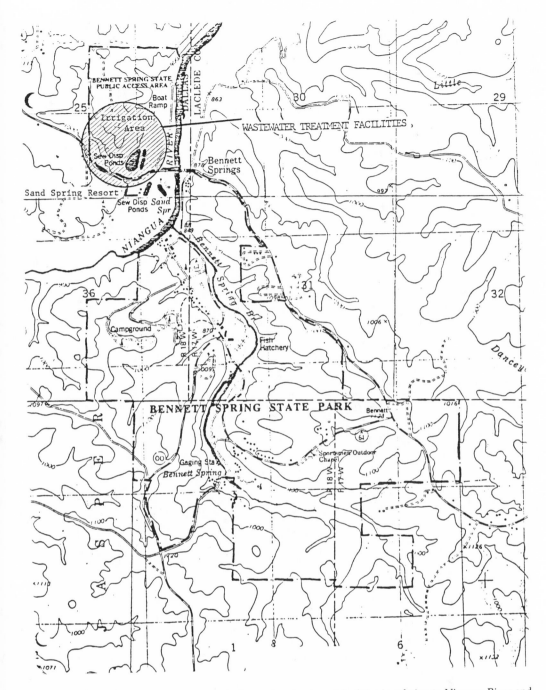

Figure 2. Map of Bennett Spring State Park wastewater treatment facilities in relation to Niangua River and surrounding area.

When selecting the type of treatment system for a park or other small- to medium-size facilities, certain factors favor land treatment: (1) Land treatment does not require highly trained operators (just common sense). (2) Very reliable treatment is provided because radically fluctuating loads do not cause treatment plant upsets. (3) Construction and operating costs can be less than other systems providing comparable treatment. (4) The

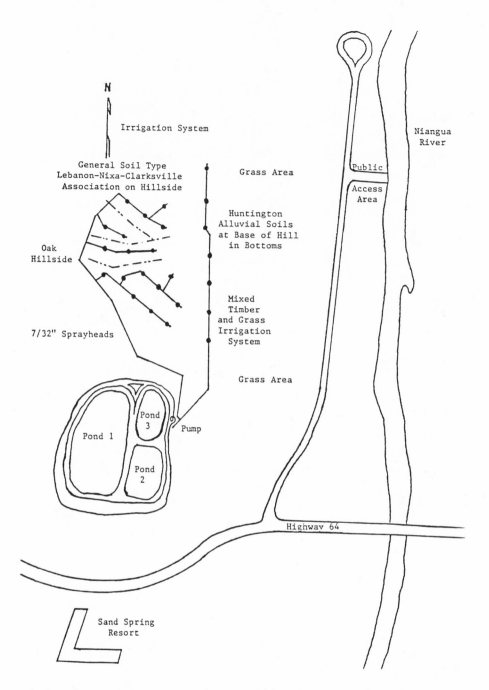

Figure 3. Plain view sketch of Bennett Spring State Park wastewater treatment facilities.

land application system provides an advanced form of treatment, and groundwater recharge meets the standard for drinking water. (5) This method can eliminate discharges to waterways.

The Bennett Spring land application system has served as a model for other Missouri parks, including Montauk, Roaring River, Crowder, Knob Noster, Thousand Hills, Stockton, Wakonda, Mark Twain, Pomme De Terre, Arrow Rock, Washington, Cuiver

NAME	COUNTY
Montauk	Dent
Roaring River	Barry
Crowder	Grundy
Knob Noster	Johnson
Thousand Hills	Adair
Stockton	Henry
Wakonda	Lewis
Mark Twain	Monroe
Pomme De Terre	Cedar
Arrow Rock	Saline
Washington	Washington
Cuiver River	Lincoln
Wappello	Wayne

Figure 4. Wastewater irrigation sites at state parks in Missouri.

River, and Wappello (Figure 4). Many of these systems are also timber or combined timber and grassland irrigation systems. The national park system also has installed similar land application facilities at Alley Spring, Round Spring, Big Spring, and Long Branch National Parks. This treatment concept has also been applied to industrial wastes at Richland and Cadet, Missouri, as well as numerous commercial facilities near the Lake of the Ozarks.

The largest facility in Missouri to be modeled after the park's system will be the new municipal wastewater treatment facility for the city of Ava. A tour of the Bennett Spring facility by city council members was a deciding factor in Ava's choice of a land treatment system.

The Bennett Spring experience was instrumental in development of the current Missouri Department of Natural Resources regulations requiring feasibility analysis of no-discharge land treatment systems before approving NPDES permits for traditional treatment and discharge options. Missouri has approximately 4.9 million hectares (12 million acres) of forest land, mostly in the Missouri Ozarks. In these areas, timber irrigation can be an economically feasible alternative for the advanced wastewater treatment required to protect sensitive environmental resource areas.

REFERENCES

Department of Natural Resources. 1985. Design of wastewater land application facilities. Water Pollution Control Program, Jefferson City, Missouri.

Sopper, W. E., and C. J. Sagmuller. 1966. Forest vegetation growth responses to irrigation with municipal sewage effluent. p. 639–647. *In* Proceedings, First Pan American Soil Conservation Congress, Sao Paulo, Brazil.

———. 1971. Effects of trees and forests in neutralizing waste. *In* Trees and forests in an urbanizing environment. Cooperative Extension Service, University of Massachusetts, Amherst.

Turner, J. A. 1973. Soil-water quality and tree growth under stabilization lagoon sewage effluent irrigation in southwest Missouri. M.S. thesis, University of Missouri, Columbia.

U.S. Department of Agriculture. 1978. Missouri resources appraisal. Soil Conservation Service, Columbia, Missouri.

Case Studies: Municipal Sludge

Pack Forest Sludge Demonstration Program: History and Current Activities

CHARLES L. HENRY and DALE W. COLE

ABSTRACT   The University of Washington College of Forest Resources has been involved with the Municipality of Metropolitan Seattle (Metro) in a major research program investigating the feasibility of applying dewatered municipal sewage sludge to forest lands. Research at Pack Forest has included application technology, environmental effects, and growth response. The research in sludge applications during the initial ten years at Pack Forest has had a number of positive results. The program has shown that, properly managed, sludge application to forest land is an environmentally sound practice in which excellent tree growth response may be expected. The success of the program has led to greater acceptance of this alternative and increased interest from municipalities and forest landowners. Federal, state, and local guidelines for sludge use in forest lands have relied on information developed in this program. Although the program and others around the country have greatly increased the interest in forest sludge applications, this is still a transitional period in the evolution from research to full-scale operations. Recognizing the critical nature of this period, the University and Metro have recently entered a second phase, in which the emphasis at Pack Forest is on demonstration rather than research. The major program elements are (1) operational scale projects, (2) equipment development, and (3) public information.

Since 1973 the University of Washington College of Forest Resources has been involved in a major research program investigating the feasibility of applying dewatered municipal sewage sludge to forest lands. This program was and continues to be conducted at the University's Charles Lathrop Pack Demonstration Forest near Eatonville, Washington. Funding and sludge for research have been provided by the Municipality of Metropolitan Seattle (Metro).

Although sludge applications had for some time been made agriculturally, in 1973 the newness of forest applications raised some interesting questions: (1) How could sludge be placed in forest environments and what would these operations cost? (2) Could the beneficial aspects of sludge such as nutrient content and soil conditioning properties enhance tree growth as it had for agricultural crops? (3) What were the environmental concerns specific for forest land applications? Research emphasis since 1973 has centered on these three questions.

Initial research employed relatively deep sludge applications to recently clearcut sites for a number of reasons. At this time sludge management was governed by the "disposal philosophy" that favored maximum applications to a minimum land base; and heavy applications would make it easier to identify public health issues. It was also not known what application rates would be needed to promote accelerated tree growth.

Clearcut sites were chosen because technology was not available at the beginning of the program to apply a dewatered sludge to forested areas.

Early research clearly demonstrated dramatic growth responses, but also revealed problems associated with heavy applications. Nitrate leaching was found to be the limiting criteria for application rates, and problems in plantation establishment required alternatives to clearcut sites. In subsequent years new application techniques were developed enabling sludge applications to existing forest stands, and appropriate rates were established for protection of the environment.

Several positive results came from this research during the initial ten years at Pack Forest. First, the program has shown that, properly managed, sludge application to forest land is an environmentally sound practice in which excellent tree growth response may be expected. Recommendations for forest applications have been developed which currently are the basis for the forest application sections in EPA's Process Design Manual (U.S. EPA 1983) and the Washington Department of Ecology's Best Management Practices (WDOE 1982). The success of this program has led to greater acceptance of this alternative and increased interest from municipalities and forest landowners. To continue increased acceptance, a major six-year demonstration program at Pack Forest was embarked on in 1983.

APPLICATION TECHNOLOGY AND ECONOMIC CONSIDERATIONS

Two sludge application techniques were used at the Pack Forest project: (1) sludge was spread directly on clearcut forest land after being transported by long-haul vehicles from Seattle Metro's Westpoint Treatment Plant, and (2) sludge was sprayed under the canopy of an existing older stand or over the canopy of a young plantation.

Early research in the project focused on the problems involved with these two methods. With the equipment technology available at the time and the goal for year-round sludge management, it immediately became evident that temporary sludge storage would be required. An old gravel pit on Pack Forest land was investigated as to suitability for sludge storage. Nearby springs used for domestic drinking water were found to be well protected from possible leachates from the gravel pit, because of extensive clay lenses occurring in the bottom of the pit below the gravel. The basin was improved and enlarged by adding dikes. The basin has been an integral working portion of the program for over ten years, and has been a part of many studies in equipment technology development for sludge transfer.

Following agricultural practices, the first method of sludge application involved direct application to clearcuts. Stumps and debris were piled and burned, thus facilitating access of long-haul vehicles. Heavy applications (10 to 25 cm of 18% dewatered sludge) were made to force questions of plant phytotoxicity, health effects, and soil improvement and growth response. Tree seedlings planted in freshly delivered sludge obviously could not survive in the anaerobic conditions caused by the high moisture and organic levels. Sludge was left to dry during summer months, disked in, then planted in rye grass, oats, or both. These grasses were very effective in dewatering the sludge to more acceptable moisture content levels, and were also able to take up much of the available nutrients provided by the sludge. Disking at planting time the next year left a very nutrient-rich and acceptable medium for tree planting. Since little site preparation was required and large amounts of sludge were used, applications made in this method appeared very attractive economically.

Unfortunately, problems immediately began to arise. In areas where 25 cm of dewatered sludge (470 dry Mg/ha) had been applied, the total nitrogen loading was approximately 20,000 kg/ha, of which about 7,500 kg/ha of nitrogen were available during the first year. This had two major effects (Edmonds and Cole 1976, 1977, 1980): (1) The increased nutrient levels in this "soil," the soil conditioning properties of the sludge, and the introduction of a grass seed source dramatically increased grass competition. The vigorous grass growth created an excellent habitat for small mammals, such as voles, which feed on the bark of seedlings. In addition, a severe deer browse problem arose. These three factors have caused very high seedling mortality, and have greatly added to the cost of plantation establishment. (2) The tremendous input of nitrogen exceeded the capacity of the site to utilize and store it. Excess available nitrogen was transformed into nitrate, which was free to move through the soil profile and leach into groundwaters. Nitrate levels in monitoring wells were well in excess of EPA drinking water standards.

Whereas these two problems did arise, many questions were answered which indicated the value of the sludge. Excellent growth responses were recorded for a variety of tree species, particularly Douglas-fir (Henry and Cole 1983). Even at high application rates, no plant toxicity occurred. Public health related questions of pathogens and heavy metals were also investigated (Edmonds and Cole 1976, 1977, 1980, Bledsoe 1981, Henry and Cole 1983). Pathogens were found to die off fairly rapidly, and approached background levels within a year. Very little heavy metal movement was found, including movement into the soil, into the plants, and up into the food chain (i.e., wildlife). These positive aspects encouraged the continuation of the program and new approaches for application methods and application rates.

Because it was assumed that established stands would not have survival problems from competition, vole damage, or deer browse, a mobile application system capable of negotiating forest trails was designed to spray apply sludge (Nichols 1980). An old army 4 by 4 was reconditioned, and a metal tank was installed with a pump and cannon. The tanker with spraying cannon was capable of shooting sludge at 18% solids 45 m into or over a stand of trees. Existing or newly constructed trails were used, spaced at a maximum of 90 m for complete coverage. This system proved very promising, and present application vehicles have been designed around this concept.

Two application rate and timing philosophies were evaluated: (1) annual applications designed to meet only the annual uptake requirements of the trees, and (2) a heavy application rate one year, followed by a number of years in which no applications are made. The second technique had some definite advantages. It was felt that application costs would be lower because entry into a site would be made less frequently, and the public could use the site for recreation in the nonsludge years. It was also not known what rate was required to promote accelerated growth. Initial applications were 5 cm of 18% solids sludge. In some of the original stands receiving the 5 cm applications, growth response has not slowed after eight years. It is felt that a 2.5 cm (47 dry Mg/ha) application will show continued maximum response for a period up to five years.

Application of sludge to mature stands where the sludge was sprayed under the canopy took place year round, except where soil conditions limited vehicle access. Over-the-canopy applications to younger stands were generally made during the nongrowing season, increasing the possibility of the sludge being washed from the foliage by rainfall.

It was also felt that sludge covering new foliage could either cause mortality or significantly retard growth. However, in small areas, applications have been made over the canopy during the growing season without any noticeable detrimental effects.

GROWTH RESPONSE

Douglas-fir is, of course, the most commercially important tree species in the Pacific Northwest, so its response to sludge additions was crucial. Most of the applications at Pack Forest have been made to Douglas-fir stands, although many others have been tried in the species trials plantations.

Growth response has also been documented on 60-year-old stands (Cole et al. 1984). A six-year average diameter growth response of 93% for unthinned and 48% for thinned stands was measured for sludge-treated trees over untreated trees. Some of the results from one of the young plantations established on gravelly outwash soils are shown in Table 1 (Henry and Cole 1983).

TABLE 1. Growth response and mortality for a young plantation established in sludge-amended soil.

Species	Increase over Control (%)		Total Mortality (%)
	Height	Diameter	
Douglas-fir	64	152	41
Western hemlock	369	693	99
Sitka spruce	140	225	49
Western redcedar	65	114	86
Hybrid cottonwood	590	893	54
Grand fir	200	1,250	22
Sequoia	286	1,033	14

Although the growth response of plantations established in sludge-amended soils has been dramatic, survival is a major problem. Table 1 gives the percentage of mortality for species in this plantation. Although almost all species have shown high mortality, it is particularly recommended that Western redcedar and Western hemlock not be planted in a sludge-amended clearcut. In addition, this problem with plantation establishment has led to the recommendation that, where possible, sludge be applied to existing stands rather than clearcuts. The minimum age recommended is seven years, at which time the trees are generally not subject to grass competition, vole damage, or deer browse.

The accelerated wood growth has led to the concern by many that wood quality has decreased. A 10 to 15% lower specific gravity for the sludge-grown wood was found (Cole et al. 1984). This apparently reflects the change in forest site quality produced by the sludge treatment and is within the specific gravity range found in Douglas-fir on higher sites.

The economical effectiveness of forest sludge applications depends greatly on both the accelerated growth benefits and the effect on wood quality. Although some work has been completed in this area (Edmonds and Cole 1976, 1977, 1980, Bledsoe 1981, Henry and Cole 1983), results depend greatly on the long-term effect of the sludge on response and quality. At this time, results in both areas are preliminary.

ENVIRONMENTAL CONSIDERATIONS

Because sludge is derived from human and industrial residues, public health concerns exist. In particular, four groups of constituents have been identified: (1) heavy metals, (2) pathogens, (3) toxic organics, and (4) nitrates.

Many studies have been conducted at Pack Forest concerning these constituents (Edmonds and Cole 1976, 1977, 1980, Bledsoe 1981, Henry and Cole 1983). The research has shown that in most cases the limiting criteria for applications is nitrogen. Pathogens and toxic organics die off or degrade in a fairly short time, and heavy metals are not generally a concern in a nonfood-chain crop.

Municipal sewage sludge contains large amounts of nitrogen. Nitrogen available immediately from sludge additions or shortly afterward, through decomposition of the organics, can follow a number of pathways, including gaseous losses from volatilization of ammonia or denitrification, uptake by the trees and understory, storage in the soil, or loss through nitrate leaching. In order, then, to have both maximum utilization of the nitrogen from the sludge and minimal impact on the environment, one must have adequate knowledge of the concentrations of nitrogen compounds as the sludge is applied, and must proceed through calculations estimating the above processes. Table 2 presents a conceptual model for a nitrogen balance, based on the literature and studies at Pack Forest.

Another important consideration is the loss of dissolved and suspended sludge-related materials from the site through runoff. Although some preliminary work has been conducted in this area (Henry and Cole 1983), more extensive research is required to define adequate buffers and recommendations for maximum slopes.

TABLE 2. First-year conceptual nitrogen balance for sludge application to forest soils.

Sludge Application			Nitrogen Transformations and Leaching				
Dry Mg/ha	Organic N (kg/ha)	Ammonia N (kg/ha)	Mineralized (kg/ha)	Volatilized, Denitrified (kg/ha)	Stored (kg/ha)	Uptake (kg/ha)	Leached (kg/ha)
24	784	224	+157	−179	−90	−112	= 0
47	1,568	448	+314	−358	−224	−112	= 68
94	3,137	896	+627	−717	−336	−112	= 358
188	6,274	1,792	+1,254	−1,434	−448	−112	=1,052

Assumptions:
1. 3.4% organic N, 0.9% ammonia-N.
2. 20% mineralization rate in year 1.
3. Volatilization of 50% initially available nitrogen.
4. Denitrification of 25% of remaining available and mineralized nitrogen.
5. Soil storage is immobilized nitrogen plus available nitrogen held by exchange sites.

CURRENT ACTIVITIES

Although two years ago this program and others around the country had greatly increased the interest in forest sludge applications, it was still a transitional period between research and full-scale operations. Since this option is relatively new compared with agricultural sludge practices, full acceptance remains dependent on continued "fine tuning" of application rates and technology, and on increasing confidence among regulators and the public. Because this is such a critical period, the Pack Forest Sludge Demonstration Program was developed in 1983 as a joint effort between Metro, the Washington Department of Ecology, and the College of Forest Resources. The emphasis of this second phase at Pack Forest is on demonstration rather than research.

In addition to the municipal sludge program, studies have been initiated in the use of pulp and paper sludges as soil amendments. A nursery bed demonstration was installed in April 1984 with four species of trees planted in eight treatments of primary and secondary pulp and paper sludges and municipal sludges (Henry, this volume). This demonstration was funded by Boise Cascade, Crown Zellerbach, and the city of Tacoma. Another study of the nitrogen balance for primary and secondary pulp and paper sludges was initiated early in 1985. Funding for this project is from Crown Zellerbach and the Washington Department of Ecology.

The Pack Forest Sludge Demonstration Program has three major elements: (1) operational scale projects, (2) equipment development, and (3) public information.

Operational Scale Projects

Seven sites at Pack Forest will receive municipal sewage sludge during the six years of the program. Three older Douglas-fir sites will have sludge applied under the canopy; the major site, referred to as the Silvicultural Demonstration Site, is discussed in more detail below. Two young Douglas-fir plantations will receive sludge over the canopy. Two recently planted hybrid cottonwood plantations will have annual sludge applications over the canopy; the larger site, referred to as the Fiber Plantation, is discussed later. The eighth site is the Right of Way Demonstration, where sludge is being used to promote a grass habitat, also discussed later.

Silvicultural Demonstration Site. The purpose of the Silvicultural Demonstration Site is to demonstrate large-scale operational application of municipal sludge under the canopy of an older Douglas-fir stand. The application of sludge to smaller experimental plots of existing stands at Pack Forest has met with excellent success in terms of growth response, while negative environmental effects have been minimized through proper application rates. The success of these research plots at Pack Forest has had a major impact on sludge management alternatives available for municipalities in the Pacific Northwest, particularly Metro. As a "demonstration site" this project will help bridge the gap between research and fully operational programs for municipalities. The emphasis is on development of site design and management techniques, development or modification of transfer, storage, and application equipment, and evaluation of site impacts.

A 100 hectare site at Pack Forest was chosen for this demonstration. In the summer and fall of 1983 a preliminary site plan and a site management plan were completed. The preliminary site plan was used to solicit comments from the sponsors of the project, local county health officials responsible for approving the site (Tacoma–Pierce County Health Department), and state regulatory people (Washington Department of Ecology). Field work included soils mapping, vegetation inventory, groundwater assessment, topographic information, and location of buffers and areas generally unacceptable for sludge application. The preliminary site plan included summaries of these items plus a preliminary schedule for applications, preliminary application trails, and a sludge routing plan.

The site management plan was a more detailed and complete design for the development and operations of the site. Application area boundaries and areas unacceptable for sludge applications were mapped, which included excessively steep slopes (over 30%), waterways with buffers, and groves of inappropriate tree species (red alder). Through this process the area acceptable for applications was reduced to about 50 hectares. All 50 usable hectares will be applied with 47 dry Mg/ha in 1985. An additional application of 47 dry Mg/ha will be made in five years to 25 hectares of lowest productivity. A plan was made for thinning the site to facilitate even and complete sludge application, and to maximize tree growth response. Designs were made for road improvement for transfer vehicles and trail construction for application vehicles.

In the spring of 1984 an intensive thinning began on portions of the 50 hectares, and has continued on the remaining portions through mid-1985. Road and application trail construction began in the fall of 1984, and was halted during the wet season. Construction continued on additional trails and other facilities in the spring of 1985. Applications of sludge began in mid-June 1985.

The site received a lot of care in design and preparation. This level of effort may not be required for all sites, but there were a number of reasons why it was desirable at this one. First, it was meant to be a showcase for tours of regulators, sludge managers, landowners, and the general public. Second, this project is the basis of many other studies. It provided a site in which to develop design and management techniques, where equipment modifications will be identified and tested, and where operational techniques will be further developed. In addition, the following site impacts will be studied.

1. *Growth response.* Nine growth plots have been installed with three replications of thinned with sludge application, thinned only, and control. In addition, growth response will be measured by inventory plots, and by coring and destructive sampling at the end of the six-year program.

2. *Groundwater, soil water, and soil monitoring.* Two wells and ten lysimeters have been installed to monitor, in particular, nitrate leaching. Soils are monitored for heavy metals.

3. *Runoff.* Weirs have been installed at the outlets of two small, side-by-side watersheds. One watershed will remain a control while the other will have sludge applied to within 15 m of the stream channel in the fall of 1985, the beginning of the rainy season. Streamflow quantity and quality will be monitored to assess the effectiveness of the 15 m buffers.

4. *Mushroom plots.* Several mushroom plots have been established to assess the effect of sludge on quantity and quality of edible mushrooms.

5. *Operational evaluations.* These studies include sludge spray trajectory and evenness of application, slopes and soil conditions appropriate for vehicle access, soil compaction and erosion from application-vehicle travel on different soil types, and optimal scheduling and access.

Fiber Plantation. The purpose of the Fiber Plantation is to demonstrate the use of municipal sludge as a soil amendment and nutrient source for increased growth of woody biomass, and to gain experience in management of sludge-treated cottonwood plantations. On appropriate sites, cottonwood shows unmatched early growth compared with other Pacific Northwest species. In addition, cottonwoods are easily established by cuttings, and successive rotations quickly reestablish through coppicing. A hybrid cottonwood clone (*Populus trichocarpa* x *P. deltoides*) has been planted at Pack Forest in some of the species trials in sludge-treated areas, and has shown excellent growth response. The success of this species on soil amended with sludge set the stage for design of a large-scale plantation of hybrid cottonwoods as part of the Sludge Demonstration Program.

An old pastureland at Pack Forest was chose for this demonstration. In the summer of 1983 a site plan was prepared. This plan was, as in the Silvicultural Demonstration Site, used to solicit comments from those involved in the project. The site plan included a site description and map, topography, soils, a plantation establishment design, and sludge application schedule.

Four hectares of pastureland were disked and prepared for planting in the fall of 1983. Cuttings were planted in two hectares in the spring of 1984, and the remaining two hectares in the spring of 1985. Municipal sludge (13%) was sprayed over the one-year-old cuttings and the recently planted cuttings in April 1985 just before leaf-out. Annual applications will be made to this plantation to match the nitrogen uptake rate of the cottonwoods and supporting vegetation.

Also studied at this site are: (1) *Growth response*. Control and sludge plots have been established in both the 1984 and 1985 sections of the plantation, and will be measured on an annual basis. (2) *Soil water and soils monitoring*. Four lysimeters have been installed to monitor nitrate leaching. Soils are monitored for heavy metals. (3) *Sludge appplication timing*. Different timings have been used for the sludge applications. Sludge has been applied after a full year's growth and to new cuttings, both before leaf-out. Sludge has also been applied after leaf-out to portions of the plantation with and without a following wash with water. This is to assess whether significant damage will occur to the young trees and result in a retarded growth or mortality.

Right of Way Demonstration. The purpose of the Right of Way Demonstration is to show how municipal sludge can be used as a means of biological maintenance. The addition of sludge to cleared forest land has been shown to promote a habitat of herbaceous vegetation in past work at Pack Forest. Tree generation can be reduced or prevented on cleared sites treated with sludge and seeded with select grasses and other herbs, because of increased competition for water, light, and nutrients. This habitat can also increase populations of wildlife such as voles and larger herbivores by providing greater forage, increased nutrient content of forage plants, and greater cover for small animals. More wildlife also leads to greater mortality from girdling and browse damage of tree seedlings trying to get established in this habitat. If this sludge management method proves effective, sludge application could replace herbicides or mechanical means of vegetation control under powerlines or similar rights of way, while at the same time creating a productive wildlife habitat.

A powerline right of way near Pack Forest was chosen for this demonstration. In the summer of 1983 a site plan was prepared. This plan was, again, used to solicit comments from those involved in the project. The site plan included a site description and map, topography, soils, a drainage analysis, a vegetation and wildlife inventory, and design and schedule for sludge application and treatment.

The right of way is a 915 by 27 m wide strip maintained by Tacoma City Light. Clearing of vegetation took place in August 1984, and sludge was applied to 610 m at a rate of 94 dry Mg/ha. A 305 m control was maintained. The sludge was then evenly spread and disked into the soil. Seeding of the area took place in September 1984 with three rye grasses, Kentucky bluegrass, and bird's-foot trefoil.

Grass establishment began immediately following the first rain. Differences between the control and the treated area began to be evident after a few months. A marked difference in the color and density of the grass was noticeable in the early spring. The treated area showed heavy browse on virtually every blade of grass. By April the grass in the treated area exploded with thick, lush growth, while the control remained sparse.

This site is studied for (1) vegetation establishment and production on both the control and the sludge-treated area, (2) wildlife use on both areas, (3) groundwater, soil water, and soils monitoring. Two wells and four lysimeters have been installed to monitor, in particular, nitrate leaching. Soils are monitored for heavy metals.

Equipment Development

During the initial ten years of sludge use at Pack Forest, an appropriate application vehicle evolved, capable of applying sludge from skid trails up to 45 m into existing stands. Application vehicles were filled by a manure-type chopper pump pulling sludge from an earthen basin. Although this method has worked well during the life of the research programs at Pack Forest, in-ground storage is not appropriate or desirable at all sites. Basins are becoming expensive because of increasing regulatory scrutiny over leachate control, and can be a nuisance and a liability. With these thoughts in mind, Metro was interested in developing a portable storage system, and this task became part of the Pack Forest Sludge Demonstration Program.

A feasibility study was performed to assess the alternatives available for a portable system. Initially the following objectives were identified: (1) the storage requirements were 115 cubic meters; (2) the system could be moved by a crew of three; (3) moving to a new area would require less than four hours; (4) the system would not require in-ground facilities; and (5) dilution of the sludge from 20-25% to 13-14% was required. An immediate problem arose—how to transfer a 20-25% solids material, since end dumping was not possible because there were no in-ground facilities.

From this initial study, it was evident that not all the objectives could be met. However, a concept was later developed to utilize the long-haul vehicle for dilution and mixing, thus allowing sludge to be pumped into a storage vessel. A 115 cubic meter portable (on wheels) tank similar to that used to store cement was incorporated into the concept. Work on the preliminary and final designs was a joint effort by Metro, the Pack Program personnel, and SCS Engineers, Inc. The system is now fully operational and an integral part of Metro's operations. It is capable of rewatering and emptying 23 cubic meters of 25% solids sludge in about 20 minutes. Set-up time for the storage and transfer facility has decreased to just over one day.

Public Information

In the 1983 Workshop on Utilization of Municipal Wastewater and Sludge on Land (Page et al. 1983), public acceptance was mentioned over and over as a key issue for success in implementing a land application program. This is certainly true of forest applications of sludge, because of relative newness compared with agricultural applications. Several technically sound forest sludge programs proposed in Washington did not receive public support and subsequently died. In response to this, a major element included in the Pack Forest Sludge Demonstration Program is public information.

Because Pack Forest has been the site for extensive research in sludge applications to forest lands, and because new operational projects are under way, Pack Forest has been the site of numerous tours. Participants in these tours include the general public, regulators, designers, landowners, and sludge generators. During the last two years, an average of two tours per month have been given. In addition, requests for presentations to civic groups, city councils, and professional societies have occurred on an average of one a month.

Another method used to help increase acceptance of this alternative was the information newsletter. The goal of this newsletter was to compile a list of all known publications on sludge and wastewater application to forest lands, and to include information on those involved. The publication, *Silvicycle* (Henry et al. 1983), was sent out quarterly from October 1983 to December 1984, and the plan is to continue to publish it in an annual edition.

The Forest Land Applications Symposium has also been an element in this public information effort. And local sludge and wastewater application workshops have been held.

To tie together and continue these efforts, and to identify and conduct future research, it is proposed to establish a regional forest land application cooperative for municipal and industrial sludges and wastewaters, with specific objectives such as the following ones: (1) to be a central organization responsive to all parties, interests, and concerns in a forest application alternative, (2) to assimilate and distribute information on the cooperative's work, through the means of a newsletter, annual report, research reports, committee meeting minutes, tours, workshops, and symposia, (3) to provide a centralized library for related reference material, and (4) to define and coordinate sludge and wastewater research as requested by the cooperative members. Through the cooperative it would be possible to provide continuity and consistency in monitoring areas that have received sludge or wastewater, and to develop research or demonstrations specific to the issues brought forth by cooperative members.

SUMMARY

The work done at Pack Forest, and by others across the country, supports the position that sludge is indeed a resource, and sludge utilization can be environmentally compatible in properly designed and managed sites. Although public health concerns exist, if the proper precautions are taken, public health risks are minimal. The addition of sludge provides nutrients and soil conditioning that have led to excellent growth responses.

REFERENCES

Bledsoe, C. S. (ed.) 1981. Municipal sludge application to Pacific Northwest forest lands. Institute of Forest Resources Contribution 41. College of Forest Resources, University of Washington, Seattle. 155 p.

Cole, D. W., M. L. Rinehart, D. G. Briggs, C. L. Henry, and F. Mecifi. 1984. Response of Douglas-fir to sludge application: Volume growth and specific gravity. p. 77–84. *In* 1984 TAPPI Research and Development Conference, Appleton, Wisconsin. Technical Association of the Pulp and Paper Industry, Technology Park, Atlanta, Georgia.

Edmonds, R. L., and D. W. Cole (eds.) 1976. Use of dewatered sludge as an amendment for forest growth. Vol. 1. Center for Ecosystem Studies, College of Forest Resources, University of Washington, Seattle. 112 p.

———. 1977. Use of dewatered sludge as an amendment for forest growth: Management and biological assessments. Vol. 2. Center for Ecosystem Studies, College of Forest Resources, University of Washington, Seattle. 120 p.

———. 1980. Use of dewatered sludge as an amendment for forest growth. Vol. 3. Institute of Forest Resources, University of Washington, Seattle. 120 p.

Henry, C. L., and D. W. Cole (eds.) 1983. Use of dewatered sludge as an amendment for forest growth. Vol. 4. Institute of Forest Resources, University of Washington, Seattle. 110 p.

Nichols, C. G. 1980. Engineering aspects of dewatered sewage sludge land application to forest soils. M.S. thesis, University of Washington, Seattle. 84 p.

Page, A. L., T. L. Gleason III, J. E. Smith, Jr., I. K. Iskandar, and L. E. Sommers (eds.) 1983. Proceedings of the 1983 Workshop on Utilization of Municipal Wastewater and Sludge on Land. University of California, Riverside. 480 p.

U.S. Environmental Protection Agency. 1983. Process design manual: Land application of municipal sludge. EPA 625/1-83-016.

Washington Department of Ecology. 1982. Best management practices for use of municipal sewage sludge. WDOE 82-12. 98 p.

Chapter 44

Silvigrow: Metro's Forest Sludge Application Program

JOHN SPENCER and PETER S. MACHNO

ABSTRACT   In the last two decades sludge has evolved from a waste by-product of the sewage treatment process to a marketable resource. In 1972 the Council of the Municipality of Metropolitan Seattle (Metro) made a major, long-range policy decision that sludge is a resource that should be put to beneficial uses. That decision moved the agency away from "waste disposal" and into recycling, marketing, and management of a resource. The decision was also a commitment to a decade of environmental, technological, and management research, resulting in realistic management regulations and establishing a base for developing public confidence in modern sludge utilization practices. After thirteen years of effort, sludge application to forest land in the Seattle metropolitan area is beyond the research and demonstration stage: operation Silvigrow—the application of sludge to forest lands—began in May 1985.

Thirteen years ago, the Municipality of Metropolitan Seattle decided it was time to transform its policy of sludge disposal into one of sludge recycling. The Metro Council rejected bids for short-term disposal of its 23,000 cubic meters of sludge and committed to a long-range proposal by the University of Washington to use it in a reforestation research project. The Council was aware that this decision was a decade-long commitment to an experiment fraught with potential pitfalls and problems. There were no federal, state, or local standards for sludge disposal or reuse, and the public perceived sewage as a waste to be gotten rid of. The decision was significant not only because it perpetuated Metro's reputation as a pioneer in innovative concepts but also because it altered the entire focus of the agency. Metro moved out of the disposal business and into the recycling business long before recycling was fashionable.

During the 1960s, sludge was considered a waste that needed to be eliminated as expeditiously as possible. In that decade, Metro was discharging sludge into Puget Sound. While such a practice was perfectly acceptable in those years, it would be deplored today in the Northwest. The disposal method still is used, however, in some parts of the country.

Metro began experimenting with alternative uses for sludge in 1965. The agency constructed a sludge holding basin at West Point and started examining the idea of dewatering sludge for use in greenbelt development. In 1970, after Metro made a firm decision to quit discharging sludge into Puget Sound, the agency began planning and constructing a sludge dewatering facility. The last sludge was discharged into the water in 1972, when the Metro Council adopted a recycling policy. An analysis of land-based options for sludge use began in earnest. Early in 1973, Metro sought competitive bids for the disposal of approximately 23,000 cubic meters of sludge. Of the six bids received, three were competitive in the $13 per cubic meter range, amounting to approximately $300,000 per year.

This land-use application of sludge was as new to the bidders as it was to Metro, and all bids had certain disadvantages. All had proposed different disinfection methods and included processing and storage sites that had to be reviewed and approved by the state Department of Ecology. At the same time, the University of Washington came forward with a proposal to use Metro's entire sludge supply in a reforestation project at approximately half the price of the low bidder—$150,000 a year.

It would be good to be able to say at this point that the rest is history, but that is far from the case. Metro's decision changed sludge from a waste to a resource, but to make that resource valuable it had to be marketable. Before a market could be created it had to have public acceptance, and before it could hope to gain public acceptance it was necessary to know more about the product. Hence, a decade of research began on public health and environmental questions. Questions were raised about the heavy metals in sludge fertilizer and the possibility of those metals entering the food chain through plant-eating animals such as deer.

There were questions about technology: How would sludge be applied? And there were public health questions: When could a forest that had been applied with sludge be open for traditional uses? It was also important to know how different varieties of trees would respond to sludge application. If the Douglas-fir did not respond well, for instance, then sludge fertilizer would not be a marketable commodity in the Northwest.

The initial forestry approach at the University of Washington Pack Forest was to determine if sludge could be used to renovate clearcut forest lands in western Washington. Sludge fertilizer application to those areas created such rapid growth of plants and grass that the acreage became a haven for wildlife such as field mice, deer, and elk that forage on young trees. Project managers concluded that sludge application would have to be limited to existing forests to avoid such destruction by wildlife.

This move away from clearcut to existing forests created problems of access and spreading. How do you get into the forested area, and what method should be used to apply the sludge fertilizer? The new approach led to the development of a forest application vehicle capable of pumping a mixture of 87% water and 13% solids and spraying it 45 meters.

Between 1973 and 1982, more than 80 hectares at Pack Forest received sludge applications. Tree growth was dramatic. Douglas-fir, the bread-and-butter tree in the Northwest, and Sitka spruce thrived, while the redcedar did not.

At the same time that Metro was conducting its research, regulatory agencies began developing rules for sludge handling, management, and disposal. The regulatory course—which progressed from inattention to ambiguity to prohibition to resource management—developed nearly in tandem with environmental research and technology development. It wasn't until 1978 that the U.S. Environmental Protection Agency created the Resource Conservation and Recovery Act and Code of Federal Regulations 257 to guide sludge management. The EPA action was complicated by the fact that sludge was regulated by both the Solid Waste Act and the Clean Water Act. At the same time, CFR 257 was criticized because it dealt only with land application and didn't provide for a trade-off between air and water quality.

Finally, in 1981 the Washington State Department of Ecology (WDOE) took the lead and developed regulations, guidelines, and Best Management Practices for the state.

This allowed local health departments to issue permits in consultation with WDOE, ending the "ping pong" effect between solid waste and water quality laws.

By 1980, the Metro Council thought that the Pack Forest research project had provided sufficient answers to the environmental and public health questions and decided to begin an operational project. It was here, while following EPA public involvement regulations, that a strategic mistake was made. The degree of public acceptance of the projects was overestimated, especially outside Metro's home base of King County. Specifically, projects had been proposed in conjunction with the Washington State Department of Natural Resources in Mason County, St. Regis Corporation in Pierce County, a private property owner in Thurston County, and the Pilchuck Tree Farm in Snohomish County. EPA public involvement regulations required public hearings even though no specific project or site had been identified.

One such meeting was held in Shelton, a logging community in Mason County. To say that the citizenry was less than enthusiastic is an understatement. The reaction bordered on open rebellion. Citizens were critical of the Department of Natural Resources, the private timber companies involved, and particularly of Metro, which they perceived as an agency from another county trying to dispose of its sludge in their backyard. If the program was such a good one, they asked, why wasn't Metro implementing it in King County forests? Metro faced similar reactions from citizens in Snohomish County, where the agency had proposed to apply sludge at the Pilchuck Tree Farm.

At the same time controversy was brewing over Metro's sludge application plans, the media and the environmental community were scrutinizing the entire timber industry for herbicide use, clearcutting, and other environmentally sensitive issues. The publicity of the sludge program pushed the black hat more firmly on the head of the timber industry. Timber companies backed away from promoting sludge application in forests until the idea became more acceptable to the public. The message was to slow down and use small demonstration projects to prove that forest applications could be managed without significant risk to the general public or the environment.

Strong citizen reaction to the sludge program taught Metro several lessons. Perhaps the most significant was the need to limit sludge application sites to remote forest lands, far away from populated areas. Metro now approaches sludge fertilization projects with the idea of being a good neighbor. The agency has become more sensitive to the concerns of citizens located in the vicinity of proposed sludge application sites and has worked to instill their confidence in such projects. Metro also has been successful in gaining valuable support for its sludge projects from community leaders and elected officials.

Education and communication are key elements in developing support for any project, and Metro has provided the public with a variety of information related to sludge—brochures, documents, fact sheets, newsletters. As part of the agency's public information campaign, project managers work to maintain rapport with newspaper editors and reporters from communities located near sludge application sites.

In 1983, the Metro Council adopted a long-range sludge management plan that reaffirmed the 1972 commitment to recycling sludge. To develop the plan, Metro evaluated eight major categories of potential sludge management methods. These included agricultural application, compost production and marketing, manufacture of dried sludge products similar to Milorganite, incineration, sanitary landfill disposal, ocean disposal, silvicultural or forestry application, and soil improvement and reclamation.

Nineteen variations of these eight alternatives were evaluated according to economic, social, and environmental factors, such as energy, air emissions, soil, water, wildlife, public health, land availability, land use, public acceptance, agency acceptance, proven experience, flexibility, and regulations. Of the eight alternatives, the management plan recommended three as the most beneficial: forestry application, soil improvement and reclamation, and compost production.

During the same year that the plan was approved, Metro decided to initiate silvicultural demonstration projects in King County before pursuing projects in other counties. To accomplish this goal, the agency purchased 930 hectares of remote forest land in King County. The year 1983 was pivotal for Metro's silviculture program, and progress has been rapid in the last two years. Along with the demonstration project on 70 hectares at Pack Forest and the project on Metro property in east King County, agreements have been negotiated for demonstration projects with Weyerhaeuser, Boise Cascade, and the Washington State Department of Natural Resources.

In addition, Metro has made strides in technology with the development and use of transfer tankers as an alternative to on-site holding basins. Such a development is significant in gaining public acceptance. Holding basins are perceived as "black lagoons" and are disliked by regulatory agencies and timber companies. Citizens, meanwhile, approach them with a "not in my backyard" attitude.

After more than a decade, Metro feels there is reason to believe it has turned the corner on sludge management. Silvigrow, the term Metro developed to describe its use of sludge fertilizer in forests, is a registered trademark. And Metro is in the enviable position of having demand exceed production by 250% and having to establish priorities for silviculture projects. The agency is gaining the confidence of area communities, timber companies, and environmental groups.

Word of the project's success is getting out. Staff members receive inquiries daily from across the United States and Canada and recently provided information on Silvigrow to the producers of the "Ripley's Believe It Or Not" television show. Silvigrow also was featured on the national Cable News Network (CNN).

Metro doesn't limit its sludge application to forests. When the agency began recycling sludge twelve years ago, the resulting fertilizer was used primarily on soil improvement projects. When sludge is applied to infertile land, the organic content acts as a soil builder, creating a humuslike topsoil where none existed. This organic base, supported by the fertilizer components and other necessary trace elements in sludge, provides a growth medium capable of supporting a wide range of vegetation.

Metro sludge fertilizer has been used to improve soil at several sites in eastern and western Washington, including Gas Works and Myrtle Edwards parks in Seattle, Edmonds and South Seattle community colleges, a strip mining operation near Centralia, and a landfill in Kittitas County. A portion of Metro's sludge is used yet another way—in a composted sludge-sawdust product called GroCo. Composted sludge can be used as a mulch, as a soil amendment to provide nutrients and improve water retention, and as a substitute for manure composts, peat moss, and other elements used in soil mixes. The Sawdust Supply Company of Kent produces GroCo, which has been used for public and private landscaping projects. The cities of Carnation and Snoqualmie have applied GroCo in local parks to improve soil for grass planting and landscaping. After 1985, Metro's need for composting services probably will exceed the capacity of GroCo's facilities. At that time, the agency will decide whether to engage another private contractor for composting services.

Overall, the past thirteen years have provided Metro with a foundation for the future. It has replaced "waste disposal" with goal-oriented "commodity management." This will allow Metro to capture the value of sludge through a broad-based marketing strategy, recognizing that sludge fertilizer is now a valuable addition to the energy, forestry, and fertilizer markets. This base is vital in facing the challenges of the future. Secondary treatment will more than double Metro's sludge production by 1991 from 62,000 Mg (tonnes) to more than 157,000 Mg per year. This means Metro will have to be even more aggressive in management and marketing techniques.

Metro will need to continue to develop community, timber company, and regulatory agency confidence in the silviculture strategy while at the same time refining the necessary technology. Metro also needs to analyze new technology in areas such as dehydration, wet oxidation, and composting—not necessarily as an alternative to forest land application but to broaden the options for sludge use and the potential market. Finally, Metro, the state, and the federal government need to explore more fully the concept of privatization—the transformation of public sector operations into private, profit-making enterprises. It could well be that sludge, once the useless by-product of the wastewater treatment process, may become the catalyst for further development of the privatization concept. This would happen because sludge has value and marketing potential that could lead to a positive return on investments.

The potential benefits of privatization include access to financial markets, lower user rates, more rapid project implementation, and lower construction costs. But caution is necessary to protect the integrity of public service requirements, such as competitive bidding and Women and Minority Business Enterprise regulations, and to avoid monopolistic situations.

These are difficult and complex problems. When confronted with such challenges, both the public and private sectors tend to avoid them and continue business as usual. But if the Metro Council and the University of Washington had not taken some risk with sludge management thirteen years ago, Metro probably still would be discharging sludge into Puget Sound or some landfill, or looking for abandoned coal mines to dump it into. Metro has come a long way in a short period, but it can continue only with aggressive and creative exploration of new concepts and technologies.

Chapter 45

Municipal Sludge Application in Forests of Northern Michigan: A Case Study

DALE G. BROCKWAY and PHU V. NGUYEN

ABSTRACT   A large-scale operational demonstration and research project was cooperatively established by the U.S. Environmental Protection Agency, Michigan Department of Natural Resources, and Michigan State University to evaluate the practice of forest land application as an option for sludge utilization. Project objectives included completing (1) a logistic and economic assessment and demonstration of the technology available for conducting sludge applications in forest stands, and (2) several research studies that would augment knowledge in the areas of public involvement and acceptance, wildlife populations, food-chain transmission of potential toxicants, groundwater quality, nutrient cycling, and vegetation growth. Field trials in four forest types (aspen, oak, pine, northern hardwoods) were of a completely random design covering 54 ha, of which 18 ha were treated with nearly 4 million liters of anaerobically digested sludge. Average solids loading ranged from 8 to 10 Mg/ha, resulting in total nitrogen levels of 400 to 800 kg/ha. Differences in loading levels of nutrients, heavy metals, and trace elements were generally not significant among treated plots. Sludge was transported by truck a distance of 80 km to the study sites and sprayed by an all-terrain tanker on the forest floor at a cost of $48,576. The resulting unit cost of 1.3 cents per liter was comparable to typical operational costs for sludge application to farmland, considering the greater transport distance in this study. Preliminary findings indicate an enhanced nutritive quality of forage on fertilized plots and a resulting increase in use by both deer and elk. Increases in plant growth were related to elevated levels of soil nitrogen, phosphorus, calcium, and magnesium. Slight increases of nitrate-N were observed in soil percolate within one year of application, but these rapidly returned to near background concentrations. Analysis of sociological data provided new insights into public concerns and attitudes and outlined a process for constructive citizen involvement in program planning.

Implementation of secondary treatment standards for wastewater discharges in the Great Lakes Region has resulted in increased volumes of wastewater sludge requiring removal from treatment facilities. In Michigan, combined residential and industrial water use and mandatory effluent phosphorus removal have resulted in the annual generation of 202,500 dry Mg (223,218 tons) of sludge by municipal wastewater treatment plants (MDNR 1983). While traditional strategies for dealing with this residual waste have emphasized options for disposal such as incineration and use as landfill, sludge management programs developed since 1978 have increasingly turned to nutrient and organic matter utilization through the practice of land application.

Numerous concerns have been raised about the hazards of land application of sludge. The concerns of state residents center on public health and environmental quality (Peyton et al. 1983). The potential presence of pathogens, heavy metals, and toxic organic compounds in sludge is a leading health concern. Nutrient enrichment of groundwater and contamination of wildlife, soil, and groundwater by toxic metals and organics are the major environmental quality concerns of land application programs.

BACKGROUND

In Michigan during the last decade, fertilization of farmland has become the most frequently selected option for the land application of wastewater sludge. Literature devoted to agricultural application research is far more voluminous than that dealing with studies of silvicultural use. However, numerous municipalities in the northern two-thirds of the state do not have cropland available for sludge recycling. In this locale are millions of hectares of forest land that could serve as sites where the constituents in sludge would be assimilated in a manner that could stimulate forest productivity while posing minimal danger to environmental quality or public health.

Forest crops (wood products) are generally nonedible, thereby diminishing the risk of human exposure to hazardous elements in the food chain. The long-term accumulation of biomass on a forest site provides substantial storage capacity for elements applied in sludge over the length of a crop rotation. The harvest of tree boles and whole trees offers a means of removing sludge-applied elements from the treated forest site. Forest soils are generally porous, resulting in minimal surface runoff of applied nutrients, and usually nutritionally impoverished, providing opportunity to increase soil organic matter and nutrient levels substantially through sludge additions. Native forest plants, though adapted to low ambient nutrient levels in forest soils, have demonstrated their ability to respond with nutrient and biomass increases following fertilization with sludge (Brockway 1983). Forest sites are usually remote from large population centers and are used for recreational activities of a dispersed nature, minimizing the opportunity for human contact with recently applied sludge.

PROJECT AND OBJECTIVES

Early USDA Forest Service research provided evidence that forest ecosystems could accommodate sludge constituents in a manner that would enhance nutrient cycling and biological productivity while ensuring protection of groundwater in the phreatic aquifer (Urie et al. 1978, Brockway 1979, Brockway and Urie 1983). In building on these small-plot studies, a demonstration and research project was cooperatively established by the U.S. Environmental Protection Agency, Michigan Department of Natural Resources, and Michigan State University to evaluate the practice of forest land application as a viable option for sludge utilization. Project objectives included completing (1) a logistic and economic assessment and demonstration of the technology available for conducting sludge applications in forest stands and (2) several research studies that would augment the base of knowledge in the areas of public involvement and acceptance, wildlife populations, food-chain transmission of potential toxicants, nutrient availability and groundwater quality, nutrient cycling, and vegetation growth and nutrition.

This project was seen as a means of bridging the gap between small-plot research and eventual large-scale program operation by municipalities and industries. The anticipated products are to include (1) numerous summaries of scientific findings, (2) operational guidelines based on project and previous research and management experience, (3) materials for a public education program, and (4) a model for encouraging constructive public involvement in future program planning decisions. Each of these will address the information needs of the public, municipal and industrial planners and managers, government agency officials, and the scientific community. These documents will aid state agency specialists in providing sludge generators with information to assist them in establishing properly planned and managed forest land application programs in accord with the statewide strategy for land recycling of wastewater sludge.

MATERIALS AND METHODS

Demonstration Sites

Wastewater sludge fertilization trials were conducted in Montmorency County on the Atlanta Forest Area (Figure 1) of the Mackinac State Forest in northeastern lower Michigan (45°N, 84° 10'W). Vegetation on each site was representative of the upland forest types of major commercial importance in the northern portion of the state. Permeable glacial drift materials formed the parent material for the soils, which are low in native fertility and allow rapid infiltration of excess precipitation falling on all four of the forest sites (Hart et al. 1983). Annual precipitation averages 735 mm (29 inches) in this area

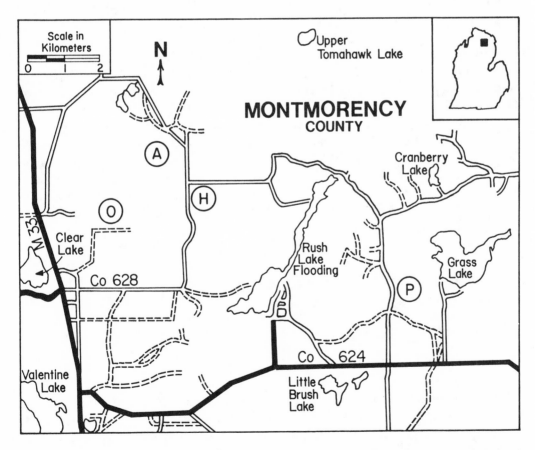

Figure 1. Sludge fertilization sites in northern Michigan. A = aspen; O = oak; P = pine; H = northern hardwoods.

(Strommen 1967). The sites are underlain by a phreatic aquifer which is contiguous with the regional groundwater system (Urie et al. 1983).

Aspen Site. The aspen site was occupied by a 10-year-old stand of regeneration which is predominantly bigtooth aspen (*Populus grandidentata* Michx.) containing a secondary component of quaking aspen (*Populus tremuloides* Michx.), northern pin oak (*Quercus ellipsoidallis* L.), cherry (*Prunus* spp. L.), and other species. Soils on this site generally belong to the Grayling series (Spodic Udipsamment) and the Rubicon series (Entic Haplorthod). Grayling soils are excessively drained and developed on deep glacial outwash sands. Rubicon soils are deep, excessively drained and formed in sandy glacio-fluvial deposits. Depth to groundwater was 5 to 8 m (16 to 26 ft).

Oak Site. The oak site was occupied by a 70-year-old stand that was a mixture of red oak (*Quercus rubra* L.) and white oak (*Q. alba* L.) with scattered pines (*Pinus* spp. L.) and aspen. The stand contained 388 trees/ha (157/acre) and an average combined basal area of more than 20 m²/ha (87 ft²/acre). Soils were predominantly of the Graycalm series (Alfic Udipsamment) with smaller areas of the Rubicon series. Graycalm soils are somewhat excessively drained and formed in deep glacio-fluvial sands. Depth to groundwater was in excess of 25 m as determined from on-site drilling.

Pine Site. The pine site was occupied by a 50-year-old plantation that was a mixture of jack pine (*Pinus banksiana* Lamb.) and red pine (*P. resinosa* Ait.). The stand contained 557 trees/ha with a combined basal area of 17.3 m²/ha. Soils on the site were of the Grayling series with a smaller area of the Montcalm series (Eutric Glossoboralf). Montcalm soils are deep, well drained, and formed in sandy and loamy glacio-fluvial deposits. Depth to groundwater was 6 to 7 m.

Northern Hardwoods Site. The northern hardwoods site was occupied by a 50-year-old stand that was predominantly red maple (*Acer rubrum* L.) and sugar maple (*A. saccharum* Marsh.) with remnants of American beech (*Fagus grandifolia* Ehrh.), yellow birch (*Betula alleghaniensis* Britton), and white birch (*B. papyrifera* Marsh.) and a minor number of red oak, American basswood (*Tilia americana* L.), white ash (*Fraxinus americana* L.), and eastern hemlock (*Tsuga canadensis* [L.] Carr.). The stand contained 288 trees/ha and an average combined basal area of 14 m²/ha. Soils were primarily of the Mancelona series, Melita series, and Menominee series (Alfic Haplorthods) with minor areas of the Kawkawlin series (Aquic Eutroboralf) and Sims series (Mollic Haplaquept). Mancelona soils are deep, excessively drained, and formed in sandy and gravelly glacio-fluvial upland deposits. Melita soils are deep, somewhat excessively drained, and formed in sandy materials overlying loamy deposits. Menominee soils are moderately well to well drained and formed in sandy material overlying loamy deposits at 50 to 100 cm. Kawkawlin soils are deep, somewhat poorly drained, and formed in moderately fine-textured glacial tills and ground moraines. Sims soils are deep, poorly to somewhat poorly drained, and formed in fine-textured glacial tills and ground moraines. Depth to groundwater ranged from 1 to 15 m.

Experimental Design

Three replications of three experimental treatments were assigned to completely randomized plots within each study site. The treatments consisted of (1) a control group of plots left undisturbed, (2) a group that underwent access trail development but received no sludge application, and (3) a group that underwent access trail development and received a single application of liquid sludge. Experimental plots were each 1.5 ha (3.7

acres) in area and of a rectangular shape approximately 100 by 150 m. The study plots covered a total of 54 ha, of which 18 ha were treated with nearly 4 million liters (1 million gallons) of wastewater sludge. The sludge application rate averaged 9 Mg of dry solids per ha (4 tons/acre). This design was sufficient to evaluate large-scale operational procedures, costs, and limitations, while affording adequate area for the conduct of a diverse array of research studies.

Site Preparation and Sludge Application

Prior to sludge application, a grid of parallel trails at 20 m (66 ft) intervals was prepared to facilitate application vehicle access and more uniform sludge distribution. Trees harvested from the oak, pine, and northern hardwood sites were felled and removed as whole trees from the stand using a rubber-tired skidder. Because of their small unmerchantable size, trees on the aspen site were removed at the groundline with a bulldozer blade.

Anaerobically digested sludges from the municipal wastewater treatment facilities in Alpena and Rogers City, Michigan, were transported by tank truck to the demonstration sites, where single applications of liquid were sprayed on the forest floor. Applications were conducted in October and November 1981 on the oak and aspen sites and in June and July 1982 on the pine and northern hardwood sites. An all-terrain vehicle, equipped with high flotation tires, a standard pressure-vacuum pump, and a modified three-nozzle spray system, was used for sludge applications on each site (Figure 2).

SLUDGE LOADING AND DISTRIBUTION

Because of the variation in site characteristics, such as microtopography and vegetation structure, and that encountered in operation of application equipment, such as vehicle speed, discharge rate, and tank pressure, a substantial amount of variation in solids, nutrient, and trace element loading can be anticipated on any sludge-treated forest site. An overall assessment indicated that this variation in loading and distribution of sludge constituents was less than expected. However, certain factors did influence the uniformity of sludge applications both among and within treated plots.

Loading Rate Variation Among Plots

Review of past studies of sludge application to forest land revealed that precise estimates of solids, nutrient, and trace element loading rates could not be reported because of the extreme variation in composition of sludge materials. Although only ranges of nutrient and trace element loadings are generally documented, relatively precise measures of these loadings can be obtained from analysis of samples collected from each truckload of sludge and from records of application, which include source of sludge, plot where applied, and number of liters delivered to each plot. Such data were recorded for the aspen and oak sites, while on the pine and hardwood sites more precise loading rates were computed because records of sludge source and volume applied to each strip within a treated plot were included in the loading calculation.

Aspen Site. The aspen site was treated with 1,112,878 liters (294,412 gal) of Alpena wastewater sludge. Variation in the distribution of this liquid is shown in Figure 3. The average dry solids content of the material was 3.2%, resulting in a mean sludge loading rate of approximately 10 Mg/ha (Table 1). The loading rates of nutrients and trace ele-

Figure 2. Sludge application on aspen (above) and oak sites.

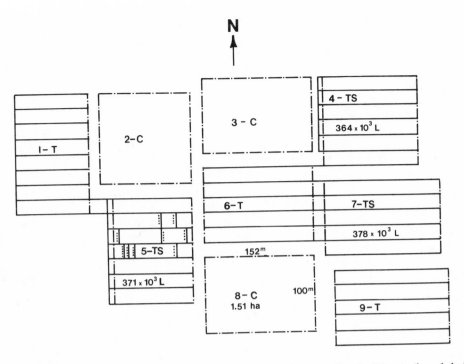

Figure 3. Sludge loading and distribution, aspen site. C = control; T = trails only; TS = trails and sludge; —— = access trail; —— · —— = plot boundary; ······· = sampling transect.

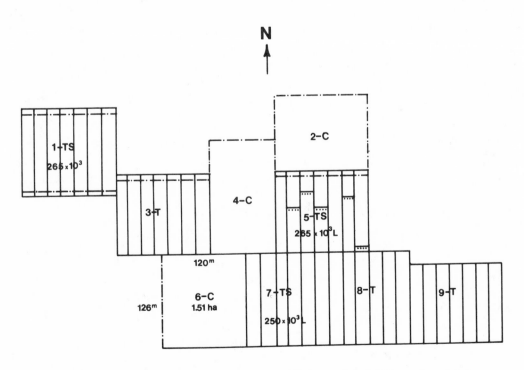

Figure 4. Sludge loading and distribution, oak site. C = control; T = trails only; TS = trails and sludge; —— = access trail; —— · —— = plot boundary; ····· = sampling transect.

ments were computed from data on area of application, volume of sludge applied, and chemical analysis of sludge samples collected during the application period. Loading rates for nitrogen and phosphorus averaged 561 and 291 kg/ha (500 and 260 lb/acre), respectively. Differences in concentrations and loading rates for most major elements were generally not statistically significant between plots.

TABLE 1. Solids, nutrient, and trace element loading, aspen site.

Component	Plot 4 (1.13 ha)	Plot 5 (1.20 ha)	Plot 7 (1.24 ha)	Mean (1.19 ha)
	(kilograms/hectare)			
Solids	8,586a	11,190b	10,165b	9,980
Nitrogen	514.5a	587.3b	580.0b	560.6
Phosphorus	257.2a	309.1b	305.2b	290.5
Potassium	24.82a	27.07b	26.73ab	26.21
Magnesium	39.92a	47.57b	46.58b	44.36
Calcium	402.4a	419.8a	431.7a	418.0
Sodium	30.51a	31.30a	32.53a	31.45
Aluminum	257.2a	343.3b	311.6b	304.0
Iron	470.5a	612.5b	588.6b	557.2
Manganese	6.27a	7.79b	7.06ab	7.04
Copper	4.52a	6.02b	6.49b	5.68
Zinc	9.92a	13.75b	13.20b	12.29
Cadmium	0.54a	0.22b	0.08b	0.28
Boron	0.47a	0.51a	0.34a	0.44
Nickel	0.39a	0.39a	0.49a	0.42
Chromium	1.54a	2.00b	1.88ab	1.81

Plot means followed by the same letter are not significantly different at the 0.05 level (Duncan's multiple range test).

TABLE 2. Solids, nutrient, and trace element loading, oak site.

Component	Plot 1 (1.11 ha)	Plot 5 (1.17 ha)	Plot 7 (1.10 ha)	Mean (1.13 ha)
	(kilograms/hectare)			
Solids	13,964a	4,461b	5,632b	8,019
Nitrogen	453.5a	430.3a	317.9b	400.6
Phosphorus	453.5a	181.2b	181.7b	272.1
Potassium	33.35a	14.33b	16.37b	21.35
Magnesium	80.44a	36.66b	35.56b	50.89
Calcium	1,205.0a	300.3b	351.6b	619.0
Sodium	32.58a	21.58b	21.48b	25.21
Aluminum	275.5a	59.6b	103.8c	146.3
Iron	787.0a	338.8b	349.3b	491.7
Manganese	14.97a	1.16b	3.19b	6.44
Copper	6.06a	6.78a	5.54a	6.13
Zinc	15.62a	5.70b	6.44b	9.25
Cadmium	0.11a	0.47b	0.69b	0.42
Boron	0.06a	0.66b	0.57b	0.43
Nickel	0.58a	0.15b	0.21b	0.31
Chromium	1.53a	0.38b	0.65c	0.85

Plot means followed by the same letter are not significantly different at the 0.05 level (Duncan's multiple range test).

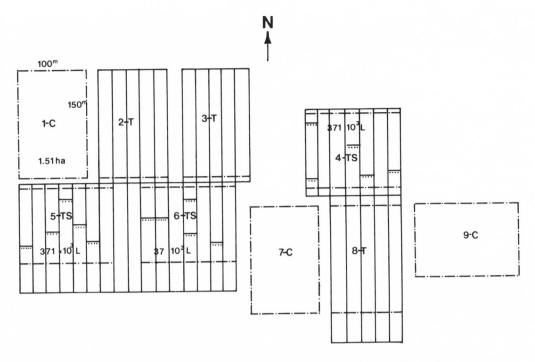

Figure 5. Sludge loading and distribution, pine site. C = control; T = trails only; TS = trails and sludge; ——— = access trail; ——— · ——— = plot boundary; ⋯⋯⋯ = sampling transect.

Oak Site. The oak site was treated with 264,971 liters of wastewater sludge from Alpena (plot 1) and 514,801 liters of wastewater sludge from Rogers City (plots 5 and 7). Variation in the distribution of these liquids is shown in Figure 4. The average dry solids content of these materials was 3.4%, resulting in a mean sludge loading rate of approximately 8 Mg/ha (Table 2). Plot 1 received the highest application rate (14 Mg/ha). Over the entire site, the nitrogen loading rate averaged 401 kg/ha, while that for phosphorus was 272 kg/ha. Nutrient loadings for plot 1 were much higher than those of other plots. Because of the different chemical characteristics of the two sludges, significant differences were found between plot 1 and plots 5 and 7 for most major elements, except nitrogen, copper, and boron.

Pine Site. The pine site was treated with 1,112,878 liters of Alpena wastewater sludge. Variation in the distribution of this liquid is shown in Figure 5. The average dry solids content was 2.6%, resulting in a mean sludge loading rate of approximately 8 Mg/ha (Table 3). The nitrogen loading rate averaged 379 kg/ha and that of phosphorus 253 kg/ha. Differences in the loading rates of most elements were generally not statistically significant between plots.

Northern Hardwoods Site. The northern hardwoods site was treated with 673,783 liters of Rogers City wastewater sludge. Variation in the distribution of this liquid is shown in Figure 6. The average dry solids content was 5.1%, resulting in a mean sludge loading rate of approximately 9 Mg/ha (Table 4). Because of the higher solids content of this sludge, nutrient additions to these plots were higher than those on other sites. The nitrogen loading rate averaged 783 kg/ha and that of phosphorus 384 kg/ha. Trace element additions were lower on this site than on the other sites. Differences in the loading rates of nutrients and trace elements were not statistically significant between plots.

TABLE 3. Solids, nutrient, and trace element loading, pine site.

Component	Plot 4 (1.25 ha)	Plot 5 (1.13 ha)	Plot 6 (1.15 ha)	Mean (1.18 ha)
	(kilograms/hectare)			
Solids	10,119a	7,058a	7,419a	8,119
Nitrogen	356.1a	459.6a	322.6a	379.4
Phosphorus	237.4a	295.4a	225.8a	252.9
Potassium	23.24a	22.13a	21.00a	22.12
Magnesium	37.09a	31.81a	27.84a	32.25
Calcium	451.9a	355.5ba	313.2b	373.5
Sodium	28.64a	31.61a	30.29a	30.18
Aluminum	176.3a	123.2ba	113.9b	137.8
Iron	592.6a	468.8a	441.3a	500.9
Manganese	5.19a	3.46a	2.76a	3.80
Copper	5.40a	3.66ba	3.61b	4.22
Zinc	10.36a	6.50b	5.97b	7.61
Cadmium	0.14a	0.42a	0.51a	0.36
Boron	0.68a	0.76a	0.69a	0.71
Nickel	0.39a	0.26a	0.40a	0.35
Chromium	1.11a	0.77ba	0.71b	0.86

Plot means followed by the same letter are not significantly different at the 0.05 level (Duncan's multiple range test).

TABLE 4. Solids, nutrient, and trace element loading, northern hardwoods site.

Component	Plot 4 (1.24 ha)	Plot 5 (1.25 ha)	Plot 9 (1.19 ha)	Mean (1.23 ha)
	(kilograms/hectare)			
Solids	8,851a	8,895a	9,885a	9,210
Nitrogen	659.3a	697.7a	992.4a	783.1
Phosphorus	362.6a	387.6a	400.8a	383.7
Potassium	12.46a	12.11a	11.09a	11.89
Magnesium	55.38a	53.68a	40.46a	49.84
Calcium	574.1a	487.0a	447.9a	503.0
Sodium	20.08a	18.33a	17.31a	18.57
Aluminum	90.0a	78.3a	71.0a	79.8
Iron	507.0a	487.4a	403.4a	456.9
Manganese	1.95a	1.63a	1.40a	1.66
Copper	12.05a	11.36a	9.06a	10.82
Zinc	10.12a	8.55a	7.14a	8.60
Cadmium	0.08a	0.08a	0.07a	0.08
Boron	0.30a	0.30a	0.22a	0.27
Nickel	0.26a	0.20a	0.17a	0.21
Chromium	0.68a	0.57a	0.50a	0.58

Plot means followed by the same letter are not significantly different at the 0.05 level (Duncan's multiple range test).

Loading Rate Variation Within Plots

Since variation was apparent in the terrain and vegetation density and structure within treated plots of such large size, it was anticipated that significant variation in sludge loading rates might result. Such variability in nutrient and trace element application rates could result in a differential response of the ecological components to treatment. Therefore, 280 catchment samplers were set out in a series of 28 transects covering all treated plots to quantify the variation in sludge loading rates within plot application areas.

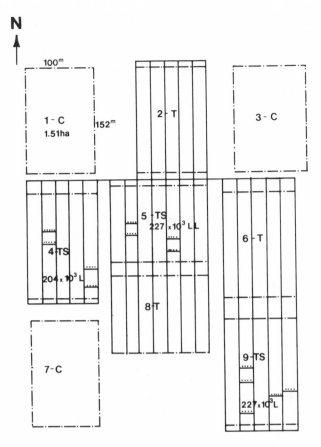

Figure 6. Sludge loading and distribution, northern hardwoods site. C = control; T = trails only; TS = trails and sludge; ———— = access trail; ——— · ——— = plot boundary; ‥‥‥ = sampling transect.

Data from sludge samples collected were compared with factors that could affect the uniformity of sludge application. Sludge loading rates along and across application strips were affected by variation in application vehicle speed and tank pressure, pit and mound microtopography (including tree stumps), vegetation structure and density, wind speed, and distance from application vehicle. Among these, distance from the application vehicle and vegetation density and structure appeared to produce the most prominent effects.

Distance Effects. The variation of sludge loading rate with distance from application vehicle is shown in Figures 7, 8, 9, and 10 for the aspen, oak, pine, and northern hardwoods sites, respectively. Generally, there were nonsignificant variations in sludge application on the sites. However, a trend of decreasing solids loading with increasing distance from the point of discharge was noted on the aspen site (Figure 7). This measurable but nonsignificant trend resulted from mechanical difficulties with the application vehicle that were encountered while treating this site.

Basal Area Effects. The variation of sludge loading rate with stand basal area is shown in Figures 11 and 12 for northern hardwoods and pine, respectively. No significant variations in sludge applications attributable to basal area were found on the aspen, oak, or northern hardwoods sites. However, a trend of decreasing solids loading with increasing basal area was measured on the pine site. This trend was believed to be the result of

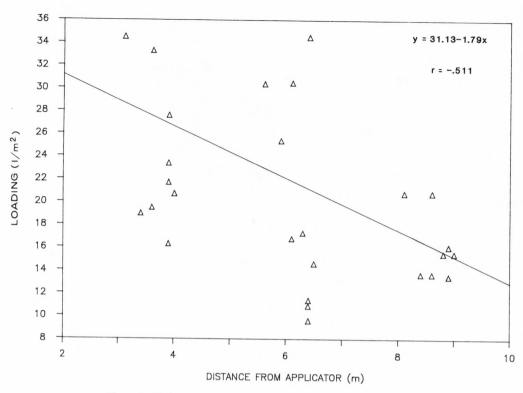

Figure 7. Sludge loading rate versus discharge distance, aspen site.

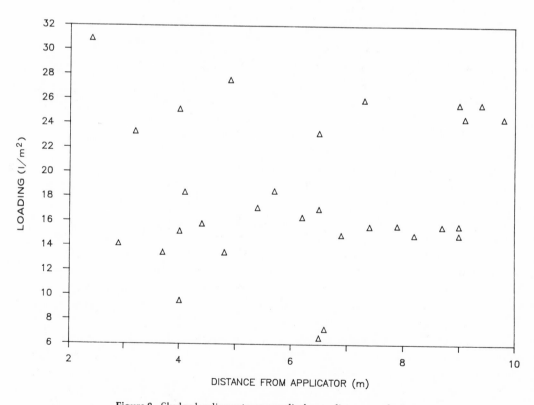

Figure 8. Sludge loading rate versus discharge distance, oak site.

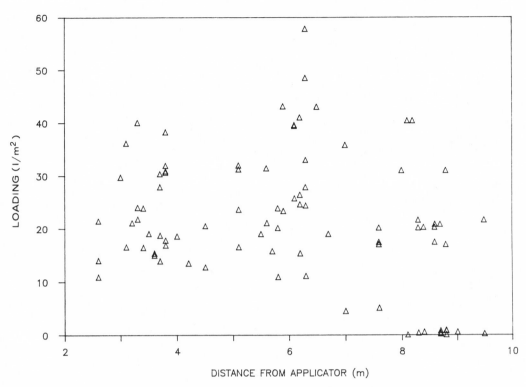

Figure 9. Sludge loading rate versus discharge distance, pine site.

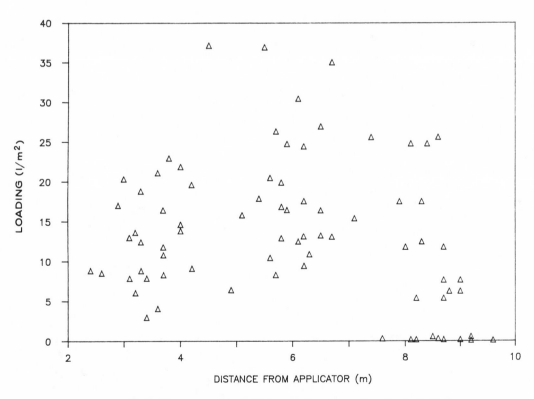

Figure 10. Sludge loading rate versus discharge distance, northern hardwoods site.

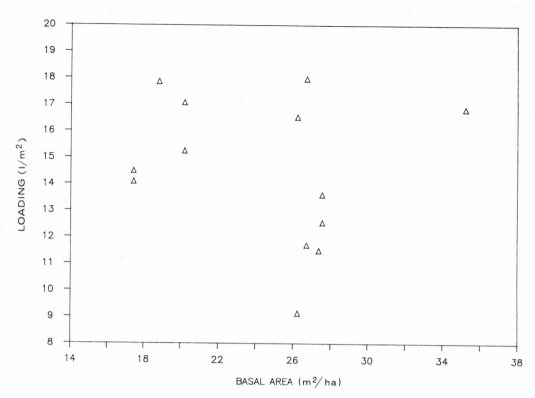

Figure 11. Sludge loading rate versus stand basal area, northern hardwoods site.

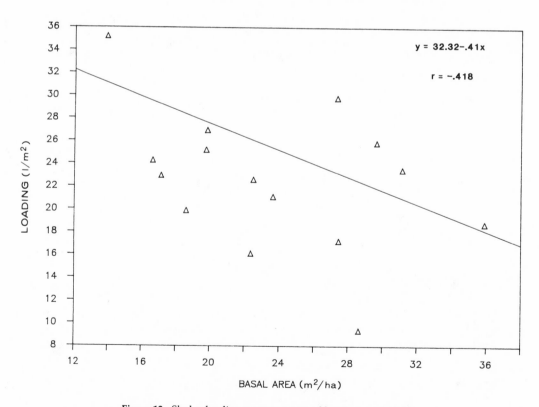

Figure 12. Sludge loading rate versus stand basal area, pine site.

uniform size and distribution of trees in the pole-sized plantation, providing a greater barrier to the movement of sludge discharged from the application vehicle. The more irregular distribution of trees in the naturally regenerating stands provided a lesser degree of physical impedence in this regard.

LOGISTICS AND COSTS

Site preparation to provide vehicle access in the stand is a major initial consideration in planning a forest land application program for wastewater sludge. If stands consist of young, unmerchantable age classes, site access may need be developed at a net cost to the manager. Such was the case with the aspen stand, in which trails were cleared at a cost of $1,485 ($163.91/ha) using a bulldozer. In contrast, a net income may be generated by harvest of timber growing in proposed access trails when trees are of sufficient size and quality. Following development of access trails on the pine, oak, and northern hardwoods sites, net respective returns from sale of timber were $340 ($37.53/ha), $158 ($17.44/ha), and $140 ($15.45/ha). Where the services of consulting foresters were required in site preparation, a rate of $21 per hour resulted in a total fee of $3,973 ($109.63/ha) for the project.

Using one 32,000 liter and two 23,000 liter tank trucks, sludge was transported from the municipal wastewater treatment plants at Rogers City and Alpena, a distance of 80 km (50 miles) to each of the forest sites. Loading time at each treatment plant varied from 45 to 60 minutes for each truck, and one-way transport time on the highway was approximately one hour. On-site unloading for each truck ranged from 30 to 40 minutes, resulting in a total delivery cycle of three to four hours per load. During a working day without mishap, each truck could complete three to four deliveries. More typically, because of operational delays, daily sludge delivery rates averaged 147,615 liters, requiring a travel distance of 950 km and 18 man-hours during the 26 days on which sludge was transported.

Sludge application was conducted using an Ag-Gator 2004, manufactured by Ag Chem Equipment Company of Minneapolis, Minnesota. This application vehicle was equipped with a standard pressure-vacuum pump that was used to fill and empty its 8,300 liter tank. Liquid sludge could be laterally discharged distances up to 10 m from one side of the vehicle through a modified spray system of three nozzles arranged to evenly cover near, intermediate, and distant bands of the forest floor.

Contractual costs for transport and application of 3,679,311 liters (972,074 gal) of liquid sludge totaled $48,576. This amount was equally apportioned by the contractor for transportation, application, and administration (Table 5). Had this procedure been a sludge reapplication to a previously used site, the contractor estimated a reapportionment of costs to 40% for transportation and 30% each for application and administration. The resultant lower total cost would be a product of less time needed in planning and greater efficiency in reapplication based on previous on-site experience.

While trafficability was satisfactory on most forest sites, pit and mound microtopography and high stumps remaining in trails at the completion of whole-tree skidding on the northern hardwoods site complicated application vehicle operation. Stumps caused the puncture of one high flotation tire on the application vehicle, and the generally rough terrain contributed to the eventual rupture of the hydraulic unit on its articulated steering mechanism. Repair costs for these breakdowns totaled $4,070.

TABLE 5. Contractor cost breakdown for transport and application.

	Initial Application	Subsequent Application
Transportation	$16,515.84 (34%)	$16,515.84 (40%)
Labor	1,651.58 (10%)	1,651.58 (10%)
Equipment	11,561.09 (70%)	11,561.09 (70%)
Fuel	3,303.17 (20%)	3,303.17 (20%)
Application	16,030.08 (33%)	12,144.00 (30%)
Labor	1,603.01 (10%)	1,214.40 (10%)
Equipment	12,824.06 (80%)	9,715.20 (80%)
Fuel	1,603.01 (10%)	1,214.40 (10%)
Administration	16,030.08 (33%)	12,144.00 (30%)
Totals	$48,576.00	$40,803.84

The cost of initial sludge transport and application to the four forest sites averaged 1.3 cents per liter (4.8 cents per gallon). If the expenditures for equipment repair are added, the total unit cost increases to 1.4 cents per liter. When care in site selection, stand preparation, and equipment operation are exercised, this cost increment for repairs can be minimized. If the expenditures for site preparation and retaining consulting foresters are also added, the total unit cost increases to 1.5 cents per liter of sludge applied. When care is taken to select sites containing merchantable timber that will be harvested and sold in the course of developing access trails, this cost increase can also be abated. Had the procedure been a sludge reapplication to forest sites receiving periodic operational use, the transport and application contractor cost estimate would have approximated 1.1 cents per liter.

These costs are comparable to those for sludge transport and application to farmland. Because the expenditures reported are for a demonstration and research project established to meet precise scientific criteria, the forest sites were located 80 km from the sludge source. Typical haul distances for operational sludge fertilization programs would more likely approximate 16 to 32 km, proportionally reducing transportation costs. This further decrease in program costs below those quoted above would make sludge application to forest land a highly attractive recycling alternative.

Finally, the costs related to creating stand access trails and those for repairing equipment subject to travel over residual stumps could be eliminated by careful planning during the establishment of a plantation scheduled to receive fertilizer applications of sludge at some future time. This could be accomplished by leaving one pair of unplanted seedling rows at 20 m intervals when a forest site is planted. The resultant system of parallel access trails would enable the stand to easily accommodate sludge application vehicles in the future and facilitate entry for intermediate silvicultural operations throughout the rotation.

SOCIAL CONSIDERATIONS

During the initial phases of project planning, numerous hours were devoted to evaluation of the logistic, physical, and biological characteristics of several candidate forest sites for land application of sludge. Of equal importance was assessment of the social climate for conduct of sludge application on publicly owned lands in each locale. Recognizing that forest land application of sludge is a relatively unfamiliar practice to a large segment of the population, it was essential to approach local elected officials in an open atmosphere where available information and concerns could be discussed.

Meetings held early in the course of program planning with representatives of the Huron Pines Resource Conservation and Development Council, the Northeastern Michigan Council of Governments, the Montmorency County Planning and Zoning Commission, and Montmorency Township were vital to the success of the project. Subsequent annual field tours have been conducted to update federal and state agency specialists, municipal officials, local groups, and individuals on progress of the project. During late summer of 1985 an information sharing conference and field workshop was held in the locale to summarize project findings and afford discussion opportunity for those groups and individuals who have an interest in land application of sludge on northern forest sites. This meeting also initiated a program of public education, wherein state agency specialists will disseminate information that objectively discusses the benefits and risks of wastewater sludge application on forest land.

PRELIMINARY FINDINGS

As indicated earlier, several research studies were conducted during the five years of the project to enhance the base of knowledge concerning the physical, chemical, biological, and social aspects of forest land application of sludge. Without offering the degree of detail provided in individual papers appearing elsewhere in this volume, the major highlights of initial findings are summarized.

Sociological

In northern Michigan, a survey of public opinion revealed that residents were generally undecided about the practice of sludge applications on forest land (Peyton et al. 1983). A lack of available information concerning this practice largely accounted for the absence of strongly held opinions. Forest application was perceived as being of benefit to forest growth and long-term environmental quality; however, short-term public health and environmental quality concerns were also noted. While university and state agency specialists generally enjoyed highest credibility as accurate information sources, local residents who mistrust state agency intentions were most likely to hold negative opinions about forest land sludge applications.

The low level of available information concerning forest application of sludge represents an important opportunity for public agencies to provide information that will allow state residents to develop accurate opinions about sludge utilization options. The basic aim of a public education program should be to provide local citizens with correct information, which would allow them to reach decisions that simultaneously evaluate the various options for sludge use and select the alternative that maximizes benefits and minimizes risks. In this effort there exists an opportunity to improve relations between the state agency and numerous groups and individuals.

Wildlife

Browse utilization by white-tailed deer (*Odocoileus virginianus* Zimmermann) and elk (*Cervus canadensis* Erxleben) was found to increase significantly on plots that had been provided with trails or with trails and sludge (Haufler and Campa 1983). Trails appeared to encourage intermediate levels of ungulate use, and sludge addition resulted in the highest levels. These results are thought to be population responses to the greater ease of access provided by trails and higher nutritional value of browse plants growing on sludge-fertilized plots. Although no changes in digestibility or the content of crude fat and fiber were detected, increases in crude protein and hemicellulose were measured in wildlife food plants (Haufler et al. 1983). Heavy metal concentrations in forage plants growing on sludge-fertilized plots did not significantly differ from those in species found on control areas.

Small mammals were observed in greater numbers on the sludge-fertilized plots of the aspen, pine, and northern hardwoods sites one and two years following treatment (Haufler et al. 1983). A slight, but nonsignificant, decrease was measured on the fertilized plots of the oak site. Herbaceous and woody plant species composition and structure on the aspen, oak, and pine sites varied little as a result of sludge application. Sludge addition appears to have decreased herbaceous and woody plant cover on the northern hardwoods site, a result of applications conducted during the growing season. Sludge fertilization increased the annual production of primary wildlife food plants substantially on the aspen and oak sites and slightly on the northern hardwoods site, but resulted in no change on the pine site. While sludge addition and access trail development generally stimulated annual production, applications conducted during the growing season were less effective, because the understory species were smothered by the loading of sludge solids.

Nutrient Availability and Water Quality

Analyses for major nutrients and heavy metals in soil leachate and groundwater have established that concentration fluctuations occurred in a manner generally unrelated to sludge treatment (Urie et al. 1983). Slight elevations in nitrate-N and sodium levels were observed in water samples collected on the oak site during the 1982 growing season. These concentrations soon thereafter declined to near background. It was apparent from the data that adequate protection for the phreatic aquifer existed in these forest ecosystems when sludge was applied at recommended rates.

Vegetation and Soils

On the aspen site no changes in understory cover or diversity were observed one year following sludge application (Hart et al. 1983). The number of tree seedlings present also could not be related to treatment. Aspen mortality was increased to a significant degree by fungal infections caused by *Cytospora chrysosperma*, *Fusarium* spp., and *Armillaria mellea*. The increased activity of these pathogens was related to the method of access trail preparation and the increased browsing activity of ungulates on sludge-fertilized plots.

Exchangeable bases applied with the sludge were largely retained in the forest floor and the upper 45 cm of mineral soil (Hart et al. 1983). Phosphorus and trace element levels significantly increased in the forest floor following sludge application. One year after treatment, forest floor weight had decreased 14% as a result of microbial activity stimulated by nitrogen and phosphorus addition. While subsoil was minimally affected by sludge application, surface soil nitrogen, phosphorus, calcium, magnesium, and aluminum levels were increased. The retention of nutrients in the forest floor was approximately 45% of the total applied.

SUMMARY

Nearly 4 million liters of anaerobically digested wastewater sludge were applied to the forest floor of four forest types of major commercial importance. The large areas receiving the treatment of approximately 9 Mg of dry solids per hectare represented an important intermediate step between earlier small-plot research and eventual full-scale operation of municipal and industrial programs. The technology currently available to conduct liquid sludge application on forest land was demonstrated to be efficient in providing uniform distribution and loading of nutrients and trace elements on each site. Costs of transportation and application compared quite favorably with those of farmland applications, making forest land application a truly viable sludge recycling alternative. In addition, the opportunity to conduct a diverse array of land application research studies was provided. Preliminary findings of this research offer encouraging evidence of improved plant growth, enhanced wildlife habitat, and adequate protection for groundwater quality and the public health. Findings have also provided new insights into public concerns and attitudes and outlined a process for constructive citizen involvement in program planning. It is hoped that this demonstration project will serve as a model for future land application programs developed in the forests of northern Michigan and similar environments.

ACKNOWLEDGMENTS

Although the information in this document has been funded in part by the U.S. Environmental Protection Agency under assistance agreement No. S005551-01 to the Michigan Department of Natural Resources and Michigan State University, it has not been subjected to the Agency's publication review process and therefore may not necessarily reflect the views of the Agency, and no official endorsement should be inferred. Mention of trade names or commercial products does not constitute endorsement or recommendation for use.

Michigan Agricultural Experiment Station journal article number 11866.

REFERENCES

Brockway, D. G. 1979. Evaluation of northern pine plantations as disposal sites for municipal and industrial sludge. Ph.D. diss., Michigan State University, East Lansing. University Microfilms, Ann Arbor, Mich. (Diss. Abstr. 40-2919B).

_____. 1983. Forest floor, soil, and vegetation responses to sludge fertilization in red and white pine plantations. Soil Sci. Soc. Am. J. 47:776–784.

Brockway, D. G., and D. H. Urie. 1983. Determining sludge fertilization rates for forests from nitrate-N in leachate and groundwater. J. Environ. Qual. 12:487–492.

Hart, J. B., P. V. Nguyen, J. H. Hart, C. W. Ramm, D. M. Merkel, and C. Thomas. 1983. Ecological monitoring of sludge fertilization on state forest lands in northern lower Michigan. Annual progress report. Department of Forestry, Michigan State University, East Lansing. 101 p.

Haufler, J. B., and H. Campa. 1983. Deer and elk use of forages treated with municipal sewage sludge. Annual progress report. Department of Fisheries and Wildlife, Michigan State University, East Lansing. 35 p.

Haufler, J. B., E. Seon, and D. Woodyard. 1983. Influences on wildlife populations of the application of sewage sludge to upland forest types. Annual progress report. Department of Fisheries and Wildlife, Michigan State University, East Lansing. 136 p.

Michigan Department of Natural Resources. 1983. A summary of current municipal sludge production and use in Michigan. In-house Paper, Groundwater Quality Division, Lansing. 9 p.

Peyton, R. B., L. Gigliotti, and T. Lagerstrom. 1983. A survey of public perception and acceptance of sludge application to state forest lands in

Michigan. Phase I summary report. Department of Fisheries and Wildlife, Michigan State University, East Lansing. 134 p.

Strommen, N. D. 1967. Monthly precipitation probabilities for climatic divisions in Michigan. ESSA-Weather Bureau, Michigan Weather Service in Cooperation with Michigan Department of Agriculture. 22 p.

Urie, D. H., A. R. Harris, and J. H. Cooley. 1978. Municipal and industrial sludge fertilization of forests and wildlife openings. p. 467–480. *In* Proceedings of First Annual Conference of Applied Research and Practice of Industrial and Municipal Waste, Department of Engineering and Applied Science, University of Wisconsin, Madison.

Urie, D. H., J. B. Hart, P. V. Nguyen, and A. Burton. 1983. Hydrologic and water quality effects from sludge application to forests in northern lower Michigan. Annual progress report. Department of Forestry, Michigan State University, East Lansing. 35 p.

Chapter 46

Reclamation of Severely Devastated Sites with Dried Sewage Sludge in the Southeast

CHARLES R. BERRY

ABSTRACT   The potential of dried sewage sludge for use in reclamation in the Southeast has been demonstrated in field experiments on severely eroded sites in the Tennessee Copper Basin, borrow pits in South Carolina, and a kaolin spoil in Georgia. In the Copper Basin, ten-year-old loblolly, shortleaf, and Virginia pines growing in plots amended with 34 Mg/ha of dried sewage sludge produced 45% or more additional height growth and 50% or more additional diameter growth than in plots that received 896 kg/ha of 10-10-10 fertilizer and 1,417 kg/ha unslaked lime. Loblolly pine growing in borrow pit plots amended with 34 Mg/ha of dried sewage sludge produced a 15-fold increase in biomass compared with growth on plots amended with 560 kg/ha of 10-10-10 fertilizer plus 2,240 kg/ha of dolomitic lime. On another borrow pit, better growth of several hardwood species as well as loblolly pine was achieved on plots amended with sludge at 68 Mg/ha compared with growth on plots amended with 10-10-10 fertilizer at 1,120 kg/ha and lime at 2,240 kg/ha. Volume increases induced by sludge compared with fertilizer after two years are yellow poplar, 661%; green ash, 278%; sweetgum, 123%; and sycamore, 148%. Sweetgum planted on kaolin spoil plots amended with sewage sludge produced 101% more volume growth than on plots amended with 560 kg/ha of 10-10-10 fertilizer plus 2,240 kg/ha of dolomitic lime. Subsoiling at a depth of either 46 or 92 cm proved to be a better site preparation procedure on a borrow pit than disking. On plots amended with sewage sludge, subsoiling 92 cm deep induced better growth on loblolly pine than subsoiling 46 cm deep. On plots amended with fertilizer, however, subsoiling to 92 cm promoted washing and leaching of fertilizer, which resulted in slower growth than subsoiling to 46 cm.

Dried municipal sewage sludge is an effective, long-lasting amendment for use in reclamation of devastated sites. Numerous studies and pilot programs have demonstrated the nutritive and soil conditioning value of sewage sludge for the establishment of cover crops and various tree species (Lejcher and Kunkle 1973, Sutton and Vimmerstedt 1973, Roth et al. 1979, Kerr et al. 1979, Cole 1982). A guide for revegetation of mined land in eastern United States using municipal sludge was also published recently (Sopper and Seaker 1983). These reports, however, relate work and methods carried out in reclamation of surface coal mines in the northeastern and midwestern states. Studies of sewage sludge in the Southeast have been made by Wells et al. (1984) on established stands, by Hinkle (1982) on reclamation of abandoned pyrite mines in Virginia, and on other disturbed sites by Berry (1977, 1979, 1985), Berry and Marx (1977, 1980), Ruehle (1980), and Kormanik and Schultz (1985).

Disturbed lands in the southeastern United States include strip-mine coal spoils in Alabama, Tennessee, and Virginia; phosphate mine spoils in Florida, North Carolina, and Tennessee; kaolin mine spoils in Georgia, and North and South Carolina; and borrow pits in all states. Other problem sites are severely eroded former agricultural lands, including over 2 million hectares (5 million acres) in the Piedmont Plateau in Alabama, Georgia, Virginia, and the Carolinas, and 2 to 3 million hectares (5 to 7.5 million acres) of entisols, soils with sand two or more meters deep in Alabama, Florida, Georgia, and South Carolina. Also in need of reforestation is the Copper Basin, Tennessee, which was badly eroded following the loss of vegetation brought about by air pollution from primitive smelting methods.

This paper summarizes several studies demonstrating the value of sewage sludge as a soil amendment for reclamation of devastated sites in the southeastern United States. Study sites were the Copper Basin, kaolin spoils, and borrow pits, none of which contained substances toxic to plants, but all were extremely low in plant nutrients and in massive structure, and were generally poorly aerated.

MATERIALS AND METHODS

The sewage sludges for these studies were obtained from plants employing secondary treatment with anaerobic digestion and sand bed drying. Analysis reveals nutrient concentrations of 1.5 to 2% nitrogen, 0.5 to 1.0% phosphorus, and 0.025 to 0.5% potassium, with 5 to 10 μg/g cadmium and 50 to 250 μg/g zinc.

Experimental plots were prepared by first grading and removing all woody vegetation, then thoroughly incorporating amendments, either sludge or fertilizer, by disking into the surface material to a depth of 15 to 20 cm (6 to 8 inches). Control plots were similarly disked. After disking, the sites were subsoiled to a depth of up to about 92 cm (36 inches). Spacing between subsoiled furrows and whether all furrows ran in one direction (parallel lines) or two directions (creating a grid pattern) varied with the experiment. In the more recently installed experiments, subsoiling was done in early fall to allow furrows to settle before planting. Seedlings were then planted in furrows.

Soil samples were composited from five subsamples per plot at a depth of 0 to 15 cm, air dried at room temperature for 10 days, and chemically analyzed after double acid extraction (0.05N HC1 plus 0.025N H_2SO_4). Phosphorus was determined colorimetrically and cations by atomic absorption. Total N was determined by Kjeldahl, organic matter by wet oxidation chromic acid digestion, CEC by saturation with ammonium and replacement with K^+, and pH by a glass electrode in a mixture of two parts water and one part soil. One foliage sample per plot, consisting of a subsample from each tree on a plot, was dried at 75°C, ground in a Wiley mill, and chemically analyzed. Total N was determined by Kjeldahl and other elements by dry-ash methods (Wells et al. 1973).

Treatment plots contained from 16 to 36 trees when installed, and in most experiments treatments were replicated five times. Treatments were evaluated primarily by comparing height and ground-line diameter or diameter at breast height (dbh). In some cases the plots have been thinned with little or no mortality since thinning. Early survival and growth have been described in previous reports (Berry 1982, 1985, Berry and Marx 1977, Kormanik and Schultz 1985). In some cases, D^2H (ground-line diameter squared times height) was also computed. Hatchell et al. (1985), studying loblolly pine in the Southeast, obtained R^2 values of 0.910 to 0.951 between D^2H and stem weight and 0.908 to 0.956 between D^2H and stem volume. R^2 values for log regressions were even higher (Hatchell et al. 1985). Data were analyzed by analysis of variance, and means were separated by Duncan's multiple range test.

Tennessee Copper Basin

Starting in the early 1840s and continuing into the 1900s, air pollution from the processing of copper ore killed nearly all natural vegetation on several thousand acres in the Tennessee Copper Basin. With the vegetation gone, erosion became severe, and in time most of the A and B horizon soil throughout the basin had eroded. Analyses of soil in this area revealed levels of total N and exchangeable P and K as low as 60, 1, and 5 μg/g, respectively. In recent years, tree seedlings planted there barely survived unless nutrients were applied (Berry 1979). Tree seedlings in routine plantings in the area are commonly fertilized with a commercial 9 gram starter tablet (Sierra Chemical Co., Milpitas, California). Attempts to reclaim the basin, however, have met with only marginal success, except near the perimeter, where air pollution and erosion were not so severe.

A single application of dried sewage sludge 1.3 cm (0.5 inch) deep amounting to 34 Mg/ha (15 tons/acre) was broadcast and incorporated into the soil before planting in 1975 (Table 1). Thirty-six trees of each species, spaced 0.9 by 0.9 m, were planted in each plot,

TABLE 1. Soil chemical properties of sewage sludge and fertilizer-amended plots in the Tennessee Copper Basin after ten years.

Treatment	Nitrogen (ppm)	Phosphorus (ppm)	Potassium (ppm)	Calcium (ppm)	Magnesium (ppm)	Organic Matter (%)	pH
Sludge*	760a	44a	35a	84a	6a	1.2a	4.2b
Fertilizer**	460b	5b	30a	65a	5a	0.5b	4.5a

Means within a column followed by the same letter do not differ significantly at p = 0.05.

*34 Mg/ha (15 tons/acre).
**896 kg/ha (800 lb/acre) of 10-10-10 fertilizer plus 1,417 kg/ha (1,265 lb/acre) of CaO.

but after five years plots were thinned to a uniform stocking of 10 to 12 trees. Growth of loblolly (*Pinus taeda* L.), shortleaf (*P. echinata* Mill.), and Virginia (*P. virginiana* Mill.) pines was significantly greater after ten years in sludge plots than in plots receiving 896 kg/ha (800 lb/acre) of 10-10-10 fertilizer applied in combination with 1,417 kg/ha (1,265 lb/acre) of unslaked lime (Table 2). Pines in the sludge-treated plots continue to grow vigorously, and are producing a thick layer of duff. Erosion is no longer a problem on treated plots.

While fair growth of all pines was achieved on fertilizer plots, growth on sludge plots is much better. Soil analyses reveal significantly more nitrogen, phosphorus, and organic matter on sludge plots (Table 1), accounting for the superior growth of trees on those plots.

Borrow Pits

The use of landfill for construction projects results in the creation of borrow pits. Typically, borrow pits in the southeastern United States consist of exposed hard substratum material that has poor internal drainage and is low in nutrients and organic matter (Table 3). Several experiments designed to study revegetation of borrow pits were installed at the Savannah River Forest Station, Aiken, South Carolina. Four experiments described here—two with loblolly pine, one with sweetgum (*Liquidambar styraciflua* L.), and one with several hardwood species—have shown the value of dried sewage sludge as an amendment to improve tree growth on these devastated sites.

Loblolly Pine. This species was planted on plots with (1) 560 kg/ha (500 lb/acre) of 10-10-10 fertilizer and 2,240 kg/ha (2,000 lb/acre) of dolomitic lime, (2) 34 Mg/ha (15 tons/acre) of dried sludge, or (3) nothing (Berry and Marx 1980). Twenty-five trees, spaced 1.2 by 1.2 m, were planted in each plot but were thinned to 10 to 12 uniformly spaced trees after five years. After three years, soil analysis showed that the sludge application maintained respective levels of nitrogen, phosphorus, calcium, and organic matter at 5, 11, 4,

TABLE 2. Growth of loblolly, shortleaf, and Virginia pines in the Tennessee Copper Basin amended with sewage sludge or fertilizer after ten years.

Treatment	Height (cm)	dbh (cm)
Loblolly Pine		
Sludge*	702a	11.1a
Fertilizer**	450b	6.6b
Increased growth induced by sludge	56%	66%
Shortleaf Pine		
Sludge*	476a	6.1a
Fertilizer**	327b	4.1b
Increased growth induced by sludge	46%	51%
Virginia Pine		
Sludge*	608a	9.1a
Fertilizer**	363b	4.5b
Increased growth induced by sludge	67%	101%

Means within a column and within a species followed by the same letter do not differ significantly at p = 0.05.

*34 Mg/ha (15 tons/acre).
**896 kg/ha (800 lb/acre) of 10-10-10 fertilizer plus 1,417 kg/ha (1,265 lb/acre) of CaO.

TABLE 3. Soil properties of a borrow pit as influenced by sewage sludge or fertilizer and lime after three years.

Treatment	Nitrogen (ppm)	Phosphorus (ppm)	Potassium (ppm)	Calcium (ppm)	Magnesium (ppm)	Organic Matter (%)	pH
Sewage sludge*	595a	84a	7a	22a	18b	1.6a	4.2b
Fertilizer and lime**	153b	13b	7a	16a	65a	0.6b	4.9a
Control (no amendment)	112b	7b	6a	4b	11b	0.4b	4.2b

Means within a column followed by the same letter do not differ significantly at p = 0.05.

*34 Mg/ha (15 tons/acre).
**560 kg/ha (500 lb/acre) of 10-10-10 fertilizer plus 2,240 kg/ha (2,000 lb/acre) of dolomitic lime.

and 3 times higher than on control plots and that the fertilizer-lime application increased calcium threefold and magnesium nearly fivefold, and raised pH from 4.2 to 4.9 (Table 3). While the fertilizer and lime treatment increased certain soil nutrients and pH, it did not increase significantly the growth of loblolly pine and only slightly improved growth of grass. Sewage sludge, on the other hand, provided a 15-fold increase in seedling volume (D^2H) over five years compared with the fertilizer and lime treatment (Table 4). Similar results were obtained by Ruehle (1980) in an adjacent study with container-grown loblolly pine seedlings. He reported that seedlings after two years had twenty times more volume on plots receiving 34 Mg/ha (15 tons/acre) sewage sludge than on plots receiving 560 kg/ha (500 lb/acre) of 10-10-10 fertilizer with 2,240 kg/ha dolomitic lime. Ruehle also found that ectomycorrhizal treatments interacted with soil amendments; the ectomycorrhizal fungus *Pisolithus tinctorius* stimulated 300% more volume growth than *Thelephora terrestris*, but only on plots amended with sewage sludge. A second loblolly pine experiment, also on a borrow pit, will be discussed later in connection with subsoiling.

Sweetgum. This species grows well on a variety of forest sites and also grew well on a borrow pit amended with sewage sludge. Concentrations of soil nitrogen and soil phosphorus were increased after application of sewage sludge up to 68 Mg/ha (30 tons/acre) (Table 5). Twenty-five trees, spaced 1.2 by 1.2 m, were planted in each plot, but were thinned to 10 to 12 uniformly spaced trees after five years. Growth of trees increased with increasing amounts of sludge up to 34 Mg/ha (15 tons/acre), but beyond that level no additional benefit was obtained (Kormanik and Schultz 1985). Total heights aver-

TABLE 4. Growth of loblolly pine on a borrow pit amended with sewage sludge and fertilizer after five years.

Treatment	Height (cm)	Diameter (cm)	D^2H (cm^3)
Sewage sludge*	438a	11.8a	60,948a
Fertilizer and lime**	156b	4.4b	4,030b
Control	99b	3.2b	2,047b

Means within a column followed by the same letter do not differ significantly at p = 0.05.

*34 Mg/ha (15 tons/acre).
**560 kg/ha (500 lb/acre) of 10-10-10 fertilizer plus 2,240 kg/ha (2,000 lb/acre) of dolomitic lime.

TABLE 5. Effects of sludge application on soil nitrogen, soil phosphorus, and growth of sweetgum after five years.

Treatment (Mg/ha)	Soil Nitrogen (ppm)	Soil Phosphorus (ppm)	Height (cm)	Root-collar Diameter (cm)
0	140d	4d	74c	1.5c
17	566c	45c	275b	5.2b
34	819b	57b	362a	6.8a
68	1,838a	75a	365a	6.8a

Means followed by the same letter within a column do not differ significantly at p = 0.05.

aging 3.6 m, after five years on plots amended with either 34 Mg/ha (15 tons/acre) or 68 Mg/ha (30 tons/acre),were equal to or greater than five-year heights reported for sweetgum on reforestation sites.

Additional Species. After determining that growth of loblolly pine and sweetgum on a borrow pit could be increased dramatically by applications of sewage sludge, an additional study to determine the benefits that might be obtained for growth by additional hardwood species was desired. For this study, a split-plot design with sludge or fertilizer in major plots was utilized to compare the relative adaptibility of several hardwoods to a sludge reclaimed borrow pit. Species compared in this experiment were sycamore (*Platanus occidentalis* L.), green ash (*Fraxinus pennsylvanica* Marsh), and yellow poplar (*Liriodendron tulipifera* L.) as well as loblolly pine and sweetgum. Twenty-five trees, spaced 2.4 by 2.4 m, were planted in each treatment plot.

A highly significant species × fertility interaction in terms of D^2H was observed after the second growing season (Table 6). After two years on sludge plots, sycamore was growing significantly faster than all other species; while on fertilizer plots, sycamore and loblolly pine did not differ significantly. Similarly, ash grew faster than sweetgum on sludge plots, but the two species did not differ significantly on fertilizer plots. All species appear to be growing faster on sludge plots than on fertilizer plots. It is interesting, however, to note the relative gain in growth brought about by sludge compared with fertilizer by species, which ranged from 123 to 661%. Growth, however, of several, if not all, these species in this experiment was retarded by weed competition, which has been exceptionally heavy in sludge plots.

TABLE 6.   Relative D^2H of five tree species on sludge plots and fertilizer plots on a borrow pit after two years.

Species	Sludge*	Fertilizer**	Increased Growth Induced by Sludge
	D^2H (cm^3)	D^2H (cm^3)	(%)
Sycamore	6,004a	2,417a	148
Loblolly	3,708b	2,614a	42
Ash	2,921bc	772b	278
Yellow poplar	2,383cd	313b	661
Sweetgum	1,364d	613b	123

Means followed by the same letter within a column do not differ significantly at p = 0.05.

*68 Mg/ha (30 tons/acre).

**1,120 kg/ha (1,000 lb/acre) of 10-10-10 fertilizer plus 2,240 kg/ha (2,000 lb/acre) of dolomitic lime.

Kaolin Spoil

About 8,498 hectares (21,000 acres) in Georgia had been surface mined for kaolin clay by 1973, and it was estimated that up to 120,000 additional hectares (297,000 acres) of land will eventually be mined (May 1977). Berry and Marx (1977) found that loblolly pine seedlings grew well in kaolin spoil when amended with either fertilizer or sewage sludge. Chemical analysis of the kaolin spoils studied in this experiment proved to be quite low in major nutrients: nitrogen, 83 µg/g; phosphorus, 2µg/g; potassium, 10 µg/g;

calcium, 67 μg/g; and magnesium, 19 μg/g. CEC was 1.0, and pH was 4.7. Fertilizer applied at 1,120 kg/ha (1,000 lb/acre) significantly increased potassium, but none of the other nutrients. Application of sewage sludge at a rate of 34 Mg/ha (15 tons/acre) increased nitrogen, phosphorus, calcium, and CEC compared with the fertilizer or control treatments (Table 7). Twenty-five trees, spaced 2.4 by 2.4 m, were planted in each treatment plot. After three years a significant twofold growth response of sweetgum was obtained when 34 Mg/ha (15 tons/acre) of sewage sludge were applied instead of 1,120 kg/ha (1,000 lb/acre) of 10-10-10 fertilizer with 2,240 kg/ha (2,000 lb/acre) of dolomitic lime (Table 8).

May (1977), testing several species for survival and growth, found that loblolly pine and sycamore were the most adaptable, but European black alder (*Alnus glutinosa* L.), bicolor lespedeza (*Lespedeza bicolor* Turez.), slash pine (*Pinus elliottii* Englem. var. *elliottii*), and Virginia pine were also good. Sweetgum was not considered a good species. In spite of this report by May, it was felt that sweetgum seedlings with abundant endomycorrhizae could make good growth on kaolin spoil amended with sewage sludge.

TABLE 7. Soil properties of a kaolin spoil amended with sewage sludge or fertilizer and lime.

Treatment	Nitrogen (ppm)	Phosphorus (ppm)	Potassium (ppm)	Calcium (ppm)	Magnesium (ppm)	CEC (meg/100g)	pH
Sewage sludge*	1,070a	77a	21ab	1,333a	72a	9.0a	5.8a
Fertilizer**	120b	17b	44a	250b	39ab	2.6b	5.1b
Control†	73b	3b	5b	283b	47ab	2.8b	5.1b

Means followed by the same letter within a column do not differ significantly at p = 0.05.

*34 Mg/ha (15 tons/acre) of dried sewage sludge.
**1,120 kg/ha (1,000 lb/acre) of 10-10-10 fertilizer plus 2,240 kg/ha (2,000 lb/acre) of dolomitic lime.
†2,240 kg/ha (2,000 lb/acre) of dolomitic lime.

TABLE 8. Growth and survival of sweetgum seedlings on a kaolin spoil amended with sewage sludge or fertilizer after three years.

Treatment	Survival (%)	Height (cm)	Root-collar Diameter (cm)	D^2H (cm^3)
Sludge*	93.0a	117.7a	2.05a	671a
Fertilizer**	98.6a	100.1b	1.85a	334b
Control† (limed)	95.3a	79.8c	1.76a	323b

Means followed by the same letter within a column do not differ significantly at p = 0.05.

*34 Mg/ha (15 tons/acre) of dried sewage sludge.
**1,120 kg/ha (1,000 lb/acre) of 10-10-10 fertilizer plus 2,240 kg/ha (2,000 lb/acre) of dolomitic lime.
†2,240 kg/ha (2,000 lb/acre) of dolomitic lime.

Subsoiling

Surface materials on all devastated areas referred to in this paper were massive in structure and were too hard for adequate seedling root penetration without extensive site preparation. As a routine site preparation procedure, all study areas described above were loosened by subsoiling in addition to the disking required to incorporate amendments. The value of subsoiling was shown by growth of loblolly pine on borrow pit plots subsoiled to different intensities. Twenty-five trees, spaced 2.2 by 2.2 m, were planted in each treatment plot. This study revealed that after four years, depth of subsoiling significantly affected rate of growth. Neither spacing of furrows nor furrow pattern—that is, whether all furrows were parallel or in a grid pattern (Berry 1985)—affected rate of growth. Subsoiling to either 46 or 92 cm (36 inches) was better than disking alone on either fertilizer and lime or sewage sludge plots (Table 9). Subsoiling to 92 cm was better than 46 cm when sewage sludge was used, but when fertilizer was used, better growth was obtained by subsoiling to a 46 cm depth. More intensive subsoiling enhances washing and leaching and therefore loss of inorganic nutrients. Since nutrients in organic form, as in sewage sludge, are not as prone to washing or leaching, the 92 cm depth of subsoiling was more beneficial than shallow subsoiling because of improved moisture relations.

TABLE 9. Influence of subsoiling depth on growth of loblolly pine on a borrow pit amended with sewage sludge or fertilizer after four years.

Subsoiling Depth (cm)	Sludge*		Fertilizer**	
	Height (cm)	dbh (cm)	Height (cm)	dbh (cm)
15 (disk-control)	288c	3.0c	217c	1.7c
46	359b	4.3b	279a	2.7a
92	375a	4.6a	255b	2.3b

Means followed by the same letter within a column do not differ significantly at p = 0.05.

*17 Mg/ha (7.5 tons/acre).
**1,120 kg/ha (1,000 lb/acre) of 10-10-10 fertilizer plus 2,240 kg/ha (2,000 lb/acre) of dolomitic lime.

DISCUSSION

The studies summarized in this paper show that dried sewage sludge is useful for reclaiming disturbed sites in the Southeast. The sludges used, while relatively low in total N, and exchangeable P and K, were excellent sources of nutrients and organic matter at the levels tested. Most impoverished sites in the Southeast could be reclaimed by incorporating 34 Mg/ha (15 tons/acre) of a similar sludge. In no case did the relatively light sludge applications result in environmental problems.

Metz et al. (1970) found 670 to 900 kg/ha (598 to 803 lb/acre) of nitrogen, 15 to 22 kg/ha (13 to 20 lb/acre) of phosphorus, and 62 to 108 kg/ha (55 to 96 lb/acre) of potassium in the forest floor and the upper 7.6 cm (3 inches) of mineral soil in 20-year-old southern pine plantations. A sludge application rate of 34 Mg/ha (15 tons/acre) at 2% N, 1% P, and 0.5% K is equivalent to 680 kg/ha (600 lb/acre) of N, 340 kg/ha (300 lb/acre) of P, and 170 kg/ha (150 lb/acre) of K, and theoretically would transform the nutrient status of the most barren site to a level equivalent to or better than an average southern pine plantation. In the Copper Basin study, soil nitrogen and phosphorus remain high after ten years with no indication of a nutrient depletion or slowing of tree growth. Indications are that the benefits of sludge application will last at least as long as a rotation and perhaps indefinitely, assuming good soil conservation practices. Inorganic fertilizer, on the other hand, if not incorporated in vegetation soon after application will be washed or leached away and lost from the site. As growth data from the subsoiling experiment show, sewage sludge is not nearly as prone to be washed or leached as inorganic fertilizer, a fact that helps explain the superior growth of trees on sludge plots. Thus, such a treatment appears to be more than adequate for mere reclamation, and promises to enable restoration of a devastated site to a fully productive forest.

The value of subsoiling hard material or compacted soil cannot be overstated. In observations made during the subsoiling study, roots of loblolly pine were found to penetrate 92 cm (36 inches) in a subsoiled trench in two years; but without subsoiling, roots would not penetrate the massive material on this site.

Timing of physical treatment and amendment application is important and ideally should be in the following sequence: (1) application of amendment prior to incorporation, (2) incorporation of amendment prior to subsoiling, (3) subsoiling in August or September when ground is dry, and (4) planting in February or March. As much time as possible, up to six months, should elapse between subsoiling and planting to allow furrows to close naturally without compaction.

Planting good quality, heavily mycorrhizal seedlings after following the above procedure should virtually ensure the quick establishment of a healthy, rapidly growing stand, even on a site as adverse as a borrow pit. Survival has been excellent in these studies. Most mortality has been the result of disease—fusiform rust (*Cronartium fusiforme* Hedgc. and Hunt ex Cumm.)—or weed competition.

Competition by naturally occurring weeds did not occur on sites that were barren before reclamation was attempted. On sites supporting weeds and shrubs, however, amendments and site preparation stimulated luxuriant weed growth, resulting in severe competition for moisture. Unfortunately, the use of grasses to obtain a quick ground cover often results in high mortality and slow initial growth for trees (Berry 1977). When trees are to be used for reclamation, consideration should be given to initial erosion control by subsoiling or terracing on contour instead of planting grasses so competitive to tree seedlings.

Since trees in the oldest study reported here are barely ten years old, their yield cannot be determined by volume tables. It would be premature therefore to predict economic benefits. With a few broad assumptions, however, one can make a conservative estimate of the economic benefit of applying sludge compared with no treatment at all on a devastated site such as a borrow pit. With an application rate of 34 Mg/ha there is an abundance of nutrients, and therefore growth is limited only by species characteristics, climate, soil, weed competition, and other environmental factors.

Two facts are central in an economic analysis: (1) On these devastated sites, stands will probably never reach merchantable size without the addition of fertilizer or sludge. (2) Trees on plots that received 17 to 34 Mg/ha of sludge have grown as well or better than trees on the best nearby undisturbed sites. The indications, therefore, are that sludge-amended sites will be as productive as the best adjacent sites, whereas without the addition of sludge or fertilizer the yield will be extremely low or nothing.

The studies summarized in this paper show the value of dried sewage sludge for reclaiming devastated sites in the Southeast. The sludges used, though low in total N and exchangeable P and K, were excellent soil amendments at the rates tested, and applications of 34 Mg/ha (15 tons/acre) gave excellent results. Combining subsoiling with sludge application is encouraged in order to provide for quick, deep root penetration to obtain better moisture relations and anchoring of trees.

ACKNOWLEDGMENT

This report is based on studies carried out in cooperation with the Cities Service Company, Copper Hill, Tennessee, Yara Engineering Corporation, Sandersville, Georgia, and the U.S. Department of Energy (Contract DOE-A109-76-SRO-870), Aiken, South Carolina.

REFERENCES

Berry, C. R. 1977. Initial response of pine seedlings and weeds to dried sewage sludge in rehabilitation of an eroded forest site. USDA For. Serv. Res. Note SE-249.

——. 1979. Slit application of fertilizer tablets and sewage sludge improve growth of loblolly pine seedlings in the Tennessee Copper Basin. Reclam. Rev. 2:33–38.

——. 1982. Dried sewage sludge improves growth of pines in the Tennessee Copper Basin. Reclam. Reveg. Res. 1:195–201.

——. 1985. Subsoiling and sewage sludge aid loblolly pine establishment on adverse sites. Reclam. Reveg. Res. 3:301–311.

Berry, C. R., and D. H. Marx. 1977. Growth of loblolly pine seedlings in strip-mined kaolin spoil as influenced by sewage sludge. J. Environ. Qual. 379–381.

——. 1980. Significance of various soil amendments to borrow pit reclamation with loblolly pine and fescue after 3 years. Reclam. Rev. 3:87–94.

Cole, D. W. 1982. Response of forest ecosystems to sludge and wastewater applications: A case study in western Washington. p. 274–291. In W. E. Sopper, E. M. Seaker, and R. K. Bastian (eds.) Land reclamation and biomass production with municipal wastewater and sludge. Pennsylvania State University Press, University Park.

Hatchell, G. E., C. R. Berry, and H. D. Muse. 1985. Nondestructive indices related to aboveground biomass of young loblolly and sand pines on ectomycorrhizal and fertilizer plots. For. Sci. 31:417–425.

Hinkle, K. R. 1982. Use of municipal sludge in the reclamation of abandoned pyrite mines in Virginia. p. 421–432. In W. E. Sopper, E. M. Seaker, and R. K. Bastian (eds.) Land reclamation and biomass production with municipal wastewater and sludge. Pennsylvania State University Press, University Park.

Kerr, S. N., W. E. Sopper, and B. R. Edgerton. 1979. Reclaiming anthracite refuse banks with heat-dried sewage sludge. p. 333–351. In W. E.

Sopper and S. N. Kerr (eds.) Utilization of municipal sewage effluent and sludge on forest and disturbed land. Pennsylvania State University Press, University Park.

Kormanik, Paul P., and Richard C. Schultz. 1985. Significance of sewage sludge amendments to borrow pit reclamation with sweetgum and fescue. USDA For. Serv. Res. Note SE-329.

Lejcher, T. R., and S. H. Kunkle. 1973. Restoration of acid spoil banks with treated sewage sludge. p. 184–199. In W. E. Sopper and L. T. Kardos (eds.) Recycling treated municipal wastewater and sludge through forest and cropland. Pennsylvania State University Press, University Park.

May, J. T. 1977. Highlights of a decade of research and reclamation on kaolin clay strip mining spoil, 1966–1976. Georgia For. Res. Council Report 37.

Metz, L. J., C. G. Wells, and P. P. Kormanik. 1970. Comparing the forest floor and surface soil beneath four pine species in the Virginia Piedmont. USDA For. Serv. Res. Paper SE-55.

Roth, P. L., B. D. Jayko, and G. T. Weaver. 1979. Initial survival and performance of woody plant species on sludge-treated spoils of the Palzo mine. p. 389–394. In W. E. Sopper and S. N. Kerr (eds.) Utilization of municipal sewage effluent and sludge on forest and disturbed land. Pennsylvania State University Press, University Park.

Ruehle, J. L. 1980. Growth of containerized loblolly pine with specific ectomycorrhizae after two years on an amended borrow pit. Reclam. Rev. 3:95–101.

Sopper, W. E., and E. M. Seaker. 1983. A guide for revegetation of mined land in eastern United States using municipal sludge. School of Forest Resources and Institute for Research on Land and Water Resources, Pennsylvania State University, University Park. 93 p.

Sutton, P., and J. P. Vimmerstedt. 1973. Treat strip-mine spoils with sewage sludge. Ohio Report 58:121–123.

Wells, C. G., D. M. Crutchfield, M. M. Berenyi, and D. B. Davey. 1973. Soil and foliar guidelines for phosphorus fertilization of loblolly pine. USDA For. Serv. Res. Paper SE-110.

Wells, C. G., K. W. McLeod, C. E. Murphy, J. R. Jensen, J. C. Corey, W. H. McKee, and E. J. Christensen. 1984. Response of loblolly pine plantations to two sources of sewage sludge. p. 85–94 *In* 1984 TAPPI Research and Development Conference, Appleton, Wisconsin. Technical Association of the Pulp and Paper Industry, Technology Park, Atlanta, Georgia.

Chapter 47

Land Treatment of Sludge in a Tropical Forest in Puerto Rico

WADE L. NUTTER and ROBERTO TORRES

ABSTRACT   The Barceloneta Wastewater Treatment Plant was one of the first regional plants in Puerto Rico. In operation since 1981, it serves two communities and about twenty-five industries, including a number of fermentation type operations. The current daily flow averages 0.26 m³/sec (6 mgd) and is discharged by ocean outfall. The sludge production is high, approximately 16 dry Mg/day (18 dry tons) and is high in essential plant nutrients. After lengthy study, land treatment of the sludge was determined to be the most viable alternative. An ancient marine slough, the Cano Tiburones, adjacent to the treatment plant was identified as a potential land treatment site. The Cano has been under drainage and agricultural management (primarily sugar cane) since the early part of this century. The Cano is a groundwater discharge zone, and drainage has enhanced this discharge as well as saltwater intrusion. A study of the Cano and the sludge characteristics indicated that sludge injection and a forest management program could provide treatment of the sludge and ensure protection of groundwater quality. A prototype study has been established, and Caribbean pine and several tropical hardwoods have been planted to determine biomass production and suitability for application of sludge. Management of the 800 hectare land treatment system presents unique forest, water resource, and waste management problems that are being addressed through application of land treatment technology and operation of a 30 hectare prototype system.

The Barceloneta Wastewater Treatment Plant began operation in August 1981 as a secondary treatment facility serving the industries and municipalities of Barceloneta and Manati. The plant was constructed under a unique arrangement between industries and the Puerto Rico Aqueduct and Sewer Authority (PRASA). The secondary treatment facilities were specially designed to treat industrial wastewaters discharged by the signatory industries of the Facility Agreement. Under the conditions of the agreement, PRASA operates the plant with assistance from an Advisory Council composed of the signatory industries. The joint industrial-municipal secondary treatment facility was designed to eliminate the need for industrial pretreatment by the individual signatory industries.

The plant is located on a bluff overlooking the Atlantic Ocean to the north and the Cano Tiburones and the municipality of Barceloneta to the south. The unit processes include grit removal, primary clarification and activated sludge for treatment of the wastewater and flotation thickening, and aerobic digestion and belt-filter pressing for treatment of the sludge. The plant has a design capacity of 0.36 m³/sec (8.3 million gallons per day, mgd). At present, flow to the plant averages about 0.26 m³/sec (6 mgd). The secondary treated wastewater is discharged to the ocean in compliance with the regulations of the Puerto Rico Environmental Quality Board and the U.S. Environmental

Protection Agency. The sludge production of approximately 16 dry Mg/day (18 dry tons) was placed in a landfill adjacent to the plant for over four years.

Alternative sludge disposal and treatment methods considered were incineration, ocean dumping, ocean discharge, expansion of the landfill at a new site, and land treatment. Except for land treatment, the other alternatives pose legal or environmental problems. Incineration would create a major air emission source and problems of disposal of the concentrated ash. Ocean based alternatives—dumping or discharge—posed environmental and legal constraints. Landfilling provided a short-term alternative by concentrating the sludge at the expense of dedicating land in perpetuity to a single use. Land treatment is, on the other hand, not a disposal method but rather a treatment method that utilizes the sludge as a resource for growing crops or trees. Not only does land treatment offer a low environmental risk, the sludge is treated and land productivity improved. Based on analysis of the Barceloneta sludge, over 1,100 kg of nitrogen and 65 kg of phosphorus are produced each day. Equally important to plant nutrition are the other essential plant nutrients also present. Thus the sludge is a fertilizer resource that is available for converting present idle land into highly productive land. A more intensive land use will also create opportunities for increased employment, an important consideration in local economic development.

LAND TREATMENT OVERVIEW

Land application of municipal and industrial sludges has been practiced worldwide for centuries. The U.S. EPA (1983) design manual for land application of municipal sludge indicates that as much as 40% of the 5,770,000 dry Mg (6,350,000 dry tons) of municipal sludge produced in the United States in 1982 was applied to the land for utilization by food-chain and nonfood-chain crops. Sludge production is increasing each year as more wastewater treatment plants are coming on line as a result of the Federal Water Pollution Control Act Amendments of 1972 (Public Law 92-500). Traditional methods of sludge disposal such as landfills and ocean dumping are no longer environmentally acceptable in many locations, or present landfills cannot accommodate the massive quantities produced.

Sludge contains many of the essential nutrients required by plants as well as other constituents such as heavy metals and organic compounds. Since much of the organic material in land-applied sludge decomposes slowly over a period of one to several years, the nutrients and other constituents are released slowly to plant roots and soil and, as a result, are utilized by the plant, transformed by biological or chemical action, adsorbed by the soil, or leached to groundwater. To determine the environmentally safe levels of sludge that can be applied to the land, a detailed assessment of the site, sludge, and management operations was undertaken and design criteria developed. The assessment drew upon existing and site specific information in the fields of soil science, geology, agronomy, forestry, engineering, and groundwater and surface water hydrology.

The basic precept that constrains the design and operation of a sludge land treatment system is generally known as the nondegradation constraint (Overcash and Pal 1979). This constraint in effect states that sludge constituents may be applied to the land in quantities no greater than that assimilated (i.e., treated) by the plant and soil system, and that any accumulation of sludge constituents in the soil cannot exceed accepted safe levels. In other words, sludge may be applied to the land at a loading that is not harmful to the environment, including plants, animals, and humans; nor is there restriction on any future intended use of the site. This constraint differs from the one that applies to a landfill: the site must be dedicated in perpetuity to a single use. Little treatment of the waste occurs while it is in the landfill.

Most sludge land treatment systems in the United States to date (there are no municipal sludge land treatment systems in Puerto Rico) have been agronomic systems producing crops such as small grains and forage grass. Forage grass has been the most popular system because it is easy to manage over a wide variety of climatic and soil conditions. Crops grown on sludge-amended soils are harvested and processed for animal feed or human consumption. In recent years the application of sludge to forest lands has become more common (Sopper and Kerr 1979). Forest soils often have low natural fertility, and dramatic gains in tree growth have been experienced. Sludge has also proved to be an important soil building material and source of nutrients in disturbed land reclamation programs.

LAND TREATMENT OF BARCELONETA SLUDGE

Industrial input to the Barceloneta plant represents approximately 75% of the total flow and accounts for the high production of sludge compared with the volume of wastewater. The bulk of the industrial input is fermentation wastes from pharmaceutical plants. The sludge is therefore atypical of municipal sludge: nitrogen concentrations are two to three times greater and the metal concentrations such as cadmium, zinc, lead, and chromium are lower than the average concentration expected in most municipal sludges. Concentrations of organics are within the range reported for municipal sludges (U.S. EPA 1979 and Sommers 1977). The characterization of a 1:1 ratio mixture of the primary and secondary sludge is presented in Table 1.

Many studies (as reviewed by U.S. EPA 1983) have shown that when sludge is stabilized by a digestion process, as is the case at the Barceloneta plant, bacterial and viral pathogens are at environmentally acceptable levels for land treatment, and no threat to public health exists when the systems are properly managed.

A research study undertaken by the Engineering Research Center, Mayaguez Campus, University of Puerto Rico (de Ramirez 1982) determined the suitability of land treating the Barceloneta sludge. Four crops were planted in the test soil pots: rice, pepper, stargrass, and *Dracaena*. All crops showed increases in growth and were suitable to be consumed by animals or humans. Sludge was applied to soils common to the Barceloneta area at loadings equivalent to and twice the recommended agronomic nitrogen rates. These rates resulted in loadings between 227 and 1,360 kg/ha nitrogen, depending on the crop. A high volume of water was added to the waste and soil mixtures in the pots to simulate extreme leaching conditions. Under these conditions, metals in the sludge showed little migration in the soil and did not pose a threat to groundwater. The study concluded that the stabilized sludge is suitable for land treatment and that adverse environmental impacts are not likely to occur.

The area available for land treatment of the Barceloneta sludge is the former sugar cane production area of the Cano Tiburones. The site is an ancient marine slough that filled with organic swamp deposits. The swamp deposits, and later mineral alluvial deposits in parts of the area, are as much as 33 to 43 m thick and overlie the Aymamon limestone. The area is bounded to the south by limestone hills and to the north by a

TABLE 1. Characterization of the Barceloneta sludge (dry basis). The means represent the results of seven analyses over a fifteen-month period.

Constituent	Range	Mean
	(ppm)	
Cadmium	<3.8 - <19.4	<10.2
Chromium	112 - 172	142
Copper	162 - 442	264
Lead	36 - <78	<50
Mercury	<0.04 - 2.29	<1.14
Nickel	29 - <44	<37
Zinc	357 - 618	481
Nitrate-N	<20 - 4,887	<1,661
Boron	2 - <53	<37
Arsenic	0.8 - <6.5	<2.7
Total phenol	28 - <49	<39
Total cyanide	<75 - 457	<320
Methylene chloride	<0.2 - 150	<55
Chloroform	<0.2 - <1.3	<0.6
Toluene	<0.2 - 13	<4.5
Di-n-butyl phthalate	<0.4 - <11	<4.4
Bis(2-ethylhexyl)phthalate	<0.4 - 2	<8.1
	(percent)	
Total Kjeldahl-N	1.15 - 11.6	6.3
Ammonia-N	0.9 - 1.54	1.29
Phosphorus	<0.002 - 1.06	<0.36
TOC	17.2 - 43.7	26.3
Chloride	1.52 - 3.36	<2.29
Sulfate	1.13 - 3.76	2.25
Calcium	2.93 - 4.93	3.70
Magnesium	0.16 - 1.46	0.65
Potassium	0.22 - 0.57	0.41
Sodium	2.84 - 3.68	3.18
COD	33 - 131	84
Total solids	2.34 - 2.66	2.52
pH	6.8 - 7.9	7.4

The sludge analyzed is a 1:1 mixture of primary and secondary sludge.

narrow strip of cemented sand dunes and occasional limestone outcrops along the Atlantic Ocean shore. The eastern boundary is the Manati River, which over time has created a natural levee of mineral soil with elevations of 1 to 3 meters above mean sea level. Approximately 4 kilometers to the west the elevations are for the most part at or below sea level and the soil is predominantly organic. Diversion canals have been constructed to intercept surface and subsurface discharge to the area. The eastern portion of the area (adjacent to the Manati River) is drained by a series of canals that discharge to the ocean through a tidal gate. All other canals for the most part are directed to the west and are discharged to the ocean by pumps.

The U.S. Geological Survey (Zack and Class-Cacho 1984) reported that the Cano Tiburones area is a groundwater discharge region. In other words, groundwater from the limestone region to the south flows to and upwells in the Cano. In addition, there is seawater intrusion and mixing with the groundwater along the northern boundary. Since the area is a groundwater discharge rather than a recharge area, there is no potential for polluting deep groundwater supply aquifers if by chance the sludge land treatment system suffered an upset.

The total land area of the potential land treatment site is some 800 hectares; this area does not include organic soils. The area has been mapped by the USDA Soil Conservation Service, and the principal soil series are the Toa (Fluventic Hapludoll), Bajura (Vertic Tropaquept), and Coloso (Aeric Tropic Fluvaquent). The soil textures range from clay to silty clay and the Toa is well drained, the Coloso somewhat poorly drained, and the Bajura poorly drained. The drainage canals result in sufficient drainage to make the soils suitable for production of crops that do not require frequent tillage. Background soil chemical analyses of the Coloso series in the prototype study area are presented in Table 2. It is interesting to note the background levels of the organic compounds in the soil. These levels can, in part, be attributed to residues from natural soil decomposition processes; however, others cannot at this time be reconciled.

The site is subject to periodic flooding by overflow of the Manati River. The frequency of flooding is, on the average, greater than once every several years, and the period of inundation does not last more than two to three days. This type of flooding does not hinder current cattle grazing and forage crop farming practices on soils adjacent to the Manati River.

Vegetation management is a critical factor in the successful operation of a sludge land treatment system. Since nitrogen is the major nutrient in the sludge, crops must be managed to utilize and remove through harvest a substantial portion of the nitrogen applied. Forage (hay) crops and forests are the two principal crops that have a potential market in Puerto Rico. Sludge application to hay has resulted in a 100 to 200% increase in dry matter production, and to trees a 200 to 300% increase in diameter growth (U.S. EPA 1983).

TABLE 2. Mean concentrations of selected parameters in the A and B horizons at the prototype study area.

Parameter	A Horizon		B Horizon	
	Mean	SE	Mean	SE
pH	6.5	0.16	6.7	0.11
Nitrate-N (mg/kg)	<1.17	0.39	<1.81	0.67
Ammonia-N (mg/kg)	4.5	1.12	7.3	2.46
Total Kjeldahl-N (mg/kg)	41.1	2.43	31.6	2.35
Sulfate (mg/kg)	379	53.6	558	95.1
TOC (mg/kg)	299	10.7	287	16.2
Organic matter (%)	5.2	0.52	6.3	0.46
Total cyanide (mg/kg)	<0.10	--	<0.10	--
Amenable cyanide (mg/kg)	<0.10	--	<0.10	--
Total phosphorus (mg/kg)	105	3.8	76.9	10.08
Sodium (mg/kg)	8.2	1.24	10.5	0.99
Cadmium (mg/kg)	<0.10	--	<0.10	--
Zinc (mg/kg)	0.61	0.01	0.62	0.01
Nickel (mg/kg)	<0.20	0.03	0.20	0.04
Lead (mg/kg)	<2.0	--	<2.0	--
Copper (mg/kg)	0.78	0.02	0.85	0.03
Calcium (mg/kg)	1.9	0.09	2.0	0.09
Magnesium (mg/kg)	675	16.5	697	24.7
Potassium (mg/kg)	18.2	1.40	17.4	1.54
Toluene (ppb)	27.0	6.79	44.0	24.89
Methylene chloride (ppb)	21.5	6.44	26.5	11.98
Methyl chloride (ppb)	<10.0	--	<10.5	0.47
Chloroform (ppb)	<5.0	--	<5.5	0.47
Di-n-butyl phthalate (ppb)	<4,825	3,286	<4,873	3,712
Bis(2-ethylhexyl)phthalate (ppb)	<81.9	67.20	<195	86.7

Local tree nurseries can produce only enough seedlings (in addition to other commitments) to plant 40 to 80 hectares per year. Thus a forage crop will be established initially on most of the area, and it will be replaced with an expansion of the forest plantings each year. The forage crop will supplement the local demand for cattle feed, and the trees, when harvested, will serve to meet a critical islandwide lumber shortage.

PROJECT IMPLEMENTATION

The land treatment system will be managed by a Puerto Rico agency, the Solid Waste Management Authority (SWMA), with funding from the industry Advisory Council. SWMA will supervise all implementation and supervisory activities of land treatment of the Barceloneta sludge.

The implementation plan calls for the establishment of a prototype land treatment system on approximately 33 hectares to be used to test crop and forest establishment procedures, sludge injection techniques under different climatic and soil conditions, and other operational practices. In addition, and perhaps most important, the prototype site will be intensively monitored to determine interactions between the sludge and the soil, plants, and groundwater. Although there will be a monitoring program to cover the complete system when implemented, the prototype area will continue to be intensively monitored to provide a continuous data source. Sludge will not be applied to the soil during excessively wet periods. Therefore, a storage facility of approximately seven to ten days will be available at the plant site. The storage capability will also permit shutdown for equipment repair and maintenance.

The sludge will be injected to a depth of 20 to 25 cm using a commercially available sludge injector. The sludge injector will transport 17,000 liters per injection trip on flotation tires to minimize soil compaction. Sludge will be injected once or several times per year over the entire area depending on the type of crop and frequency of harvest. For instance, sludge will be injected three times each year in the hay crop area to coincide with harvest and only once each year in the forest. The injector will be equipped with a roller mechanism to close the injection slit. Although the sludge is stabilized by the digestion process and has little or no odor, injection in the soil eliminates any possibility of offensive odor production at the site and the potential for washoff of sludge if the area is flooded.

Nitrogen controls the amount of sludge that may be applied in any one year. The metal loading on an annual basis is considerably lower than U.S. EPA recommendations. The tree species planted as part of the prototype study are bigleaf mahogany, hybrid (St. Croix) mahogany, mahoe, *Eucalyptus robusta*, teak, kenaf, and Caribbean pine. The forage crop planted is Pangola grass. The annual loading to the Pangola grass is 8.2 dry Mg/yr (9.0 dry tons) and to the forest 3.86 dry Mg/yr (4.25 dry tons). This is equivalent to an annual loading of 585 and 275 kg N/ha per year, respectively.

The total land area required is dependent on the mix of hay and forest production. It is expected that the tree harvest age for the Caribbean pine will be 12 to 14 years. Therefore, the initial plantings must be for hay, with each year a reduction in hay production as trees are planted. It is anticipated that the final crop mix will be 70% forest and 30% hay. The land required for this mix is approximately 600 hectares. Considering unusable areas such as canals, buffer strips, and roads, approximately 800 hectares are required to treat all the sludge produced by the Barceloneta plant.

REFERENCES

de Ramirez, L. M. 1982. Agricultural utilization of sludge from the Barceloneta Waste Treatment Plant. Research report. Engineering Research Center, School of Engineering, University of Puerto Rico, Mayaquez. 58 p.

Overcash, M. R., and D. Pal. 1979. Design of land treatment systems for industrial wastes: Theory and practice. Ann Arbor Science Publishers, Ann Arbor, Michigan. 684 p.

Sommers, L. E. 1977. Chemical composition of sewage sludges and analysis of their potential use as fertilizers. J. Environ. Qual. 6:225–232.

Sopper, W. E., and S. N. Kerr (eds.) 1979. Utilization of municipal sewage effluent and sludge on forest and disturbed land. Pennsylvania State University Press, University Park. 537 p.

U.S. Environmental Protection Agency. 1979. Fate of priority pollutants in publicly owned treatment works: Pilot study. EPA 440/1-79-300.

————. 1983. Process design manual: Land application of municipal sludge. EPA 625/1-83-016.

Zack, Allen L., and Angel Class-Cacho. 1984. Restoration of freshwater in the Cano Tiburones area, Puerto Rico. U.S. Geological Survey Water Resources Investigations Report 83-4071. 33 p.

Chapter 48

The Sludge Application Program at the Savannah River Plant

J. C. COREY, M. W. LOWER, and C. E. DAVIS

ABSTRACT   Since 1980 a research program has been conducted at the Savannah River Plant to evaluate the use of domestic sewage sludge to enhance forest productivity. The objective of the program has been to determine the environmental impact of using sewage sludge as a soil conditioner and slow-release fertilizer. The potential impacts of sludge application on nutrient cycling, organic carbon budgets, and biomass production have been studied. Soil, soil water, groundwater, and stand biomass samples have been analyzed to monitor the availability and movement of nutrients and metals.

Studies have been under way at the Savannah River Plant (SRP) in southwestern South Carolina since 1980 to evaluate the environmental effects of applying sewage sludge for forest productivity. The Savannah River Plant, encompassing 77,701 hectares (192,000 acres), is a Department of Energy installation in southwestern South Carolina which produces fissionable nuclear materials for national defense weapons programs. Currently some 70,882 hectares of SRP lands are classified as suitable for growing pine or hardwood trees. Over 107 million pine trees, largely longleaf and loblolly pine, have been planted at the site since 1952. At present, about 500 million board feet of sawtimber and one million cords of pulpwood exist on site, with an inventory value exceeding $53 million.

Forests at the SRP are ideally suited for evaluating sewage sludge management: (1) there is a diversity of soils representative of the coastal plain area; (2) there are extensive, protected forests on the 77,701 hectare site, with limited public access; and (3) the forests are primarily managed for sustained yield timber production, which provides a variety of stand age classes and conditions ideal for forest research.

The implementation of environmentally sound waste management technologies for disposal or use of sewage sludges was mandated under the Federal Water Pollution Control Act Amendments of 1972 and the Clean Water Act Amendment of 1977. As a consequence of improved wastewater treatment, the quantities of sewage sludge generated and requiring disposal have increased. The most common means of disposal are through use of landfills and lagoons, by incineration and dumping in the ocean, and by application to land.

THE PROGRAM

The forestry management expertise provided by the U.S. Forest Service (USFS) at SRP, the Southeastern Forest Experimental Station (SEFES) of Charleston, South Carolina and Research Triangle Park, North Carolina, the Savannah River Ecology Laboratory (SREL), and the Savannah River Laboratory (SRL) organizations was beneficial to programmatic design and implementation.

The primary organizational roles of the integrated study effort under way since 1980 have been as follows: SRL, research coordination and administration, site preparation, plot sampling, groundwater monitoring, and biomass estimation and assessment correlation; USFS, oversight and consultation, pine forest management, and site preparation; SEFES, experimental design, sample collection and analysis, and interpretation of results; and SREL, nutrient cycling (stemflow, throughfall, litterfall, understory) and wildlife habitat.

Sludge Characteristics

The sludge used in the forest productivity research came from two off-site sources, the Horse Creek Pollution Control Facility in North Augusta, South Carolina, and the Augusta Wastewater Treatment Plant in Augusta, Georgia. Because the two plants treat wastewater differently, the resultant sludge provided a range in nutrient and heavy metal composition. A compilation of average concentrations and application rates for the Horse Creek and Augusta sludges is given in Table 1. Analyses of the Horse Creek sludge indicated lower nitrogen levels and higher heavy metal concentrations compared with the Augusta sludge (Table 1). These differences were attributed to different treatment processes and influent compositions at the two facilities. Both sludges had pH ranges of 7.3 to 7.4. Prior to land application, analyses of both sludges were performed on a routine basis for nutrients and metals.

The Horse Creek Pollution Control Facility receives 80% of its wastewater from industrial sources, primarily textile mills. Sewage is aerobically digested, and solids are heated for 0.3 hour at 375°C under 250 psi. The heat-treated sludge is then dewatered and contains 30% solids. These processes condition and stabilize the sludge and destroy pathogenic organisms without extensive chemical use. The Augusta Wastewater Treatment Plant receives 80% of its influent wastewater from domestic sources. Sewage treatment is by both conventional aerobic and anaerobic processes, producing a "liquid" sludge with a 2 to 3% solid suspension.

The Horse Creek sludge was broadcast on pine forest experimental sites with a manure spreader at a rate equivalent to 632 kg/ha. The Augusta sludge was sprayed on the forest floor with a pressurized tanker. Application rates for the Augusta sludge were equivalent to 402 and 804 kg/ha. Control plots were established at all forestry research sites. Control plots received no sludge amendments so that biomass production and environmental impacts could be compared with those on experimental sites receiving sludge amendments.

Objectives

The principal objective of the forest productivity program has been to evaluate optimum methods and alternative techniques for the land disposal of sewage sludge, with the goal of increasing biomass production of the pine forests without environmental degradation. This objective was accomplished by examining (1) the growth and survival responses of pine trees at various ages following sludge application at different rates on both sandy and sandy clay soils, (2) the necessary management decisions to minimize the potential for adverse environmental and health impacts following application of

TABLE 1. Average concentrations and application rates for Augusta and Horse Creek sludge.

Sludge Property	Average Concentrations				Application Rates		
	Augusta	Horse Creek	Augusta	Horse Creek	Augusta Low	Augusta High	Horse Creek Low
pH	7.32	7.37					
	(g/100 g)				(kilograms/hectare)		
Oven-dry	2.48	55.72			5,555	11,100	49,925
	(g/100 g, oven-dry)						
Ash	46.02	36.11			2,556	5,113	18,027
	(ug/g, wet)		(ug/g, oven-dry)				
Kjeldahl-N	1,795	7,066	72,374	12,676	402	804	632
Ammonium-N	753	134	30,361	240	169	338	12
Nitrate-N	3.7	0	149	0	0.82	1.64	0
Phosphorus	402	4,206	16,209	7,546	90	180	377
Potassium	66	170	2,661	305	15	30	15
Calcium	361	11,649	14,556	20,898	81	162	1,043
Magnesium	61	1,943	2,460	3,486	14	28	174
Sodium	519	1,638	20,926	2,939	116	232	147
Manganese	3.4	167	137	300	0.76	1.52	15
Zinc	33	1,203	1,330	2,158	7.39	14.78	108
Copper	7.9	696.0	318	1,249	0.77	1.54	62
Lead	5.9	98.0	238	176	1.32	2.64	8.79
Nickel	1.1	23.9	44	43	0.24	0.48	2.15
Cadmium	1.1	3.3	44	6	0.24	0.48	0.29
Sulfur	148	2,031	5,967	3,644	33	66	181
Iron	2,525	23,662	10,808	42,449	565	1,130	2,119
Boron	6.3	149.5	254	268	1.41	2.82	13
Chromium*	4.3	772.0	173	1,386	0.96	1.92	69
Antimony*	0.2	0.9	8	2	0.04	0.08	0.09
Selenium*	2.0	42.3	81	76	0.45	0.90	3.79
Arsenic*	2.6	19.5	105	35	0.58	1.16	1.74
Tin*	2.5	22.3	101	40	0.56	1.12	2.00
Cobalt*	0.2	4.6	8	8	0.04	0.08	0.39
Mercury*	0.1	17.2	4	31	0.02	0.04	1.55

*Analyzed by inductive coupled argon plasma spectrophotometer.

sewage sludge as a soil conditioner and slow-release fertilizer, (3) the cycling and fate of nutrients and heavy metals in sewage sludge applied to various soil systems, and resultant impacts on biomass and understory associated with loblolly pine stands, and (4) the responses of pine plantations amended with sewage sludge to inorganic fertilizers and/ or herbicides.

During 1980-81, 0.2 ha experimental plots were installed covering 18 ha (45 acres) of four different age loblolly pine stands on two soil types, and the physical and chemical characteristics of the trees and soils were determined. Following a single sludge application at these sites, the monitoring program, continuing today, was initiated to evaluate environment impacts and modifications of nutrient and elemental cycling on the treated plots.

Seven experimental loblolly pine sites were selected at five locations at SRP, in plantations planted in 1953, 1973, 1978, and 1981. The ages of the stands were selected on the basis of stages of growth that represented times of near maximum stemwood formation, crown closure, maximum hardwood competition, and seedling establishment. The 1953, 1973, and 1978 plantings amended with sewage sludge are in areas having two different soil classes—a sandy soil and a heavier sandy clay.

Lucy Site

The Lucy site is a loblolly pine forest in northwestern SRP typified by thick loamy sands known as the Fuquay series. The trees were planted in 1953. This experimental site received a single sludge amendment in 1981 of both the solid (Horse Creek) and liquid (Augusta) sludges. This site was selected in order to evaluate the effects of sludge amendments on low fertility sandy soils and to determine whether the organic matter in the sludges would improve soil moisture holding properties, thus enhancing biomass production by the retention of nutrients.

For practical purposes, the Lucy sand is part of the Fuquay series, with 46 to 76 cm of sand over sandy clay subsoils. The soil is considered highly permeable, acid, and infertile by agronomic standards. The experimental research design consisted of randomized gross plots 0.2 ha in size, with rows 2.4 m (8 ft) apart. Plots measured about 44 by 46 m. Liquid sludge was applied in 1981 at rates of 0, 233,000, and 466,000 liters per hectare (equivalent to 0, 402, and 804 kg N/ha). Solid sludge was applied at rates of 0 and 89,668 wet kg/ha (equivalent to 0 and 632 kg N/ha).

Stands at the Lucy site had been periodically thinned by the U.S. Forest Service at SRP. Understory vegetation at the time of sludge application was minimal. Baseline stand measurements of pine diameter at breast height (dbh) at specific plots within the gross plots were collected in 1981 prior to the sludge application. Soil, soil water, litterfall, throughfall, stemflow, and groundwater samples have been collected and analyzed periodically since the sludge treatment, to monitor the effects on nutrient and heavy metal cycling.

Orangeburg Site

The Orangeburg site is in central SRP and encompasses about 3.2 ha (8 acres) of loblolly pine stands planted in 1953. Sludge applications at this site were performed in 1981 similar to the regime performed at the Lucy site; soils at the Orangeburg site, however, were a heavier sandy clay.

The study at the Orangeburg site entailed the design of three randomized blocks with three replications at the site. Gross plots measured 0.2 ha. Liquid and solid sludges from Augusta and Horse Creek, respectively, were applied at rates equivalent to 0, 402, and 804 kg N/ha for the liquid and 0 and 632 kg N/ha for the solid material.

Prior to sludge application, tree identification and dbh measurements were performed for baseline purposes. Incidence of fusiform rust disease was documented. Following the sludge application, soil, soil water, litterfall, and groundwater samples were collected periodically to ascertain environmental effects. Biomass production measurements, stem measurements, and stem analyses were made in 1984.

Road F Site

The Road F site is a loblolly pine plantation in northwestern SRP. Loblolly pines, planted in 1973, were grown on both a sandy soil and heavier sandy clay. Soils at the Road F site received a sewage sludge amendment in 1980 to determine the cycling and assimilation of nutrients and heavy metals.

Sewage sludge from the Augusta Wastewater Treatment Plant was sprayed in a liquid form at three levels equal to 0, 402, and 804 kg N/ha at the Road F pine forest. The treatment design involved three replications of the three treatments imposed in a block design on the Wagram-Lucy (sand) and Fuquay-Dothan (sandy clay) soils.

Tree spacing at this site was about 1.8 by 3 m. Prior to sludge application, the stand had an average height of 5.61 m and dbh of 9.1 cm. Control of competition, both hardwood and herbaceous species, was not of concern, since pines at the Road F site represent crown closure and have overtopped the hardwoods.

Gross plots were designed for an area 46 by 46 m (0.2 ha). Each gross plot contained 15 rows about 46 m long. Measurement plots were established within gross plots of 34 by 34 m or 0.11 ha. Soil water, groundwater, litterfall, and biomass samples were periodically collected and analyzed.

Lower Kato Road Site

The Lower Kato Road site is a 20 ha plantation of loblolly pines planted in 1978 on two soils, a sandy (>90%) soil (Lucy) and a heavier sandy clay (10 to 20% clay in surface horizons, Orangeburg). Sludge amendments were imposed on three-year-old loblolly pine stands in 1981. Liquid and solid sludges were applied at rates equivalent to 0, 402, and 804 kg N/ha for the liquid sludge and 0 and 632 kg N/ha for the solid sludge.

The experimental design consisted of a randomized block design with three replications installed on the two soils. Competing vegetation of a variety of hardwood and herbaceous species existed at this site. Split-plot treatment applications of granular Velpar® was applied to reduce hardwood competition. Plot sizes at the Lower Kato Road site were identical to the 0.2 ha plots established at the other research sites. Gross plots at this site contained 15 rows 46 m long.

Kato Road Site

The response of loblolly pine seedlings at establishment to two sources of sewage sludge at three levels applied with and without disking in combination with tip moth control and herbicides to control weed competition was evaluated at the Kato Road site. This establishment site is a 14 ha plantation that was treated in February and March 1981 with sewage sludge from both the Augusta Wastewater Treatment Plant and the Horse Creek Pollution Control Facility. Site soils were characteristic of the Orangeburg series throughout the site.

The establishment site encompassed a complex experimental design. Three replications of twelve combinations of fertilizer sources were designed, along with varying disking and tip moth control treatments (Furadan®). Imposed on the twelve treatments were three herbicide treatments consisting of no herbicide application, Velpar® (Du Pont) applied at 0.78 kg/ha, and Goal® (Rohm and Haas) applied at 0.56 kg/ha. Sludge and disking treatments were applied prior to planting, and herbicides the spring of the first growing season.

The Kato Road site had two sources of sludge applied at rates equivalent to 0, 402, 632, and 804 kg N/ha compared with 28 and 55 kg N/ha applied as diammonium phosphate. Following amendments and establishment plantings in 1981, soil, soil water, groundwater, and growth measurements have been made annually to determine the effectiveness of sewage sludge as a fertilizer and soil conditioner for loblolly pines at establishment.

COLLECTION, ANALYSIS, AND INTERPRETATION
OF ENVIRONMENTAL SAMPLES

Prior to and following sludge amendments at the experimental forest research sites, soil, soil water, and groundwater samples were collected and analyzed for all parameters of interest, so that the environmental effects of sludge amendments could be determined. Tree height and/or dbh measurements were made at the time of the sludge amendment and annually thereafter to gauge biomass production and growth response. This section highlights the environmental and growth response monitoring data obtained prior to and following the sludge application through 1984.

Soil Analyses

Release of nutrients in forest soils is the primary uptake pathway of nutrient sources to trees and understory. Chemical characteristics of forest soils govern the rate of mobility and availability of the nutrients. Low levels of available nitrogen in forest soils have been found to limit loblolly pine production in the southeastern United States (Wells 1970).

TABLE 2. Soil properties (0-10 cm) on the 1953 stand with sandy soil, Fuquay series (Lucy site).

Soil Property	No Sludge	Augusta Sludge (402 kg N/ha)	Augusta Sludge (804 kg N/ha)	Horse Creek Sludge (632 kg N/ha)
pH	5.3	5.0	4.8	5.5
		(ppm)		
Total N	343	335	390	330
Available P	35	21	55	32
		(meq/100 g)		
Exchangeable Ca	0.88	0.43	0.33	0.67
Exchangeable Mg	0.14	0.07	0.05	0.15
Exchangeable K	0.04	0.02	0.02	0.02
Exchangeable Na	0.01	0.01	0.01	0.02
		(ppm)		
Total potassium	35	29	52	22
Total carbon	128	83	5,621	138
Total magnesium	74	71	269	60
Total sodium	0	0	97	1
Total lead	1	4	0	1
Total copper	1	1	1	1
Total zinc	2	3	3	2
Total iron	1,444	1,661	1,729	1,401
Total nickel	3	5	9	2
Total cadmium	1	1	1	1
Total chromium	1	2	6	1
Total manganese	124	72	95	75
		(percent)		
Organic matter	1.78	1.48	1.44	1.48
Clay	8	9	8	8

Inorganic nitrogen (ammonium and nitrate) in sludges added to forest soils becomes immediately available for tree and understory uptake. The organic nitrogen is not immediately available and must be converted to inorganic forms via nitrogen mineralization to render it available for uptake. During the first year after sludge treatments, nitrogen mineralization rates range from about 8 to 20%, with reduced rates occurring in subsequent years (SCS 1982).

An extensive soil sampling and analysis program was conducted at the experimental forestry sites preceding and following the sludge application. Soil analyses included nutrients, cations, and metals.

Soil properties were found to be quite variable following the sludge application, reflecting both the nature of the sludge product and the low cation exchange capacity of the soils (Wells et al. 1982). The most extensive and comprehensive soil analyses were conducted at the 1953 stands. Growth data collected through 1984 indicated that the trees were still responding to the one sludge application in 1981.

Physical and chemical characteristics of the sandy and sandy clay soils at the 1953 stands six months after sludge treatment are presented in Tables 2 and 3. General trends, noted by Wells et al. (1982), were as follows: (1) Soil pH decreased as much as 0.5

TABLE 3. Soil properties (0-10 cm) on the 1953 stand with sandy clay soil, Orangeburg series (Orangeburg site).

Soil Property	No Sludge	Augusta Sludge (402 kg N/ha)	Augusta Sludge (804 kg N/ha)	Horse Creek Sludge (632 kg N/ha)
pH	5.3	4.5	4.9	5.3
(ppm)				
Total N	245	255	195	285
Available P	10.0	24	19	13
(meq/100 g)				
Exchangeable Ca	0.63	0.49	0.40	0.70
Exchangeable Mg	0.16	0.12	0.07	0.25
Exchangeable K	0.03	0.03	0.02	0.03
Exchangeable Na	0.05	0.04	0.03	0.02
(ppm)				
Total potassium	53	32	49	58
Total carbon	88	75	63	95
Total magnesium	138	91	109	134
Total sodium	1.5	1	1	2
Total lead	9.0	3	5	12
Total copper	5	2	2	4
Total zinc	6	4	4	6
Total iron	2,023	1,575	1,705	1,995
Total nickel	5	1	1	4
Total cadmium	0	1	1	0
Total chromium	1	1	0	1
Total manganese	87	120	125	215
(percent)				
Organic matter	0.81	0.91	0.77	1.16
Clay	11	8	9	10

units with the Augusta sludge, but the Horse Creek sludge had no effect on soil pH. (2) Available phosphorus levels increased with application of the Augusta sludge. (3) Exchangeable bases (calcium, magnesium, potassium, sodium) in soils were reduced from prior levels, a condition attributed to nitrate production near the surface which leached these exchangeable constituents deeper into the soil. (4) Organic matter in the soils was reduced at the sandy site and increased at the sandy clay site. Retention of organic matter was attributed to the lower permeability of clay materials in soils at the Orangeburg site. In all, degradation of site soils attributable to sludge amendments was clearly not evidenced.

Soil Water Analyses

Soil water samples were collected at the experimental sites following sludge applications, using fritted glass lysimeters, so that quantities of nutrients and cations leaching in the soils could be measured. These lysimeters, installed at 0.5 and 1 m depths, were constructed in such a way that solution contacted only glass and polyethylene tubing (Wells et al. 1984). Two pairs of lysimeters at both depths were installed in each of the 0.2 ha experimental plots at the stands planted in 1953. In the 1973, 1978, and 1981 stands, one lysimeter was installed at each depth, except for the 1973 stand, which had lysimeters in only the heavier clay soil (Wells et al. 1984). Lysimeter samples were collected over a 24-hour period and analyzed off site for nutrients, cations, and heavy metals.

Results of soil water analyses through 1983 indicated that the age of the pine plantation and the nitrogen concentration in the sludge amendments were the major controlling variables for nitrate-N in soil water (Wells et al. 1982). One full year after the sludge application, the order of nitrate concentration in soil water at the 0.5 and 1 m depths was 1-year-old > 27-year-old > 8-year-old = 3-year-old (Wells et al. 1982). Following the first full year of soil water data, nitrate-N levels in soil water for all age classes generally declined. Results of soil water analyses showed the following additional trends.

1. The 1953 plots on which sludge was applied equivalent to 804 kg N/ha indicated overloading of nitrogen to the system (Figures 1 and 2) at both lysimeter depths. Elevated nitrate-N levels at the 0.5 m depth persisted at the Lucy site for a full year following sludge application (Figure 2).

2. High levels of nitrate-N at the 1953 stands within a year after treatment indicated extremely rapid nitrification, which was generally unexpected in soils of pH 5.0 (Wells et al. 1982).

3. Nitrate-N levels in soil water at both depths in the 1953, 1973, and 1978 stands for control plots and plots treated with Horse Creek (632 kg N/ha) and Augusta (402 kg N/ha) sludges were all comparable.

4. Some nitrate leaching was evidenced from lysimeter samples at the 1973 stands one year after treatment (Figure 3).

5. Wells and McKee (1983) documented that no relation was noted at the 1973 stand between the amount of precipitation and soil water nitrate-N concentration.

6. Nitrate-N concentrations in soil water peaked at 27 mg/l at the 0.5 m depth in April 1982 at the 1978 stand applied with the high rate of nitrogen in liquid sludge (Figure 4).

7. More efficient uptake of nitrates by the stands with little or no nutrient leaching to greater root depths was evidenced at the 1978 stand compared with the 1973 stand on similar soils and equivalent sludge treatments.

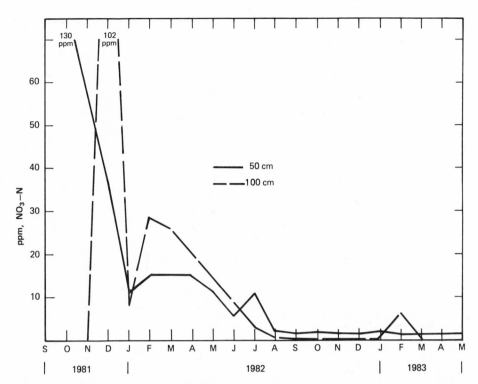

Figure 1. Nitrate-N concentrations in soil water at 0.5 and 1.0 m depths on 1953 stand clay (Orangeburg) site treated with high rate of Augusta sludge (800 lb N/acre) applied July 1981.

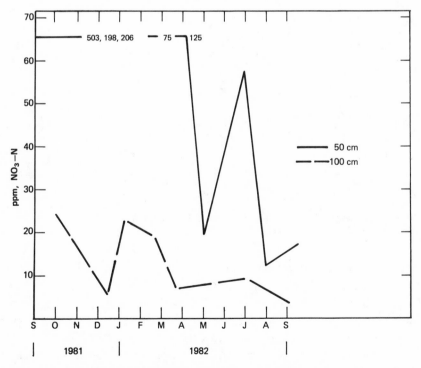

Figure 2. Nitrate-N concentrations in soil water at 0.5 and 1.0 m depths on 1953 stand sandy site (Fuquay series) treated with high rate of Augusta sludge (800 lb N/acre) applied July 1981.

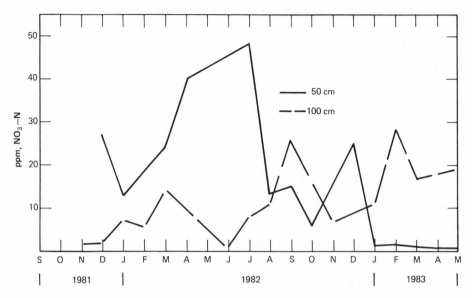

Figure 3. Nitrate-N concentrations in soil water at 0.5 and 1.0 m depths on 1973 stand clay (Road F) site with high rate of Augusta sludge (800 lb N/acre) applied September 1981.

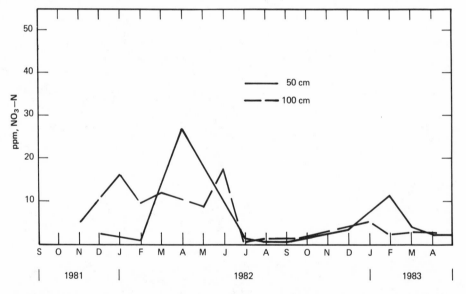

Figure 4. Nitrate-N concentrations in soil water at 0.5 and 1.0 m depths on 1978 stand clay (Lower Kato Road) site with high rate of Augusta sludge (800 lb N/acre) applied November 1981.

8. At the establishment stand, average values of the replicates for nitrate-N on each treatment indicated that the high rate of application of Augusta sludge at both the 0.5 and 1 m depths resulted in highest soil water nitrate-N values (Figures 5 and 6).

9. Solid sludge applied at 632 kg N/ha at the establishment stand produced nitrate-N concentrations only slightly greater than those determined in control plots.

10. Typical ammonium-N concentrations at all plots on all age stands ranged from 0.1 to 0.5 mg/l; losses of ammonium-N from plots on all age stands and sludge treatments were infinitesimal (Wells et al. 1982).

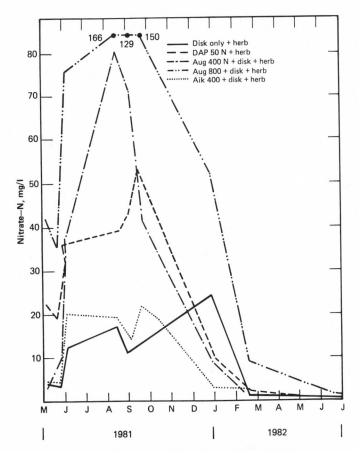

Figure 5. Nitrate-N concentrations in soil water at 0.5 m depths at establishment (Kato Road) site amended with varying fertilizations.

11. Treatment effects on phosphorus concentrations at both soil water depths on all age stands were found to be insignificant; losses of phosphorus were found to be less than 1.1 kg P/ha per year (Wells et al. 1982).

12. Calcium was the major cation in soil water solution followed in order by sodium, potassium, and magnesium; all cation concentrations generally appeared to be directly related to nitrate-N concentrations in soil water.

13. Sludge amendments did not increase heavy metal concentrations in soil water at either the 0.5 or 1 m depth (Wells and McKee 1983).

Groundwater Analyses

The terms and conditions of the Industrial Waste Permit issued to SRP for the land application of sewage sludge by the South Carolina Department of Health and Environmental Control (SCDHEC) included extensive groundwater monitoring requirements. SCDHEC mandated that three groundwater wells be installed at each of the experimental forest site pits receiving sludge amendments. Following well installation, the permit required that groundwater samples be collected quarterly and analyzed for water depth, pH, conductivity, and concentrations of nitrates (as N), total dissolved solids, sodium, and chlorides. Additional well samples were to be collected annually and analyzed for

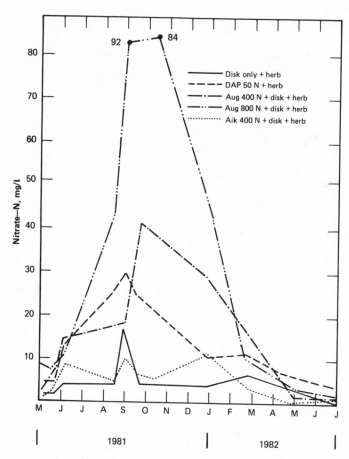

Figure 6. Nitrate-N concentrations in soil water at 1.0 m depths at establishment (Kato Road) site amended with varying fertilizations.

orthophosphates, total Kjeldahl nitrogen, and a variety of cations and heavy metals. A complete list of required groundwater monitoring parameters is given in Table 4.

In the 4.5 years since the inception of this monitoring effort, groundwater samples from each well have been collected and analyzed seventeen times for quarterly parameters and four times for annual parameters (Table 4). Results of all analyses indicate no

TABLE 4. Required groundwater monitoring parameters.

Quarterly Monitoring	Annual Monitoring
Water depth	Orthophosphates
pH	Total Kjeldahl nitrogen
Nitrate-N	Cadmium
Total dissolved solids	Copper
Sodium	Iron
Chlorides	Nickel
Conductivity	Lead
	Calcium
	Magnesium
	Manganese
	Potassium

groundwater degradation as a consequence of the sludge treatments. A compilation of mean values for quarterly parameters at all well locations is given in Table 5, which clearly indicates that mean nitrate-N concentrations are well below the 10 mg/l maximum contaminant level. All other quarterly parameters are within ranges typically associated with other ambient, upgradient shallow groundwaters at SRP. No leaching of heavy metals into local groundwater as a consequence of sludge application has been noted from annual analyses.

TABLE 5. Results of groundwater monitoring at the sludge application sites, 1980-85.

			Mean Value*				
Site	Depth (m)	pH (range)	Nitrate-N (mg/1)	TDS (mg/1)	Sodium (mg/1)	Chloride (mg/1)	Conductivity (µmhos/cm)
Lower Kato Road	20.00	3.70-5.55	0.51	18	1.58	2.56	17(6)
Lower Kato Road	14.36	4.28-5.75	1.10	27	2.14	3.31	23(6)
Lower Kato Road	9.39	4.10-5.67	0.65	18	1.37	2.25	16(6)
Orangeburg	18.71	4.11-5.22	1.56	32	2.19	3.13	26(6)
Orangeburg	14.30	3.90-5.60	0.88	22	1.34	1.76	15(6)
Orangeburg	14.27	3.98-5.28	0.82	19	1.09	1.72	15(6)
Lucy	22.30	4.20-5.60	3.39	46	8.11	8.54	49(6)
Lucy	22.08	4.00-5.71	1.02(14)	23(13)	3.27(14)	3.76(14)	26(6)
Lucy	20.76	4.00-5.21	0.81(15)	21(13)	2.47(15)	3.47(15)	23(6)
Kato Road	14.54	4.40-6.45	1.10(14)	39(14)	3.00(14)	3.31(15)	25(6)
Kato Road	22.74	3.90-6.28	1.68(16)	34(16)	1.98(16)	2.60(15)	26(6)
Kato Road	28.40	4.29-6.47	0.29(16)	30(16)	2.56(16)	6.15(15)	26(6)
Road F	14.89	4.35-5.72	1.78(13)	29(13)	2.40(13)	1.71(13)	26(6)
Road F	14.32	4.45-5.57	0.69(13)	22(13)	1.36(13)	1.83(13)	19(6)
Road F	18.23	4.32-5.62	0.80(13)	19(13)	1.58(13)	1.55(13)	15(6)

*Seventeen analyses, except as noted in parentheses.

TDS = total dissolved solids.

Growth Response of Loblolly Pines

Loblolly pine stands planted in 1953 and 1973 responded to the liquid sludge treatment equivalent to 402 kg N/ha, but gave no further growth response to the higher nitrogen loading of 804 kg N/ha. Basal area growth was increased more rapidly by sludge treatments on the 1973 stand; however, the total stem volume growth per hectare response was larger on the 1953 stands. Weed and hardwood competition was found to prevent growth response on the 1.2 m high, three-year-old (1978) stand (Lower Kato Road site). On the establishment stand, the supply of nutrients was sufficient for growth without sludge treatments (Wells et al. 1984). Growth response measurements collected to date on the 1953 and 1973 stands indicate that the trees are still responding to the single sludge application in 1985, four years after treatment. Detailed results are found in the paper by McKee et al. in this volume.

Nutrient Cycling

The nutrient cycle in a forest ecosystem may be described as the flow of nutrients into, within, and from the system. A simplified schematic of the major cycling components— mineral soil, forest floor, and vegetation—is given in Figure 7. In this generalized model,

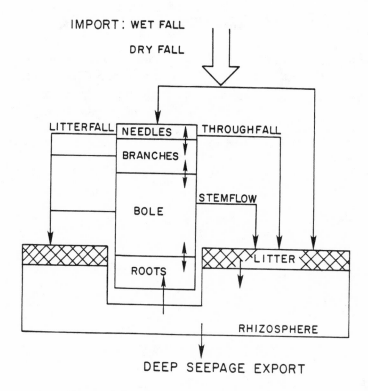

Figure 7. Typical nutrient cycle in forest ecosystem.

boxes represent elemental pools or compartments while arrows represent fluxes of elements between compartments.

Nutrient components or pools change rapidly when a stand is first established but tend to reach an equilibrium as the stand matures. After an initial accumulation of nutrients in the trees, a subsequent increase of nutrients in the forest floor occurs at the expense of nutrients in the mineral soil.

Data have shown that the nutrient cycle of undisturbed pine stands in the southeastern United States is tight (Waide and Swank 1975). Nutrient-tight pine systems are generally associated with soils of low nitrification rates, low fertility, and active annual root growth. As a result, losses from the system are typically restored via input, although component sizes and rates of transfer change as the stand develops. Extensive discussion of the nitrogen, phosphorus, and potassium cycles is found in the experimental research of Wells (1970), Wells and Jorgensen (1975), and Wells, Nicholas, and Buol (1975), respectively.

Extensive macronutrient and pH analyses in throughfall, stemflow, and vegetation were made at the 1953 and 1973 experimental stands from 1982 until late 1983, so that estimates of nutrient flux and system inventory could be made. Wells et al. (1984) documented that liquid sludge applied at rates equivalent to 402 and 804 kg N/ha increased litterfall weight some 20 to 30% over that of both the control plot litterfall at both age stands and the solid sludge-amended sites (632 kg N/ha) at the 1953 stands. Enhanced summer needlefall above rates associated with control plots was noted at plots amended with liquid sewage sludge. Mean needle length, and nitrogen and phosphorus concentrations in needle litterfall, were increased in liquid sludge plots compared with control plots (Wells et al. 1984). Understory biomass of grass, herb, shrub, and vine were enhanced on the liquid sludge plots at both the 1953 and 1973 stands. Detailed results of the nutrient cycling studies from these plots are discussed by McLeod et al., in this volume.

CONCLUSION

Biomass production was increased by liquid sludge application on forest stands age eight years and older at an equivalent nitrogen loading rate of 402 kg N/ha. At this loading rate, no adverse environmental impacts occurred in either soil, soil water, or groundwater concentrations of nutrients or metals at the experimental forest sites. Nutrients in the sludge were assimilated by cover vegetation, and metals in the sludges were retained in the upper soil layer so as not to have changed soil concentrations from normal background levels.

ACKNOWLEDGMENT

The information contained in this paper was developed during the course of work under contract No. DE-AC09-76SR00001 with the U.S. Department of Energy.

REFERENCES

SCS Engineers. 1982. Utilization of municipal sewage sludge on land. Vol. 3. Prepared for U.S. Environmental Protection Agency by SCS Engineers, Long Beach, California.

Waide, J. B., and W. T. Swank. 1975. Nutrient recycling and the stability of ecosystems: Implications for forest management in the southeastern U.S. Proceedings, SAF Annual Meeting, Washington, D.C.

Wells, C. G. 1970. Nitrogen and potassium fertilization of loblolly pine in S.C. Piedmont soil. Forest Service 16:172–176.

Wells, C. G., and J. R. Jorgensen. 1975. Nutrient cycling in loblolly pine plantations. p. 137–157. In B. Bernier and C. H. Winget (eds.) Forest soils and forest land. Proceedings, Fourth North America Forest Soils Conference, Laval University, Quebec.

Wells, C. G., A. K. Nicholas, and S. W. Buol. 1975. Some effects of fertilization on mineral cycling in loblolly pine. p. 754–764. In F. G. Howell, J. B. Gentry, and M. H. Smith (eds.) Mineral cycling in southeastern ecosystems. ERDA Symposium Series (CONF-740513), Augusta, Georgia.

Wells, C. G., W. H. McKee, Jr., E. J. Christensen, G. J. Hollod, and C. E. Davis. 1982. Elemental cycling in southern pine plantations with nutrient and heavy metal amendments. Annual report FY-1982. USDA, Southeastern Forest Experimental Station, Asheville, North Carolina.

Wells, C. G., and W. H. McKee, Jr. 1983. Elemental cycling in southern pine plantations with nutrient and heavy metal amendments. Annual report FY-1983. USDA, Southeastern Forest Experimental Station, Research Triangle Park, North Carolina.

Wells, C. G., K. W. McLeod, C. E. Murphy, J. R. Jensen, J. C. Corey, W. H. McKee, Jr., and E. J. Christensen. 1984. Response of loblolly pine plantations to two sources of sewage sludge. p. 85–94. In 1984 TAPPI Research and Development Conference, Appleton, Wisconsin. Technical Association of the Pulp and Paper Industry, Technology Park, Atlanta, Georgia.

Case Studies: Industrial and Pulp and Paper

Pulp Mill Sludge Application to a Cottonwood Plantation

WALTER J. SHIELDS, JR., MICHAEL D. HUDDY, and SHELDON G. SOMERS

ABSTRACT   Crown Zellerbach's Wauna (Oregon) pulp and paper mill will begin land application of all secondary treatment plant sludge in 1986 to alluvial land that will be planted with hybrid cottonwoods. The sludge will be pumped as a slurry through a 5 km (3 mile) pressurized pipeline and applied through a traveling gun spray irrigation system. Engineering tests will be conducted after the pipeline is functional to determine the feasibility of applying a portion of Wauna's primary sludge as well. This paper discusses a number of research experiments that preceded the decision to go operational, including study of cottonwood response to sludge treatments, effects on groundwater and soil solution chemistry, sludge decomposition, effects of leaf coating, and application equipment suitability.

Crown Zellerbach's mill at Wauna, Oregon produces about 352,000 metric tons (Mg) of groundwood newsprint and bleached kraft tissue paper annually. Two types of sludges are produced: primary sludge consisting mainly of cellulosic fibers, and waste-activated secondary sludge consisting mainly of microbial biomass. Currently, Wauna dewaters and uses as landfill about 9,504 Mg of primary and 4,800 Mg of secondary sludge annually. However, the secondary sludge landfill has only a few years' capacity remaining. Since the Oregon Department of Environmental Quality indicated that permitting a new disposal site would be difficult, if not impossible, an alternative method to dispose of secondary sludge was needed.

In 1980 Crown began research on short-rotation intensively cultured (SRIC) cottonwood plantations on diked floodplain land near the Wauna mill. In addition to research on culture, genetics, propagation, and spacing, sludge trials were also installed. Young (1983) and Thacker (1984) have reviewed the literature and paper industry experiences in land application of paper mill sludges. The use of Wauna mill sludge as a soil amendment for SRIC plantations was investigated for several reasons: (1) it was assumed that pulp mill sludges would be an excellent soil amendment, improving soil tilth and providing nitrogen for the nutrient-demanding hybrid cottonwood trees; (2) suitable land for both the SRIC plantations and sludge application was located within 5 to 8 km (3 to 5 miles) from the mill; and (3) the mill needed a long-term, cost-effective method to dispose of secondary sludge.

Results of these experiments and operational trials led to a decision by the Wauna mill to install a 5 km pipeline from the treatment plant to an SRIC cottonwood plantation in 1986. Although the pipeline will have the capacity to handle both primary and secondary sludges, only secondary sludge will be applied initially.

In this paper, the results of research experiments and operational trials are summarized and components of the operational design are discussed.

RESEARCH EXPERIMENTS

Sludge Characterization

Analyses of Wauna primary and secondary sludges over the last several years are summarized in Table 1. Metals were extracted following a sludge digestion in nitric and hydrochloric acid and concentrations measured with a Jarrell-Ash Model 975 inductively coupled argon plasma emission spectrophotometer (ICP). All analyses followed standard procedures (APHA 1976). Using the assumption that volatile solids content (mass lost at 105°C) is about double the organic C content of paper mill sludges (Dolar et al. 1972), the C:N ratio of Wauna primary sludge is about 213, whereas the nitrogenous secondary sludge has a C:N ratio of approximately 7.5.

TABLE 1. Analysis of Wauna sludges, oven-dry basis.

Parameters	Primary Sludge		Secondary Sludge	
	Mean	SD	Mean	SD
Volatile solids (%)	76.7		88.7	
Ash (%)	16	5	13	1
pH	8.0		7.3	
	(micrograms/gram)			
Total N	1,800	290	59,000	3,100
Total P	340	75	7,800	470
Aluminum	910	360	1,600	210
Arsenic	<4.5		<4.7	
Barium	31	13	40	4
Cadmium	<1.4		<1.4	
Calcium	6,500	3,400	4,400	300
Chromium	54	37	32	4
Copper	40	22	23	1
Iron	1,200	460	930	72
Potassium	210	70	4,800	180
Lead	<9.3		<11	
Magnesium	1,300	750	2,800	140
Manganese	180	54	440	20
Nickel	11	4	8	1
Selenium	<5.2		<6.1	
Silver	<1.2		<1.4	
Sodium	4,800	73	6,000	1,000
Zinc	40	19	98	14

SD = standard deviation.

Tree Response

Cottonwood and Alder Study. In 1980, Wauna primary and secondary sludges were applied to 0.006 ha (25 by 25 ft) research plots which were then planted with black cottonwood (*Populus trichocarpa*), red alder (*Alnus rubra*), and a 1:1 mixture of the two species. The plots were located at the CZ Westport Research Station near the Wauna mill. The soil type was a Locoda silt loam (Fluvaquentic Humaquept, fine-silty, mixed, acid), a very poorly drained alluvial soil. The treatments were no sludge (control), primary sludge, secondary sludge, and a 1:1 mixture of the two sludges. The application rates for the three sludge treatments were 114, 228, and 456 Mg/ha (50, 100, and 200 dry tons/acre). Each species-sludge combination was replicated three times for a total of 90 plots (Table 2).

Sludge was spread on tilled soil in October 1980. The secondary sludge treatments were rototilled, and primary sludge was left on the surface as a mulch. Dormant cottonwood cuttings and red alder container seedlings were planted at a 1.5 by 1.5 m (5 by 5 ft) spacing the following spring (nine measurement trees per plot with two buffer rows between plots).

Owing to a very wet spring and the failure of the tidal gate that drains the experimental area, it was not possible to control weeds in the research plots. The secondary sludge stimulated rapid weed growth, which killed most plot trees before it was dry enough for weed cultivation. The top dressing of primary sludge, however, was an effective mulch, resulting in significantly greater tree survival in those plots (Table 3).

All plots were rototilled during the summer of 1981. The trees were taken out, but the plots were maintained to study the effects of the sludge on soil properties (discussed below).

Cottonwood Study. The flooding and resultant weed control problems were corrected in the summer of 1981, and a second tree response study was installed at the Westport

TABLE 2. Experimental design for test of Wauna sludges on growth and survival of cottonwood and red alder.

Factor	N	Description
Species	3	Cottonwood, alder, mixture
Sludge treatment	3	Primary, secondary, 1:1 mixture
Sludge rate	3	114, 228, 456 dry Mg/ha
Control	1	No sludge, no fertilizer
Replication	3	Completely random
Total	90	

TABLE 3. Tree survival following the first growing season of the alder and cottonwood study (each value represents a mean of nine plots with nine measurement trees per plot).

Treatment	Survival (%)		
	Cottonwood	Mixed	Red alder
Control	58	52	55
Primary: 114 dry Mg/ha	62	75	20
Primary: 228 dry Mg/ha	85	50	33
Primary: 456 dry Mg/ha	92	73	30
Combined: 114 dry Mg/ha	20	30	20
Combined: 228 dry Mg/ha	7	13	10
Combined: 456 dry Mg/ha	12	8	0
Secondary: 114 dry Mg/ha	7	10	0
Secondary: 228 dry Mg/ha	10	8	0
Secondary: 456 dry Mg/ha	5	0	0

Research Station in the fall of 1981. The treatments were 82 dry Mg/ha (36 tons/acre) of primary sludge, 141 dry Mg/ha (62 tons/acre) mixed 2:1 primary and secondary sludge, and 28 kg/ha (150 lb/acre) balanced fertilizer (N:P:K:S = 14:24:14:4). Each treatment was replicated three times for a total of nine plots. Plot size was 232 m² (2,500 ft²) with a 6.1 m (20 ft) buffer between plots. The sludge was incorporated in the fall, and dormant cuttings from a local Columbia River cottonwood clone were planted at a 3.1 by 3.1 m spacing the following spring. Fertilizer was applied to the control plots after planting. This experiment was also located on the Locoda silt loam soil.

The first-year growth response of trees growing on the primary sludge plots averaged 41% higher than the fertilized control plots; the combined sludge plots showed a 68% increase over control. The treatment differences were evident, although not as pronounced, during the following two growing seasons (Table 4).

TABLE 4. Mean cottonwood height growth (cm) and standard deviation on plots treated with fertilizer, primary sludge, and combined primary and secondary sludge.

Tree Age	Fertilizer		Primary Sludge		Combined Sludge	
	Mean	SD	Mean	SD	Mean	SD
1	82	20.4	116	5.3	138	19.1
2	160	5.3	224	32.7	250	40.6
3	232	15.3	303	56.8	321	60.2

Sludge Slurry Study. The above experiments, as well as most previous studies of plant response to pulp mill sludge, simulated ground spreading of dewatered sludge. Ongoing feasibility studies at Wauna indicated that pipeline transport of undewatered clarifier solids to an irrigation system for application to the cottonwood plantations would be the most cost-effective delivery system. The pipeline slurry would contain about 1.2% solids and would have a higher BOD (biochemical oxygen demand) and salt content than sludge that has been dewatered and then reconstituted.

This study was designed to determine the effects of spray irrigation of sludge slurry on young cottonwood plants. Dormant cuttings of one cottonwood clone ("Hybrid 5," *P. trichocarpa* × *deltoides*) were planted in eighty 15 liter (4 gal) pots in a peat, sand, and vermiculite mixture in April 1983. The treatments, applied six times at regular intervals throughout the summer, are described below and summarized in Table 5.

1. *Foliar application.* The aerial portion of each tree was dipped into a sludge slurry (approximately 2:1 primary to secondary sludge mixture) taken fresh from the Wauna treatment plant. After leaf coating, half of the dipped trees were washed with a water spray to simulate rainfall (or water irrigation) following sludge application.

2. *Root application.* In order to separate the possible detrimental effects of leaf coating from sludge impact on root growth, sludge slurry was applied to the potting medium without foliar contact. There were three application rates: 1, 2, and 4 liters per pot. The cumulative dry weight application after six treatments was 34, 36, and 136 dry Mg/ha, respectively (15, 30, and 60 tons/acre). The root applications were not incorporated into the potting medium. Half of the pots receiving the 1 and 2 liter root application were also

TABLE 5. Height growth (adjusted for diameter of dormant cutting) of potted hybrid poplar receiving various foliar and root applications of mixed primary and secondary sludge slurry.

Foliar Application	Foliage Washed	Root Application (liters)	Sample Size	Tree Height (cm)	
				Mean	SD
No	No	0	10	67.3	3.0
Yes	No	0	10	78.8	3.2
Yes	Yes	0	10	64.9	3.0
Yes	No	1	5	74.5	4.3
Yes	No	2	5	75.2	4.2
Yes	Yes	1	5	70.9	4.2
Yes	Yes	2	5	57.5	4.7
No	Yes	1	5	67.0	4.2
No	Yes	2	5	71.5	4.3
No	No	1	10	70.2	4.2
No	No	2	10	63.0	4.2
No	No	4	10	51.0	3.5

treated with foliar application, water spray, and the combined foliar and wash treatments.

Since height growth was significantly correlated to the diameter of the dormant cottonwood cuttings, the height means listed in Table 5 are adjusted for cutting diameter.

To test the effects of the various treatments, F-values were computed for selected contrasts of adjusted treatment means. The results are as follows: (1) Foliar application increased height growth over control (F = 5.88*). This was rather surprising since the leaves were so thoroughly coated with fibrous sludge. The slurry dried rapidly on the leaf surfaces and tended to peel away after several days. The increase in height growth may have been due to sludge nutrients, since the control pots were not fertilized. (2) Root applications at the 1 and 2 liter rates did not affect tree growth compared with control. However, the 2 liter treatment caused significant growth reduction (F = 12.83**). This was clearly due to "drowning" the trees with a deep cover of sludge, which severely inhibited gas exchange for root metabolism. The lighter treatments dried between treatment intervals, allowing gas exchange through shrinkage cracks. (3) Mean height of trees that were washed after foliar application was not different from control and actually lower than for trees that were dipped and not washed (F = 9.13**).

Effects on Soil

Methods. A composite of five soil samples from the 0 to 15 cm depth was collected on each of the ninety alder-cottonwood plots treated with sludge in 1980 and on the nine cottonwood plots that were established in 1980 and 1981. The soils were sampled twice, in June 1983 and June 1984. Soil samples were air dried and passed through a 2 mm sieve. Exchangeable calcium, magnesium, potassium, and sodium were extracted with neutral $1N$ ammonium acetate and concentrations determined by ICP. Phosphorus was extracted with $0.002N$ sulfuric acid and concentrations determined by the molybdophosphoric blue colorimetric method. Other soil analyses included total nitrogen by the semimicro-Kjeldahl procedure, easily oxidizable organic matter by the Walkley-Black method, and soil pH by glass electrode in a 1:2 soil-calcium chloride ($0.05N$) solution (Black 1965).

TABLE 6. Soil analysis of red alder and cottonwood experimental plots that received sludge in the fall of 1980 (each value is a mean of nine plots).

Sample	Rate (dry Mg/ha)	TKN (%)	P (ppm)	K	Ca (meq/100 g soil)	Mg	Na	SAR*	Organic Matter (%)	pH
1983 Sample										
Control	0	0.32	25.42	0.35	9.02	4.02	0.28	0.16	4.37	4.17
Primary sludge	114	0.29	7.47	0.24	9.62	4.10	0.35	0.05	5.77	4.22
Primary sludge	228	0.32	10.76	0.25	9.86	4.10	0.37	0.05	6.24	4.23
Primary sludge	456	0.32	15.90	0.25	10.88	4.10	0.53	0.03	7.28	4.62
Primary plus secondary	114	0.36	15.12	0.12	9.25	4.03	0.33	1.11	7.78	4.18
Primary plus secondary	228	0.30	18.52	0.23	9.82	4.20	0.36	0.01	6.12	4.34
Primary plus secondary	456	0.39	18.98	0.28	10.00	4.01	0.34	1.49	6.21	4.28
Secondary sludge	114	0.39	13.46	0.27	8.83	3.93	0.34	0.29	7.24	4.04
Secondary sludge	228	0.34	39.75	0.28	8.83	3.83	0.29	0.01	7.23	4.06
Secondary sludge	456	0.40	30.89	0.31	8.50	3.74	0.35	0.15	7.02	4.06
Least sig. difference (p = 0.05)**		0.10	28.76	0.08	0.83	0.42	0.08	1.46	3.33	0.18
1984 Sample										
Control	0	0.42	37.44	0.44	7.51	3.32	0.20	0.09	9.09	4.00
Primary sludge	114	0.49	43.38	0.30	8.23	3.34	0.28	0.17	9.90	4.20
Primary sludge	228	0.43	34.56	0.35	8.11	3.28	0.29	0.12	12.14	4.11
Primary sludge	456	0.44	27.81	0.30	9.92	3.74	0.41	0.16	10.88	4.28
Combined sludge	114	0.48	29.91	0.33	7.51	3.22	0.27	0.12	11.54	3.99
Combined sludge	228	0.48	38.03	0.33	8.10	3.36	0.32	0.13	12.13	4.08
Combined sludge	456	0.42	31.59	0.40	8.67	3.29	0.26	0.11	10.87	4.11
Secondary sludge	114	0.50	33.26	0.31	7.32	3.16	0.30	0.13	12.70	3.76
Secondary sludge	228	0.53	63.40	0.43	7.40	3.21	0.31	0.13	12.08	3.78
Secondary sludge	456	0.60	48.87	0.47	7.15	3.08	0.30	0.13	13.52	3.76
Least sig. difference (p = 0.05)**		0.10	17.89	0.12	1.62	0.70	0.13	0.12	3.33	0.21

*SAR (sodium adsorption ratio) = $(Na)/((Ca + Mg)/2)**0.5$.
**Treatment means that differ by more than this value are significantly different at p = 0.05 (Snedecor and Cochran 1967).

Results. The presence of red alder had no effect on soil nitrogen or any other measured soil variable, so all the species treatments in the 1980 red alder-cottonwood study were grouped according to sludge treatment (Table 6).

1. *Nitrogen.* Secondary sludge, particularly at the highest rate, caused a prolonged increase in total nitrogen levels in the soil. The 1981 cottonwood study also showed an increased nitrogen level with the combined sludges, although the effect was not significant (Table 7).

2. *Phosphorus.* Even though Wauna secondary sludge is relatively rich in phosphorus (7,800 μg/g, oven-dry basis), sludge treatments did not consistently raise plant available phosphorus levels in the soil (Tables 6 and 7). The lack of treatment response may be due to the apparent acidification effect of secondary sludge. In strongly acid soils, soluble iron and aluminum ions react with $H_2PO_4^-$ ions to form insoluble hydroxy phosphates (Cole and Jackson 1950).

3. *Exchangeable bases.* Primary sludge treatments clearly raised exchangeable calcium levels over those in control and secondary sludge only plots (Tables 6 and 7). Sludge effects on levels of potassium, magnesium, and sodium were not as apparent.

High sodium levels in land-applied kraft mill sludges sometimes draw questions about degradation of soil structure by replacement of calcium and magnesium with sodium on colloid surfaces. Critical values of the sodium adsorption ratio (SAR), an index of sodium imbalance, for fine-textured soils range from 5 to 15 (Overcash and Pal 1979). SAR values in the sludge-treated plots averaged less than 0.2 (Tables 6 and 7). The highly mobile sodium ions were undoubtedly preferentially leached in these soils with a shallow, rapidly fluctuating groundwater table.

4. *Soil pH.* Apparently, the production of nitric acid by mineralization of organic nitrogen in secondary sludge caused a marked depression in soil pH. When primary sludge was combined with secondary, thereby decreasing the nitrogen mineralization rates, soils were not acidified in relation to control plots.

OPERATIONAL TRIALS

Trucking Dewatered Sludge

Application. All the dewatered primary and secondary sludge produced at Wauna between March and July 1983—about 34,000 dry Mg—was hauled to the Westland District and applied to 26 ha of pastureland (Figure 1) at an average rate of about 192 dry Mg/ha (84 tons/acre). Primary sludge dewatered to about 18% solids on vacuum coil filters and secondary sludge dewatered to about 11% on a Reitz belt press were diluted to about 3% solids in a 800,000 liter capacity mixing lagoon. The sludge cake was mixed by water jets and a hydraulically driven stinger propeller. The reconstituted sludge was pumped by a Cornell 4NHDH centrifugal pump through 25.4 cm (10 inch) diameter aluminum mainline to a traveling spray gun applicator.

The inclusion of primary sludge caused a number of problems. First of all, the bearing seals and packing for the sludge pump were inadequate. The bearings "burned out" on the first pump twice, resulting in significant downtime. A better sludge pump was installed in June: a Cornell 4×4×14T with a 35 cm (14 inch) two-port enclosed impeller and a single mechanical seal lubricated with fresh water and packed with standard graphite packing. Corrosion of the bearing packing still occurred, although not as severely as before. The stress on pump bearings was probably due to inadequate mixing of the

TABLE 7. Soil analysis of cottonwood experimental plots that received sludge in the fall of 1981 (each value is a mean of three plots).

Sample	TKN (%)	P (ppm)	K	Ca (meq/100 g soil)	Mg	Na	SAR*	Organic Matter (%)	pH
1983 Sample									
Control (with NPKS fertilizer)	0.21	22.3	0.29	9.78	4.05	0.23	0.09	6.90	4.29
Primary sludge (82 Mg/ha)	0.24	41.6	0.27	11.07	4.55	0.31	0.11	8.43	5.10
Primary plus secondary (141 Mg/ha)	0.29	28.1	0.33	11.38	4.34	0.31	0.11	8.67	4.83
Least sig. difference (p = 0.05)**	0.09	17.0	0.10	1.82	0.88	0.13	0.04	0.90	0.53
1984 Sample									
Control (with NPKS fertilizer)	0.32	10.7	0.31	10.79	4.28	0.23	0.08	6.47	4.26
Primary sludge (82 Mg/ha)	0.32	12.0	0.30	12.65	4.94	0.32	0.11	7.47	4.94
Primary plus secondary (141 Mg/ha)	0.41	12.0	0.43	12.81	4.46	0.31	0.11	10.47	4.72
Least sig. difference (p = 0.05)**	0.11	16.2	0.07	1.59	0.61	0.15	0.05	3.47	0.40

*SAR (sodium adsorption ratio) = $(Na)/((Ca + Mg)/2)^{**}0.5$.
**Treatment means that differ by more than this value are significantly different at p = 0.05 (Snedecor and Cochran 1967).

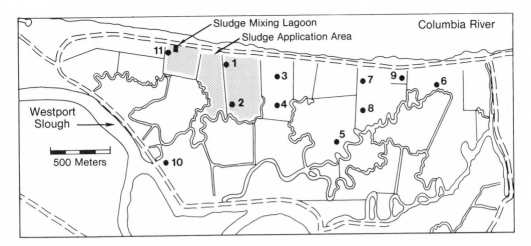

Figure 1. Westland Drainage District with groundwater monitoring wells and 1983 sludge application area.

dense filter cake. A chopper pump (Parma Model 34), installed in the mixing lagoon in July, significantly improved sludge mixing and sludge pump performance.

Primary sludge filter cake and solid objects frequently plugged the irrigation system. Large knots, rocks, and pieces of metal and plastic that somehow passed the traveling screens in the main sewer line at the mill commonly plugged elbow joints, risers, and spray nozzles. When joints ruptured following plugging, very heavy sludge deposits collected in depressions on the application site and sometimes flowed into adjacent drainage ditches. These problems were largely eliminated after the chopper pump was installed.

Groundwater Sampling. Three monitoring wells were installed within the sludge application area. An additional eight "control" wells were located in fields separated from the sludge treatment area by actively flowing drainage ditches. Monitoring wells were encased with schedule 40 PVC pipe—5.1 cm (2 inch) inside diameter—with screens that were slotted to 0.5 mm (0.02 inch) and packed with filter sand. Average depth of the wells was about 2 m.

Wells were sampled six times between March 1983 and October 1984: once prior to sludge application; twice during the application period; and 3, 8, and 15 months after the project ended. Measurements included pH, conductivity, alkalinity, color, tannin, total Kjeldahl N (TKN), total P, total organic C (TOC), cations (sodium, potassium, iron, and manganese) and anions (nitrate, chloride, and sulfate) (APHA 1976).

The well next to the sludge mixing lagoon (well no. 11), damaged when the lagoon was expanded, was relocated approximately 30 m from the sludge pit (Figure 1). Conductivity samples from the relocated well indicated seepage from the mixing lagoon. Salt concentration, as indicated by conductivity, was not influenced by sludge application away from the immediate vicinity of the pit (Figure 2).

Tannin, sodium, and iron showed slight increases in the sludge-treated areas and then declined rapidly to control levels. Sludge application had no apparent effect on color, TKN, phosphorus, TOC, potassium, manganese, and anions measured.

On the basis of periodic analyses of sludge nitrogen content, approximately 100 Mg of nitrogen were applied during the trial. This was equivalent to an application rate of 3,846

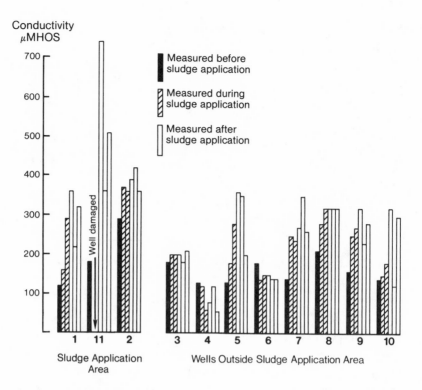

Figure 2. Conductivity of groundwater measured six times before, during, and after sludge application.

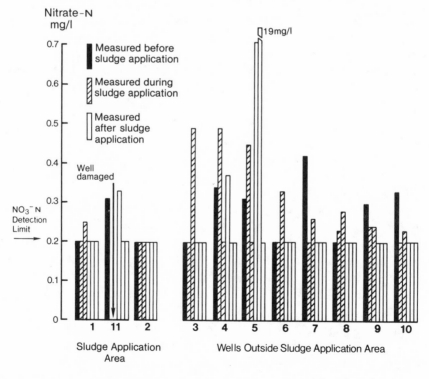

Figure 3. Nitrate-N concentrations in groundwater measured six times before, during, and after sludge application.

kg/ha (3,435 lb/acre). Even with this extremely high loading, nitrate-N concentrations in well samples were well below the EPA drinking water standards of 10 mg/l (Figure 3). Nitrate-N concentrations for well no. 5 in a control field peaked at 19 mg/l in October 1983. This was probably due to grazing cattle, which were frequently seen gathering around this well.

The low nitrate-N concentration in the monitoring wells was surprising. Ammonia volatilization, and particularly denitrification, accounted for some loss; however, the bulk of the applied nitrogen was undoubtedly immobilized because of high C:N ratios (40 to 230) of the sludge mixture. The nitrate that was slowly mineralized and leached through the soil profile was diluted by the vast amount of tidally influenced groundwater moving past the well screens.

Public Relations. The major public relations problem associated with this operational trial resulted from heavy truck traffic. Neighbors complained about noise, dust, traffic, small spills, and safety. A strong effort was made to inform local residents and elected officials about the concept of using pulp mill clarifier solids as a beneficial soil amendment. Local newspaper articles, public hearings, newsletters, and personal visits to local residents were generally effective in convincing the community that sludge utilization was *not* sludge disposal.

Sludge Slurry Operational Trial

Introduction and Methods. Because of the public objections to trucking and the problems with reconstituting dewatered sludge, we recommended that the Wauna mill consider pipeline transport of sludge to the Westland District. We conducted a controlled application-equipment trial in 1984 using undewatered sludge slurry (1) to test the effectiveness of the Parma chopper pump, (2) to test two types of traveling gun applicators, an externally powered winch and a hydraulically (slurry) powered trip valve ("water jack"), and (3) to determine the distribution pattern of different sludge mixtures. Three types of sludge slurry were tested: primary sludge, secondary sludge, and a 1:1 mixture of the two sludges.

Clarifier underflow was transported in tanker trucks and pumped into a storage tank at the Westport Research Station by a Parma Model 34 chopper pump. The stored slurry could be either recirculated or pumped directly to the field applicators using a nonclog sludge pump (Farmstar Model 2400), which maintained a pressure of 4.9 kg/cm^2 (70 lb/inch^2) at the nozzle head.

Primary sludge and mixed (1:1) primary and secondary sludge slurries were applied at a nominal rate of 3.1 cm depth/ha (approximately 0.5 acre-inch), or 3.6 dry Mg/ha (1.6 tons/acre) as determined by traveler speed and nozzle pressure. Secondary sludge slurry was applied at about 6.2 cm depth/ha (one acre-inch). Dry weight conversions were based on an average oven-dry solids content of 2.8% for all three slurries.

Uniformity of application was measured by placing 3.8 liter (1 gal) capacity pots at 7.6 m (25 ft) intervals to cover an application diameter of 76 m (250 ft) and a traveler path of 46 m (150 ft). Each of the three slurry types was sampled in the collectors for volume and percentage of solids. Ten "control" pots were also set out to calculate a correction factor for rainfall.

Results and Discussion. The chopper pump allowed primary sludge solids (i.e., knots and bark) less than about 2.5 cm (1 inch) to pass through to the field applicator, resulting

TABLE 8. Dry weight of sludge applied as estimated by volume and consistency
measurements from collector pots.

Sludge Type: Distance (m)	Secondary (dry Mg/ha)	Combined (dry Mg/ha)	Primary (dry Mg/ha)
7.8	6.8	3.0	3.9
15.6	5.9	2.5	4.1
22.9	7.5	2.5	4.1
30.5	8.0	1.6	0.5
38.1	1.8	1.1	0.2
Mean	5.2	2.1	2.1
Nominal rate	7.3	3.6	3.6

in frequent plugging of the 2.5 cm (1 inch) diameter irrigation nozzle. We eliminated this problem by installing a 3.3 cm (1.3 inch) diameter nozzle.

The water jack method of winching the applicator worked well with secondary sludge but not at all with primary sludge or mixed sludge. Wood fibers and small pieces of undigested wood immediately plugged the water jack's hydraulic system. An external motorized portable winch with positive direct drive will be needed to pull applicators when primary sludge is being pumped.

The application rate calculated from volume and consistency data matched the nominal rate of 7.3 dry Mg/ha for secondary sludge as set by pump pressure and traveler speed (Table 8). The combined sludge rate was lower than the nominal rate of 3.6 dry Mg/ha, but the distribution was relatively uniform. Primary sludge measurements approximated the nominal rate within 23 m of the applicator and then declined rapidly (Table 8). The poor distribution of primary sludge may have been caused by (1) inadequate mixing in the storage tank or (2) separation of the larger solids in the spray irrigation stream by gravity or wind.

OPERATIONAL DESIGN

Sludge Delivery System

In the spring of 1986, a 4.8 km (3 mile) pipeline will be built to transport secondary sludge from Wauna to the Westland District. Initially, only secondary sludge will be pumped; however, the pipeline will have the capacity to handle primary sludge as well.

Underflow from the secondary clarifiers will be pumped to a 2.7 million liter (720,000 gal) buffer storage pond at the treatment plant. Two pumps, each capable of pumping 37.8 l/sec (600 gpm) against a head of 7.620 kg/cm² (250 ft of water), will alternate in pumping a 1.2% consistency slurry through the buried 25.4 cm (10 inch) diameter polyethylene pipe to the field pump station. A 95,000 liter (25,000 gal) buffer tank with float switches will enable the operator to maintain a uniform flow to the field applicators. Sludge will be pumped through buried mainline throughout the application area (Figure 4) by a sludge pump with a 37.8 l/sec (600 gpm) capacity. A number of shutoff switches and shunt valves will be built into the system to handle surges.

Traveling gun applicators will be fed by an 8.9 cm (3.5 inch) diameter flexible hose attached to the buried pipeline at hydrants spaced at about 68 m (223 ft) (Figure 5). Each applicator will be pulled across the field by a 402 m (0.25 mile) cable, operating from an

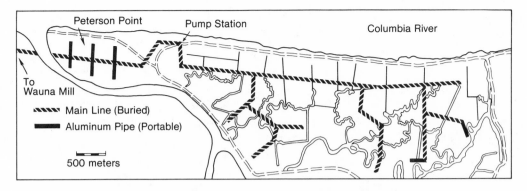

Figure 4. Field pipeline layout for Westland District and Peterson Point.

anchored, motor-powered winch. The applicators will be lightweight and equipped with flotation tires for easy traveling over soft, wet terrain.

The average daily secondary sludge production of 11 dry Mg (12 tons), or 908,000 liters (240,000 gal) of 1.2% slurry, can be handled by two applicators, each spraying at a rate of 18.9 l/sec (300 gpm), during one 8-hour shift. A third standby applicator will be positioned to start a run as soon as a functioning applicator finishes a 402 m (0.25 mile) set or when sludge production exceeds the capacity of two applicators. A hose reel and purge

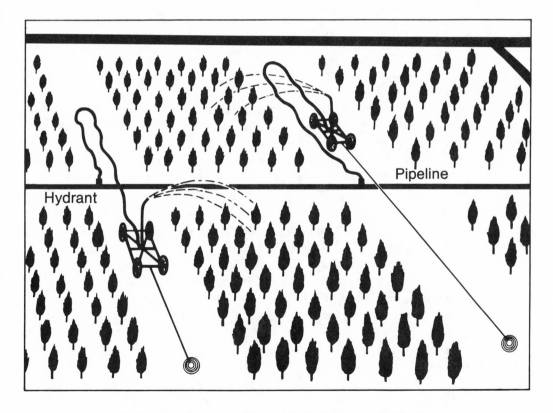

Figure 5. Traveling gun applicators with flexible hose and cable winch.

pump will be mounted on a tracked all-terrain vehicle for moving the feeder hoses and applicators to each new set location. This system can be operated by one full-time field operator and a part-time manager.

Application Rates and Schedule

Application rates are limited by nitrate leaching and soil infiltration capacity. The maximum annual application was based on N loading using the following assumptions:

Total Kjeldahl N of secondary sludge: 5.8%

Ammonia-N content: 0.2%

Volatilization loss of N:

1. 50% of ammonia-N lost during application
2. 25% of available N lost annually via denitrification

Mineralization rate:

1. 20% of applied N during first year
2. 8% of residual N during second year
3. 3% of residual N during third plus years

Plant N uptake rate:

1. Dense grass pasture: 267 kg/ha (300 lb/acre)
2. Hybrid poplar plantation (2,152 trees/ha):
      89 kg/ha (100 lb/acre) first year
      178 kg/ha (200 lb/acre) second year
      267 kg/ha (300 lb/acre) third year

The above assumptions are based on values from the literature (U.S. EPA 1983). Experiments were established in 1985 to obtain better estimates of ammonia volatilization, denitrification, and mineralization rates. Nitrogen uptake estimates will be refined through annual foliar analyses. The average annual sludge loading rate based on these initial assumptions is about 27 dry Mg/ha (12 tons/acre) for all but the first two years following plantation establishment.

When the ground is wet during the rainy season (November through May), no more than 3.1 cm depth/ha (0.5 acre-inch) of slurry (about 1.5 dry Mg/ha) will be applied during one applicator run to avoid runoff and puddling. Slurry volume will be limited to 6.2 cm depth/ha per applicator run at other times to ensure percolation and equipment operability. Subsurface drainage pipes have been installed and ditches have been dredged to improve drainage of the Locoda silt loam soil that is predominant in the Westland District.

The Westland District has been divided into four management blocks of about 80 ha (200 acres) each. Each block will be planted in sequential years to provide flexibility in scheduling sludge applications. The trees will be planted at a spacing of 1.5 by 3 m (5 by 10 ft). Fifteen m (50 ft) wide clearings will be left for the traveler paths, so that sludge can be sprayed over the canopy when the trees get large (Figure 5).

Sludge will be tilled into the soil during the first three years after planting in conjunction with normal weed cultivation. Following harvest in the sixth growing season, harvested plantation blocks will receive sludge and then be prepared for planting the following spring.

Composite sludge samples will be analyzed weekly for nitrogen and percentage of oven-dry solids and measured monthly for metals, volatile organics, pH, and ash content. The eleven monitoring wells will be sampled every three months during the first year and annually thereafter. Soils and foliage samples will be analyzed annually.

Inclusion of primary sludge into the system will require an evaluation of required changes at the treatment plant, including improved traveling screens, rock wells, and chopper pump(s).

SUMMARY

1. Crown Zellerbach's Wauna (Oregon) mill produces 9,504 Mg of cellulosic primary sludge and 4,800 Mg of waste-activated secondary sludge annually. The C:N ratios of primary and secondary sludges are about 213 and 7.5, respectively. Heavy metal content is very low.

2. Application of primary sludge to cottonwood plots increased height growth 41% over the fertilized control plots. Mixed sludge treatments resulted in a 68% height growth increase over control.

3. The aerial portion of potted cottonwood trees was dipped into a slurry of mixed sludge to simulate an extreme case of leaf coating from spray irrigation of sludge. Leaf coating had no detrimental effect on cottonwood height growth.

4. Secondary sludge application caused a prolonged increase in soil total nitrogen concentrations. Available soil phosphorus was not increased with secondary sludge treatments, probably because of the acidifying effects of nitrification. Primary sludge increased exchangeable calcium levels. All sludge treatments had very low sodium adsorption ratios in the soil plow layer.

5. During the 1983 operational trial, 34,000 dry Mg of primary and secondary sludge were hauled dewatered to a floodplain application site. The sludge was reconstituted to 3% oven-dry solids in a mixing lagoon and applied to 26 hectares of land by traveling spray irrigators. Inadequate mixing and large solid objects in primary sludge caused pump failures and plugs in pipelines and applicators. Installation of a chopper pump eliminated most of these problems.

6. Monitoring wells away from the immediate vicinity of the mixing lagoon showed a slight and short-lived increase in tannin, sodium, and iron. The well next to the mixing lagoon showed high conductivity, indicating seepage. Nitrate concentrations were very low—probably because of nitrogen immobilization in the high C:N ratio mixed sludge, denitrification, and dilution by the tidally influenced groundwater.

7. The hydraulic jack used to winch the applicator worked well with secondary sludge but not at all with primary sludge.

8. The spray pattern of secondary sludge was relatively uniform within an application diameter of about 67 m (200 ft). The spray pattern of primary sludge was not as well distributed.

9. In 1986 a 5 km pipeline will be built to transport secondary sludge to a short-rotation, intensively cultured cottonwood plantation. Traveling gun applicators will be fed by flexible hose attached to the buried pipeline at hydrants spaced about 68 m apart.

10. Annual application rates are limited by nitrogen loading to a maximum of about 27 dry Mg/ha (12 tons/acre) depending on plantation age.

REFERENCES

American Public Health Association. 1976. Standard methods for the examination of water and wastewater. APHA, Washington, D.C. 1193 p.

Black, C. A. (ed.) 1965. Methods of soil analysis. Part 2. American Society of Agronomy, Madison, Wisconsin.

Cole, C. V., and M. L. Jackson. 1950. Solubility equilibrium constant of dehydroxy aluminum dihydrogen phosphate relating to a mechanism of phosphate fixation in soils. Soil Sci. Soc. Am. Proc. 15:84–89.

Dolar, S. G., J. R. Boyle, and D. R. Keeney. 1972. Paper mill sludge disposal of soils: Effects on the yield and mineral nutrition of oats. J. Environ. Qual. 1:405–409.

Overcash, M. R., and D. Pal. 1979. Design of land treatment systems for industrial wastes: Theory and practice. Ann Arbor Science Publishers, Ann Arbor, Michigan.

Snedecor, G. W., and W. G. Cochran. 1967. Statistical methods. 6th ed. Iowa State University Press, Ames. 593 p.

Thacker, W. E. 1984. The land application and related utilization of pulp and paper mill sludges. NCASI Tech. Bull. 439. National Council of the Paper Industry for Air and Stream Improvement, New York.

U.S. Environmental Protection Agency. 1983. Process design manual: Land application of municipal sludge. EPA 625/1-83-016. CERI, Cincinnati, Ohio.

Young, S. R. 1983. Pulp and paper sludge treatment/disposal alternatives: Land. ESD Res. Memo. 442-9. Crown Zellerbach Environmental Service Division, Camas, Washington.

Chapter 50

Land Treatment of Chemical Manufacturing Wastewater

R. L. KENDALL, W. G. ALGIERE, and W. L. NUTTER

ABSTRACT   National Starch and Chemical Corporation produces vinyl and acrylic polymer emulsions at a plant in Woodruff, South Carolina. Wastewater generated from rinsing and washing of blending vessels is treated in a chemical coagulation/DAF system. When plant operations started in 1975, wastewater was discharged to a second order stream. In 1978 a 4.5 hectare land treatment system was installed, but operational problems caused groundwater quality to deteriorate downgradient from the irrigation site. To develop design criteria for a new system, site testing was performed to determine site characteristics. A waste assimilation study determined the wastewater application rate. The pretreatment assessment concluded that modifications were necessary to reduce the amount of sodium and chloride in the wastewater. The site testing and waste assimilation studies indicated that the COD loading was the limiting waste constituent, requiring a total of 9 hectares. The new irrigation system consists of mature hardwood and pine forest divided into seven zones. Wastewater is generally irrigated two hours each day for a week on a single zone, after which the zone is rested for six weeks while the other zones are irrigated. Storage for three days of flow, or 190,000 liters, is available for periods of inclement weather or system maintenance. Soil, soil water, and groundwater are monitored to determine the impacts of wastewater irrigation. Soil water and groundwater data have been within primary and secondary drinking water standards since operation began in March 1983. By properly designing the new land treatment system, it has been possible to achieve environmental standards while treating wastewater in a cost-effective manner.

Land treatment of industrial wastewater has gained widespread regulatory acceptance over the past several years. A large number of land treatment systems now provide operational and monitoring data that document the high level of treatment attainable by this technology. The data from these systems allow new systems to be designed that are capable of meeting standards for groundwater quality while providing a cost-effective method of wastewater treatment.

Application of industrial wastewater to forest land has a number of advantages over application to grass or cropland. Forest vegetation generally is tolerant of a larger number of industrial waste constituents. The hydrologic properties of forest soils allow waste assimilation over a wider range of climatic conditions. The major disadvantage of forest sites is that slopes tend to be steeper and soils somewhat shallower than on agricultural land. When forest sites are selected for land treatment, it is essential that a detailed site investigation be performed to identify areas unsuitable for wastewater application.

National Starch and Chemical Company (NSCC) produces vinyl and acrylic polymer emulsions at a plant in Woodruff, South Carolina, approximately 48 kilometers (30 miles) east of Greenville. The primary source of wastewater at the plant is from rinsing

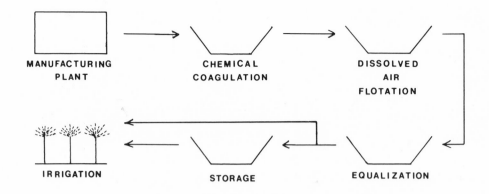

Figure 1. Wastewater treatment schematic.

and washing of blending vessels used in the manufacturing processes. The rate of wastewater generation averages 490,000 liters/week (130,000 gal/wk), but production may vary up or down by as much as 50% over the course of the year. Wastewater is treated in a chemical coagulation and dissolved air flotation (DAF) system to remove polymer that remains in suspension in the raw wastewater. The wastewater is then pumped to an equalization tank before being pumped to the irrigation system. A 1.9 million liter (500,000 gallon) storage pond is used when irrigation is not possible because of climatic conditions. Figure 1 is a schematic of the wastewater treatment system.

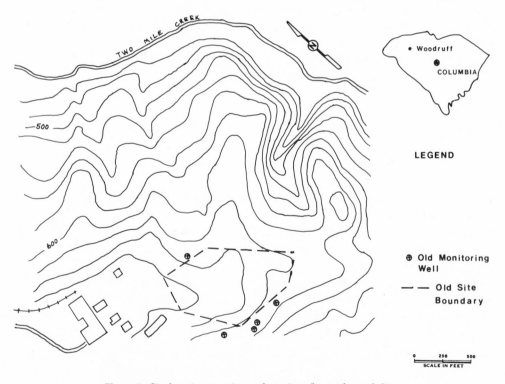

Figure 2. Site location map (manufacturing plant at lower left).

When the plant operations started in 1975, treated wastewater was discharged to Two Mile Creek, a second order stream that drains to the Atlantic Ocean. The high cost of treatment to meet the NPDES discharge standards led NSCC to investigate land treatment as a waste treatment alternative. A limited site investigation and waste characterization was undertaken during the land treatment design process. In 1978 a 4.5 hectare (11 acre) land treatment system was installed in a thirty-year-old loblolly pine plantation with the approval of the South Carolina Department of Health and Environmental Control (DHEC). Figure 2 shows the location of the site relative to the manufacturing plant.

Minor problems with runoff during periods of intense rainfall were encountered from the outset of operations on areas that had been disturbed during installation of the underground laterals. The occurrence of runoff was considered by DHEC to be a violation of the operating procedures, and NSCC was subsequently required to construct berms at the lower limits of the land treatment site to retain the runoff. As runoff collected, a small impoundment began to extend back into the irrigation field, creating wet soil conditions for much of the year.

Shortly after the berms were constructed, elevated levels of chemical oxygen demand (COD), sodium, chloride, and conductivity were detected in wells located downgradient from the impoundment. Upon closer inspection it was evident that an old roadbed was located in the lowest part of the irrigation field where the berms were constructed. Wastewater impounded behind the berms was percolating through the roadbed directly into the groundwater without passing through the soil profile.

After the berms were constructed, conditions on the irrigation field deteriorated to such a degree that the only feasible alternative was to abandon the site. Before committing to developing a new land treatment system, NSCC decided to have a new site and waste evaluation performed based on the state-of-the-art approach described by Overcash and Pal (1979).

SITE INVESTIGATION

A reconnaissance of the 40 hectare (100 acre) NSCC property was performed to determine if suitable land was available for relocating the land treatment system. The available land consisted of natural mixed hardwood and planted loblolly pine forest with slopes ranging from 5% to 15%. The USDA Soil Conservation Service had mapped the area as two soil series—Cecil and Pacolet—both of which are deep, well-drained soils located on uplands in the Piedmont physiographic province (SCS 1978). From the site reconnaissance and review of the Soil Survey, there appeared to be no limitations for developing a new land treatment system.

To verify the preliminary conclusions of the site reconnaissance, a detailed site investigation was performed. An area of about 20 hectares (50 acres) was selected for investigation, based on projections of land area requirements made during the previous phase of the investigation. A total of twenty-three hand auger borings were advanced to a depth of 120 cm (48 inches) to describe the soil profile according to SCS terminology.

The soil profiles described during the investigation were similar to the typical SCS description for the soil series present. The A horizon was typically brown sandy loam from 0 to 12 cm in depth. The B1 horizon, extending from 12 cm to 20 cm, was a reddish brown clay. Below this was the B21 horizon, consisting of a brownish red clay from 20 cm to 30 cm in depth. The B22 horizon consisted of a red clay loam with few mica flakes from 30 cm to 75 cm in depth. From 75 to 105 cm the B3 horizon consisted of red clay loam with common mica flakes and few small quartz fragments. Saprolite or decomposed mica schist bedrock was encountered below 105 cm.

The area under investigation had been farmed until the 1940s and considerable erosion had occurred. Two deep gullies extended from Two Mile Creek to near the top of the slopes in the study site. In several locations saprolite was encountered at depths as shallow as 30 cm. Approximately 30% of the study site had been planted to loblolly pine under the Land Bank program during the 1950s. The remainder of the site developed a mixed hardwood forest through natural succession. The forest vegetation had stabilized the surface soil and allowed an A horizon to begin to develop over the entire site.

In addition to hand auger borings, a series of mechanical borings were advanced to 7.5 m to describe the shallow geology of the site and determine depths to groundwater. Auger refusal was encountered at 3 m in one of the seven borings, but the remaining borings did not hit hard rock. Sandy micaceous silt typical of the saprolite on the site was encountered at all boring locations. Groundwater was not encountered in any of the borings.

Soil permeability tests were performed on undisturbed samples collected at each mechanical boring. Samples were collected in the upper 40 cm, where the horizon with the lowest hydraulic conductivity is typically located. Tests were performed in the laboratory using the saturated falling head method. The average value for the clay horizon was 1.0 cm/hr, which is slightly below the SCS predicted permeability of 1.5 cm/hr.

The mean hydraulic conductivity of 1.0 cm/hr was used to estimate a safe percolation value for calculation of a hydrologic budget for the site. Rainfall and evapotranspiration data were gathered and combined with the percolation value to calculate the acceptable hydraulic loading for the site. By this process the safe loading was calculated to be 16 cm/month for the most limiting month of January. This was selected as the design limit for hydraulic loading throughout the year so that storage requirements could be kept to a minimum.

WASTE CHARACTERIZATION

The absence of a thorough waste characterization was one of the major shortcomings of the initial system design. An extensive sampling and analytical program was undertaken to provide an accurate characterization of the waste, in order that proper loading rates could be determined that would minimize the impacts on the groundwater at the new site.

After the first set of analyses it became evident that concentrations of sodium, chloride, and total dissolved solids (TDS) were such that an excessive land area would be required to control leaching so that groundwater standards would be met. After a review of the chemical coagulation process in the wastewater treatment system, experiments were undertaken with different chemicals to coagulate the polymers in the wastewater. Potassium was substituted for sodium, and sulfate replaced chloride in the process. The result was an improvement, but TDS concentrations were still very high.

Experiments were undertaken using a lime slurry for chemical coagulation. These experiments were successful in reducing the TDS and in bringing COD down to an acceptable level. The final effluent quality resulting from the wastewater treatment process modifications is presented in Table 1.

TABLE 1.   Final effluent quality.

Constituent	Concentration (mg/1)
pH	8.5
Specific conductance (mmhos/cm)	8.23
Chloride	360
Total solids	7,910
Total dissolved solids	6,000
Chemical oxygen demand	5,600
Total Kjeldahl nitrogen	80
Ammonia-N	48
Nitrate-N	2
Total sulfur	1,100
Magnesium	4.3
Potassium	1,200
Calcium	62
Sodium	330
Total organic carbon	1,000

LAND TREATMENT PROCESS DESIGN

To determine the land requirement for assimilation of the NSCC wastewater, a land limiting constituent analysis was performed. The assimilative pathways for each significant waste constituent were evaluated and the assimilative capacity of the site was established. Nitrogen, COD, and dissolved solids were identified as the waste constituents that required the most detailed evaluation. The pathways contributing to assimilation of the critical waste constituents are biological decomposition or uptake and leaching of ionic compounds.

Nitrogen assimilation occurs by microbial immobilization, ammonia volatilization, denitrification, plant uptake, and nitrate leaching. To determine the land required for nitrogen assimilation, an approach similar to that recommended for land application of municipal sludge was taken (U.S. EPA 1983).

Allowance is first made for volatilization of ammonia. The ammonia-N remaining after volatilization plus the nitrate-N in the wastewater are assumed to be immediately available for plant uptake. A mineralization rate is estimated for the organic fraction of the total nitrogen in the waste stream. The mineralized organic nitrogen plus the ammonia-N and nitrate-N represent the plant available nitrogen (N_p). A nitrogen uptake rate for the vegetation proposed for the site is determined, and the N_p divided by the uptake rate yields the land requirement. This empirical approach to determining nitrogen assimilation is valid when the wastewater contains a small proportion of nitrate-N.

For the NSCC waste stream, nitrate-N represents about 2% of the total nitrogen (2 mg/1), ammonia-N represents nearly 60% (48 mg/1), and organic nitrogen slightly less than 40% (32 mg/1). Because of the high pH in the wastewater (8.5), ammonia volatilization was estimated at 60% of the applied ammonia-N. The mineralization rate for the organic nitrogen was estimated at 30% the first year, 15% of the second year, 8% the third year, and leveling out at 3% in the fifth and succeeding years.

Based on the estimates presented above and an annual wastewater volume of 7.6 million liters/year, the Np for the first year was calculated as 234 kg/yr. By the fifth year the nitrogen mineralization reaches equilibrium, and the amount of Np is 280 kg/yr. At ten years, with continued nitrogen mineralization, Np equals 300 kg/yr. The mixed hardwood and pine forests were conservatively estimated to have a nitrogen uptake and storage rate of 100 kg/ha per year. The resulting land area requirement from the calculations presented above was 3.0 ha for nitrogen assimilation after ten years.

Oxygen-demanding waste constituents, reflected by COD, require an aerobic environment for assimilation. The study site soils are well drained and capable of assimilating a large amount of COD if the soils are not hydraulically overloaded. The assimilative capacity, based on oxygen diffusion calculations, was estimated to be 15,000 kg/ha per year. The mass generation rate of COD of 135,000 kg/yr, divided by the assimilative capacity, yielded a land requirement of 9 hectares for COD assimilation.

The assimilative capacity for dissolved solids is related to two factors: (1) the effect of salts on the osmotic potential of vegetation on the site and (2) the effect of dissolved solids migrating to groundwater. If the salinity of soil water exceeds the osmotic potential, an adverse vegetation response occurs. The conductivity of the NSCC wastewater was measured at 8 mmhos/cm, which is much higher than normal irrigation water. Considering the relatively small volume of wastewater to be applied (470,000 l/wk), the average rainfall for the area (120 cm/yr), and the large land area available, the application of dissolved solids at a concentration of 6,000 mg/1 was not considered to be detrimental.

Leaching of dissolved solids from land treatment sites must be controlled so that the secondary drinking water standard of 500 mg/1 is not exceeded in receiving groundwater. Calculations were performed to estimate the impact of rainfall and mixing with ambient groundwater on the concentration of dissolved solids that leach out of the soil profile. These calculations indicated that the wastewater must be distributed over 8.4 hectares to accomplish the necessary dilution.

After evaluating the critical waste constituents, COD was identified as the land limiting constituent. To adequately assimilate the COD loading a total of 9 hectares was required. For the plans being made for plant expansion within the next five years, the decision was to install a system on 11 hectares.

DESIGN CRITERIA

The application of 470,000 liters per week to 11 hectares results in an average loading of 22 cm/yr, or less than 0.5 cm/wk over the entire site. This is well below the hydraulic loading limit of 16 cm/month. Because of the low hydraulic loading, only a portion of the site is irrigated each week. The site is divided into seven zones, with a single zone irrigated two or three times during a week with an average application of 3 cm/wk. Irrigation is moved to another zone for the next week until all seven zones are irrigated.

Experience on the earlier land treatment system demonstrated the problems associated with installing irrigation distribution systems underground in forests. Because of this experience, an aboveground system consisting of aluminum irrigation pipe was selected. Self-draining gaskets between sections of pipe allow wastewater to drain from the pipe and risers, eliminating freeze damage during the winter. Sprinklers are the impact type rated at 19 1/min (5 gpm), 3.4 atm (50 psi) with 30 m (100 ft) diameter. A new pump rated at 900 1/min (240 gpm) with a 20 hp motor was installed.

MONITORING

A monitoring system consisting of three downgradient and one upgradient groundwater wells set at approximately 8 meters below ground surface was implemented. Four pairs of lysimeters were also installed; each pair included a shallow lysimeter at 30 cm and a deeper lysimeter at 90 cm. Monitoring was scheduled on a quarterly basis in accordance with DHEC requirements. Representative results of the monitoring data are summarized in Table 2. All groundwater data have consistently been within primary

TABLE 2. Representative monitoring results.

Parameter	Upgradient Wells	Downgradient Wells	Lysimeters
Nitrate (mg/l)	1.3	0.2	0.3
Total dissolved solids (mg/l)	160	100	500
Chemical oxygen demand (mg/l)	10	40	80
pH	7.2	5.1	7.3
Specific conductance (μmhos)	180	50	600

and secondary drinking water standards. As expected, lysimeter monitoring data indicate that there is a higher concentration of dissolved solids and COD in the soil water than in the groundwater. The low concentration of nitrate in the lysimeters supports the nitrogen calculations performed during the process design phase.

CONCLUSION

Proper design is critical to achieving water quality standards for receiving groundwater at land treatment sites. When design and operation are optimized, it is possible to achieve a high level of wastewater treatment in the soil before recharge to groundwater occurs. The earlier forest land treatment system operated by NSCC highlighted some of the problems that may be encountered when proper design procedures are not followed. The new system, designed using a state-of-the-art approach, has operated well with little more than routine maintenance.

REFERENCES

Overcash, M. R., and D. Pal. 1979. Design of land treatment systems for industrial wastes: Theory and practice. Ann Arbor Science Publishers, Ann Arbor, Michigan.

Soil Conservation Service. 1978. Soil Survey of Spartanburg County. U.S. Department of Agriculture.

U.S. Government Printing Office, Washington, D.C.

U.S. Environmental Protection Agency. 1983. Process design manual: Land application of municipal sludge. EPA 625/1-83-016.

Chapter 51

Applying a Resin Based Sludge
to Forested Areas: A Case Study

GLENN L. TAYLOR and RON D. PRESLEY

ABSTRACT   In the past, most resin plants in the wood products industry have placed waste sludge in lagoons for long-term storage. The Georgia-Pacific facility in Conway, North Carolina had approximately 16,320 cubic meters (21,340 cubic yards) of resin sludge stored in lagoons. The sludge contained phenol, formaldehyde, 90,000 ppm COD, 6,930 ppm TKN, 20% total solids, and high levels of calcium, sodium, and sulfate (3,500 ppm, 4,300 ppm, and 6,790 ppm, respectively). Of the 52 acres (21.2 ha) investigated for potential land treatment of sludge, 32 acres (12.9 ha) were in a loblolly pine plantation about ten years old. The sludge characteristics that determined the amount of land area needed to treat the waste were organic matter, nitrogen, and sulfate. All 52 acres were utilized. A decision was made to make three separate applications of sludge, at least 90 days apart. After considering several application techniques, spray irrigation was selected. The sludge was diluted to about 10% solids and pumped to a booster pump near the application point. The booster pump fed a traveling irrigation unit. The first application occurred during October 1984, with satisfactory results. The traveling gun performed better than expected in the forested area, with no damage occurring to the trees. Groundwater wells are being monitored, with no indications of any impacts from the sludge application.

BACKGROUND OF THE PROBLEM

Georgia-Pacific Corporation, at its resin facility in Conway, North Carolina, manufactures wood and paper product resins. The plant is a major supplier of wood product adhesives for the United States upper eastern coast. In addition to the adhesives, several other resins and formaldehyde are manufactured on site. In the late 1960s, three above-ground, lined lagoons were constructed to temporarily store urea-formaldehyde and phenol-formaldehyde based resin wastes. In addition, tank and sump cleanings, truck washouts, and customers' excess glue wastewater were stored in the lagoons. The intended use of the lagoons was to hold the wastewater until the material could be recycled to the reactors for reuse.

In 1981, two concerns became evident. First, the synthetic liners were deteriorating; second, a thick resin sludge had accumulated in all three lagoons, greatly reducing the wastewater storage volume. A decision was made to switch to steel tanks for the storage of wastewater prior to reuse. The problem was how to handle the lagoons in an environmentally acceptable manner.

Since the advent of RCRA and more stringent solid waste regulations, many wood products plants and resins plants are being faced with disposal difficulties and rapidly rising costs. The sludge at the Conway facility contained enough free phenol and formaldehyde to prohibit disposal in a sanitary landfill, but not enough to be classified as a hazardous waste. Considering the sludge volume of 16,320 cubic meters (21,340 cubic yards), it became evident that disposal could be extremely expensive.

ANALYSIS OF OPTIONS

An engineering evaluation of the known disposal alternatives was undertaken in 1981 by the Georgia-Pacific Central Engineering staff. Extensive laboratory tests were performed on sludge composites from each lagoon to determine constituents and concentrations present. Dewatering, filtration, chemical fixation, oxidation, solidification, as well as sealing and capping in place, were all explored. Each was eliminated because of expense or lack of effectiveness on resin wastes. During an in-house review of options in late 1981, the idea of utilizing land treatment technology emerged.

Preliminary discussions were held with the North Carolina Department of Natural Resources and Community Development, Division of Environmental Management (DEM), in September 1982. The discussions revealed that the state agency would consider issuing a land treatment permit, provided proper studies were performed.

Additional sludge testing, performed in December 1982, is summarized in Table 1. These tests indicated a drop in both urea-formaldehyde and phenol-formaldehyde concentrations from the tests conducted in 1981. This drop was attributed to biological activity and other oxidation within the lagoons.

During early 1983, the Georgia-Pacific Central Engineering staff performed an extensive international data search, which revealed that a land treatment project for this type of waste had never been documented. Therefore, no historical data concerning applica-

TABLE 1. Sludge characterization.

Constituent	Composite Concentration (ppm)	Mass (kg)	(lb)
Phenol	3.77	61.5	135.6
Formaldehyde	0.35	5.7	12.6
BOD	90.0	1,468	3,237
COD	90,000	1,468,392	3,237,250
Total Kjeldahl-N	6,930	158,425	349,268
Ammonia N	204	3,328	7,338
Nitrate and nitrate N	0.41	6.7	14.7
Chloride	147	2,399	5,288
Calcium	3,500	57,104	125,893
Sodium	4,300	70,157	154,669
Phosphorus	9.78	160	352
Sulfate	6,790	110,782	244,232
Potassium	118	1,925	4,244
Chromium	1.47	24.0	52.9
Mercury	0.009	0.1	0.3
Zinc	9.87	161	355

Methylene chloride extract		EP extract	
Phenol content	36 ppm	Extractable phenol	25 ppm
TOC content	250,000 ppm	Extractable TOC	2,300 ppm

BOD = biochemical oxygen demand.   COD = chemical oxygen demand.
TOC = total organic carbon.

tion rates, permit limits, or results were available. For this reason, it was decided to retain outside land treatment consultants to assist in determining the feasibility of the project. In early 1983, ERM-Southeast, Inc. was retained to conduct a feasibility evaluation of the site and to determine loading rates.

LAND TREATMENT FEASIBILITY

The preliminary site selected consisted of four areas totaling 21.2 hectares (52 acres) of land that surrounded the plant. About 12.9 hectares (32 acres), or 62%, of the preliminary site were planted in young pines 4.0 meters (13 feet) high; the remainder of the site was in mixed grass fields. The first step was to address the soil science, geophysical, and groundwater conditions on the site. Ten soil test pits were dug on the site to characterize the shallow soil profiles and for acquiring samples for physical and chemical analysis. Five USDA Soil Conservation Service (SCS) soil series were identified: Norfolk, Goldsboro, Lynchburg, Dunbar, and Rains. The distribution of the soil series on the site is

TABLE 2. Distribution of soil types.

Site	Norfolk	Goldsboro	Lynchburg	Dunbar	Rains	Total
1	5.4	0.2	3.7	--	3.3	12.6 (5.1)
2	0.1	4.0	--	1.3	--	5.4 (2.2)
3	1.8	--	--	0.7	--	2.5 (1.0)
4	23.8	3.4	--	4.6	--	31.8 (12.9)
Total	31.1	7.6	3.7	6.6	3.3	52.3 (21.2)
	(12.6)	(3.1)	(1.5)	(2.7)	(1.3)	

Area expressed in acres with hectares in parenthesis.

shown in Table 2. Table 3 summarizes the published physical characteristics of the five soil series based on SCS information.

The Norfolk and Goldsboro soil series were considered to be well drained with the seasonally high water table at adequate depth for year-round sludge application. These two soils make up 74% of the land area being considered. The remaining three soil series make up 26% of the land area. These three soils have a high water table during the winter and early spring that would interfere with the year-round operation of a land treatment system.

A drill rig was used to bore four deeper soil borings and to install groundwater monitoring wells. The site lies within the Middle Coastal Plain Physiographic and Geologic Province of North Carolina. The region is characterized by a thick wedge of sedimentary material which overlies the ancient Precambrian basement complex. Geologic units consist of variability bedded sand, clay, limestone, and marl. The sediments are divided into hydrologic characteristics. The monitoring wells were sampled to establish a background level for reference evaluation and future monitoring. The results are presented in Table 4.

The land area requirements are based on the capacity of the land to assimilate the individual waste constituents such that the land is not irreversibly converted to an unproductive state. The Georgia-Pacific situation at the Conway plant is different from that

TABLE 3. SCS published soil characteristics.

Soil Series	Horizon	Texture (USDA)	Permeability (in./hr)	Permeability (cm/hr)	pH
Norfolk	A	sandy loam	2.6-6.0	5.1-15.2	5.5
	B1	sandy loam	0.6-2.0	1.5- 5.1	5.5
	B2	sandy clay loam	0.6-2.0	1.5- 5.1	5.0
	B3	--	--	--	5.0
Goldsboro	A	loamy sand	2.0-6.0	5.1-15.2	5.5
	B	sandy loam	0.6-2.0	1.5- 5.1	5.0
Lynchburg	A	sandy loam	2.0-6.0	5.1-15.2	4.5
	B	sandy clay loam	0.6-2.0	1.5- 5.1	4.5
	C	clay loam	0.6-2.0	1.5- 5.1	4.5
Dunbar	A	sandy loam	2.0-6.0	5.1-15.2	5.0
	B	sandy clay	0.2-0.6	0.5- 1.5	5.0
Rains	A	loamy sand	2.0-6.0	5.1-15.2	5.5
	B1	sandy loam	0.6-2.0	1.5- 5.1	5.0
	B2	sandy clay loam	0.6-2.0	1.5- 5.1	5.0
	B3	sandy loam	0.6-2.0	1.5- 5.1	5.0

of most land treatment systems, since sludge will not be continuously applied. The waste management practices have been modified and the lagoons will not be further utilized for sludge holding. Once emptied, the lagoons will be permanently closed. Each constituent in the sludge was evaluated and compared with the various soil types on site. It was found that organic matter, nitrogen, and sulfate were the limiting constituents. From the loading rate calculations, it was determined that 5,438 cubic meters (7,113 cubic yards) of sludge could be applied to 15.4 hectares (38 acres) every three months for the three application periods without environmental degradation. This equates to 3.6 centimeters (1.4 inches) of sludge over the 15.4 hectares.

The completed study, test data, and loading rates were submitted to the North Carolina DEM in March 1984. Owing to the precedent-setting nature of the project, DEM performed a very intensive review of the application. Experts in waste treatment, agriculture, and geology from North Carolina State University were called in to review the report and project. On June 6, 1984, a land treatment permit was issued.

TABLE 4. Average background groundwater quality before sludge application.

Constituent	Concentration
Phenol	<0.3 mg/l
Formaldehyde	<0.11 mg/l
Chromium	<0.03 mg/l
Zinc	0.02 mg/l
Chloride	4.68 mg/l
Phosphate	0.07 mg/l
Nitrate-N	4.8 mg/l
pH	5.6

APPLICATION METHOD

After study and discussion with sludge contractors, a Bush mobil spray system fed by a Guzzler sludge pump was determined to be the most cost-effective means to remove, transport, and uniformly apply the sludge. Bryson Environmental was selected to perform the contract pumping. The first phase of the application took eighteen days in October 1984. A modified Chaney-Bush traveling spray irrigation unit moved through the fields to apply the sludge. The 18 to 20% solids sludge was diluted to about 10% solids and pumped from the lagoon using a Guzzler hydraulic pump. The sludge was pumped for as much as 762 meters (2,500 feet) to an irrigation booster pump and from the booster pump, another 198 meters (650 feet) to the traveler irrigation unit. The traveler was supplied with a telescoping orchard mount to spray above the young pine trees. A Nelson "Big Gun" with a 5.1 centimeter (2 inch) nozzle was used to apply the sludge at 3,785 liters per minute (1,000 gallons per minute), at 125 psi.

RESULTS

Results from the monitoring program indicate better than anticipated results. Within two days after spraying, grasses began to turn darker green. Anaerobic soil conditions did not occur, and a light disking, which was anticipated, was not necessary. Pines sprayed with sludge were washed clean by the first rain. Close inspection of the needles by Georgia-Pacific scientists revealed no adverse effects from spray impact or sludge constituents. Soil sampling by both DEM and Georgia-Pacific, in December 1984, confirmed excellent soil conditions after the sludge application. Groundwater monitoring has not shown any increase above drinking water standards, though there have been some slight changes in nitrate-N, pH, and chloride concentrations. Table 5 presents a summary of the groundwater monitoring results since the first application. The last phase of sludge application was completed in October and November 1985, and the ponds have been closed.

TABLE 5.   Average groundwater quality after first sludge application.

Constituent	Concentration
Phenol	<0.4 mg/ml
Formaldehyde	0.11 mg/l
Chromium	0.03 mg/l
Zinc	0.02 mg/l
Chloride	6.87 mg/l
Phosphate	0.08 mg/l
Nitrate-N	8.9 mg/l
pH	6.0

COSTS

Research and development costs were approximately $20,000. Equipment costs were approximately $15,000. Implementation costs for each application were approximately

$45,000. Engineering costs to monitor the application procedures and soils were approximately $14,000. The equipment used on this site will be moved to other Georgia-Pacific sites as the need arises, so a portion of the equipment costs will be recoverable. The approximate $10 per cubic yard costs ($13 per cubic meter), including the research and development costs, were significantly less than the $60 to $70 per cubic yard costs ($78 to $92 per cubic meter), plus transportation in some cases, that were projected for the alternatives.

AWARD

This project won the national 1984 Environmental and Energy Achievement Award for Solid Waste Management, presented by the American Paper Institute and the National Forest Products Association.

Contributors

W. G. Algiere
National Starch and Chemical Corporation
P.O. Box 578
Woodruff, SC 29388

E. Allender
Manager
Land Energy Pty. Ltd.
P.O. Box 1
Macclesfield, South Australia 5153

Ken Arnold
Unit Chief
Land Application Unit
Missouri Department of Natural Resources
Jefferson City, MO 65101

Don Barnett, P.E.
Environmental Engineer
Land Application Unit
Missouri Department of Natural Resources
Jefferson City, MO 65101

Robert K. Bastian
Office of Municipal Pollution Control
U.S. Environmental Protection Agency
401 M Street, S.W.
Washington, DC 20460

Charles R. Berry
Institute for Mycorrhizal Research and Development
Forestry Sciences Laboratory
Athens, GA 30602

D. G. Briggs
Assistant Professor
College of Forest Resources
University of Washington
Seattle, WA 98195

Dale G. Brockway
Department of Natural Resources
P.O. Box 30028
Lansing, MI 48909

Robert S. Burd
Director, Water Division
Region 10
U.S. Environmental Protection Agency
Seattle, WA 98101

Andrew J. Burton
Department of Forestry
Michigan State University
East Lansing, MI 48824

Henry Campa III
Department of Fisheries and Wildlife
Michigan State University
East Lansing, MI 48824

Debra Carey
Biological Technician
USDA Forest Service
Southeastern Forest Experiment Station
Research Triangle Park, NC 27709

A. C. Chang
Professor of Agricultural Engineering
University of California
Riverside, CA 92521

Terrill J. Chang
SCS Engineers
1008 140th N.E.
Bellevue, WA 98005

Roberta Chapman-King
Municipality of Metropolitan Seattle
821 Second Avenue
Seattle, WA 98104
(formerly with the College of Forest Resources,
University of Washington, Seattle)

Dale W. Cole
College of Forest Resources
University of Washington
Seattle, WA 98195

J. C. Corey
Environmental Sciences Division
E. I. du Pont de Nemours & Co., Inc.
Savannah River Laboratory
Aiken, SC 29808

Ronald W. Crites
George S. Nolte & Associates
1700 L Street
Sacramento, CA 95814

C. E. Davis
Forest Research Consultant
E. I. du Pont de Nemours & Co., Inc.
Savannah River Laboratory
Aiken, SC 29808

Lisa A. Donovan
Savannah River Ecology Laboratory
University of Georgia
P.O. Drawer E
Aiken, SC 29801

R. L. Edmonds
College of Forest Resources
University of Washington
Seattle, WA 98195

Larry M. Gigliotti
Department of Fisheries and Wildlife
Michigan State University
East Lansing, MI 48823

Charles C. Grier
School of Forestry
Northern Arizona University
Flagstaff, AZ 86011
(formerly with the College of Forest Resources,
University of Washington, Seattle)

A. Ray Harris
Principal Soil Scientist
Forestry Sciences Laboratory
Grand Rapids, MN 55744

James B. Hart, Jr.
Department of Forestry
Michigan State University
East Lansing, MI 48824

John H. Hart
Department of Forestry
Michigan State University
East Lansing, MI 48824

Jonathan B. Haufler
Department of Fisheries and Wildlife
Michigan State University
East Lansing, MI 48824

Charles L. Henry
College of Forest Resources
University of Washington
Seattle, WA 98195

Thomas M. Hinckley
College of Forest Resources
University of Washington
Seattle, WA 98195

Michael E. Hodgson
Geography Department
University of South Carolina
Columbia, SC 29208

G. J. Hollod
E. I. du Pont de Nemours & Co., Inc.
Wilmington, DE 19805

Michael D. Huddy
Northwest Hardwoods
4640 S. W. McAdams
Portland, OR 97201
(formerly with Crown Zellerbach Forestry Research
Division)

John R. Jensen
Geography Department
University of South Carolina
Columbia, SC 29208

R. L. Kendall
Earth Systems Associates
2563 Hearthwood Place
Marietta, GA 30064

P. Kube
Research Forester
Conservation Commission of the Northern Territory
P.O. Box 1046
Alice Springs, Northern Territory 5750
Australia

T. J. Logan
Professor of Soil Chemistry
The Ohio State University
Columbus, OH 43210

Elliot D. Lomnitz
U.S. Environmental Protection Agency
401 M Street, S.W. (WH-585)
Washington, DC 20460

M. W. Lower
E. I. du Pont de Nemours & Co., Inc.
Savannah River Laboratory
Aiken, SC 29808

Gary S. MacConnell
Environment Engineer
Camp Dresser & McKee Inc.
Maitland, FL 32751

Peter S. Machno
Manager
Sludge Management Program
Municipality of Metropolitan Seattle
821 Second Avenue
Seattle, WA 98104

Halkard E. Mackey, Jr.
Savannah River Laboratory
Aiken, SC 29808

W. H. McKee, Jr.
Soil Scientist
USDA Forest Service
Southeastern Forest Experiment Station
Research Triangle Park, NC 27709

M. R. McKevlin
Biologist
USDA Forest Service
Southeastern Forest Experiment Station
Research Triangle Park, NC 27709

Kenneth W. McLeod
Associate Research Ecologist
Savannah River Ecology Laboratory
Institute of Ecology
University of Georgia
P.O. Drawer E
Aiken, SC 29801

R. C. McNeil
Biological Technician
USDA Forest Service
Southeastern Forest Experiment Station
Research Triangle Park, NC 27709

F. Mecifi
Research Assistant
College of Forest Resources
University of Washington
Seattle, WA 98195

Dennis M. Merkel
Department of Forestry
Michigan State University
East Lansing, MI 48824

Louis J. Metz (retired)
Principal Soil Scientist
USDA Forest Service
Southeastern Forest Experiment Station
Research Triangle Park, NC 27709

Barbara E. Moore
Department of Civil Engineering
The University of Texas at Austin
Austin, TX 78712

Sydney Munger
Senior Microbiologist
Municipality of Metropolitan Seattle
821 Second Avenue
Seattle, WA 98104

C. E. Murphy
Research Ecologist
Savannah River Laboratory
E. I. du Pont de Nemours & Co., Inc.
Aiken, SC 29801

Phu V. Nguyen
Department of Forestry
Michigan State University
East Lansing, MI 48824

Charles G. Nichols
Municipality of Metropolitan Seattle
821 Second Avenue
Seattle, WA 98104

Wade L. Nutter
School of Forest Resources
University of Georgia
Athens, GA 30602

Michael Overcash
Professor of Chemical Engineering
North Carolina State University
Raleigh, NC 27695

A. L. Page
Professor of Soil Science
University of California
Riverside, CA 92521

R. Ben Peyton
Department of Fisheries and Wildlife
Michigan State University
East Lansing, MI 48823

Ron D. Presley
Senior Environmental Engineer
Georgia-Pacific Corporation
P.O. Box 105605
Atlanta, GA 30348

Carl W. Ramm
Department of Forestry
Michigan State University
East Lansing, MI 48824

Jane T. Red
School of Forest Resources
University of Georgia
Athens, GA 30602

Sherwood C. Reed
U.S. Army Cold Regions Research and Engineering
Laboratory
Hanover, NH 03755

Gail L. Ridgeway
University of New Hampshire
Botany Department
Durham, NH 03824
(formerly with Lake Superior State College)

P. Sandell
Forester-in-charge (Southern Region)
Conservation Commission of the Northern Territory
P.O. Box 1046
Alice Springs, Northern Territory 5750
Australia

K. C. Sherrod
Savannah River Ecology Laboratory
Institute of Ecology
University of Georgia
P.O. Drawer E.
Aiken, SC 29801

Walter J. Shields, Jr.
CH2M Hill, Inc.
1500 114th S.E.
Bellevue, WA 98004
(formerly with Crown Zellerbach Forestry Research
Division)

W. R. Smith
Associate Professor
College of Forest Resources
University of Washington
Seattle, WA 98195

Sheldon G. Somers
University of Washington
Pack Forest
Eatonville, WA 98328
(formerly with Crown Zellerbach Forestry Research
Division)

William E. Sopper
Professor of Forest Hydrology
Institute for Research on Land and Water Resources
The Pennsylvania State University
University Park, PA 16802

Charles A. Sorber
Department of Civil Engineering
The University of Texas at Austin
Austin, TX 78712

John Spencer
Director
Water Pollution Control Department
Municipality of Metropolitan Seattle
821 Second Avenue
Seattle, WA 98104

H. T. L. Stewart
Scientific Officer
State Forests and Lands Service
Department of Conservation
Forests and Lands
G.P.O. 4018
Melbourne, Victoria 3001
Australia

D. M. Stone
Research Forester
USDA Forest Service
North Central Forest Experiment Station
Grand Rapids, MN 55744

Glenn L. Taylor
Principal
ERM-Southeast, Inc.
Suite 201
2623 Sandy Plains Road
Marietta, GA 30066

William E. Thacker
Research Engineer
National Council of the Paper Industry for Air and
Stream Improvement, Inc. (NCASI)
Western Michigan University
Kalamazoo, MI 49008

H. A. Thomas
Entomologist
USDA Forest Service
Southeastern Forest Experiment Station
Research Triangle Park, NC 27709

Roberto Torres
Merck Sharp & Dohme Quimica de Puerto Rico, Inc.
Barceloneta, PR 00617

Dean H. Urie
Research Associate
Department of Forestry
Michigan State University
East Lansing, MI 48823
(formerly Principal Hydrologist, USDA Forest Service)

Jerome P. Weber
North Carolina State University
Weed Science Center
Box 7627
Raleigh, NC 27695

Carol G. Wells
Principal Soil Scientist
USDA Forest Service
Southeastern Forest Experiment Station
Research Triangle Park, NC 27709

Stephen D. West
College of Forest Resources
University of Washington
Seattle, WA 98195

David K. Woodyard
Department of Fisheries and Wildlife
Michigan State University
East Lansing, MI 48824

R. J. Zasoski
College of Forest Resources
University of Washington
Seattle, WA 98195

Index

Abies procera. See Fir

Acacia aneura, 437

Acer spp. *See* Maple

Acidity of soil, 50, 66, 385, 403; and trace elements, 87, 88; and cadmium regulations, 89; cropland, 91; and solubility of trace elements, 97; and leaching of metals, 107, 108; and aluminum, 108; and nitrification and leaching study, 133, 136; in nitrogen transformations study, 145, 148; in element flux study, 155, 156; in heavy metal study, 171, 172–174; in oak studies, 288, 289; in Southeast, 309; in Penn State study, 408, 409, 416; in Australia, 439; in Georgia study, 448, 450; in Copper Basin, 499; in borrow pit, 500; in kaolin spoil, 503; of Barceloneta sludge and soil, 511, 512; at Savannah River Plant, 520, 521–522, 527; in Wauna experiments, 539

Adelaide, Australia, 435, 437

Adsorption: retention of viruses in soil, 81; and leaching of metals, 107; of nitrates, 107; SAR, 43–44, 538, 539, 540, 547

Aerial photography, 325

Aerosols, 121, 122

Aesthetic concerns of public, 375, 376

Agriculture: sludge used in, 12, 20, 168; wastewater used in, 28, 56; and maintenance, 37; use of pulp and paper sludge for, 44, 46, 259–260; hydraulic loading capacity, 57; and trace elements, 87–90; cadmium effects, 89; contrasts between cropland and forest land, 90–94, 125–126, 178, 356; old fields, 91; soil profile, 91; and organic priority pollutant study, 126–131; and nitrification, 132; heavy metal buildup, 168; public acceptance of land application, 341–342, 372; sludge guidelines for, 384; tests in Puerto Rico, 510

Agronomic rate, 385–386

Aiken, South Carolina, 179, 325, 326. *See also* Savannah River Plant

Air pollution, 23, 498, 499, 509

Alabama, 497, 498

Alder, 50; red, 467, 534–535, 537–539; black, 503

Alice Springs, Australia, 433, 434, 437–439

Alpena, Michigan, 145, 190, 266, 283, 481, 485, 491

Aluminum, 225; in pulp and paper sludge, 45, 46; toxicity in acid soils, 108; in element flux study, 162, 163, 166; in oak growth studies, 284, 287, 288; at northern Michigan sites, 484, 486, 484; in Wauna sludge and soil, 534, 539

Amelanchier. See Serviceberry

Ammonia, 35, 50, 225, 228, 232, 233, 236, 237, 261, 264, 289, 399, 402, 511, 512, 546, 553, 557; volatilization, 34, 142, 225, 228, 230, 234, 240, 387, 465, 543, 546; in nitrification study, 133, 134, 136, 140; in nitrogen transformations study, 148, 151, 152; "ammonia stripping," 225; nitrification, 234–235; in regional sludges, 239, 240

Ammonium: nitrification and leaching in soil treated with sulfuric acid and nitrification inhibitor, 132–141; ammonium-N concentrations, 137, 139, 147, 149, 151, 156, 159–162, 163, 165, 273, 408, 445, 454, 517; in nitrogen transformations study, 142–152; in element flux study, 159–166; in loblolly growth study, 273

Animals. *See* Wildlife

Anions, 228, 232, 241. *See also* Leaching

Antimony, 273, 517

Application methods, 21, 67, 156, 169, 274–275, 283, 481, 482; effect on wildlife, 111; technology for wastewater use, 349–354, 554–555; technology for sludge use, 356–366, 463, 481, 482; sprayers, 143, 156, 357, 362, 365, 407, 417, 422–423, 424, 462, 463, 481, 482, 491, 516, 560; sprinklers, 349–354, 399–405, 417, 421, 422–424, 426, 433, 434, 443, 453, 554–555; traveling gun, 350, 354, 362–363, 539, 543, 544, 560; manure spreaders, 357, 358, 362, 364, 469, 516; over-the-canopy, 357, 462, 463, 463–464, 560; sludge hauling, 359–360, 469, 481, 491, 492, 539, 560; transfer, 360–362, 462, 469, 475, 481; sludge systems, 362–364; comparison of systems, 365; in Australia, 436; uniformity of application, 481; distance from application vehicle, 487–489; sludge injection, 513; pipeline, 533, 536, 543, 544–546, 554–555; foliar and root applications, 536–537; chopper pump, 541, 543; water jack, 543, 544; mobile spray system, 560; for resin based sludge, 560; hydraulic pump, 560. *See also* Irrigation; Technology

Application rates. *See* Design criteria

Arizona, 27

Armillaria mellea, 269–271, 494

Arsenic, 511; in pulp and paper sludge, 45, 46; in sewage sludge, 86, 87; in loblolly pine study, 273; in Savannah River sludge, 517; in Wauna sludge, 534

Ascaris, 75, 76, 78, 79, 81, 82

Ash, 401, 454; white, 143, 200, 201, 203, 422, 480; green, 502

Aspen (*Populus* spp.), 34, 95, 111, 112, 113, 115, 143, 209, 350; in nitrogen transformations study, 143, 144, 145, 147, 148, 149, 150, 151; heavy metal storage in soil, 168–175; deer and elk forage, 188–197; nitrogen uptake, 236, 237; sludge application rates, 239, 388; nitrogen transformation and utilization (conceptual model), 240; mortality study, 266–271; in short-term growth study, 282–283; in long-term growth study, 294; in northern Michigan, 480, 481, 482, 483, 484, 487, 488, 494. *See also* Bigtooth aspen; Quaking aspen

Atlanta. *See* Clayton County, Georgia

Attitudes. *See* Public opinion

Augusta, Georgia, 233. *See also* Savannah River Plant

Augusta Wastewater Treatment Plant, 179, 516–525

Australia, 28, 209; irrigation of tree plantations, 431–440; research, 432–433; case studies, 433–440; effluent characteristics, 434, 437; growth response, 435, 438; health concerns, 440

Automobiles, 96

Ava, Missouri, 457

Bacteria, 74–76, 79–80; size, 82; die-off rate, 82; fecal coliforms, 81, 101, 103, 104

Balsam fir, 34, 350

Barceloneta Wastewater Treatment Plant, 508–513

Barium, 45, 534

Basal area effects, 487–491

Baseflow, 403–404

Basswood, 143, 200, 480

Beech, 422, 424, 480; American, 143, 200

Benzo(a)pyrene, 120–121, 123–124

Bennett Spring State Park, 27, 452–457

Berms, 551

Berries, 121, 122, 123, 200. *See also individual species*

Best management practices (BMPs), 23, 340, 462, 473

Betula spp. *See* Birch

Big Cotton Indian Creek, 395

Bigtooth aspen, 112, 143, 200; in nitrogen transformations study, 143, 144, 145, 147, 148, 149, 150, 151; heavy metal storage in soils, 168–175; deer forage, 189, 191, 192, 193, 194, 195, 196; in vegetation and small mammal study, 201–204; mortality study, 266–271; in northern Michigan, 480, 481, 482, 483, 484, 487, 488, 494

Biochemical oxygen demand (BOD), 42–43, 48, 399, 536, 557

Biological constituents in sludge, 18, 19

Biomass: trace elements in, 93–94; effect of interaction between site and fertilization on, 215–216; effect of nitrogen on, 237; remote sensing using airborne data, 324–333; harvesting, 401; prescribed burning, 450; fiber plantation, 467–468. *See also* Vegetation

Birch, 143, 350, 422, 424, 480

Bis(2-ethylhexyl)phthalate, 18, 120, 511, 512

Boise Cascade, 260–264, 466, 475

Boron, 31, 86; in pulp and paper sludge, 45; in sewage sludge, 86, 87; toxicity, 88; effect on wildlife, 114; in nitrification and leaching study, 135; in element flux study, 156; in loblolly pine study, 273; in Clayton County, 399; in Penn State study, 408; at northern Michigan sites, 484, 485, 486; in Barceloneta sludge and soil, 511; in Savannah River sludge, 517

Borrow pits, 497–498, 499–502

Bracken fern, 189, 200, 201, 203

Brambles, 189, 200

Bremerton, Washington, 356, 364

Brockton, Massachusetts, 3

Bryson Environmental, 560

Buffer zones, 66, 67, 118, 189, 200, 396, 401, 465, 466, 467, 551

Burning, 443–450

Cable News Network, 475

Cadillac, Michigan, 169, 233

Cadmium: in pulp and paper sludge, 45, 46, 259; pretreatment, 86; in sewage sludge, 86, 87; toxicity, 88–89, 114, 320, 321; regulation by EPA, 89; recommended limits, 90; in herbivores and carnivores, 92, 320; in deer tissue, 193, 197; in forest soil, 93, 95; from smelter, 94, 95; leaf tissue concentrations, 95; and drinking water guidelines, 101; in leachates, 106–107; in forages, 114–115; in nitrification and leaching study, 135; in element flux study, 156, 162, 163, 166; in heavy metal study, 169, 170, 171, 172, 174; in vegetation and small mammals, 202, 203, 204; in oak growth studies, 284, 287, 288; in loblolly pine studies, 273, 315, 316, 317, 318, 319, 320, 321; in Penn State study, 408, 409, 413, 414, 415, 416; at northern Michigan sites, 484, 486; in Barceloneta sludge and soil, 510, 511, 512; in Savannah River sludge and soil, 517, 520, 521; in Wauna sludge, 534

Calcium: in pulp and paper sludge, 45; in pulp and paper wastewater, 48, 49; in forest biomass and soil, 91, 224, 241, 450; in nitrification and leaching study, 134, 135, 136, 138, 139; leaching of, 154, 234; in element flux study, 155, 156, 161, 162, 163, 164, 165, 166; and pH, 172; renovation of, 228; in sludge, 232, 233, 234; application rates, 233; in loblolly pine studies, 273, 315, 316, 317, 318, 319; in oak growth studies, 284, 287, 288, 289; in Clayton County, 399; in Penn State study, 408, 409, 415, 416; in Australian irrigation, 437; at Georgia site, 445, 446, 447, 448, 449; at northern Michigan site, 484, 486, 494; in Copper Basin soil, 499; in borrow pit, 500; in kaolin spoil, 503; in Barceloneta sludge and soil, 511, 512; in Savannah River sludge and soil, 517, 520, 521, 522, 525; in Wauna sludge and soil, 534, 538, 539, 540, 547; in NSCC effluent, 553; in resin based sludge, 557

California, 292, 350, 352, 354

Cancer, 119, 122, 123

Cano Tiburones, 508–518

Carbon: C:N ratio, 46, 142–143, 145, 148, 156, 157, 235, 238, 258, 259, 261–262, 263, 264, 278, 534, 543, 547; organic, 49, 145; total organic carbon (TOC), 104, 105, 511, 512, 541, 557; net primary productivity, 217; in Pack Forest study, 261; and denitrification, 448; in Savannah River Plant soil, 520, 521

Carbon 14 analysis, 126–131

Carcinogens, 119, 122, 123

Carex spp. *See* Sedge

Carnation, Washington, 475

Carpinus caroliniana. See Ironwood

Carya tomentosa. See Hickory

Casuarina, 432

Cation exchange capacity (CEC): and cadmium in soil, 89, 90; and trace elements, 90; and forest soils, 91, 301, 302, 521; in kaolin spoil, 503

Cations, 228, 232, 241

Cattle, 435
Cedar, 388
Center for Disease Control, 74, 76
Cervus elaphus canadensis. See Elk
Chapman Forestry Foundation, 354
Chemical oxygen demand (COD), 551, 553, 554, 555, 557
Chemical substances in sludge, 18, 19. *See also* Organic substances; Toxic substances
Cherry (*Prunus* spp.), 143, 480; black, 266. *See also* Pin cherry
Chipmunks, 113, 201, 204
Chloride, 49, 228, 241, 402, 511, 512, 527, 551, 552, 553, 557, 559, 560
Chlorinated lignin derivatives, 47
Chlorine, 399, 408, 427, 437, 445, 446; in pulp and paper wastewater, 44; in pulp and paper sludge, 45, 47
Chloroform, 18, 120, 511, 512; in pulp and paper sludge, 47
Cholera, 74, 75, 82
Chopper pump, 541, 543
Christmas tree production, 33–34, 210, 354
Chromium: in pulp and paper sludge, 45, 46; pretreatment, 86; in sewage sludge, 86, 87; in soil, 88, 95; recommended limits, 90; in nitrification and leaching study, 135; in element flux study, 156, 162, 163, 166; in heavy metal study, 169, 171, 172, 174; in deer tissue, 193, 197; in vegetation and small mammals, 202, 203, 204; in loblolly pine studies, 273, 315, 316, 317, 318, 319, 320; in oak growth studies, 284, 287, 288; in Penn State study, 408, 409, 413, 414, 415, 416; at northern Michigan sites, 484, 486; in Barceloneta sludge, 510, 511; in Savannah River sludge and soil, 517, 520, 521; in Wauna sludge, 534; in resin based sludge, 557; groundwater content before and after sludge application, 559, 560
Citellus tridecemlineatus. See Thirteen-lined ground squirrel
Clay in pulp and paper sludge, 44, 46
Clayton County, Georgia, 27, 350, 352; wastewater use in, 393–405; Clayton County Water Authority, 394, 396; costs of system, 395–396, 401; public involvement, 396; computer use, 401; site disturbance, 401–402; rehabilitation, 402
Clean Air Act Amendments, 23
Clean Water Act (CWA), 3–4, 7–8, 22–23, 338, 340, 473, 515
Clearcuts, 357, 461–462, 464; in heavy metal study, 168–175; and deer and elk forage, 188–197
Clethrionomys gapperi, 113
Coagulation/DAF system, 552–555
Coal mines, 497
Coastal plain (Southeast), 272–280, 301–306, 308–323, 515–529
Cobalt: in pulp and paper sludge, 45, 46; pretreatment, 86; in sewage sludge, 86, 87, 88; in element flux study, 162, 163, 166; in loblolly pine study, 273; in Penn State study, 408, 409, 413–414, 415, 416; in Savannah River sludge, 517

Code of Federal Regulations, 257, 473
Collembola, 178, 180, 184, 185, 186
Colorado, 270
Columbian black-tailed deer, 113–114
Compost, 12, 79, 80; Milorganite, 50; GroCo, 475
Comptonia. See Sweetfern
Computers, 401
Conceptual models: forest tree growth, 211–219; nitrogen dynamics, 239, 240; aspen ecosystem, 270–271
Conference on Recycling Municipal Sludges and Effluents on Land, 55, 59
Conifers, 350, 421; growth response, 209, 216; nitrogen levels, 228, 229. *See also individual species*
Cooperative State Research Service, 88, 89
Copper: in pulp and paper sludge, 45, 46; pretreatment, 86; in sewage sludge, 86, 87; in animal feed, 88; recommended limits, 89, 90; in forest soil, 93, 95; from smelter, 94, 95; movement in soil, 95; leaf tissue concentrations, 95; fractionation, 96–97; effect on wildlife, 114; in nitrification and leaching study, 134, 135, 136, 139; in element flux study, 156, 162, 163, 166; in heavy metal study, 170, 171, 172, 174; in deer tissue, 193, 197; in vegetation and small mammals, 202, 203, 204; in loblolly pine studies, 273, 316, 317, 318, 319, 320; in oak growth studies, 284, 287, 288, 289; in Clayton County, 399; in Penn State study, 408, 409, 413, 414, 415, 416; at northern Michigan sites, 484, 485, 486; in Barceloneta sludge and soil, 511, 512; in Savannah River sludge and soil, 517, 520, 521; in Wauna sludge, 534
Copper Basin, Tennessee, 498, 499, 500, 504
Corn, 126–131, 260
Costs: of pulp and paper sludge application, 49; of wastewater irrigation, 58–59; of sludge application, 64–65, 67, 356–366, 491–492, 560–561; operation and maintenance, 351–354, 364–365, 424–425, 491–492; in future, 353–354, 492; sprinkler systems, 354, 424; of Clayton County system, 395–396, 401; in Michigan operations, 491–492; of trail construction, 491; of transport and application, 491–492; of Georgia-Pacific operation, 560–561
Cottonwood, 31, 48, 50, 454; black (*Populus trichocarpa*), 212, 534–536, 537–540; eastern (*P. deltoides*), 212, 400, 434; hybrid, 260–264, 464, 466, 467–468, 536–537; in Australia, 434; short-rotation intensively cultured, 533
Covington, Georgia, 27, 350
Cropland. *See* Agriculture
Crown density, 34
Crown Zellerbach Corporation, 51, 260–264, 466, 533–547
Cultural practices: effects on wood characteristics, 250–256; pruning, 251, 254, 435. *See also* Fertilization; Thinning
Cyanide, 511, 512
Cycling of forest nutrients. *See* Nutrients
Cypress, 454; bald, 212
Cypress domes, 30

Cytospora chrysosperma, 268–271, 494

Dalton, Georgia, 27, 350
Danthonia spicata. See Poverty grass
Deciduous forests, 219
Deer, 113–114; BaP and PCBs in deer fat, 121, 122, 123; in human food chain, 123–124, 188; trace metals in, 197; browsing on sludge-treated vegetation, 188–197, 267–271, 286, 463, 464, 493, 494
Deer mice, 113, 114, 201, 202, 204
Defiance, Ohio, 86
Demonstration projects, 67–68, 461–470, 474, 475
Design criteria: EPA design manuals, 55, 56, 420; for wastewater systems, 57–58, 398–399, 404; for sludge systems, 65–67; to avoid groundwater contamination, 82, 321, 398, 404; EPA guidelines, 339, 427; application rate and timing, 358; agronomic guidelines, 384–386; site specific, 387; proposed guidelines, 385–389; in Clayton County, 398–399, 404; for living filter system, 408–409, 417–418; for New England systems, 420–429; design limits, 427–428; at Kings Bay, Georgia, 442–445; at Pack Forest, 467; site preparation, 491, 492; for chemical coagulation/DAF system, 554–555. *See also* Treatment of sludge and wastewater
Detroit Lakes, Minnesota, 27
Devastated sites, 497–506
Dewberry, 411
Diammonium phosphate (DAP), 156, 158, 160, 161, 163, 165, 310
Dichlorobenzene. *See* Substituted monocyclic aromatics
Die-off rate, 79, 82, 102
Di-n-butyl phthalate (DnBP), 18, 126–131, 511, 512
Disease, 75, 76, 77. *See also* Infectious elements; Public health
Disinfection, 79, 80, 421, 422, 426, 429, 473
Distance effects, 487–489
Dissolved solids, 552–553, 555
Disturbed soil, 92, 497–506
Ditches, 425, 443
Douglas-fir, 16, 30, 31, 32, 209, 213, 350, 473; cumulative biomass, 33; growth response to sludge treatment, 64, 210, 215, 232, 289; fecal coliforms in sludge, 103; understory plants, 112, 113; wildlife studies, 114; nitrate leaching, 154, 237; drought resistance, 212, 216; foliar nitrogen, 216, 236; net primary productivity, 217; nitrogen renovation, 227; nitrogen uptake of vegetation, 229; nitrogen transformation and utilization, 231, 235, 236, 237, 240, 241; leaching, 238; sludge application rates, 238; sludge ammonia content, 239; conceptual model, 240; effect of sludge on 60-year-old stand, 246–257; juvenile wood, 249; specific gravity, 249; Pack Forest sludge study, 260–264, 464, 466; Snoqualmie Pass system, 350, 351
Dracaena, 510
Drainage. *See* Water relations
Drinking water. *See* Groundwater
Drought resistance, 212, 216
Dunbar soil, 558, 559

Earthworms, 37, 92, 93, 114, 115, 204
East Lansing, Michigan, 226
Economic concerns of public, 375, 376, 381
Economic factors: wastewater use, 58–59; sludge use, 64–65, 67; revenue from timber sale, 491. *See also* Costs
Ectomycorrhizae, 501, 503
Effluent. *See* Wastewater
Effluent storage. *See* Lagoon storage; Storage systems
EG&G, Inc., 325, 333
Elemental cycling. *See* Nutrients
Elk, 113, 188–197; browsing on sludge-treated vegetation, 188–197, 267–271, 493
Elm, 48
Energy consumption, 425
Energy supply, 38, 437
Environmental considerations, 373, 374, 375, 376, 380, 465; need for trained operators, 60; and Metro, 474. *See also* Pathogens; Public health; Risk management
Equilibrium solution modeling, 96
ERM-Southeast, Inc., 558
Erosion on severely devastated sites, 497–506
Eucalyptus, 48, 354, 513; Australian treatments, 432, 435–437, 437–439
Evaporation, 157
Evapotranspiration, 92, 240, 398, 400, 445

Fagus grandifolia. See Beech
Fecal coliforms, 81, 101, 103, 104, 427
Federal Water Pollution Control Act, 22–23
Federal Water Pollution Control Act Amendments, 221, 394, 509, 515
Fertilization: interaction with site quality, 215–216; effects on forest ecosystems, 224–225; nitrogen, 224; phosphorus, 224; effect on wood properties, 251, 253, 254, 256; in loblolly pine study, 274, 275; understory response, 308–323; of devastated sites, 499–506. *See also* Sewage sludge; Sludge; Wastewater
Fertilizer, 474; sludge and wastewater as, 11–12; sludge compared with urea, 64, 218; time-released, 127; diammonium phosphate, 156, 158, 160, 161, 163, 165, 310, 519; with pulp and paper sludge, 259; sludge pellets, 396, 401, 405; Milorganite, 50, 92, 111, 474; Silvigrow, 121, 475–476; GroCo, 475; for devastated areas, 499, 500, 501
Fescue, 126–131
Fiber plantation, 467–468
Fir, 388, 424; balsam, 34, 350; noble, 260–264; grand, 464
Fire: prescribed burning, 443–450; prevention, 321
Firewood, 354. *See also* Fuelwood
Flint River Wastewater Facilities Plan, 394, 396
Flooded gum, 436
Flooding, 512
Florida, 29, 497, 498
Fluorine, 86, 87
Foliage: leaf area, 213–217, 305; necrosis, 31, 464; potassium in, 228; foliar nitrogen levels in regional vegetation, 229; nitrogen in, 228, 229, 236, 241,

260, 263, 403; phosphorus in, 236, 403; in pulp and paper and municipal sludge study, 260; premature needle drop, 280, 301, 305; in oak studies, 283–284, 286; in loblolly pine study, 301–306; needle xylem pressure potential, 302–306; pine, 403; hardwood, 403; analyses in Penn State study, 412–414; and prescribed burning, 449; needlefall, 528. *See also* Litter

Food-chain risks, 36

Forages, 112–114, 468; toxicity, 114–115, 320; deer and elk use of sludge-treated, 118–197, 266–271, 286, 494; nutrients in, 112–113, 114, 190, 193, 319–320; cadmium in, 197, 320, 321; metals in vegetation and small mammals, 199–204; sludge guidelines, 384; near trails, 494; forage grass, 510; crops in Puerto Rico, 512, 513. *See also* Grasses; Vegetation

Forest ecosystem nutrient cycling. *See* Nutrients

Forest irrigation. *See* Irrigation

Forest land: commercial land area in Pacific Northwest, 5; acreage by region; 14–15; compared with cropland, 90–94; old fields, 91; soil horizon, 91

Forest land application: potentials, 13–22, 52; federal and nonfederal forest land by region, 14–15; current use, 27–28; slow-rate systems in humid states, 29; reluctance to use forests, 51. *See also* Pulp and paper industry; Sewage sludge; Sludge; Soils; Treatment of sludge and wastewater; Wastewater

Forestry Sciences Laboratory, 157

Forest soils. *See* Soils

Formaldehyde, 556–557, 559, 560

Fractionation, 96

Fragaria virginiana. *See* Strawberry, wild

Fragipan, 425, 453

Fraxinus. *See* Ash

Fremont, Michigan, 27

Fruit processing waste, 436

Fuelwood, 354, 436, 437, 437–439

Fungi, 268–271, 494, 501

Fuquay soil, 272, 273, 302–306, 309, 313, 314, 315, 518, 519, 520, 523

Furadan®, 274, 275, 276, 278, 280, 519

Fusarium spp., 494

Fusiform rust, 505, 518

Gallberry, 443, 448, 449, 450

Gaultheria. *See* Wintergreen

Georgia, 157, 350, 498; wastewater use in, 27, 29, 31, 55, 350, 352, 393–405, 442–450; foliar nitrogen, 229; nitrogen uptake of vegetation, 229; nutrients in sludge, 233; application rates, 233; Environmental Protection Division, 394, 404; kaolin mine spoils, 497, 502–503. *See also* Clayton County, Georgia

Georgia-Pacific Corporation, 556–561

Germany, 106, 107

Giardia lamblia, 75, 76, 78, 82

Goal®, 274, 275, 519

Goddard, Maurice, 37

Goldenrod, 411, 412, 413, 414

Goldsboro soil, 558, 559

Grand Traverse County, Michigan, 368, 372

Grasses, 260, 264, 357, 463, 464, 505; panic grass, 189, 194, 195, 200, 201, 203; nitrogen uptake, 229, 230; rye grass, 260, 462, 468; in loblollly pine forests, 310–312; heavy metals in, 320; poverty grass, 411; for right of way demonstration, 468; forage, 510, 513; in resin based sludge-treated plots, 558, 560

Grazing, 435, 439

Great Lakes region, 477; nitrogen assimilation, 231, 235, 236, 237, 239, 240; nutrients in sludge, 232, 233; sludge application rates, 238. *See also* Lake States; Michigan

Greenbelts, 32

Greenhouse studies, 126–131, 260–264, 466. *See also* Pot studies

Greenville, Maine, 421

Greenville, South Carolina, 549

Griffin, Georgia, 394

GroCo, 475

Ground cover, 230, 411. *See also* Grasses; Vegetation

Groundwater: nitrates in, 36, 50, 57, 65, 107–108, 132, 234, 235, 237, 384, 494, 542–543; trace chemicals and pathogens in, 36, 77, 103; and color of wastewater, 43; drinking water standards, 50, 65, 83, 100–101, 107–108, 242, 394, 402, 404, 416, 417, 427, 463, 554; pathways of microbial transport, 77, 81–82; die-off rate of microorganisms, 82; infiltration, 103, 105, 439; organics, 104–105; metals, 106–107; risk assessment, 119; BaP and PCBs in, 121, 122, 123; in nitrogen transformation study, 151; in Douglas-fir stand, 237; and site assimilation capacity, 241; and sludge storage basins, 359; public concerns about, 373; regulations in Michigan, 383–389; monitoring, 385, 402, 417, 418, 423, 439, 444, 453, 463, 467, 525, 541, 555, 560; table elevations, 404; living filter system, 407, 416, 417, 452; fragipan barrier, 425, 453; percolate quality, 427, 446; discharge in Barceloneta area, 511; at Savannah River Plant, 525–527; conductivity, 542, 554; and polymer wastes, 554; and resin based wastes, 559, 560. *See also* Leaching; Nitrate

Growth response: data needed, 57, 58; from use of sludge, 64–65, 261; and trace elements, 95; of forest trees to wastewater and sludge application, 209–219, 400–401, 463; factors limiting growth, 211–213; growth model, 213–219; net primary productivity, 217; ways of measuring growth, 247, 276, 293; in Pack Forest, 261, 463, 464, 467; of four ages of loblolly pine to sludge, 272–280; volume of wood produced, 279–280; short-term response of oak, 282–291; long-term response of oak, 292–300; mean annual increment, 293, 297–299; in Clayton County, 400–401; of plantations in Australia, 435, 438; at Missouri site, 454; at Savannah River Plant, 527; Wauna experiments, 536–537

Gum, 436, 438

Habitat. *See* Wildlife

Halogenated short chain aliphatics, 119, 120

Hamamelis virginiana. *See* Witch hazel

Harbor Springs, Michigan, 27, 34, 226

Hardwood forests, 350, 351, 352, 354, 398, 400; and

inorganic nitrogen leached to groundwater, 33; sprout growth, 34; and pulp and paper sludge fertilization, 49; northern, 111–112, 113, 143, 227, 237, 293, 388, 480, 485, 489, 490, 491; in nitrogen transformations study, 143, 144, 145, 147, 148, 149, 150, 151; mixed hardwood, 174, 200–204, 227, 228, 231, 237, 238, 293, 350, 351, 352, 407, 412, 421, 422, 551; potassium in soil, 227; foliar nitrogen levels, 229, 403; nitrogen uptake by vegetation, 229; nitrogen transformation and utilization, 231, 236, 237, 412; nitrogen leaching, 237, 238; nitrogen application rates, 388; phosphorus, 403; in northern Michigan, 480, 481, 485, 487, 489, 490, 491, 494; on devastated sites, 502; in South Carolina, 551. *See also individual species*

Harvest scheduling model, 401

Hawkweed, 194, 195, 201

Hay, 512, 513

Hazardous and Solid Waste Amendments, 23

Hazardous substance, 5, 46

Health risks. *See* Public health

Heavy metals: in pulp and paper sludge, 45, 46, 50, 259; in sludge, 66, 102, 170, 293, 463; in wastewater, 102; in forages, 114, 319–320; and nitrification and leaching study, 135, 136; element flux, 156; storage in soils of aspen forest, 168–175; recycling, 178; in grasses, 320; guidelines for, 385. *See also* Trace elements; *individual elements*

Helen, Georgia, 27

Helminths. *See* Parasites

Hemlock, 350, 422; western (*Tsuga heterophylla*), 213, 464; eastern (*T. canadensis*), 143, 480

Hepatitis, 75, 76

Herbicides, 158, 162, 210, 468, 519; in loblolly pine studies, 274, 275, 276, 309, 311, 312, 313, 321, 400; used in Australia, 432, 438

Herbivores, 112, 113, 188, 197, 203, 270, 468. *See also* Deer; Wildlife; *individual species*

Hickory, 388, 412

High resolution airborne remote sensor data, 324–333

Hop hornbeam, 200, 201, 203

Horse Creek Pollution Control Facility, 179, 516–525

Hublersburg soil, 407–408, 409–412, 414

Huie Land Treatment System, E. L., 350, 395, 396

Humid regions, 32

Humus: in heavy metal study, 169–174; importance of mesofauna, 178; decomposition of, 228, 235; in oak forest studies, 288, 295

Huron-Manistee National Forest, 294

Hydrology, 92; baseflow, 403–404; conductivity, 82, 428, 552, 553, 554, 555; loading capacity, 57, 91, 399, 428, 443, 446, 453, 552, 554; and industrial waste, 549. *See also* Groundwater; Water quality; Water relations

Idaho, 5

Ilex glabra. See Gallberry

Illinois, 197

Immobilization: of nitrogen, 142–143, 145, 148, 159, 236, 238, 262, 297; of heavy metals, 168–175

Incineration, 10, 372, 474, 509

Indiana, 89, 134, 136

Industrial sludges and wastewater. *See* Fruit processing waste; Pulp and paper industry; Sludge; Textile mill waste; Treatment of sludge and wastewater; Wastewater

Infectious elements: in wastewater, 73–74; pathways to humans, 77; minimal infectious dose, 78, 80. *See also* Bacteria; Disinfection; Parasites; Public health; Viruses

Injection process, 513

Inorganic compounds in soil, 96

Insect control, 127, 274, 275, 276, 278, 280, 519

Insectivores, 114, 203

International Agency for Research on Cancer, 122

Invertebrates. *See* Mesofauna

Ion exchange capacity and virus adsorption, 81

Iron, 225; in pulp and paper sludge, 45; in sewage sludge, 87, 233; in element flux study, 156, 162, 163, 166; in heavy metal study, 170, 171, 172, 174; in loblolly pine study, 273; in oak growth studies, 284, 287, 288, 289; in Clayton County, 399; in Penn State study, 408; deficiency, 432; at northern Michigan sites, 484, 486; in Savannah River sludge and soil, 517, 520, 521; in Wauna sludge and soil, 534, 539, 547

Ironwood, 283

Irrigation, 400–405; treated effluents, 12; current status of wastewater use, 26–39; list of projects, 27; slow-rate systems, 29, 34, 354, 432, 442; nitrogen removal, 34; costs of, 58–59, 350–354, 365, 424–425; site disturbance, 110–111, 400, 401–402, 421; and forest floor decomposition, 241; effect on streamflow, 403–404; runoff, 423, 426; of tree plantations in Australia, 431–440; flood irrigation of pastures, 432, 438; coastal plain, slash pine system in Georgia, 442–450; burned and nonburned areas compared, 443–450; at Bennett Spring State Park, 452–457; Wauna mill operation, 536–547; National Starch and Chemical system, 549–555; resin based sludge, 560. *See also* Application methods; Wastewater

Italian rye grass, 112

Jack pine, 111, 112, 113, 143, 144, 200, 388, 480

Japan, 88

Juniper, 213, 215

Kaolin spoils, 497, 498, 502–503

Kalkaska County, Michigan, 343, 344, 376, 380

Kenaf, 513

Kennett Square, Pennsylvania, 350, 353

Killington, Vermont, 27

King County, Washington, 474, 475

Kings Bay Submarine Base, Georgia, 27, 350, 442–450

Laboratory incubation, 132–141, 142–152

Lagoon storage, 9–10, 225, 351, 352, 421, 422, 423, 434, 442, 475; overflow, 437; three-cell system, 453; resin wastes, 556–557, 559

Lake of the Pines, California, 350, 352

Lake States, 31, 32, 270. *See also* Great Lakes area

Land application. *See* Application methods; Sewage sludge; Sludge; Wastewater

Land Energy Pty Ltd., 436

Landfill, 51, 264, 372, 474, 475, 509, 510, 533. *See also* Borrow pits

Landsat MSS data, 324, 325

Larch, 31

Leaching: of nitrate, 35, 107–108, 132–141, 151, 152, 154, 163, 165, 178, 228, 230, 231, 232, 235, 237–239, 289, 290, 321, 385, 446; from pulp and paper wastewater, 43; of metals, 66, 106–107; of trace elements, 91, 164, 166; in forests, 101, 321; process of, 102–103; pulses, 103; total organic carbon, 105; toxic organics, 105; and soil acidity, 107; element flux in loblolly pine plantation, 154–166; movement of heavy metals, 168–175; of potassium, 227; of calcium, 228; of magnesium, 228; and ground cover, 230; in regions of U.S., 231; of phosphorus, 233; in oak studies, 289; and root mass, 321; of dissolved solids, 554. *See also* Groundwater

Lead: in pulp and paper sludge, 45, 46; pretreatment, 86; in sewage sludge, 86, 87; recommended limits on, 89, 90; in calcareous tissues, 92; from smelter, 94, 95; leaf tissue concentrations, 95; in forest soil, 95, 107; and soil acidity, 107; effect on wildlife, 114; in nitrification and leaching study, 135, 136; in element flux study, 156, 162, 163, 166; in heavy metal study, 170, 171, 172, 174; in loblolly pine studies, 273, 315, 316, 317, 318, 319; in Penn State study, 408, 409, 413, 414, 415, 416; in Barceloneta sludge and soil, 510, 511, 512; in Savannah River sludge and soil, 517, 520, 521; in Wauna sludge, 534

Leaf area, 213–217

Leelanau County, Michigan, 368, 372

Legionellosis, 74, 75

Legislation: federal laws concerning sludge and wastewater use, 22–24. *See also* Clean Water Act; U.S. Environmental Protection Agency

Lespedeza, 503

Liebig, J., 211, 212

Lime, 499, 500, 501, 503, 504

Limestone region, 510–511

Liquidambar styraciflua. See Sweetgum

Litter, 178–179; decomposition of, 93, 228; effect of smelter on, 94–95; organic substances in, 105, 106, 121, 122, 123; and volatilization, 148; in oak forest studies, 287, 288, 290, 294–295; early litterfall, 306; Australian study, 433; litterfall rate, 528–529. *See also* Humus; Soils

Livestock, 87

"Living filter" system, 406–418, 452

Loblolly pine, 48, 209, 289, 290, 350, 352, 398, 400, 515; and nitrification, 133, 134–135, 528; element flux in plantations, 154–166; influence of sludge on mesofauna, 177–186; phosphorus, 233, 403; nitrogen, 235–236, 237, 241, 303, 402; leaching, 237, 238; nitrogen transformation and utilization (conceptual model), 240; study of growth response of four ages to sludge, 272–280; water relations, 301–306; understory response to fertilization,

308–323; remote sensing of biomass, 324–333; harvesting, 400; in Copper Basin, 499–501; in borrow pit, 502, 504; in kaolin spoil, 502, 503; and subsoiling, 505; at Savannah River Plant, 517–519, 520, 527; nutrient cycling, 528; at NSCC site, 551

Logistics, 491–492

Lolium multiflorum, 112

Lombardy poplar, 30, 31, 32

Loxton, Australia, 433, 434, 435–437

Lucy soil, 273, 309, 312, 313, 518, 519, 520, 522, 527

Lynchburg soil, 558, 559

Mackinac State Forest, 199, 479

Mackinaw City, Michigan, 27, 38, 350

Macroorganisms, 93

Magnesium: in pulp and paper sludge, 45; in pulp and paper wastewater, 48, 49; in nitrification and leaching study, 134, 135, 139; in element flux study, 155, 156, 160, 161, 162, 163, 164, 165, 166; in forest ecosystem, 224; renovation of, 228; in loblolly pine studies, 273, 315, 316, 317, 318, 319; in oak growth studies, 284, 287, 288; in Clayton County, 399; in Penn State study, 408, 409, 415, 416; in Australian irrigation, 437; at Georgia site, 445, 446, 447, 448, 449, 450; at northern Michigan sites, 484, 486, 494; in Copper Basin soil, 499; in borrow pit, 500; in kaolin spoil, 503; in Barceloneta sludge and soil, 511, 512, in Savannah River sludge and soil, 517, 520, 521, 522, 525; in Wauna sludge and soil, 534, 538, 539, 540; in NSCC effluent, 553

Mahoe, 513

Mahogany, 513

Maine, 27, 420, 421

Maintenance, 37

Mammals, small: effect of sludge on, 111, 113, 114, 199–204, 494; in study of metal concentration, 199–204; species collected for trace element monitoring, 201, 204

Management needs, 37–38

Manati River, 510, 512

Manganese: in nitrification and leaching study, 134, 135, 136; in element flux study, 156, 162, 163, 164; in heavy metal study, 170, 171, 172, 174; in loblolly pine study, 273, 316, 317, 318, 319, 320; in oak growth studies, 284, 287, 288, 289; in Penn State study, 408; at northern Michigan sites, 484, 486; in Savannah River sludge and soil, 517, 520, 521; in Wauna sludge and soil, 534, 541

Manistee National Forest, 169, 290, 294, 295

Manure spreaders, 357, 358, 362, 364, 365

Maple, 31, 424, 454; red, 112, 143, 200, 201, 203, 282, 283, 285, 294, 412, 422, 480; sugar, 112, 143, 200, 201, 422, 480; striped, 200; nitrogen renovation, 227; in short-term growth study, 282, 285

Marine Protection, Research and Sanctuaries Act, 8, 23

Marquette County, Michigan, 368

Massachusetts, 3

Meadow jumping mice, 113, 201, 202, 204

Meadow voles, 92, 113, 114, 201, 204

Mean annual increment, 293, 297–299
Menominee County, Michigan, 368
Mercury, 273; in pulp and paper sludge, 45, 46; in sewage sludge, 86, 87; effect on wildlife, 114; in Penn State study, 408; in Barceloneta sludge, 511; in Savannah River sludge, 517; in resin based sludge, 557
Mesofauna, 177–186
Metals: in sludge and wastewater, 18, 19, 65–66, 106–107; in pulp and paper compared with sewage sludge, 45; in pulp and paper sludge, 50; loading guidelines, 89, 90; and soil acidity, 107; movement of, 168–175. See also Heavy metals; Trace elements
Methylene chloride, 120, 511, 512, 557
Metro. See Municipality of Metropolitan Seattle
Mice, 113, 114
Michigan, 27, 29, 31, 32, 33, 35, 55, 111, 112, 113, 154, 189, 209, 350; nitrogen transformation study, 142–152; heavy metal study, 169; metals in vegetation and small mammals, 199; Department of Natural Resources, 200, 342–343, 344, 370, 374, 380, 383–389, 478; potassium in soil, 227; foliar nitrogen, 229, 236; nitrogen uptake of vegetation, 229; nitrogen leaching, 237–238; nutrients in sludge, 233; application rates, 233; aspen mortality study, 266–271; oak studies, 282–291, 292–300; public acceptance surveys, 343–347, 367–381; Department of Agriculture, 380; Department of Public Health, 380; United Conservation Clubs, 380; research results related to sludge guidelines, 383–389; sludge application in forests of northern Michigan, 477–495; local organizations and officials, 493. See also Great Lakes area
Michigan State University, 143, 146, 151, 293, 296, 478–479; studies on public acceptance, 342–343
Microbiological aspects, 73–83
Microflora, 80
Microorganisms: of concern, 74–77; and treatment effectiveness, 78–82; survival of, 80–81; and soil decomposition, 93; effect on pulp and paper sludge, 259
Microtus pennsylvanicus, 92, 113
Microtus townsendii, 111
Middleville, Michigan, 27, 32, 226
Mildura Wines Ltd., 439
Mill wastes. See Pulp and paper industry
Milorganite, 50, 92, 111, 474
Mine sites, 96, 475, 497, 502–503
Minnesota, 27, 29
Missaukee County, Michigan, 368
Missouri, 27, 452–457
Mites, 178, 180, 182, 184, 185, 186
Model for growth of forest trees, 211–219
Model of aspen ecosystem, 270–271
Models of nitrogen dynamics, 239, 240, 465
Molybdenum, 45, 46, 86, 87, 88
Monocyclic aromatics, 119, 120
Montana, 29, 197
Montezuma School, 352
Montmorency County, Michigan, 143, 189, 200, 266,

282, 294, 295, 343, 344, 479
Moore Brothers Pty Ltd., 436, 437
Morrison soil, 412
Mount Lemmon, Arizona, 27
Mount Sunapee, New Hampshire, 27, 350, 420, 421
Mule deer, 197
Mulga, 437
Municipality of Metropolitan Seattle (Metro), 67, 68, 358, 364, 461, 466; risk assessment, 117, 118–124; Silvigrow, 121, 475–476
Municipal sludge. See Sewage sludge
Murray River, 434, 435–436
Mushroom plots, 467

Napaeozapus insignis. See Woodland jumping mouse
Naphthalene, 47
National Council of the Paper Industry for Air and Stream Improvement, Inc., 41
National forests, 13
National Pollution Discharge Elimination System (NPDES), 23, 384, 457, 551
National Starch and Chemical Corporation, 549–555
Naval submarine base, 27, 350, 442–450
Needle xylem pressure potential, 302–306
Net primary productivity, 217–218
New England, 420–429
New Hampshire, 27, 29, 237, 350, 351, 354, 420, 421–424
New Jersey, 27
New York, 292, 293
Niangua River, 452–456
Nickel: in pulp and paper sludge, 45, 46; pretreatment, 86; in sewage sludge, 86, 87; recommended limits on, 89, 90; leaf tissue concentrations, 95; in forest soil, 95; in leachates, 106–107; effect on wildlife, 114; in nitrification and leaching study, 135, 136; in element flux study, 156, 162, 163, 166; in heavy metal study, 170, 171, 172, 174; in loblolly pine studies, 273, 316, 317, 318, 319, 320; in deer tissue, 193, 197; in vegetation and small mammals, 202; in oak growth studies, 284, 287, 288; in Penn State study, 408, 409, 413, 414, 415, 416; at northern Michigan sites, 484, 486; in Barceloneta sludge and soil, 511, 512; in Savannah River sludge and soil, 520, 521; in Wauna sludge, 534
Nitrate, 35–36, 232, 241–242, 399, 522; and pulp and paper wastewater, 43; and pulp and paper sludge, 49, 50; drinking water standards, 57, 101, 102, 230, 412, 416, 463; assessing levels of in soil, 68, 228; nitrification and leaching in soil treated with sulfuric acid and nitrification inhibitor, 132–141; nitrate-N concentrations, 137, 139, 147, 149, 151, 156, 159–162, 163, 165, 233, 240, 273, 289, 290, 402, 408, 410–413, 417, 445, 446, 454, 463, 494, 511, 512, 517, 522, 527, 542–543, 553, 555, 557, 559, 560; in element flux study, 157–166; denitrification, 228; in regional conceptual models, 240. See also Groundwater; Leaching
Nitrogen: in sludge and wastewater, 11–12, 19, 221, 225–226, 232–239, 284, 465; assimilation by plants, 31–32, 33, 34, 35, 112, 229–230, 237, 240, 385, 546;

accumulation in soil, 32, 108, 217–218, 240, 292, 447; renovation capacity, 33, 227, 228–229, 240, 410–411, 446; and vegetation removed, 33; and water quality, 34; in pulp and paper wastewater, 43, 49; in pulp and paper sludge, 44–46, 50, 259, 261; information needed, 57, 66; and drinking water guidelines, 101; in organic pollutants study, 126; nitrification and leaching in soil treated with sulfuric acid and nitrification inhibitor, 132–141; nitrogen transformations study, 142–152; immobilization, 142–143, 145, 148, 159, 236, 238, 262, 297; total inorganic N, 148, 152; in element flux study, 155, 156, 159; in soil mesofauna study, 180; in forages, 190; effect on growth and water uptake, 211, 212; effect on photosynthesis, 216, 217; fertilizer, 224–225; concentrations in wastewater in regions of U.S., 226, 231, 239, 241, 408; nitrification and denitrification, 228, 234–235, 448; foliar levels, 228, 229, 236, 241, 263, 286; application rates, 145, 156, 180, 230, 232, 233, 240, 241, 388, 465; transformation and utilization in regions of U.S., 231, 240; concentrations in sludge in regions of U.S., 233; mineralization, 234, 235, 239, 240, 385, 387, 521, 546, 553; uptake rates in various forests in regions of U.S., 236; storage in overstory, 237; conceptual models of nitrogen dynamics, 239, 240, 465; irrigation-applied, 241; most effective uptake of, 241, 465; short-term transfer rates, 242; in Pack Forest, 261, 263, 463, 465; in loblolly pine growth studies, 273, 315, 316, 317, 318, 319, 528; in oak growth studies, 284, 286, 287, 288, 289, 296, 297, 298; in remote sensing study, 325, 330, 331, 332; recommended application rates, 388, 465; in Clayton County, 399, 402, 403; in Penn State study, 408, 409–412; at New Hampshire site, 423, 428, 429; at Vermont site, 426, 427–428, 429; at Georgia site, 445, 446, 447, 448, 449; at northern Michigan sites, 484, 485, 486, 494; in Copper Basin soil, 499, 504; in borrow pit, 500; in kaolin spoil, 502, 503; in Barceloneta sludge and soil, 510, 511, 512; in Savannah River sludge and soil, 517, 520, 521, 522; in Wauna sludge and soil, 534, 538, 539, 540, 541, 546; in NSCC effluent, 553–554; in resin based sludge, 557, 559. *See also* Ammonia; Ammonium; Nitrate

NOAA Advanced High Resolution Radiometer data, 324

Norfolk soil, 558, 559

North Carolina, 162, 497, 498; Department of Natural Resources and Community Development, 557, 559, 560

North Carolina State University, 126, 559

Northeastern Regional Research Project (NE-96), 88, 89

Nursery studies, 126–131, 260–264, 466. *See also* Pot studies

Nutrients: in sludge and wastewater, 11–12, 232, 233, 241; in pulp and paper sludge, 44–45; nutrient cycling in forests, 93–94, 178, 221–242, 527–528; nutrient cycling information needed, 57; protein, 112, 114; in forages, 112–113, 114, 190, 193; effect on mesofauna, 184; nutrient loss, 224, 446; renovation capacity, 226–230, 446; retention rates, 298–299. *See also* Fertilization; *individual nutrients*

Oak (*Quercus* spp.), 31, 111, 143–144, 454; red, 112, 143, 200, 201, 203, 282, 283, 285, 286, 294, 412, 422, 480; white, 143, 200, 201, 212, 282, 283, 285, 286, 294, 412, 422, 480; swamp white, 212; black, 412; scarlet, 412; northern pin, 143, 480; and nitrification, 134; in nitrogen transformations study, 143, 144, 145, 146, 147, 148, 149, 150, 151; in vegetation and small mammal study, 199–204; in mortality study, 266; in Michigan short-term growth study, 282–291; long-term growth study, 292–300; nitrogen application rates, 388, 484, 485; in northern Michigan, 480, 481, 482, 483, 484, 485, 487, 488, 491, 494

Oats, 260, 462

Ocean disposal of wastes, 7–8, 10, 23–24, 221–222, 472, 474, 508

Odocoileus columbianus. See Columbian black-tailed deer

Odocoileus hemionus. See Mule deer

Odocoileus virginianus. See White-tailed deer

Office of Municipal Pollution Control, 338

Office of Water Regulations and Standards, 338

Ohio, 86, 89

Old fields, 33, 91, 409–412, 413, 414–417; nitrogen uptake of vegetation, 229

Operation and maintenance. *See* Costs

Operations: need for trained operators, 60; importance of demonstration studies, 67–68; problems, 401–402; evaluations, 467

Orangeburg soil, 272, 273, 309, 312, 313, 314, 315, 518, 519, 522, 523, 527

Oregon, 5, 12, 213, 216, 533; Department of Environmental Quality, 533

Organic substances: in sludge, 18, 19, 66, 104–105; in pulp and paper wastewater, 44; in pulp and paper sludge, 44–46, 47; decomposition of, 102, 104–105, 127; risk assessment, 118–124; priority pollutants, 44, 47, 118–124, 125–131; in land treatment systems, 125–131; in municipal sludge, 510. *See also* Toxic substances

Orthophosphate, 226, 232, 454, 526

Osceola County, Michigan, 368

Osmocote®, 127

Ostrya virginiana. See Hop hornbeam

Pacific Northwest: use of fertilizer, 224; denitrification, 228; nitrogen assimilation, 229, 230, 231, 236, 237, 239, 240; growth response to use of sludge, 232; nutrients in sludge, 232, 233; soil nitrogen, 235; leaching, 237; sludge application rates, 238; ammonia content of sludge, 239, 240; technology and costs of sludge application, 356–366. *See also* Oregon; Washington State; Pack Forest

Pack Forest, Washington, 27, 226; growth response, 64; application rates, 65–66; total organic carbon in leachates, 105; study using sewage and pulp and paper sludge, 258–264; demonstration program,

461–470, 475; wildlife, 473

Palmetto, 50; saw palmetto, 443, 448, 449

Pangola grass, 513

Panic grass, 189, 194, 195, 200, 201, 203

Paradichlorobenzene (pDCB), 126–131

Parasites, 74–76, 78, 79, 83; and wastewater treatment, 79–80; survival, 81; transport, 82; size, 82; die-off rate, 82

Parks: irrigation with wastewater, 452–457; sludge use, 475

Pates Creek, 395–396, 397, 402–404

Pathogens, 17, 18, 19, 20, 21, 36, 74–78; in pulp and paper wastewater, 43; in pulp and paper sludge, 47; die-off, 79, 82, 102, 463; and treatment, 79–80, 429, 510; in groundwater, 103; fungi, 268–271; in snow containing wastewater, 426. *See also* Infectious elements; Disinfection

Pathways of environmental concern, 339–340

Pathways of microbial transport, 77

Pathways of organic transport, 122

Pellets, sludge, 396, 401, 405

Pennsylvania, 27, 29, 33, 34, 35, 37, 55, 209, 350, 352, 353, 354; potassium in soil, 227; foliar nitrogen, 229; nitrogen uptake of vegetation, 229; leaching, 238

Pennsylvania State University, 27, 29, 30, 226, 349, 432; living filter system, 406–418

Pepper, 510

Peromyscus spp. *See* Deer mice

Pesticides, 127, 280; tip moth, 274, 275, 276, 278, 280

pH. *See* Acidity of soil

Pharmaceutical plant wastes, 510

Phenol, 18, 511, 556, 557, 559, 560

Phosphorus: in sludge and wastewater, 11–12, 19, 221, 225–226, 232–233; renovation, 35, 43, 49, 226–227; in pulp and paper wastewater, 43, 49; in pulp and paper sludge, 45, 50, 261; in vegetation, 112–113, 233; in nitrification and leaching study, 135; in element flux study, 155, 156, 163, 164, 165; in forages, 190, 193, 194, 195, 196; fertilizer, 224–225; orthophosphate, 226, 232, 454, 526; leaching, 233; application rates, 233; foliar levels, 236, 286; retention in soil, 241, 428, 429, 450; short-term transfer rates, 242; in Pack Forest study, 261; in loblolly pine studies, 273, 315, 316, 317, 381, 319; in oak growth studies, 284, 286, 287, 288, 289, 296, 297, 298; existing high level in soil, 384; in Clayton County, 399, 402, 403; in Penn State study, 408, 409, 415, 416; at New Hampshire site, 423, 429; at Vermont site, 426, 428, 429; equation for site assessment, 428; in Australian irrigation, 437; at Georgia site, 445, 446, 447, 448, 449; phosphate concentrations, 454, 559, 560; mandatory removal, 477; at northern Michigan sites, 484, 485, 486, 494; in Copper Basin soil, 499, 504; in borrow pit, 500; in kaolin spoil, 502, 503; in Barceloneta sludge and soil, 511, 512; in Savannah River sludge and soil, 517, 520, 521, 522, 525; in Wauna sludge and soil, 534, 538, 539, 540, 541, 547; in resin based sludge, 557

Photosynthesis, 216, 217, 236, 305, 327, 356, 433

Phthalates, 119, 120, 511, 512; in pulp and paper sludge, 47; di-n-butyl phthalate study, 126–131

Picea sitchensis. See Spruce

Pilchuck Tree Farm, 474

Pin cherry, 189, 200; deer forage, 191, 192, 193, 194, 195, 196; in vegetation and small mammal study, 201, 203

Pine, 111, 113, 350, 351, 352, 398, 454; Caribbean, 513; jack, 111, 112, 113, 143, 144, 200, 388, 480; longleaf, 303, 304, 515; Monterey, 209; pitch, 31; ponderosa, 213, 214, 292; radiata, 432–433; red, 29, 30–31, 35, 50, 143, 144, 200, 227, 228, 229, 231, 236, 237, 260, 292, 293, 350, 388, 407, 409, 412, 480; Scotch, 34, 209; shortleaf, 212, 499, 500; slash, 48, 350, 442–450, 503; southern, 504; Virginia, 499, 500 503; white, 31, 209, 236, 260–264, 292, 350, 388, 422; and nitrification, 134, 528; in nitrogen transformations study, 143, 144, 145, 146, 148, 149, 150, 151, 152; in vegetation and small mammal study, 199–204; foliar nitrogen, 229, 260; nitrogen uptake of vegetation, 229; nitrogen transformation and utilization, 231, 236, 409–412; nitrogen leaching, 237, 238, 290, 410; sludge application rates, 239; nitrogen and phosphorus rates, 242, 403, 485, 486; juvenile wood, 249; in short-term growth study, 282; in long-term growth study, 294; nitrogen application rates, 388; in Australia, 432; in northern Michigan, 480, 481, 485, 486, 487, 489, 490, 491, 494; in Copper Basin, 499, 500; nutrient cycling, 528; in resin based sludge-treated plots, 558, 560. *See also* Loblolly pine

Pine River Experiment Forest, 169

Pinus spp. *See* Pine

Pinus taeda. See Loblolly pine

Pipeline, 533, 536, 543, 544–546

Pisolithus tinctorius, 501

Plant growth. *See* Growth response

Plant pathogens, 268–271

Plants. *See* Vegetation

Plating waste, 169

Pokeweed, 36–37

Poland, 28

Pollutants: risk management, 4–5; organic priority pollutants, 44, 47, 118–124, 125–131; classification of pollutant levels, 96; EPA regulations, 338–340. *See also* Air pollution; Risk assessment; Toxic substances; Trace elements

Polychlorinated biphenyls (PCBs), 24; in pulp and paper sludge, 47; risk assessment, 119–124

Polynuclear aromatic hydrocarbons (PAHs), 119–120

Pond pretreatment, 29, 34

Ponds, 395, 425. *See also* Lagoon storage; Storage systems

Poplar, 30, 31, 32, 48; cumulative biomass, 33; disease, 38; in Pacific Northwest, 227, 231; in Great Lakes area, 227, 231; soil phosphorus, 227; nitrogen renovation, 227; foliar nitrogen, 229; nitrogen uptake of vegetation, 229; harvesting, 230; nitrogen transformation and utilization, 231, 236, 241; hybrid, 354; yellow, 401, 502; in Australia, 434–435, 439. *See also* Lombardy poplar

Populus deltoides. See Cottonwood

Populus grandidentata. See Bigtooth aspen

Populus spp. *See* Aspen

Populus tremuloides. See Quaking aspen

Populus trichocarpa. See Cottonwood

Potassium: in sludge and wastewater, 11–12, 19, 221, 232, 233; in pulp and paper sludge, 45, 261; in pulp and paper wastewater, 48; in nitrification and leaching study, 134, 135, 136; in element flux study, 155, 156, 160, 161, 162, 163, 164, 165, 166; renovation of, 227–228; leaching, 227, 233; application rates, 233; foliar levels, 236; retention in soil, 241, 292, 450; in Pack Forest study, 261; in oak growth studies, 284, 287, 288, 289; in loblolly pine studies, 273, 315, 316, 317, 318, 319; in Clayton County, 399; in Penn State study, 408, 409, 415, 416; in Australia irrigation, 437; at Georgia site, 445, 446, 447–448, 449; mobility of, 448; at northern Michigan site, 484, 486; in Copper Basin soil, 499; in borrow pit, 500; in kaolin spoil, 502, 503; in Barceloneta sludge and soil, 511, 512; in Savannah River sludge and soil, 517, 520, 521, 522, 525; in Wauna sludge and soil, 534, 538, 539, 540, 541; used to coagulate polymers, 552; in NSCC effluent, 553; in resin based sludge, 557

Pot studies, 126–131, 510, 536–537, 544

Poverty grass, 411

Prescribed burning, 443–450

Preservationists, 373, 375

Pretreatment, 86, 171, 175

Priority pollutants. *See* Pollutants

Program for Effective Residuals Management (PERM), 384

Protein content of forages, 112, 114, 188, 190, 193, 194, 195, 196, 435, 494

Protozoans. *See* Parasites

Pruning, 251, 253, 435

Prunus pennsylvanica. See Pin cherry

Prunus spp. *See* Cherry

Pseudotsuga menziesii. See Douglas-fir

Pteridium aquilinum. See Bracken fern

Public acceptance, 4–5, 17–18, 20, 38, 59–60, 63–64, 396; planning for the public dimension of land application, 341–347, 469–470; major objections, 342, 373; Michigan studies, 342–343, 493; preferred uses of sludge and wastewater, 343, 372; attitude of public officials, 343–344, 368, 371, 380–381, 493; public acceptance survey, 367–381; environmental concerns, 373, 374, 375, 376; health concerns, 373, 375, 376; economic concerns, 375, 376, 381; aesthetic concerns, 375, 376; preferred forest activities, 377; information sources, 379–380; in Georgia planning, 396; newsletter, 470; public hearings, 474; Metro's program, 474; Wauna operation, 543

Public health: risk assessment, 4, 36, 118–124; sludge characteristics and constituents of concern, 18, 66; organic elements, 66, 119–124; infectious elements, 73–74; microorganisms of concern, 74–77; minimal infectious dose, 78; effectiveness of wastewater treatment, 79–82; cancer risk, 119, 122–123; public concerns about land application, 373, 375, 376; in Australia, 440. *See also* Infectious elements

Publicly owned and operated wastewater treatment works (POTWs), 8–10, 23; improvements in treatment, 20

Public surveys, 343–344, 367–381, 493

Puerto Rico, 508–513

Puget Sound, 472

Pulp and paper industry: land application of wastewater and sludge, 41–52; profile of the industry, 41–42; wastewater characteristics, 42–44; sludge characteristics, 44–47, 259; priority pollutants, 44, 47; wastewater application programs, 47–49, 51; sludge application, 49–51, 51–52, 466, 533–547; current status, 51–52; sludge used with municipal sewage sludge, 258–264, 466; sludge production, 259; undesirable wood for, 252; tracheid dimensions, 255; sludge from Wauna mill, 533–547; resin wastes, 556–561

Quaking aspen, 189, 200; deer forage, 191, 192, 193, 194, 195, 196; in vegetation and small mammal study, 201–204; mortality study, 266–271; in northern Michigan, 480, 481, 482, 483, 484, 487, 488, 494

Quercus spp. *See* Oak

Rains soil, 558, 559

Raleigh, North Carolina, 127

Raverdeaux poplar, 32

Reclamation of devastated land, 497–506

Recreational use of forests, 373, 377

Recycled water. *See* Wastewater

Red alder, 50

Red-backed voles, 113

Redcedar, 464, 473

Regeneration of site, 59

Regression analysis, 296–297, 298, 373, 375, 498

Regulations and guidelines for sludge use, 20, 23, 63, 337–340, 383–389; on trace elements, 89–90

Regulatory agencies and forest systems, 59, 63–64

Remote sensing of forest biomass, 324–333

Research results related to sludge guidelines, 383–389

Research Triangle Park, North Carolina, 157

Resin wastes, 556–561

Resource Conservation and Recovery Act of 1976 (RCRA), 5, 7, 23, 473, 556

Revenue from timber sales, 491

Rice, 510

Right of way demonstration, 468–469

"Ripley's Believe It Or Not," 475

Risk assessment, 58, 68, 117–124, 338–339. *See also* Public health

Risk management, 4–5

River red gum, 436, 438

Rodents, 37, 357. *See also* Voles

Rogers City, Michigan, 283, 481, 485, 491

Root rot, 269

Root systems, 210; and net primary productivity, 217–218; and leaching, 321; Australian studies, 432; and subsoiling, 505, 506

Rotation length, 439; short-rotation intensively cul-

tured (SRIC), 533
Round Spring Park, 27
Rubus flagellaris. *See* Dewberry
Rubus spp. *See* Brambles
Ruminant animals, 87, 88
Runoff, 92, 400, 401, 423, 425, 426, 428, 435, 465, 467, 551

St. Helens, Oregon, 260
St. Marys, Georgia, 350
St. Regis Corporation, 474
Salinity: of pulp and paper wastewater, 43, 48, 49; of pulp and paper sludge, 46–47, 50; of Murray River, 435–436; in Australian soil, 439
Salmonellosis, 74: *Salmonella*, 75, 76, 78, 80, 81, 82
Savannah River Ecology Laboratory, 516
Savannah River Laboratory, 516
Savannah River Plant, 133, 155, 179, 309; loblolly pine growth response study, 272–280; loblolly pine water relations, 301–306; loblolly pine understory fertilization, 308–323; borrow pits, 499; sludge application program, 515–529
Sawdust Supply Company, 475
Seabrook Farms, New Jersey, 27, 349
Seattle, Washington, 233, 356, 364. *See also* Municipality of Metropolitan Seattle
Sedge, 112, 113, 189, 200, 201, 203
Selenium: in pulp and paper sludge, 45, 46; in sewage sludge, 86, 87; in crops, 88; in loblolly pine study, 273; in Savannah River sludge, 517; in Wauna sludge, 534
Sequoia, 464
Serenoa repens. *See* Palmetto
Serviceberry, 200
Sewage sludge: history of use of, 3; disposal alternatives, 7–10; U.S. sewage and septage, 8; municipal sludge production, 9, 10, 509; land application practices, 10–11, 12; nutrients in, 11–12, 225–230, 284, 298–299; forest land application potentials, 13–22; common sludge constituents, 17–19; finding application sites, 21; compared to pulp and paper sludge, 45; future directions, 62–69, 366; pathogens in, 76; treatment effectiveness, 79–80; and virus survival, 81–82; and trace elements, 85–87; effect on water quality, 100–108; and wildlife, 110–116, 188–197; risk assessment, 117–124; behavior of organic compounds in land treatment systems, 125–131; nitrification and leaching in soil treated with sulfuric acid and nitrification inhibitor, 132–141; nitrogen transformations study, 142–152; spraying liquid sludge, 143, 156, 357; application rates, 145, 156, 180, 230, 232, 233, 238, 284, 321, 462, 463, 481–491, 504; element flux in loblolly pine plantation, 154–166; heavy metal storage in soils of an aspen forest, 168–175; clumps of dried sludge, 170, 171, 172; pretreatment, 86, 171, 175; influence on soil mesofauna, 177–186; metals in vegetation and small mammals, 199–204; growth response of forest trees to, 209–219; production and discharge figures, 221; nitrogen, 232; phosphorus, 232–233, 477; effect on

60-year-old Douglas-fir stand, 246–257; used with pulp and paper sludge, 258–264; annual production, 259; aspen mortality, 266–271; growth response of loblolly pine study, 271–280; short-term growth response of oak forests, 282–291; long-term growth response of oak forests, 292–300; nutrient retention rates, 298–299; response of loblolly pine, water relations, 301–306; and loblolly pine understory, 308–323; regulations and guidelines, 7–8, 22–24, 100–101, 337–340, 383–389; public acceptance, 341–347, 367–381; technology and costs of application, 356–366; application rate and timing, 358, 463, 468, 481, 505; percentage of solids, 358; storage, 358–360, 462; transfer equipment, 360–362, 452, 475, 481, 491; rewatering procedure, 361, 469; application systems, 362–364; comparison of storage and application systems, 364–365; research results related to guidelines, 383–389; stabilization, 388; pelletizing, 396, 401; dewatered, 462, 463, Pack Forest demonstration program, 461–470; heavy applications, 462; tree species not recommended, 464, 467, 473; Silvigrow, 121, 475–476; Metro's program, 472–476; privatization of production, 476; use in northern Michigan, 477–495; variation in application rates, 481, 487; distance effects, 487–489; basal area effects, 487–491; site preparation costs, 491; transport, 491–492; dried sludge used on devastated sites, 497–506; nondegradation constraint, 509–510; at Savannah River Plant, 515–529. *See also* Application methods; Treatment of sludge and wastewater; Wastewater
Sheep, 435
Shelton, Washington, 474
Shigellosis, 74; *Shigella*, 75, 76, 78, 81, 82
Shoal Creek System, 350, 395
Silver, 45, 534
Silvicultural demonstration site, 466–467
Silviculture. *See* Cultural practices
Silvicycle, 470
Silvigrow, 121, 475–476
Site disturbance, 59, 110–111, 401–402, 405, 467, 551
Site preparation, 491; subsoiling, 498, 503–504. *See also* Trail construction
Site productivity research and guidelines, 383–389
Site quality, 213, 215–216, 217, 232, 233, 239, 241, 242, 292, 551
Site selection. *See* Design criteria
Sitka spruce, 213, 215
Ski areas, 351, 422, 424
Slopes: design criteria for, 57–58, 66, 67, 384, 465, 467; risk assessment, 119; in Clayton County system, 397; in New Hampshire, 423; in Vermont, 424, 425; and industrial waste, 549
Sludge, 204; from pulp and paper industry, 42, 44–47, 49–51, 51–52, 258–264, 466, 533–547, 556–561; content of pulp and paper sludge, 45; costs of operation, 49, 64–65; growth response to, 64–65; textile mill waste, 156, 179, 273, 309, 516; in Puerto Rico, 510–513; from pharmaceutical plants, 510; injection process, 513; at Savannah River Plant,

515–529; slurry experiments, 536–537; polymer wastes, 549–555; resin wastes, 556–561. *See also* Sewage sludge; Treatment of sludge and wastewater; Wastewater

Slurry, 536–537

Smelters, 94–95

Snoqualmie, Washington, 475

Snoqualmie Pass, Washington, 350, 351, 354

Snow-making operations, 426

Snowmelt, 426, 427, 429

Sodium: in pulp and paper wastewater, 43, 48, 49; SAR, 43–44, 538, 539, 540, 547; in pulp and paper sludge, 45, 46–47, 50; salt toxicity, 50; in nitrification and leaching study, 135; in element flux study, 156; renovation of, 228; retention in soil, 241; in loblolly pine studies, 273, 316, 317, 318, 319; in oak studies, 289; in Clayton County, 399; in Penn State study, 408; in Australian irrigation, 437; at Georgia site, 445, 446; at northern Michigan sites, 484, 486, 494; in Barceloneta sludge and soil, 511, 512; in Savannah River sludge and soil, 517, 520, 521, 522, 525, 527; in Wauna sludge and soil, 534, 538, 539, 540, 547; at NSCC site, 551, 552, 553; in resin based sludge, 557

Soil organisms, 92

Soils: accumulation of nitrogen in, 32, 34–36, 132–141, 235, 240; sandy, 33, 35, 44, 49, 50, 166, 238, 260, 290, 292, 301, 326, 328, 329–330, 331, 518; clay, 44, 166, 226, 326, 328, 329–330, 331; and survival of microorganisms, 80–81; microbial transport, 81–82; adsorption and virus retention, 81; trace elements in, 86–90, 94–95; comparison of cropland and forest land, 90–94; composition of litter, 93; elemental cycling, 93–94, 222–224; inorganic compounds in, 96; total organic carbon in leachates, 105; effects of organic priority pollutants and sludge on, 129–131; nitrogen transformations study, 142–152; element flux in loblolly pine plantations, 154–166; storage of heavy metals, 168–175, 467; influence of sludge on mesofauna, 177–186; renovation capacity, 226–230, 241, 446; and leaching, 238; nitrogen storage capacity, 238, 240; in oak studies, 284, 286–289, 294–295, 297, 298; in loblolly pine understory study, 309, 323; in remote sensing of biomass study, 326, 328, 329–330, 331; phosphorus in, 384, 428, 447, 494; in Penn State study, 407–408, 409, 414–416; Australian studies, 432; wind erosion, 436; in burned and nonburned irrigated areas, 445–450; permeability, 453, 479, 552; fragipan, 425, 453; forest floor weight and nutrient retention, 494; severely devastated sites, 497–506; subsoiling, 498, 503–504, 505; in Copper Basin, 499; borrow pits, 497–498, 499–502; kaolin spoils, 497, 498, 502–503; in southern pine plantation, 504; in Barceloneta area, 512; at Savannah River Plant, 520–522; at Wauna operation, 538–540; site investigation method, 551–552; at resin based sludge-treated areas, 558–559. *See also* Acidity of soil; Buffer zones; Fertilization; Humus; Leaching; Litter; Metals; Runoff; Slopes; Trace elements; Water relations; *individual nutrients or soil types*

Solar radiation measurement, 302, 306

Solidago spp. *See* Goldenrod

Solid-phase mineralogy, 96

Solid Waste Disposal Act, 23, 473

Sorghum, 260

South Carolina, 134, 179, 272, 498; foliar nitrogen, 236; remote sensing, 324; mine spoils, 497; borrow pits, 499–502; NSCC wastewater, 549–551; Department of Health and Environmental Control, 525. *See also* Savannah River Plant

Southeast: use of fertilizer, 224; nitrogen assimilation, 228, 231, 236, 237, 239, 240; nutrients in sludge, 232, 233; wastewater use in, 27, 29; soil acidity in, 133; soil needs, 301; severely devastated land, 497. *See also* Savannah River Plant; Coastal plain

Southeastern Forest Experimental Station, 515

Soybeans, 126–131

Specific gravity, 249, 253, 254–255, 464

Spray application methods, 143, 156, 357, 362–363, 365, 407, 417, 422–423, 424, 463, 481, 482, 491, 516, 560

Spring Creek, Pennsylvania, 406

Sprinkler irrigation, 349–354, 399–405, 417, 421, 422–424, 426, 433, 434, 443, 453, 554–555

Spruce, 31, 34, 350, 388, 424; Sitka, 213, 215, 464, 473; foliar nitrogen, 229; white, 231, 407, 409, 411–412, 413, 414, 415, 416, 424; nitrogen transformation and utilization, 231, 411–412; foliar analysis, 412–413

Stargrass, 510

State College, Pennsylvania, 350, 352, 354, 406

Statistical analysis, 146, 157, 296, 329–330, 445, 498

Storage systems: wastewater, 353–354, 396, 401, 550; Vermont requirement, 353; costs of, 354; sludge, 358–360, 462, 469, 513; comparison of systems, 365; portable system, 469; for resin wastes, 556–557. *See also* Lagoon storage

Storms, 410, 412

Strawberry, wild, 194, 195, 201, 412, 413, 414

Stream contamination, 423, 426, 452. *See also* Buffer zones; Runoff

Streamflow, 403–404, 467

Strip mine spoils, 44, 475, 497

Submarine base, 27, 350, 442–450

Subsoiling, 498, 503–504, 505

Substituted monocyclic aromatics, 119, 120

Sugarloaf Mountain, Maine, 27, 421

Sulfate, 49; Sulfate-S, 160, 161, 162, 163, 164, 165; in wastewater, 228; leaching, 241; in Barceloneta sludge and soil, 511, 512; used to coagulate polymers, 552; in resin based sludge, 557, 559

Sulfite, 49

Sulfur: in pulp and paper sludge, 45; in nitrification and leaching study, 135; in element flux study, 156; in loblolly pine study, 273; in Savannah River sludge, 517; in NSCC effluent, 553

Sunapee, New Hampshire, 27, 350, 420, 421

Sunscald, 267, 268, 270, 271

Survey research, 341–347, 367–381, 493

Survival of microorganisms, 80–81

Sweetfern, 189, 200
Sweetgum, 48, 401, 501–502, 503
Sycamore, 48, 401, 454, 502

Tacoma–Pierce County Health Department, 466
Tacoma, Washington, 260–264, 356, 466; Tacoma City
    Light, 468
Tamias striatus, 113
Tanker-sprayer systems, 357, 362, 365
Tannin, 541
Taxodium distichum. See Cypress
Teak, 513
Technology: for waste disposal, 8; methods for incor-
    poration and transfer of, 60; for sludge applica-
    tion, 67–68, 356–366; for wastewater application,
    349–354. *See also* Application methods; Costs
Temperature effect, 132, 140, 146, 150
Tennessee, 497, 498; Copper Basin, 498, 499, 500, 504
Terracing, 505
Terrain. *See* Buffer zones; Slopes
Textile mill waste, 156, 179, 273, 309, 516
Thelephora terrestris, 501
Thinning, 251–253, 254, 255, 256, 467, 498, 499, 501,
    518
Thirteen-lined ground squirrels, 113, 201, 204
Tilia americana. See Basswood
Tin, 45, 46, 273, 517
Tip moth treatment, 274, 275, 276, 278, 280, 519
Titanium, 45
Toluene, 18, 511, 512
Total dissolved solids (TDS), 552–553, 555
Total organic halogen (TOX), 47
Total suspended solids (TSS), 42–43
Toxic substances, 17, 18, 19, 20; risk assessment, 4–5,
    118–124; trace elements, 87, 94–96, 320; organics,
    105; in wildlife forage, 114–115, 188–197, 320. *See
    also* Heavy metals; Organic substances; Pollutants;
    Trace elements
Toxic Substances Control Act, 24
Trace chemicals: and wildlife, 36, 37; in pulp and pa-
    per wastewater and sludge, 44, 50
Trace elements, in municipal sewage sludge, 85–87;
    toxicity of, 87–89, 94–95; recommended limits on,
    89–90; in cropland compared with forest land,
    90–94; leaching, 91; in forest biomass and soil,
    93–94; in cropland, 93; anthropogenic sources,
    94–97; smelting, 94–95, 96; effect on invertebrates
    and small mammals, 94, 114; mining, 96, solubil-
    ity and bioavailability, 96–97; in element flux
    study, 162–164; in deer, 197; in vegetation and
    small mammals, 199–204; in loblolly pine studies,
    273, 315–319; in Penn State study, 408, 409,
    412–417; in northern Michigan sites, 484, 486; vari-
    ation in loading rates, 486–487. *See also* Heavy me-
    tals; leaching
Trail construction, 111, 189–190, 286, 467, 491; in
    Michigan, 200, 481, 491; for tanker-sprayer sys-
    tems, 357; for manure spreaders, 357; and wildlife
    use, 494
Traveling gun application, 350, 354, 362–363, 365,
    539, 543, 544, 560

Treatment of sludge and wastewater: disinfection, 79,
    80, 421, 422, 426, 429; effectiveness, 79–80; and
    trace elements, 85–87; pretreatment, 86, 171, 175;
    for Savannah River Plant, 179, 184; pulp and pa-
    per sludge, 259; sludge hauling, 359–360; in Clay-
    ton County, 394–405; Penn State's system,
    406–418; trickling filter, 408; energy consumption,
    425; in Australia, 434, 436; in Puerto Rico, 508–513;
    nondegradation constraint, 509–510; Savannah
    River sludges, 516; at Crown Zellerbach's Wauna
    mill, 533–534; National Starch and Chemical,
    549–555; chemical coagulation/DAF system,
    552–555. *See also* Costs; Irrigation; Lagoon storage;
    Sewage sludge; Sludge; Storage systems; Waste-
    water
Tree growth, 30–34; from pulp and paper sludge ap-
    plication, 50–51; models needed, 58. *See also*
    Growth response
Trickling filter plant, 408
Trident submarine base, 442, 445
Tropical storm Agnes, 410, 412
Tsuga canadensis. See Hemlock
Tsuga heterophylla. See Hemlock

Understory vegetation. *See* Vegetation
Unicoi State Park, 27, 29, 226, 350
U.S. Department of Agriculture, 88, 89, 478
U.S. Department of Energy, 325, 515
U.S. Department of the Navy, 442
U.S. Environmental Protection Agency: and risk
    management, 4–5, 7, 119; Region 10, 5; waste
    management guidelines, 7–8, 508–509; Policy on
    Municipal Sludge Management, 22; design manu-
    als, 55, 56, 420, 462, 509; trace element regulation,
    89, 202; drinking water guidelines, 100–101; poli-
    cies and guidelines for sludge reuse and disposal,
    337–340, 388, 473, 474; Sludge Task Force,
    337–340; state sludge management regulations,
    340; sponsorship of research, 343, 478; accuracy as
    an information source, 380; review of Clayton
    County system, 404
U.S. Forest Service, 515, 516
U.S. Forest Service and Michigan State University
    Forest Soil and Water Co-op laboratory, 170, 202
University of Georgia, 157, 349, 442
University of Minnesota Analytical Laboratory,
    169–170
University of Missouri, 453
University of Puerto Rico, 510
University of Washington, 51, 67, 293, 349, 364, 461,
    472, 473, 476. *See also* Pack Forest
Upper coastal plain, 272–280, 301–306, 308–323
Urea fertilizer, 64, 218

Vaccinium, 200
Vegetation: nitrogen uptake in, 31, 32, 35, 228, 236,
    237, 411, 427; cumulative biomass, 33, 311–312;
    nutrient cycling, 33; on strip mine spoil, 44; and
    trace elements, 95, 202, 413; changes in species
    composition, 111; in wildlife habitat, 111, 188–197,
    199–204; nutritive quality, 112–113; effects of or-

ganic priority pollutants on, 126–131; competition with trees, 214–215; analysis of samples, 190; metals in, 199–204; in loblolly pine forests, 301, 308–323; categories, 310; wildfire, 321; and remote sensing of biomass, 330; role in living filter system, 407, 410–414; qualitative understory survey, 444–445, 448–450; in Georgia study, 448–450; and prescribed burning, 449, 450; in Pack Forest, 463; in northern Michigan, 494; revegetation of devastated land, 497; in Puerto Rico, 512–513; and industrial wastes, 549; osmotic potential, 554. See also Biomass; Forages; Grasses; Herbicides; Weeds

Velpar®, 274, 275, 276, 309, 519

Vermont, 27, 350, 351, 420, 421, 424–427; storage requirement, 353

Victoria, Australia, 432, 433–435, 436, 440

Vinyl and acrylic polymer wastes, 549–550, 552–553

Virginia, 497, 498

Viruses, 18, 19, 75, 76–77, 78; and wastewater treatment, 79–80; survival of, 80–81; adsorption, 81; size, 82; die-off rate, 82; in groundwater, 103, 108

Volatiles, 120

Voles, 92, 111, 113, 114, 357, 463, 464, 468

Wagrum soil, 273, 302–306, 309, 313, 314, 317, 519

Walnut, 454

Wangaratta, Australia, 433–435

Washington State, 5, 27, 29, 30, 31, 55, 111, 112, 113, 114, 209, 216, 469; Department of Ecology, 118, 462, 466, 473, 474; Department of Natural Resources, 474, 475; nitrogen uptake of vegetation, 229; nutrients in sludge, 233; application rates, 233; wastewater irrigation, 350, 351, 354; technology and costs of sludge application, 356–366. See also Pacific Northwest; Pack Forest; University of Washington

Waste disposal sheet, 384–385, 386

Wastewater: POTWs, 8–10, 20, 23; nutrients in, 11–12, 225–230, 241, 445; current status of use for forest application, 26–39, 350; future directions, 55–61; list of projects, 27; tree growth, 30–34; from pulp and paper industry, 42–44, 47–49, 51; location of systems in U.S., 55, 350; design criteria for systems, 57–58, 553–555; recycling, 58–59; pathogens in, 76; treatment effectiveness, 79–80; form of trace metals in, 97; effect on water quality, 100–108; effect on soil mesofauna, 179; growth response of forest trees to, 209–219; renovation capacity, 226–230, 241; nitrogen, 225, 226, 227, 228–230, 426, 434; phosphorus, 225, 226–227, 426, 434; potassium, 227–228; public acceptance, 341–347, 367–381, 396; technology and costs of application, 349–354; storage, 9–10, 225, 351, 352, 353, 354, 396; use in Clayton County, 393–405; Penn State's living filter, 406–418; in New England, 420–429; in Australia, 431–440; from fruit processing, 436; coastal plain, slash pine system, 442–450; from Georgia submarine base, 442–450; industrial waste, 549; National Starch and Chemical, 549–555; vinyl and acrylic polymer wastes, 549–550, 552–553. See also Application methods; Ir-

rigation; Lagoon Storage; Sewage Sludge; Sludge; Storage systems; Treatment of sludge and wastewater

Water jack, 543, 544

Water pollution control, 22–23, 77. See also Clean Water Act; Groundwater; Pollutants

Water quality, 100–108, 394, 403, 494; nitrogen removal, 34–36; and pathways of microorganisms, 77; flow rate of leachate, 103; pulses, 103; pathogens, 103; surface water, 121, 122, 123; Georgia standards, 394, 404; West Dover, Vermont, 426, 427. See also Groundwater

Water reclamation. See Wastewater

Water relations: information needed, 57–58; water balance estimates, 157, 401; drainage, 157, 159, 447, 448; drought resistance, 212; in model of forest tree growth, 213–216; in loblolly pine forest, 301–306; water stress, 301. See also Hydrology; Leaching; Runoff

Watersheds, 403, 404, 467

Water table, 384, 387, 404, 443, 444; Australian studies, 432

Wauna, Oregon, 260, 533–547

Weeds, 37, 158, 210, 260, 264, 357, 505, 535; in Australia, 435, 436, 438, 439. See also Herbicides

West Dover, Vermont, 27, 350, 351, 353, 420, 421, 424–427, 428

Wetlands irrigation, 30

Wexford County, Michigan, 368

Weyerhaeuser Company, 475

Wheat, 126–131

White-tailed deer, 133, 188–197, 493

Wildfire, 321, 450

Wildlife: public concern about, 18, 345, 373; effects of wastewater on, 36–37; risk assessment, 38; and Giardia lamblia, 76; trace elements in, 92; responses to application of sewage sludge, 110–116, 266–271, 468, 473, 493–494; responses to habitat changes, 113–114; and toxic substances, 114–115, 320; vegetation and small mammals study, 199–204; and sludge guidelines, 384. See also Deer; Forages; Mammals, small; individual species

Willow, 48

Windthrow, 210, 211, 280

Winery irrigation, 439

Wintergreen, 200

Wisconsin, 89

Witch hazel, 200

Wolfeboro, New Hampshire, 27, 350, 351, 354, 420, 421, 422–424, 428

Women and Minority Business Enterprise, 476

Woodcock, 37

Woodland jumping mouse, 201, 202, 204

Wood properties: effect of sludge on Douglas-fir stand, 246–257; undesirable features, 247, 250–253, 255, 257; hormone concentrations, 248, 254, 256; cultural practices, 250–256; Australian studies, 432

Wood quality: data needed, 58, 65; for pulp production, 178, 252, 255; specific gravity, 249, 254–255, 464; terminology, 249; knots, 250–252; juvenile wood, 249, 251, 252; rings, 251, 252; compression wood, 251, 252–253; taper, 251, 253; spiral grain, 251, 253, crookedness, 251, 253; veneer, 252

Woodruff, South Carolina, 549

Workshop on Utilization of Municipal Wastewater and Sludge on Land, 56, 62

Xylem pressure potential, 302–306

Yulara, Australia, 439

Zapus hudsonius, 113

Zinc: in pulp and paper sludge, 45, 46; from galvanized pipe, 86; pretreatment, 86; in sewage sludge, 86, 87; recommended limits on, 89, 90; in animals, 92, 114; in deer tissue, 193, 197; in forest soil, 93, 95; from smelter, 94, 95; leaf tissue concentrations, 95; in leachates, 107; in nitrification and leaching study, 134, 135, 136, 139; in element flux study, 156, 162, 163, 166; effect on wildlife, 114; in vegetation and small mammals, 202, 203, 204; in oak growth studies, 284, 287, 288, 289; in heavy metal study, 170, 171, 172, 174, 175; in loblolly pine studies, 273, 316, 317, 318, 319, 320; in Clayton County, 399; in Penn State study, 408, 409, 413, 414, 415, 416; at northern Michigan sites, 484, 486; in Barceloneta sludge and soil, 510, 511, 512; in Savannah River sludge and soil, 517, 520, 521; in Wauna sludge, 534; in resin based sludge, 557; groundwater quality before and after sludge application, 559, 560